T0248588

Handbook of

Food and Bioprocess Modeling Techniques

FOOD SCIENCE AND TECHNOLOGY

Editorial Advisory Board

Handbook of
Food and Bioprocess Modeling Techniques

edited by

Shyam S. Sablani   M. Shafiur Rahman
Ashim K. Datta   Arun S. Mujumdar

CRC Press
Taylor & Francis Group
Boca Raton   London   New York

CRC Press is an imprint of the
Taylor & Francis Group, an informa business

CRC Press
Taylor & Francis Group
6000 Broken Sound Parkway NW, Suite 300
Boca Raton, FL 33487-2742

ISBN-13: 978-0-367-45326-8 (pbk)
ISBN-13: 978-0-8247-2671-3 (hbk)

**Visit the Taylor & Francis Web site at
http://www.taylorandfrancis.com**

**and the CRC Press Web site at
http://www.crcpress.com**

Library of Congress Cataloging-in-Publication Data

Handbook of food and bioprocess modeling techniques / editors, Shyam S. Sablani ... [et al.].
    p. ; cm. -- (Food science and technology ; 166)
  "A CRC title."
  Includes bibliographical references and index.
  ISBN-13: 978-0-8247-2671-3 (hardcover: alk. paper)
  ISBN-10: 0-8247-2671-5 (hardcover : alk. paper)
  1. Biochemical engineering--Mathematical models. 2. Biochemical engineering--Mathematical models. 3. Food industry and trade--Mathematical models. 4. Food--Biotechnology--Mathematical models.   I. Sablani, Shyam S. II. Series: Food science and technology (Taylor & Francis) ; 166.
  [DNLM: 1. Food Technology. 2. Biotechnology. 3. Food Handling--methods. 4. Models, Theoretical.   W1 FO509P v.166 2006 / WA 695 H2355 2006]

TP248.25.M39H36 2006
664'.02015118--dc22
                                                             2006021065

Dedication

To our families

Preface

Models can reduce the number of experiments; thereby reducing time and expenses and providing process optimization, predictive capability, improved process automation, and control possibilities. A physics-based model can also provide insight into a process for which experimentation may not be practical. Food and bioproducts go through complex physical, chemical, and biological changes during their processing and storage, making modeling an important tool in understanding and controlling these changes. With the advancement of computers, the use of models has become integral to most studies in food science and engineering.

In addition to modeling techniques popular in the past, newer ones have become commonplace due to the enhanced ease of their development using computers. As the choices in modeling techniques have increased, user awareness of alternatives has become increasingly important. Although books are available on any one modeling technique, a user is faced with the daunting task of combing through a number of these books to decide on the most appropriate model. Our goal for this handbook is to aid this process by providing, in one volume, succinct descriptions of a large number of modeling techniques, followed by examples of applications in the food context. This presentation differs from the many theoretical details of a model that the reader can follow-up in texts on individual modeling techniques. The chapters in this handbook are structured, generally, to begin with a short introduction to the modeling technique followed by details on how that technique can be utilized in specific food and bioprocess applications.

The concise introduction and the context of food for many different models alone make this handbook unique. Having access to many models at once, the reader can see at a glance which models have been most successful in specific applications. By collecting the various modeling possibilities into one volume, the book can lead someone unfamiliar with modeling to look into the benefits of this useful tool. The book allows the reader to conveniently shop for the model that best suits, not only the application in question, but also his or her own technical expertise. Including both physics- and observation-based models provides a unique perspective to the user, in terms of both modeling alternatives and ways to combine various types of models to describe the same process. The book emphasizes problem formulation, particularly in the case of physics-based models, from an application point of view and thereby makes it more accessible to the end user.

This handbook is intended as a reference book for food and bioprocess researchers in academia and also for professionals engaged in process and product development in the food and bioprocess industries. It is also quite appropriate as a text or reference for courses on modeling for graduate or upper level undergraduate students in engineering related to food and bioprocesses (chemical, biological, agricultural, and food engineering) and some food science disciplines. The highly international (fifteen countries) authorship and reviews of the book ensure both the quality of individual chapters and their usefulness around the globe.

This book is the work of its contributing authors. The editors would like to thank them for their willingness, time, and effort as the manuscripts went through many revisions. The editors also gratefully acknowledge the reviews and insightful comments by the many individuals from 14 countries that greatly improved the quality of the book. These individuals include: José M. Aguilera (Universidad Católica de Chile, Chile), Ioannis Arvanitoyannis (University of Thessaly, Greece), Julio R. Banga (Spanish Council for Scientific Research, Spain), Pilar Buera (Facultad de Ciencias Exactas y Naturales, Argentina), Fred Eisenberg (Eisenberg Research, USA), Hao Feng (University of Illinois, USA), Amit Halder (Cornell University, USA), Ashish Dhall (Cornell University, USA), James P. Gleeson (University College Cork, Ireland), Lihan Huang (USDA Eastern Regional Research Center, USA), V. K. Jindal (Asian Institute of Technology, Thailand), Elena Maestri (University of Parma, Italy), S. Nakai (University of British Columbia, Canada), Bart Nicolai (Catholic University of Leuven, Belgium), John Nieber (University of Minnesota at Twin Cities, USA), Jitendra Paliwal (University of Manitoba, Canada), Vineet Rakesh (Cornell University,

USA), H. S. Ramaswamy (McGill University, Canada), Ajay K. Ray (National University of Singapore, Singapore), Phil Richardson (Campden & Chorleywood Food Research Association, UK), Arnab Sarkar (Masterfoods, USA), Bhaskar Sen Gupta (Queen's University, Belfast, UK), Christian Trägårdh (Lund University, Sweden), Gilles Trystram (ENSIA, France), Rolf Verberg (University of Pittsburgh, USA), and Weibiao Zhou (National University of Singapore, Singapore). We thank the editors at Taylor & Francis for their support and help. We also wish to express our gratitude to our respective universities for giving us the opportunity and the facilities to complete such an important project.

**Shyam S. Sablani, Ashim K. Datta,
Mohammad Shafiur Rahman, and Arun S. Mujumdar**

The Editors

Shyam S. Sablani, PhD, is an associate professor of food and bioprocess engineering at Sultan Qaboos University, Muscat, Oman. He obtained a BE (1986) in mechanical engineering from Ravishanker University, an MS (1989) in mechanical engineering from the Indian Institute of Technology, Madras, India, and a PhD (1996) in food and process engineering from McGill University, Montreal. Following a brief period as a research associate at Agriculture and Agri-Food Canada, St. Hycinthe, he moved to the Sultan Qaboos University, Oman, in 1997. Dr. Sablani is a member of the American Society of Mechanical Engineers, Institute of Food Technologists, American Society of Agricultural and Biological Engineers, and the Indian Society of Heat and Mass Transfer. He received a certificate of merit from the Institute of Food Technologists (1994) and two Outstanding Researcher Awards from Sultan Qaboos University (2002 and 2004). His research interests are heat and mass transport relevant to food processing, artificial neural network modeling, and food properties. His current research focuses on physical and chemical changes in biomaterials during processing and storage. He has authored and co-authored more than 70 refereed journal articles and 15 book chapters. He serves as the associate editor of the *International Journal of Food Properties*.

Ashim K. Datta, PhD, is a professor in the Department of Biological and Environmental Engineering, Cornell University, Ithaca, New York. His research area is computer-aided food process engineering and he is currently working on developing software for integrating food properties with engineering analysis of complex food processes. The author and co-author of over 80 journal articles and book chapters, two co-edited books and one textbook related to food process modeling, Dr. Datta is a member of the American Institute of Chemical Engineers, American Society of Agricultural and Biological Engineers, Institute of Food Technologists, and the Institute of Biological Engineering. He received a BTech (1979) from the Indian Institute of Technology, Kharagpur, India, an MS (1982) from the University of Illinois at Urbana-Champaign and a PhD (1985) from the University of Florida, Gainesville, all in agricultural engineering.

Mohammad Shafiur Rahman, PhD, is an associate professor at the Sultan Qaboos University, Oman. He has authored or co-authored over 200 technical articles, including 66 refereed journal papers, 68 conference papers, 33 reports, 8 popular articles, and 3 books. He has authored and edited two books, *Food Properties Handbook* and *Handbook of Food Preservation*, the former having been translated into Spanish. He served as associate editor for the *Handbook of Food Science, Technology, and Engineering*. Dr. Rahman is founding editor of the *International Journal of Food Properties* and serves on the editorial boards of several journals. Dr. Rahman is member of the New Zealand Institute of Food Science and Technology, the Institute of Food Technologists, the American Society of Agricultural Engineers, and the American Institute of Chemical Engineers. He received a BSc Eng (chemical) (1983), an MSc Eng (chemical) (1984) from Bangladesh University of Engineering and Technology, Dhaka, an MS (1985) in food engineering from Leeds University, England, and a PhD (1992) in food engineering from the University of New South Wales, Sydney. Dr. Rahman has received numerous awards and fellowships in recognition of his research and teaching achievements, including the Hort Research Chairman's Award, the Bilateral Research Activities Program (BRAP) Award, an Outstanding Researcher Award, CAMS, Sultan Qaboos University 2003, and the British Council Fellowship.

Arun S. Mujumdar, PhD, is currently a professor of mechanical engineering at the National University of Singapore. He was a member of the Department of Chemical Engineering faculty of McGill University, Canada prior to joining NUS. Dr. Mujumdar has over 350 journal papers, 250 conference papers, 60 edited works, two co-authored books, and over 100 book chapters to his credit. Winner of numerous international awards for his distinguished contributions to drying technology, heat and mass transfer, and chemical engineering in general, he is editor-in-chief of the international journal *Drying Technology* and the widely acclaimed *Handbook of Industrial Drying*, now in a third enhanced edition. He holds a BChem Eng from the University of Mumbai, and a MEng, and PhD from McGill University.

Contributors

Ioannis S. Arvanitoyannis
Laboratory of Food Quality and Technology
School of Agricultural Sciences
University of Thessaly
Nea Ionia Magnesias, Volos, Greece

Christopher G. J. Baker
Department of Chemical Engineering
College of Engineering and Petroleum
Kuwait University
Safat, Kuwait

Xiao Dong Chen
Department of Chemical Engineering
Monash University
Clayton Campus, Victoria, Australia

Kevin Cronin
Department of Process Engineering
University College Cork
Cork, Ireland

Zhanfeng Cui
Department of Engineering Science
Oxford University
Oxford, England

Ashim K. Datta
Department of Biological and
  Environmental Engineering
Cornell University
Ithaca, New York

Ilhan Dilber
Fluent Inc.
Evanston, Illinois

Jinglie Dou
Department of Food Science
University of British Columbia
Vancouver, British Columbia, Canada

Eli Feinerman
Faculty of Agriculture
The Center for Agricultural Economic
  Research
The Hebrew University of Jerusalem
Rehovot, Israel

Maria C. Giannakourou
Laboratory of Food Chemistry and Technology
School of Chemical Engineering
National Technical University of Athens
Athens, Greece

James P. Gleeson
School of Mathematical Sciences
University College Cork
Cork, Ireland

Maarten L. A. T. M. Hertog
Flanders Centre/Laboratory of
  Postharvest Technology
Catholic University of Leuven
Leuven, Belgium

Geoffrey Holmes
Department of Computer Science
University of Waikato
Hamilton, New Zealand

David Hughes
Department of Engineering Science
Oxford University
Oxford, England

Ashwini Kumar
Fluent Inc.
Evanston, Illinois

Haitham M. S. Lababidi
Department of Chemical Engineering
College of Engineering and Petroleum
Kuwait University
Safat, Kuwait

Eunice C. Y. Li-Chan
Department of Food Science
University of British Columbia
Vancouver, British Columbia, Canada

Law C. Lim
School of Chemical and Environmental
Engineering, Faculty of Engineering
  and Computer Science,
The University of Nottingham,
Malaysia Campus, Jalan Broga
Selangor, Malaysia

T. Morimoto
Faculty of Agriculture
Department of Biomechanical Systems
Ehime University
Matsuyama, Japan

Arun S. Mujumdar
Department of Mechnical and
 Production Engineering
National University of Singapore
Singapore

Shuryo Nakai
Department of Food Science
University of British Columbia
Vancouver, British Columbia, Canada

Bart M. Nicolaï
Flanders Centre/Laboratory of
 Postharvest Technology
Catholic University of Leuven
Leuven, Belgium

Mohammad Shafiur Rahman
Department of Food Science and Nutrition
Sultan Qaboos University
Muscat, Oman

Shyam S. Sablani
Department of Food Science and Nutrition
Sultan Qaboos University
Muscat, Oman

Sam Saguy
Faculty of Agricultural, Food and
 Environmental Sciences
The Institute of Biochemistry
 Food Science and Nutrition
The Hebrew University of Jerusalem
Rehovot, Israel

Nico Scheerlinck
Flanders Centre/Laboratory of
 Postharvest Technology
Catholic University of Leuven
Leuven, Belgium

R. G. M. van der Sman
Food and Bioprocess Engineering
University of Wageningen
Wageningen, The Netherlands

Taha Taha
Department of Engineering Science
Oxford University
Oxford, England

Petros S. Taoukis
Laboratory of Food Chemistry
 and Technology
School of Chemical Engineering
National Technical University of Athens
Athens, Greece

Contents

CHAPTER 1

Mathematical Modeling Techniques in Food and Bioprocesses: An Overview

Ashim K. Datta and Shyam S. Sablani

CONTENTS

1.1   MATHEMATICAL MODELING

A model is an analog of a physical reality, typically simpler and idealized. Models can be physical or mathematical and are created with the goal to gain insight into the reality in a more convenient way. A physical model can be a miniature, such as a benchtop version of an industrial scale piece of equipment. A mathematical model is a mathematical analog of the physical reality, describing the properties and features of a real system in terms of mathematical variables and operations. The phenomenal growth in the computing power and its associated user-friendliness

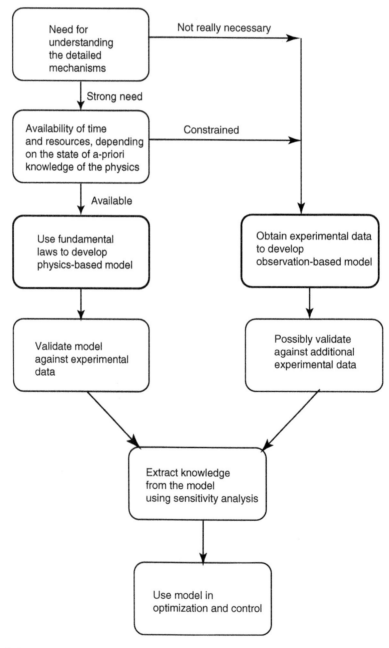

Figure 1.1   A simple overview of model development and use.

have allowed models to be more realistic and have fueled rapid growth in the use of models in product, process, and equipment design and research. Many advantages of a model include (1) reduction of the number of experiments, thus reducing time and expenses; (2) providing great insight into the process (in case of a physics-based model) that may not even be possible with experimentation; (3) process optimization; (4) predictive capability, i.e., ways of performing "what if" scenarios; and (5) providing improved process automation and control capabilities.

Mathematical models can be classified somewhat loosely depending on the starting point in making a model. In observation-based models, the starting point is the experimental data from which a model is built. It is primarily empirical in nature. In contrast, the starting point for physics-based models is the universal physical laws that should describe the presumed physical phenomena. Physics-based models are also validated against experimental data, but in physics-based models the experimental data do not have to exist before the model. The decision on whether to build an observation-based or a physics-based model depends on a number of factors, including the need and available resources, as shown in Figure 1.1. After a model is built, its parameters can be varied to see their effects—this process is termed *parametric sensitivity analysis*. A model can also be used to control a process. These conceptual steps are also shown in Figure 1.1.

1.2 CLASSIFICATION OF MATHEMATICAL MODELING TECHNIQUES

Classification of mathematical models can be in many different dimensions (Gershenfeld 1999), as shown in Figure 1.2. As implied in this figure, there is a continuum between the two extremes for any particular dimension noted in this figure. For example, it can be argued that even a model that is obviously physics-based, such as a fluid flow in a porous media, has permeability as a parameter that is experimentally measured and is made up of many different parameters characterizing the porous matrix and the fluid. It is possible to use a lattice Boltzmann simulation for the same physical process that will not need most of these matrix and fluid parameters and, therefore, can be perceived as more fundamental.

The chapters in this text cover much of the range shown in Figure 1.2 for any particular dimension. Physics-based (first-principle-based) vs. data-driven models is the primary dimension along which the chapters are grouped. Scale of models is another dimension covered here. The lattice Boltzmann simulation in Chapter 2, for example, is at a smaller scale than the macroscale

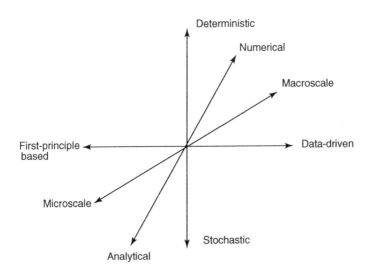

Figure 1.2 Various dimensions of a model. This is not an exhaustive list.

continuum models in Chapter 3 through Chapter 6. Another dimension is deterministic vs. stochastic. For example, the deterministic models in Chapter 3 through Chapter 6 can be made stochastic by following the discussion in Chapter 7. Analytical vs. numerical method of solution is another dimension of models. Numerical models have major advantages over analytical solution techniques in terms of being able to model more realistic situations. Thus, Chapter 2 through Chapter 5 cover mostly numerical solutions, although some references to analytical solutions are provided as well.

1.3 SCOPE OF THE HANDBOOK

Each chapter in this book describes a particular modeling technique in the context of food and bioprocessing applications. Entire books have been written on each of the chapters in this handbook. However, these books are frequently not with food and bioprocess as the main focus. Also, no one book covers the breadth of modeling techniques included here. The motivation behind this handbook was to bring many different modeling techniques, as varied as physics-based and observation-based models, under one umbrella with food and bioprocess applications as the focus. Because the end goal of even very different modeling approaches, such as physics-based and observation-based models, can be the same (e.g., to understand and optimize the system), any two modeling techniques can be conceptually thought of as competing alternatives. This is more so in food and bioprocess applications in which the processes are complex enough that the superiority of any one type of modeling technique in an industrial scenario that demands quick answer is far from obvious. Another reason for discussing various models under one roof is that different types of models can be pooled to obtain models that combine the respective advantages. Succinct discussion of each model in the same context of food and bioprocess can help trigger such possibilities. The modeling techniques selected in the handbook are either already being used or have a great potential in food and bioprocess applications. Emphasis has been placed on how to formulate food and bioprocess problems using a particular modeling technique, away from the theory behind the technique. Thus, the chapters are generally structured to have a short introduction to the modeling technique, followed by the details on how that technique can be used in specific food and bioprocess applications. Although optimization is often one of the major goals in modeling, optimization itself is a broad topic that could not be included (with the exception of linear programming) in this text because of its extensive coverage of modeling, and the reader is referred to the excellent article by Banga et al. (2003).

1.4 SHORT OVERVIEW OF MODELS PRESENTED IN THIS HANDBOOK

A short description of each type of model presented in this handbook is presented in this section. There is no such thing as the best model because the choice of a model depends on a number of factors, the most obvious ones being the goal (whether to know the detailed physics), the modeler's background (statistics vs. engineering or physics), and the time available (physics-based models typically take longer). Some of this is also noted in the schematic in Figure 1.1.

1.4.1 Physics-Based Models (Chapter 2 through Chapter 8)

Physics-based models follow from fundamental physical laws such as conservation of mass and energy and Newton's laws of motion; however, empirical (but fairly universal) rate laws are needed to apply the conservation laws at the macroscopic scale. For example, to obtain temperatures using a physics-based model, combine conservation of energy with Fourier's law (which is empirical) of heat conduction. The biggest advantages of physics-based models are that they provide insight into the physical process in a manner that is more precise and more trustable (because we start from universal conservation laws), and the parameters in such models are measurable, often using available techniques.

Physics-based models can be divided into three scales: molecular, macro, and meso (between molecular and macro). An example of a model at the molecular scale is the molecular dynamic model discussed later. Models such as the lattice Boltzmann model discussed in this book are in the mesoscale. Macroscopic models are the most common among physics-based models in food. Examples of macroscopic models are the commonly used continuum models of fluid flow, heat transfer, and mass transfer. As we expand food and biological applications at micro- or nanoscale, such as in detection of microorganisms in a microfluidic biosensor, scales will be approached where the continuum models in Chapter 2 through Chapter 5 will break down (Gad-el-Hak 2005). Similarly, in very short time scales, continuum assumption breaks down, and mesoscale or molecular scale models become necessary (Mitra et al. 1995). General discussion of models when continuum assumption breaks down can be seen in Tien et al. (1998).

Physics-based models today are less common in food and bioprocessing product, process, and equipment design than in some manufacturing, such as automobile and aerospace. This can be primarily attributed to variability in biomaterials and the complexities of transformations that food and biomaterials undergo during processing; however, this scenario is changing as the appropriate computational tools are being developed. In fact, the physics-based model (such as computational fluid dynamics, or CFD) is one of the areas in food process engineering experiencing rapid growth.

1.4.1.1 Molecular Dynamic Models

Molecular dynamic (MD) models are physics-based models at the smallest scale. In its most rudimentary version, repelling force between pairs of atoms at close range and attractive force between them over a range of separations are represented in a potential function (such as Lennard–Jones), for which there are many choices (Rapaport 2004). The spatial derivative of this potential function provides the corresponding force. Forces between one atom and a number of its neighbors are then added to obtain the combined force, and Newton's second law of motion is then used to obtain the acceleration from the force. This acceleration is then numerically integrated to obtain the trajectory describing the way the molecule would move. Physical properties of the system can be calculated as the appropriate time average over the trajectory, if it is of sufficient length. Although applications of molecular dynamics relevant to food processing (such as protein functionality and solution properties of carbohydrates) have been reported (Schmidt et al. 1994; Ueda et al. 1998), there appears to be very little ongoing work in applying MD to systems of direct relevance to food processing. Thus, MD has been excluded from this handbook.

1.4.1.2 Lattice Boltzmann Models (Chapter 2)

The lattice Boltzmann (LB) method is physics-based, but at an intermediate scale (referred to as mesoscale) between the molecular dynamic model mentioned above that is at the microscale and continuum models mentioned below that are at the macroscale, where physical quantities are assumed to be continuous. LB is based on kinetic theory describing the dynamics of a large system of particles. The continuum assumption breaks down at some point going from the macroscale toward the microscale. Examples of such systems can be colloidal suspensions, polymer solutions, and flow-through porous media. This is where the lattice Boltzmann model is useful and is currently being pursued in relation to food processes.

Other mesoscale simulations are also being used in food. For example, in Pugnaloni et al. (2005), large compression and expansion of viscoelastic protein films are studied in relation to stability of foams and emulsions during formation and storage.

1.4.1.3 Continuum Models (Chapter 3 through Chapter 6)

Continuum models presented in Chapter 3 through Chapter 6 primarily deal with transport phenomena, i.e., fluid flow, heat transfer, and mass transfer. These physics-based models are based on fundamental physical laws. Typically, these models consist of a governing equation that describes the physics of the process along with equations that describe the condition at the boundary of the system. The conditions at the boundary determine how the system interacts with the surroundings. Mathematically, they are needed to obtain particular solutions of the governing equation. The solution of the combined governing equation-boundary condition system can be made as exact an analog of the physical system as desired by including as much detail of the physical processes as necessary.

Physics-based models have several advantages over observation-based models: (1) they can be exact analogs of the physical process; (2) they allow in-depth understanding of the physical process as opposed to treating it as a black box; (3) they allow us to see the effect of changing parameters more easily; and (4) models of two different processes can share the same basic parameter (such as mass diffusivity and permeability measured for one process can be useful for other processes). The disadvantages of a physics-based model are as follows: (1) high level of specialized technical background is required; (2) generally more work is required to apply to real-life problems; and (3) often longer development time and more resources are needed.

In the past 10 years or so, physics-based continuum models have really picked up because of the available powerful and user-friendly software. These software programs do have limitations, however, that apply to food related problems because of complexities in the process and significant changes in the material due to processing. For example, rapid evaporation, as is true in baking, frying, and some drying operations, is hard to implement in most of these software. Also, these continuum models rely heavily on properties data that are only sparsely available for food systems.

There are other physics-based continuum models for which more details could not be included because of the scope of this handbook. For example, electromagnetic heating of food such as microwave and radio frequency heating is modeled using the governing Maxwell's equations, some details of which are provided in Chapter 3. Likewise, solid mechanics problems in food, such as during chewing, puffing, texture development, etc., are governed by the equations of solid mechanics, which also are not included in the book.

1.4.1.4 Kinetic Models (Chapter 7)

Kinetic models mathematically describe rates of chemical or microbiological reactions. They generally can be considered to be physics-based. However, in complex chemical and microbiological processes, as is true for food and bioprocesses, the mechanisms are generally hard to obtain and are not always available. The kinetic models for such systems are more data-driven than fundamental (as could be true for simple systems).

1.4.1.5 Stochastic Models (Chapter 8)

The physics-based continuum models have material properties that are typically measured. These models are often treated as deterministic ones, i.e., the parameter values are considered fixed. However, due to biological and other sources of variability, these measured parameters can have random variations. For example, viscosity of a sample can have random variation because of its biological variability. In a fluid flow model that uses viscosity, the final answer of interest, such as pressure drop, would also have the random fluctuations corresponding to the random variations in viscosity. Inclusion of such random variations makes the physics-based models more realistic. Techniques to include such uncertainty are presented in Chapter 6 and Chapter 8.

1.4.2 Observation-Based Models (Chapter 9 through Chapter 15)

The physics-based modeling process described in part I assumes that a model is known, which is frequently difficult to achieve in complex processes. Although a physics-based model may also be adjusted based on measured data, observation-based models (see Figure 1.3) are inferred primarily from measured data. Observational models are black box models to different degrees in relation to the physics of the process. The classical statistical models can have a model in mind (often based on some understanding of the process) before obtaining the measured data. This makes them less of a black box than models such as neural network or genetic algorithm that are frequently completely data driven; no prior assumption is made about the model and no attempt is made to physically interpret the model parameters once the model is built. Loosely speaking, though, all observational models are referred to as data-driven models. For this handbook (Figure 1.3), we separate the classical statistical models from the rest of the observation-based models and refer to the rest as data-driven models.

There are many practical situations in which time and resources do not permit a complete physics-based understanding of a process. Physics-based models often require more specialized training and/or longer development time. In some applications, detailed understanding provided by the physics-based model may not be necessary. For example, in process control, detailed physics-based models often are not needed, and observation-based models can suffice. Observation-based models can be extremely powerful in providing a practical, useful relationship between input and output parameters for complex processes. The types of data available and the purpose of modeling usually influence the kind of observation models to be used. General information on how to choose a model for a particular situation is hard to locate. An excellent Internet source guiding data-driven model choice and development can be seen in NIST (2005). Because observation-based models are built from data without necessarily considering the physics involved, use of such models beyond the range of data used (extrapolation) is more difficult than in the case of physics-based models.

1.4.2.1 Response Surface Methodology (Chapter 9)

This is a statistical technique that uses regression analysis to develop a relationship between the input and output parameters by treating it as an optimization problem. The principle of experimental design is used to plan the experiments to obtain information in the most efficient manner. Using experimental design, the most significant factors are found before doing the response surface and finding the optimum. This method is quite popular in food applications. It is important to note that finding the optimum using response surface is not limited to experimental data. Physics-based models can also be used to generate data that can be optimized using the response surface methodology similar to the method for experimental data (Qian and Zhang 2005).

1.4.2.2 Multivariate Analysis (Chapter 10)

Multivariate analysis (MVA) is a collection of statistical procedures that involve observation and analysis of multiple measurements made on one or several samples of items. MVA techniques are classified in two categories: dependence and interdependence methods. In a dependence technique, the dependent variable is predicted or explained by independent variables. Interdependence methods are not used for prediction purposes and are aimed at interpreting the analysis output to opt

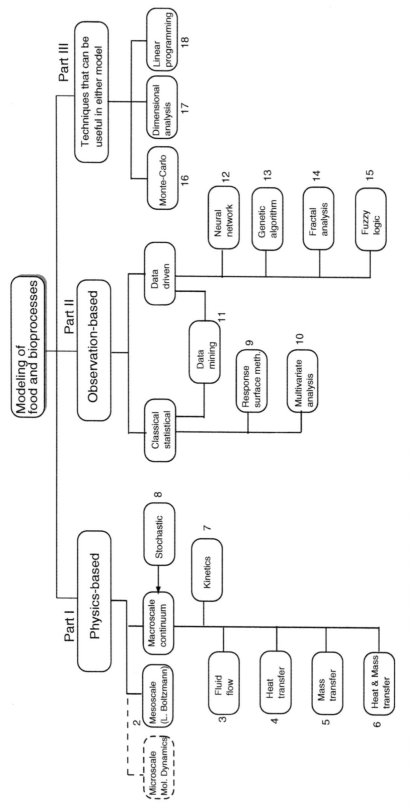

Figure 1.3   Various models presented in this handbook and their relationships.

for the best and most representative model. MVA is likely to be used in situations when one is not sure of the significant factors and how they interact in a complex process. It is also a popular modeling process in food.

1.4.2.3 Data Mining (Chapter 11)

Data mining refers to automatic searching of large volumes of data to establish relationships and identify patterns. To do this, data mining uses statistical techniques and other computing methodology such as machine learning and pattern recognition. Data mining techniques can also include neural network analysis and genetic algorithms. Thus, it can be seen as a meta tool that can combine a number of modeling tools.

1.4.2.4 Neural Network (Chapter 12)

An artificial neural network model (as opposed to a biological neural network) is an interconnected group of functions (equivalent to neurons or nerve cells in a biological system) that can represent complex input–output relationships. The power of neural networks lies in their ability to represent both linear and nonlinear relationships and in their ability to learn these relationships directly from the modeled data. Generally, large amounts of data are needed in the learning process.

1.4.2.5 Genetic Algorithms (Chapter 13)

Genetic algorithms are search algorithms in a combinational optimization problem that mimick the mechanics of the biological evolution process based on genetic operators. Unlike other optimization techniques such as linear programming, genetic algorithms require little knowledge of the process itself.

1.4.2.6 Fractal Analysis (Chapter 14)

Fractal analysis uses the concepts from fractal geometry. It has been primarily used to characterize surface microstructure (such as roughness) in foods and to relate properties such as texture, oil absorption in frying, or the Darcy permeability of a gel to the microstructure. Although fractal analysis may use some concepts from physics, the models developed are not first principle-based. Processes governed by nonlinear dynamics can exhibit a chaotic behavior that can also be modeled by this procedure. Applications to food have been only sporadic.

1.4.2.7 Fuzzy Logic (Chapter 15)

Fuzzy logic is derived from the fuzzy set theory that permits the gradual assessment of the membership of elements in relation to a set in contrast to the classical situation where an element strictly belongs or does not belong to a set. It seems to be most successful for the following: (1) complex models where understanding is strictly limited or quite judgmental; and (2) processes in which human reasoning and perception are involved. In food processing, the applications have been in computer vision to evaluate food quality, in process control, and in equipment selection.

1.4.3 Some Generic Modeling Techniques (Chapter 16 through Chapter 18)

Included in this part of the book are three generic modeling techniques that are somewhat universal and can be used in either physics-based or observation-based model building or for optimization once a model is built.

1.4.3.1 *Monte-Carlo Technique (Chapter 16)*

Monte Carlo refers to a generic approach whereby a probabilistic analog is set up for a mathematical problem, and the analog is solved by stochastic sampling. Chapter 7 shows the application of this technique to physics-based models.

1.4.3.2 *Dimensional Analysis (Chapter 17)*

This is typically an intermediate step before developing mostly physics-based (but can be data-driven) models that is used to reduce the number of variables in a complex problem. This can reduce the computational or experimental complexity of a problem.

1.4.3.3 *Linear Programming (Chapter 18)*

This is a well-known technique that is used for the optimization of linear models. It can be used in the context of a physics-based or a data-driven model.

1.4.4 Combining Models

Various modeling approaches can be combined to develop models that are even closer to reality and that have greater predictive power. For example, a physics-based model can be combined with an observation-based model by treating the output from the physics-based model as analogous to experimental data. See, for example, Eisenberg (2001) or work in a different application (Sudharsan and Ng 2000). Such a combined model is useful when only a portion of the system can be represented using a physics-based model or when the parameters in the physics-based model are uncertain. Two or more observation-based modeling techniques can also be combined (e.g., Panigrahi 1998), which is sometimes referred to as a hybrid model. A challenge, however, is to combine diverse methods in a seamless manner to provide a model that is easy to use.

1.5 CHARACTERISTICS OF FOOD AND BIOPROCESSES

Some characteristics of food and bioprocesses are as follows: (1) they often involve drastic physical, chemical, and biological transformation of the material, during processing. Many of these transformations have not been characterized, primarily because of the following: (1) such a large variety of possible materials; (2) their biological origin, variabilities are significant, even in the same material; (3) because the material contains large amounts of water, unless temperatures are low, there is always evaporation in the food matrix. This evaporation is hard to handle in physics-based models and increases complexity of the process; and (4) many food processes involve coupling of different physics (e.g., microwave heating involves heat transfer and electromagnetics), thus compounding complexities. As novel processing technologies are introduced and combination technologies such as hurdle technology become more popular, complexities will only increase in the future.

The industry in this area is characterized by a lower profit level and less room for drastic changes, than, for instance, automotive and aerospace industries. This translates to lower investment in research and development, which in turn leads to the generally lower level of technical sophistication as compared to other industries. Modeling, particularly physics-based modeling, often requires time and resources that are not available in the food industry. Consequently, with the exception of a handful of large multinational companies, modeling in general and physics-based modeling in particular are viewed as less critical and somewhat esoteric. It is expected that as the computer technology continues to advance, modeling (particularly physics-based modeling) will become easier and perhaps more of a viable alternative in the industry.

ACKNOWLEDGMENTS

Author Datta greatly acknowledges discussions with Professor James Booth of the Department of Biological Statistics and Computational Biology, Professor John Brady of the Department of Food Science, Professor Jean Hunter of the Department of Biological and Environmental Engineering, and Mr. Parthanil Roy of the School of Operations Research and Industrial Engineering, all of Cornell University.

REFERENCES

Banga, J. R., Balsa-Canto, E., Moles, C. G., and Alonso, A. A., Improving food processing using modern optimization methods, *Trends in Food Science and Technology*, 14, 131–144, 2003.

Eisenberg, F. G., Virtual experiments using computational fluid dynamics. *Proceedings of 7th Conference on Food Engineering*, American Institute of Chemical Engineers, New York, 2001.

Gad-el-Hak, M., Liquids: The holy grail of microfluidic modeling, *Physics of Fluids*, 17, 1–13, 2005.

Gershenfeld, N., *The Nature of Mathematical Modeling*, Cambridge: Cambridge University Press, 1999.

Mitra, K., Kumar, S., Vedavarz, A., and Moallemi, M. K., Experimental evidence of hyperbolic heat conduction in processed meat, *Journal of Heat Transfer, Transactions of the ASME*, 117(3), 568–573, 1995.

NIST. 2005. NIST/SEMATECH e-Handbook of Statistical Methods, http://www.itl.nist.gov/div898/handbook/pmd/pmd.htm.

Panigrahi, S., Neuro-fuzzy systems: Applications and potential in biology and agriculture, *AI Applications*, 12(1–3), 83–95, 1998.

Pugnaloni, L. A., Ettelaie, R., and Dickinson, E., Brownian dynamics simulation of adsorbed layers of interacting particles subjected to large extensional deformation, *Journal of Colloid and Interface Science*, 287, 401–414, 2005.

Qian, F. P. and Zhang, M. Y., Study of the natural vortex length of a cyclone with response surface methodology, *Computers and Chemical Engineering*, 29(10), 2155–2162, 2005.

Rapaport, D. C., *The Art of Molecular Dynamics Simulation*, Cambridge: Cambridge University Press, 2004.

Schmidt, R. K., Tasaki, K., and Brady, J. W., Computer modeling studies of the interaction of water with carbohydrates, *Journal of Food Engineering*, 22(1–4), 43–57, 1994.

Sudharsan, N. M. and Ng, E. Y. K., Parametric optimization for tumor identification bioheat equation using ANOVA and the Taguchi method. Proceedings of the IMechE, Part H, *Journal of Engineering in Medicine*, 214(H5), 505–512, 2000.

Tien, C.-L., Majumdar, A., and Gerner, F. M., *Microscale Energy Transport*, Washington, DC: Taylor & Francis, 1998.

Ueda, K., Imamura, A., and Brady, J. W., Molecular dynamics simulation of a double-helical b-Carrageenan hexamer fragment in water, *The Journal of Physical Chemistry A*, 102(17), 2749–2758, 1998.

PART I

Physics-Based Models

CHAPTER 2

Lattice Boltzmann Simulation of Microstructures

R. G. M. van der Sman

CONTENTS

2.1   INTRODUCTION

The lattice Boltzmann method is a recently developed computer modeling methodology that is gaining attention in the academic world, especially for the simulation of complex fluid phenomena at the mesoscopic scale.[1,2] The mesoscop ic scale lies between the molecular (micro) scale, and the macroscopic scale, where physical quantities are assumed to be continuous. Somewhere between the micro and macroscale, the continuum approach breaks down and some parts of physical systems cannot be assumed to be continuous. Examples of these mesoscale systems are emulsions, colloidal suspensions, flow in porous media, and polymer solutions.

Many processed foods can be viewed as such complex fluids. Often during processing, one wants to control the mesoscale structure of the foods because it significantly influences the texture of the food. Lattice Boltzmann is a powerful simulation tool to probe microstructural development during processing. This chapter will discuss (1) the basics of lattice Boltzmann, (2) the principles of lattice Boltzmann models for complex fluids as emulsions and suspensions, and (3) applications of these models in research projects in our lab, focusing on the use of microtechnology to control food microstructure.[3-5]

2.2   THE BASICS OF LATTICE BOLTZMANN

2.2.1   Discretising Kinetic Theory

The lattice Boltzmann method started as a novel modeling methodology for the description of physical transport phenomena such as fluid flow[6] and (convection) diffusion,[7,8] the domain of traditional finite-volume or finite-element modeling methods. These traditional methods are based on the continuum approach, and can be viewed as the discretization of partial differential equations such as the Navier–Stokes equation or the Fourier equation. In contrast, the lattice Boltzmann method is based on kinetic theory—the physical theory describing the dynamics of large systems of particles.

In principle, all equations governing physical transport phenomena at the macroscopic (continuum) scale can be derived from classical kinetic theory. However, if applied to real-sized molecules, the classical kinetic theory can only be applied to small-sized systems. Currently, computing power is still insufficient to investigate problems at the process engineering scale. However, by taking a discrete version of kinetic theory (and in particular the Boltzmann equation) the lattice Boltzmann method has been proven to be capable of simulating physical transport phenomena at the macroscale, even with moderate computing power.

The lattice Boltzmann method discretizes kinetic theory as follows:

- (Fluid) mass is collected in discrete (lattice-gas) particles.
- Particles are located on points of a regular lattice.
- Particles move according to a finite, discrete set of velocities, taking them to adjacent lattice points in discrete time steps.

Therefore, in the world of lattice Boltzmann space, time and *particle velocity* are discrete variables. Particles encountering each other at a lattice point will collide with each other, after which they will propagate to an adjacent lattice site. This two-step process of streaming and collision is depicted in Figure 2.1. This very simple view of physics can model real physical phenomena, if one adheres to physical conservation laws and if the lattice has sufficient symmetry.

In analogy with the Boltzmann equation of the classical kinetic theory, the system of particles is described by a particle velocity distribution functions $f(\mathbf{x},c_i,t)=f_i(\mathbf{x},t)$, which represents the mass density of lattice gas particles located at lattice point $\mathbf{x}$ at time t and moving with velocity c_i. At a lattice site, particles can have different velocities that are chosen from a finite set $\{c_i\}$. These are such that the particles move to adjacent lattice sites in exactly one time step.

The connection of this particle view to the macroscopic view of continuous fields is obtained by taking "moments" of the particle velocity distribution function. The total mass density of particles at lattice point $\mathbf{x}$, $\rho(\mathbf{x},t)$, is obtained by summing over all directions i:

$$\rho(\mathbf{x},t) = \sum_i f_i(\mathbf{x},t). \tag{2.1}$$

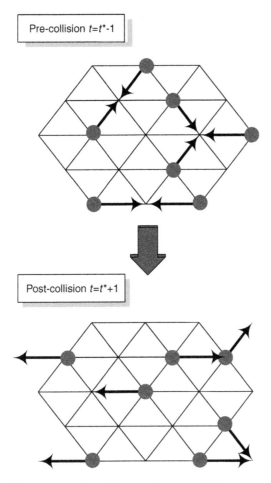

Figure 2.1   Lattice gas particles moving and colliding on a hexagonal lattice.

This is the zeroth order moment of the particle distribution function. In the case of lattice Boltzmann schemes for fluid flow, the average fluid velocity u is obtained from the first-order moment of the particle distribution function. Therefore, the particle momenta are summed over all directions i:

$$\rho(\mathbf{x},t)u(\mathbf{x},t) = \sum_i c_i f_i(\mathbf{x},t). \tag{2.2}$$

Note that the average fluid velocity, which in kinetic theory is called the *drift velocity*, is the net result of particles moving in various directions. On average, the particles are moving in a certain direction, which becomes the fluid flow direction.

As lattice Boltzmann originates from the field of statistical physics, most of the scientific literature assumes a physics background.[2] In general, food engineers lack a sound training in statistical physics. In this section, the basics of the lattice Boltzmann method are reviewed with the assumption that the readers have only an engineering background. First, kinetic theory is introduced using a lattice Boltzmann scheme for the simplest physical transport phenomenon: one-dimensional diffusion. To demystify the statistical mechanics picture of lattice Boltzmann, the equivalence of a special type of lattice Boltzmann scheme with finite-volume schemes is shown for one-dimensional convection-diffusion. This section is concluded with the description of lattice Boltzmann for fluid flow, and its related boundary conditions.

2.2.2 1-D Diffusion

The basics of the lattice Boltzmann scheme are best illustrated with the simplest phenomenon it can describe: one-dimensional (1-D) diffusion. At the macroscopic (or continuum) scale, 1-D diffusion is described by the following partial differential equation:

$$\frac{\partial \rho}{\partial t} = D\frac{\partial^2 \rho}{\partial x^2} \tag{2.3}$$

In this simple case, it is sufficient to assume that the lattice gas particles have either positive ($c_1 = +c$) or negative velocity ($c_2 = -c$), moving them either to the right or left neighboring lattice points (see Figure 2.2). Their respective particle mass densities are denoted as f_1 and f_2. The magnitude of their velocities, c, is such that $c = \Delta x/\Delta t$, where Δx is the lattice spacing and Δt is the time step. This type of lattice is denoted as D1Q2. In the following sections, lattice types will be denoted as DdQq,[6] where d is the dimension and q is the number of velocities in the velocity set $\{c_i\}$.

During collision, a fraction (α) of the lattice gas particles collides and reverses velocity. One should imagine that the collisions occur with a background fluid in which the diffusing particles are dissolved. The background fluid does not need to be modeled explicitly. Using the above defined scattering rate α, the collision rules can be written to express the post-collision distribution functions f_i' in terms of the pre-collision ones (f_i):

$$\begin{aligned}
f_1' &= (1-\alpha)f_1 + \alpha f_2, \\
f_2' &= (1-\alpha)f_2 + \alpha f_1.
\end{aligned} \tag{2.4}$$

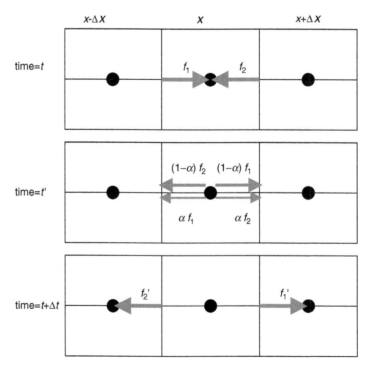

Figure 2.2 Collision and streaming on a D1Q2 lattice.

Note that this collision conserves the number of particles: $f_1 + f_2 = f_1' + f_2'$; this is a prerequisite for diffusion phenomena.

The collision rules can also be written in matrix notation, where Λ represents the scattering matrix:

$$f_i'(x,t) = \sum_j \Lambda_{ij} f_j(x,t). \tag{2.5}$$

Here, $\Lambda_{ij} = 1 - \alpha$ for $i = j$, whereas $\Lambda_{ij} = \alpha$ for $i \neq j$.

The velocity distribution that is invariant under collision, is by definition the equilibrium velocity distribution f_i^{eq}:

$$f_i^{eq} = \sum_j \Lambda_{ij} f_j^{eq}. \tag{2.6}$$

For the D1Q2 lattice, the equilibrium distribution function is simply a weighted function of the density:

$$f_i^{eq} = \frac{1}{2}\rho = \frac{1}{2}(f_1 + f_2), \tag{2.7}$$

where $\rho = \Sigma_i f_i$ is the density of lattice gas particles at the corresponding lattice point.

As the equilibrium distribution is invariant under collision, it is an eigenvector of the scattering matrix Λ_{ij} with eigenvalue $\lambda_0 = 1$. For the general case, it holds that in lattice Boltzmann schemes the distribution functions can always be decomposed in a linear combination of eigenvectors of Λ_{ij}. The number of eigenvectors is always equal to the number of velocities in the set $\{c_i\}$. Hence, for the simple case of diffusion on a D1Q2 lattice, the distribution function f_i may be decomposed into an equilibrium distribution, f_i^{eq}, and a nonequilibrium component, f_i^{neq}:

$$f_i(x,t) = f_i^{eq}(x,t) + f_i^{neq}(x,t), \tag{2.8}$$

where the nonequilibrium distribution function is given by:

$$f_i^{neq} = \frac{1}{2c}(f_1 - f_2)c_i = \pm\frac{1}{2}(f_1 - f_2). \tag{2.9}$$

Multiplication of the nonequilibrium distribution with the scattering matrix shows that it is indeed an eigenvector:

$$\sum_j \Lambda_{ij} f_j^{neq} = (1 - 2\alpha)f_i^{neq}, \tag{2.10}$$

with an eigenvalue of $\lambda_1 = (1 - 2\alpha)$. The fact that the absolute value ($|\lambda_1|$) is smaller than unity shows that the collision process tends to decrease the perturbations from equilibrium; in other words, the process tends to drive the system towards equilibrium.

After the collision step, the particles propagate (stream) to the neighbouring lattice points:

$$f_i(x + \Delta x_i, t + \Delta t) = f_i'(x,t). \tag{2.11}$$

Here, the lattice spacing is linked to the particle velocity: $\Delta x_i = c_i \Delta t$.

The two-step process of collision and propagation can be combined into a single equation:

$$f_i(x + \Delta x_i, t + \Delta t) = \sum_j \Lambda_{ij} f_j(x,t). \tag{2.12}$$

Normally the governing equation of the lattice Boltzmann scheme is written in another form. First, the pre-collision distribution is split into equilibrium and nonequilibrium components. Using the fact that the equilibrium is invariant under collision (see Equation 2.6),

$$f_i(x + \Delta x_i, t + \Delta t) = f_i^{eq}(x,t) + \sum_j \Lambda_{ij} f_j^{neq}(x,t). \tag{2.13}$$

Second, the scattering matrix is replaced by the operator: $\Omega = I - \Lambda$, i.e., $\Omega_{ij} = \delta_{ij} - \Lambda_{ij}$. Here, δ_{ij} is the Kronecker delta, with $\delta_{ij} = 1$ for $i = j$, and $\delta_{ij} = 0$ for $i \neq j$. The governing equation of the lattice Boltzmann scheme then becomes:

$$f_i(x + \Delta x_i, t + \Delta t) - f_i(x,t) = -\sum_j \Omega_{ij} f_j^{neq}(x,t). \tag{2.14}$$

Linear algebra states that the eigenvectors of Λ_{ij} are also eigenvectors of Ω_{ij}, with eigenvalues $\omega_n = 1 - \lambda_n$. Note that the eigenvalue of the equilibrium distribution function becomes $\omega_0 = 0$, whereas that of the nonequilibrium part of the distribution function becomes $\omega_1 = 2\alpha = \omega$. Hence, the scheme can simply be described by

$$f_i(x + \Delta x_i, t + \Delta t) - f_i(x,t) = -\omega f_i^{neq}(x,t) \tag{2.15}$$

This last equation has a form similar to the governing equation of the classical kinetic theory—the Boltzmann equation—hence the name *lattice Boltzmann method*. Equation 2.15 is consequently called the *lattice Boltzmann equation*.

Not only for the simple case of 1-D diffusion, but also for the more general cases, one finds that the governing equations of the lattice Boltzmann method are discretized versions of the corresponding continuum Boltzmann equations. This is made evident by comparing the above lattice Boltzmann equation with a special (single relaxation-time) approximation of the classical Boltzmann equation:

$$\frac{\partial f(x,t)}{\partial t} + c \cdot \nabla f = -\omega[f(x,t) - f^{eq}(x,t)]. \tag{2.16}$$

This approximation is due to Bhatnagar, Groos, and Krook, and the lattice Boltzmann equivalent is named the *lattice BGK equation*:

$$f_i(x + \Delta x_i, t + \Delta t) - f_i(x,t) = -\omega[f_i(x,t) - f_i^{eq}(x,t)]. \tag{2.17}$$

The left-hand side of the lattice BGK equation is a discretization of the left-hand side of the Boltzmann equation using Euler forward time discretization and upwind spatial discretization.

In classical kinetic theory, the equilibrium distribution function (of an ideal gas) is the well-known Maxwell–Boltzmann distribution:

$$f^{eq}(c) = \frac{\rho}{\left(2\pi c_s^2\right)^{D/2}} \exp\left[-\frac{(c-u)^2}{2c_s^2}\right] \tag{2.18}$$

where D the dimension of the system.

Notice that this is a Gaussian distribution of the particle velocities, with average velocity u, and with variance c_s, which has the physical meaning of the speed of sound of an ideal gas and relates pressure to density: $p = \rho c_s^2$, (the ideal gas law). The Maxwell–Boltzmann distribution is normalized such that the total density of particles is equal to ρ.

As in kinetic theory, the resulting physics in the lattice Boltzmann method is governed by the relaxation parameters Ω_{ij} (determining the values of the transport coefficients like diffusivity or viscosity), and by the explicit form of the equilibrium distribution (determining the type of transport phenomenon). It is shown,[9] that the lattice Boltzmann method describes the proper physics if the moments of the equilibrium distribution equal those of the classical Maxwell–Boltzmann distribution (up to a certain order).

Theory shows that, for (convection) diffusion, the moments up to second order must be satisfied.[10,7,8] In 1-D, these moments are

$$M_0 = \sum_i f_i^{eq} = \int f^{eq}(c)dc = \rho,$$

$$M_1 = \sum_i c_i f_i^{eq} = \int c f^{eq}(c)dc = \rho u, \qquad (2.19)$$

$$M_2 = \sum_i c_i^2 f_i^{eq} = \int c^2 f^{eq}(c)dc = \rho c_s^2 + \rho u^2.$$

For diffusion, the average velocity, u, is zero, and the variance of the equilibrium distribution is c_s. For convection diffusion, u is externally imposed. For (convection) diffusion, the variance, c_s, does not have any physical meaning, and can be used as a free parameter to optimize lattice Boltzmann schemes. For fluid dynamics, it has the meaning of the speed of sound.

Using the formalism of matching the moments of the equilibrium distribution, f_i^{eq}, to those of the Maxwell–Boltzmann distribution, one can take a very natural and gradual build-up of complexity of physical transport phenonema to be modelled with lattice Boltzmann. The example discussed here started with diffusion; the next step is convection-diffusion. For convection diffusion, moments up to second order must be satisfied, but now they are also velocity dependent. From convection diffusion, it is a small step towards fluid flow. Then, in addition to the second-order moments that are identical to those of convection-diffusion, the third-order moments must also be satisfied.

The relations for the transport coefficients (i.e., diffusivity or viscosity) are derived using the Chapman–Enskog expansion,[2,8] which is an advanced mathematical procedure originating from kinetic theory. For the simple case of diffusion, the diffusivity is

$$D = c_s^2 \left(\frac{1}{\omega} - \frac{1}{2} \right) \Delta t. \qquad (2.20)$$

For fluid flow (see Section 2.4), the expression for the kinematic viscosity is very similar: $\eta = c_s^2 \left(\frac{1}{\omega} - \frac{1}{2} \right) \Delta t.$

2.2.3 Equivalence with Finite-Volume Schemes

The lattice Boltzmann method can also be viewed as a special kind of finite-volume scheme.[11–13] For fluid flow, this perspective does not follow quite immediately. However, for more simple phenomena such as diffusion and convection diffusion, this relation is very evident.[12,13] Here, this is illustrated for a lattice Boltzmann scheme for convection diffusion. Instead of the D1Q2 lattice introduced above, the D1Q3 lattice—a 1-D lattice with rest particle particles in addition to particles moving to the right or left—is now needed. Therefore, the velocity set is $c_i = \{0, +1, -1\}$. Rest particles will be indicated with index $i = 0$.

One finds a straightforward equivalence with a finite-volume scheme for a special choice of the relaxation parameter: $\omega = 1$. This value is inserted into the lattice BGK scheme:

$$f_i(x + \Delta x_i, t + \Delta t) = f_i^{eq}(x,t). \qquad (2.21)$$

The equilibrium distribution follows directly from the above constraints on its moments (Equation 2.19), giving

$$f_i^{eq} = w_i\rho\left[1 + \frac{c_i u}{c_s^2} + \frac{u^2}{c_s^2}\right], \tag{2.22}$$

where $w_i = c_s^2/2c^2$ for $i=1, 2$, and $w_0 = 1 - w_1 - w_2$. Note that the equilibrium distribution for only diffusion is obtained by setting $u=0$.

By applying the Gauss theorem, a finite-difference equation is derived for the mass density that may be compared to traditional discretization schemes. The Gauss theorem relates the change in mass over time to the net mass flux, integrated over the surface of the volume element (i.e., the lattice cell) that is being considered. Hence, it is a formulation of the mass balance:

$$\rho(x,t) - \rho(x,t - \Delta t) = f_2^{eq}(x + \Delta x,t) - f_1^{eq}(x,t) + f_1^{eq}(x - \Delta x,t) - f_2^{eq}(x,t). \tag{2.23}$$

Here, $\Delta x = c\Delta t$ has been used.

Substitution of the expression for the local equilibrium distribution function results in

$$\rho(x,t) - \rho(x,t - \Delta t) = w_1[\rho(x + \Delta x,t) - \rho(x,t) + \rho(x - \Delta x,t) - \rho(x,t)] - w_1\frac{cu}{c_s^2}[\rho(x + \Delta x,t)$$

$$- \rho(x - \Delta x,t)] + w_1\frac{u^2}{c_s^2}[\rho(x + \Delta x,t) - \rho(x,t) + \rho(x - \Delta x,t) - \rho(x,t)]. \tag{2.24}$$

Recognize the finite difference stencils in the above equation, for the Euler forward time derivative, $\mathcal{D}_t\rho(x,t)$, and the first- and second-order spatial derivatives following central differencing, $\mathcal{D}_x\rho(x,t)$ and $\mathcal{D}_x^2\rho(x,t)$. They read:

$$\mathcal{D}_t\rho(x,t) = \frac{\rho(x,t) - \rho(x,t - \Delta t)}{\Delta t} \tag{2.25}$$

$$\mathcal{D}_x\rho(x,t) = \frac{\rho(x + \Delta x,t) - \rho(x - \Delta x,t)}{2\Delta x} \tag{2.26}$$

$$\mathcal{D}_x^2\rho(x,t) = \frac{\rho(x + \Delta x,t) + \rho(x - \Delta x,t) - 2\rho(x,t)}{\Delta x^2}. \tag{2.27}$$

Using $w_1 = c_s^2/2c^2$ and $c = \Delta x/\Delta t$, $D = c_s^2\Delta t/2$ results in:

$$\mathcal{D}_t\rho(x,t) + u\mathcal{D}_x\rho(x,t) = \left(D + \frac{u^2}{2}\right)\mathcal{D}_x^2\rho(x,t). \tag{2.28}$$

It follows that the above finite-difference equation for the evolution of the density, Equation 2.28, is identical to the Lax–Wendroff finite-volume scheme.[12] This scheme is obtained by central differencing of first- and second-order spatial derivatives, and a second order Taylor expansion of the Euler forward time derivative. In this expansion, the second-order time derivative is replaced with second-order spatial derivatives (as follows from the convection-diffusion equation), thereby eliminating numerical diffusion. Without this correction of the time derivative, the finite-volume scheme would be overly dispersive, leading to significant, and even unstable, numerical oscillations. Evidently, the lattice Boltzmann scheme automatically corrects for numerical diffusion as a consequence of the constraints for the equilibrium distribution, Equation 2.19.

In the more general case of $\omega > 1$, the Chapman–Enskog expansion shows that the nonequilibrium component of the distribution function is related to the density gradient, $\partial\rho/\partial x$.[10,8] From a finite-volume viewpoint, one can state that in the lattice Boltzmann method, the density gradient is added as an extra state variable at a grid cell, in addition to the density, which is normally the only state variable that is used in traditional finite-volume methods. It is probable that lattice Boltzmann is related to a special finite-volume scheme that uses Hermite interpolation. However, such finite-volume schemes are very rare. Finite-volume schemes are usually based on Lagrangian interpolation.

The use of gradients as state variables makes lattice Boltzmann effectively a higher-order scheme. For $\omega > 1$ and $c_s^2 = c^2/3$, the lattice Boltzmann scheme for convection-diffusion is third-order accurate in spatial dimensions. Furthermore, it is virtually without numerical diffusion or dispersion (which plague either upwind or Lax–Wendroff finite-volume schemes), and is stable even for high grid-Peclet numbers ($Pe^* \approx 1000$, provided that density gradients are moderate).

2.2.4 Fluid Flow

The lattice Boltzmann scheme for fluid flow also follows from the constraint that the moments of the equilibrium distribution are equal to those of the classical Maxwell–Boltzmann distribution for an ideal gas.[9] Lattice Boltzmann normally operates in the weakly-compressible limit (simulating incompressible flow with an ideal gas at low Mach numbers, $Ma = u/c_s \ll 1$), and the moments must be satisfied up to third order. The constraints on the moments are:[9]

$$\sum_i f_i^{eq} = \rho,$$

$$\sum_i c_{i,\alpha} f_i^{eq} = \rho u_\alpha,$$

$$\sum_i c_{i,\alpha} c_{i,\beta} f_i^{eq} = \rho c_s^2 \delta_{\alpha\beta} + \rho u_\alpha u_\beta, \tag{2.29}$$

$$\sum_i c_{i,\alpha} c_{i,\beta} c_{i,\gamma} f_i^{eq} = \rho u_\alpha u_\beta u_\gamma + \rho c_s^2 (u_\alpha \delta_{\beta\gamma} + u_\beta \delta_{\alpha\gamma} + u_\gamma \delta_{\alpha\beta}).$$

Here, α, β, and γ indicate the Cartesian components of the particle velocities, c_i, and the fluid flow velocity, u.

Note that the zeroth-order moment is the density, ρ, the first-order moments are the components of the momentum density, ρu_α, and the second-order moments are the components of the momentum flux tensor, $\rho c_s^2 \delta_{\alpha\beta} + \rho u_\alpha u_\beta$, with the pressure proportional to the density as stated in the ideal gas law: $p = \rho c_s^2$. Here, c_s is the speed of sound in the lattice gas. The Kronecker delta, $\delta_{\alpha\beta}$, again indicates the components of the unit tensor. The third-order moments are related to the (kinetic) energy fluxes. Third-order moments only have to be satisfied for fluid flow. However, in most lattice Boltzmann schemes, the third-order term in the velocity, $\rho u_\alpha u_\beta u_\gamma$ is omitted, as the scheme is operated at low Mach numbers and the third-order term is negligible. Further note that the constraints for (convection)-diffusion are a subset of the constraints for fluid flow.

The equilibrium distribution satisfying the above constraints, Equation 2.29, can be expressed (using the Einstein convention of summation over repeated indices) as:[6]

$$f_i^{eq} = w_i \rho \left[1 + \frac{c_{i,\alpha} u_\alpha}{c_s^2} + \frac{(c_{i,\alpha} u_\alpha)^2}{2c_s^4} - \frac{u_\alpha u_\alpha}{2c_s^2} \right]. \tag{2.30}$$

The D2Q9 and D3Q19 lattices are most commonly used; their velocity sets are depicted in Figure 2.3. From the isotropy of viscosity, it follows that the speed of sound should be equal to $c_s^2 = c^2/3$. Note that for fluid flow, particles propagating to next-nearest neighbor sites are also needed. The weight factors are listed in Table 2.1. Note that the value of the weight factors only depends on the magnitude of the particle velocity, $\|c_i\|$.

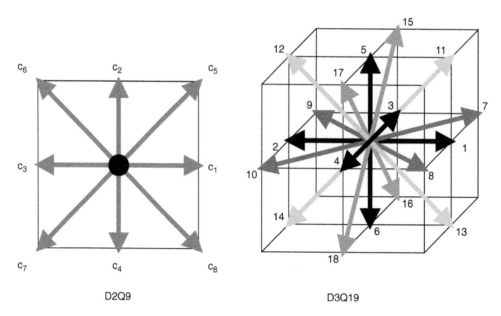

Figure 2.3 2-D and 3-D lattices commonly used for fluid flow.

2.2.5 Boundary Conditions

Fluid flow can be driven either by (1) moving boundaries, or (2) pressure differences. The fluid flow can also meet stationary (nonmoving) boundaries, for which the no-slip boundary condition must be applied. The implementation of these boundary conditions for the case of a D2Q9 lattice will be briefly discussed.

Pressure Boundary Conditions. In the case of a system with one inlet and outlet, both of the same size, it is very convenient to apply pressure-periodic boundary conditions, as introduced by Inamuro.[14] Contrary to pure periodic boundary conditions, a fraction of the lattice gas particles will be reflected, thereby creating a pressure difference $\Delta p = \Delta \rho c_s^2$. For a problem with the inlet and outlet at $x=1$ and $x=NX$, respectively, the boundary conditions are:

$$
\begin{aligned}
f_1(x = 1,t + \Delta t) &= f_1'(x = NX,t) + w_1\Delta\rho \\
f_5(x = 1,t + \Delta t) &= f_5'(x = NX,t) + w_5\Delta\rho \\
f_8(x = 1,t + \Delta t) &= f_8'(x = NX,t) + w_8\Delta\rho \\
f_3(x = NX,t + \Delta t) &= f_3'(x = 1,t) - w_3\Delta\rho \\
f_6(x = NX,t + \Delta t) &= f_6'(x = 1,t) - w_6\Delta\rho \\
f_7(x = NX,t + \Delta t) &= f_7'(x = 1,t) - w_7\Delta\rho.
\end{aligned}
\tag{2.31}
$$

Table 2.1   Weight Factors, w_h for D2Q9 and D3Q19 Lattices

$\|c_i\|^2$	D2Q9	D3Q19
0	4/9	1/3
1	1/9	1/18
2	1/36	1/36

For systems that are not periodic, other types of pressure boundary conditions must be formulated. Such conditions are developed by Zou and He.[15]

As the pressure boundary condition is a Dirichlet boundary condition, it is more convenient to prescibe it at the lattice site. Where the density and the momentum density are always fully determined, i.e.,

$$\rho = \sum_i f_i,$$

$$\rho u_x = \sum_i f_i c_{i,x}, \tag{2.32}$$

$$\rho u_x = \sum_i f_i c_{i,x},$$

the inlet and outlet pressures are given via the boundary conditions, and determine the respective densities ρ_0 and ρ_1. Furthermore, a zero vertical-velocity component at inlet and outlet must be enforced.

At the inlet, f_1, f_5, and f_8 are unknown. The velocity can be obtained from the known particle distribution functions by elimination of $f_1 + f_5 + f_8$ from Equation 2.32:

$$\rho_0 u_x = \rho_0 c_x - c_x(f_0 + f_2 + f_4 + 2(f_3 + f_6 + f_7)). \tag{2.33}$$

Note that f_6 and f_7 contain information on the viscous stress tensor, $S_{\alpha\beta} = \nu(\partial_\alpha u_\beta + \partial_\beta u_\alpha)$, which should be preserved. The particle distribution normal to the inlet does not contain that information. Therefore,

$$f_1 - f_3 = \frac{2w_1 c_x}{c_s^2} \rho_0 u_x \tag{2.34}$$

may be applied. Now the unknown populations are fully defined. Using the values of w_i gives

$$f_1 = f_3 + \frac{2\rho_0 u_x}{3c},$$

$$f_5 = f_7 - (f_2 - f_4)/2 + \frac{\rho_0 u_x}{6c}, \tag{2.35}$$

$$f_8 = f_6 + (f_2 - f_4)/2 + \frac{\rho_0 u_x}{6c}.$$

At the outlet, the followed may be obtained in a similar fashion:

$$f_3 = f_1 - \frac{2\rho_1 u_x}{3c},$$

$$f_6 = f_8 - (f_2 - f_4)/2 + \frac{\rho_1 u_x}{6c}, \tag{2.36}$$

$$f_7 = f_5 + (f_2 - f_4)/2 + \frac{\rho_1 u_x}{6c}.$$

Velocity Boundary Conditions at Inlet and Outlet. Prescribing a velocity at the inlet or outlet is quite similar to prescribing the pressure at the boundary. The inlet and outlet are again thought to be located at the lattice sites, and the velocity field at inlet or outlet has only a horizontal component. Therefore, as in the case of pressure boundary conditions, three constraints are observed: $\rho = \sum_i f_i$, $\rho u_x = \sum_i f_i c_{ix}$, $\rho u_y = \sum_i f_i c_{iy} = 0$. However, now the density, ρ, is unknown, and u_x is known from the prescribed velocity profile (a parabolic one in the case of Poisseuille flow). As for the pressure boundary conditions, the above constraints are not sufficient to close the set of equations. Consequently, the density is also taken to be equal to the standard (or average) density, ρ_0, because we work in the weakly compressible limit. At the inlet, the boundary conditions again become:

$$f_1 = f_3 + \frac{2\rho_0 u_x}{3c},$$

$$f_5 = f_7 - (f_2 - f_4)/2 + \frac{\rho_0 u_x}{6c}, \qquad (2.37)$$

$$f_8 = f_6 + (f_2 - f_4)/2 + \frac{\rho_0 u_x}{6c}.$$

At the outlet, the boundary conditions are identical to Equation 2.36.

No-Slip Boundary Conditions. Following the same line of reasoning, no-slip boundary conditions at (moving) walls, as introduced by Ladd,[16] may be derived. Consider Couette flow, with top and bottom walls moving at velocity $\pm u_x$. In lattice Boltzmann, a natural boundary between fluid and solid is halfway lattice sites, coinciding with the boundary of the Wigner–Seitz cell of the lattice. In the example of Couette flow, no-slip boundary conditions should be enforced at *halfway* lattice sites at top and bottom. At the top wall, after propagation, f_2, f_5, and f_6 have left the fluid node, and via the boundary condition, f_4, f_7, and f_8 have to be injected into the fluid node. The constraint of zero mass flux through the wall demands that

$$f_2 + f_5 + f_6 = f_4 + f_7 + f_8. \qquad (2.38)$$

The moving wall adds momentum to the ejected particles. To obtain a wall velocity u_{wall}, one needs

$$c(f_5 + f_8) - c(f_6 + f_7) = \frac{\rho u_{\text{wall}}}{6}. \qquad (2.39)$$

Again, assuming bounce-back for a particle propagating normal to the wall, then $f_2 = f_4$. Consequently, the boundary conditions at the top wall become:

$$f_4 = f_2,$$

$$f_7 = f_5 - \frac{\rho_0 u_{\text{wall}}}{6c}, \qquad (2.40)$$

$$f_8 = f_6 + \frac{\rho_0 u_{\text{wall}}}{6c}.$$

In the case of $u_{\text{wall}} = 0$, the wall is not moving, and the boundary conditions are therefore:

$$f_4 = f_2; \quad f_7 = f_5; \quad f_8 = f_6. \qquad (2.41)$$

These boundary conditions are the often-used "bounce-back" boundary conditions, that impose the no-slip boundary conditions for stationary walls.[2]

2.2.6   What Makes Lattice Boltzmann Special?

Despite the fact that the lattice Boltzmann scheme has been derived by discretization of kinetic theory, it does not seem significantly different from traditional finite-volume schemes. However, as demonstrated above, the lattice Boltzmann can be distinguished from traditional finite volumes by a number of favorable attributes. In summary:

- In the case of $\omega > 1$, the lattice Boltzmann schemes make use of gradients in conserved quantities, i.e., density gradient or the symmetric part of the deformation rate tensor, as state variables, making it effectively a higher-order scheme.
- Fluid flow is solved in the weakly compressible limit using the ideal gas equation of state. This makes it possible to model fluid flow through complex geometries with relatively coarse grids. Traditional schemes based on incompressible fluids have to solve the Poisson equation for the pressure, requiring refined meshes that result in longer computation times. The weakly compressible limit, however, does imply that the use of lattice Boltzmann schemes should be limited to low-Mach-number flows, $u/c_s \ll 1$ (or rather low Courant numbers, $u/c \ll 1$).
- Lattice Boltzmann schemes are implemented in a stream-and-collide fashion, where the most demanding computations (collision step) needs only local information. This makes it very simple to code and, above all, ideal for parallel computing. Practice shows that the performance scales almost linearly with the number of processors. Consequently, numerous implementations of lattice Boltzmann schemes are found on PC clusters, parallel-computing architectures, or even grid environments.
- Via its link to kinetic theory, it is relatively simple to extend the lattice Boltzmann to complex fluids that are often described on a thermodynamic basis or on a particulate basis. Examples of these complex fluids to which lattice Boltzmann has been applied are immiscible fluids, surfactant stabilised emulsions, polymer melts, microemulsions, and suspensions.

2.3 LATTICE BOLTZMANN SCHEMES FOR COMPLEX FLUIDS

The last property mentioned in the previous paragraph makes lattice Boltzmann very attractive to foods, which are one of the most notorious complex fluids. This "notorious complexity" can be probed using the "simplicity" of lattice Boltzmann and the "ease" with which models from the field of soft condensed matter can be incorporated. Soft condensed matter is a relatively new field in physics. It has been recognized as being of highly significant value by major food manufacturers for designing new food products with innovative textures and functional ingredients.[18]

Below, the principles of lattice Boltzmann schemes for complex fluids will be described. As in the field of soft condensed matter, the microstructure of complex fluids can be represented in two fundamentally different ways: (1) continuous fields of order parameters, governed by free-energy functionals, and (2) in a particulate sense, governed by a force model entering Newton's law. Among the methods that follow the first approach are self-consistent field theory,[19] density functional theory,[20] and phase field theory.[26,25] Among those following the second approach are Brownian dynamics, molecular dynamics, Stokesian dynamics,[21] and stochastic rotation dynamics.[22] Lattice Boltzmann schemes for complex fluids mostly follow either one of the approaches, although there are a few examples where the methods are combined.

Two cases are presented below: (1) a lattice Boltzmann scheme for emulsions, based on the free-energy functional approach, and (2) a lattice Boltzmann scheme for suspension, with the dispersed phase represented by particles moving on a fixed lattice.

2.3.1 Lattice Boltzmann Scheme for Emulsions

Lattice Boltzmann schemes for emulsions (or immiscible fluids, in general) use the diffuse-interface concept, where the dispersed or continuous phase is indicated with a color function or order parameter.[23,28] At the interface, the order parameter changes value in a region of several grid spacings, hence the term *diffuse interface*. Gradients in the color function or order parameter give rise to a capillary pressure that is incorporated as a body force in the Navier–Stokes equation. The color function evolves more or less as a passive scalar that is transported by the fluid flow. Consequently, the interface does not have to be tracked explicitly. Therefore, there is no need for adaptive grid refinements at interfaces as is required in finite-element and boundary-element methods. In many CFD packages, one finds a diffuse-interface method based on volume of fluid.[24,33] The disadvantage of the VOF method is that it is not based on a physical theory, making it difficult to incorporate complex physical processes such as surfactant adsorption.

Another numerical method based on the diffuse interface is the phase field method, which is based on the Cahn–Hilliard theory of phase separation in fluids.[25–28] The Cahn–Hilliard theory is based on a free-energy functional of the order parameter. From the functional, two thermodynamic quantities are derived: (1) the chemical potential driving the diffusion of the order parameter, and (2) the pressure tensor to be inserted in the Navier–Stokes equation. The latter describes the velocity field in both the dispersed and continuous phases. The capillary pressure due to the interface is incorporated into the pressure tensor and is related to gradients in the order parameter. The use of the Cahn–Hilliard theory in the context of lattice Boltzmann schemes is pioneered by Julia Yeomans and coworkers,[28] and among other multiphase flow phenomena, it has been shown to accurately describe droplet breakup.[29] Recently, we have extended the model of Yeomans and coworkers with surfactant adsorption.[32] Below, the main concepts of the Cahn–Hilliard theory, as used in the context of lattice Boltzmann, are described.

For a homogeneous immiscible binary fluid (e.g., an oil–water mixture) with phase densities ρ_a and ρ_b, the bulk free-energy density, F_0, is given by regular solution theory:

$$F_0 = kT[\rho_a \ln \rho_a + \rho_b \ln \rho_b + \chi_{ab}\rho_a\rho_b]. \tag{2.42}$$

Instead of densities ρ_a and ρ_b, the free energy can also be described by the total density, $\rho_0 = \rho_a + \rho_b$, and the density difference, $\phi = (\rho_a - \rho_b)/\rho$, also known as the *order parameter*. A frequently used approximation of the regular solution free energy (under the assumption of a constant density ρ_0) is

$$F_0 \approx -\frac{A}{2}\phi^2 + \frac{B}{4}\phi^4. \tag{2.43}$$

For an inhomogeneous fluid (having interfaces between the immiscible phases), a gradient term must be included in the free energy in the spirit of van der Waals, cf.:[28]

$$F = F_0 + \frac{\kappa}{2}(\nabla\phi)^2, \tag{2.44}$$

where κ is a parameter that is linked to the surface tension.

The chemical potential is then:

$$\mu = \frac{\delta F}{\delta\phi} = -A\phi + B\phi^3 - \kappa\nabla^2\phi. \tag{2.45}$$

This is the driving force for diffusion of the order parameter.

In addition to the chemical potential, the capillary pressure tensor must be defined to be inserted in the Navier–Stokes equation. The scalar part of the stress tensor (the hydrostatic pressure), p_0, follows from its thermodynamic definition:[30]

$$p_0 = \phi\mu - F. \tag{2.46}$$

Substitution of the expressions of the free-energy density and chemical potential then gives:

$$p_0 = \rho c_s^2 - \frac{A}{2}\phi^2 + \frac{3B}{4}\phi^4 - \kappa\phi\nabla^2\phi - \frac{\kappa}{2}(\nabla\phi)^2. \tag{2.47}$$

For the complete pressure tensor, $P_{\alpha\beta}$, one must ensure that it obeys the condition of mechanical equilibrium, specifically that it is divergence-free: $\partial_\beta P_{\alpha\beta}=0$. A suitable choice is[28]

$$P_{\alpha\beta} = p_0\delta_{\alpha\beta} + \kappa(\partial_\alpha\phi)(\partial_\beta\phi). \tag{2.48}$$

The evolution of the momentum density, ρu, and the order parameter, ϕ, are described, respectively, by the Navier–Stokes Equation (with the capillary stress tensor) and the convection-diffusion Equation (with the diffusive flux proportional to the gradient in the chemical potential):

$$\partial_t\phi + \partial_\alpha\phi u_\alpha = \partial_\alpha M\partial_\alpha\mu, \tag{2.49}$$

$$\partial_t\rho u_\alpha + \partial_\beta\rho u_\alpha u_\beta = \partial_\beta P_{\alpha\beta} + \partial_\alpha\rho\nu\partial_\alpha u_\alpha, \tag{2.50}$$

where ν is the local viscosity of the fluid, which can be made dependent on ϕ in case of viscosity differences between the fluid phases, and M is the mobility.

The total density, ρ, and the order parameter, ϕ, are modelled with two particle velocity distribution functions, f_i and g_i, respectively. Both distributions evolve according to the lattice Boltzmann equation:

$$f_i(x + \Delta x_i, t + \Delta t) - f_i(x,t) = -\omega_f[f_i(x,t) - f_i^{eq}(x,t)],$$
$$g_i(x + \Delta x_i, t + \Delta t) - g_i(x,t) = -\omega_g[g_i(x,t) - g_i^{eq}(x,t)]. \tag{2.51}$$

To obtain the correct governing physics, the moments of equilibrium distributions must follow:[28]

$$\begin{aligned}
\sum_i f_i^{eq} &= \rho, \\
\sum_i c_{i,\alpha} f_i^{eq} &= \rho u_\alpha, \\
\sum_i c_{i,\alpha} c_{i,\beta} f_i^{eq} &= P_{\alpha\beta} + \rho u_\alpha u_\beta, \\
\sum_i c_{i,\alpha} c_{i,\beta} c_{i,\gamma} f_{i\,eq} &= \rho c_s^2(u_\alpha\delta_{\beta\gamma} + u_\beta\delta_{\alpha\gamma} + u_\gamma\delta_{\alpha\beta}).
\end{aligned} \tag{2.52}$$

and

$$\begin{aligned}
\sum_i g_i^{eq} &= \phi, \\
\sum_i c_{i,\alpha} g_i^{eq} &= \phi u_\alpha, \\
\sum_i c_{i,\alpha} c_{i,\beta} g_i^{eq} &= \Gamma\mu\delta_{\alpha\beta} + \phi u_\alpha u_\beta,
\end{aligned} \tag{2.53}$$

where Γ is a parameter involved in the definition of the mobility: $M=\Gamma(1/\omega_g - 1/2)\Delta t$.

Figure 2.4 shows an example of the application of this model to droplet breakup in shear flow. The D2Q9 lattice with Reynolds number $Re=1.0$ and capillary number $Ca=1.0$ have been used here.

2.3.2 Lattice Boltzmann Scheme for Suspensions

Lattice Boltzmann schemes simulate particle suspensions by mapping the suspended particles onto a fixed grid on which the Navier–Stokes equation is solved.[16] The interaction between the fluid and the particles is by special fluid–solid boundary conditions (see below). These boundary

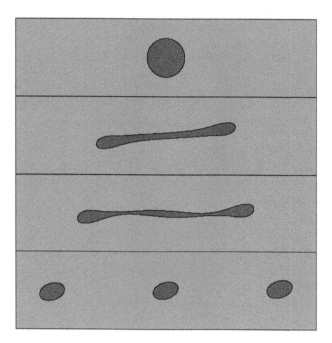

Figure 2.4 Droplet breakup in shear flow.

conditions prescribe the velocity on the fluid–solid boundary. They also determine the force and torque exerted by the fluid on the particle. The dynamics of the colloidal particles are solved by a separate algorithm that solves Newton's equation of motion using the forces and torques as obtained from the fluid–solid boundary conditions. To solve the particle dynamics, one can use efficient schemes that are developed in the field of molecular dynamics.

Particles having a defined radius, R, are mapped onto the lattice by simple digitization. Lattice sites with a distance to the particle's center that is smaller than the radius are marked as *solid*. All other unoccupied sites are consequently *fluid*. For conservation of mass, the solid lattice sites are occupied with lattice gas particles to which the collision and propagation steps are applied. Via application of the boundary conditions, this fluid remains inside the particle and flows according to the translational and angular velocities.

Such a digitized particle is shown in Figure 2.5. Observe that via digitization the round shape of the particle is transformed to a staircase representation. The arrows indicate lattice gas particles from the external fluid that are bounced back to the fluid after having collided with the suspended particle. Via this bounce-back process, the appropriate (no-slip) boundary conditions are applied on the surface of the particle.

As described in Section 2.5, the no-slip boundary conditions are imposed halfway between the solid lattice site (located at $\mathbf{x}$) and the fluid lattice site (located at $\mathbf{x}+c_i\Delta t$). The fluid velocity, u_b, at this position is given by

$$u_b = \mathbf{U}_p + \mathbf{\Omega}_p \times \left(\mathbf{x} + \frac{1}{2}c_i\Delta t - \mathbf{X}_p\right). \tag{2.54}$$

Here $\mathbf{U}_p$ is the translational velocity, $\mathbf{\Omega}_p$ is the angular speed, and $\mathbf{X}_p$ is the location of the particle's center.

For moving solid–fluid boundaries, momentum must be exchanged between the incoming particles from the solid and fluid sides of the boundary. The incoming distribution functions are

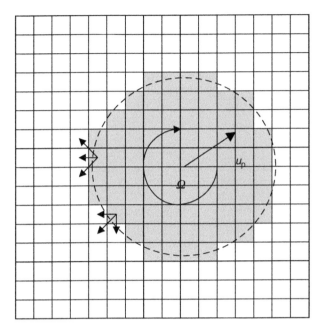

Figure 2.5 Digitization of a suspended particle on the lattice. Here, arrows indicate lattice gas particles that bounce back from the suspended particle.

the post-collisional distribution functions $f_i'(\mathbf{x},t)$ and $f_{i*}'(\mathbf{x}+\Delta\mathbf{x}_i,t)$, having defined $c_{i*}=-c_i$. When the boundary conditions are applied, the reflected particles arrive back at the lattice site at time $t+\Delta t$, ready for the next collision step.

The following rule imposes the required no-slip boundary condition:[16]

$$f_{i*}(\mathbf{x},t+\Delta t) = f_i'(\mathbf{x},t) + \Delta f,$$

$$f_i(\mathbf{x}+\Delta\mathbf{x}_i,t+\Delta t) = f_{i*}'(\mathbf{x},t) - \Delta f,$$ (2.55)

$$\Delta f = w_i\rho\frac{c_i\cdot u_b}{c_s^2}.$$

Notice that the no-slip boundary condition for flat moving walls, derived in the previous section, is a subset of the one given above.

The exchanged momentum, Δfc, is proportional with ρu_b, the velocity of the solid-fluid boundary. The momentum exchange is given by

$$\Delta\mathbf{p}\left(\mathbf{x}+\frac{1}{2}\Delta\mathbf{x}_i\right) = c_i f_i(\mathbf{x}+\Delta\mathbf{x}_i,t+\Delta t) + c_{i*}f_{i*}(\mathbf{x},t+\Delta t) - c_i f_i'(\mathbf{x},t) - c_{i*}f_{i*}'(\mathbf{x},t).$$ (2.56)

This momentum exchange induces a local force,

$$\Delta\mathbf{F}_p\left(\mathbf{x}+\frac{1}{2}\Delta\mathbf{x}_i\right) = -\frac{\Delta\mathbf{p}\left(\mathbf{x}+\frac{1}{2}\Delta\mathbf{x}_i\right)}{\Delta t},$$ (2.57)

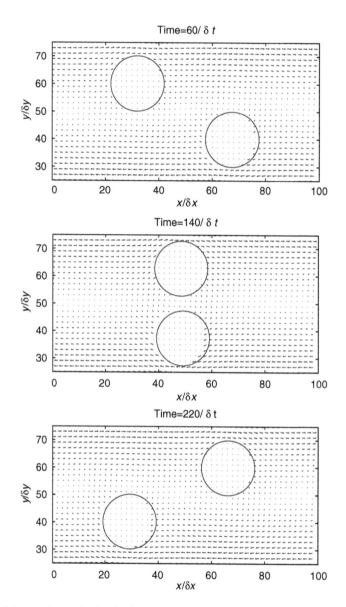

Figure 2.6   Two disks passing each other in shear flow.

and torque,

$$\Delta \mathbf{T}_p\left(\mathbf{x} + \frac{1}{2}\Delta \mathbf{x}_i\right) = \left(\mathbf{x} + \frac{1}{2}\Delta \mathbf{x}_i\right) \times \Delta F\left(\mathbf{x} + \frac{1}{2}\Delta x_i\right), \tag{2.58}$$

on the solid particle, indicated by the index p. The total force $\mathbf{F}_p$ and torque $\mathbf{T}_p$ on this colloidal particle is obtained by summing the $\Delta \mathbf{F}_p(\mathbf{x} + (1/2)\Delta \mathbf{x}_i)$ and $\Delta \mathbf{T}_p(\mathbf{x} + (1/2)\Delta \mathbf{x}_i)$ over all lattice links between fluid lattice sites and solid lattice sites corresponding to the current particle p.

After computing the total force and torque that is exerted on the particle, the kinematic properties of the suspended particle are updated by solving Newton's law using the Verlett scheme:[16]

$$u_p(t + \Delta t) = u_p(t) + \frac{\mathbf{F}_p}{m_p} \Delta t,$$

$$\mathbf{X}_p(t + \Delta t) = \mathbf{X}_p(t) + u_p(t + \Delta t)\Delta t + \frac{1}{2} \frac{\mathbf{F}_p}{m_p} \Delta t^2, \tag{2.59}$$

$$\mathbf{\Omega}_p(t + \Delta t) = \mathbf{\Omega}_p(t) + \frac{\mathbf{T}_p}{I_p} \Delta t,$$

where m_p and I_p are the mass and the inertia, respectively, of the suspended particle.

Figure 2.6 shows a typical example of flow computed with this type of scheme: two suspended discs that pass each other in shear flow. Simulation is performed with the D2Q9 lattice, with particle Reynolds number $Re = 0.1$.

2.4   APPLICATIONS

2.4.1   Emulsification in Microchannel T-Junctions

During the last decade, a large volume of research has been dedicated toward the manufacture of monodisperse emulsions using membranes or microfluidic devices.[33,34,39,37] These monodisperse emulsions have positive attributes as stability against Ostwald ripening and creaming. By mixing monodisperse emulsions, food producers obtain full control over the textural properties of emulsions. Furthermore, monodisperse emulsions are very advantageous as a pre-emulsion in the process of making double emulsions.[36]

Despite ten years of research, the physics of droplet formation in confined systems, such as membranes or microfluidic devices, is still far from understood. We have embarked on two projects that focus on preparing monodisperse emulsions using microsieves. To obtain a better understanding of the governing physics, we have developed lattice Boltzmann schemes describing the process of droplet formation in a confined geometry. Our model is based on the diffuse-interface scheme as developed by Yeomans and coworkers[28] that is explained in the previous section. Below, results obtained from modelling droplet formation in microchannel T-junctions, which are thought to act as a model system for membranes and microsieves, are presented.

Figure 2.7 shows snapshots of a droplet formed during an experiment and in simulation under the same flow conditions. Comparing experimental and numerical data for droplet diameter shows that our model accurate predicts the experimental data, without any parameter estimation.[5] An important phenomenon that is essential for droplet breakup is *contact-line dynamics*.[41] This is evident in both experiment and simulation by the intrusion of the continuous phase (water) into the side branch, which is hydrophilic (as is the main channel). Wetting boundary conditions are

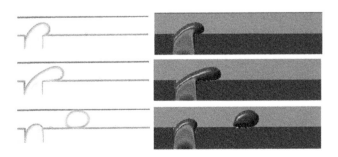

Figure 2.7   Comparison of experiment and simulation. (From van der Graaf, S., Nisisako, T., Schroën, C. G. P. H., van der Sman, R. G. M., and Boom, R. M., *Langmuir*, 22(9), 4144–4152, 2006.)

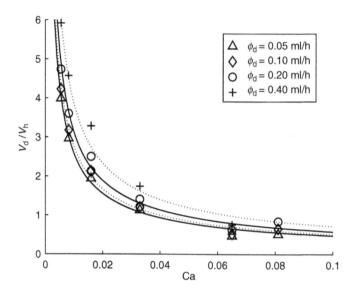

Figure 2.8 Scaling between droplet diameter and capillary number. Droplet diameter is scaled by $V_h = \pi d_h^3/6$, where d_h is the hydraulic diameter of the main flow channel. (From van der Graaf, S., Nisisako, T., Schroën, C. G. P. H., van der Sman, R. G. M., and Boom, R. M., *Langmuir*, 22(9), 4144–4152, 2006.)

implemented with the Cahn boundary condition,[38,5] from which contact-line dynamics are an emergent property.

After having validated the model against experimental data, we have performed a parameter study on how the droplet's volume, V_p, depends on the capillary number, Ca, and the dispersed flow rate, ϕ_d (see Figure 2.8). Via this parameter study, we have obtained the following scaling rule:

$$V_p = (V_{p,0} + \tau_0 \phi_d) Ca^{-0.75}, \tag{2.60}$$

where $V_{p,0}$ and τ_0 are constants that depend on the geometry of the microfluidic device. This relation holds for the regime $Ca < 0.1$, where the droplet diameter is of comparable order or smaller than the hydraulic diameter of the main channel. This scaling is quite different from the scaling found in the regime $Ca \gg 0.1$, where the confining walls have little influence and the droplet breakup is governed by a balance between the shear force on the droplet and the surface tension force.[40,39]

Recently, we have extended this model to encompass surfactants, which can diffuse and absorb on the droplet interface and thereby lower the interfacial tension.[32] Surfactants are important for emulsion droplet stabilization and they also lower the interfacial tension. This allows the emulsification process to operate in the low-shear regime. Low shear rates are beneficial for the conservation of energy and for protecting some shear-sensitive ingredients such as the internal phase of double emulsions.[36] The correct surfactant dynamics emerge from a free-energy functional that is based on the original formulation by Diamant and Andelman,[31] but converted from a sharp-interface to a diffuse-interface formulation.[32] The free-energy functional is simply added to the Cahn–Hilliard free energy from which the chemical potential for surfactants and an extended pressure tensor is derived.

2.4.2 Shear-Induced Diffusion in Microfiltration Processes

The above-mentioned microsieves are also very promising for fractionation of polydisperse suspensions such as milk. A sieve designed for fractionation is depicted in Figure 2.9.

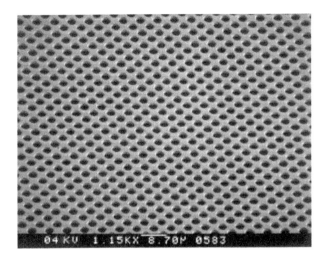

Figure 2.9   Silicon–nitride microsieve with monodisperse pores of 1-μm diameter made with standard silicium lithography technology. (From Kuiper, S., van Rijn, C. J. M., Nijdam, W., Krijnen. G. J. M., and Elwenspoek, M. C., *J. Membr. Sci.*, 180, 15–28, 2000.)

The microsieve is operated as a microfiltration device. However, as is generally true for filtration devices, microfiltration processes are hindered by fouling, i.e., the formation of a concentration polarization layer or even a cake layer on top of the microsieve. This fouling layer must be minimal for the microsieve to be successful in practice. For monodisperse emulsions in Stokes flow, the physics behind fouling-layer formation, i.e., shear-induced diffusion, is well understood.[21] However, for milk fractionation, one must deal with polydisperse suspensions and nonzero Reynolds number flows. To understand shear-induced diffusion in these practical systems, we have developed lattice Boltzmann schemes based on the scheme developed by Ladd,[16] as discussed in the previous section.

First, we have performed simulations of suspensions with fully resolved particles, having diameters significantly larger than the lattice spacing. The results show significant deviations from the monodisperse/Stokes flow case.[42,44] For finite Re-number flows, we observed a higher shear-induced diffusivity than in the case of Stokes flow, and in a bidisperse suspension, the shear induced diffusivity of the smaller particles was largely determined by the larger particles (Figure 2.10).

To perform process optimisation, we have also developed a continuum model describing fouling-layer formation on a microsieve. This continuum model consists of a Navier–Stokes equation coupled to a convection-diffusion equation.[3] Here, the Navier–Stokes equation describes the flow of the fluid as a whole, whereas the convection-diffusion equation describes the evolution of the density of the suspended particles. The shear-induced diffusivity, D_{eff}, that enters the convection-diffusion equation depends on the local shear-rate, γ, which is directly obtained from the Navier–Stokes equation. In the Navier–Stokes equation, the effective fluid viscosity, μ_{eff}, depends on the suspended particle density, ϕ. Therefore, there is a two-way coupling between the convection-diffusion and the Navier–Stokes equation. For this application, the continuum model is based on the lattice Boltzmann scheme because it has favorable properties for the simulation of convection-diffusion (because it shows little to no numerical diffusion or dispersion), and because the shear rate can be directly obtained from the *local* nonequilibrium part of the velocity distribution function.

The Navier–Stokes equation and the convection-diffusion equation are simulated with two lattice gases with distributions f_i and g_i, with the macroscopic quantities represented by hydrodynamic moments of the distribution functions: $\rho = \Sigma_i f_i$ is the density field of the fluid,

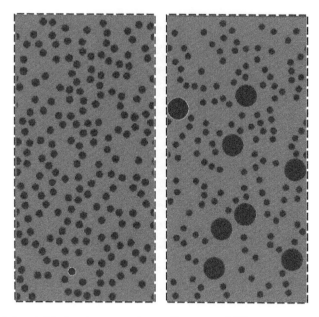

Figure 2.10 Shear-induced diffusion phenomena in mondisperse and bidisperse suspensions in shear flow.

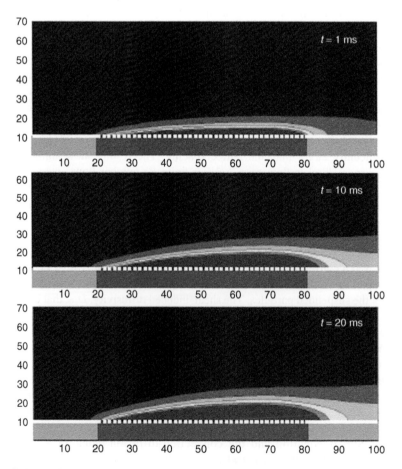

Figure 2.11 Build up of the fouling layer above a microsieve membrane.

$\rho u = \Sigma_i f_i c_i$ is the momentum density of the fluid, $p = \rho c_s^2$ is the pressure field, and $\phi = \Sigma_i g_i$ is the density field of the suspended particles. These macroscopic quantities are governed by the following set of partial differential equations:

$$\partial_t \rho + \nabla \cdot \rho u = 0, \tag{2.61}$$

$$\partial_t \rho u + \nabla \cdot \rho u u = -\nabla p + \nabla \mu_{\text{eff}}(\nabla u + \nabla u^T) + \mathbf{F}_{\text{drag}}, \tag{2.62}$$

$$\partial_t \phi + \nabla \cdot \phi u = \nabla D_{\text{eff}} \nabla \phi. \tag{2.63}$$

For monodisperse suspensions in a Stokes-flow regime $(Re=0)$, analytical expressions are available for the effective transport coefficients:[45,3]

$$\mu_{\text{eff}} = \mu_0 \left[1 + 1.5 \frac{\phi}{(1 - \phi/\phi_{\max})} \right]^2, \tag{2.64}$$

$$D_{\text{eff}} = 0.33 \gamma a^2 \phi^2 [1 + 0.5 \exp (8.8\phi)].$$

Our simulation results are in good agreement with an analytical theory of membrane fouling.[45] Figure 2.11 shows some snapshots of the growth of the fouling layer on a microsieve membrane.[3]

For polydisperse suspensions and finite Reynolds-number flows $(Re>0.1)$, other closure relations for the effective transport coefficients are needed. Our simulations on fully resolved particles[42,44] provided a first insight in these closure relations, but more work is required.

2.5 CONCLUSIONS

In this chapter, we have introduced the basics of the lattice Boltzmann scheme. Due to its simplicity, it is a powerful method for simulating macroscopic transport phenomena such as fluid flow and convection-diffusion in complex geometries such as porous media.[2] Due to the fact that the collision step is performed locally at each lattice site, the technique is very efficient with respect to parallel computing, showing almost linear scaling of the performance with the number of processors.

With respect to food processing, the value of the lattice Boltzmann lies more in the fact that it can very efficiently model complex fluids with fully resolved structural elements, such as emulsions and suspensions. We have succesfully applied lattice Boltzmann to the modelling of emulsification in microdevices and the fouling of microsieves in applications of microfiltration of dairy suspensions.

In recent years, the lattice Boltzmann method has been applied to other complex fluids such as foams,[46] liquid crystals,[47] polymer blends with nanofillers,[48] and viscoelastic fluids.[49] This demonstrates that the lattice Boltzmann method has the potential to model many other structured foods.

REFERENCES

1. Benzi, R., Succi, S., and Vergassola, M., The lattice Boltzmann equation: Theory and applications, *Phys. Rep.*, 222(3), 145–197, 1992.
2. Chen, S. and Doolen, G. D., Lattice Boltzmann method for fluid flows, *Ann. Rev. Fluid Mech.*, 30, 329–364, 1998.

3. Kromkamp, J., Bastiaanse, A., Swarts, J., Brans, G., van der Sman, R. G. M., and Boom, R. M., A suspension flow model for hydrodynamics and concentration polarisation in crossflow microfiltration, *J. Membr. Sci.*, 253(1–2), 67–79, 2005.

4. Brans, G., Kromkamp, J., Schroën, C. G. P. H., van der Sman, R. G. M., and Boom, R. M., Filtration with microsieves: Influence of surface properties and sieve design, *J. Membr. Sci.*, in press.

5. van der Graaf, S., Nisisako, T., Schroën, C. G. P. H., van der Sman, R. G. M., and Boom, R. M., Lattice Boltzmann simulations of droplet formation in a T-shaped microchannel, *Langmuir*, 22(9), 4144–4152, 2006.

6. Qian, Y. H., d'Humieres, D., and Lallemand, P., Lattice BGK models for Navier–Stokes equation, *Europhys. Lett.*, 17(6), 479–484, 1992.

7. van der Sman, R. G. M. and Ernst, M. H., Convection diffusion lattice Boltzmann scheme for irregular lattices, *J. Comput. Phys.*, 160, 1–17, 2000.

8. van der Sman, R. G. M., Lattice Boltzmann scheme for natural convection on square and rectangular grids, *Phys. Rev. E*, 74, 026705, 2006.

9. McNamara, G. and Alder, B., Analysis of the lattice Boltzmann treatment of hydrodynamics, *Physica A*, 194, 218–228, 1993.

10. van der Sman, R. G. M. and Ernst, M. H., Diffusion lattice Boltzmann scheme on an orthorhombic lattice, *J. Stat. Phys.*, 94(1/2), 203–216, 1999.

11. Junk, M., A finite difference interpretation of the lattice Boltzmann method, *Numer. Meth. Part. Diffl. Equat.*, 17(4), 383–402, 2001.

12. van der Sman, R. G. M., Finite Boltzmann schemes, *Comput. Fluids*, 35, 59–64, 2006.

13. van der Sman, R. G. M., Diffusion on unstructured triangular grids using lattice Boltzmann, *Future Generation Comp. Sys.*, 20(6), 965–971, 2004.

14. Inamuro, T., Yoshino, M., and Ogino, F., Non-slip boundary condition for lattice Boltzmann simulations, *Phys. Fluids*, 7(12), 2928–2930, 1995.

15. Zou, Q. S. and He, X. Y., On pressure and velocity boundary conditions for the lattice Boltzmann BGK model, *Phys. Fluids*, 9(6), 1591–1598, 1997.

16. Ladd, A. J. C. and Verberg, R., Lattice-Boltzmann simulations of particle-fluid suspensions, *J. Stat. Phys.*, 104(5–6), 1191–1251, 2001.

17. Hill, R. J., Koch, D. L., and Ladd, A. J. C., Moderate-Reynolds-number flows in ordered and random arrays of spheres, *J. Fluid Mech.*, 448, 243–278, 2001.

18. Mezzenga, R., Schurtenberger, P., Burbidge, A., and Michel, M., Understanding foods as soft materials, *Nat. Mater.*, 4, 729–740, 2005.

19. Fredrickson, G. H., Ganesan, V., and Drolet, F., Field-theoretic computer simulation methods for polymers and complex fluids, *Macromolecules*, 35(1), 16–39, 2002.

20. Fraaije, J. G. E. M., Zvelindovsky, A. V., and Sevink, G. J. A., Computational soft nanotechnology with Mesodyn, *Mol. Simul.*, 30(4), 225–238, 2004.

21. Brady, J. F. and Bossis, G., Stokesian dynamics, *Ann. Rev. Fluid Mech.*, 20, 111–157, 1988.

22. Malevanets, A. and Kapral, R., Mesoscopic model for solvent dynamics, *J. Chem. Phys.*, 110(17), 8605–8613, 1999.

23. Gunstensen, A. K., Rothman, D. H., Zaleski, S., and Zanetti, G., Lattice Boltzmann model of immiscible fluids, *Phys. Rev. A*, 43(8), 4320–4327, 1991.

24. Brackbill, J. U., Kothe, D. B., and Zemach, C., A continuum method for modelling surface tension, *J. Comput. Phys.*, 100(2), 335–354, 1992.

25. Jacqmin, D., Calculation of two-phase Navier–Stokes flows using phase-field modeling, *J. Comput. Phys.*, 155(1), 96–127, 1999.

26. Anderson, D. M., McFadden, G. B., and Wheeler, A. A., Diffusive-interface methods in fluid mechanics, *Ann. Rev. Fluid Mech.*, 30, 165–193, 1998.

27. Cahn, J. W. and Hilliard, J. E., Free energy of a non-uniform system I. Interfacial energy, *J. Chem. Phys.*, 28, 258–267, 1958.

28. Swift, M. R., Orlandini, E., Osborne, W. R., and Yeomans, J. M., Lattice Boltzmann simulations of liquid–gas and binary fluid mixtures, *Phys. Rev. E*, 54, 5041–5052, 1996.

29. Wagner, A. J. and Yeomans, J. M., Effect of shear on droplets in a binary mixture, *Int. J. Mod. Phys. C*, 8(4), 773–782, 1997.

30. Lamura, A., Gonella, G., and Yeomans, J. M., A lattice Boltzmann model of ternary fluid mixtures, *Europhys. Lett.*, 45(3), 314–320, 1999.
31. Diamant, H., Ariel, G., and Andelman, D., Kinetics of surfactant adsorption: The free energy approach, *Colloids Surf. A*, 183–185, 259–276, 2001.
32. van der Sman, R. G. M. and van der Graaf, S., Diffuse interface model for surfactant adsorption onto evolving interface, *Rheologica Acta*, 2006.
33. Abrahamse, A. J., van der Padt, A., Boom, R. M., and de Heij, W. B. C., Process fundamentals of membrane emulsification-simulation with CFD, *AIChE J.*, 47(6), 1285–1291, 2001.
34. Sugiura, S., Nakajima, M., Iwamoto, S., and Seki, M., Interfacial tension driven monodispersed droplet formation from microfabricated channel array, *Langmuir*, 17, 5562–5566, 2001.
35. Kuiper, S., van Rijn, C. J. M., Nijdam, W., Krijnen, G. J. M., and Elwenspoek, M. C., Determination of particle-release conditions in microfiltration: A simple single-particle model tested on a model membrane, *J. Membr. Sci.*, 180, 15–28, 2000.
36. van der Graaf, S., Schroen, C. G. P. H., and Boom, R. M., Preparation of double emulsions by membrane emulsification-a review, *J. Membr. Sci.*, 251, 7–15, 2005.
37. Verberg, R., Pooley, C. M., Yeomans, J. M., and Balazs, A. C., Pattern formation in binary fluids confined between rough, chemically heterogeneous surfaces, *Phys. Rev. Lett.*, 93(18), 184501, 2004.
38. Briant, A. J., Papatzacos, P., and Yeomans, J. M., Lattice Boltzmann simulations of contact line motion in a liquid–gas system, *Philos. Trans. R. Soc. London*, 360, 485–495, 2002.
39. Garstecki, P., Stone, H. A., and Whitesides, G. M., Mechanism for flow-rate controlled breakup in confined geometries: A route to monodisperse emulsions, *Phys. Rev. Lett.*, 94, 164501, 2005.
40. Tice, J. D., Song, H., Lyon, A. D., and Ismagilov, R. F., Formation of droplets and mixing in multi-phase microfluidics at low values of the Reynolds and the capillary numbers, *Langmuir*, 19, 9127–9133, 2003.
41. Christov, N. C., Ganchev, D. N., Vassileva, N. D., Denkov, N. D., Danov, K. D., and Kralchevesky, P. A., Capillary mechanisms in membrane emulsification: Oil-in-water emulsions stabilised by Tween 20 and milk proteins, *Colloids Surf. A*, 209, 83–104, 2000.
42. Kromkamp, J., van den Ende, D. T. M., Khandai, D., van der Sman, R. G. M., and Boom, R. M., Shear-induced self-diffusion and microstructure in non-Brownian suspensions at non-zero Reynolds numbers, *J. Fluid Mech.*, 529, 253–278, 2005.
43. Brans, G., van der Sman, R. G. M., and Schroën, C. G. P. H., Optimization of the membrane and pore geometry for micro-machined membranes, *J. Membr. Sci.*, 278(1–2), 239–250, 2006.
44. Kromkamp, J., van den Ende, D. T. M., Kandhai, D., van der Sman, R. G. M., and Boom, R. M., Lattice Boltzmann simulation of 2D and 3D non-Brownian suspensions in Couette flow, *Chem. Eng. Sci.*, 61(2), 858–873, 2006.
45. Romero, C. A. and Davies, R. H., Global model of crossflow microfiltration based on hydrodynamic particle diffusion, *J. Membr. Sci.*, 39, 157–185, 1988.
46. Korner, C., Thies, M., and Singer, R. F., Modeling of metal foaming with lattice Boltzmann automata, *Adv. Eng. Mater.*, 4(10), 765–769, 2002.
47. Denniston, C., Orlandini, E., and Yeomans, J. M., Simulations of liquid crystal hydrodynamics in the isotropic and nematic phases, *Europhys. Lett.*, 52(4), 481–487, 2000.
48. Verberg, R., Yeomans, J. M., and Balazs, A. C., Modeling the flow of fluid/particle mixtures in microchannels: Encapsulating nanoparticles within monodisperse droplets, *J. Chem. Phys.*, 123(22), 224706, 2005.
49. Onishi, J., Chen, Y., and Ohash, H., A lattice Boltzmann model for polymeric liquids, *Prog. Comput. Fluid Dyn.*, 5(1–2), 75–84, 2005.

CHAPTER 3

Fluid Flow and Its Modeling Using Computational Fluid Dynamics

Ashwini Kumar and Ilhan Dilber

CONTENTS

3.1   INTRODUCTION TO FLUID FLOW MODELING

As described in Chapter 1, fluid flow analysis is based on physics-based models frequently encountered in many current engineering problems. It has a very wide range of application, and it has many aspects such as the flow regime (laminar or turbulent), type of fluid (gas or liquid), and interaction with its surroundings (heat transfer, moving boundaries, and mixtures). The fluid flow problem is challenging enough, yet it is often combined with several other phenomena that are tightly coupled to the fluid that makes solution of most fluid problems only possible through complex numerical simulations or series of experiments. The flow phenomenon involves many models with each being its own discipline, and many textbooks are written to cover the individual disciplines in great lengths. It is impossible to cover all the details of every one of the sub-models of fluid flow in one chapter. Yet, every engineer who has a fluid problem to solve needs to be aware of these models in order to make correct decisions. The goal of this chapter is not to show all the theories behind these physical models, but it is to give the reader a reference to the models and terminologies frequently encountered as well as some useful tips on how to solve a fluid flow problem, specifically targeting the food industry.

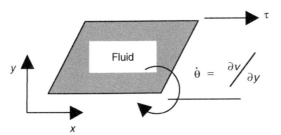

Figure 3.1 Fluid volume under shear.

Food processing operations involve several steps where engineering analysis or modeling can be extremely helpful. A thorough application of physical principles will lead to a better design tool that has the potential to change the dynamics of food engineering research toward right by design rather than build and test. Numerical modeling can lead to virtual experimentation where initial designs can be tested before a pilot plant testing or full industrial testing can take place.[1-2]

A brief review of some basic fundamentals of fluid dynamics and fluid flow is presented followed by a discussion of the modeling of these processes. A fluid is a substance that continuously deforms under shear forces, and fluid mechanics is the study of flow while it is at rest (fluid statics) or in motion (fluid dynamics). A shearing stress (τ in Figure 3.1), defined as shear force per unit area, is created whenever a tangential force acts on a surface. The fluid parcel's deformation rate ($\dot{\theta}$ in Figure 3.1) is a function of this shearing stress.

Fluids such as liquids and gases keep deforming or flowing under this stress. Some fluids may not deform unless this stress is greater than some yield stress. This yield stress plays an important role in the physical properties for many substances in the food industry, as will be examined later in Section 3.4.2.

3.2 DYNAMICS OF FLUIDS

The fluid motion is governed by physical laws that, when mathematically expressed, allow all fluid parameters (e.g., pressure, temperature, velocity) to be determined at every point in space. These physical laws are expressed in terms of complex non-linear partial differential equations. Although the fundamental equations of fluid flow have been known for almost two centuries, their solutions were limited to simplest flow problems until the advent of computers. With computers' increasing powers and the continued development of new numerical methods in the past three decades, these equations can now be solved for increasingly more complex and realistic flow problems.

In most physical phenomena, the familiar reference frame (or the coordinate axes) is fixed in space (stationary). This type of reference frame is called *Eulerian frame*. In this reference frame, the fluid would flow through a stationary coordinate system. However, in fluid mechanics, there are cases where it is desirable to have the reference frame move either with the fluid body or with a particle moving inside the fluid. This moving frame of reference is called *Lagrangian frame*. Because fluid dynamics involves fluid in motion, many equations have additional terms that arise from the moving frame of reference. For simplicity, sometimes these terms are grouped with the temporal derivative. This combined derivative term is defined as the operator

$$\frac{D}{Dt}(.) \equiv \frac{\partial(.)}{\partial t} + \left(v_x \frac{\partial}{\partial x} + v_y \frac{\partial}{\partial y} + v_z \frac{\partial}{\partial z} \right)(.) = \frac{\partial(.)}{\partial t} + (\mathbf{v} \cdot \nabla)(.), \tag{3.1}$$

where **v** is the velocity of the fluid or the reference frame in motion. The derivative notation in the equation above is often called the *Lagrangian Derivative*, or *material derivative*, or *substantial derivative*, or *derivative following the fluid*. The first term on the right-hand side is the ordinary time derivative in fixed reference frame (Eulerian derivative), and the second term is the result of changes because of a body in motion or the reference frame's being in motion (Lagrangian frame). These changes are referred to as *advection*, and the second term is often called the *advective term* in the equation.

3.3 GOVERNING EQUATIONS

Physics-based models in fluid dynamics involve solving governing equations for fluid flow, heat transfer, chemical species transport, and related phenomena. These fundamental equations are based on conservation principles. In simple terms, this can be expressed as:

$$\text{Rate} = \{\text{rate of influx} - \text{rate of outflux}\} + \{\text{rate of generation} - \text{rate of consumption}\}$$

There are three major laws of conservation. The first, conservation of mass, says that the mass is neither created nor destroyed in the fluid parcel. The second, conservation of momentum, says that the rate of momentum's change is the sum of the forces acting on the fluid. Third, the conservation of energy, says that the change in total energy is net heat transfer minus net work done by the fluid. These equations can become highly nonlinear, and only a few simple problems have exact solutions. The numerical techniques are used to obtain approximate solutions to these problems, and this field is termed as the computational fluid dynamics (CFD) that is explained in more detail in a later section of this chapter. Although *fluid dynamics* implies the study of the dynamics of fluids, many applications not only deal with just the motion of fluids, but they also involve heat transfer, mass transfer, and chemical reactions. They often involve solid parts as well. However, this chapter will focus on the study of fluids. The heat and mass transfer are explained in their respective chapters in this handbook.

The following sections describe the governing equations for these various phenomena. For simplicity, the equations presented here are in Cartesian coordinates. Complete derivation of these equations and expressions for other coordinate systems (cylindrical and spherical) can be found in Bird et al.[3]

3.3.1 Conservation of Mass

The conservation of mass is expressed by the continuity equation that states that in a given volume, mass can neither be created nor destroyed, and the net mass flow through all its surfaces must be equal to the rate of accumulation

$$\frac{D\rho}{Dt} + \rho \nabla \cdot \mathbf{v} = 0. \tag{3.2}$$

For incompressible fluids, as is the case in a majority of food applications, the density is constant, and the above equation reduces to

$$\nabla \cdot \mathbf{v} = 0. \tag{3.3}$$

that is referred to as the incompressibility constraint. This equation applies for single-fluid systems.

For multiple component systems, each individual component in the mixture must satisfy the principle of mass conservation as well as the mixture itself as a whole. The mass conservation for the individual component is given by

$$\rho\left(\frac{Dc_n}{Dt} + c_n \nabla \cdot \mathbf{v}\right) = -\nabla \cdot \mathbf{j}^{(n)} + q_{c_n} + R_n,\tag{3.4}$$

where c_n is the mass fraction of specie n, $\mathbf{j}^{(n)}$ is the diffusive mass flux of species n, R_n is the rate of production of c_n because of chemical reaction, and q_{c_n} is a general source term. In principal, the mass flux can have contributions from thermal gradients. However, in most cases, the diffusive mass flux is assumed to be the result of only molecular diffusion (concentration gradient). In this simplified case, the diffusive mass flux reduces to

$$\mathbf{j}^{(n)} = -\nabla(\rho \alpha_n c_n).\tag{3.5}$$

Here, α_n is the mass diffusivity. This reduces the species equation to

$$\rho\left(\frac{Dc_n}{Dt} + c_n \nabla \cdot \mathbf{v}\right) = \nabla \cdot (\rho \alpha_n \nabla c_n) + q_{c_n} + R_n.\tag{3.6}$$

Mass transport is discussed in more detail in Chapter 5 and Chapter 6 of this handbook.

Table 3.1 Simplified Forms of the Momentum Equation

Navier–Stokes equation without external forces: Momentum equation for an incompressible fluid with constant viscosity and in the absence of body forces or external forces. Typically used for simple fluids in simple flow conditions driven by pressure gradients and fluid inertia	$\rho\frac{D\mathbf{v}}{Dt} = -\nabla P + \mu\nabla^2\mathbf{v}$
Euler equation: Momentum equation where viscous term is negligible relative to inertial term (inviscid flows), neglecting body forces. This is typically used in aerospace applications and is not encountered in food industry but is given here for completeness	$\rho\frac{D\mathbf{v}}{Dt} = -\nabla P$
Stokes flow: Momentum equation where inertial forces are negligible relative to viscous terms (highly viscous flows, creeping flows), neglecting body forces. This is frequently encountered in very viscous fluids (such as honey, peanut butter) that are subject to very slow motion	$\nabla P = \mu\nabla^2\mathbf{v}$
Bernoulli equation: Momentum equation when the flow is inviscid, steady, incompressible, and with no heat transfer. Bernoulli equation has many practical applications in fluid mechanics (specifically, fluid statics) where the flow is strictly driven by pressure gradient or gravity, and friction is negligible	$P + 1/2\rho v^2 + \rho gh = \text{Constant}$
Poiseuille flow: Momentum equation for viscous, laminar, steady state flows without any external forces. Such flows occur in circular pipes or channels with constant cross-sections (pressure can only change in the direction of flow). This equation can be integrated to yield a parabolic velocity profile. f is the pressure drop per unit length, and v is the velocity in the direction of flow	$\nabla^2 v = f$
Laplace equation: Simplest form of momentum equation. It assumes a perfect fluid (no viscosity), no external forces, away from boundaries (zero vorticity), and the flow is solely governed by Reynolds number. The velocity can then be expressed as the gradient of a scalar field ($\mathbf{u} = \nabla\phi$), and continuity ($\nabla \cdot \mathbf{u} = 0$) yields Laplace's equation. The velocity field can be reconstructed from the scalar quantity ϕ	$\nabla^2\phi = 0$

3.3.2 Conservation of Momentum

Conservation of momentum is derived from Newton's second law, stating that the acceleration of a body is equal to the sum of all external forces per unit mass

$$\rho \frac{D\mathbf{v}}{Dt} = -\nabla P - \nabla \cdot \tau + \rho \mathbf{g}, \tag{3.7}$$

where $\mathbf{v}$ is the velocity, τ is the stress tensor, $\mathbf{g}$ is the acceleration due to gravity, and P is the pressure. This is the most general form of the momentum equation. The full expression of the stress tensor is outside the scope of this chapter, but its relationship to viscosity is briefly discussed in Section 3.4.2. Assuming constant viscosity, the momentum equation reduces to

$$\rho \frac{D\mathbf{v}}{Dt} = -\nabla P + \mu \nabla^2 \mathbf{v} + \rho \mathbf{g}. \tag{3.8}$$

The term on the equation's left side includes the temporal and the advection terms, and the right side contains the pressure gradient, viscous diffusion, and gravitational force. Equation 3.8 is the form most commonly referred to as the Navier–Stokes equation (along with the continuity equation), though many articles in literature refer to either the full representation, using the stress tensor or to further simplified versions of this equation. There are many other simplified versions of this equation that have significant practical applications, and each has been individually studied extensively in literature. Table 3.1 shows a few of these simplified versions of the equation.

3.3.3 Conservation of Energy

Although this chapter's main focus is fluid motion, many food processes often involve heat transfer, and many properties are a function of temperature that directly affect the fluid flow problem. Therefore, the conservation of energy equation is often coupled with the Navier–Stokes equation, and, for an incompressible fluid with constant conductivity, k, is given by

$$\rho c_p \left(\frac{\partial T}{\partial t} + \mathbf{v} \cdot \nabla T \right) = k(\nabla^2 T) + Q + \Phi. \tag{3.9}$$

In the above equations, T is temperature, c_p is specific heat, Q is heat source, and Φ is the mechanical or viscous dissipation term. The details of the energy equation are described in next chapter of this book. However, as seen from the above equation, there is a tight coupling between the velocity field and the temperature field. Therefore, coupling the energy equation with the momentum equation is worth mentioning. When it comes to solving the energy equation (or specie equation) along with momentum, there are four classes of flows:

- *Isothermal Flows*. In this case, there is no need to solve the energy equation at all as the entire system is at a constant temperature, or the temperature variations are of no concern (and have no affect on properties). Typically, these flows involve simple fluids such as air and water, and they occur in neutral environment (e.g., room temperature) where the focus is the flow itself.
- *Advection–Diffusion Problems*. This class of problems assumes the flow field is known, constant in time, and completely independent of the energy field. Then the energy equation can be solved independent from the momentum equation. One can obtain the velocity field from the momentum equation and use these velocities in the energy equation to obtain the full solution. This class of problems is often encountered when

the fluid properties are independent of temperature, but one is interested in finding the temperature variations (or more typically used for species concentration variations) as a result of fluid flow. These types of flows also involve simple fluids that tend to have properties independent of temperature (or changes are negligible for the problem at hand).

- *Weakly-Coupled Flows*. In this model, the buoyancy term is ignored, leaving only the convection terms as the coupling between the equations. This allows for solving of the energy and the momentum equations decoupled from each other. Typically, the flow field does not feel the presence of the temperature field because fluid density and viscosity are not dependent on temperature, but the temperature field is dependent on the flow field through the convective term (heat transfer properties can be a function of temperature). These flows may also include shear heating as long as the flow properties are constant.
- *Strongly-Coupled Flows*. If buoyancy cannot be ignored in the flow (such as natural convection flow) or if there are non-linear boundary conditions or temperature dependent flow properties, then a full set of equations must be solved fully-coupled. If the fluid viscosity is dependent on temperature, the flow and thermal equations are also strongly coupled. This is a common case in many food applications where temperature is present because most food properties are temperature dependent (besides the buoyancy effects of temperature dependent density, temperature dependent viscosity often has dramatic impact on the flow).

Depending on the particular problem being solved, other governing equations may be considered. For example, high temperature problems will require the solution of the appropriate equations for radiation heat transfer. Turbulent flows necessitate the solving of the turbulence modeling equations (for example, the so-called k–ε model involves convection–diffusion equations for the kinetic energy of turbulence, k, and the turbulent dissipation, ε). In addition to the governing equations, an equation of state that relates the fluid density to the local temperature, pressure, and composition is also required.

3.4   PHYSICAL PROPERTIES OF THE FLUID

Most of the complications with solving flow problems in foods are related to variability in material properties. As a result, having either good property data or reliable property models for the material is often mandatory for successful modeling. This chapter focuses on the flow process itself, and as a result, this section concentrates on flow properties. Although they have significant impact on fully coupled flows, heat and mass transfer material properties including thermal conductivity, specific heat, and diffusivity are discussed in the respective chapters in this handbook.

3.4.1   Density

Density is defined as mass per unit volume and is one of the two most important physical properties (the other being viscosity) in fluid flow analysis. The fluid's density can be dependent on temperature (e.g., buoyant flows) or pressure (e.g., compressible flows). Typically, the food industry deals with incompressible fluids (except for high pressure processing of food) or fluids that may only have temperature dependency. Therefore, the fluid's compressibility is not considered in this chapter. However, temperature or species dependency of the fluid properties such as density, viscosity, conductivity, etc. often need to be considered.

The constant density assumption states that the density has no variation and can be expressed as a constant, ρ_0:

$$\rho = \rho_0.$$

Substituting this into the Navier–Stokes equations would simplify them by eliminating variations on density. However, density's temperature or species dependence results in buoyancy effects that significantly alter the flow and need to be considered. This is often accomplished by making use of the so-called *Boussinesq approximation* that assumes the density's variations of density only dependent on temperature or species, and they only affect the gravitational force term. With this assumption, the density can be split into a constant ρ_0 and a temperature (or species) dependent term, ρ:

$$(\rho - \rho_0)\mathbf{g},$$

where the temperature dependence of the density can be expressed by the relationship

$$\frac{\partial \rho}{\rho} = -\alpha \partial T, \tag{3.10}$$

where α is the thermal expansion coefficient. Substitution of this expression into the set of flow equations will allow modeling of incompressible flows with variable density model.

While searching for density, two other definitions may be found: *specific weight* and *specific gravity*. Specific weight is defined as weight per unit volume, whereas the density is defined as mass per unit volume.

$$\text{Specific weight} = \rho \times g.$$

Specific gravity is slightly different. It is the ratio the fluid's density to the density of water at the standard temperature and pressure. At 4°C, the density of water is 1000 kg/m³.

3.4.2 Viscosity

Along with the density, viscosity is an important property that characterizes the flow resistances in fluid flow problems, and it is the property that determines the rheology of the material. Viscosity relates the shear stress in the fluid to the rate of deformation of the fluid, and it often has a complex, non-linear behavior.

A fluid with linear relationship between stress and rate of deformation is called Newtonian fluid, and the rate of proportionality is referred to as viscosity. For a simple 1D case, the Newtonian law of viscosity can be written as

$$\tau = -\mu \frac{du}{dy} \tag{3.11}$$

where τ is the stress exerted by the fluid on the wall, du/dy is the velocity gradient perpendicular to plane of shear, and μ is the viscosity.

The ratio of absolute viscosity to density of fluid is referred to as kinematic viscosity, ν,

$$\nu = \frac{\mu}{\rho}. \tag{3.12}$$

In the SI system, μ has units of N s/m² (or Pa s), and it includes a measure of force. Kinematic viscosity has units of m²/s and does not have a dependence on force.

A fluid that does not obey the relationship in Equation 3.11 is called a non-Newtonian fluid. A material that has a time-dependent stress response to both the strain applied and the strain rate at which it was applied is called a viscoelastic material. There are many viscosity models, depending on whether the fluid is inelastic, elastic, or viscoelastic; shear rate dependent or temperature dependent, etc. Most of these models are based on heuristic or empirical data and are applicable for only some types of fluid or some range of values. The success of a simulation involving complex fluids may often depend on good selection of a model that describes its viscosity (often called the apparent viscosity). Some commonly used non-Newtonian viscosity models for shear-rate dependent fluids are given in Table 3.2. For many food materials, the rheology is unknown, and these models would only provide an approximation. In most cases, the viscosity models need to be adjusted by selecting suitable coefficients or a characteristic yield stress to determine the applicability of a model over the range of operating conditions. These parameters are often obtained by trial and error from experiments or from a series of simple numerical simulations that can be compared to some experiment. Time spent obtaining reliable rheological data for the food material is extremely valuable for successful CFD simulation and can, in some cases, be the bulk of the effort.

When temperature dependence is to be considered, it should be combined with the shear-rate dependency of viscosity that is often expressed as

Table 3.2 Commonly Used Viscosity Models for Non-Newtonian Fluids

Viscosity Model	Formulation
Power law: Commonly used for food materials with high shear rates; commonly used for shear thinning foods	$\mu = K(\dot{\gamma})^{n-1}$
Bird–Carreau: Commonly used for low-shear-rate dependency of viscosity such as doughs	$\mu = \mu_\infty + (\mu_0 - \mu_\infty)(1 + \lambda^2 \dot{\gamma}^2)^{(n-1)/2}$
Carreau–Yasuda: A variation of Bird–Carreau with an additional exponent	$\mu = \mu_\infty + (\mu_0 - \mu_\infty)[1 + (\lambda\dot{\gamma})^a]^{(n-1)/a}$
Cross law: Commonly used for low shear rate dependency of viscosity (similar to Bird–Carreau)	$\mu = \frac{\mu_0}{1+(\lambda\dot{\gamma})^m}$
Bingham: Commonly used for food materials such as yogurt where a constant yield stress is required (sometimes, these equations are written in terms of yield stress rather than shear rate)	$\mu = \begin{cases} \mu_0 + \dfrac{\tau_0}{\gamma} & \dot{\gamma} \geq \dot{\gamma}_c \\ \mu_0 + \tau_0 \dfrac{(2-(\dot{\gamma}/\dot{\gamma}_c))}{\dot{\gamma}_c} & \dot{\gamma} < \dot{\gamma}_c \end{cases}$
Modified Bingham: An analytical form of Bingham law that may be easier to calculate and provide more stable solutions in numerical computations	$\mu = \mu_0 + \tau_0(1 - \exp(-m\dot{\gamma})/\dot{\gamma})$, where $m = 3/\dot{\gamma}_c$, such that the standard and modified Bingham laws exhibit the same behavior above the critical shear rate, $\dot{\gamma}_c$
Herschel–Bulkley: Used for similar materials as Bingham law but incorporates shear-thinning behavior as well	$\mu = \begin{cases} \dfrac{\tau_0}{\gamma} + K\left(\dfrac{\dot{\gamma}}{\dot{\gamma}_c}\right) & \dot{\gamma} > \dot{\gamma}_c \\ \dfrac{\tau_0\left(2 - \dfrac{\dot{\gamma}}{\dot{\gamma}_c}\right)}{\dot{\gamma}_c} + K\left[(2-n)+(n-1)\dfrac{\dot{\gamma}}{\dot{\gamma}_c}\right] & \dot{\gamma} \leq \dot{\gamma}_c \end{cases}$
Log–Log: Purely empirical law that sometimes provides better fit for experimental data	$\mu = \mu_0 10^{a_0 + a_1[\log(\dot{\gamma}/\dot{\gamma}_c)] + a_{11}[\log(\dot{\gamma}/\dot{\gamma}_c)^2]}$, $a_i =$ the coefficients of the polynomial expression

μ, viscosity; μ_0, zero-shear-rate viscosity; λ, natural time (i.e., inverse of the shear rate where the fluid changes from Newtonian to power-law behavior); γ, shear rate; γ_c, critical shear rate; τ_0, yield stress; K, consistency; n, power-law index.

$$\mu = H(T)\mu_0(\dot{\gamma}), \tag{3.13}$$

where $H(T)$ is the Arrhenius law or other applicable laws describing temperature dependence, and $\mu_0(\dot{\gamma})$ is a shear-rate dependent viscosity (as described above) at some reference temperature.

3.4.3 Vapor Pressure

Vapor pressure is the pressure exerted by a substance's vapor at equilibrium with its liquid and solid phases at any given temperature. This is important when studying evaporation and condensation from foods.

3.4.4 Surface Tension

A fluid's surface is defined as the interface between the fluid body and its surroundings that can be another immiscible fluid, a different phase of the fluid itself, or an open surface to atmosphere. Surface tension is a force that exists on that interface and that causes that surface layer to act like a film or sheet, wrapping around the volume and separating it from its surroundings. Although this force is always in the tangential direction of the surface, its net influence is always in the normal direction. It is an important property in flows with two or more immiscible fluids. Temperature dependence of this property is an important factor in determining the shape of the interface or the flow direction.

3.5 FLOW TYPES

Fluid flows can be classified in many different ways. Whereas some classifications are based on the flow regime (e.g., laminar or turbulent), others are based on the number of phases present (e.g., mixture of immiscible fluids or solid, liquid, and gas phases). The flow type not only determines the nature of the flow but also whether or not additional sets of equations are needed.

3.5.1 Flow Regime

The flow regime is determined by one of the most well-known nondimensional numbers in engineering: the Reynolds number, Re. The Reynolds number relates the inertial forces to viscous forces in the flow and is defined as

$$Re = \frac{\rho u L}{\mu} = \frac{uL}{\nu}, \tag{3.14}$$

where L is characteristic dimension, and u is the characteristic velocity. For example, for a circular pipe, L is the diameter, and u is the average velocity. For a noncircular pipe, a hydraulic diameter is used as characteristic dimension and is given by

$$\text{Hydraulic Diameter} = \frac{4A}{P}, \tag{3.15}$$

where A is the cross section area of the pipe, and P is the wetted perimeter.

Very low values of Re ($Re \ll 1$) characterizes flows with extremely small velocity or very high viscosity. Such flows are referred to as creeping flows. In these flows, inertial effects can be ignored, meaning that density will not be an important variable. If the Reynolds number is very large and there are no wall effects, viscous effects can be ignored, and the flow is solved as nonviscous flow (called *inviscid flow*). Moderately low levels of Re characterize laminar flows. For pipe flow, the laminar flow regime exists when Re is less than 2000–2200. As Reynolds number increases, the flow goes through transitional into the turbulent regime. Turbulent flows require additional modeling techniques or additional set of equations to be solved and coupled with the flow equations. Turbulence has significant impacts on energy dissipation (both inertial and heat).

To successfully model turbulent flows, an adequate knowledge of such flows and of turbulence models is needed to properly choose the appropriate turbulence model. Such knowledge is also important to determine if computed results are realistic. The purpose of these models is to quantify effects of turbulence to the flow. This is often done by calculating a new flow property, turbulent viscosity. This turbulent viscosity is a calculated quantity. There are several models based on the number of additional equations solved for this. However, invoking additional equations entails the solution of additional transport equations that can significantly increase the CPU requirements of the numerical solution. Several models are available, including mixing length model, standard k–ε, and k–ω. In most food applications, the standard k–ε model will be satisfactory and is the default model in most commercial software products.

3.5.2 Flows with Multiple Phases

There are three phases of matter: solid, liquid, and gas. However, in fluid flow, a much broader definition is used to distinguish different materials as well as phases of the materials. Therefore, there may be multiple liquids as well as solid, liquid, or gas phases of different materials coexisting in the flow. Flows involving different phases of matter can be categorized in two major groups: the so-called free-surface flows and multiphase flows. Free surface flows are those where there is a clear and continuous interface between phases. In this type of flows, interface location and its dynamics are of typical interest. However, there are many applications where the flow involves a mixture of particulate matter with liquids or gases (as in slurries, food particles suspended in liquids, cyclones, dryers) where the interface is not clearly defined or is of a much smaller scale to be explicitly defined. Such flows are termed as multiphase flows and need a different kind of formulation. In this type of flows, fluid particle interactions are of typical interest.

Most food products are essentially a mixture of multiple ingredients, including emulsions and suspensions. These products have been simplified as homogeneous mixtures, and the behavior has been captured in their rheology. New models such as population balance in conjunction with CFD offer more tools to engineers to more accurately handle complex physics. The following sections briefly describe various models related to fluid flow involving different phases of matter.

3.5.2.1 Free-Surface Flows

If the flow involves free surfaces or an interface between different fluids or different phases of matter (such as filling processes, extrusion, bubbly flows, droplet laden gas flows, flows with melting or freezing interface), good data on the surface tension of the fluid is needed. Surface tension determines the behavior of the interface between the two phases and will have a great impact in accurate prediction of the bubble's or droplet's size or shape or the filling's behavior (splashing, separation, or break up). The position or motion of the free surface is governed by the balance of forces; in steady state flows, the net force on the interface must be zero. The forces that act on the interface are the pressure and shear forces on either side of the interface and the surface tension force. Surface tension is a force acting in the tangential direction of the interface, but its net influence on the interface is in the normal direction (therefore, it defines the shape and motion of the interface). There are various classes of free surface problems such as seen in Figure 3.2, each requiring a different setup.

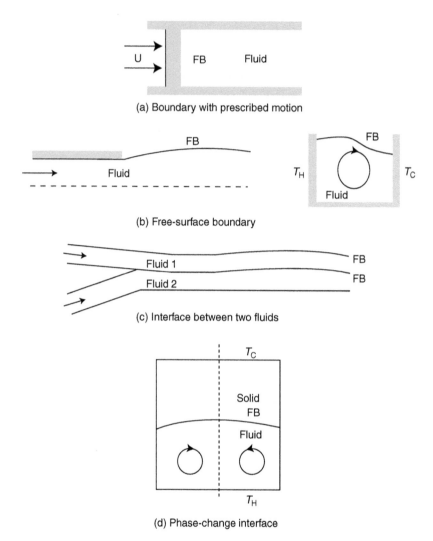

Figure 3.2   Different types of free surface flows (FB indicates free-surface boundary and T_C and T_H refer to cold and hot temperatures).

3.5.2.2 Discrete Particle Model

The discrete particle model is often used when the dispersed phase is relatively low in concentration relative to the main carrier fluid. A particle can be solid, droplet, or bubble in liquid or gas phase. This is characterized by a low volume fraction of the particle phase (that may have a high mass fraction relative to the fluid). In this limit, the particle–particle interactions can be ignored, and the only interaction considered is the particle–fluid interaction. The particles can exchange momentum, mass, and heat with the surrounding fluid. The model is based on solving the equations of motion on individual particles in Lagrangian (moving) frame of refer-. ence, and the fluid is solved in the Eulerian (fixed) frame of reference. And at any point in time, the conservation principles are applied for the transfer of momentum, mass, and heat between the two phases.

The forces acting on a particle can be the sum of pressure forces, buoyancy, and external forces. Once the forces are defined, the motion and trajectory of the particles can be predicted. Even for steady state flows, the particle trajectory needs to be integrated over time. If the particles are very light relative to the fluid or the concentration of particles is very low, the presence of the particulate phase will not disturb the flow, and the only interaction is from the fluid to the particulate phase. In this case, the equations for the dispersed phase can be solved independent of the flow equations. However, if the particles do impact the flow field, the equation of motion needs to be coupled with the fluid equations.

3.5.2.3 Multiphase Models

If the concentration of the particulate phase is high and particle–particle interactions cannot be ignored or if large volumes of immiscible fluids are present, the discrete particle model assumption fails, and the multiphase model should be used. Multiphase flows can be categorized in four regimes.

- Gas–liquid or liquid–liquid flows (e.g., bubbly flow, droplets, slug flow, immiscible fluids with clear interface)
- Gas–solid flows (e.g., particle laden gas flow, fluidized beds)
- Liquid–solid flows (e.g., soups, slurries, sedimentation, hydrotransport)
- Three phase flows (a combination of any of the above)

Three different models are commonly used to treat multiphase flows: the Volume Of Fluid (VOF) model, the Mixture model, and the Eulerian multiphase model.

VOF Model. The VOF model is essentially a surface tracking technique that is used when the shape of the interface between phases is important, yet the deformations of the interface are too large to be tracked explicitly as in the Free-Surface model or if the interface breaks up or coalesces that cannot be modeled by the Free Surface technique. VOF allows for break-ups, agglomerations, and large displacements (such as in jet break up and deformations, filling process, sloshing). Instead of tracking the exact interface, in VOF, the volume fraction of the secondary phase tracks and helps analytically reconstitute the interface.

Mixture Model. This can be viewed as a simplified Eulerian model. It is used for two or more phases where the momentum equation is solved for the mixture, and it prescribes the relative velocities of the phases. It is applicable when the loading of the secondary phase is light such as in bubbly flows, sedimentation, or cyclone separators. The mixture momentum equation is obtained by summing up individual momentum equations whereby the physical properties and solution parameters become mixture properties and mixture parameters.

Eulerian Model. In multiphase models, because the volume cannot be occupied by more than one phase at a time, the concept of volume fraction for each phase is introduced. Each volume fraction is assumed to be continuous in space and time, and they must all add up to one. Conservation principles are applied to each phase, and the equations are often closed by empirical relationships between pressure and interface exchange coefficients, or in some cases, by kinetic theory. In this model, each particulate phase is treated as a separate continuum with its own set of continuum equations and associated properties. For this approximation to be realistic, the particles in the continuum must behave similarly. Therefore, not only should they have the same physical properties and composition, but the size of the particles must also be relatively similar so that they will behave the same under a given force field. As a result, the particulate phase is typically broken down into sets of similar material and size, each different size of the same material being treated as a different material. This is the most complex and general model for the simulation of multiphase flows, and it requires additional computational resources to solve the large set of partial differential equations (PDEs) that result. Typical applications of this model include bubble columns and particle suspension.

3.6 TYPICAL BOUNDARY CONDITIONS

The solution of the governing equations requires appropriate boundary conditions and initial conditions for the fluid domain. The equations are valid for most flows, and their unique solution depends on the specification of flow conditions at the domain boundaries. The boundary conditions can be thought of as operating conditions. Depending on the problem, either a degree of freedom can be constrained (e.g., defined velocity component or pressure), or surface forces or fluxes can be applied (e.g., mass flux) on any boundary. The former is also referred to as Dirichlet boundary conditions and the latter as Neumann boundary condition. A third type of boundary condition is called Robin boundary condition where a linear combination of the solution and its normal derivative is specified.

Pressure, often specified as a boundary condition at the inlet or outlet, is a critical flow parameter that directly impacts the solution. Therefore, it is important to understand what is meant by pressure when setting up a problem as it has more than one representation. *Pressure* is defined as the normal force per unit area exerted on a surface immersed in a fluid. There are several ways this pressure can be expressed:

- *Atmospheric pressure (static pressure)* is the pressure exerted at the surface of a body by a column of air in the atmosphere. *Standard pressure* is the average atmospheric pressure at sea level; it is defined as 1 atm on Earth, which is equal to 760 mm Hg or 101,325 Pa.
- *Dynamic pressure* is the pressure that represents the fluid kinetic energy,

$$P_{\text{dynamic}} \equiv 1/2\rho v^2. \tag{3.16}$$

- *Total pressure* is the sum of dynamic and static or hydrostatic pressure,

$$P_{\text{total}} = P_{\text{static}} + P_{\text{dynamic}}. \tag{3.17}$$

When specifying pressure boundary condition, it is very important to understand which definition is appropriate and is needed for the solution. Table 3.3 lists commonly used boundary conditions. One should be careful not to over constrain the system of equations by specifying too many boundary conditions (for example, specifying velocity at both inlet and outlet).

3.7 FLOW MODELING

The traditional engineering approach to process and equipment design is experimentation: build laboratory scale prototypes; take samples or measurements; rely on heuristic or experimental data available. These methodologies have inherent difficulties.

Table 3.3 Common Boundary Conditions

Boundary Zone	Boundary Condition
Inlets/Outlets	Velocity components (plug flow or constant viscosity or fully developed profile)
	Pressure drop and flow direction
	Scalar variables, e.g., temperature or species concentration
	Kinetic energy and eddy dissipation for turbulent flows
Walls (solid surfaces)	Fluid sticks to the walls (no-slip condition)
	Moving walls
	Flow through the walls (porous surface)
	Surface reactions at wall boundary
Other Zones	Symmetry where normal gradient is zero
	Periodic where only part of the domain is modeled

- Prototyping is expensive; therefore, the number of prototypes that built should be limited.
- Laboratory-scale prototypes do not reflect the behavior of full-scale models; therefore, scale-up models should also be performed.
- Building prototypes can also take a long time for a very short test.
- Collecting accurate and relevant data can be difficult. Often, the process of interest is inaccessible for such intrusion and severely limits the amount of data that can be collected.
- Taking measurements during the process can often alter the process itself thereby reducing the reliability of the experiment.

It is almost impossible to test each and every scenario. Therefore, several techniques are used to extend the experimental results to untested situations. Some of these techniques include similitude, dimensional analysis, and modeling. As elucidated in the previous chapters, food processing operations involve several steps where engineering analysis or modeling can be extremely helpful. A thorough application of physical principles leads to a better design tool that has the potential to change the dynamics of food engineering research toward right by design rather than build and test. Modeling can lead to virtual experimentation where initial designs can be tested before a pilot plant testing or full industrial testing can take place at the end.[3-4] Software simulations have many advantages over traditional methods:

- *Time Saving.* Typically, computer models can be built much faster than prototypes, and they can be executed faster than running an experiment.
- *Cost Saving.* Computer software and the hardware necessary to run it typically costs far less than building a prototype or running an experiment in the lab. By reducing the number of actual prototypes, simulations can reduce the cost of research and development for product design and improvement.
- *Nonintrusive.* Computer software deals with a virtual model and eliminates intrusion into the process or any hazardous conditions that may exist.
- *Extensive Information.* The software will generate any data of interest throughout the whole domain of interest. It is like putting thousands of thermocouples in the experimental unit to measure the temperature.
- *Parametric Study.* With software, many "what-if" scenarios can be executed to gain more insight about the design or the process at hand.

Software simulations have been extensively used in many computer-aided design (CAD) and computer-aided engineering (CAE) applications in a wide range of industries. They have been widely accepted in aerospace and automotive industries as a powerful engineering design tool and are now finding their way into the food and beverage industry. Computational fluid dynamics (CFD) is one of the tools used in this mix of computational tools available to today's food engineers. Realizing their benefits and returns on investment (ROI), many leading food companies have been using these tools for many years for many kinds of industrial problems. The applications range from aerodynamics of a potato chip in a dryer[5] to oil flow in industry scale fryers to studying the detailed flow and thermal field in a mixing tank. They can be used as aids to scale-up processes from lab or pilot scale to full production. Troubleshooting the performance of existing equipment is often carried out with the help of CFD. Determining how an existing piece of process equipment will operate under new conditions or with new input materials is also a common task for CFD.

CFD is a form of numerical experiment that can elucidate flow and thermal fields in a manner not achievable in a real food engineering experiment. Such numerical experiments carried out parallel with the physical experiments can be used to help interpret these physical results and to ascertain a basic physical and phenomenological aspect that is not evident or achievable in physical

Table 3.4   Applications of CFD in Food and Beverage Industry

Process	Associated Physics/Models
Aseptic processing	Forced convection, nutrient retention, microbial kill, voltage field, and joule heating for ohmic heating
Baking ovens	Natural or forced convection and radiation
Beer and wine processes	Settling tanks or mixing tanks with multiphase
Can and bottle filling	Volume of fluids
Clean rooms, fume hoods, and ventilation	Species transfer
Cold storage, refrigerators, freezers, cooling tunnels, and food storage cabinets	Heat transfer and particle tracking
Candy making	Solidification, free surfaces for coating and enrobing, and VOF for dipping
Cyclone separators	Discrete particle methods
Deep-fat frying	Reactions
Dipping processes	VOF
Dough sheeting	Free surface
Drying	Fluidized beds, forced convection, and moisture content predictions
Equipment design (flow meters, heat exchangers, pumps, fans, etc.)	Multiple physics
Extrusion (single or twin-screw)	Complex geometry, viscous dissipation, complex rheology, screw design, and die design
Ice cream freezing, thawing, and coating	Solidification, freezing, and enrobing using VOF or free surface
Mixers (draft tube mixing, static mixers, mixing tanks)	Mixing tank and multiple phase
Packaging	Thermoforming and blow molding models
Pasteurization	Natural convection
Soda dispensers	VOF, multiphase, and mixing
Sprays and spray drying	Multiphase, spray models, forced convection, species transport, and moisture content predictions
Sterilization and canning	Natural convection, microbial kill, and nutrient retention

experiments. CFD is only another tool in the design and analysis of the processes of interest. It is not a substitute for measurements and experiments, but it complements other tools available to the engineer. Its purpose is to supplement knowledge about the process, and it is a tool to see through the walls and into the regions of interest without intrusion or where it may be impossible to reach by physical means. It helps eliminate the repetitive task of prototyping and parametric studies, scale-up, and the like. This is very important to understand as all CFD models are based rely on experimentation, either to provide input values such as properties and conditions or to verify the results of the simulation. CFD can be effectively used as a predictive tool if there is enough confidence in the models that is based on some validation or previous experience.

It is important to understand the methodology by which the analysis process should be started to prepare for simulations. This is specifically important in the food industry as in many cases, the properties of the materials handled are unknown, or they consist of mixtures of various materials and consistencies. The processes and the physics behind them may not be well understood and they may rely on past experience or historical data.

Flow modeling is not a new concept. It has been used for several decades in many industries. Engelman and Sani[6] were the first to apply a general-purpose CFD code to a food application. In this case, they successfully simulated natural convection profiles, including the pasteurization of beer in glass bottles that, in fact, is a time-dependent problem as beer bottles travel through different heating and cooling sections. Datta[7] developed his own code for natural convection and conduction heat transfer in a water-like food in a can undergoing sterilization. Kumar[8] extended this work to thick and thin soups, obeying non-Newtonian rheology both for canning and for aseptic processing while using the same code as Engelman and Sani.[6]

There are several review papers in the literature including those by Puri and Anantheswaran,[9] Scott and Richardson,[10] and Wang and Sun.[11] Scott and Richardson[10] outlined the trends in food processes' modeling. Flow modeling has been used in analyzing pasteurization of beer,[6] thermal processing,[7,8,12–17] aseptic processing,[18–21] dairy applications,[22] extrusion,[23,24] mixing,[25] spray drying,[26] cooling and refrigeration,[27,28] thin film UV reactor,[29] and even sucrose crystallization.[30] Table 3.4 lists some applications and associated physical models where CFD is either being used or can be used in the industry. This list is not exhaustive as engineers are continuously simulating new and challenging problems. In this section, a brief overview of CFD techniques is presented. Then the power and utility of these techniques for solving practical problems, via a series of examples, are illustrated.

CFD involves transforming the governing equations to a set of algebraic equations (called discretization) and breaking up the solution domain into many small cells or elements (called meshing) over which those algebraic equations can be numerically solved. The results are then visualized in terms of graphical displays called post-processing. The following sections describe these steps in more detail.

It is not necessary to write one's own software program to solve the flow equations. In today's marketplace, several commercial CFD packages are available, including FLUENT™, FiDAP™, POLYFLOW™, FloWizard™, and ANSYS-CFX™ from ANSYS, Inc.; and STAR-CD™ and STAR-CCM+™ from CD-Adapco. A complete and up-to-date list can be found at http://www.cfd-online.com. These commercial software programs should give similar results for a given problem provided the necessary capabilities are available in each. The following sections describe the steps involved in setting up a CFD problem in more detail.

3.8 DISCRETIZATION

The full conservation equations' numerical solution involves converting the continuous domain of the set of partial differential equations (PDEs) to a system of algebraic equations. This is accomplished by approximating the flow variables by simpler algebraic functions defined over a small arbitrary volume, typically referred to as a control volume, and substituting these functions in the original equations. This leads to a set of simultaneous equations where the unknowns are values of the solution variables at discrete locations throughout the domain. The approximation method used defines a specific discretization method such as finite difference, finite volume, finite element, spectral element, and boundary element. In most commercial software packages, the discretization method is transparent to the end user and is automatically done by the software; however, understanding the concept is still valuable for the engineer.

The main ingredients of the finite volume method will be briefly described here. First, the domain of interest is divided into a number of small cells or volumes. The governing equations are then integrated over each volume. The derivatives with respect to coordinates and time are replaced with difference equations. The result is a series of simultaneous algebraic equations of the form

$$a_P \phi_P = \sum a_{nb} \phi_{nb} + b, \tag{3.18}$$

where ϕ_P is the value of the unknown quantity (e.g., velocity component, energy, or species concentration) in cell P; ϕ_{nb} are the values in the neighboring cells; a_P and a_{nb} are the coefficients; and b is any source term. This approach yields a method that is inherently conservative.

Finite element or spectral methods differ in the formulation of the approximating functions and are based on a variational principle. The unknown functions are expressed in terms of well-defined linear or higher-order polynomial functions, and they are integrated over a given volume. This yields the unknowns as the coefficients of the approximating polynomials. The derivation of these equations is somewhat involved and outside the scope of this chapter. The reader is referred to finite element method reference books for further detail.[31]

The primary variables in most differencing techniques such as finite volume are evaluated at the vertices, face center, or cell center. Whereas in the case of variational techniques, the variables are obtained at integration points that lie somewhere within the volume. In variational techniques, the solution is exact at integration points, and the values are interpolated to the vertices or nodes of the element for post-processing and continuity purposes.

The algebraic equations obtained can then be solved on a computer directly or iteratively by use of one or more of many well-established numerical solution methods, either by direct integration or by iterative solution techniques. Direct solvers yield more accurate solutions; however, they require very large memory. Iterative solvers require relatively small memory, but they may lead to convergence and numerical stability issues for complex flows or complex rheologies of the fluid. The set of equations can be solved in a fully coupled manner (if the variables are strongly dependent on each other), completely decoupled (if the variables have negligible interdependency), or partially coupled.

3.8.1   Meshing

Now that the PDEs are converted to a set of discretized algebraic equations, the locations where the discrete values will be computed need to be defined. This requires subdividing the region of interest into small, discrete volumes, and the approximation functions can be evaluated either at the center of the volume or the vertices. This method of subdividing the original geometry into small discrete volumes is commonly referred to as meshing. Meshing is one of the most important and time consuming aspects of CFD modeling. Yet, steady advances in meshing technology and modern software tools make this task easier and more transparent. The subdivision of the geometrical domain into smaller volumes allows mapping the discretized equations to these small domains. Therefore, the discretized equations are no longer applied to arbitrary volumes but to the control volumes obtained by meshing technique as shown in Figure 3.3 below.

As shown in Figure 3.4, there are many ways of meshing a simple circle. They range from mapped mesh (structured, curvilinear rows and columns of mesh) to completely paved mesh (unstructured distribution of cells of various shapes and sizes) to a mesh with boundary layers to resolve gradients near the wall. A mapped mesh gives more control to the user whereas an

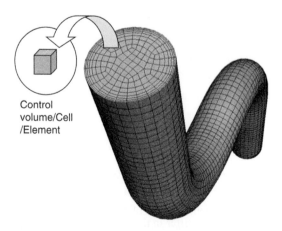

Control
volume/Cell
/Element

Figure 3.3   Meshing of a pipe into control volumes.

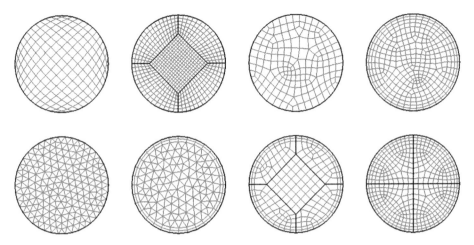

Figure 3.4 Different choices of meshes in a simple circular cross-section of a pipe.

automatic paver creates the mesh based on the algorithm it is based on. The choice of meshing technique depends on the problem and the physical properties at hand. Figure 3.5 illustrates the application of multiblocked mapped mesh to a cyclone separator. The finer the mesh, the better the approximation. However, the mesh's density also directly correlates with the number of unknowns to be solved and has a direct impact on the computational effort to solve the problem.

The challenge is to know the type and size of the mesh that are appropriate to capture the details of the problem without exceeding the available time and computer resources. Furthermore, when solving coupled phenomena such as flow and heat or mass transfer, it is very common for the different phenomena to have different dimensional scales, requiring different mesh sizes, at different locations of the domain. Ideally, the final solution should be independent of grid size, meaning the solution should not change if mesh is further refined.

3.8.1.1 Modeling: Steps Involved in Setting Up a CFD Problem

One of the main goals of using CFD, as opposed to experimentation, is the time and cost savings it provides. Yet, many food problems involve complex physics and complex geometries that require

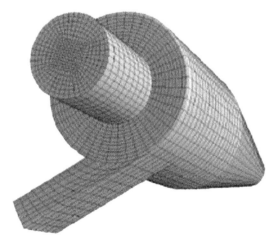

Figure 3.5 Mesh for a cyclone separator.

the solution of complex set of non-linear equations. Numerical solutions of non-linear partial differential equations require an iterative solution approach. Such CFD problems are quite large and require a substantial amount of computer time (they can vary anywhere from minutes to days of simulation, depending on the complexity of the problem). To be successful with CFD, simple steps must be taken starting with the simplest problem and then adding additional physics to build-up to the complex model. Often the simple models provide enough engineering information to solve the underlying problem as the full model. Therefore, simplification of either or both of the geometry and the physics of the problem is an important first step of any CFD study. Some of these simplifications are listed in Table 3.5 and also are elaborated in the case studies in Section 3.9.

Although it is not necessary to know the details used to solve the complex equations of CFD, it is advantageous to be aware of the basic concepts involved when setting up a problem to be solved with CFD software packages. When solving complex problems (problems with multiple physics and fluids with complex properties), the techniques involved and their limitations to achieve acceptable results must be considered. Many modern CFD software packages will hide most of the complexity from the end user. As a result, it is easy to make mistakes because of many steps

Table 3.5   Some Possible Simplifications to Simulation Problems

Domain selection	Isolate areas of interest and model them
Simulation goals	Conceptual design to study basic flow and heat transfer final design for validation
Geometry	CAD geometry vs. creating the model from scratch
Model simplification	Full 3D model vs. 2D, axisymmetric, or periodicity even for a full 3D model, start with a simple 2D or axisymmetric porous body instead of modeling tube banks, vents, and filters
Physics simplification	Transient vs. steady
	Isothermal vs. non-isothermal (conduction, convection, radiation)
	Laminar vs. turbulent
	Turbulent methods—simple models such as mixing length to standard k–ε to a large eddy simulation (LES)
	Single species vs. multiple species
	Single phase vs. multiphase
	Single physics vs. multiple physics
Material properties	Constant properties vs. complex models, describing their dependence on temperature, pressure, etc.
	If certain properties are unknown, one can still model it to find the operational window by changing this property by $\pm 10\%$ and performing parametric modeling
Mesh creation	Generate mesh with own resources
	Commercial mesh generation software
	Simplified automated mesh generation tools
	Mesh grading schemes
	Mesh sensitivity analysis using simple 2D models
Solver selection	Writing own code or commercial CFD code
	Using wizard-based CFD tool
	Full-featured CFD code
Postprocessing	Commercial code's built-in postprocessing tool
	Creating eye-catching animations using rendering software such as Fieldview™ (Intelligent Light, Inc.) or EnSight™ (CEI, Inc.)
Parametric study, optimization, and validation	Parameterizing geometries and/or operating conditions and/or material properties
	Design optimization
	Parametric study: cycling through the above steps
	When to go to prototyping or lab modeling to verify the CFD predictions
	Experimental validation
	Analytical validation
	Literature survey

being automated by the software. This section will outline the steps involved in setting up, solving, and analyzing the results for successful use of CFD in design and analysis of flow problems.

The solution of a flow problem using CFD involves the following steps:

- *Geometry Creation*—a geometric model of the simulation domain is built or transferred from a suitable CAD system.
- *Grid Generation*—the solid model is divided into many small, finite volumes (also referred to as grids, elements, or cells, depending on the discretization method).
- *Problem Specification*—the problem can be completely described by specifying which equations need to be solved, operating conditions, and material properties.
- *Solution*—the governing equations are solved to obtain the desired flow parameters.
- *Postprocessing*—the results are visualized via color contours, velocity vectors, flow pathlines, XY plots, and other qualitative and quantitative means.
- *Validation*—results are compared to available data, making sure that they are reasonable and acceptable.

In the following sections, these steps will be discussed to show some of the important things to watch for in each step.

Geometry Creation. The very first step in a CFD simulation process is identifying the geometry to use. The CFD engineer not only needs to define the shape involved, but he or she also needs to determine whether a two-dimensional approximation or symmetry along an axis (axisymmetric model) assumption applies to the flow involved or if the problem has any periodicity (it is important to distinguish the symmetry or periodicity of the geometry from that of the solution). Taking advantage of any such symmetry property avoids building a full 3-dimensional solution domain and reduces the model creation and computational time significantly. Some examples of geometric simplification are illustrated in Figure 3.6.

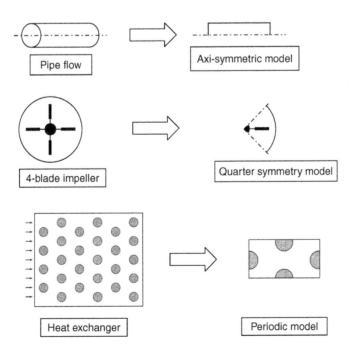

Figure 3.6 Geometry simplifications.

Once that decision is made, there are typically two choices for creating a computer model of the geometry available:

- Import drawings from a computer-aided design (CAD) program.
- Use the tools provided by the CFD application (or other preprocessor software) to build the geometry from scratch.

If the CAD model is already available, it is advisable to use this CAD model rather than recreating it in the preprocessing software. Whatever method is used to create the geometry, it is important to keep the geometry clean, meaning all boundaries of the domain must be connected with continuous lines. If there are regions in the domain that really do not significantly impact flow field, the geometry should be simplified by excluding them. Many CAD packages will tolerate gaps and holes in their model. Although those are acceptable for manufacturing purposes, CFD requires well-defined and continuous air-tight boundaries for the discretization and mesh generation to work, for boundary conditions to be applied accurately, and for continuity to be satisfied (undefined or missing boundary segments are not allowed). The following are a few specific things to watch for in defining the model geometry:

- Inlet and outlet surfaces must be defined along with all the boundaries of the domain, and they cannot be left blank.
- One can specify fully developed velocity profile at the inlet boundary to reduce the size of the computational domain rather than modeling a long pipe leading to the region of interest.
- The outlet boundary must be located far away from the recirculation regions if zero stress boundary condition is assumed at the outlet (see Figure 3.7).
- Solid regions of the geometry may or may not need to be included in the model, depending on whether or not there is heat or mass transfer that involves the solid material.
- Although, in some cases, the goal of numerical simulation is to understand the flow in a given geometry; in many other cases, the goal is actually to come up with optimal geometry for the process. In this case, the geometry (typically some specific part of the domain) becomes part of the solution. This leads to a series of simulations, each with a slightly different geometry, to better understand the process and change certain

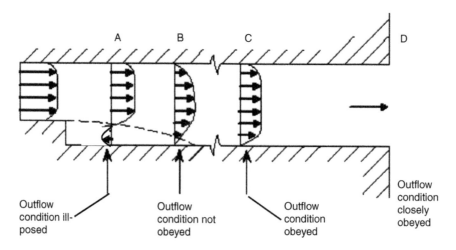

Figure 3.7   Geometry selection.

design parameters to achieve the optimal (or an improved) design. This is referred to as design optimization by parametric study, and this is where CFD can be a powerful tool because it typically requires a fraction of the cost of manufacturing the various design options. For these types of applications, the ease of changing the geometry, typically using parameters to define the boundary of interest, to allow for fast, easy modifications to the model must be considered.

Another important consideration in identifying the geometry of interest is whether or not to include all the details in the model at all. In some cases, the process involves intricate details that would be prohibitively expensive to explicitly model such as a perforated plate or a bank of tubes or a grill in a heat exchanger. In many cases, rather than explicitly defining the details of those features, an approximation can be made by using the method called *Porous Media*. This technique essentially replaces the complexity of the geometry with an effective permeability to emulate the pressure drop across the obstacle in question. If the details of the flow through this domain (e.g., the pores of the perforated plate or around the individual tubes of the heat exchanger bank) is not of primary interest and an effective pressure drop or heat transfer through this domain is satisfactory, then porous medium approximation is a very effective tool.

Mesh Generation. For simple to medium complexity problems, the automatic mesh generators available with the CFD software will do an acceptable job. However, regardless of how the mesh is created, the CFD user should have some understanding of the relationship of the mesh quality to the problem being solved. Unfortunately, in most cases, the mesh quality is a subjective concept and can only be acquired from experience on related flow simulations. However, there are guidelines one can follow to minimize the impact of bad mesh on the results.

It is not necessarily how the mesh looks on the computer screen that matters or if all the cells are nicely formed and rectangular in shape, but it is more of where the mesh density is, how it is distributed or concentrated in the geometry, and how the mesh size relates to the physical properties of the flow. For example, the flow field may be fairly regular and smooth; however, the conductivity of the fluid may be very low, requiring much finer mesh near heat sources or sinks to resolve the temperature field where the flow field would otherwise require a coarser mesh. Sometimes the location of the sharp gradients in the flow field is the result of the solution itself such as the case for shear-thinning fluids. In this case, it is difficult to predict the exact location of the high shear areas until the flow field is obtained. Typically, one would either guess the location of high shear and confirm from the solution or create fine mesh across the domain. The mesh size must be smaller than the characteristic scale of the physics being solved, or one must be able to resolve the rate of change of the solution variable. Determining where to refine the mesh is especially difficult for fluids with variable and non-linear physical properties as the location of gradients would be determined by the solution and can be difficult to predict.

There are two possible outcomes of bad mesh: a deformed mesh or an inappropriate mesh size may lead to convergence difficulties, or the program will converge but to a wrong solution. This is why it is wiser to err on the safe side and use as fine a mesh as can be afforded. Alternatively, one can perform a series of simulations with increasingly finer mesh until the change in mesh density no longer has an impact on the solution. This technique is called *mesh sensitivity analysis*. Some commercial products also have the so-called *adaptive meshing technique* that means the software will remesh the domain based on the solution, refining the mesh near high gradients.

In some cases, there is a choice of whether to use triangular mesh (triangles, tets, prisms) or rectangular mesh (rectangles, bricks, hexagonal). Typically, triangular mesh is easier to generate, but it is harder to control the number of cells generated. Rectangular mesh may be more difficult to generate for complex geometries, but it gives better control on the size and quality of the mesh. For complicated geometries, a hybrid scheme using both tetrahedra and hexahedra is good solution. Many automatic mesh generation tools make this task somewhat transparent to the user. Some sample elements or cell types are shown in Table 3.6.

Table 3.6   Types of Basic Elements/Cells

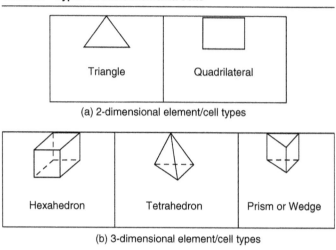

(a) 2-dimensional element/cell types

(b) 3-dimensional element/cell types

Problem Definition. Just as meshing is important for the accuracy of the numerical solution, the definition of the problem is essential for the accuracy of the physical model. The solution can only be as accurate, but not more so, than the problem definition. Problem definition includes definition of boundary conditions, definition of physical properties, coupling with other phenomena (such as heat transfer, electric or magnetic fields, body forces), and definition of other physical models to be used. Typically, this stage of the modeling offers most options for simplification by making various assumptions, and as a result, is the stage that requires most familiarity with the process and the physics involved. Leaving out an important phenomenon from the simulation will yield completely useless results. One must understand the physics and the implications of making certain assumptions or simplifications to the model, and one must be willing to accept the risks and the consequences. For example, ignoring heat transfer computation for a fluid that has significant temperature dependency of the viscosity or density will yield unrealistic flow patterns. One needs to consider questions such as is the flow transient; is the flow turbulent, and if so, which turbulence model is most appropriate; is there a mass transfer or chemical reaction that would impact the flow; what viscosity model is most appropriate for the fluid; etc.

In the food industry, the most challenging part of setting up a fluid flow problem is probably obtaining accurate physical properties and data for validation.[32] Most fluids in the food industry have either very complex rheology or unknown rheological properties. Furthermore, recipes in the food industry and other fast moving consumer goods frequently change. Variability of material properties is also very common (e.g., moisture content, fat content). Obtaining accurate data for the material of interest if often challenging and requires its own study, simulation, or experimentation. However, this is probably a very good investment in time if one wishes to get good results from the flow simulation. The lack of physical data is typically addressed by performing a series of parametric studies called numerical experimentation. Similar to parametric experimentation for design, parametric study of fluid properties ($\pm 10\%$) or physical models is a very common use of CFD to better understand the process involved and to identify the process window.

Solution. The solution phase of CFD simulation is the most compute-intensive part. Once the mesh is defined and the physical model is specified, the only control left is selection of numerical solution options. Most commercial software will provide more than one solver, e.g., segregated solver, direct solver, and coupled solver. Different physical models would benefit from different solvers. The choice of solver is often dependent on the models being used in the simulation. Therefore, the reader is referred to the documentation of the software being utilized.

3.8.1.2 Postprocessing

This is the most rewarding part of a CFD simulation process. It should be mentioned that although it is very easy to get colorful graphics or animations from CFD, it is imperative that these results are realistic. This requires some knowledge of the process being studied. A careful examination of certain flow behavior can give valuable insight into if the solution is realistic or not. A CFD analysis provides tremendous information about the flow field. Graphical plots can be obtained for numerous quantities, including velocity field, pressure distribution, temperature distribution, heat flux calculation, shear rate or shear stress distribution, recirculation zones, trajectories of massless particles (similar to injecting a dye in the model) or particles with mass, derived quantities such as drag force, streamlines, etc.

In some cases, a quantity's values at some specified locations are of primary interest. Possibly, some experimental data are available at those specified locations that can be compared to the simulation results. In most cases, especially in design work, it is not the value at a certain location, but the overall flow pattern, pressure drop (relates to power savings for example), recirculation zones, and similar global features that are of interest. This is where postprocessing becomes a powerful tool. It lets one see inside the fluid and inside the walls—the invisible and the inaccessible. It is similar to an MRI or CAT Scan of the system, yielding tremendous information for further analysis.

3.8.2 Modeling: Steps Involved in Engineering Analysis Using CFD

3.8.2.1 Data Collection

The very first step before attempting any simulation is to collect data about the materials of interest. The results obtained out of a simulation are as good as the data put into it. It is crucial to have good data about the process and the material properties before attempting the actual simulation. Often, the information needed is not available, and one needs to actually run experiments or simulations to obtain the data needed for the process simulation. Reliable physical property data is essential for reliable results. If the property information is not readily available, it can be obtained from experimental techniques. If experiments are not feasible, a series of simple numerical simulations can be performed specifically for the purpose of predicting the behavior of the fluid at hand.

3.8.2.2 Experimental Validation

The next step before starting a simulation is to define a process or an experiment to compare the results against. This may involve defining a test case to experimentally reproduce the process as well as by simulation. This test case helps verify the property data collected in step one, and it increases confidence in the simulation (both the simulation software and the physical models utilized). The validation could also be directly against the process being simulated. If specific, reliable data can be collected from the process that can be compared directly to the simulation results. This latter approach is typically preferred as it reduces the cost of intermediate steps, and it verifies the simulation directly against the process. However, not all processes allow collection of measurements or data, either because the data collection may significantly interfere with the process, or the process environment can be hostile and may not allow access.

3.8.2.3 Refer to the Code's Validation

Most commercial codes include validation examples. Some of these validation examples are based on historical benchmarks that every CFD code has to perform. Other validations may be based on the experimental data. Unfortunately, most of the industrial data is proprietary. Most companies perform their internal benchmarks to their own specifications before accepting the results of a simulation. Without validation, the colorful results may look realistic, but they can be inaccurate because of improper selection or oversimplification of geometry, physical models, boundary conditions, or properties.

3.8.2.4 Parametric Studies and Design Optimization Using CFD

The next step is to start the simulation. Very few simulations will yield the result looked for in one attempt. The purpose of doing the simulation is to better understand the system of interest, to improve the process, or to study the different conditions it may operate under. All of these cases suggest an iterative process. One would typically start with a base case scenario, compare the results to existing conditions and verify, then start varying certain process parameters to see how a change in them affects the end product. In most cases, there are numerous parameters available for the designer to play with in different combinations. These parameters can be geometrical, physical property, environmental operating conditions, and the like. This series of simulations to experiment with various parameters is referred to as *parametric study*. Each parametric study has a range of operating conditions that the process engineer is interested in that is called the process window. Defining a good and focused process window will help reach the simulation goal much faster. Therefore, good planning and good understanding of the process window is essential to successful simulation.

The traditional way to optimize a given geometry is by trial and error. The same concept extends to CFD: re-draw a component in CAD, then re-mesh and re-apply the various boundary conditions required to solve the problem. This process is tedious and difficult to automate. New tools are coming into the market where they integrate some of the optimization steps in the CFD, and these tools work with CFD side-by-side. One such tool is Sculptor™ (Optimal Solutions Software, LLC, Idaho Falls, Idaho) that allows parametric shape deformation of the CFD mesh, vastly reducing the man-hours and computational requirements for design optimization. Some CFD software includes some form of shape optimization. For example, inverse die design features can compute the shape of the outlet of an extrusion die needed to produce the desired product shape (therefore taking into account die swell effects).[4] Because of the swelling of the material at the die exit, the actual shape of the die can be quite different than the final extrudate shape. As a result, estimating the actual die shape that yields the desired extrudate is often a challenging task. Such inverse die design features in the CFD software can be a very useful tool in such applications.

3.8.2.5 Limitations

CFD has been around for the past few decades, and it is still in the process of evolving, adding new physical models, being able to handle more complex geometries, and solving larger problems thanks to the increased speed and capacity of modern computers. It can help the engineer solve more problems with more physical models. The dynamics of fluids, coupled with heat transfer and other complex phenomena such as particulate matter and electromagnetic heating, make modeling of the whole process extremely complex, especially in the food industry. As with any numerical modeling approach, CFD has certain inherent limitations. These are described next.

3.8.2.6 Assumptions

Certain assumptions about the process are needed to bring the problem at hand to a reasonable size. The assumptions are typically simplifications to the model. For example, one may assume constant property where the property may, in reality, change as a result of temperature, or one may assume the fluid is similar to a better known material that allows the use of its well-known properties. These assumptions are extremely important to get realistic results. With incorrect assumptions, the solution will give wrong results accurately.

3.8.2.7 Accuracy of Property Data

Getting a good description of the material is the next important factor in the process. As previously mentioned, the quality of the material property will determine how realistic the simulation is. A model with good assumptions and bad property data is like driving a good car with the wrong directions. It is important to spend the time doing research, experiments, or parametric studies to get data as realistic as possible. This builds confidence on the simulation.

3.8.2.8 Physical Models

Once the material is identified, the next steps are to identify what type of physics is involved and to select the relevant physical models. Is there heat transfer? Is there more than one phase? Are there chemical reactions or non-linear properties? Is the flow laminar or turbulent? If it is turbulent, which turbulence model is appropriate? Note that the physics may be present in the process; however, they may not be relevant in the simulation and can be excluded for the flow phenomena.

3.9 STEP-BY-STEP CASE STUDIES

This section illustrates how one can solve industrial problems with CFD using the lessons learned in the preceding sections. The first example illustrates the basic concepts of setting up a CFD problem using a simple axisymmetric pipe flow, showing the various steps of the modeling. Even for complicated geometries, it is often recommended to start with a simple two-dimensional model with a simple physics and to gradually increase the complexity of the problem. This approach will help an inexperienced engineer gain some insight into and build confidence in the model and the tools available in the software before attempting the full model.

The second example illustrates the application of CFD by a small manufacturing company in trouble shooting as well as developing recommendations for a new design that solves the problems in the equipment.

The third example illustrates the use of CFD in understanding the flow of dough in an extrusion die head. This example combines complex geometry and pertinent meshing techniques, as well as physics to include heat transfer, viscous dissipation, and food rheology.

3.9.1 Case Study 1: Flow through a Pipe

3.9.1.1 Problem Description

This is a common application where liquid food such as soups flow in a tubular heat exchanger for aseptic processing. The objectives of this case study are to apply recommendations from the previous sections, make reasonable assumptions (refer to Table 3.5), postprocess the results, and then validate these results against an analytical solution. Such a simplified model is often used before starting a complex model, typically to validate the input data and physical models of the software. Table 3.7 shows a representative checklist specific to this simulation. It is advised that the engineer creates a similar checklist for new models.

This case involves a viscous fluid (such as soup) flowing through an infinitely long pipe of circular cross-section with radius, R. Initially, the fluid is at rest. At the instant $t=0$, a constant pressure gradient, dP/dz, is imposed along the pipe. The fluid begins to move under the influence of viscous and inertia forces, and the velocity profile asymptotically approaches the parabolic profile of steady Poiseuille flow. The same profile is also referred to as fully developed velocity profile. Instead of applying a constant pressure gradient in the pipe, often a plug flow velocity (calculated from volumetric flow rate and cross-sectional area) is used to simplify the problem. After a certain axial distance from the inlet, the flow becomes fully developed. To keep the analysis tractable, only steady state analysis with constant inlet velocity will be carried out. Using the properties and boundary conditions outlined in Figure 3.8 and in Figure 3.9, the computed Reynolds number is 88. Because it is well below the critical Reynolds number of 2200 for pipe flows, the flow is laminar.

One should always try to take advantage of any symmetry present in the model. If the pipe is straight and horizontal, as is the case here, the gravity effects can be ignored. This assumption is important as it leads to a major geometry simplification by solving only an axisymmetric case. Mathematically, it is referred to as assuming rotational symmetry, i.e., the tangential and radial

Table 3.7   CFD Model Check List for Case Study One

Item	Information	Values
Geometry	Overall dimensions	x: 1 m (length), y: 0.02 m (radius), z: N/A
	Dimensions	Axisymmetric
Grid	Number of cells	400
	Types of cells	4-node quads
Time dependence		Steady
Viscous models	Laminar/turbulence	Laminar
	Wall functions	None
Heat transfer	Heat transfer models	None
	Radiation	None
Species		None
Materials	Soup	Density=1000 kg/m³, Viscosity=0.01 Pa s
	Steel pipe	Not modeled
Body forces	Gravity	None as pipe is horizontal
Boundary conditions	Velocity and pressure BCs	Symmetry/axis (with un=0)
		Inlet: $v_x=0.022$ m/s and $v_y=0.00$ m/s
		Outlet: pressure outlet with zero gauge pressure
	Temperature	Not modeled
Initial conditions	Velocities	Zero velocities
Under-relaxation factors	Solver default	Pressure=0.3, Momentum=0.7
Solver	Multigrid	Default
Numerics	Pressure velocity coupling	SIMPLE
	Pressure	PRESTO
	Momentum	First order and then second-order upwind with tighter convergence
	Turbulence	None
	Species	None
Monitors		Convergence check on
		Monitor axial velocity at the exit
Convergence criteria		Allow residuals to level off
		Ensure several orders of magnitude reduction
Additional checks		Check material properties look okay (especially density)
Post processing	Plots	Convergence
		Grid
		Velocity vectors
		Line plots of velocities, pressure drop
	Computations	Maximum axial velocity at the exit

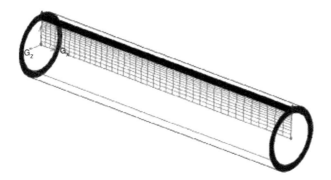

Figure 3.8 Geometry simplification from 3D to axisymmetric.

velocities are zero, and the velocity component parallel to flow direction (Z-axis) is a function of radial direction, r, only. Now with this assumption, one can work with a simple 2D slice of the pipe from the centerline to the outer wall that needs to be modeled (Figure 3.8). The main advantage of this assumption is that a relatively small mesh in 2D is needed instead of resolving a full 3D field. Furthermore, the velocities in the solids are zero, and one does not need to model the pipe wall. The boundary condition on this pipe's inner wall can be specified as zero to give the same effect (Figure 3.9).

To summarize, the following assumptions have been made:

- Flow is steady, laminar, and isothermal
- Gravitational effects are ignored
- 2D axisymmetric case
- Properties of soup are constant (i.e., not dependent on temperature or shear)

3.9.1.2 Governing Equations

Rewriting the conservation equations from Section 3.3

$$\text{Continuity}: \quad \frac{D\rho}{Dt} + \rho\nabla\cdot\mathbf{v} = 0;$$

$$\text{Momentum}: \quad \rho\frac{D\mathbf{v}}{Dt} = -\nabla P + \mu\nabla^2\mathbf{v}.$$

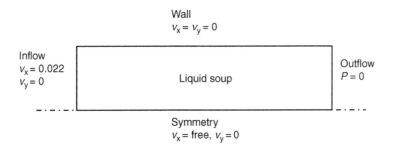

Figure 3.9 Entities and boundary conditions for the model.

Or, in expanded form,

$$\text{Continuity}: \qquad \frac{\partial \rho}{\partial t} + \mathbf{v} \cdot \nabla \rho + \rho \nabla \cdot \mathbf{v} = 0;$$

$$\text{Momentum}: \quad \rho \left(\frac{\partial \mathbf{v}}{\partial t} + \mathbf{v} \cdot \nabla \mathbf{v} \right) = -\nabla P + \mu \nabla^2 \mathbf{v}.$$

With the assumptions listed in the previous section, one can rewrite the conservation of mass and motion for this problem as follows:

$$\text{Continuity}: \qquad \nabla \cdot \mathbf{v} = 0;$$

$$\text{Momentum}: \quad \rho \mathbf{v} \cdot \nabla \mathbf{v} = -\nabla P + \mu \nabla^2 \mathbf{v}.$$

3.9.1.3 Boundary Conditions

Because an axisymmetric analysis is being performed, a symmetry boundary condition must be applied at the pipe centerline—that is the radial velocity component is constrained to be zero while the axial component of velocity is left free. At the pipe boundary, a no-slip velocity boundary condition is applied.

Depending on the objective, one can specify a number of different boundary conditions at the inlet (Table 3.3). The simplest boundary condition for the inlet is a plug flow or constant velocity at the inlet. This can be computed from the volumetric flow rate and cross-sectional area. Other choices include specifying inlet mass flow rate, volumetric flow rate, or pressure drop in the system. In cases where the assumed inlet is located at the end of long pipe, one can specify the fully developed profile. At the outer wall, a no-slip boundary condition is specified.

3.9.1.4 Geometry and Mesh Creation

Any commercial mesh generator can be used to create the mesh shown in Figure 3.10. The domain is a rectangle with length equal to 25 times the diameter. One can always create a real fine mesh in the entire domain. Please note that the mesh size directly translates to computing resources needed. Even though this is a simple 2D case where one can afford to have a fine mesh, it is always a good practice to use resources wisely. It is anticipated that this model will have two areas of high velocity gradients. One is near the inlet as the flow changes from a constant velocity to a parabolic profile, and the other is near the wall where the velocities will be reduced to zero. A proper mesh with finer resolution near the inlet and the wall will yield a small mesh sufficient to resolve these gradients. In this particular case, the model has been created using GAMBIT™ (ANSYS, Inc., canonsburg, PA), and it consists of 400 quadrilateral cells. It is also recommendes to keep the aspect ratio (length divided by width of a cell) reasonable. In this mesh, it ranges from is about 2.5 near the inlet to 50 near the outlet.

Figure 3.10 Mesh showing the grading near the wall and near the inlet.

3.9.1.5 *Results and Discussion*

A number of different plots are presented to show the results of the simulation. First, one should make sure that the solution is fully converged. The residual errors should be lower than the predetermined criteria. In this case, the solution has been converged to a default convergence criterion of 0.1% in 27 iterations (Figure 3.11). Then, to achieve further accuracy, a second-order upwinding is used to get to much tighter tolerances. The final solution has residual errors of less than 1×10^{-6}. When a second-order upwinding is invoked, the residual errors initially shoot up and then come down.

Figure 3.12 shows the axial velocity component at different distances from the inlet of the pipe ($x = 0.05$ m, 0.10 m, 0.20 m, and outlet at $x = 1.00$ m). A plug flow velocity has been specified at the inlet. As soon as the flow enters the pipe, the no slip condition at the wall of the pipe comes into effect and viscosity or viscous forces impose on the flow. The flow adjacent to the wall continuously decelerates until the boundary layer thickness reaches the full pipe radius. Once the flow is fully developed, the velocity profile does not vary in the flow direction (Figure 3.13). In fact, in this region, the pressure gradient and the shear stress in the flow are in balance. From Figure 3.12, it can be adjudged that the entrance length for this flow is about 0.20–0.21 m.

3.9.1.6 *Validation*

This is a simple flow problem where an analytical solution is available. The length of the pipe between the start and the point where the fully developed flow begins is called the entrance length, L_e. This length has been correlated with the Reynolds Number of the flow. For laminar flow, this distance is approximately

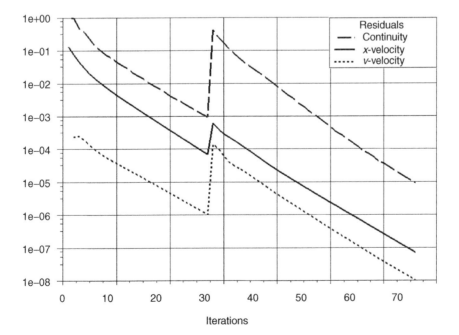

Figure 3.11 Residual monitors for Case Study 1.

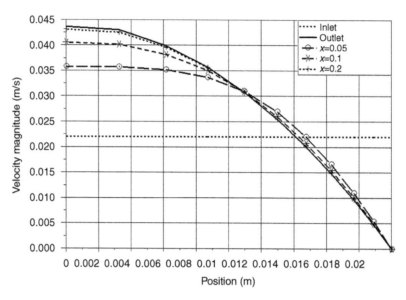

Figure 3.12   Axial velocity profiles at various distances from the inlet.

$$\text{Entrance Length} \approx 0.06 \times Re \times \text{diameter}.$$

For this case, this comes out to be 0.21 m. It correlates very well with the Figure 3.12.

After the entrance length, the velocity profile asymptotically approaches the parabolic profile of steady Poiseuille flow. The axial component at this point is given by Hagen–Poiseuille flow that is given by

$$v = -\frac{1}{4\mu}\frac{dP}{dz}(R^2 - r^2).$$

The pressure drop calculated from the code is 5 Pa. But this includes the entrance region where this equation is not valid. If one omits this section, the pressure drop in the remaining pipe is 3.508 Pa. Using this value, one can compute the axial velocity for the fully developed flow. Table 3.8 shows an excellent agreement between the analytical steady state solution and numerical solution obtained by the commercial CFD code used in this study. Just in case the match was not as expected, it could be due to several factors, including incorrect units, physical dimensions in the model, wrong material properties, incorrect boundary conditions, mesh's being too coarse to resolve the gradients, and inappropriate model selections (2D vs. axisymmetric vs. 3D; laminar vs. turbulent).

3.9.1.7   Summary

This case study illustrated how to set up a simple pipe flow problem and validate it with the analytical results. Jung and Fryer[21] validated a similar problem for a shear-thinning liquid in a tubular heat exchanger.

Figure 3.13   Velocity vectors near the outlet of the pipe.

Table 3.8 Comparison of Analytical and Numerical Results

Radial Distance from Center, m	Axial Velocity, m/s	
	Numerical Solution	Analytical Solution
0.000000	0.043584	0.043858
0.003260	0.042494	0.042693
0.006145	0.039585	0.039717
0.008699	0.035485	0.035560
0.010960	0.030661	0.030688
0.012960	0.025454	0.025441
0.014731	0.020111	0.020064
0.016298	0.014806	0.014732
0.017686	0.009659	0.009563
0.018913	0.004752	0.004637
0.020000	0.001183	0.000000

3.9.2 Case Study 2: Fluid Flow through an Industrial Scale French Fryer

3.9.2.1 Problem Description

This case study is courtesy of Gem Equipment that is based in Woodburn, Oregon. Gem is a producer and designer of custom french fryers for all the food manufacturers. The fried potatoes and hash browns sold in restaurants and fast-food places are typically precooked by the food processor in order to sterilize the product and help it stay fresh longer. When the potatoes first enter the fryer, they are heavier than oil so they sink onto the conveyor. But as they are heated by the oil, the water boils off, the potatoes' density drops below that of the oil, and the potatoes begin to float. At this point, the flow of the oil through the fryer becomes very important. The ideal flow pattern is for the oil to move in a continuous plug pattern in the same direction as the conveyor, maintaining a constant velocity across any given section of the fryer. If there are areas of high and low velocity or if the flow moves perpendicular to the conveyor, the potatoes could be pushed to one side of the fryer. This could cause the potatoes to be unevenly cooked, and it also creates handling problems when the potatoes leave the fryer. To obtain consistent product quality, it is important that all potato pieces are exposed to the same conditions when in the fryer. Fryers are equipped with a number of devices that are used to control these conditions. For example, because the oil temperature has a direct effect on the color and textural characteristics of the finished product, temperature controls are used. By maintaining different oil temperatures in separate zones of the fryer, products with different characteristics can be simultaneously produced. Smooth oil flow in the fryer kettle is critical for uniform heat distribution and first in, first out potato products without clumps (Figure 3.14).

One of their customers is complaining that some of the tater-tots and french fries are burning during the frying process. Because the oil is extremely hot, it is very difficult to figure out why this is happening. The oil enters from a header (that has also been designed using CFD to even out the flow across the width of the fryer by optimizing a set of perforated plates) and leaves at the other end. Because the fryer is about 10–15 m long (Figure 3.14 and Figure 3.15), some of the oil is removed midstream and replenished. The engineer thinks this may be related to uneven oil flow distribution in the fryer and some of the material's getting trapped, recirculated, and burned. In a typical scenario, the engineer would observe the fryer; try to visually determine where the problems were occurring; then, based on experience, try different adjustments on the fryer until the problem was solved. With a flow modeling tool (CFD), it is decided to study the flow inside the kettle without any potatoes and to try alternate designs.

Flow is assumed to be steady, isothermal, and turbulent. The properties of oil at frying temperature are assumed to be constant.

Figure 3.14 Industrial French fryer. (Courtesy of Gem Equipment.)

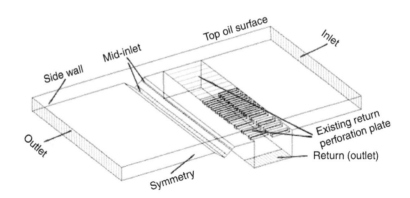

(a) Old design (only half fryer is modeled)

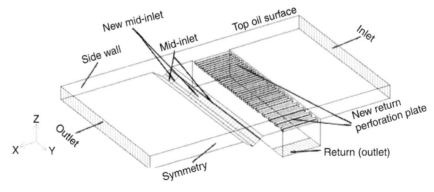

(b) New design (only half fryer is modeled)

Figure 3.15 Old and new designs of the oil frying kettle.

3.9.2.2 Governing Equations

With the above assumptions, one can rewrite the conservation of mass and motion, in vector notations, as follows:

$$\text{Continuity}: \qquad \nabla \cdot \mathbf{v} = 0$$

$$\text{Momentum}: \quad \rho \mathbf{v} \cdot \nabla \mathbf{v} = -\nabla P + \mu \nabla^2 \mathbf{v}.$$

Because the flow is turbulent, additional equations are needed to calculate turbulent viscosity. To keep the simulation manageable, a simple mixing length model is used. This mixing length can be thought of as the average length over which momentum is distributing in the channel.[33]

3.9.2.3 Boundary Conditions

For running the simulation, material properties such as density and viscosity are needed and so are boundary conditions. In this case, one needs to specify the inlet velocity and outlet velocity at the recirculation. At this time, it is needed to know if the flow is laminar or turbulent. This can be calculated from Reynolds number at the inlet. Because the cross-section of the inlet is not circular, hydraulic diameter needs to be used. The characteristic velocity can be calculated from the flow rate and cross-sectional area. The top surface of oil can be approximated with a perfect slip by specifying the normal component of velocity as zero and keeping tangential components free.

3.9.2.4 Geometry and Mesh Creation

A CAD drawing of the equipment may contain some small details like bolts and cross-bars needed for strength. These details may not add anything significant to flow analysis, but they could complicate mesh generation. At this point, one can either create the model in the preprocessor or bring in the CAD model and delete unnecessary details.

In this case, only half the fryer is considered. Once the geometry is created, it is meshed. As previously noted, the mesh has to be of an appropriate size with finer mesh where the gradients are high. In a typical case with this sized fryer, a fine quality mesh can run into a few million cells. Because of the complexity of the fryer geometry because of perforated plates, a tetrahedral mesh was created using GAMBIT™. The model contained 500,000 tetrahedral cells.

3.9.2.5 Results and Discussion

This particular model was solved using FiDAP™ on a personal computer, and results were postprocessed using the built-in postprocessor. It took overnight to get the converged results with residuals reaching a default convergence criterion.

Because the model is large, it is sometimes difficult to visualize the results even with multiple plane cuts to plot velocity vectors and speed contours. Whenever there is a problem with flow, it is a recommended to compute pathlines of massless particle (or particle tracks with density effects) in the domain. This is the line that one would get from a long exposure photograph, highlighting a single fluid particle. Depending on the software, one can inject these massless traces anywhere in the domain and study their behavior. These pathlines are extremely important in finding the areas of vortex or even a dead zone. The particles trapped in these vortices or dead zones tend to get overcooked and burned.

The first model of the new header clearly shows flow distribution problems (Figure 3.16), but it provides engineers with complete information on the flow through the device. This information helps to create possible solutions. In particular, engineers made changes to the model's part that represents the perforations used to release oil from the header. They tried four or five different designs on the computer, and they finally found one that works for a wide range of conditions. The analysis showed a considerable improvement in flow patterns (Figure 3.17). These changes were then implemented in the industrial fryer, and the original problem will burnout oil, and overcooked material was alleviated.

3.9.2.6 Summary

The new design has demonstrated the ability to provide even flow distribution in every application where it has been tried. Over the past few years when Gem Equipment has used this new design, not a single header has required adjustment in the field. Gem engineers who were initially skeptical of computer simulation's ability to predict flow through the complex fryer geometry have become believers. As a result, the company is now using CFD to optimize other elements of the fryer design such as the oil return passageways.

Although the service engineers were able to solve the problem, the cost of the service calls and the inconvenience to customers caused Gem to look at the simulating oil flow while designing these headers. After validating it with actual data in the plant, a CFD analysis became a routine even before building any new fryers. Unlike a physical prototype, the geometry of the CFD model can be quickly changed on the computer and re-analyzed to explore different design options in project design or operating conditions. Gem engineers began by modeling the oil flow through their existing header design.

This case study illustrated several key aspects of the flow modeling: keeping the problem as simple as possible to gain insight into the process and making design changes in the simulation rather than physically build.

3.9.3 Case Study 3: Modeling Flows in Extrusion Dies

Food extrusion has rapidly developed over the last fifty years with applications being continually expanded to new areas of food processing. Among the many applications of food extruder are continuous pasta presses, ready-to-eat (RTE) cereal products, expanded corn curls, texturized puffed corn meal, soup and gravy bases, pet-foods, and cookies or cracker-type shapes. Several

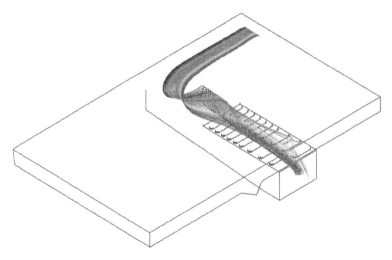

Figure 3.16   (See color insert following page 178.) Unwanted vortex generated by improper geometry over fryer pan return.

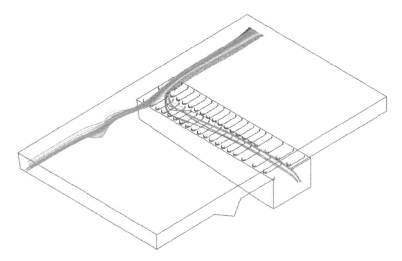

Figure 3.17 (See color insert following page 178.) Vortex removed using FiDAP™ to analyze new 3D geometry.

designs of extruders including single-screw and twin-screws are widely employed in commercial food production. With these developments, food engineers are faced with many challenges such as

- Increasing the productivity of an existing extrusion line
- Retrofitting an existing line for new products
- Designing an optimum extruder process and system
- Scaling up an extrusion process developed in the laboratory
- Specifying the control parameters for an extruder

In this process, the solid food material such as corn or soy flour with desired additives is conveyed by a single screw or twin rotating screws inside a barrel. The friction from the interface between the flowing material and the wall increases the temperature through a viscous heating process, locally melting the corn meal. After the meal has melted and has completed its flow around the screw(s), the pressure within the barrel increases because of a restriction at the discharge of the barrel. This restriction is due to one or more orifices or shaped openings called a die. Discharge pressures can be quite high as well and generally cause product to expand or swell with extensive flashing of moisture. These openings can be quite simple, i.e., circular, annual or quite complex. The design of such dies is complicated and difficult. Traditionally, dies have been designed using an expensive trial and error procedure with possibly ten or twenty trials and modifications. The main reason for this lengthy design process is that the flow patterns inside the extruder and dies are unknown. Flow modeling is an ideal choice to supplement experiments. Instead of modeling the entire process, it can be broken into several smaller studies including flow inside the extruder (depends on choices of the co-rotating screw elements and barrel size), flow through the die (shape and number of holes), and deformation of the extrudate. All these have been independently studied using CFD.[23,24,34] One can predict the shape of the extruded product and even start with the desired product shape by asking the software program to compute the die lip shape required to produce that product. The result is a considerable savings of time and expense because the amount of trial-and-error testing is reduced. For illustration purposes, only flow through extrusion die is studied here.

3.9.3.1 *Problem Description and Assumptions*

Consider a food extrusion die with eight outlets (a production die generally has many more holes) situated at the end of a twin-screw extruder. The flow of laminar, non-Newtonian dough, melts (such as corn meal) inside the die is affected by the heat transfer to the dough, temperature sensitive transport properties (such as viscosity), the shear rate dependence of viscosity and viscous heating.[35] Viscous heating can significantly raise the temperature of the dough. This increase in temperature lowers the viscosity resulting in lower pressure drop in the die head for a given flow rate. The rheological properties of the dough can be defined using the Carreau model for its dependence on shear rate and temperature.

3.9.3.2 Equations

The governing equations for a steady state laminar flow problem are as follows:

$$\text{Continuity}: \qquad \nabla \cdot \mathbf{v} = 0;$$

$$\text{Momentum}: \quad \rho \mathbf{v} \cdot \nabla \mathbf{v} = -\nabla P + \nabla \cdot \left[\mu(\dot{\gamma}, T) \nabla \times \mathbf{v} \right];$$

$$\text{Energy}: \qquad \rho c_p \mathbf{v} \cdot \nabla T = k(\nabla^2 T) + Q + \Phi.$$

Because the viscosity of dough may depend on temperature as well as shear rate, heat transfer needs to be included in the CFD model.

3.9.3.3 Boundary Conditions

The velocity profile and temperature field needs to be defined at the inlet as boundary conditions. Because the flow is coming from the end of a twin screw extruder, the velocity field across the cross-section is not readily available. Therefore, as an initial guess, a plug flow profile (constant velocity) can be used. All the exterior walls at the die plate have natural convection boundary conditions whereas the exterior walls near the inlet are assumed to adiabatic.

3.9.3.4 Geometry and Mesh Creation

In this case, a CAD model can directly be imported into a preprocessing package. Using the symmetric nature of the geometry, only half of the model is modeled. The big advantage of this is that the resulting mesh will be half the size. To simplify further, only the flow domain can be modeled, ignoring the solid metal surrounding it. Therefore, the boundary conditions are directly imposed on the fluid touching the metal surface. In some commercial codes, there is an option to add metal thickness for thermal considerations.

Since viscous dissipation is important, a good quality mesh is needed to resolve the thermal gradients near the walls. It is advisable to use hexahedral meshes for these problems. Unfortunately, because of the complexity of the geometry, it is impossible to get all hex mesh in the domain of interest. A good compromise is to put hexahedral mesh where it is possible with finer mesh near the wall and then use tetrahedral mesh in the remaining areas. Boundary layers near the walls are used to better capture gradients. A mesh created with GAMBIT™ contains about 1.63 million cells (Figure 3.18 and Figure 3.19) with about 900,000 hexahedral, 521,000 tetrahedral, 170,000 prisms, and 28,000 pyramids. Going from a simple 2D model, as in Case Study 1 (Section 3.9.1) with 400 cells, to a full 3D model with 1.63 million cells to resolve the flow patterns illustrates the complexity of the simulation as well as a need for compute power for large models.

3.9.3.5 Results and Discussion

This simulation has been solved using a commercially available CFD package, FLUENT™. The mesh created in GAMBIT™ is imported into the FLUENT™. All boundary and initial conditions,

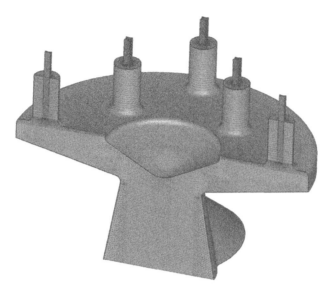

Figure 3.18 Geometry and mesh for an extrusion die (only half the domain is shown).

dough properties (density, viscosity, thermal conductivity, and specific heat) are then defined in FLUENT™. A nonstandard Carreau viscosity model is incorporated by writing a few lines of code to describe the relationship of viscosity with temperature and shear rate. For added accuracy, a double precision solution with second order is obtained. Because the mesh size is big, one can run this simulation on two or more processors (called parallel processing) to cut down the simulation time. This particular case took a few hours on a personal computer with 2 gigabytes of memory.

Again, the first thing is to check is whether the solution has been converged or not. When the physics is complicated, default values of simulation controls (such as under-relaxation values, multigrid options, pressure velocity coupling, discretization) may need to be adjusted to make the solution stable. When dealing with highly nonlinear viscous flows, it is recommended to

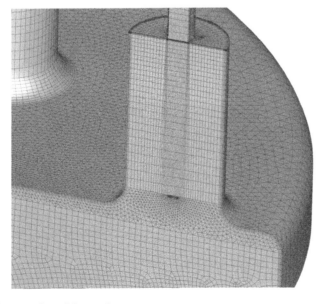

Figure 3.19 Close-up view of the mesh.

obtain the initial solution using a representative constant viscosity and to use this solution for an initial guess for a full model. Viscous dissipation should be turned on only when the solution seems to be stable. One can find these bells and whistles in the documentation of the code. In addition to residuals, it is always advised to monitor intermediate solutions of certain variables (e.g., integrated value of static pressure at the inlet).

Similar to the previous two cases, a number of different plots can be generated to visualize the results of the simulation. Figure 3.20 shows velocity vectors on two planes, one on symmetry plane and the other cutting through one of the die exits. As the flow enters the domain from the inlet, it travels into a converging channel. Shortly after that, it gets split because of the presence of a strategically placed cone. There is an area of lower velocity (blue color) near the ends of the channel. This is potentially an area of concern for recirculation or is even a dead zone. The stagnant flow in the dead spaces, if any, results in a comparatively lower temperature of the material. A plot of massless pathlines from the inlet or any other place in the computation domain can also be used to further study this recirculation. In addition to velocity vectors, one can plot velocity magnitude, pressure contours, shear rate, temperature, and variable properties. Figure 3.21 shows the velocity magnitude on one symmetry plane and non-Newtonian viscosity on the other plane. One can then correlate the effect of velocity on the non-Newtonian viscosity. The areas of high velocity experience higher shear rate that, in turn, lowers the viscosity.

Significant pressure loss occurs in the narrow section of the die orifice. Because of the narrow geometry at the die exit, the dough experiences high shear that causes a considerable rise in the production temperature of the material. Figure 3.22 shows that the flow enters the domain at given temperature (shown with green) and exits at higher temperature (shown with red). The temperature increase is often referred to as the temperature rise because of viscous dissipation. In this particular case, the viscous dissipation raises the temperature of the dough by about 40°C. As noted in the boundary conditions, the exterior walls are exposed to the room temperature and cools off the material inside as shown by the blue color near the edges.

For food engineers, the die flow balance is an important area for die design. A well-designed die usually exhibits uniform flow from all die openings. CFD modeling offers significant advantages over the physical experiments and eliminates the need to build multiple die parts and to perform time-consuming experiments.

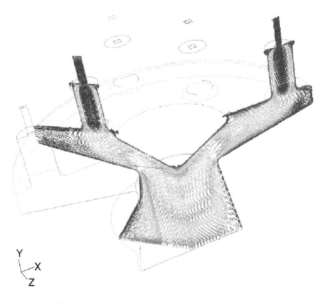

Y
X
Z

Figure 3.20   (See color insert following page 178.) Plot of velocity vectors in the die.

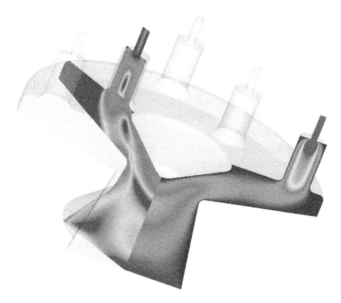

Figure 3.21 (See color insert following page 178.) Plots of velocity magnitude (right side) and non-Newtonian viscosity (left side).

3.9.3.6 Summary

This particular case study builds on the previous two case studies by modeling a more complicated geometry, including non-Newtonian viscosity and solving energy equations. In this case, a Carreau model has been used to describe the viscosity dependence on shear. Most foods exhibit viscoelastic behavior that could have a significant effect on flow.[23] CFD simulations such as shown here are regularly used by the food industry to understand the pressure and temperature distribution inside a die head and to optimize the extrusion process.

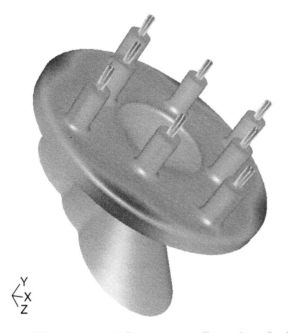

Figure 3.22 (See color insert following page 178.) Temperature profiles on the walls of die.

3.10 CHAPTER SUMMARY

This chapter's goal was to give the reader enough background into the fluid flow phenomenon so that one can illustrate the power of numerical experimentation in the field of CFD. The examples selected are designed to guide the reader to specific aspects of setting up a fluid flow problem. The many physical models available for the fluid problem may be different, but the approach and challenges to fluid flow's modeling is no different than in other physics-based modeling approaches such as one discussed in the heat transfer chapter of this handbook. The key points of interest are essentially the same, and they can be summarized as problem setup, simplification options, definition of properties and boundary conditions, solution methodology, and presentation of results. Although the physics involved in the food industry are fairly complex (either because of the complexity of materials handled or the combination of physical phenomena involved), advances in both the software industry as well as in computer hardware make it possible for these modeling techniques to be applicable to more and more realistic processes that make them attractive in more application areas. The numerical physics-based modeling of fluid flow represents another powerful tool in the mix of all other tools available to the food engineer (including experimental, analytical, and heuristic approaches). As these tools become more applicable to more food processes, it is expected that the use of these tools will continue to grow and present some challenges to food engineers in the years to come.

ACKNOWLEDGMENTS

The authors would like to take this opportunity to thank Eric Grald, Raj Venturmalli, Bill Wangard, and Joan Johnson for providing critical feedback; GEM Equipment for providing a case study; and Preeti Gupta for illustrations. Some of this chapter's content came from our training notes prepared by many engineers over the years.

REFERENCES

1. Kumar, A. and Swartzel, K. R., Selected food engineering problems and their solutions through FEM. Fed-vol. 171, In *Advances in Finite Element Analysis in Fluid Dynamics*, American Society of Mechanical Engineers, New York, pp. 107–113, 1993.
2. Eisenberg, F. G., Virtual experiments using computational fluid dynamics, Presented at the 7th Conference on Food Engineering, AIChE, 2001.
3. Bird, R. B., Stewart, W. E., and Lightfoot, E. N., *Transport Phenomena*, New York: Wiley, 1960.
4. Dhanasekharan, M., Grald, E. W., and Mathur, R. P., How flow modeling benefits the food industry. In addition to aiding the development of new products and processes, flow modeling software can help focus experimental work and pilot plant trials, *Food Technology*, 58(3), 32–35, 2003.
5. Lang, T., Keynote Speech, Fluent Users Group Meeting, Dearborn, MI, 2004.
6. Engelman, M. S. and Sani, R. L., Finite-element simulation of an in-package pasteurization process, *Numerical Heat Transfer*, 6, 41–54, 1983.
7. Datta, A. K., Numerical modeling of natural convection and conduction heat transfer in canned foods with application to on-line process control, PhD diss., University of Florida, 1985.
8. Kumar, A., Modeling thermal and aseptic processing of liquid food products, PhD diss., University of Minnesota, 1990.
9. Puri, V. M. and Anantheswaran, R. C., Finite element method in food processing—a review. Paper No. 90-6523, ASAE Winter Meeting, Chicago, IL, 1990.
10. Scott, G. and Richardson, P., The application of computational fluid dynamics in the food industry, *Trends in Food Science and Technology*, 8(April), 119–124, 1997.
11. Wang, L. and Sun, D. W., Recent developments in numerical modeling of heating & cooling processes in the food industry—a review, *Trends in Food Science and Technology*, 14, 408–423, 2003.

12. Datta, A. K. and Teixeira, A. A., Numerical modeling of natural convection heating of canned liquid foods, *Transactions of the American Society of Agricultural Engineers*, 30, 1542, 1987.

13. Kumar, A., Bhattacharya, M., and Blaylock, J., Numerical simulation of natural convection heating of canned thick viscous liquid food products, *Journal of Food Science*, 55(5), 1403–1411, 1990. 1420

14. Kumar, A. and Bhattacharya, M., Transient temperature and velocity profiles in a canned non-Newtonian liquid food during sterilization in a still-cook retort, *International Journal of Heat and Mass Transfer*, 34(4–5), 1083–1096, 1991.

15. Kumar, A., Blaylock, J., and Swartzel, K. R., Modeling Thermal and Aseptic Processes of Foods Using Fidap. 4th Fidap Users Conference, Fluid Dynamics International, April 14–16, 1991.

16. Ghani, A. G., Faird, M. M., and Zarrouk, S. J., The effect of can rotation on sterilization of liquid food using computational fluid dynamics, *Journal of Food Engineering*, 57, 9–16, 2003.

17. Alvarez-Vazques, L. J. and Martinez, A., Modeling and control of natural convection in canned foods, *Journal of Applied Mathematics*, 63, 247–265, 1999.

18. Kumar, A. and Bhattacharya, M., Numerical analysis of aseptic processing of a non-Newtonian liquid food in a tubular heat exchanger, *Chemical Engineering Progress*, 103, 27–51, 1991.

19. Sandeep, K. P. and Zuritz, C. A., Modeling the flow of non-Newtonian suspensions in straight and helical holding tubes, The American Society of Agricultural Engineers, Atlanta, GA, December 13–16, 1994.

20. Fitt, A. D. and Please, C. P., Asymptotic analysis of the flow of shear-thinning foodstuffs in annular scraped heat exchangers, *Journal of Engineering Mathematics*, 39, 345–366, 2001.

21. Jung, A. and Fryer, P. J., Optimising the quality of safe food: computational modeling of a continuous sterilization process, *Chemical Engineering Science*, 54, 717–730, 1999.

22. Grijspeerdt, K., Hazarka, B., and Vucinic, D., Application of computational fluid dynamics to model the hydrodynamics of plate heat exchangers for milk processing, *Journal of Food Engineering*, 57, 237–242, 2003.

23. Dhanasekharan, M. and Kokini, J. L., Viscoelastic flow modeling in the extrusion of a dough-like fluid, *Journal of Food Process Engineering*, 23(3), 237–247, 2000.

24. Dhanasekharan, K. M. and Kokini, J. L., Design and scaling of wheat dough extrusion by numerical simulation of flow and heat transfer, *Journal of Food Engineering*, 60, 421–430, 2003.

25. Jongen, T., Characterization of batch mixers using numerical flow simulations, *AIChE Journal*, 46(November), 2140–2150, 2000.

26. Langrish, T. A. G. and Fletcher, D. F., Spray drying of food ingredients & applications of CFD in spray drying, *Chemical Engineering and Processing*, 40, 345–354, 2001.

27. Torok, D. F., Computational thermofluid modeling in the food processing industry, 4th Annual Fidap User Conference, April 14–16, 1991.

28. Sun, E. W. and Zehua, H., CFD simulation of coupled heat and mass transfer through porous foods during vacuum cooling process, *International Journal of Refrigeration*, 26, 19–27, 2003.

29. Kucuk Unluturk, S., Arastoopour, H., and Koutchma, T., Modeling of UV dose distribution in a thin-film UV reactor for processing of apple cider, *Journal of Food Engineering*, 65, 125–136, 2004.

30. Sima, M. A. and Harris, J. A., Numerical modelling of flow in a vertical cooling crystalliser, *Journal of Fluids Engineering*, 121, 148–154, 1999.

31. Zienkiewicz, O. C., *The Fnite Element Method in Engineering Science*, London: McGraw-Hill, 1971.

32. Singh, R. P., Moving boundaries in food engineering, *Food Technology*, 54(2), 44–53, 2000.

33. Fluent, Inc., *FiDAP Theory Manual*, Lebanon, NH: Fluent Inc., 2005.

34. Ashokan, B., Fanning, L., and Kokini, J. L., Accurate prediction of the flow profile in a continuous mixer using new 3-D FEM numerical simulation techniques, Presented at the Fluent Users Group Meeting, Dearborn, MI, 2005.

35. Kumar, A., Bhattacharya, M., and Padmanabhan, M., Modeling flow in cylindrical extruder dies, *Journal of Food Science*, 54(6), 1584–1589, 1989.

CHAPTER 4

Heat Transfer

Ashim K. Datta

CONTENTS

4.1 INTRODUCTION

Physics-based models for phenomena such as fluid flow (described in Chapter 3), heat transfer (this chapter) and mass transfer (Chapter 5 and Chapter 6) are the workhorse of models used in many process-engineering applications. These models are also known as *continuum-based transport models* or simply *transport models*. Unlike the observation-based models discussed in Chapter 9– Chapter 15 of this book, these physics-based models are based on fundamental physical laws. The relative advantages and disadvantages of a physics-based model have been mentioned in Chapter 1. Because these transport models are used extensively in engineering, every branch of transport modeling is a discipline by itself. This is quite true for heat transfer, where many textbooks (e.g., Refs. 1,2) and handbooks (e.g., Ref. 3) cover the models comprehensively. This knowledge cannot be reproduced in a single chapter and there is no utility in attempting to do so. On the other hand, time and time again, the primary difficulty of a newcomer in any physics-based modeling is in simplifying a physical situation and choosing the most appropriate analysis among the many available in textbooks and handbooks. Guidance geared specifically to the newcomer in food-process modeling, given in terms of general approaches to developing heat transfer models or general classes of models that have worked in the past, should be novel and uniquely helpful. This will not repeat the general information on heat transfer available in the literature. This chapter will pursue such an approach.

4.2 DEVELOPMENT OF PHYSICS-BASED MODELS IN HEAT TRANSFER

A physics-based model of heat transfer uses two basic physical laws, as shown in Figure 4.1. The first law is that of energy conservation which states that the total energy of a system is conserved—it can only interconvert into a different form. This law of energy conservation is applied to the conservation of

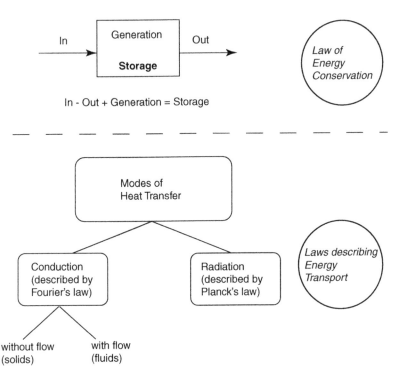

Figure 4.1 The law of *energy conservation* and the laws describing *energy transport* via the fundamental mechanisms of conduction and radiation together make up the foundation of a physics-based model of heat transfer processes.

thermal energy where conversion from another energy form to thermal energy is treated as a generation (source) term. The second law (Figure 4.1) pertains to the two fundamental mechanisms of heat transfer, conduction, and radiation. The law for conduction heat transfer is generalized from empirical observations and states that the flux (amount per unit area per unit time) of thermal energy at a location due to conduction in a material is proportional to the gradient in temperature at that location:

$$q_x = -k \frac{dT}{dx}.$$
(4.1)

Here, T is temperature at location x and q_x is the heat flux at x, along the positive x direction. The coefficient, k, is the called the *thermal conductivity* of the material—it is a material property that is almost always determined experimentally. This equation is also known as *Fourier's law of heat conduction*. The commonly known term of *convection heat transfer* typically refers to the addition of flow (as in a fluid) to conduction heat transfer. For most of the discussions in this chapter, porous media will not be considered, i.e., the discussions deal with either a pure solid or a pure liquid.

Transport of thermal energy has another fundamental mechanism besides conduction: through electromagnetic radiation from a surface. A body at any temperature above absolute zero emits radiation in all directions over a wide range of wavelengths. The quantity (amount) and quality (spectral distribution) of energy emitted by this radiative heat transfer depends on the temperature. The energy emitted by a body over all wavelengths and all directions is the radiative heat flux (E_b, in W/m^2) and is given by

$$E_b = \varepsilon \sigma T^4$$
(4.2)

where T is the absolute temperature of the surface, ε is a surface property called the *emissivity*, and σ is the Stefan–Boltzmann constant. Equation 4.2 is known as the Stefan–Boltzmann law that is derived from the more general Planck's law of radiation which provides the energy emitted as a function of temperature and wavelength. Modeling of radiative heat transfer is described in more detail in Section 4.10.

4.3 PROBLEM FORMULATION IN HEAT TRANSFER: CONDUCTION AND CONVECTION

By applying the law of conservation of energy and the Fourier's law of heat conduction on a differential volume and taking the limiting case of this as volume goes to zero, one arrives at the general equation for thermal energy transport. Such derivations can be found in a large number of undergraduate and graduate textbooks on heat transfer (e.g., Ref. 4). A very general form of the heat Equation (with constant thermal properties) is given by:

$$\rho C_p \left(\underbrace{\frac{\partial T}{\partial t}}_{\text{Transient}} + \underbrace{v_x \frac{\partial T}{\partial x} + v_y \frac{\partial T}{\partial y} + v_z \frac{\partial T}{\partial z}}_{\text{Convection}} \right) = \underbrace{k \left[\frac{\partial^2 T}{\partial x^2} + \frac{\partial^2 T}{\partial y^2} + \frac{\partial^2 T}{\partial z^2} \right]}_{\text{Conduction}} + \underbrace{Q}_{\text{Heat source}}$$

$$+ 2\mu \left\{ \left(\frac{\partial v_x}{\partial x} \right)^2 + \left(\frac{\partial v_y}{\partial y} \right)^2 + \left(\frac{\partial v_z}{\partial z} \right)^2 \right\} + \mu \left\{ \left(\frac{\partial v_x}{\partial y} + \frac{\partial v_y}{\partial x} \right)^2 + \left(\frac{\partial v_x}{\partial z} + \frac{\partial v_z}{\partial x} \right)^2 + \left(\frac{\partial v_y}{\partial z} + \frac{\partial v_z}{\partial y} \right)^2 \right\}$$
(4.3)

Here, T is temperature at position (x, y, z), v is the velocity, and Q is a heat source (discussed later). The terms k, ρ, C_p are thermal conductivity, density, and specific heat, respectively, that are called the *thermal properties*. The velocities v_x, v_y, and v_z can be obtained from solving the Navier–Stokes equation, as described in Chapter 2. The equation is given here in Cartesian coordinates and the same equation in other coordinate systems can be found in other textbooks (e.g.,Ref. 4). The equation is rarely solved with all the terms as noted. Important simplifications are made when problems are reduced to 2D or 1D and terms representing various physical processes are dropped. A common simplification is to drop the last term that represents viscous dissipation of heat. Dropping this term leads to the more familiar heat equation

$$\frac{\partial T}{\partial t} + v_x \frac{\partial T}{\partial x} + v_y \frac{\partial T}{\partial y} + v_z \frac{\partial T}{\partial z} = \alpha \left[\frac{\partial^2 T}{\partial x^2} + \frac{\partial^2 T}{\partial y^2} + \frac{\partial^2 T}{\partial z^2} \right] + \frac{Q}{\rho C_p} \tag{4.4}$$

where $\alpha = k/(\rho C_p)$, called the *thermal diffusivity*, has been used.

One of the simplest situations occurs when the diffusional resistance to heat transfer from the interior of the material to its boundary is lower than the resistance to heat transfer from the boundary to the surroundings. Under this situation, the above equation may be reformulated into what is called the *lumped parameter analysis*. This is further discussed in Section 4.7.3. When formulating a heat-transfer problem for all other situations where diffusion needs to be considered, several considerations are necessary, primarily to simplify the situation.

4.3.1 Solid, Fluid, or Conjugate Problem

One of the first steps in formulating a heat-transfer problem is to decide on the domain of analysis. In a problem involving both solids and fluids, in some cases both the solid and fluid need to be included in the analysis. For example, in a slow-freezing process involving a solid and colder surrounding air, the temperature drop in the solid that is being frozen influences the air flow around it; both the solid and surrounding air may need to be included in the analysis. Such a problem is called a *conjugate problem*; it is discussed in Section 4.11. However, in most situations, the analysis can involve either the solid or the fluid. When the analysis is carried out for the solid, the fluid is replaced by a surface heat-transfer coefficient. When the analysis is carried out for the fluid, the solid may be replaced by a surface temperature.

4.3.2 Transient

The transient term denotes rate of change of storage. Solutions to transient problems can be computer time consuming and also numerically challenging. Therefore, it is important to decide whether a transient solution is really necessary. The transient term should be retained in situations such as

- When changes with time are likely to be significant and is of interest
- The timescale of interest is short enough that transient effects are important
- In some situations in numerical solution, even though steady state is of interest, where the researcher may have to start from a transient simulation and run the simulation for long enough time to reach steady state

Examples of some situations where the transient term may be ignored occur if one expects the processes to reach steady state. This can happen in (1) heat transfer in a tube flow, or (2) heat loss through the walls of a cold-storage container when outside and inside conditions are fairly constant.

4.3.3 Convection

This term represents the transport of energy or species due to bulk flow. It should be kept

- For a fluid, when movement (flow) is expected
- When it is likely to have a strong contribution in relation to diffusion, as determined by the Peclet number ($Pe = uL/\alpha$, where L is the characteristic dimension)

The convection term can be ignored, for example, for a solid region that does not move (flow) or is not subject to bulk flow (such as in a porous media).

4.3.4 Conduction

This term represents the contribution due to conduction or diffusion. It is needed when large temperature gradients are expected in the system. The diffusion term may be ignored in situations such as

- Uniform and rapid heat generation, such as in short-time microwave or ohmic heating
- When all of the thermal resistance is outside the material, given by the condition $hL/k <$ 0.1; under this condition, a specialized, simpler, lumped parameter analysis (Section 4.7.3) may be carried out

4.3.5 Heat Source

This term represents the contribution due to volumetric generation or depletion of energy. More details of source terms are provided later.

- It is needed when energy deposition needs to be considered due to volumetric heating from electromagnetic, ultrasonic, or other sources (in heat transfer).
- It may be ignored when heat generation due to electromagnetic, pressure, or other modes of heating are insignificant compared to conduction or other modes of heat transfer.

4.3.6 Coupling with Other Physics

In many situations, heat transfer is coupled with other physics. For example, in frying, heat transfer is strongly coupled with moisture transport. In microwave heating, heat transfer is coupled with electromagnetics. In processes such as drying or bread baking, where dimensional changes are involved, heat transfer is coupled with solid mechanics. Again, inclusion of other physics to provide more details should be balanced against complexities in computation that arise due to the inclusion of such physics. This is further discussed in Section 4.11.

4.3.7 Uncertainty or Stochastic Variations

Equation 4.3 is deterministic in nature, i.e., there is no built-in uncertainty in either the properties or the other parameters. In reality, there are uncertainties in food properties (because food is a biological material that can have natural variations) and in process parameters such as temperature of the fluid being heated (because of the limitations of the process control system). Physics-based models can be made more realistic by the inclusion of such uncertainties. This is discussed at length in Chapter 8. In the absence of such detailed formulation of the stochastic variations that are unavailable in many commercial software programs, one can run the deterministic model mentioned in this chapter for a range of input parameters (such as thermal conductivity) covering the uncertainty and obtain a range of output parameters that would represent the extremities in cases of parameter uncertainty.

4.4 SOLUTION TECHNIQUES FOR THE HEAT EQUATION

Until computers found widespread use, the primary mode of solving the equations in physics-based models was analytical. Analytical solutions can provide simplicity and more insight into the physical process and make it easier to obtain the dependence of a particular parameter, but they are often restricted to rather simple situations. Numerical solutions, on the other hand, are extremely flexible. Numerical methods generally work for arbitrary initial and boundary conditions, complex geometries, thermal properties varying in space and time, etc. Thus, numerical solutions are well suited for heat-transfer problems. Many books have been written on numerical solutions to heat-transfer equations and these numerical methods have been incorporated in numerous commercial, public-domain, and personally owned software. In the past, most of the research-level problem solving in numerical heat transfer has been carried out through development of individual codes, a trend that still continues to some extent today. As general-purpose commercial codes become more powerful and easy to use, the use of these general-purpose codes is becoming more prevalent.

The use of commercial software to solve heat-transfer problems can be artificially divided into three categories: (1) small-scale teaching software, (2) large-scale industrial and research software, and (3) software that couples multiphysics. An example of small-scale teaching software is FEHT (F-Chart Software, Madison, Wisconsin). Restricted versions of large-scale industrial software are also available as a teaching tool, such as FlowLab (Fluent, Inc., New Hampshire). A number of computational fluid dynamics (CFD) software programs are available at the large industrial and research scales that can perform most of the (uncoupled) problems mentioned in this chapter. An extensive list of such codes is available at "http://www-berkeley.ansys.com/cfd"; most of the programs are commercial. Several examples of the use of CFD in food-processing problems are provided in Chapter 2. At the time of this writing, two of the commercial software programs that have substantial multiphysics capabilities are ANSYS (Canonsburg, Pennsylvania) and COMSOL Multiphysics (COMSOL, Inc., Burlington, Massachusetts). Significant activities are underway within the CFD community to couple the physics available in various standalone codes; consequently, a substantial increase in multiphysics capabilities should be expected in the future.

4.5 THERMAL PROPERTIES

Thermal properties that are needed for heat-transfer modeling include properties related to heat conduction (thermal conductivity, specific heat, and density), those related to thermal radiation (reflectance, absorptance, transmittance, and emittance) and those related to fluid flow (viscosity) when the material is flowing. Thermal properties, like other food properties, are often a strong function of processing conditions such as temperature, moisture, and even their histories. It would be especially useful if correlations of the needed properties were available that covered ranges of temperature, moisture, and other states of the food during processing. Such detailed data, as illustrated by the thermal conductivity data shown in Table 4.1, are generally unavailable. When such detailed data is not available, sources of data can be (1) reference books on food properties (e.g., Refs. 5,6,7), (2) computerized databases (e.g., Ref. 8), (3) individual research papers, and (4) measurement. If these fail, the only choice is to think of the composition of the food in terms of air,

Table 4.1   Thermal Conductivity of Individual Components, to Be Combined with Composition to Obtain Thermal Conductivity of Food Materials

Component	Conductivity Equation
Water	$0.57109 + 1.762 \times 10^{-3}T - 6.7036 \times 10^{-6}T^2$
Ice	$2.21960 - 6.2489 \times 10^{-3}T + 1.0154 \times 10^{-4}T^2$
Proteins	$0.17881 + 1.1958 \times 10^{-3}T - 2.7178 \times 10^{-6}T^2$
Fats	$0.18071 - 2.7604 \times 10^{-3}T - 1.7749 \times 10^{-7}T^2$
Carbohydrates	$0.20141 + 1.3874 \times 10^{-3}T - 4.3312 \times 10^{-6}T^2$
Fibres	$0.18331 + 1.2497 \times 10^{-3}T - 3.1683 \times 10^{-6}T^2$
Ash	$0.32961 + 1.4011 \times 10^{-3}T - 2.9069 \times 10^{-6}T^2$
Air	0.025

Source: Nesvadba, P., *Engineering Properties of Foods*, Rao, M. A., Rizvi, S. S. H., and Datta, A. K., Eds., Boca Raton, FL: Taylor & Francis, 2005.

water, and ice, each having a distinct thermal conductivity value. Using an initial guess of the property based on the composition, a sensitivity analysis of the heat-transfer model to thermal property can be accomplished by running the model for a range of thermal property values. Such sensitivity analysis will provide the likely range of the final parameter (such as temperature) from the model. The sensitivity analysis can also serve as a guide to where more accurate data is needed if computations show a process to be particularly sensitive to a given property.

4.6   INTERNAL HEATING AND SOURCE TERMS

4.6.1   What Is a Source Term?

As previously mentioned, the source or generation term, Q, in Equation 4.3 refers to an energy source or energy generation that seems to appear out of nowhere. Because energy cannot really come from nowhere, it is actually converted from other forms. The most common type of conversion is from electromagnetic energy, a typical example being microwave heating of food, where the energy of the microwaves is converted into heat inside the food material, thus appearing as a source term in the energy equation.

Thus, the focus of this section is to obtain the source term, Q, necessary to solve Equation 4.3. In general,

$$Q = f\left(\begin{array}{c} \text{food} \\ \text{parameters} \end{array}, \begin{array}{c} \text{heating} \\ \text{parameters} \end{array}\right). \tag{4.5}$$

An example of a heating parameter in microwave heating is the oven power level or the oven geometry. The physical mechanism that leads to the heating will now be discussed for a number of food heating situations, along with the formulation of Q. Note that the source term is volumetric, i.e., Q is distributed in the material (not necessarily uniformly) and its units are W/m^3.

4.6.2   Heat Source Term for Electromagnetic Heating: Microwaves

The effect of microwave heating is typically formulated as a heat source term in the heat equation. Before presenting the heat source term for microwave heating, some detailed discussions of the nature of microwave heating are now presented to provide the reader with some insight into this important mode of heating. At the end of this section, a summary is provided of the heat source term formulations for microwave heating. It is important to note that the modeling of microwave heating described below is another example of physics-based modeling.

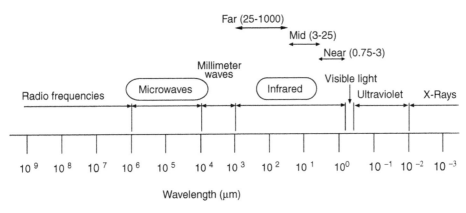

Figure 4.2 The electromagnetic spectrum showing the regions of microwave and infrared (with its sub regions) energy that are used to heat food materials. (From Datta, A. K. and Almeida, M., *Engineering Properties of Foods*, Rao, M. A., Rizvi, S. S. H., and Datta, A. K., Eds., Boca Raton, FL: Taylor & Francis, 2005.)

Microwave heating of food refers to the heating of food using electromagnetic waves in the microwave region (the electromagnetic spectrum is shown in Figure 4.2) in a resonant cavity, usually at frequencies of 915 MHz and 2450 MHz. At these frequencies, water molecules as dipoles and ions present in the food try to follow the alternating electric field of the microwaves and generate frictional heat. The general form of the volumetric heat source term, Q, in microwave heating is given by

$$Q = 2\pi f \varepsilon_0 \varepsilon'' E^2 \quad \frac{\text{W}}{\text{m}^3} \tag{4.6}$$

where f is the frequency of the microwaves, ε'' is dielectric loss (a property of the food material), and E is the strength of the electric field at any time t at a given location. The use of this equation to model microwave heating is not straightforward. Section 4.6.2.6 will provide the precise formulations of Q for various situations. However, to help the reader understand the reasoning behind these formulations and to be able to use them properly, this section will provide physical insight into the microwave heating process, including showing how to model the electric field, E.

4.6.2.1 Electromagnetic Interaction with a Food Material and Dielectric Properties

The two properties that determine a material's interaction with microwaves are the dielectric constant, ε', and dielectric loss, ε''. Air in a microwave oven can absorb very little of the microwave energy; only the food is heated by the electromagnetic waves. Dielectric properties of foods cover a significant range, as shown in Figure 4.3.

Like many other food properties, first-principle-based prediction equations for dielectric properties as a function of temperature and composition are not available. Empirical or semiempirical correlations are the only possibilities, but even these are scarce (e.g., see Ref. 11).

Dielectric properties of foods can vary strongly with temperature. As the ice in a frozen food melts, absorption of microwaves increases tremendously. This leads to one of the sharpest changes in dielectric properties. Such changes in dielectric properties can be described using procedures such as those in.[12] Above freezing, the dielectric constant decreases with temperature. The dielectric loss of an unfrozen material decreases with temperature, except for materials with high salt contents, where dielectric loss increases due to increased ionic activity at higher temperature. From the few available correlations,[11] an example of temperature variation of dielectric properties can be

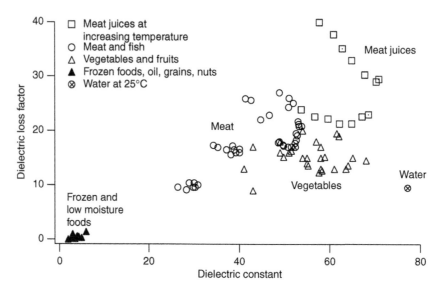

Figure 4.3 A scatter plot of some of the literature data for a variety of food materials showing some approximate grouping. Most of the data are from a frequency range of 2400–2500 MHz. Temperature varies between 5°C and 65°C for meats and meat juices, while the vegetable data is mostly at 23°C. (From Datta, A. K. and Almeida, M., *Engineering Properties of Foods*, Rao, M. A. Rizvi, S. S. H., and Datta, A. K., Eds., Boca Raton, FL: Taylor & Francis, 2005.)

seen in the data for meats:

$$\varepsilon'_{\text{meats}} = m_{\text{water}}(1.0707 - 0.0018485T) + m_{\text{ash}}(4.7947) + 8.542;$$

$$\varepsilon''_{\text{meats}} = m_{\text{water}}(3.4472 - 0.01868T + 0.000025T^2) + m_{\text{ash}}(-57.093 + 0.23109T) - 3.5985.$$

Some other correlations can be seen in Ref. 11.

Moisture content is one of the major determinants of food dielectric properties. Increases in the dielectric constant and loss factor of food systems with moisture content have been shown in various studies. However, few data or correlations are available for data as a function of moisture content, especially for lower moisture contents. In simulations of microwave drying, for example, moisture content variation is needed. In the absence of available data, approximate variations can be developed based on composition and data on similar materials, followed by a sensitivity analysis, as discussed under thermal properties.

During processing, both temperature and moisture of the food may change. Thus, to use Equation 4.6, one needs both ε'' and ε' (because E depends on it) as functions of temperature and moisture. This couples electromagnetics and heat transfer (see Section 4.11.2 for details of such modeling). The following two sections discuss how to solve for the electric field, E, thereby allowing calculation of the heat source term.

4.6.2.2   Modeling of Microwave Heating: Solutions for Idealized Plane Wave

The shape and size of the food greatly affect its spatial distribution of microwave absorption. Consider first the simplest situation of plane electromagnetic waves propagating in an infinite medium, which is different from food heating in a cavity (discussed later). Maxwell's equations describing the electromagnetic wave propagation can be simplified, assuming electric field polarization in the x direction, the magnetic field polarization in the y direction, and the wave propagation in the z direction, as

$$\frac{\partial^2 E_x}{\partial x^2} - \mu\sigma_e \frac{\partial E_x}{\partial t} - \mu\varepsilon \frac{\partial^2 E_x}{\partial t^2} = 0. \tag{4.7}$$

Assuming that a solution to this 1D electromagnetic wave propagation equation is $E_x e^{j\omega t}$, then the spatial variation given by E_x can be written as

$$E_x = E_x^+ e^{-\alpha z} e^{-j\beta z}, \tag{4.8}$$

where E_x^+ is the electric field at $z=0$. From this expression, the magnitude of the electric field is given by

$$|E_x| = E_x^+ e^{-\alpha z}. \tag{4.9}$$

This shows that the electric field decays exponentially into the material from the surface. The coefficients α and β are given by

$$\alpha = 2\pi f \sqrt{\mu\varepsilon} \left(\frac{1}{2} \left[\sqrt{1 + (\varepsilon''/\varepsilon)^2} - 1 \right] \right)^{1/2}, \tag{4.10}$$

$$\beta = 2\pi f \sqrt{\mu\varepsilon} \left(\frac{1}{2} \left[\sqrt{1 + (\varepsilon''/\varepsilon)^2} + 1 \right] \right)^{1/2}. \tag{4.11}$$

The heat generation in the material is related to the electric field by Equation 4.6. Substituting Equation 4.9 into Equation 4.6 gives

$$Q = 2\pi f \varepsilon_0 \varepsilon''_{\text{eff}} |E_x|^2 = 2\pi f \varepsilon_0 \varepsilon''_{\text{eff}} E_x^{+2} e^{-2\alpha z} = Q_0 e^{-z/\delta} \tag{4.12}$$

where Q_0 is the rate of heat generation at the surface and $\delta = 1/2\alpha$ is the power penetration depth that is often referred to as simply *penetration depth*. From Equation 4.10, penetration depth, δ, is a function of food properties. Typical ranges of penetration depths for various food groups are shown in Figure 4.4.

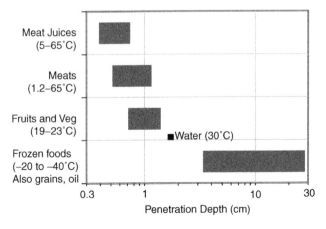

Figure 4.4 Typical ranges of penetration depths for various groups of food materials (the dielectric properties correspond to those shown in Figure 4.3). Data are at frequencies near 2.45 GHz. (From Datta, A. K., Sumnu, G., and Raghavan, G. S. V., *Engineering Properties of Foods*, Rao, M. A., Rizvi, S. S. H., and Datta, A. K., Eds., New York: Marcel Dekker, 2005.)

4.6.2.3 *Modeling of Microwave Heating: Solutions for Heating in an Oven (a Cavity)*

A microwave oven is a 3D cavity, as illustrated in Figure 4.5, with a spherical food load inside of it. The inside walls of the cavity reflect microwaves. Inside this cavity, the incoming electromagnetic waves from the magnetron (shown as excitation and waveguide) and the reflected waves from the cavity walls form resonant patterns. The qualitative pattern, as well as the magnitude of electric fields inside the food, are quite different from those in the air. In contrast to a plane wave (Equation 4.9), there is no easy and universal solution to obtain the electric field inside a food placed in such a cavity. Over the years, exponential decay inside a food has been used in many oven-heating situations, but for most of the situations this is essentially a qualitative assumption and can be completely wrong, depending on the size of the food and its dielectric properties. The proper method of obtaining the electric field patterns inside a food is to solve the Maxwell's equations of electromagnetics for the oven and obtain the volumetric rate of heating as given by Equation 4.6. This process is described below.

4.6.2.3.1 *Governing Equations*

The electromagnetic fields, such as those that are responsible for the heating of the food material inside a microwave oven (Figure 4.5), are described by the Maxwell's equations

$$\nabla \times \mathbf{E} = -\frac{\partial}{\partial t}(\mu \mathbf{H}) \tag{4.13}$$

$$\nabla \times \mathbf{H} = \frac{\partial}{\partial t}(\varepsilon' \varepsilon_0 \mathbf{E}) + \varepsilon''_{eff} \varepsilon_0 \omega \mathbf{E} \tag{4.14}$$

$$\nabla \cdot (\varepsilon \mathbf{E}) = 0 \tag{4.15}$$

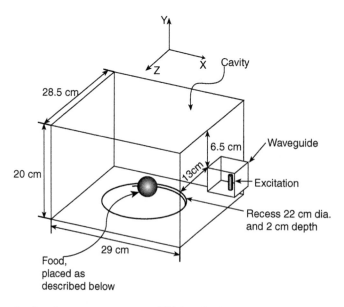

Figure 4.5 Schematic of a microwave oven system (GE, Inc., Louisville, KY, Model number JE635WW03, rated at 0.92 kW) with food placed at the center of the oven. (From Zhang, H. and Datta, A. K., *Transactions of the Institution of Chemical Engineers*, 83, 14, 2005.)

$$\nabla \cdot \mathbf{H} = 0, \tag{4.16}$$

where $\mathbf{E}$ and $\mathbf{H}$ are the electric and magnetic field vectors, respectively. In food materials, heating is carried out by the electric field, primarily through its interaction with water and ions, as noted earlier. The complex permittivity, ε, is given by

$$\varepsilon = \varepsilon' + j\varepsilon''_{\mathrm{eff}}, \tag{4.17}$$

where the properties ε' and $\varepsilon''_{\mathrm{eff}}$ are functions of locations in the food due to temperature (and moisture) variations. These properties also vary with the frequency of the microwaves; however, in food applications, the microwave frequency is usually fixed. In the above equations, ε_0 is the permittivity of free space (8.86×10^{-12} F/m) and $\omega = 2\pi f$ is the angular frequency of the microwaves. For a short discussion of Maxwell's equations and their solution in heating applications, see Ref. 13. The Maxwell's equations are to be solved to obtain the electric field, $\mathbf{E}$, as a function of position in the food and heating time. The rate of volumetric heat generation is calculated from this electric field using Equation 4.6.

4.6.2.3.2 Boundary Conditions

Boundary conditions for the electromagnetic modeling of a cavity are set on the walls of the cavity, which are considered perfect conductors. The entire cavity interior is treated as a dielectric, with appropriate dielectric properties of air and food in the regions that they occupy. Note that in modeling of the entire cavity, the food–air interface does not have to be treated in any special manner by the modeler because this is built into Maxwell's equations. In the interior of a perfect electrical conductor, the electric field is zero. This condition, together with Maxwell's equations, leads to the boundary condition at the air–wall interface as

$$E_{\mathrm{t,air}} = 0 \tag{4.18}$$

$$B_{\mathrm{n,air}} = 0. \tag{4.19}$$

Here, the subscripts t and n stand for tangential and normal directions, respectively. These conditions are necessary to determine the solution. Input parameters needed for the solution are the geometry of the food, inside geometry of the oven, the dielectric properties of the food material and the magnitude of the excitation. In most cases, the magnitude of the excitation is obtained by matching experimental data on temperature rise.

4.6.2.3.3 Numerical Solution and Experimental Verification

There are a number of commercial software programs based on various computational methods, such as the finite difference time-domain method and the finite element method (see http://www.emclab.umr.edu/csoft.html. Many of these codes are not particularly efficient for cavity-heating applications where the electromagnetics must be solved for high frequencies and coupled with heat transfer. Two of the codes that have such coupling capabilities for cavity-heating applications are ANSYS and COMSOL. An example of using ANSYS in microwave heating of food can be seen in Ref. 14. Direct experimental verification of the electric field is difficult because few, if any, sensors are available for this purpose. Typically, the temperature measurements serve as indirect validation of the electric field distributions.

4.6.2.4 *Factors Affecting Heat Generation: Food Volume*

Because a complete solution of the electromagnetics in a cavity depends strongly on oven and food parameters, only general trends are discussed here. In microwave heating, like conventional heating, the total power absorbed increases with volume, eventually leveling off at a power that depends primarily on the magnetron power level, and to a lesser extent on the dielectric properties and geometry. Thus, total power absorption with load volume is typically described by a curve similar to that shown in Figure 4.6. The curves in this figure show experimental data and numerical computations for heating of water in a cylindrical container. This relationship is sometimes described by an empirical equation of the form

$$q_{total} = a(1 - e^{-bV}), \tag{4.20}$$

that can also be written as

$$\frac{q_{total}}{V} = \frac{a}{V}(1 - e^{-bV}), \tag{4.21}$$

where a and b are empirical constants. Thus, power absorption per unit volume decreases as load volume increases. When the complete set of Maxwell's equations are solved for a cavity, as discussed above, this volume effect of the load is automatically taken into account. However, if the modeling is to be based on other simplified formulations, as described below under various heat-source formulations, the total power absorbed by a certain volume of load needs to be experimentally measured.

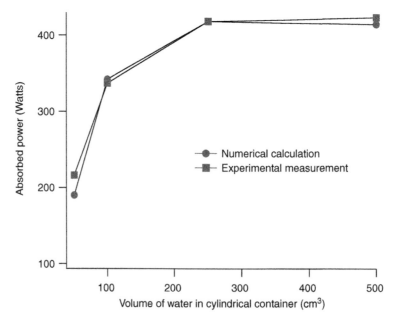

Figure 4.6   Magnitude of power absorbed in different volumes of water loads when full power of the oven is applied, obtained from experiment and electromagnetic simulations. (From Zhang, H. and Datta, A. K., *Transactions of the Institution of Chemical Engineers*, 81C, 257, 2003.)

4.6.2.5 Factors Affecting Heat Generation: Food Shape

Food shape can significantly affect both the total power absorption as well as its spatial distribution inside the food. One of the major problems in microwave heating is the high intensity of electromagnetic fields at the edges and corners of the food. Another very significant feature of microwave heating is the possibility of focusing due to curved shapes. Focusing is a function of the dielectric properties of food and the food's size, as illustrated in Figure 4.7. Such shape effects can only be modeled by solving the Maxwell's equations for the particular situation.

4.6.2.6 Summary of Heat Source Formulations for Microwave Heating

4.6.2.6.1 Constant Value of Heat Generation

For very thin or low-loss materials, where the penetration depth is expected to be very large compared to the size of the food, sometimes a reasonable assumption can be constant volumetric heat generation, i.e.,

$$Q = \text{constant.} \tag{4.22}$$

4.6.2.6.2 Exponentially Decaying Heat Generation from the Surface

This is the most used (and misused) expression for heat generation. This form actually originates from the plane electromagnetic waves propagating in an infinite media, as shown earlier, and is given by

$$Q = Q_0 e^{-z/\delta}. \tag{4.23}$$

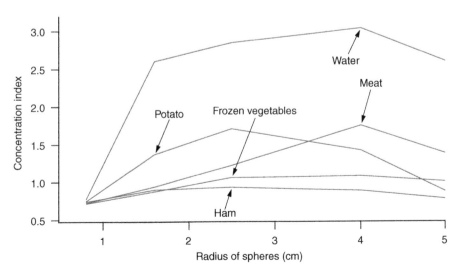

Figure 4.7 Concentration index (a measure of the intensity and the extent of focusing of energy) as a function of radius for different spherical food materials heated in a microwave cavity, obtained from electromagnetic simulations. (From Zhang, H. and Datta, A. K., *Handbook of Microwave Food Technology*, Datta A. K. and Anantheswaran, S., Eds., New York: Marcel Dekker, 2000.)

This equation was shown[18] to be valid for plane electromagnetic-wave heating of finite slabs where the slab thickness is greater than 5.4δ–0.08 cm, where δ is in cm. For cavity heating, the use of Equation 4.23 is mostly empirical because no studies have proven the equation to be true for a cavity.

4.6.2.6.3 Detailed Electromagnetic Solution

A detailed electromagnetic solution is the right method to accurately obtain the heat generation values. However, it is also a computationally challenging problem, as previously discussed. As improved software becomes available, this approach should become easier. Here, the electric field is computed from solving the Maxwell's Equation (Section 4.6.2) and the heat generation is computed from the electric field as

$$Q = 2\pi f \varepsilon_0 \varepsilon''_{eff} |E|^2, \tag{4.24}$$

where $|E|$ is the magnitude of the electric field at any location. This equation can be used to calculate heat generation for any situation where E is known.

4.6.3 Heat Source Term for Electromagnetic Heating: Infrared

4.6.3.1 Zero Penetration: Inclusion as a Boundary Condition

If the depth of penetration of infrared radiation is negligible, the heat flux at the surface, $q|_{surface}$, computed from radiative transfer calculations, is used as a boundary condition:

$$-k\frac{\partial T}{\partial x} = q|_{surface}. \tag{4.25}$$

Further discussion on heat-flux boundary condition can be seen in Section 4.10. Heat flux can be measured experimentally[14] or obtained from the radiative heat-transfer analysis mentioned in Section 4.10.

4.6.3.2 Significant Penetration: Volumetric Heating with Exponential Decay

If the depth of penetration, δ, as defined earlier, is significant, a different formulation is required. Surface radiant heat flux at the boundary given by Equation 4.25 is no longer used; instead, radiant heating is included as a volumetric heat source term given by

$$Q = -\frac{dq}{dx} = \frac{q|_{surface}}{\delta} e^{-x/\delta}, \tag{4.26}$$

where $q|_{surface}$ is obtained from experiment or radiative heat transfer analysis, as discussed previously. Some values of penetration depth for food materials can be seen in Ref. 11. Although penetration depth is a function of wavelength and moisture content, etc., such detailed information is generally unavailable (representative information for potato can be seen in Ref.11).

4.6.4 Heat Source Term for Electromagnetic Heating: Ohmic

Before providing the heat source term for ohmic heating, a brief introduction is provided to the physics of ohmic heating. A summary of the source term formulations is provided at the end of this section.

4.6.4.1 Electrical Conductivity of Foods and Interaction of Low-Frequency Electric Fields with Food Material

Ohmic or resistive heating depends on the electrical conductivity of the material. Food formulations with a moderate percentage of free water having dissolved ionic salts are reasonably good conductors and can be heated by the process. Most pumpable foods with water content exceeding 30% conduct electricity sufficiently well for ohmic heating. Covalent, nonionized fluids such as fats, oils, alcohols, and sugar syrups, and nonmetallic solids such as bone, cellulose, and crystalline structures (including ice) cannot be directly ohmically heated. Conductivity of most electrolytic solutions increases with temperature due to increased ionic mobility and decreasing viscosity.

4.6.4.2 Modeling of Ohmic Heating

The source term, Q, for ohmic heating is given by

$$Q = \sigma_{\mathrm{eff}} E^2, \tag{4.27}$$

where σ_{eff} is the conductivity of the material and E is the magnitude of the electric field. Note that Equation 4.27 and Equation 4.6 are equivalent because the effective dielectric loss can be related to effective conductivity as

$$\sigma_{\mathrm{eff}} = 2\pi f \varepsilon_0 \varepsilon''_{\mathrm{eff}}. \tag{4.28}$$

The volumetric heating term for ohmic heating can be written in terms of voltage as

$$Q = \frac{1}{\rho}\left(\frac{\Delta V}{\Delta L}\right)^2. \tag{4.29}$$

Here, a voltage drop of ΔV occurs over a length ΔL for a material of resistivity ρ. Note that this is the same as Equation 4.27 because $E = \Delta V/\Delta L$, and $\sigma = 1/\rho$ is the conductivity (units of Siemens/m).

The voltage, V, is obtained from a simplified version of the Maxwell's equation of electromagnetics described previously under microwave heating. It is given by

$$\nabla \cdot (\sigma \nabla V) = 0, \tag{4.30}$$

which in 2D Cartesian coordinates is given by

$$\frac{\partial}{\partial x}\left(\sigma \frac{\partial V}{\partial x}\right) + \frac{\partial}{\partial y}\left(\sigma \frac{\partial V}{\partial y}\right) = 0 \tag{4.31}$$

The voltage equation (Equation 4.31) is second-order in space. It therefore needs two boundary conditions for each spatial dimension x and y. A common boundary condition would be to set the voltage at one end as the applied voltage and the voltage at the other end as zero. The input parameter to this equation is typically the conductivity, σ, as a function of temperature. Notice the similarity of Equation 4.31 to the steady-state diffusion equation:

$$\frac{\partial}{\partial x}\left(D \frac{\partial c}{\partial x}\right) + \frac{\partial}{\partial y}\left(D \frac{\partial c}{\partial y}\right) = 0. \tag{4.32}$$

Thus, a common way to solve Equation 4.31 in the context of CFD software is to use the voltage as a species. The value of σ is input as the value of diffusivity for this species. This is how ohmic heating is implemented in the software FIDAP, for example.

4.6.4.3 Summary of Heat Source Formulations for Ohmic Heating

4.6.4.3.1 Constant Value of Heat Generation

When the variation of electrical conductivity with temperature and composition can be ignored, a constant value of heat generation can be used, i.e.,

$$Q = \text{constant.} \tag{4.33}$$

4.6.4.3.2 Spatial and Time Variation in Heat Generation

Electrical conductivity variation cannot often be ignored. For this situation, the heat generation will be given by

$$Q = \sigma_{\text{eff}} E^2, \tag{4.34}$$

where E is obtained from solving Equation 4.31.

4.6.5 Heat Source Term for High-Pressure Heating

In high-pressure processing, pressure buildup leads to volumetric heating from the work of compression. This is a thermodynamic effect. The volumetric heating can be modeled as

$$Q = \beta T \frac{dP}{dt} \frac{W}{m^3}, \tag{4.35}$$

where β is the thermal volumetric expansion coefficient and T and P are temperature and pressure at any time t, respectively. Some data on β values of food materials is becoming available.[19]

4.6.6 Heat Source Term for Respirative Heating

Respiration is the process by which stored organic materials (carbohydrates, proteins, and fats) are broken into simple endproducts with a release of energy. This process uses oxygen and produces carbon dioxide and water vapor. The respiration rate of a product determines its transit and post-harvest life. Stored food reserves are lost during respiration, which means less food value, loss of flavor, loss of saleable weight, and more rapid deterioration. Respiration rates of commodities are directly related to product temperature; the higher the temperature, the higher the respiration rate. Rapid cooling to the commodity's safe temperature is most critical for those commodities with inherently higher respiration rates. An example of a heat and mass transfer model that includes respirative heating can be seen in Ref. 20. Typically, the heat source term is treated as a constant.

4.6.7 Heat Source Term for Ultrasonic Heating

In ultrasonic heating, mechanical energy carried by longitudinal waves is converted into thermal energy by frictional losses. The associated frequencies are usually over 20 kHz, beyond the range of human perception. It is not associated with the electromagnetic spectrum. Although an exponential decay of energy levels from the surface into the material can sometimes be assumed, there are significant complications.[21] Modeling of ultrasonic heating in the food context is rare, but studies from biomedical applications can be quite relevant.

4.7 HEAT TRANSFER IN SOLIDS

The various approaches used to model heat transfer in food processes can be summarized as shown in Figure 4.8, in order of complexity from left to right. Although the first four situations (two resistance formulations, lumped parameter, and analytical solution to the heat conduction equation) can frequently lend themselves to simpler analytical solutions, often a real situation must be grossly simplified to achieve this. A numerical solution is the most flexible method and can accommodate almost any conduction heating situation. Such numerical solutions and relevant software are discussed in Chapter 3. The conjugate or solid–fluid coupled problems are generally much more complex and numerical solutions are usually needed. The first four situations are discussed in this section, whereas the conjugate problem is discussed under the fluids section.

4.7.1 Thermal Resistance Formulation: Steady State

Here, the discussion starts from the simplest possible situation for heat transfer that is steady state and also one-dimensional. Figure 4.9 shows two such situations: one is a thin slab where the heat transfer can be considered only along the thickness; the other is a long cylinder where the heat transfer can be considered only in the radial direction. For no heat generation and steady-state heat transfer, the general governing equation (Equation 4.3) for a slab becomes

$$\frac{d^2T}{dx^2} = 0. \tag{4.36}$$

For the simplest boundary condition where constant temperatures of T_1 and T_2 can be assumed at the two surfaces given by $T(x=0)=T_1$ and $T(x=L)=T_2$, the solution is given by

$$T = \frac{T_2 - T_1}{L} x + T_1, \tag{4.37}$$

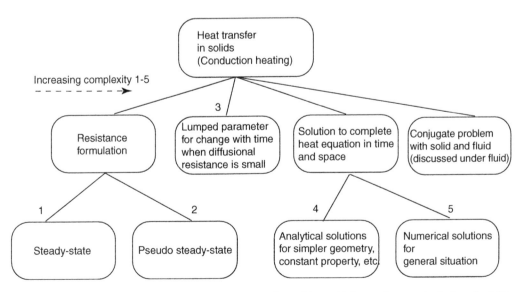

Figure 4.8 Approaches to solving a conduction heat transfer problem, with increasing complexity from the left to the right.

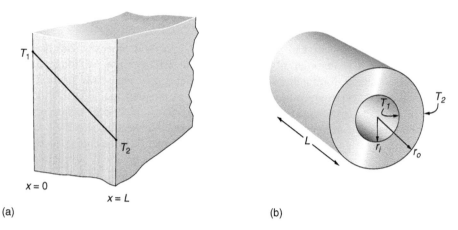

Figure 4.9 Simple slab and cylindrical geometries.

which shows a linear change in temperature from T_1 to T_2 at steady state. The heat flow in the positive x direction is given by

$$q_x = -kA\frac{\mathrm{d}T}{\mathrm{d}x} = \frac{T_1 - T_2}{L/kA} = \frac{\text{Temperature difference}}{\text{Thermal resistance}}.$$ (4.38)

The term L/kA in the above equation is the thermal resistance for a slab. Similarly, heat flow outward in the radial direction per unit length of a long cylindrical annulus (Figure 4.9b) is written as

$$q_r = \frac{T_1 - T_2}{(\ln(r_o/r_i))/2\pi k},$$ (4.39)

where r_i and r_o are the inner and outer radius of the cylinder, respectively.

In a manner similar to the steady-state slab and cylindrical annulus, the convective heat transfer equation

$$q_x = hA(T - T_\infty) = \frac{T - T_\infty}{1/hA}$$ (4.40)

can also be considered in terms of a (convective) thermal resistance $1/hA$ and a driving force of $T - T_\infty$.

In practice, two or more resistances are frequently present. For steady-state situations, such resistances can be combined in a simple manner. For example, for a slab with convection on both sides, the heat flux can be calculated by considering the thermal resistances in series (an analogy to current flow):

$$q_x = \frac{T_h - T_c}{\underbrace{\frac{1}{h_h A}}_{\text{convective}} + \underbrace{\frac{L_1}{k_1 A}}_{\text{conductive}} + \underbrace{\frac{1}{h_c A}}_{\text{convective}}}$$

$$= \frac{\text{Temperature difference}}{\sum \text{Thermal resistance}}.$$ (4.41)

Here, h_h and h_c are the heat transfer coefficients on the two sides of the slab, respectively. Equation 4.41 or its variation is quite useful in modeling heat flux, for example, through a composite wall (such as heat gained through the walls of a cold-storage building, heat loss through the insulation of a heating pipe, etc.).

4.7.2 Thermal Resistance Analysis: Pseudo Steady State

The resistance formulation developed in the previous section for a steady state has been extended to processes that are not, strictly speaking, at steady state, but are changing slowly enough that they can be considered almost steady or "pseudo" steady state. Use of this approach in modeling food processes, particularly those involving a phase change such as in freezing, has been common. For example, a freezing process is described in Section 4.9.2. As illustrated in Figure 4.10 for a freezing process, the boundary between frozen and unfrozen, or the *freezing front*, moves slow enough that the frozen region can be considered at steady state. However, the freezing front does move with time and therefore the process is not at a true steady state. This analysis method was more popular before numerical computation became commonplace.

4.7.3 Lumped Parameter Analysis

In some special cases, temperature variation in all three spatial directions can be ignored. When the temperature variations are ignored, the situation is considered lumped. Temperature would then vary only with time. This lumped parameter condition is possible when the internal resistance in the solid is small compared to the external resistance in the fluid (this is explained in greater detail later). The governing equation for such heat transfer is given by

$$\frac{dT}{dt} = -\frac{hA}{mC_p}(T - T_\infty). \tag{4.42}$$

It is possible (e.g., see Ref. 22) to start from the governing heat-transfer equation (Equation 4.4 without any heat generation or convection) and derive Equation 4.42. Because it is a first-order equation, it needs one condition that, in this case, is the initial condition

$$T(t = 0) = T_i. \tag{4.43}$$

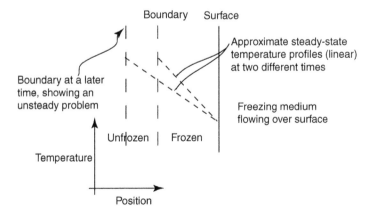

Figure 4.10   Illustration of a pseudo-steady state process where the process changes slowly enough that it can be considered to be at a steady state at any given time.

Solving Equation 4.42 gives temperature as a function of time in lumped parameter heat transfer as

$$\frac{T - T_\infty}{T_i - T_\infty} = \exp\left(-\frac{hA}{mC_p}t\right) = \exp\left(-\frac{t}{mC_p/hA}\right). \tag{4.44}$$

Equation 4.44 shows that an infinite time is required to reach the steady state or the final temperature of T_∞. As the temperature of the solid becomes close to fluid temperature, the rate of heat transfer drops; consequently, the solid can never quite reach the fluid temperature, T_∞. This example illustrates the power of analytical solutions in providing detailed physical insight for simple heat-transfer situations. The validity of this analytical solution depends on the ratio of internal to external resistance (also called the *Biot number*) being less than 0.1. For a slab, this leads to $(L/kA)/(1/hA) < 0.1$ or $hL/k < 0.1$. For an arbitrary shape, the following must be satisfied for a lumped parameter model to be used:

$$\frac{h(V/A)}{k} < 0.1, \tag{4.45}$$

where V is the volume of the solid and A is the surface area. The error in temperature calculation is less than 5% when Equation 4.44 is used and Equation 4.45 is satisfied. An example of the use of lumped parameter analysis in food can be seen in Ref. 23.

4.7.4   Analytical Solution to the Heat Equation: Application to Canning

In a more complex model, when internal diffusional resistances cannot be ignored (i.e., $h(V/A)/k \geq 0.1$), lumped parameter modeling is not possible and the complete governing equations for heat transfer in a solid must be used. For example, the governing equation for symmetric heating or cooling of an infinite slab without any heat generation can be simplified as

$$\frac{\partial T}{\partial t} = \frac{k}{\rho C_p}\frac{\partial^2 T}{\partial x^2}. \tag{4.46}$$

The boundary conditions are

$$\left.\frac{\partial T}{\partial x}\right|_{x=0,t} = 0 \quad \text{(from symmetry)} \tag{4.47}$$

$$T(L, t > 0) = T_s \quad \text{(surface temperature is specified)} \tag{4.48}$$

and the initial condition is

$$T(x, t = 0) = T_i, \tag{4.49}$$

where T_i is the constant initial temperature and T_s is the constant temperature at the two surfaces of the slab at time $t > 0$.

$$\frac{T - T_s}{T_i - T_s} = \sum_{n=0}^{\infty} \frac{4(-1)^n}{(2n + 1)\pi}\cos\frac{(2n + 1)\pi x}{2L} e^{-\alpha\left(\frac{(2n+1)\pi}{2L}\right)^2 t}, \tag{4.50}$$

where $\alpha\{=k/\rho C_p\}$ is the thermal diffusivity. The terms in the series ($n=0,1,\ldots$ in Equation 4.50) drop off rapidly for large values of time and the solution can be written as

$$\frac{T - T_s}{T_i - T_s} = \underbrace{\frac{4}{\pi} \cos \frac{\pi x}{2L}}_{\text{spatial}} \underbrace{e^{-\alpha \left(\frac{\pi}{2L}\right)^2 t}}_{\text{time}}. \tag{4.51}$$

This expression clearly shows that eventually (for times long after $t=0$), the time-temperature relationship is exponential at a given position (x value). This again shows the power of analytical solutions. A volumetric average in temperature given by

$$T_{av} = \frac{1}{L} \int_0^L T dx \tag{4.52}$$

can be calculated as

$$\frac{T_{av} - T_s}{T_i - T_s} = \frac{8}{\pi^2} e^{-\alpha \left(\frac{\pi}{2L}\right)^2 t}. \tag{4.53}$$

This implies that for a given change of average temperature (measured in terms of fractional change of the total possible change $T_i - T_s$), the time required increases with the square of the thickness, i.e., $t \propto L^2$. This observation can be generalized for other geometries by saving the time required is proportional to the square of the characteristic dimension. Expressions for temperatures as a function of position and time (such as Equation 4.50) can be developed for cylindrical, spherical, and other coordinate systems.

It is extremely important to consider the limitations of the analytical solution described in this section (Equation 4.50). These limitations include (1) uniform initial temperature; (2) constant boundary temperature; (3) slab of uniform thickness; (4) slab is much wider in the other dimensions compared to its thickness; and (5) constant thermal properties. When these conditions are not satisfied, it is prudent to use a numerical solution instead.

4.7.4.1 Example: Canning of Solid Foods

Sterilization of food involves destruction of microorganisms carried in it so that the food can be stored longer and be safe to consume. Canning, where a container is filled with food and then sealed, is one of the oldest methods for food sterilization. Later, the sealed container is heated in a closed vessel with steam or hot water long enough to kill the microorganisms and then it is cooled. Heating the container more than the required amount of time would unnecessarily degrade the food's quality while wasting energy. However, insufficient heating could lead to serious public health hazards.

Cylindrical cans are possibly the most common type of container used in the canning industry. For solids packed into such a container, the heat conduction in the food is described by the heat equation in the cylindrical geometry

$$\frac{\partial T}{\partial t} = \alpha \left(\frac{\partial^2 T}{\partial r^2} + \frac{1}{r} \frac{\partial T}{\partial r} + \frac{\partial^2 T}{\partial z^2} \right), \tag{4.54}$$

with the initial condition of constant temperature and a boundary condition of convection at the can surface. One assumption is that the thermal resistance of the metal wall is ignored in comparison to that of the food. The death of bacteria during the heating process is often considered to follow first-order kinetics with a temperature dependency given by a Z value, where $Z = 2.303RT_0^2/E_a$ is a linearization of the Arrhenius kinetics with activation energy E_a around a reference temperature of T_0. A quantity called the F_0 value is used in the food industry; F_0 stands for the time of heating at a reference temperature of $T_0 = 121°C$ that would lead to the same destruction of bacteria as in a given process. Thus, F_0 is defined as $F_0 = 1/k_0 \ln (c_i/c)$ where k_0 is the rate of reaction at temperature T_0, and c_i and c are the initial and final concentrations of bacteria, respectively, for the given process. Heating continues until F_0^{slowest}, calculated from

$$F_0^{\text{slowest}} = \int_0^t 10^{(T-T_R)/Z}\,dt, \tag{4.55}$$

where T is the temperature at the coldest point, is more than the desired value, i.e.,

$$F_0^{\text{slowest}} \geq F_0^{\text{desired}}, \tag{4.56}$$

where F_0^{desired} is the desired value that comes from industry practice and/or government regulations.

4.7.4.1.1 Analytical Solution: Ball's Formula

The analytical solution to Equation 4.54 for constant initial temperature and convection at the surface is given by

$$\frac{T - T_m}{T_i - T_m} = \sum_m \sum_p A_{mp} e^{(-\beta_m^2 + \eta_p^2)\alpha t} J_0(\beta_m r)\cos(\eta_p z) \tag{4.57}$$

where A_{mp} depends on initial temperature and β_m and η_p depend on the boundary conditions. For a convective boundary condition, the last two constants are given by

$$J_0'(\beta_m R) + \frac{h}{k} J_0(\beta_m R) = 0; \tag{4.58}$$

$$\eta_p \tan(\eta_p H/2) = \frac{h}{k}. \tag{4.59}$$

The previous simplification of the solution to the conduction equation in a slab (Equation 4.51) is well suited for applications involving thermal heating of food to destroy bacteria. During the initial times, the bacterial destruction is insignificant. Thus, it is only logical for researchers considering food heating applications to be interested in later times; the temperature during these times are given by

$$\frac{T - T_m}{T_i - T_m} = A e^{(-\beta^2 + \eta^2)\alpha t} J_0(\beta r)\cos(\eta z). \tag{4.60}$$

Here, the temperatures at the geometric center of the can are of interest because this is the slowest heating point in the can. Equation 4.60 can be rewritten as

$$\frac{T - T_m}{T_0 - T_m} = j10^{-t/f}, \tag{4.61}$$

where j and f are given in terms of position variable and heating parameters. Substituting temperature T from Equation 4.60 into the equation for F_0 (Equation 4.55) yields

$$F_0 = \int_{T_m + j(T_0 - T_m)}^{T} \frac{-f\left(10^{\frac{T-121}{Z}}\right)}{2.303(T - T_m)}\,dT,$$

which can be written as

$$\frac{1}{f/u^*} = \int\limits_{\frac{-j(T_0 - T_m)}{Z}}^{\frac{T_m - T}{Z}} \frac{10^{-x}}{2.303x} dx.$$

According to Ball and Olson (1957), it is only necessary to start from 44°C (80°F) below T_m to achieve all the significant lethalities. Consequently, the integral above can be written as

$$\frac{1}{f/u^*} \simeq \int\limits_{\frac{44}{Z}}^{\frac{T_m - T}{Z}} \frac{10^{-x}}{2.303x} dx.$$

In accordance with this expression, lethality calculations for the heating process in Ball and Olson (1957) do not depend on initial temperature or the j value. Because this integral is difficult to evaluate, it is tabulated for various values of f/u^* vs. $T_m - T$ for various values of j and Z. Using these graphs, the processing time for a particular bacterial reduction F_0 is calculated in the food industry. This procedure is general enough that the Balls' formula is true for conduction heating of foods in arbitrary shape and for forced convection heating. This is true, in part, because the solution given by Equation 4.60 is true for arbitrary shaped domains, as was shown in.[24]

4.7.5 Numerical Solution to Heat Equation: Application to Canning

The heat conduction or diffusion Equation (Equation 4.46, or its more general form) is a particularly common equation for which numerical software is ubiquitous. Any numerical software typically accommodates arbitrary geometry, boundary and initial conditions, property variations with temperature, etc. All of the formulations discussed earlier in this chapter can be routinely solved using numerical software, as opposed to the analytical solutions already described. Although most of the software can readily accommodate a heat source term, the software for solving heat conduction equations are often not coupled with other physics such as electromagnetics. Examples and issues in coupling of other physics with heat conduction are described in Section 4.11.

An example of using a numerical model in food applications is the modeling of the sterilization of solid foods,[25] where optimum time-temperature for sterilization was calculated by solving Equation 4.54 together with Equation 4.55. This work has been extended to obtaining optimum time-temperature for sterilization in other geometries and processing conditions.

A more comprehensive picture of the sterilization process, such as nutrient and bacterial profiles inside the container, can be computed by sequentially solving Equation 4.54 and a species equation for bacteria and/or other species equations for nutrients (see Section 4.11.3 for details of formulating species equations for this purpose). As will be discussed, this procedure is general enough that with appropriate replacement of the heat equation, it can be used in the case of convective heating.

4.7.6 Optimization of Conductive Heating

Optimization of conductive heating of foods has been pursued for over 30 years.[26] The factors relevant to the optimization process are the geometry of the container, property variations due to temperature or composition of the food, and the boundary (steam or hot water) temperature.

4.8   HEAT TRANSFER IN LIQUIDS

Modeling heat transfer in liquids is more complicated because, at least conceptually, the flow equations need to be solved in addition to the equation for energy transfer. Some unique aspects that need to be considered in the study of heat transfer in liquids include the following:

Momentum Equation. When detailed velocity profiles are needed, the momentum (Navier–Stokes) equation must be solved. This significantly increases the complexities over the solution of heat transfer in conduction heating.

Type of Fluid. The Newtonian vs. the non-Newtonian nature of the fluid, as discussed in Chapter 5, is important for heat transfer. The complexity of the model and difficulty of the solutions increase with a non-Newtonian fluid.

Type of Flow. The nature of the physics can be quite different depending on laminar vs. turbulent flow regimes. Computation of turbulent flow is more challenging compared to laminar flow. Turbulent flow is generally desired. Mechanical agitation is one method of achieving turbulent flow, such as in a scraped-surface heat exchanger or in an agitating retort. In continuous flow, tube dimensions, flow rates, and liquid properties are adjusted to obtain a Reynolds number in the turbulent range. Modeling the agitations is a challenging task. There are times, however, when turbulent flow is uneconomical to achieve, as in the case of many non-Newtonian liquids. These liquids exhibit high apparent viscosity, so the high pumping pressures required to obtain fully turbulent conditions are not economical for production rates of interest.[27]

Particles in the Fluid. When solid particles are present in the liquid, such as in a chunky soup, a very complex heat-transfer system results whose modeling has been rare. In sterilization, for example, the particles in the liquid would need to be sterilized in addition to the liquid itself. Conductive resistance for the inside of the solid food and convective resistance between the solid food surface and the liquid carrying it are added, making the sterilization of the food limited by the sterilization of the particles. The thermal resistance represented by the surface heat-transfer coefficient between the solid particles and the fluid carrying it is difficult to obtain experimentally or estimate from theoretical considerations.

Biochemical Reactions. When computing how temperature affects biochemical reactions in a fluid being heated, it is important to note the conceptual distinction between the time-temperature histories of location, as opposed to that of a liquid element.[28] Time-temperature history at a location in a moving liquid would not correspond to a single liquid element in general. Instead, it would correspond to all elements passing through that point over time. In a sterilization application, for example, depending on the nature of the flow, point sterilization can be different from element sterilization. From the standpoint of modeling food safety and quality, this distinction is important, and different methodologies need to be applied as compared to conduction heating, as detailed in Section 4.11.3.

4.8.1   Formulations of Convective Heat Transfer

Modeling of convective heat transfer can be divided into three different approaches, depending on the level of details desired, as shown in Figure 4.11. In the most simple formulation, **1**, only the rate of heat transfer between a surface and a fluid is described, without detailed information on the temperature profile in a fluid. In formulation **2**, the detailed temperature profile in the fluid is available at the expense of fairly intensive computations. In formulation **3**, at the expense of a still greater level of complex computations, details of the temperature profile are available not only for the fluid, but also for its immediate surroundings (usually a solid).

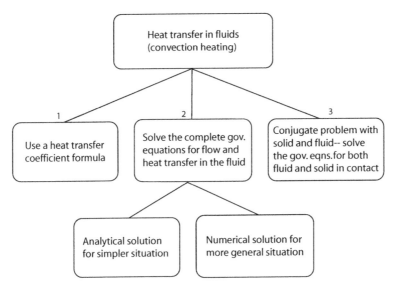

Figure 4.11 Formulations of convective heat transfer with increasing complexity from left to right.

Partly because of the complications mentioned in the earlier section and partly due to the availability of powerful CFD software, convection formulations **2** and **3** are primarily solved numerically. For formulations **1** and **2**, analytical solutions exist for simple situations. When a food-processing situation can be simplified such that analytical solutions can be used, the analytical solutions are generally encouraged because they can provide more insight into the problem with less effort.

When particles are included with the food, it becomes essentially a conjugate problem. Very few modeling studies exist for such situations. Formulation **1** has been used in sterilization to estimate the heat transfer coefficient between food pieces and the fluid for still (unagitated) heating of liquid with particulates (e.g., Ref. 29) and for agitated heating (e.g., Ref. 30,31). Using the surface heat-transfer coefficient, the temperature in the solid can be calculated. Formulation **3**, with a semianalytical solution, was developed for an axially rotating canned food with liquid particulate.[32] Formulation **3** has also been developed for solid–liquid mixture flow in a continuous heating situation.[33]

4.8.2 Modeling Using a Heat Transfer Coefficient

A great majority of practical convective heat-transfer situations can be modeled using this methodology. The goal in this type of modelling is to develop a formula for the heat-transfer coefficient, h, such that the standard equation for convective heat transfer

$$q = hA(T - T_\infty) \tag{4.62}$$

can be used. For simpler and more idealized geometries and flows, a vast array of correlations exist. An example of this is the case of fluid flow over a flat plate where the fluid temperature is different from the plate temperature. Two sets of formulas are provided—one for laminar flow and one for turbulent flow—because the heat transfer is quite different for the two situations:

$$Nu_x = 0.332 Re_x^{1/2} Pr^{1/3} \quad \text{for laminar } (Re_x < 2 \times 10^5) \tag{4.63}$$

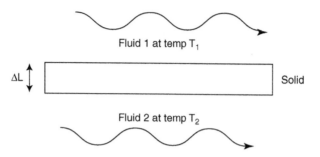

Figure 4.12 Schematic showing a solid with convective resistances on both sides.

$$Nu_L = 0.664 Pr^{1/3} Re_L^{1/2} \quad \text{for laminar } (Re_x < 2 \times 10^5) \tag{4.64}$$

$$Nu_x = 0.0288 Re_x^{4/5} Pr^{1/3} \quad \text{for turbulent } (Re_x > 3 \times 10^6) \tag{4.65}$$

$$Nu_L = 0.360 Re_L^{4/5} Pr^{1/3} \quad \text{for turbulent } (Re_x > 3 \times 10^6) \tag{4.66}$$

where $Nu_x = h_x x/k$ provides the local heat-transfer coefficient, h_x, at a location x and $Nu_L = hL/k$ provides the average heat transfer coefficient, h_L, over distance L. The quantity Pr is Prandtl number, given by $Pr = \mu C_p/k$, where μ is the viscosity, C_p is the specific heat, and k is the thermal conductivity of the fluid.

In many practical situations, heat transfer can involve fluids on each side of a solid surface, as illustrated in Figure 4.12. Here, heat transfer between fluids **1** and **2** can be written as

$$q = UA(T_1 - T_2), \tag{4.67}$$

where U is termed the *overall heat transfer coefficient* between the two fluids and is related to individual heat-transfer coefficients as

$$\frac{1}{U} = \frac{1}{h_1} + \frac{\Delta L}{k} + \frac{1}{h_2}, \tag{4.68}$$

where ΔL is the thickness of the solid. This approach in modeling is particularly useful in complex heat-transfer situations where spatial variation in temperature is either small or difficult to obtain.

Example 1: Sterilization with Agitation.   An example of heat-transfer modeling using the overall heat-transfer coefficient is the agitated batch (in-container) heating of liquids in a sterilization process, such as with canned foods. Agitation, such as that during axial rotation (see inset of Figure 4.13), increases the rate of heat transfer, thus shortening processing time and improving quality. This problem has been formulated as

$$mC_p dT = UA(T_\infty - T)dt, \tag{4.69}$$

where m is the total mass of liquid that changes its temperature by dT in time dt. Note that the definition of overall heat-transfer coefficient U uses the *mass average temperature*, T, as basis. Here, T_∞ is the heating-fluid temperature. Integrating this equation provides the time-temperature history of the fluid:

$$\frac{T - T_\infty}{T_i - T_\infty} = e^{-(UA/mC_p)t}, \tag{4.70}$$

were T_i is the initial temperature of the fluid.

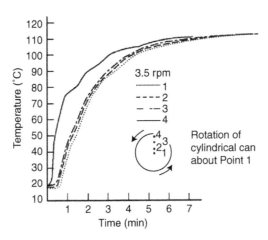

Figure 4.13 Temperature distribution at radial locations in an axially rotating horizontal can. (From Hotani, S. and Mihori, T., *Heat Sterilization of Food*, Motohiro T. and Hayakawa, K., Eds., Tokyo: KoseishaKosei-kaku, 1983.)

The value of U is obtained in practice from experiments performed to measure temperature $T(t)$ and fitting Equation 4.70 to experimental data. Correlations can be developed for U values in particular physical situations, analogous to the equations for a plate discussed previously. For example, for end-over-end (EOE) rotation in the presence of a headspace, the following dimensionless correlation exists:[35]

$$Nu = 2.9Re^{0.436}Pr^{0.287} \tag{4.71}$$

From the definition of Nusselt number, $Nu = hD/k$, the heat-transfer coefficient, h, is found from Equation 4.71 and is used to calculate temperatures from Equation 4.70. Note that only mean fluid temperature is available in this procedure. To obtain the spatial variation of the temperature profile inside the container, a more detailed analysis is required (formulation **2** of Figure 4.11), as described in the next section.

Example 2: Sterilization Without Agitation. In modeling sterilization of a canned liquid without agitation, Equation 4.61 has been used.[36] The use of this equation, which resembles Equation 4.70, for an unagitated fluid is an ad-hoc generalization without any mathematical or physical basis.[24] The f and j values needed for Equation 4.61 are found by fitting the equation to experimental temperature–time data.

From the discussions of Example 1 and Example 2 above, it is apparent that the use of h (or U) that is often obtained experimentally leads to models that are somewhere between an analytical model and an observational (nonphysics-based) model, often closer to an observational model. However, such models are very powerful in accommodating a variety of flow situations (laminar and turbulent) and fluid types (Newtonian and non-Newtonian). The limitations of these models lie in their inability to provide more details of the flow.

4.8.3 Modeling by Solving the Governing Equations of Fluid Flow Together with Heat Transfer

If more details on temperature and velocity profiles are desired, or if obtaining h values is not possible either from the simple formulas or from experimentation, then a complete solution to the governing equations for fluid flow and heat transfer in a moving fluid is necessary. Before the detailed governing equations are discussed, it is important to note that the two descriptions—use of a heat-transfer coefficient or h value, as described earlier, and the use of conservation-based governing equations and boundary conditions—are conceptually equivalent. The convective heat-transfer coefficient, h, is a simplified description of the process, and is related to a more detailed heat-transfer description by

$$-k_{\text{fluid}}\frac{\partial T}{\partial y}\bigg|_{y=0,\text{ in fluid}} = h(T_s - T_\infty) \qquad (4.72)$$

which is the defining equation for h; this can be simplified as

$$h = \frac{-k_{\text{fluid}}\frac{\partial T}{\partial y}\big|_{y=0,\text{ in fluid}}}{T_s - T_\infty}. \qquad (4.73)$$

From detailed computations described below from which $T(y)$ is obtained, Equation 4.73 can be used to obtain h, if desired. Obtaining h, however, is not necessary because the rate of heat transfer can be obtained from $-k_f \partial T/\partial y|_{y=0}$ when $T(y)$ is available. This approach of obtaining the complete solution is becoming increasingly popular as powerful commercial CFD solvers are able to easily solve for relatively complex situations (see Chapter 3 for examples). Note also that CFD programs to calculate h can be used in a way analogous to experimentation and these h values for different flow parameters can be combined to produce correlations such as Equation 4.71 that were previously developed only from experimental data.

Example 1: Sterilization Without Agitation.   An example of a solution to the complete governing equations is the modeling of a sterilization process that involves no agitation. Agitation is sometimes avoided to maintain the product or package integrity or for economical reasons in small production volumes. Without any agitation, liquid moves by natural convection due to the buoyancy induced by the change in temperature. This couples the flow and temperature fields and makes their computation a challenging task, as can be seen in a number of studies including pasteurization of beer in bottles[37], and sterilization of canned Newtonian[38,39] and non-Newtonian[40,41] fluids. The governing equations for the heating of a Newtonian fluid in a cylindrical can are given by

$$\frac{\partial T}{\partial t} + v\frac{\partial T}{\partial r} + u\frac{\partial T}{\partial z} = \frac{k}{\rho C_p}\left(\frac{1}{r}\frac{\partial}{\partial r}\left(r\frac{\partial T}{\partial r}\right) + \frac{\partial^2 T}{\partial z^2}\right) \qquad (4.74)$$

$$\rho\left(\frac{\partial u}{\partial t} + v\frac{\partial u}{\partial r} + u\frac{\partial u}{\partial z}\right) = -\frac{\partial p}{\partial z} + \mu\left(\frac{1}{r}\frac{\partial}{\partial r}\left(r\frac{\partial T}{\partial r}\right) + \frac{\partial^2 u}{\partial z^2}\right) + \rho g \qquad (4.75)$$

$$\rho\left(\frac{\partial v}{\partial t} + v\frac{\partial v}{\partial r} + u\frac{\partial v}{\partial z}\right) = -\frac{\partial p}{\partial r} + \mu\left(\frac{\partial}{\partial r}\left(\frac{1}{r}\frac{\partial(rv)}{\partial r}\right) + \frac{\partial^2 v}{\partial z^2}\right) \qquad (4.76)$$

$$\frac{\partial c}{\partial t} + v\frac{\partial c}{\partial r} + u\frac{\partial c}{\partial z} = -kc. \qquad (4.77)$$

Density is treated as a constant, except in the term (in Equation 4.75) containing gravity, g, where

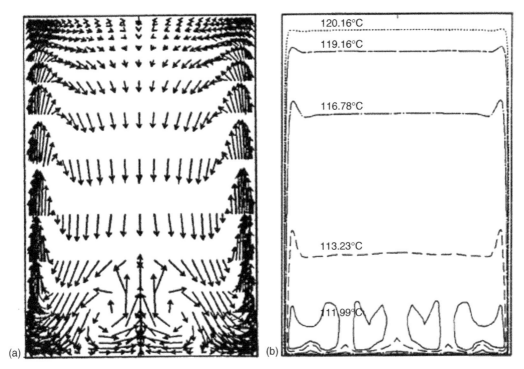

Figure 4.14 Computed velocity vectors and temperature contours after 30 min in natural convection heating of liquid in a cylindrical metal can heated from all sides with the wall temperature set at 121°C. (From Datta, A. K. and Teixeira, A., *Transactions of the ASAE*, 30, 1542, 1987.)

the following equation is used:

$$\rho = \rho_0(1 - \beta(T - T_0)).$$ (4.78)

Here, ρ_0 is the density at temperature T_0 and β is the volumetric expansion coefficient.

The boundary condition for the thermal problem is constant temperature at all boundaries. For the velocities, the boundary conditions are no-slip at all of the walls. Symmetry boundary conditions (zero gradients) for velocity and temperature are used at the centerline. The additional equation, Equation 4.77, is used to calculate the bacterial concentration, c, from which F_0 can be calculated at any position and time as $F_0 = 1/k_0 \log(c_i/c)$, whose lowest value will provide F_0^{slowest}. This formulation can be readily solved in a CFD solver that includes the solution of the Navier–Stokes equations. Major difficulties in solving this problem come from the large Rayleigh number (e.g., see Ref. 38). Figure 4.14 shows the typical flow patterns and temperature profiles in such nonagitated heating, starting from a fluid at rest at a uniform temperature. Its walls are raised to the retort temperature. Heating of a liquid (carrot-orange soup) in a pouch was also modeled using 3D rectangular Cartesian coordinate version of the above equations[42]. Starch gelatinization has been included [41,43] using an apparent viscosity model that changed with the gelatinization phase. This model could predict how the well-known broken heating curves (temperature–time curves with drastic changes in slope) develop during heating of starch-containing products.

Microwave heating was included in the study of[28] using a heat-generation term that decayed from the surface to the center as

$$Q = Q_0 \exp\left(-\frac{R-r}{\delta_p}\right).$$ (4.79)

Buoyancy-driven recirculating flows, as in conventional heating, can be seen in Figure 4.15.[44,45]

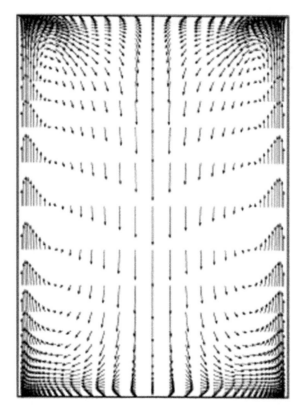

Figure 4.15 Flow patterns during microwave heating of water in a microwave transparent cylindrical container without any agitation and without any focussing effect of the microwaves due to the curved geometry of the cylinder.

The direction of circulation would depend on electric field variations inside the container. If focusing effects (discussed in Section 4.6.2) are present, as might be true in a small-diameter container, the flow pattern can be reversed, i.e., liquid can rise at the center.[46] Such details can be seen by obtaining the microwave field distribution from solving the complete electromagnetics (Equation 4.13 through Equation 4.16), instead of Equation 4.79.

Example 2: Sterilization in Agitated Heating.   Another example of detailed solutions of a convective heat-transfer problem in foods is the work on axially rotating containers.[47,48] In,[48] the governing energy equation, the momentum equation with the buoyancy term, and the continuity equations are solved. The boundary conditions for the fluid were no-slip at the can boundary, given by

$$r = R \quad 0° \leq \theta \leq 360° \quad v = \Omega R, \tag{4.80}$$

where R is the can radius, θ is the angular position of the can in degrees, and Ω is the can angular velocity ($\Omega = 2\pi N/60$) and N is the can rotational speed, in rpm, which is an intermittent rotation for the particular equipment (Steritort™), adapted from.[47] Examples of temperature data obtained from this study are shown in Figure 4.16. The figure shows that the heat-transfer rate at the slowest heating point increased dramatically when the can was intermittently rotated. The F_0 values can be found by solving an equation similar to Equation 4.77.

Example 3: Sterilization in Continuous Flow.   As an example of the continuous heating of fluids, consider sterilization of a fluid in a tubular heat exchanger where the fluid flows through the

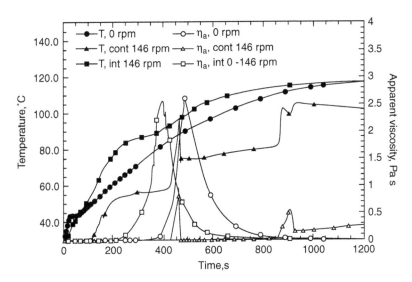

Figure 4.16 Temperature history and the change of viscosity at the slowest heating point of intermittently rotating canned 3.5% corn starch dispersion, compared to that of the stationary (0 rpm) and the continuously rotating (146 rpm) can.

tube and the heating medium is outside the tube. Continuous sterilization is often used as part of an aseptic processing system where the package is sterilized separately and the sterilized product is filled and sealed in a sterile environment. Steam is generally used as the heating medium for most continuous sterilizing systems, although microwaves and direct resistance heating have also been used.

As an example of a detailed model for a continuous heating system for fluid, consider the study of.[49] They set up the problem using the governing equations of steady laminar flow in a tube, steady-state heat transfer, and complex changes in viscosity during heating of a 4% waxy rice starch dispersion that is captured using a temperature-dependent effective viscosity. Natural convection effects are ignored in this problem. The resulting velocity, temperature, and viscosity profiles are shown in Figure 4.17. Calculation of the nutrient and bacterial concentration in a sterilization application can be readily added to the above formulation by having a species equation for each of bacteria and nutrients (similar to Equation 4.77). For a given sterilization at the cold point, the average concentration of nutrients increases with reduction in tube diameter due to a smaller spread of temperature values.[50]

4.8.4 Modeling Conjugate Problems by Considering Both the Fluid and Solid

This is formulation **3** in Figure 4.11. When solid–fluid systems in contact have comparable thermal resistances in solid and fluid phases, a conjugate heat-transfer analysis may be required where both the fluid and the solid are to be considered simultaneously. Such a problem is obviously more complex and should be considered only when necessary. The heat-transfer coefficient, used in formulation **1**, is not required here (as in formulation **2**) because detailed spatial temperature profiles are computed (see discussion surrounding Equation 4.73).

Example 1: Convective Freezing in Air.   Consider a freezing scenario that involves natural convection, as shown in Figure 4.18, where the imprecise knowledge of the heat-transfer coefficient, h, has been found to be the major error source in freezing-time prediction methods. Small deviations in the convective heat-transfer coefficient were found to result in large deviations in the

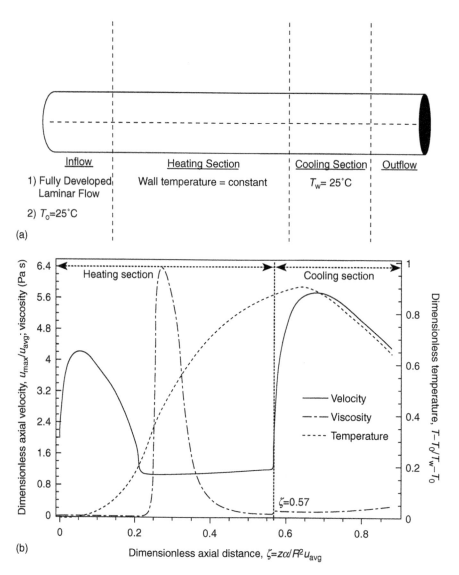

Figure 4.17 Schematic of a continuous tube heating system and computed dimensionless axial velocity and temperature near the centerline along the length of a tube of a starch solution flowing in the tube and heated from the walls. (From Liao, H. J., Rao, M. A., and Datta, A. K., *Food and Bioproducts Processing*, 78, 48, 2000.)

core temperature of food that was chilled by air.[51] Because the heat-transfer coefficient, h, changes with time, one way of obtaining this result is to solve the conjugate problem.

Example 2: Continuous Heating of Solid–Liquid Mixtures. Another example of a conjugate problem can be the continuous heating of solid–liquid mixture in aseptic processing of foods. A model proposed by[33] for continuous ohmic heating (Figure 4.19) is now described. In this formulation, three-dimensional variation of temperature within the tube as well as inside the spherical particles is considered. The solid–liquid mixture is considered homogeneous. The tube wall is surrounded by air at constant temperature. The electric field is applied longitudinally along the flow path with the tube being electrically nonconducting. The spherical particles are considered of uniform shape and size. The temperature distribution in the spherical particles is formulated using

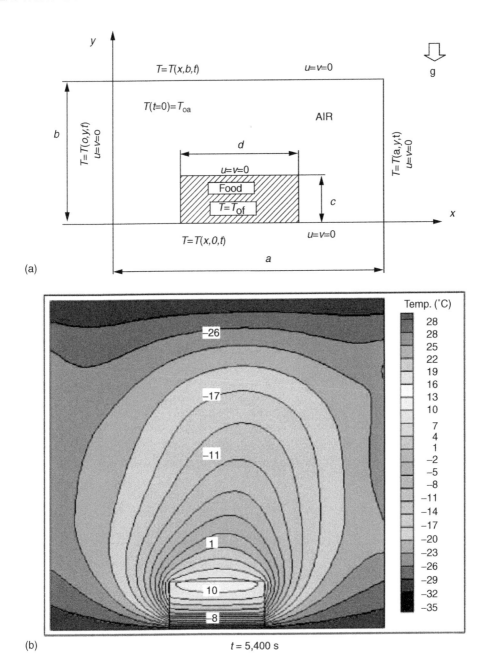

Figure 4.18 Schematic (above) of a conjugate solid–fluid heat transfer problem in freezing and the computed temperature profiles. (From Moraga, N. O. and Barraza, H. G., *Journal of Food Engineering*, 56, 17, 2003.)

$$\rho_p C_{P_p} \frac{\partial T_p}{\partial t} = \nabla \cdot (k_p \nabla T_p) + \dot{u}_p, \tag{4.81}$$

where $\dot{u}_p$ is the heat-source term for the particle that is calculated from the voltage gradient using

$$\dot{u}_p = |\nabla V|^2 \sigma_{op}(1 + m_p T_p), \tag{4.82}$$

where the voltage gradient itself is obtained by solving the voltage equation (Equation 4.30). Both

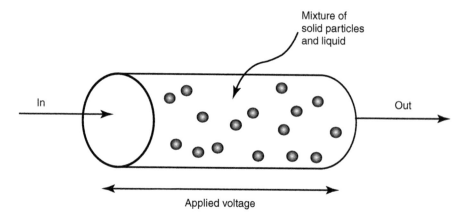

Figure 4.19 Schematic of continuous flow ohmic heating of a solid–liquid mixture in a circular tube.

axial and radial variations of the electric field are considered. The temperature distribution in the fluid is given by

$$\rho_f C_{pf} \bar{v}_z v_{ff} \frac{\partial T_f}{\partial z} \beta(v_{ff}) \nabla \cdot (k_f \nabla T_f) - n_p A_p h_{fp}(T_f - T_{ps}) + \dot{u}_f v_{ff}, \tag{4.83}$$

where $\dot{u}_f$ is the heat-source term for the fluid that is calculated in a similar manner as for the particle

$$\dot{u}_f = |\nabla V|^2 \sigma_{0f}(1 + m_f T_f). \tag{4.84}$$

The quantity v_{ff} is the volume fraction of the fluid and $\beta(v_{ff})$ is the fraction of conductive heat transfer through the mixture in the fluid phase and is approximated as the area fraction of the fluid phase, given by

$$\beta(v_{ff}) = 1 - (1 - v_{ff})^{2/3}. \tag{4.85}$$

The velocity, v_z, of fluid is taken to be that for a fully developed non-Newtonian flow in a tube, given by

$$v_z = \left(\frac{3j + 1}{j + 1}\right) v_m \left[1 - \left(\frac{r}{R}\right)^{j+1}\right]. \tag{4.86}$$

Assumption of the velocity profile avoids solving for detailed fluid motion and the detailed motion of the particles, including particle–fluid and particle–particle interactions. They considered the use of fine particle models (that treat the particles as point sources of momentum) in the literature to not be applicable because the solid particles in this study have significant dimensions. Some approaches in literature for large particles that consider Stokes flow far from boundaries are also considered to be inapplicable in this case because the solid pieces are large and confined close to boundaries. At the high viscosity of liquids used here, solids are considered to be entrained in the liquid; this allows the same average velocity for both phases and, therefore, makes it possible to treat the flow as homogeneous. However, it is noted from experimental work that significant local interphase relative velocity may exist between the solid and the liquid, even though the velocities on the average are essentially equal. The effect of the local interphase relative velocity was included by treating the heat-transfer coefficient between the fluid and the particle as an independent parameter whose sensitivity was studied. The model provided particle and fluid temperatures.

4.9   HEAT TRANSFER WITH CHANGE OF PHASE

Evaporation and freezing involve changes of phase. Evaporation is discussed in the context of simultaneous heat and mass transfer in Chapter 4 and a very brief introduction is given in Section 4.11.1. Heat transfer during freezing in food systems has been modeled in two ways: using a simple formula that is valid for freezing of a pure liquid such as water, and using a more comprehensive apparent specific-heat formulation that takes care of freezing over a temperature range, as is true for a food material. Before presenting these two models, the process of gradual freezing in food materials is discussed.

4.9.1   Modeling of Ice Formation as It Relates to Temperature

The gradual freezing process in food materials can be described by the fraction $f(T)$ of initial water frozen at any temperature T. Property variation with respect to temperature can be obtained from $f(T)$. Because of large variations in food composition, there are no general formulas available to describe $f(T)$. Generally, there are two ways of obtaining $f(T)$ and the apparent specific heat, C_{pa} (see Equation 4.90 below)

4.9.1.1   Using the Freezing-Point Depression Equation

Most of the literature uses the procedure of Heldman (1974), which is to use the freezing-point depression equation,

$$\frac{\Delta H_f M_w}{R} \left[\frac{1}{T_0} - \frac{1}{T} \right] = \ln \left[\frac{(x_w - x_b)/M_w}{(x_w - x_b)/M_w + x_s/M_s} \right],$$
(4.87)

$$f(T) = 1 - \frac{x_w}{x_0},$$
(4.88)

where x_0 is the initial water content, x_w is the water content at temperature T, x_b is the bound water content, x_s is the amount of soluble solutes in the product, M_s is the effective molecular weight of the solutes, and T_0 is the melting point of pure water. To solve Equation 4.87, bound water x_b and moles of solute x_s/M_s must be known. There is little knowledge of the amount of exact solutes dissolved and often various guesses are made. The bound water, x_b, must also be known to be able to use Equation 4.87. There are approximate empirical equations to estimate bound water (Pham 1987). The apparent specific heat, C_{pa}, would relate to the ice fraction, f, as

$$C_{pa} = w \left[(1-f)C_{p_u} + fC_{p_i} + \frac{\partial f}{\partial T} \right] + (1-w)C_{p_s}.$$
(4.89)

Note that C_{pa} relates to df/dT, which is the rate of ice formation and is the latent heat contribution to C_{pa}.

4.9.1.2   Using Measured Enthalpy Data

An alternative procedure for estimating $f(T)$ is from enthalpy data, $H(T)$, available either from experiment or approximate calculations. Enthalpy data is related to an apparent specific heat as

$$C_{pa} = \frac{dH}{dT},$$
(4.90)

where this apparent specific heat includes latent heat, as well as various specific heats. It is related to the fraction of water frozen $f(T)$ by Equation 4.89. Knowing the amount of water, w, in the sample before freezing, $f(T)$, the fraction of water frozen at any temperature T can be calculated from Equation 4.89. The equation needs one boundary condition that can be $f(T_{if})=0$, where T_{if} is the initial freezing point. Although this procedure requires the T_{if} information, it uses measured or calculated specific material data (C_{pa}) more directly and therefore is likely to provide better accuracy.

4.9.2 Simple Model for Pure Materials: Pseudo Steady-State Formulation

Because the freezing process involves latent heat in addition to sensible heat, analysis of this common process is rather complicated. A very simplified solution (known as the *Plank's solution*) that is used in practice and that preserves some of the essential physics of the process is described below.[22]

Consider symmetric (both surfaces at T_∞) freezing of a slab of pure liquid, as illustrated in Figure 4.20. The latent heat evolved at the interface of frozen and unfrozen regions is removed through the frozen layer. Although the thermal conductivity of the frozen layer is higher than that of the unfrozen layer, it is still small in absolute terms. The rate of heat transfer in the frozen layer is slow enough to be regarded as in pseudo-steady-state condition (see discussion in Section 4.7.2). Thus, although the temperature profile changes with time, this happens slowly enough that it closely approaches a steady-state profile at any time. A common boundary condition is convection at the surface. For this case, additional convective resistance $1/hA$ must be added to the conductive resistance x/kA of the frozen layer, with the total temperature difference becoming T_m-T_∞. The heat flow can be written as

$$q = \frac{T_m - T_\infty}{1/hA + x/kA}.$$

This heat flow can also be written in terms of the rate at which latent heat is given off due to the

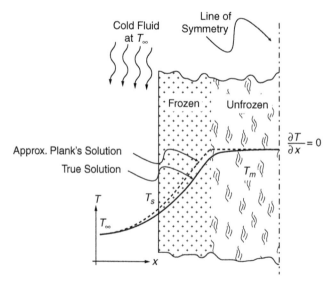

Figure 4.20 Freezing of a pure (stationary) slab of liquid showing two distinct regions, frozen and unfrozen. Plank's assumption of a linear temperature profile is compared with the "true" profile.

movement of the interface between frozen and unfrozen regions:

$$q = \Delta H_f A \rho \frac{dx}{dt}.$$

Equating the two heat flows:

$$\frac{T_m - T_\infty}{\frac{x}{kA} + \frac{1}{hA}} = \Delta H_f \rho A \frac{dx}{dt}, \tag{4.91}$$

where ΔH_f is the latent heat of fusion (in J/kg) of the material per unit mass. Freezing of the slab is complete when the freezing front reaches the midpoint:

$$t_{slab} = \frac{\Delta H_f \rho}{k(T_m - T_s)} \frac{L^2}{2}, \tag{4.92}$$

where t_{slab} is the time to freeze the slab.

4.9.3 Apparent Specific-Heat-Based Formulation for Gradual Freezing of Food Materials

The previous calculation of freezing time is for a pure liquid that freezes at one temperature, unlike the water in the food that has many solutes dissolved and therefore freezes over a temperature range. This gradual freezing can be successfully captured by following an alternative procedure for calculating freezing time. To develop this, the heat equation is rewritten as

$$\frac{\partial}{\partial x} \left(k \frac{\partial T}{\partial x} \right) = \rho C_{pa} \frac{\partial T}{\partial t}, \tag{4.93}$$

where C_{pa} is the apparent specific heat defined earlier. It is the sum of sensible and latent heats of a material per unit mass. An example of experimental data for C_{pa} for beef muscle tissue is shown in Figure 4.21. Equation 4.93 provides temperature as a function of time and therefore freezing time. Note that Equation 4.93 automatically includes the phase change effects due to the use of the apparent specific heat, C_{pa}.

4.9.4 Presence of Microwaves and Radio-Frequency Heating

Thawing and tempering (bringing the temperature up to a few degrees below complete thawing) are some of the most effective uses of the microwave heating of food. However, thermal runaway effects during phase changes can lead to severe nonuniformities. Thermal runaway refers to the situation where some regions in the food that happen to thaw first, due to higher rates of heating in these regions or composition nonuniformity there, absorbs increasing amount of energy, eventually reaching boiling temperature while other regions remain frozen. Ideally, coupled solutions of electromagnetics and energy transport should be considered (see Figure 4.30) for heating a frozen food in a microwave cavity because the food can undergo very significant changes in thermal and dielectric properties. This solution has not appeared in the literature. Thawing of food in a microwave cavity has been modeled, but the studies generally assume an exponential decay of energy from the surface (e.g., Ref. 53). An example of how thawing time varies with the power level of the microwave oven is seen in Figure 4.22.

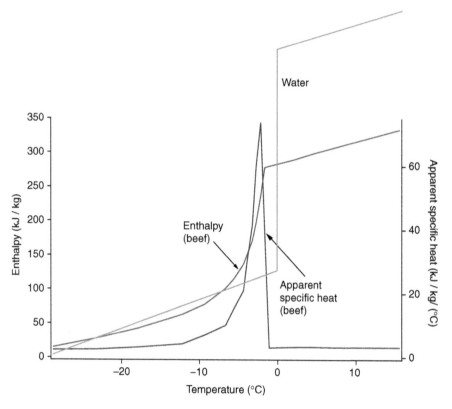

Figure 4.21 Experimental enthalpy (H) vs. temperature (T) relationship for beef muscle tissue around freezing temperature. The apparent specific heat ($c_{pa} = dH/dT$) is calculated from this data. Enthalpy for pure water/ice is superimposed for comparison.

4.9.5 Presence of High Pressure

Freezing and thawing under high pressure are being investigated for obtaining significantly improved quality of frozen and thawed foods.[54] The apparent specific-heat formulation (Equation 4.93) can be used to model such situations. The difficulty in applying this equation at high pressure lies in the lack of information on the apparent specific heat, C_{pa}, and thermal conductivity, k, at high pressures. The most logical step, used in,[56] is to use analogous relationships as those in atmospheric pressure but with adjustments for higher pressure. This is referred to as "shifting" in the work of.[56] Thus, examination of Equation 4.89 for apparent specific heat reveals two parameters—latent heat, λ, and fraction frozen, f—that depend on the initial freezing point. It appears in[56] that the shape of the f curve was assumed to stay the same at any pressure. The initial freezing point is "shifted" following the equation[55]

$$T_{if} = -0.07142P - 1.4795 \times 10^{-4}P^2, \qquad (4.94)$$

where P is the pressure in MPa and T_{if} is the initial freezing point (in °C). One may think of the entire frozen fraction curve as being shifted to the new value of T_{if} that corresponds to a new pressure. This is shown from the work of[56] in Figure 4.23. The latent heat, λ, is a function of the state diagram. The thermal conductivity of the partially frozen material at various pressures is calculated following the same equation as that for atmospheric pressure, but the frozen fraction curve in its prediction is "shifted" according to the previous discussion.

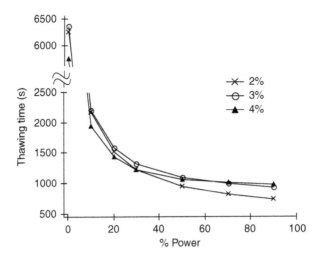

Figure 4.22   Computed thawing time for rectangular blocks (5 cm×5 cm×5 cm) of tylose (with % salt as noted), as a function of microwave oven power levels. Thawing with conventional heat (0% power) at the same surrounding temperature of 26°C and $h=30$ W/m²°C is added for comparison. (From Chamchong, M. and Datta, A. K., *Journal of Microwave Power and Electromagnetic Energy*, 34, 9, 1999.)

4.9.6   Presence of Ohmic Heating

The use of ohmic heating to thaw has been reported in the literature, but no modeling studies exist. The electrical properties of foods are a strong function of the amount of frozen vs. unfrozen water that exists in the system, coupling the ohmic heat generation equation with the heat equation. The apparent specific-heat formulation (Equation 4.93) should work well for ohmic heating.

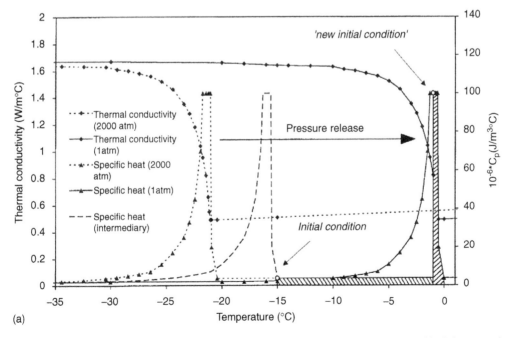

(a)

Figure 4.23   Prediction of the apparent specific heat curve at different pressures by "shifting" the curve for atmospheric pressure. (From Denys, S., Van Loey, A. M., Hendrickx, M. E., and Tobback, M. E., *Biotechnology Progress*, 13, 416, 1997.)

4.10 PROBLEM FORMULATION IN HEAT TRANSFER: RADIATION

Thermal radiation is a part of electromagnetic radiation. In the context of food processing, the entire range of infrared electromagnetic waves is typically further divided into near-infrared (0.75–3 µm), mid-infrared (3–25 µm) and far-infrared (25–1000 µm) regions. Infrared heating of foods is used in processes such as drying, baking, roasting, blanching, and surface pasteurization. When electromagnetic radiation such as infrared strikes a surface, part of it is reflected, part of it is absorbed, and the remaining, if any, is transmitted. Detailed property data for various food surfaces are generally not available. Some of these data have been discussed in.[10,57]

The wavelength of radiation incident on the food depends on the emission characteristics of the source of the radiation. Thus, it is important to know the characteristics of common sources (emitters) used for thermal radiation. Infrared emitters can be made of various materials such as quartz glass, ceramic, or metal. Figure 4.24 shows the typical spectral distribution of radiation from such emitters. Solar radiation is superimposed on this figure for comparison because solar radiation is also used in food processes such as drying. This spectral dependence of the source radiation is needed when such dependence is considered in modeling.

Radiative heat transfer in foods can typically be formulated as radiative exchange at the surface of the food combined with conduction or convection inside the food material. Thus, the modeling of radiative heat transfer typically involves a surface heat-flux boundary condition that can be written as

$$-k\frac{\partial T}{\partial n}\Big| = h(T - T_\infty) \qquad \underbrace{-q_r} \qquad . \qquad (4.95)$$
$$\text{Radiative flux}$$
$$\text{at surface}$$

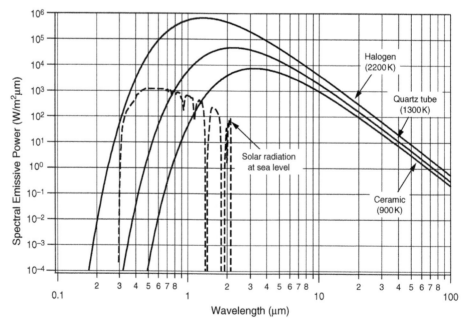

Figure 4.24 Spectral emissive powers of three classes of emitters at their typical temperatures (blackbody radiation at the noted temperatures) in the range of thermal radiation (0.1–100 µm). Measured average solar radiation at sea level is superimposed.

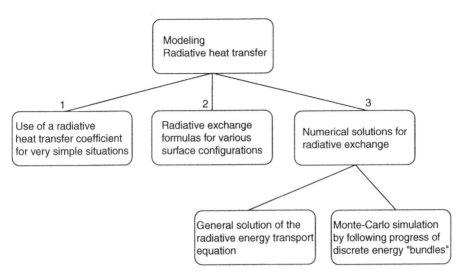

Figure 4.25 Approaches to modeling of radiative heat transfer, with increasing complexities to the right. The Monte Carlo simulation is an alternative to obtaining solution to the radiative energy transport equation.

The term q_r is a result of radiative exchange between the surface (e.g., of the food) and its surroundings. The exchange of radiative energy between two or more bodies is often a complex problem. Complexities arise from geometric issues (e.g., orientation of the bodies relative to each other) as well as the fact that surface radiative properties such as emissivity can depend on the wavelength range of interest and the direction of emission from the surface.

Modeling of radiative heat transfer can be grouped into three different approaches, as shown in Figure 4.25. The complexity of modeling increases from **1** to **3** as more details of the solution become available and with increasing accuracy. For food applications, detailed radiative heat-exchange calculations, as mentioned in formulation **3** in Figure 4.25 have only recently been reported. These formulations are now discussed.

4.10.1 Radiative Modeling in Simple Situations

In perhaps the simplest situation of radiative heat transfer, one surface is completely enclosed by the other. An example of this is a small food object placed inside an oven such that the food is completely enclosed in the oven; assume that the oven has a uniform surface temperature. The net radiative exchange is given by

$$q_{r_{1-2}} = \varepsilon_1 \sigma A_1 (T_1^4 - T_2^4), \qquad (4.96)$$

where $q_{r_{1-2}}$ is the net radiative energy transfer between bodies 1 and 2 (in W), ε_1 is the emissivity of the food surface, A_1 is the surface area of the food and T_1 and T_2 are the temperatures of the food and the oven surface, respectively. Here, emissivity is considered to be independent of direction or wavelength (also called a gray surface). The flux $q_{r_{1-2}}$ calculated this way is to be used in Equation 4.95. Other simple configurations that a food processing situation may be able to adapt to can be seen in heat-transfer textbooks.

For the special case when the surface temperatures T_1 and T_2 are similar, Equation 4.96 can be written as:

$$q_{r_{1-2}} = \varepsilon_1 \sigma A_1 4 T_1^3 (T_1 - T_2) \qquad (4.97)$$

$$= A_1 h_r (T_1 - T_2), \tag{4.98}$$

where h_r, given by

$$h_r = 4\sigma T_1^3 \varepsilon_1 \tag{4.99}$$

can be termed as a *radiative heat-transfer coefficient* analogous to a convective heat-transfer coefficient. Note that Equation 4.99 is only an approximation.

Experimental data are often collected such that several heat-transfer coefficients, such as convective, radiative, and evaporative cooling, are lumped together into what can be termed an *effective heat-transfer coefficient*, h_{eff}, and used as a boundary condition, replacing h in Equation 4.95. Of course, if h_{eff} already includes radiation effects, the q_r term in Equation 4.95 must be dropped.

4.10.2   Radiative Modeling Using Available Configuration Factors

In general, when two bodies exchange radiation, the radiative exchange also depends on their size and shape, as well as the relative orientation of their respective surfaces. The size, shape, and orientation factors are lumped in a parameter called the *configuration factor* or the *view factor*. In terms of the view factors, net radiative exchange between two bodies is given by

$$q_{r_{1-2}} = \varepsilon_1 \sigma A_1 F_{1-2}(T_1^4 - T_2^4) \tag{4.100}$$

where F_{1-2} is the view factor that stands for the fraction of radiation leaving surface 1 that is intercepted by surface 2. For a large number of geometric configurations, F_{1-2} can be found from either textbooks (e.g.,[58]) or other sources ("http://www.me.utexas.edu/howell/"). The quantity $q_{r_{1-2}}$ computed this way will be used in the boundary condition (Equation 4.95). When surface properties depend on wavelength, Equation 4.100 can be used for small ranges of wavelength and added together to obtain the total quantity, $q_{r_{1-2}}$.

4.10.3   Numerical Solution for More Complex Geometries and for Spectrally Dependent Properties: Using the Radiative Transport Equation

For a more general radiative exchange situation, the governing radiative heat-transfer equation is obtained by combining the radiative surface energy balance and Kirchoff's law, i.e., absorptance equals the emissivity for zero transmittance. Consider a surface shown in Figure 4.26. Writing in terms of the position vector, $\mathbf{r}$, the total blackbody emission, $E(\mathbf{r})$, depending on the temperature at point $\mathbf{r}$, the total radiation incident on the surface, irradiation, $G(\mathbf{r})$, and radiosity, $J(\mathbf{r})$, are related by

$$J(\mathbf{r}) = \varepsilon(\mathbf{r})E(\mathbf{r}) + \rho(\mathbf{r})G(\mathbf{r}). \tag{4.101}$$

Here, ρ is the reflectance of the surface. For an opaque surface, i.e., no energy passing through the surface, the energy balance can be written as

$$q(\mathbf{r}) = J(\mathbf{r}) - G(\mathbf{r}) = \varepsilon(\mathbf{r})E(\mathbf{r}) - \alpha(\mathbf{r})G(\mathbf{r}). \tag{4.102}$$

Here, $q(\mathbf{r})$ is the radiative heat flux at $\mathbf{r}$ and α is the absorptance of the surface. The irradiation G is now written in terms of radiosity over the total surface A, using the definition of view factor $F_{dA'-dA}$, as

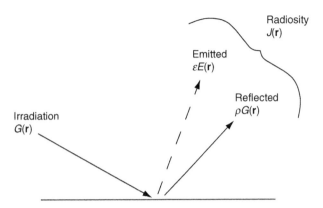

Figure 4.26   Schematic showing radiative exchange over an infinitesimal surface.

$$G(\mathbf{r})dA = \int_A J(\mathbf{r}')dF_{dA'-dA}dA'. \qquad (4.103)$$

Inserting Kirchoff's law into Equation 4.102, solving for J, and substituting into Equation 4.103 results in the following integral equation that relates temperature and radiative heat flux, q, at each location, $\mathbf{r}$, in the enclosure:

$$\frac{q(\mathbf{r})}{\varepsilon(\mathbf{r})} - \int_A \left(\frac{1}{\varepsilon(\mathbf{r}')} - 1\right)q(\mathbf{r}')dF_{dA-dA'} = E(\mathbf{r}) - \int_A E(\mathbf{r}')dF_{dA-dA'}. \qquad (4.104)$$

This integral equation is solved numerically to obtain heat flux on a food surface, $q(\mathbf{r})$, and is substituted for q_r in Equation 4.95. A number of heat-transfer packages are available to solve Equation 4.104, such as the software Fluent (Fluent, Inc., New Hampshire).

An example of a numerical computation of radiative heat transfer is now presented for radiation exchange in an oven-food system (Figure 4.27). The commercial finite element package FIDAP is used for this problem. The air in the oven is assumed transparent to the radiation. Heat conduction is assumed in the entire oven (food and air) for the short duration. The quantity $q(\mathbf{r})$ from Equation 4.104 is used as the boundary condition for the heat-conduction equation. Measured wavelength-dependent emissivity values are used. The computed radiative heat flux and temperatures at the top surface of the food in Figure 4.27 are shown in Figure 4.28.

4.10.4   Numerical Solution for More Complex Geometries and Spectrally Dependent Properties: Using Monte Carlo

Monte Carlo is another approach to modeling radiative heat transfer; it is considered particularly useful for complex configurations where this method may be the only reasonable way to attack the problem. In food applications, Monte Carlo has been used in modeling surface heating of strawberries for decontamination.[61] In a Monte Carlo simulation of radiation heat transfer, the energy emitted from a surface is simulated by the propagation of a large number of photons, which are massless units of energy. The photon is followed as it proceeds from one interaction to another, which are described as random events. This continues until the photon is absorbed or leaves the computational domain. A large number of trajectories are required to ensure that the variation due to random events is small, which can require relatively long running times. The results are used to determine the fraction of energy that has been absorbed on each surface in the geometry. This

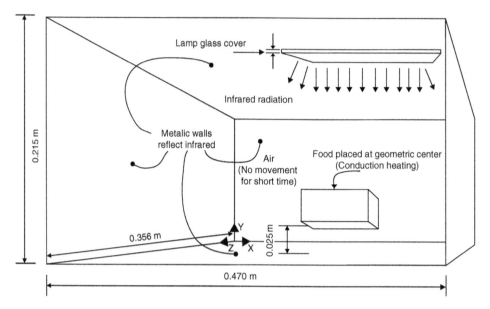

Figure 4.27 Schematic of a radiation dominant problem. The geometry of the oven is rectangular, of size
0.470 m×0.356 m×0.215 m. The food inside the oven is a potato slab of geometry 0.0470 m×
0.0356 m×0.0215 m that has a volume of $3.6×10^{-5}$ m³. Food is placed at 2.5 cm above the
geometric center of the oven's bottom surface, resting on a quartz glass tray, parallel to the
0.470 m×0.356 m oven surface. (From Almeida, M. F., Modeling infrared and combination infrared-microwave heating of foods in an oven. PhD diss., Cornell University, 2005).

method is relatively easy to set up for complex problems that involve spectral effects and/or
directional surfaces. It is implemented in commercial software such as ANSYS CFX (ANSYS
Corporation Canonsburg, Pennsylvania) which also allows the radiation heat transfer to be coupled
with conduction and/or convection.

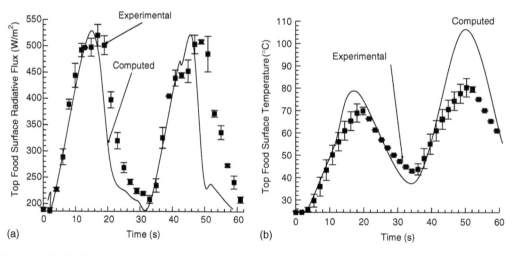

Figure 4.28 Radiative flux and temperatures from radiative analysis inside the oven shown in Figure 4.27. The
computations are compared with experimental data. (From Almeida, M. F., Modeling infrared and
combination infrared-microwave heating of foods in an oven, PhD diss., Cornell University, 2005.)

4.11 HEAT TRANSFER COUPLED WITH OTHER PHYSICS

Whether intended or unintended, temperature changes lead to other physical changes in the food material. These changes, in turn, can affect heat transfer. Thus, heat transfer can be coupled with other physics, as illustrated in Figure 4.29. One important aspect of coupling is whether the coupling exists as one-way or both ways. Example of one-way coupling can be microwave heating where the dielectric properties do not change with temperature. Whereas electromagnetic heating leads to temperature changes and therefore is coupled with heat transfer, if the dielectric properties are constant, temperature changes do not affect the electromagnetics. Another example of one-way coupling can be biochemical and microbiological changes that are affected by temperature, but these changes do not influence the heat transfer process. On the other hand, in microwave heating, if the dielectric properties change with temperature, the electromagnetics are affected and the electromagnetic model may need to be re-solved, depending on the extent of the changes.

Another important decision to make in modeling a coupled process is whether the model equations belonging to all of the physics need to be solved at every time step in a transient situation. This obviously depends on how strongly the processes are coupled, i.e., how rapidly one process changes the other. If one process does not change as rapidly as another one occurring simultaneously, the slower process does not have to be computed as often; this can provide major savings in computation time. In case of coupled electromagnetics-heat-transfer modeling, of course, the heat transfer changes the dielectric properties. But questions arise as to how much the dielectric properties change and how significant these changes are to the electromagnetic

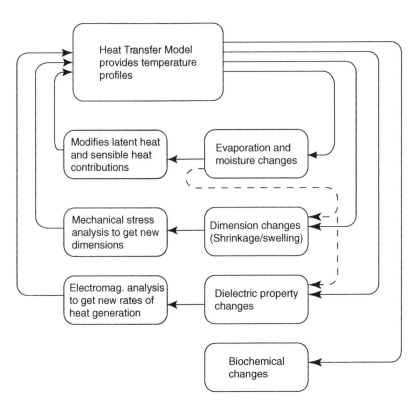

Figure 4.29 Schematic showing coupling of different types of physics with heat transfer. The connecting solid lines stand for coupling due to temperature itself whereas the dashed lines stand for additional coupling that can arise in a heating process such as moisture loss.

analysis. Answers to these questions for practical situations can be found realistically through trial and error, i.e., by running the code for various degrees of coupling.

Solutions to coupled processes have been achieved in the past mostly by using custom-developed codes. In recent years, coupled codes have become more available. An example of a commercial software program that can do multiphysics is ANSYS. It combines structural, thermal, CFD, acoustic and electromagnetic simulation capabilities. Likewise, another commercial product with flexible multiphysics capability is COMSOL Multiphysics. Another approach to coupling is the implementation of co-simulation using two different commercial software programs, such as using FLUENT and ABAQUS together to obtain fluid-structure interactions. Combining two commercial software packages may also be possible by the user, as was done for electromagnetics and heat transfer,[63] but this obviously requires more work. In the future, more software with integrated multiphysics capabilities should be available with more seamless integration between two separate commercial software programs.

4.11.1 Heat Transfer Coupled with Evaporation and Moisture Transport

Heating of food almost always is accompanied by the evaporation of water in it. Evaporation changes the heat-transfer process quite significantly and requires significant reformulation that also considers water and vapor transport. The presence of evaporation during heating and the resulting moisture transport are discussed in detail in Chapter 4. To illustrate how evaporation can be included in heat transfer, consider the following set of equations for energy, liquid water, and vapor conservation:

$$\rho C_{\mathrm{p}} \frac{\partial T}{\partial t} = \frac{\partial}{\partial x}\left(k \frac{\partial T}{\partial x}\right) \underbrace{-\lambda I}_{\substack{\text{coupling with} \\ \text{evaporation}}} \tag{4.105}$$

$$\frac{\partial W}{\partial t} = \frac{\partial}{\partial x}\left(D_{\mathrm{w}} \frac{\partial W}{\partial x}\right) - I \tag{4.106}$$

$$\frac{\partial V}{\partial t} = \frac{\partial}{\partial x}\left(D_{\mathrm{v}} \frac{\partial V}{\partial x}\right) + I, \tag{4.107}$$

where I denotes the rate of evaporation, W and V are liquid and vapor contents, and D_{w} and D_{v} are liquid and vapor diffusivities, respectively. Here, I varies with position and time in the food and *couples* heat transfer with moisture transfer. These equations are quite simplified and more complex formulations that couple heat transfer with moisture transport are presented in Chapter 4.

4.11.2 Heat Transfer Coupled with Microwaves

Coupling of microwaves with heat transfer is becoming more common as microwaves are introduced in newer processes, often in combination with other heating mechanisms. Two examples are discussed here: microwave sterilization and microwave combination heating in domestic ovens. The reader is referred to[59] for additional applications.

Example 1: Microwave Sterilization. In sterilization, microwaves offer a unique opportunity to raise the temperature quickly without heat diffusion limitations, thus lowering thermal destruction of components during the come-up time. However, the nonuniformities in the spatial distribution of energy deposition during microwave heating and the changes in distribution during the heating process can be distinctly different from conventional retort heating and can limit the implementation of a microwave sterilization process. Here, the heating of a cylindrical sample in a pressurized microwave transparent container inside a laboratory microwave oven that is somewhat similar to a domestic microwave oven is considered.[62]

This process is described by the transient heat equation:

$$\rho C_p \frac{\partial T}{\partial t} = \nabla \cdot (k \nabla T) + Q(\mathbf{r}, T) \tag{4.108}$$

where $Q(\mathbf{r}, T)$ is the spatially varying microwave heat generation obtained from the solution of Maxwell's equations describing microwave propagation (Equation 4.13 to Equation 4.16), given by

$$Q(\mathbf{r}, T) = \frac{1}{2} \omega \varepsilon_0 \varepsilon''_{\text{eff}} \mathbf{E}^2 \tag{4.109}$$

Because dielectric properties of foods, specially the salty ones, can vary considerably with temperature, Equation 4.108 and Equation 4.109 (with $\mathbf{E}$ coming from the Maxwell's equations) need to be coupled, as illustrated in Figure 4.30. In,[63] the coupled governing equations were solved using finite element methods. Two commercial FEM software packages were used: EMAS (Ansoft Corporation, Pittsburgh, PA) for electromagnetic fields and NASTRAN (MacNeal–Schwendler Corporation, Los Angeles, California) for temperature distributions. Coupling is not built into the software, and system level codes were written in C language to develop the two-way coupling as shown in Figure 4.30. Additional details of the coupling process can be seen in[17].

The effect of nonuniformity in heating can be characterized using information on volume fraction that reached a particular value of sterilization given by $F_0 = 1/k_0 \ln (c_i/c)$. This F_0 value can also be referred to as *thermal time*[64]. Figure 4.31 shows the volume fraction curve for thermal time, where the vertical axis represents the volume fraction of material below a certain thermal time or F_0 value. For the low-loss material (0.7% salt ham samples), large volume fraction (about 60%) of material falls in the low thermal-time range (below $F_0 = 6$ min), indicating that the material is

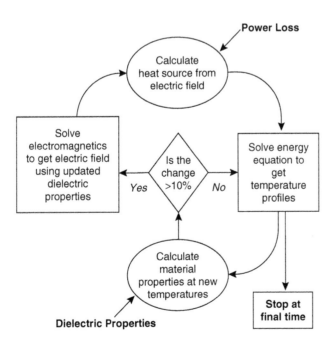

Figure 4.30 Schematic of the coupling of electromagnetic and thermal calculations. (From Zhang, H., Datta, A. K., Taub, I. A., and Doona, C., *American Institute of Chemical Engineers Journal*, 47, 1957, 2001.)

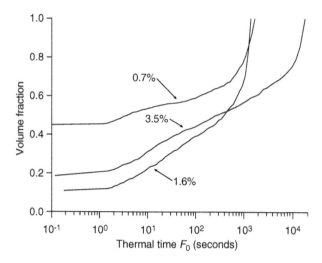

Figure 4.31   Calculated volume fraction curves for thermal time F_0 (bacteria) in ham with salt contents of 0.7%, 1.6% and 3.5% after 500 s of microwave heating. (From Zhang, H., Datta, A. K., Taub, I. A., and Doona, C., *American Institute of Chemical Engineers Journal*, 47, 1957, 2001.)

underheated, i.e., the thermal time is below the desired minimum of $F_0 = 6$ min. For a high-loss material (ham with 3.5% salt content), over 40% of material exceeds 2000-s thermal time (above $F_0 = 33$ min), representing significant overheating (i.e., thermal time much above the desired $F_0 = 6$ min) in the outer ring. Intermediate lossy (ham with 1.6% salt) material gives a better thermal time distribution and shows the smallest range of thermal time, even though the range, from 0.1 to 1200 s, is still quite large due to the nature of the cavity heating.

4.11.2.1   Need for Coupled Solutions

Because the dielectric properties are temperature sensitive, the distribution of electric fields, defined as *heating potential*, changes with the heating time. The initial distribution of heating potential, as shown in Figure 4.32a, changes as heating progresses (Figure 4.32b). The distribution of initial heating potential is different from that during the later heating time. The focusing effect (i.e., hotter locations are inside), which is initially present, is not significant after the material reaches the sterilizing temperature (when the hotter locations move to the surface). This qualitative and significant change in heating pattern is obtained by coupling the electromagnetics with energy transfer in microwave sterilization.

Example 2: Combination Microwave and Infrared Heating in a Domestic Combination Oven.   Combination microwave and infrared heating has been modeled for a microwave-infrared combination oven.[14] In this study, microwaves in a cavity with food, radiative exchange between food and oven, and conductive heat transfer inside the food are combined. Commercial modeling software ANSYS is used to model the electromagnetics and obtain the spatial distribution of microwave power deposition inside the food. The power deposition values are input as heat sources to the commercial CFD software FIDAP. In FIDAP, the surface radiative flux for infrared heating is also computed using radiative exchange analysis described under radiation heat transfer. The microwave volumetric heat generation and the radiative surface heat fluxes are combined in the governing equation and boundary condition as

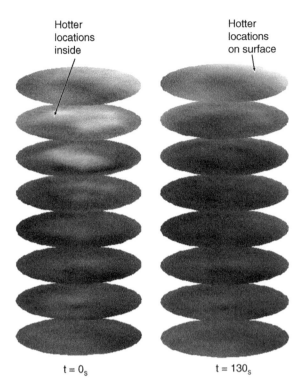

Hotter
locations
inside

Hotter
locations
on surface

t = 0_s   t = 130_s

Figure 4.32   Horizontal sections of the cylindrical food at equal vertical intervals showing the change in heating potential from initial time to after 130 s of continuous heating. Lighter shades of gray represent increased magnitude of heating potential. (From Zhang, H., Datta, A. K., Taub, I. A., and Doona, C., *American Institute of Chemical Engineers Journal*, 47, 1957, 2001.)

$$\rho_0 C_p \frac{\partial T}{\partial t} = k\nabla^2 T + q_{\text{gen}}, \tag{4.110}$$

$$-k\nabla T = -q_r + h_c(T - T_\infty) \tag{4.111}$$

where q_{gen} is microwave volumetric heating, q_r is heat flux from infrared radiative heating and h_c is the surface convective heat-transfer coefficient for the surrounding air. Both ANSYS and FIDAP are finite-element based; thus, heat source values for each Gaussian point from ANSYS were input into FIDAP using a custom-developed subroutine. Example of computed surface temperature profiles are shown in Figure 4.33.

4.11.3   Heat Transfer Coupled with Biochemical Reactions: Modeling of Safety and Quality

Heat transfer is generally a prelude to the calculation of quality and safety. If kinetics of the biochemical reactions that relate to microbiological and chemical safety of a food process or the quality resulting from the food process are known, it is a relatively simple matter to couple the reaction kinetics with temperature calculated from the heat-transfer model. Such coupling is illustrated in Figure 4.34 as a one-way coupling of heat transfer with the kinetics of biochemical changes during processing of food.

The main idea in modeling quality and safety is to consider the relevant biochemical reaction in the context of a transport equation. For example, if species A is being generated from a first-order

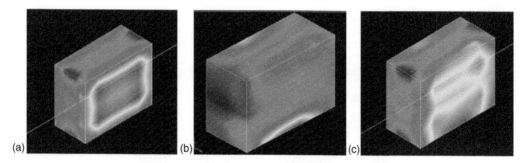

Figure 4.33   (See color insert following page 178.) Computed temperature contours showing the food surface for (a) infrared only, (b) microwave only, and (c) combined microwave and infrared heating.

reaction given by

$$\frac{Dc_A}{Dt} = -kc_A, \tag{4.112}$$

where k is temperature dependent, given typically by the Arrhenius equation

$$k = k_0 e^{(-E_a/RT)}. \tag{4.113}$$

Equation 4.112 can be expanded to obtain the species-transport Equation (see Chapter 4), as

$$\frac{\partial c_A}{\partial t} + \frac{\partial}{\partial x}(c_A u) = \underbrace{-k_0 e^{(-E_a/RT)}}_{\substack{\text{coupling with} \\ \text{heat transfer}}} c_A. \tag{4.114}$$

Equation 4.114 shows the coupling the species c_A resulting from the biochemical reaction with temperature. If the convective term is kept in the above equation, it can be used to accurately calculate the concentration field, c_A, even in a flowing system. Note that an arbitrary reaction can be included in place of Equation 4.112 to accommodate many different types of reactions and their temperature dependencies. Equation 4.114 being a transport equation, its solution is straightforward in a commercial solver. Because the equation corresponds to a diffusivity value of zero in a transport equation and a zero diffusivity may not be allowed in some computational software, a small enough value of diffusivity can be assumed that does not change the final computations of concentration values c_A.

A typical transient distribution of sterilization (F_0 values) in a cylindrical container[39] is shown in Figure 4.35. This was calculated by implicitly following the liquid elements[28] throughout the

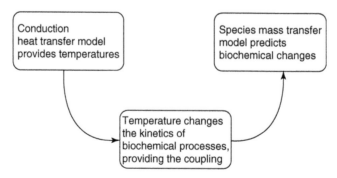

Figure 4.34   Schematic of one way coupling in modeling biochemical and physical changes that relate to quality and safety.

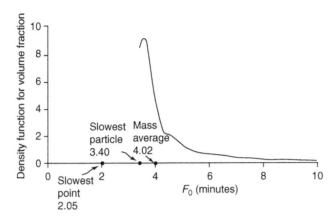

Figure 4.35 Computed distribution of sterilization (F_0 values) in liquid for the conditions in Figure 4.14 after 10 min of heating.

duration of heating and solving Equation 4.114. The volume-average (or mass-average) sterilization lies within the range of the distribution, as expected. However, the sterilization calculated based on the temperature at the slowest heating point lies completely outside the distribution. This is because all physical fluid particles stay only some of the time in the slowest heating zone and no particle stays all of the time in the slowest heating zone. Thus, all particles in the system obtain a sterilization more than what is calculated based on the slowest heating zone. Because sterilization at the slowest heating zone is easily measured by a thermocouple, this is the temperature used in practice to calculate sterilization. The slowest-point value in Figure 4.35 demonstrates that use of such slowest-heating thermocouple data provides additional overprocessing (and safety) beyond the true least sterilization of a fluid. It is interesting to note that physics-based modeling makes it easy to obtain such insight.

4.11.4 Heat Transfer Coupled with Mechanics: Thermomechanics

Temperature (and moisture) changes can lead to dimensional changes in the food and, therefore, mechanical stress. Modeling such a process involves coupling of heat (and moisture) transfer with solid mechanics.[65,66] As an example, modeling of how a material can crack during a rapid freezing process[66] is now presented using a one-way coupling of heat transfer with phase change and solid mechanics, as illustrated in Figure 4.36. In this example, coupling is a one-way heat transfer leading to mechanical changes. Water in a food material expands during the phase change process from water to ice and the amount of expansion at a location in the food depends on the extent of ice formation at that location, which in turn depends on temperature. The temperature from the thermal model is coupled with stress analysis in the material due to the expansion. A quasi-static, continuum formulation is developed where the output from the thermal model is input to the model for mechanical stress analysis. A long cylindrical geometry is used with axisymmetric freezing, making the temperatures vary only in the radial direction while the stresses vary in radial and circumferential directions.

Heat-transfer analysis of the phase-change problem will use Equation 4.93 described earlier. The boundary and initial conditions are

$$T|_{surface} = T_\infty, \tag{4.115}$$

$$T = T_i \tag{4.116}$$

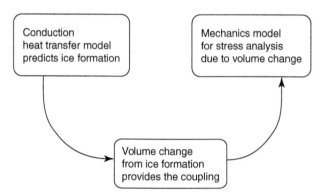

Figure 4.36 Schematic of one way coupling in studying thermal stresses and cracking during a rapid freezing process.

where T_∞ is the constant bulk temperature of the fluid being used for freezing, and T_i is the constant initial temperature of the material being frozen.

The thermal stress analysis portion is formulated as a quasi-static, linear, axisymmetric, plane-strain problem. The equilibrium equation is

$$\sigma_{ij,j} + F_i = 0 \tag{4.117}$$

where the σ_{ij} is the stress tensor, F_i is the body force per unit volume, and the comma in $\sigma_{ij,j}$ denotes the derivative with respect to the spatial variable corresponding to direction j. The range of indices is r and θ for this two-dimensional problem. The compatibility conditions are given by:

$$\varepsilon_{ij,kl} + \varepsilon_{kl,ij} - \varepsilon_{lj,ki} - \varepsilon_{ki,lj} = 0, \tag{4.118}$$

where $\varepsilon_{ij,kl}$ denotes the second derivative of ε_{ij} with respect to the two spatial variables.

It is assumed here that each component of the strain tensor can be additively decomposed into a viscoelastic strain $\varepsilon_{ij}^{(v)}$ and a thermal strain, $\varepsilon_{ij}^{(T)}$ i.e.,

$$\varepsilon_{ij} = \varepsilon_{ij}^{(v)} + \varepsilon_{ij}^{(T)}, \tag{4.119}$$

where the thermal strain is given by

$$\varepsilon_{ij}^{(T)} = \delta_{ij} \int_{T_R}^{T} \alpha dT, \tag{4.120}$$

where α is the temperature-dependent coefficient of linear thermal expansion, δ_{ij} is the Kronecker delta, and T_R is a reference temperature at which thermal strains are zero. For this study, T_R was set equal to the initial temperature T_i.

The hereditary integral formulation for linear viscoelasticity,

$$\sigma_{ij} = \delta_{ij} \int_0^t \lambda(t-\tau) \frac{\partial \varepsilon_{kk}^{(v)}}{\partial \tau} d\tau + 2 \int_0^t G(t-\tau) \frac{\partial \varepsilon_{ij}^{(v)}}{\partial \tau} d\tau, \tag{4.121}$$

in terms of the Lamé function, $\lambda(t)$, and the shear modulus function, $G(t)$, of the material. The strain deviator, $e_{ij}^{(v)}$, and bulk modulus, $K(t)$, are given, respectively, by

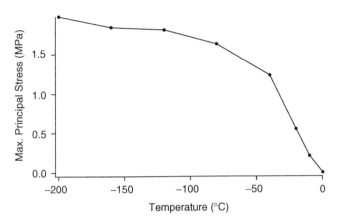

Figure 4.37 Increase in maximum principal stress as function of boundary temperature (lower boundary temperature amounts to higher cooling rate).

$$e_{ij}^{(v)} = \varepsilon_{ij}^{(v)} - \delta_{ij}\varepsilon_{kk}^{(v)}/3, \qquad (4.122)$$

$$K(t) = \lambda(t) + \frac{2}{3}G(t). \qquad (4.123)$$

An unconstrained boundary condition has been used at the surface. This is given by

$$\sigma_{ij}n_j = 0, \qquad (4.124)$$

where n_j denotes components of the unit vector normal to the surface. The governing equations were solved using the finite-element computer code ABAQUS (ABAQUS, Inc., Providence, Rhode Island). The variations in thermal and mechanical properties with temperature make both the energy and mechanics equations nonlinear.

The cooling rate was varied by changing the boundary temperature. Figure 4.37 shows the effect of the cooling rate on the maximum principal tensile stress. Two major factors influence the maximum principal stress developed: the thermal expansion with temperature and the time available for viscous decay of mechanical properties.

The most significant effect of lowering the boundary temperature in the phase change range of 0°C to −40°C is through the thermal expansion. Inside this temperature range, a decrease in temperature causes increased thermal expansion due to an increase of the frozen fraction. This increases the maximum tensile principal stress almost linearly (Figure 4.37). The second effect on the maximum principal tensile stress comes from the time available for the decay of viscoelastic properties. The faster the freezing, the less time is available for decay of properties, effectively making the material behave as a harder material (higher values of modulus).

As shown in Figure 4.37, the maximum tensile principal stresses during the phase-change period increase as the boundary temperature is lowered. These higher stresses at the boundary temperature of −200°C lead to much higher energy release rate (crack extension force), i.e., the energy available for a crack to propagate, at that temperature. Thus, the sample is more likely to crack for a boundary temperature of −200°C as compared to a boundary temperature of −40°C.

4.12 CONCLUDING REMARKS

Sterilization is the primary context in which many conduction and convection heat-transfer models have been developed. Although conduction heat-transfer modeling of food processes has been underway for more than 50 years, many important convection heat-transfer (both natural and forced) models have also been developed in the last 20 years because computing power has became more readily available. Radiation heat-transfer models are relatively rare in food processes, where only a few studies exist. Increased coupling of heat transfer with other simultaneous processes will be the major trend in modeling in the coming years. Such coupling includes internal heating of the microwaves and radio frequency—processes that have already been modeled. Coupling of heat transfer with biochemical changes and microbiological growth and decay are being pursued in the context of modeling food quality and safety. Coupling of heat transfer with evaporation and transport of moisture, as in drying and analogous processes, is also a very active area of modeling. Finally, we expect to see more modeling work relating heat transfer to mechanical changes such as those in freezing and cooking. Integration of the coupled physics with food-specific property and parameter data in a user-friendly interface will make modeling more accessible to everyone.

The many examples shown in this chapter illustrate the power of physics-based models in providing in-depth understanding of food processes. Although the chapter provides examples primarily from heat transfer, the same is true for other physical processes such as fluid flow. Of course, the physics-based approach requires the mastery of the knowledge of the physical processes to be able to mathematically formulate a problem starting from a physical process by making appropriate simplifications. A practical physics-based approach to complex processes also requires the availability of powerful and user-friendly solvers (software). As major improvements in software continue, the effort required to implement physics-based models should continually decrease, making them more and more attractive.

ACKNOWLEDGMENTS

The author appreciates the critical comments and proofreading of portions of the document by A. Dhall, A. Halder, V. Rakesh, A. Pelletier and V. Craven of the Department of Biological and Environmental Engineering at Cornell University.

REFERENCES

1. Incropera, F. P. and DeWitt, D. P., *Introduction to Heat Transfer*, New York: Wiley, 1985.
2. Ozisik, M. N., *Heat Conduction*, New York: Wiley, 1980.
3. Rohsenow, W. M., Hartnett, J. P., and Cho, Y. I., *Handbook of Heat Transfer*, New York: McGraw-Hill, 1998.
4. Bird, R. B., Stewart, W. E., and Lightfoot, E. N., *Transport Phenomena*, New York: Wiley, 1960.
5. Rao, M. A., Rizvi, S. S. H., and Datta, A. K., Eds., *Engineering Properties of Foods*, New York: CRC Press, 2005.
6. Rahman, S., *Food Properties Handbook*, Boca Raton, FL: CRC Press, 1995.
7. Saravacos, G. D. and Maroulis, Z. B., *Transport Properties of Foods*, New York: Marcel Dekker, 2001.
8. TUV NEL. The physical properties of food database, http://www.nelfood.com.
9. Nesvadba, P., Thermal properties in unfrozen foods, In *Engineering Properties of Foods*, Rao, M. A., Rizvi, S. S. H., and Datta, A. K., Eds., Boca Raton, FL: Taylor & Francis, 2005.
10. Datta, A. K. and Almeida, M., Properties relevant to infrared heating of foods, In *Engineering Properties of Foods*, Rao, M. A., Rizvi, S. S. H., and Datta, A. K., Eds., Boca Raton, FL: Taylor & Francis, 2005.
11. Datta, A. K., Sumnu, G., and Raghavan, G. S. V., Dielectric properties of foods, In *Engineering Properties of Foods*, Rao, M. A., Rizvi, S. S. H., and Datta, A. K., Eds., New York: Marcel Dekker, 2005.
12. Chamchong, M. and Datta, A. K., Thawing of foods in a microwave oven: I. Effect of power levels and power cycling, *The Journal of Microwave Power and Electromagnetic Energy*, 34, 9, 1999.

13. Saltiel, C. and Datta, A. K., Heat and mass transfer in microwave processin, *Advances in Heat Transfer*, 32, 1998.

14. Datta, A. K., Geedipalli, S. S. G., and Almeida, M., Microwave combination heating, *Food Technology*, 9, 36, 2005.

15. Zhang, H. and Datta, A. K., Microwave power absorption in single and multi-compartment foods, *Transactions of the Institution of Chemical Engineers*, 81C, 257, 2003.

16. Zhang, H. and Datta, A. K., Heating concentrations of microwaves in spherical and cylindrical foods.II: In a cavity, *Transactions of the Institution of Chemical Engineers.*, 83, 14, 2005.

17. Zhang, H. and Datta, A. K., Electromagnetics of microwave oven heating: Magnitude and uniformity of energy absorption in an oven, In *Handbook of Microwave Food Technology*, Datta, A. K. and Anantheswaran, S., Eds., New York: Marcel Dekker, 2000.

18. Ayappa, K. G., Modelling transport processes during microwave heating: A review, *Reviews in Chemical Engineering*, 13, 1, 1997.

19. Rasanyagam, V., Balasubramaniam, V. M., Ting, E., Sizer, C. E., Bush, C., and Anderson, C., *Journal of Food Science*, 68, 254, 2005.

20. Hoang, M. L., Verboven, P., Baelmans, M., and Nicolai, B. M., Sensitivity of temperature and weight loss in the bulk of chicory roots with respect to process and product parameters, *Journal of Food Engineering*, 62, 233, 2004.

21. Humphrey, V. F., Nonlinear propagation in ultrasonic fields: Measurements, modelling and harmonic imaging, *Ultrasonics*, 38, 267, 2000.

22. Datta, A. K., *Biological and Bioenvironmental Heat and Mass Transfer*, Boca Raton, FL: CRC Press, 2002.

23. Dolan, K. D., Stoeckle, A. W., and Beck, M. E., Softening kinetics of cooked dry beans at temperatures below 100C, *Journal of Texture Studies.*, 36, 351, 2005.

24. Datta, A. K., On the theoretical basis of the asymptotic semilogarithmic heat penetration curves used in food processing, *Journal of Food Engineering*, 12, 177, 1990.

25. Teixeira, A. A., Zinsmeister, G. E., and Zahradnik, J. W., Computer-simulation of variable retort control and container geometry as a possible means of improving thiamine retention in thermally processed foods, *Journal of Food Science*, 40, 656, 1975.

26. Banga, J. R., Balsa-Canto, F., Moles, C. G., and Alonso, A. A., Improving food processing using modern optimization methods, *Trends in Food Science and Technology*, 14, 131, 2003.

27. Skelland, A. H. P., *Non-Newtonian Fluid Flow and Heat Transfer*, New York: Wiley, 1967.

28. Datta, A. K., Integrated thermokinetic modeling of processed liquid food quality, In *Physical Properties and Process Control, Vol. 1 of Engineering and Food*, Spiess, W. E. L. and Schubert, H., Eds., New York: Elsevier Applied Science, 1989.

29. Sastry, S. K., Convective heat transfer coefficients for canned mushrooms processed in still retorts, Paper no. 846517, presented at the Winter Meeting of the ASAE at New Orleans, LA, 1984.

30. Sablani, S. S. and Ramaswamy, H. S., Heat transfer to particles in cans with end-over-end rotation: Influence of particle size and concentration (% v/v), *Journal of Food Process Engineering*, 20, 265, 1997.

31. Rao, M. A. and Anantheswaran, R. C., Convective heat transfer to fluid foods in cans, *Advances in Food Research*, 32, 39, 1988.

32. Stoforos, N. G. and Merson, R. L., Estimating heat transfer coefficients in liquid/particulate canned foods using only liquid temperature data, *Journal of Food Science*, 55, 478, 1990.

33. Orangi, S., Sastry, S., and Li, Q., A numerical investigation of electroconductive heating in solid–liquid mixtures, *International Journal of Heat and Mass Transfer*, 41, 2211, 1998.

34. Hotani, S. and Mihori, T., Some thermal engineering aspects of the rotation method in sterilization, In *Heat Sterilization of Food*, Motohiro, T. and Hayakawa, K., Eds., Tokyo, Japan: KoseishaKoseikaku, 1983.

35. Anantheswaran, R. C. and Rao, M. A., Heat transfer to model non-newtonian liquid foods in cans during EOE rotation, *Journal of Food Engineering*, 4, 21, 1985.

36. Ball, C. O. and Olson, F. C. W., *Sterilization in Food Technology*, New York: McGraw-Hill, 1957.

37. Engelman, M. S. and Sani, R. L., Finite-element simulation of an in-package pasteurization process, *Numerical Heat Transfer*, 6, 41, 1983.

38. Datta, A. K. and Teixeira, A., Numerical modeling of natural convection heating in canned liquid foods, *Transactions of the ASAE*, 30, 1542, 1987.
39. Datta, A. K. and Teixeira, A., Numerically predicted transient temperature and velocity profiles during natural convection heating of canned liquid foods, *Journal of Food Science*, 53, 191, 1988.
40. Kumar, A., Bhattacharya, M., and Blaylock, J., Numerical simulation of natural convection heating of canned thick viscous liquid products, *Journal of Food Science*, 55, 1403, 1990.
41. Yang, W. H. and Rao, M. A., Numerical study of parameters affecting broken heating curve, *Journal of Food Engineering*, 37, 43, 1998.
42. Abdul Ghani, A. G., Farid, M. M., and Chen, X. D., A computational and experimental study of heating and colling cycles during thermal sterilization of liquid foods in pouches using CFD, Proceedings of the Institution of Mechanical Engineers, Part E, *Journal of Process Mechanical Engineering*, 217, 1, 2003.
43. Yang, W. H. and Rao, M. A., Transient natural convection heat transfer to starch dispersion in a cylindrical container: Numerical solution and experiment, *Journal of Food Engineering*, 36, 395, 1998.
44. Prosetya, H. and Datta, A. K., Batch microwave heating of liquids: An experimental study, *The Journal of Microwave Power and Electromagnetic Energy*, 26, 215, 1991.
45. Datta, A. K., Prosetya, H., and Hu, W., Mathematical modeling of batch heating of liquids in a microwave cavity, *The Journal of Microwave Power and Electromagnetic Energy*, 27, 38, 1992.
46. Anantheswaran, R. C. and Liu, L., Effect of viscosity and salt concentration on microwave heating of model non-newtonian liquid foods in a cylindrical container, *Journal of Microwave Power and Electromagnetic Energy*, 29, 119, 1994.
47. Ladiende, F., Studies on thermal convection in self-grauitating and rotating horizontal cylinders in a vertical external gravity field. PhD diss., Ithaca, NY: Cornell University, 1988.
48. Tattiyakul, J., Rao, M. A., and Datta, A. K., Heat transfer to three canned fluids of different thermo-rheological behavior under intermittent agitation, *Transactions of the Institution of Chemical Engineers*, 80, 19, 2002.
49. Liao, H. J., Rao, M. A., and Datta, A. K., Role of thermo-rheological behaviour in simulation of continuous sterilization of a starch dispersion, *Food and Bioproducts Processing*, 78, 48, 2000.
50. Simpson, S. G. and Williams, M. C., Analysis of high temperature short time sterilization during laminar flow, *Journal of Food Science*, 39, 1047, 1974.
51. Nicolai, B. M. and DeBaerdemaeker, J., Sensitivity analysis with respect to the surface heat transfer coefficient as applied to thermal process calculations, *Journal of Food Engineering*, 28, 21, 1996.
52. Moraga, N. O. and Barraza, H. G., Predicting heat conduction during solidification of food inside a freezer due to natural convection, *Journal of Food Engineering*, 56, 17, 2003.
53. Chamchong, M. and Datta, A. K., Thawing of foods in a microwave oven: II. Effect of load geometry and dielectric properties, *Journal of Microwave Power and Electromagnetic Energy*, 34, 22, 1999.
54. LeBail, A., Chevalier, D., Mussa, D. M., and Ghoul, M., High pressure freezing and thawing of foods: A review, *International Journal of Refrigeration*, 25, 504, 2002.
55. Bridgman, P. W., Water, in the liquid and five solid forms, under pressure, *Proceedings of the American Academy of Arts and Sciences*, 47(13), 439, 1912.
56. Denys, S., Van Loey, A. M., Hendrickx, M. E., and Tobback, M. E., Modeling heat transfer during high-pressure freezing and thawing, *Biotechnology Progress*, 13, 416, 1997.
57. Il'yasov, S. G. and Krasnikov, V. V., *Physical Principles of Infrared Irradiation of Foods*, New York: Hemisphere Publishing, 1991.
58. Siegel, R. and Howell, J. R., *Thermal Radiation Heat Transfer*, New York: Hemisphere Publishing, 2002.
59. Datta, A. K., Fundamentals of heat and moisture transport for microwaveable food product and process development, In *Handbook of Microwave Food Technology*, Datta, A. K. and Anantheswaran, S., Eds., New York: Marcel Dekker, p. 99, 2000.
60. Almeida, M. F., Modeling infrared and combination infrared-microwave heating of foods in an oven. PhD diss., Cornell University, 2005.
61. Scheerlinck, N., Marquenie, D., Jancsok, P. T., Verboven, P., Moles, C. G., Banga, J. R., and Nicolai, B. M., A model-based approach to develop periodic thermal treatments for surface decontamination of strawberries, *Postharvest Biology and Technology*, 34, 39, 2004.

62. Zhang, H., Datta, A. K., Taub, I. A., and Doona, C., Electromagnetics, heat transfer, and thermo-kinetics in microwave sterilization, *American Institute of Chemical Engineers Journal*, 47, 1957, 2001.

63. Zhang, H. and Datta, A. K., Coupled electromagnetic and thermal modeling of microwave oven heating of foods, *Journal of Microwave Power and Electromagnetic Energy*, 35, 71, 2000.

64. Nauman, E. B., Nonisothermal reactors: Theory and applications of thermal time distributions, *Chemical Engineering Science*, 32, 359, 1977.

65. Akiyama, T., Liu, H., and Hayakawa, K. I., Hygrostress-multicrack formation and propagation in cylindrical viscoelastic food undergoing heat and moisture transfer processes, *International Journal of Heat and Mass Transfer*, 40, 1601, 1997.

66. Zhang, J., Datta, A. K., and Mukherjee, S., Transport processes and large deformation during baking of bread, *American Institute of Chemical Engineers Journal*, 51, 2569, 2005.

CHAPTER 5

Mass Transfer: Membrane Processes

David Hughes, Taha Taha, and Zhanfeng Cui

CONTENTS

5.1   INTRODUCTION

5.1.1   Scope of This Chapter

Membrane processes are now commonplace in the dairy, food, chemical, pharmaceutical, and water-treatment industries. Successful design of a membrane process relies on the ability to predict filtration performance; for these predictions, the modelling of mass-transfer processes is essential. After a brief introduction to the fundamentals of membranes and membrane processes, modelling of pressure-driven membrane processes will be discussed in detail. Section 5.2 will review the currently available models and Section 5.3 is a case study on the prediction of permeate flux in ultrafiltration using computational fluid dynamics.

5.1.2   Membrane Processes

A membrane is a selective barrier that allows some species to permeate the barrier while retaining others.[1,2] In food and biotechnology applications, the goal is typically to concentrate, recover, or remove particles, macromolecules, or low-molecular-weight solutes from an aqueous feed stream.[3] Membrane filtration can be categorized into four major pressure-driven membrane processes: microfiltration (MF), ultrafiltration (UF), nanofiltration (NF), reverse osmosis (RO), and direct osmotic concentration (DOC) where chemical potential is the driving force.

Membranes are porous materials,[1] with pore sizes that decrease from MF, through UF and NF, to RO. MF is principally used for separation of micron-sized species that are usually particles or large macromolecules. UF is used for the separation of macromolecules (e.g., proteins) and NF and RO are used for low-molecular-weight solutes (e.g., salts). DOC uses RO membranes and a difference in chemical potential between the feed and permeate to concentrate solutions containing low-molecular-weight solutes.

The separation mechanism for membrane processes is based upon either size exclusion or solution diffusion. The pore size and the structure of the membrane determine which mechanism is dominant. MF and UF membranes have well-defined pores and the separation is based upon size exclusion. If the size of a species is larger than the size of the membrane pores, it cannot pass through the membrane and is retained. For NF and RO membranes, which have smaller pores, there still exists some debate concerning the separation mechanism. The suggested mechanisms are size exclusion similar to MF and UF, or solution diffusion. In the solution-diffusion mechanism, species are absorbed into the membrane, diffuse through the membrane structure, and are then desorbed. The relative rates of the absorption, desorption, and diffusion of the species control the separation. Although classified as porous, NF and RO membranes are often considered to be intermediates between truly porous membranes such as those found in MF, and dense, nonporous membranes used in gas separation and pervaporation.

In a membrane process, the pressure applied across the membrane to the drive the separation is the transmembrane pressure (TMP). The portion of the feed that passes through the membrane is termed the filtrate or permeate. The volumetric flow rate of permeate per unit area of membrane is the *permeate flux*, usually denoted by J,

$$J = \frac{1}{A}\frac{dV}{dt}. \tag{5.1}$$

In this equation, V is the total volume that has permeated through the membrane at time t, and A is the area of the membrane. The SI units of flux are m s^{-1}, however, flux is often stated in Lm^{-2}h^{-1}.

The permeate may contain a fraction of the particles, macromolecules, or low-molecular-weight components from the feed stream. This transmission of a species across the membrane is conventionally described in terms of either the apparent sieving coefficient or observed rejection. The apparent sieving coefficient (S_0) is defined as the concentration of a species in the permeate (C_P) as a fraction of its concentration in the feed (C_B).[4]

$$S_a = \frac{C_p}{C_B}, \tag{5.2}$$

whereas the observed rejection (R_0) is defined as

$$R_0 = 1 - S_a. \tag{5.3}$$

In DOC, a membrane (usually RO) separates two compartments, one containing the feed solution to be concentrated, and the other containing an osmotic solution. The chemical potential of the osmotic solution (usually brine) is lower than that of the species to be concentrated; consequently, there exists an osmotic pressure difference driving liquid across the membrane from the feed solution into the osmotic solution. As a result, the feed is concentrated.

In general, membrane processes are used to either concentrate or fractionate a feed stream. In concentration, the aim is to increase the concentration of the rejected species in the retentate. This is achieved by selecting a membrane with pores much smaller than the species to be concentrated. Permeation through the membrane leaves the retentate more concentrated in the rejected species. In fractionation, the goal is to separate a feed stream containing two or more species into retentate and permeate streams that are rich in different species. In general, the membrane must have pores that are larger than the size of the species that must cross the membrane, but smaller than the size of the species that remains in the retentate to be rejected. Fractionation can be imprecise and difficult to achieve.

5.1.3 Microfiltration and Ultrafiltration

MF and UF operate on a size-exclusion principle. The membranes have a well-defined porous structure. MF membranes have the largest pores, with mean pore diameters ranging from 0.1 to 10 μm. Their porosity is generally high and this leads to a low resistance to permeate flow. Therefore, acceptable fluxes can be achieved at low TMP, typically below 2 bar.

The membranes can be produced from a wide range of both organic and inorganic materials. Organic MF membranes are made from polymers such as polysulphone, polycarbonate, and polyvinylidene fluoride. Polymer membranes tend to offer a low resistance to permeate flow and a low cost per unit area but are unable to handle high temperatures, a wide range of pHs or harsh chemical cleaning conditions. Hence for some MF applications inorganic membranes made from ceramics such as alumina (Al_2O_3) and zirconia (ZrO_2) are preferred. These membranes can resist harsh operating conditions but are expensive per unit area and may also be brittle.

Structurally, MF membranes differ from all others as they maybe either symmetric or asymmetric. Symmetric membranes have a homogenous structure and the entire structure contributes to the separation, whereas asymmetric membranes have a thin top layer supported by a more porous lower layer. The separation is governed by the thin top layer. All types of MF membranes are widely used in the food and biotechnology industries. Some examples of applications of MF include:

- Clarification of cheese whey in the dairy industry
- Production of high quality water
- Downstream processing of fermentation broths for cell recovery
- Membrane bioreactors for wastewater treatment

The pores of UF membranes are smaller than those of MF membranes. The pore size is specified by the molecular-weight cutoff (MWCO) of the apparent. The MWCO is defined as the molecular weight of the macromolecules at which the apparent sieving coefficient falls in the range of 5–10%. UF membranes have MWCOs of between 1 and 300 kDa, which requires mean pore diameters in the range 2–100 nm. As UF membranes have smaller pores—and generally a lower porosity—a higher TMP (of up to 10 bar) is required to achieve acceptable fluxes. Like MF membranes, UF membranes are made from a wide range of polymers and inorganics. The polymers include polysulphone, polyvinylidene fluoride, polyamide, and cellulose acetate, whereas the inorganics are limited to the ceramics, alumina, and zirconia. UF membranes are exclusively asymmetric in structure because of the small pore size. The advent of robust asymmetric UF membranes has lead to a growth in food and biotechnology applications. Some examples include:

- Fractionation of milk for cheese making
- Concentration of a wide range of fruit juices
- Production of high quality water
- Protein fractionation

Although membrane selection is important in MF and UF processes, the actual performance of the process is often not governed by the membrane itself but by concentration polarization and fouling (see Section 5.1.7 and Section 5.1.8). Therefore, when modeling these processes, the accurate description of fouling and concentration polarization is essential.

5.1.4 Nanofiltration and Reverse Osmosis

In NF and RO, the selection of the membrane and, in particular, the material, is much more critical. Here, fouling and concentration polarization must still be considered but the separation is usually governed by the intrinsic properties of the membrane. In fact, the intrinsic selectivity of a material towards a given solute is often the determining factor in membrane selection.[1] The range of materials used for NF and RO membranes is much smaller than that used for MF and UF, and is limited to polymers. For example, cellulose acetate membranes are often used in RO processes for the desalination of sea water because they have a relatively high permeability to water and a very low salt permeability. Polyamides are also commonly used. Although polyamides exhibit a lower water permeability, they can be operated over a wider range of pH.

All NF and RO membranes are asymmetric in structure.[5] As with UF membranes, but to a much greater extent, the separation layer must be kept thin as the pores are extremely small. It is estimated that NF and RO membranes have pores with diameters less than 2 nm. Membranes are usually classified according to their retention of monovalent (e.g., Na^+, Cl^-) and bivalent (e.g., Ca^{2+}) ions. RO membranes have high retention of both monovalent and bivalent ions, typically more than 98%, whereas NF membranes have high retention of bivalent ions, usually over 90%, but much lower retention of monovalent ions.[1] Whereas RO membranes offer high retentions, the more open structure of NF membranes leads to a lower resistance to permeate flow. NF is frequently called "loose" or "leaky" RO.

The low-molecular-weight species to be separated in NF and RO exhibit high osmotic pressures. The pressure depends upon the species and its concentration. For example, seawater has an osmotic pressure of approximately 25 bar; thus, this pressure must be overcome before water can be made to permeate through the membrane from the feed where the salt concentration is high, to the permeate where the salt concentration is low. The combination of high osmotic pressure and high resistance to permeate flow leads to typical operating pressures of 10–100 bar. Given the high pressures, pumping costs—and therefore energy costs—of NF and RO processes are much higher than those for MF and UF. The high energy costs can be tolerated because NF and RO are competing with separation technologies such as evaporation.

RO, in particular, has found many applications, the most common of which is the desalination of seawater. In the food industry, RO is very useful as a concentration step. RO applications include:

- Concentration of milk for bulk transport
- Down stream processing, particularly desalting
- Concentration of wastewater streams
- Production of ultrapure water

The applications of NF are more limited. A number of specialized applications exist, such as concentration of whey in the diary industry and downstream processing in the corn and wine making industries.

5.1.5 Direct Osmotic Concentration

In DOC, RO membranes are frequently used. There is no substantial hydraulic pressure applied across the membrane; therefore, the driving pressure is almost equal to the osmotic pressure difference across the membrane. The modelling of DOC processes is limited, but because the process shares many similarities with pressure-driven membrane processes, models developed in this area may well be applicable. It should be noted that one difference between DOC and pressure-driven separations is that there maybe a significant resistance to mass transfer on both sides of the membrane in DOC. Possible applications of DOC include the concentration of grape juice and tomato juice. The focus of the remainder of this chapter will be pressure-driven membrane separations; however, with suitable modification the models presented could be applied to DOC.

5.1.6 Mode of Operation and Module Design

There are two modes of membrane operation: dead-end and crossflow. The original mode of operation for membrane processes was dead-end, where the feed stream is normal to the membrane. The applied pressure across the membrane causes the liquid portion of the feed to permeate. Depending on their size particles, macromolecules and low-molecular-weight components in the feed either permeate or are retained on or within the membrane. Dead-end filtration is still used in a number of applications, such as cell harvesting in the biotechnology sector. In recent years cross-flow operation has become popular. In crossflow operation, the feed stream flows tangentially across the surface of the membrane. The pressure applied across the membrane forces typically 0.1–10% of the liquid portion of the feed stream to permeate through the membrane. As in dead-end operation the permeate may contain some of the species present in the feed. The majority of the feed does not permeate through the membrane and is swept out of the module, this is termed the retentate.

Module design is linked to the choice of operating mode. Membranes are supplied as either tubes of varying diameter, or as flat sheets that must then be mounted within a suitable module.

Factors affecting the choice of module design include feed properties, operating conditions, ease of cleaning, surface area of membrane per unit volume of the module (membrane packing density), and cost. Flat sheet membranes are often mounted in spiral-wound modules. Membranes supplied as tubes can be broadly classified either as capillaries, hollow fibres, or tubular membranes. Capillaries are less than 1 mm in diameter, hollow fibers a few millimetres in diameter, and tubular membranes that may be tens of millimetres in diameter. For dead-end filtration, single-use cartridges containing either hollow fibres or flat sheet membranes are commonly used. A wider range of modules are used for crossflow operation.

5.1.7 Concentration Polarization

In membrane processes, the retained or rejected species accumulate near the membrane surface. This leads to two phenomena that reduce the filtration performance of membrane systems: concentration polarization and fouling. Concentration polarization is a natural consequence of preferential transport of some species through the membrane. The applied pressure across the membrane causes convection of species towards the membrane. The concentration of rejected species increases at the membrane surface. This creates a concentration gradient leading to back-transport of the rejected species to the bulk. This is a particular problem when separating macromolecules and low-molecular-weight solutes. These species can generate significant osmotic pressure at the membrane wall; this reduces the effective applied pressure across the membrane. A reduction in the effective applied pressure leads to a reduction in the permeate flux. Concentration polarization is linked to the permeate flux through the film mass-transfer equation:

$$J = \frac{D}{\delta} \ln\left(\frac{C_{\mathrm{w}} - C_{\mathrm{P}}}{C_{\mathrm{B}} - C_{\mathrm{P}}}\right) = k \ln\left(\frac{C_{\mathrm{w}} - C_{\mathrm{P}}}{C_{\mathrm{B}} - C_{\mathrm{P}}}\right), \tag{5.4}$$

where k is the mass-transfer coefficient usually given in the SI unit m s^{-1}, and C_{w}, C_{B}, and C_{P} are the concentrations of the species at the membrane, in the bulk, and in the permeate, respectively. To increase the permeate flux, either the mass-transfer coefficient must be increased, or the bulk concentration of the retained species must be decreased.

Concentration polarization is a reversible phenomenon that does not itself affect the intrinsic properties of the membrane. If a pure water or clean liquid stream is fed to the membrane, then the permeate flux is determined by the TMP and the properties of the membrane. When a feed stream containing a species that is rejected by the membrane is filtered under the same conditions, concentration polarization will occur and the permeate flux will fall below the level for a pure-water feed stream. If the feed is switched back to a pure-water stream, the permeate flux will recover. Hence, the effect of concentration polarization is reversible.

Concentration polarization is a particular problem when filtering macromolecules and low-molecular-weight solutes that have significant osmotic pressures. Careful consideration must be given to the module design and choice of operational conditions for UF, NF, and RO processes to limit the phenomenon. MF is unaffected because particulate feeds do not exert large osmotic pressures.

5.1.8 Fouling

Membrane fouling results in a decrease in the permeate flux over time in constant TMP operation. It is a more complex phenomenon than concentration polarization and may be both reversible and irreversible. For porous membranes, fouling may occur as in-pore blocking of the membrane pores, as adsorption onto the surface of the membrane, or as the formation of an external fouling layer on the membrane surface.[1] The relative importance of the three types of fouling depends on operating conditions, the feed stream, and membrane properties. In-pore blocking and adsorption onto the membrane surface both reduce the open area of pores through which permeate can flow, thereby reducing the permeate flux. This is a common problem when filtering macromolecules such as proteins and is irreversible. The performance of the membrane is reduced and will not recover without chemical cleaning. An external fouling layer occurs when rejected species build up on the membrane surface. This causes an additional resistance to the flow of permeate. In MF applications with particulate feed streams, an external fouling layer of rejected particles, known as a *cake*, is common. In UF applications, particularly at high TMP, an external gel layer of macromolecules may form on the membrane surface.

Fouling and concentration polarization are related phenomena but should not be confused. Concentration polarization is a natural consequence of mass transport through a membrane which affects the performance of the system without actually altering the membrane itself. Fouling, in contrast, may cause changes to the membrane or the formation of an external fouling layer that acts to increase the resistance to permeate flow. In broad terms, fouling is a greater problem in MF and UF; fouling in MF can be very severe and can lead to decreases in permeate flux of more than 95%.[1] The fouling of RO and NF processes tends to be a lesser problem because usually some form of pretreatment of the feed is performed before it reaches the membranes.

5.1.9 Operating Conditions

The choice of operating conditions is important in limiting the effects of concentration polarization and fouling. The majority of membrane processes are operated in crossflow mode because the tangential flow of the feed stream across the membrane increases the mass-transfer coefficient and reduces fouling. An important consideration is whether to operate with a constant applied pressure across the membrane (constant TMP) or with a constant permeate flux. Originally, most crossflow processes were operated under high constant TMP to achieve high permeate fluxes, but this in turn led to severe fouling. In recent years, many processes have been shifted to operate at low to moderate constant TMP or constant permeate flux. Particularly in constant permeate flux operation, where the convective force towards the membrane is controlled, the effects of fouling can be reduced and even eliminated.[6] Here, examples will be given in which it is assumed that the processes are operated under constant TMP.

Feed streams in the food and biotechnology industries are aqueous but tend to be very complex. It is common for feeds to contain a mixture of particles, macromolecules, and low-molecular-weight components. For example, whole milk contains fat globules, proteins, and carbohydrates, whereas fermentation broths from the brewing industry contain yeast cells, cells debris, alcohol, and low-molecular-weight flavor components. The membrane processes and operating conditions must be carefully chosen to ensure the desired species are retained and an acceptable permeate flux is achieved.

5.2 MODELLING OF PRESSURE-DRIVEN MEMBRANE PROCESSES

5.2.1 Classification of Models

The goal of modelling in membrane processes is usually to predict the variation of permeate flux and transmission of one or more species with time under a given set of operating conditions. The inputs to a model typically include operating conditions (e.g., TMP and crossflow velocity) and feed-stream properties (e.g., particle size and feed concentration). Of particular interest is the steady-state permeate flux that is achieved after extended operation under crossflow. Membrane models are best categorized by the physical phenomena that govern the process. The four major categories of model are concentration polarization, fouling, membrane, and force balance. Each will be considered in turn and selected examples will be given. There also exist entirely empirical membrane models, but because they have few links to any underlying physical phenomena, these models are highly application-specific and will not be considered here. Emphasis will be placed on models for crossflow filtration under constant TMP given the prevalence of this operational mode in industry. However, it is interesting to note that many crossflow models are merely extensions of earlier models developed for dead-end filtration. For MF, the models commonly used are based upon force balance, concentration polarization, and fouling. Force balance models are unique to MF, in which an individual entity, usually a particle, can be readily identified. Models based on the membrane itself are rarely used in MF because the process is often dominated by fouling; the actual membrane properties are not the controlling factor in the filtration process. No category of model has been established as giving superior results, and some models use a combination of the concepts.

For UF, as for MF, few membrane models exist because fouling and concentration polarization are often significant. With UF systems, it is helpful to make a distinction as to which phenomena—fouling or concentration polarization—is controlling the process. For some UF processes, fouling is low or negligible and the permeate flux is controlled by concentration polarization. In these situations, the osmotic pressure and gel models, which are adaptations of the film mass-transfer equation, are most appropriate. The case study in Section 5.3 outlines such a process using the osmotic-pressure model. In situations where fouling dominates in UF, both concentration polarization and fouling models can be employed.

In general for NF and RO, concentration polarization and membrane models are employed. The prediction of species rejection is of particular interest for these processes and, therefore, membrane models are important. If concentration polarization is important, then a concentration-polarization-based model may be used alone or linked to an intrinsic membrane-property model. That is not to say that fouling models do not exist, they are merely less common.

5.2.2 Concentration Polarization Models

Concentration polarization models are based on a mass balance of material in the mass-transfer boundary layer. At steady state, the convective flux of species to the membrane must be balanced by a diffusive flux of species back to the bulk and a permeation of species through the membrane. Control-volume analysis leads to the film mass-transfer or concentration-polarization equation:

$$J = k \ln\left(\frac{C_w - C_P}{C_B - C_P}\right).$$
(5.5)

The permeate flux can be found if the mass-transfer coefficient (k) can be predicted and the concentrations in the bulk (C_B), in the permeate (C_P), and at the membrane (C_w) are known. The bulk concentration is a fixed operational parameter, and the concentration in the permeate may be easily measured or sensibly assumed, given knowledge of the membrane. More problematic are the prediction of the mass-transfer coefficient and the accurate determination of the concentration at the membrane.

There exist many equations and correlations to predict mass-transfer coefficients, most of which were developed for conventional mass-transfer processes where the wall is not semipermeable. The mass-transfer coefficient in the film mass-transfer equation has the same dimensions as conventional or engineering mass-transfer coefficients. However, the semipermeable nature of the wall in membrane processes alters the concentration profile in the mass-transfer boundary layer. In conventional mass-transfer processes, there is no convective flux towards the wall. The concentration profiles of species vary linearly between the wall and the bulk. However, in membrane processes, the wall is porous and there is a convective flux to the membrane. The effect of this convective flux is to curve the concentration profile between the membrane and the bulk. Consequently, care must be taken when predicting mass-transfer coefficients for membrane processes using equations or correlations developed for conventional or nonmembrane processes.

The requirement to predict or measure the concentration at the membrane is perhaps the most problematic aspect of concentration polarization models. Direct measurement of the concentration at the wall under filtration conditions is extremely difficult. The prediction or determination of concentration at the membrane is handled differently, depending on the membrane process.

Concentration polarization models have been applied to all membrane processes. The best results are achieved when fouling is limited and TMP is high, such that the steady-state flux has become independent of TMP. Initial attempts to use concentration polarization models (based on the film mass-transfer equation) to predict permeate fluxes for the MF of micron-sized particles proved unsuccessful. The predicted fluxes were much lower than those seen experimentally, owing to the Brownian diffusion coefficient being used to determine the mass-transfer coefficient. For macro-molecules and submicron particles, Brownian diffusion is the dominant back-transport mechanism.[7] However, a different mechanism clearly controls the back-diffusion of larger particles. Original workers termed this the *flux paradox*.[8]

An alterative back-transport mechanism known as *shear-induced diffusion*[9] was proposed as a solution to the flux paradox. Shear-induced diffusion results from particles in a flow undergoing random displacements from their streamlines as they interact with other particles. In a concentration gradient, a greater number of random displacements will occur on the high-concentration side of the particle, thereby causing it to move down the concentration gradient. Colton and Zydney[9] used shear-induced diffusivity instead of Brownian diffusivity to predict the mass-transfer coefficient in the film mass-transfer equation. Their work led to two solutions for flux:

$$J = 0.078\dot{\gamma}_w \left(\frac{d_p^4}{L}\right)^{1/3} \ln\left(\frac{\phi_w}{\phi_B}\right) \tag{5.6}$$

for $\phi_w - \phi_B \ll \phi_w$, and

$$J = 0.126\dot{\gamma}_w \left(\frac{\phi_w d_p^4}{\phi_B L}\right)^{1/3} \tag{5.7}$$

for $\phi_B \ll \phi_w$, where $\dot{\gamma}_w$ is the shear rate at the membrane, d_p is the particle diameter, L is the membrane length, and ϕ_w and ϕ_B are the volume concentrations at the membrane and in the bulk, respectively. The SI units for flux are m s^{-1}. In MF processes, there is often a cake fouling layer formed on the membrane surface. The concentration at the membrane is often taken to be the concentration at the top of the cake. Because most cakes are composed of approximately spherical particles, the concentration at the top of the cake is taken to be the concentration of close-packed spheres. Further investigation found that, in MF, these models are relatively insensitive to the exact value of the concentration at the membrane.[10]

The osmotic-pressure model is suitable for UF processes where the rejected species exert significant osmotic pressure. As the concentration of the rejected species at the membrane increases, the osmotic pressure of the species also increases. This leads to a reduction in the effective driving pressure across the membrane and, consequently, a reduction in permeate flux. For a UF processes with negligible fouling and total solute rejection, Darcy's law maybe written as

$$J = \frac{\text{TMP} - \Delta\pi}{\mu R_m}, \tag{5.8}$$

where μ is the permeate viscosity (Pas), R_m is the resistance of the membrane (m^{-1}), and $\Delta\pi$ and TMP are the osmotic and transmembrane pressures, respectively (Pa).

The concentration-polarization equation for total solute rejection is

$$J = k \ln\left(\frac{C_w}{C_B}\right). \tag{5.9}$$

The osmotic pressure exerted by macromolecules has been measured experimentally and is a function of concentration. Here, the concentration that determines the osmotic pressure is that at the membrane (C_w), and the relationship between concentration and osmotic pressure is often expressed in the form:

$$\Delta\pi = a_0 + a_1 C_w + a_2 C_w^2 + a_3 C_w^3 + \cdots \tag{5.10}$$

The constants (a_0, a_1, etc.) must be determined experimentally. The solution of Equation 5.8 through Equation 5.10 allows the flux and the concentration at the membrane to be determined.

Whether using the film mass-transfer equation, gel model, or osmotic-pressure model, a value for the mass-transfer coefficient must be determined. The mass-transfer coefficient may either be predicted or measured experimentally. The analogy between heat and mass transfer in conventional or engineering mass-transfer processes has long been accepted. Thus, modified semiempirical heat-transfer correlations are used to estimate the mass-transfer coefficient for membrane processes.[11]

The correlations are expressed in terms of standard dimensionless numbers that are used throughout work on mass transfer, and are not unique to membrane processes. These numbers include the variables tube diameter (d), species diffusivity (D), mass-transfer coefficient (k), crossflow velocity (U), solution density (ρ), and solution viscosity (μ):

$$\text{Sherwood number}: \quad Sh = \frac{kd}{D}; \tag{5.11}$$

$$\text{Reynolds number}: \quad Re = \frac{\rho U d}{\mu}; \tag{5.12}$$

$$\text{Schmidt number}: \quad Sc = \frac{\mu}{\rho D}. \tag{5.13}$$

These are normally expressed as:

$$Sh = aRe^b Sc^c \left(\frac{d}{L}\right)^d. \tag{5.14}$$

The coefficients a, b, c, and d depend upon the correlation selected, which in turn depends on module design. It should be noted that a consistent set of units for all the variables must be used. One such semiempirical solution for laminar flow in a tube is the Leveque solution that is used when the velocity profile is fully developed but the concentration profile is still developing:[2]

$$Sh = 1.86 \, Re^{0.33} Sc^{0.33} \left(\frac{d}{L}\right)^{0.33}. \tag{5.15}$$

The lead coefficient is also reported as 1.62.[11] The Leveque solution can be expressed in terms of mass-transfer coefficient and shear rate:

$$k = 0.94 \left(\frac{\gamma_w D^2}{L}\right)^{0.33}, \tag{5.16}$$

where the shear rate (γ_w) at the membrane is given by

$$\gamma_w = \frac{8U}{d}. \tag{5.17}$$

The Leveque solution predicts that the mass-transfer coefficient is only a function of the shear rate at the membrane (γ_w), the diffusivity of the species (D), and membrane length (L). Because species diffusivity is fixed, only the membrane length and shear rate at the membrane may realistically be changed. Given that most modules and membranes are designed to be standard sizes, the variation of membrane length is also difficult. Therefore, the only realistic method of increasing the mass-transfer coefficient is to increase shear rate.

Alternatively, if the flow is laminar but the velocity and concentration profiles are fully developed, then the Grober correlation may be used:[11]

$$Sh = 0.664 \, Re^{0.5} Sc^{0.33} \left(\frac{d}{L}\right)^{0.5}. \tag{5.18}$$

For definitions of the dimensionless numbers Sh, Re, and Sc, the reader is referred to Chapter 6. For the frequently encountered situation of turbulent flow in a tube, the Dittus–Boelter or Chilton–Colburn correlation is often used:[12]

$$Sh = 0.023 \, Re^{0.8} Sc^{0.33}. \tag{5.19}$$

Other correlations have been produced to suit specific module designs. For example, in spiral-wound modules, spacers are often added in the feed channel to cause locally high shear stresses, improve mixing, and therefore increase the mass-transfer coefficient. Schock and Miquel[13] produced a correlation for this situation; the flow in such a module with a spacer would be turbulent:

$$Sh = 0.065 \, Re^{0.065} Sc^{0.25}. \tag{5.20}$$

It is interesting to note that the dependence on the Schmidt number has changed only slightly from the Dittus–Boelter correlation, but the dependence on the Reynolds number has changed by more than an order of magnitude.

Whether analytical solutions or correlations are used, parameters such as species diffusivity and crossflow velocity or wall shear stress must be determined for use in the model. The diffusivity of many common species encountered in UF has been experimentally measured and correlations are available. If module geometry is simple and the flow is steady then prediction of crossflow velocity or wall shear stress is straightforward. However when module geometry is complex, for example spiral-wound modules, or the flow is unsteady then prediction becomes much more difficult. In these cases it is often necessary to use methods such as computational fluid dynamics.

Alternatively the mass-transfer coefficient maybe found experimentally. For MF and UF processes, this is achieved by performing experiments at a range of bulk concentrations. A rearrangement of the film mass-transfer equation may then be used to determine the mass-transfer coefficient by plotting flux vs. natural log of bulk concentration. Here, the film mass-transfer equation for a membrane with total retention is used:

$$J = -k \ln C_B + k \ln C_w. \tag{5.21}$$

The mass-transfer coefficient is found from the gradient of the resulting straight line, and the gel concentration is found from the intercept on the y-axis.

5.2.2.1 Food and Bioprocess Examples

Constenia and Lozano[14] studied the UF of apple juice through hollow fibres at constant volume-concentration ratio. Reynolds and Schmidt numbers were calculated taking into account the effect of increasing viscosity with feed concentration using the well-known rheological power law:

$$\mu = \mu_0 \gamma_w^{n-1},\tag{5.22}$$

where μ_0 is the consistency coefficient and n is the flow behavior index; these values were determined from experiment. Linear regression of log Re and log Sc vs. log Sh established the relationship:

$$Sh = 0.149\ Re^{0.336} Sc^{0.329}.\tag{5.23}$$

Both Re and Sc show a very similar dependence to the Leveque solution (Equation 5.15), however the lead coefficient is an order of magnitude lower, and no dependence on the ratio of membrane length to diameter is discussed.

5.2.2.2 Worked Example

Question: In a pilot-scale UF process, a nonfouling macromolecule is filtered under constant TMP. Use the osmotic-pressure model to predict the permeate flux. The relationship between the osmotic pressure in bar ($\Delta\pi$) and the concentration of the macromolecule at the membrane in kg m^{-3} (C_w) can be taken as

$$\log \Delta\pi = 0.25 + 0.1 C_w^{0.35}.$$

Operational Data:

TMP: 10 bar
Pure-water flux at 10 bar TMP: 7.5×10^{-5} m s^{-1}
Concentration at the membrane (C_w):   100 kg m^{-3}
Permeate Viscosity (μ): 1×10^{-3} Pas

Solution: The osmotic-pressure model (Equation 5.8) is

$$J = \frac{\text{TMP} - \Delta\pi}{\mu R_m}.$$

To find the flux, J, the osmotic pressure, $\Delta\pi$, and the resistance of the membrane, R_m must be calculated. The resistance of the membrane can be found from the pure-water flux. Because pure-water has zero osmotic pressure ($\Delta\pi = 0$), then the membrane resistance is

$$R_m = \frac{\text{TMP}}{\mu J_0} = \frac{10 \times 10^5}{1 \times 10^{-3} \times 7.5 \times 10^{-5}} = 1.33 \times 10^{13}\ \text{m}^{-1}.$$

The osmotic pressure can be calculated from the equation, given that

$$\log \Delta\pi = 0.25 + (0.1 \times 100^{0.35}) = 0.751,$$

$$\Delta\pi = 5.64\ \text{bar}.$$

The flux is then

$$J = \frac{\text{TMP} - \Delta\pi}{\mu R_m} = \frac{(10 - 5.64) \times 10^5}{1 \times 10^{-3} \times 1.33 \times 10^{13}} = 3.28 \times 10^{-5}\ \text{m s}^{-1}.$$

5.2.3 Membrane Fouling Models

Membrane fouling models are common in MF and UF, and apply equally well to all processes if fouling is occurring. Under constant TMP operation when fouling occurs the permeate flux decreases with time. The effects of fouling are often considered with respect to Darcy's law. When filtering pure water prior to exposing the membrane to any foulant, the resistance to the flow of permeate is a minimum and is only a function of the membrane and the TMP:

$$J_0 = \frac{\text{TMP}}{\mu R_m}, \tag{5.24}$$

where μ is the permeate viscosity (Pas), R_m is the resistance of the membrane (m^{-1}).

In the general case of a membrane exposed to foulant species Darcy's law becomes:

$$J = \frac{\text{TMP}}{\mu(R_m + R_{if} + R_c)}. \tag{5.25}$$

After exposure to a foulant species, the flux will decrease $(J < J_0)$ due to either internal fouling of the membrane (R_{if}) or the presence of a cake layer that causes a cake resistance (R_c). Internal fouling may occur due to in-pore fouling at the openings or within the pores or the foulant species being adsorbed onto the membrane surface. Alternatively, a fouling layer may form above the membrane, providing an additional resistance to permeate flow that is quantified by the cake resistance.[1] There are four common models that account for the different physical phenomena involved in membrane fouling: complete pore blocking, standard blocking, intermediate blocking, and cake filtration.

Both complete pore blocking and standard blocking are internal fouling mechanisms because the fouling occurs at the pore entrance or within the pores. There is no additional resistance due to any fouling layer on the surface of the membrane.[15] Standard pore blocking assumes a uniform deposition of material on the pore walls, causing a uniform decrease in the diameter of the pores and thus a reduction in the open pore area available for permeate flow. Complete pore blocking occurs when a pore is completely sealed or plugged by a particle or macromolecule. After the pore is plugged, no further flow through the pore is possible and the number of open pores available for permeation is reduced. The number of plugged pores is assumed to increase in proportion to the volume of permeate that has passed through the membrane.

Intermediate fouling is based upon a similar physical mechanism to standard pore blocking. Depositing species completely block pores, but may deposit over an already blocked pore, leading to an external fouling layer. Therefore, intermediate fouling leads to both internal and cake resistances.

Cake filtration is an external fouling mechanism and its physical basis is different to complete pore blocking or standard blocking. With cake filtration, there is no internal fouling and all additional resistance to permeate flow is provided by a fouling layer on the membrane surface. This is a common fouling mechanism in MF, particularly when particulate feeds and a membrane with pores much smaller than the size of the particles in the feed are present. The resistance of the cake (R_c) is proportional to M, the mass of cake deposited per unit area $(kg\ m^{-2})$:

$$R_c = \alpha M, \tag{5.26}$$

where α is the specific cake resistance $(m\ kg^{-1})$ and is dependent upon the species forming the cake. For rigid, nonadhesive spheres, the specific cake resistance is described well by the Carman–Kozney equation for flow through a bed of packed spheres. There are numerous expressions of this equation:[16]

$$\alpha = \frac{36 K_0 \tau^2 (1 - \varepsilon)}{\rho_c d_p^2 \varepsilon^3},$$ (5.27)

where ε is the porosity of the cake $(-)$, ρ_c is the density of the deposited particles in kg m^{-3} and d_p is the diameter of the deposited particles in meters. Here, K_0 is the Kozney coefficient rather than the mass-transfer coefficient. The tortuosity (τ) can be used to account for beds of particles that are nonspherical. In membrane applications, $36 K_0 \tau^2$ is often taken to equal 180. However, if the cake is formed from particles or species that are adhesive, or if it is formed under conditions where surface charge is important, then Carman–Kozney tends to under-predict the specific cake resistance. In these situations, the specific cake resistance must be determined experimentally. A further complication arises if compressible species such as microbial cells[7] or yeast cells[17] form the cake. The specific cake resistance then becomes a function of the applied TMP. This is often accounted for using a power law:

$$\alpha = \alpha_0 \mathrm{TMP}^s.$$ (5.28)

Again, the specific cake resistance under zero applied pressure, α_0, and the cake compressibility, s, must be determined experimentally.

The four major fouling models were brought together by Hermia[18] in a unified model for fouling in dead-end systems expressed in terms of the total permeate volume, V (m^3), and time, t (s):

$$\left(\frac{d^2 t}{dV^2}\right) = K \left(\frac{dt}{dV}\right)^n.$$ (5.29)

Each of the four fouling mechanisms is represented by a different value of n. The constant K must be determined by fitting the equations to experimental flux decline curves. The units of K and n vary with the fouling mechanism. There are many other fouling models, but the majority are essentially derivatives of Hermia's equation. Although Hermia's equation is strictly only valid for dead-end operation, it is often applied to crossflow operation by making the assumption that the convective flow of material to the membrane far exceeds the removal of material by crossflow action. Under these conditions, the effect of the crossflow is minimal and the system closely approximates dead-end filtration.

Hermia's unified model was expanded with the inclusion of a crossflow removal term to allow rigorously accurate modelling of crossflow systems:[6]

$$-\frac{dJ}{dt} J^{n-2} = K(J - J^*).$$ (5.30)

Again, n and K are constants that depend on the fouling mechanism, whereas J and J^* are, respectively, the flux at any time (m s^{-1}) and the critical flux (m s^{-1}) that is considered the flux below which no fouling occurs. In recent years, there has been considerable interest in operating MF and UF processes close to this point. There exists a subset of models that deal with prediction of the first point at which fouling occurs; most of these models focus on the first point of deposition of particulate species in MF.

The Table 5.1 summarizes the basic modes of fouling, their effects on the resistance terms in Darcy's equation, and an integrated form of Hermia's dead-end constant-pressure model.

An alternative fouling model used for UF applications is the gel model. It does not fit into Hermia's framework, but like Hermia's models it is underpinned by physical phenomena. Experimentally, it is often noted that the flux in a UF processes is only a function of hydrodynamics

Table 5.1 Summary of Hermia's Blocking Laws

Mode	Description	Darcy's Law	Hermia's Model	
Complete pore blocking	Each particle that reaches a pore, completely seals that pore	$R_{if} > 0$ and $R_c = 0$	$n = 2; J = J_0 \exp(-Kt)$	$K(s^{-1})$
Standard blocking	Diameter of pores is uniformly reduced	$R_{if} > 0$ and $R_c = 0$	$n = 1.5; J = (J_0^{-0.5} + Kt)^{-2}$	$K(m^{-0.5}s^{-0.5})$
Intermediate fouling	Particles may seal a pore, or may deposited over an already sealed pore	$R_{if} > 0$ and $R_c > 0$	$n = 1; J = (J_0^{-1} + Kt)^{-1}$	$K(m^{-1})$
Cake fouling	A layer of rejected particles builds up on the membrane surface providing an extra resistance	$R_{if} = 0$ and $R_c > 0$	$n = 0; J = (J_0^{-2} + Kt)^{-0.5}$	$K(sm^{-2})$

above a certain TMP. In the gel model, it is assumed that after the concentration at the surface increases above a certain level, the rejected species will precipitate and form a gel layer at the surface. The concentration of this gel, C_g, is fixed, and any further increase in TMP only results in a thickening of the gel layer; therefore, the flux remains constant. The formation of a gel layer is a fouling phenomenon; however, the gel model stems from a modification of the concentration-polarization equation. Here, the gel model is given for a membrane with total rejection:

$$J = k \ln\left(\frac{C_g}{C_B}\right). \tag{5.31}$$

The only factor now affecting the process is the mass-transfer coefficient. Improvement of the mass-transfer coefficient thins the gel layer and increases the flux. The mass-transfer coefficient can be determined using correlations or determined experimentally.

As a predictive tool in MF and UF, fouling models must be treated with care. Preliminary experiments must be performed to determine the dominant fouling mechanism and to allow determination of model parameters. Given the very complex nature of membrane fouling, the parameters determined in small-scale preliminary trials frequently do not scale-up well to industrial plants. An illustration of the complexity of fouling is provided by processes such as MF of solutions containing proteins. One such process is the separation of cell debris from a fermentation broth. The goal is to retain the cell debris while allowing the proteins or other products to pass with the permeate. In spite of native proteins being much smaller than the pores of MF membranes, they can cause very severe fouling. The fouling mode is found to change from internal to external over time. Thus, a model must be derived that incorporates both modes of fouling. Dead-end MF of proteins has been extensively modelled.[15,19] The models are complex, but a physical meaning can still be ascribed to many of the parameters.

5.2.3.1 Food and Bioprocess Examples

Chiang and Cheryan[20] used a modified form of the gel model to predict the steady-state permeate flux for the UF in hollow fibers. They accounted for the increase in diffusivity with feed concentration using a linear relationship:

$$D = D_0(1 + \varphi C), \tag{5.32}$$

where D_0 is the reference diffusivity ($m^2 s^{-1}$) and φ is an empirical constant with units dependant on those used for concentration. With this concentration dependence, the film mass-transfer equation is integrated and combined with a dimensionless correlation. Assuming the Schmidt number to exhibit a one-third power dependence, the following relationship was established:

$$J = 0.087 \frac{D_0(1 + 0.157C)}{d} Re^{0.64} Sc^{0.33} \left(\ln \frac{C_G}{C_B} - 0.157(C_G - C_B) \right). \qquad (5.33)$$

Both the lead coefficient and power dependence of Reynolds number are significantly different from the Leveque solution (Equation 5.15).

De Bruijn and Borquez[21] analysed the fouling mechanisms for apple juice filtration through a ceramic UF membrane using a combination of the fouling laws proposed by Hermia[18] and Field et al.[6] A characteristic relationship between total permeate volume and time was determined and parameters were fitted using experimental data. Both cake formation and internal pore blocking were found to contribute significantly to the fouling.

Bhattacharya et al.[22] used a two-stage model for the flux decline in the UF of limed sugarcane juice in stirred-cell experiments. Initially, an osmotic-pressure model was used to describe the rapid fouling in the early stages. This initial period of fouling lasted only a few seconds. The more significant long-term fouling was modelled using Darcy's law (Equation 5.25) with the decreasing flux attributed to a growing polarization layer. An empirical equation is fitted to the data to give the variation of the resistance of the polarized layer.

5.2.3.2 Worked Example

Question: UF of a macromolecule was undertaken using a single tubular membrane, which is assumed to have 100% rejection. Operation is at high TMP and the permeation rate through the membrane is found to be independent of the TMP. Using the gel model, estimate the concentration of the gel layer, given that the measured permeation rate (F) through the membrane is 22 ml min^{-1}?

Operational Data:

Membrane Diameter (d):   12.5 mm
Membrane Length (L):   1.9 m
Feed Flowrate (Q):   1.75 L min^{-1}
Feed Concentration (C_B):   0.6 wt%
Feed Kinematic Viscosity (v): 1×10^{-6} m^2 s^{-1}
Macromolecule Diffusivity (D): 8×10^{-11} m^2 s^{-1}

Solution: The gel model was used, given by Equation 5.31:

$$J = k \ln \left(\frac{C_g}{C_B} \right).$$

The model contains four variables; the feed concentration, C_B, is given. To find the gel concentration, C_g; the flux, J, and mass-transfer coefficient, k, must be determined.

The flux is defined as the permeation rate per unit area of membrane. For use in the model, the flux must be in the same units as the mass-transfer coefficient. Here, the SI units of m s^{-1} are used.
Conversion to SI units: $F = 22$ m L min$^{-1} = 3.67 \times 10^{-7}$ m s^{-1}.
The flux is given by

$$J = \frac{F}{\pi \, dL} = \frac{3.67 \times 10^{-7}}{0.0746} = 4.91 \times 10^{-6} \text{ m s}^{-1}.$$

The mass-transfer coefficient must be found from a correlation. To choose the appropriate correlation, one must determine if the flow is laminar or turbulent. This is found from the

Reynolds number. To find Reynolds number, the crossflow velocity must be calculated from the feed flowrate.

Conversion to SI units: $Q = 1.75$ L min^{-1} $= 2.92 \times 10^{-5}$ m^3 s^{-1}.

The crossflow velocity is determined by:

$$U = \frac{Q}{(1/4)\pi d^2} = \frac{2.92 \times 10^{-5}}{1.23 \times 10^{-4}} = 0.238 \text{ m s}^{-1}.$$

The Reynolds number is

$$Re = \frac{Ud}{\nu} = \frac{0.238 \times 0.0125}{1 \times 10^{-6}} = 2950.$$

The Reynolds number is greater than 2000. Therefore, flow is considered turbulent and the Dittus–Boelter correlation for the Sherwood number will be used (Equation 5.19):

$$Sh = 0.023 \, Re^{0.8} Sc^{0.33}.$$

The Schimdt number is:

$$Sc = \frac{\nu}{D} = \frac{1 \times 10^{-6}}{8 \times 10^{-11}} = 12,500.$$

The Dittus–Boelter correlation is:

$$Sh = \frac{kd}{D} = 0.023 \times 2950^{0.8} \times 12,500^{0.33} = 320.$$

From the Sherwood number, the mass-transfer coefficient can be determined by:

$$k = \frac{ShD}{d} = \frac{320 \times 8 \times 10^{-11}}{0.0125} = 2.05 \times 10^{-6} \text{ m s}^{-1}.$$

Rearranging the gel model yields:

$$C_g = C_f e^{J/k} = 0.6 \times e^{4.91 \times 10^{-6}/2.05 \times 10^{-5}} = 6.59 \text{ wt\%}.$$

5.2.4 Force Balance Models

Force based models are most readily applicable to MF because they require a particle or well-defined entity on which all of the acting forces can be assessed. In general, this is a particle that is either present in the fluid or deposited on the membrane. For particles in the fluid, the only two forces that tend to be of significance are the drag towards the membrane caused by the convective flux of solvent, and a lift force away from the membrane. The forces acting upon a deposited particle include: drag, friction, and electrostatic forces. Depending upon the likely release mechanism for the deposited particle, either a force or a torque balance may be used.

An early example of a force-based model for a particle in the flow was inertial lift.[8] Inertial lift occurs because of nonlinear interactions between particles and the surrounding flow. Inertial lift becomes important if the Reynolds number (based on particle diameter) is large enough to cause the nonlinear inertia terms in the Navier–Stokes equation to be significant. At steady state, the inertial lift velocity is expected to balance the permeation velocity. Drew, Schonberg, and Belfort[23] proposed the following expression for fast laminar flow with a thin fouling layer:

$$J = 0.036 \left(\frac{\rho}{\mu}\right) \dot{\gamma}_w^2 d_p^3. \tag{5.34}$$

The solution density, ρ, viscosity, μ, wall shear rate, γ, and particle diameter, d_p, have the usual units. Models that balance the forces on a deposited particle are more common. These models often incorporate a term to calculate the resistance of the cake formed by deposited particles and therefore involve components from fouling models. Altmann and Ripperger[24] balanced the drag force of the permeate flow, the drag force of the crossflow, the lift force, and the frictional force on a single particle resting on the membrane. As more particles were deposited, the resistance of the cake layer was estimated using the Carman–Kozeny equation. This produced a series of equations that could be solved to give the permeate flux and cake layer height.

Alternatively, a torque balance may be taken about the centre of a deposited particle to determine whether the particles will roll along the membrane or cake surface. This subset of force based models are known as *surface transport models*. Sherwood[25] considered the drag and contact forces on a sphere deposited at the top of the cake. Taking a torque balance at the point of contact, the permeate flux was predicted to be:

$$J = 2.4 d_p \dot{\gamma}_w (d_p^2 \hat{R}_c)^{2/5} \cot \theta, \tag{5.35}$$

where $\cot \theta$ is a parameter that depends on the surface morphology $(-)$ and $\hat{R}_c$ is the resistance per unit depth of the cake (m^{-2}).

Although force based models are frequently encountered in MF, they are rarely used in food and bioprocess applications because of the complex nature of the feeds. Although useful for determining the deposition of micron-sized particles, when dealing with complex feeds, the fouling of macromolecules, rather than particulate deposition, may control the steady-state flux.

5.2.5 Membrane Models

Intrinsic membrane property models are available for all of the four major membrane processes. The Karman–Cozney and Hagen–Poiseuille models can be used to predict the membrane resistance, R_m, of clean MF and UF membranes.[1] However, because MF processes are usually dominated by fouling, these models are of little practical use.

In NF and RO, when the rejection of a species is to be a determined, intrinsic membrane property models become important. The three major starting points for modelling the membrane are solution-diffusion, irreversible thermodynamics, or the extended Nerst–Plank equation. The simplest and most common model is solution-diffusion:[1]

$$\text{Solvent flux}: \quad J = L_p(\text{TMP} - \Delta\pi); \tag{5.36}$$

$$\text{Solute flux}: \quad J_s = B(C_B - C_P), \tag{5.37}$$

where L_p (m s^{-1} Pa^{-1}) and B (m^4 kg^{-1} s^{-1}) are the water and solute permeability coefficients respectively and characterise the membrane:

$$L_p = \frac{D_w c_w V_w}{RT\Delta x}, \tag{5.38}$$

$$B = \frac{D_s K_s}{\Delta x}, \tag{5.39}$$

where D_s and D_w are the diffusivities of the solute and water, respectively, in the membrane ($m^2\ s^{-1}$), c_w is the concentration of water (mol L^{-1}), V_w is the partial molar volume of water (L mol^{-1}) and K_s is the distribution of the solute in the membrane (−). The flux is inversely proportional to the thickness of the active separation layer of the membrane (Δx); this emphasises the need to minimize membrane thickness. Both L_p or B must be determined experimentally.

More complex models use the extended Nerst–Plank equation[26] to predict the flux of individual solute components, j_i, through the membrane given that the solvent flux, J, is known:

$$j_i = -D_{ip}\frac{dc_i}{dx} - \frac{z_i c_i D_{ip}}{RT}F\frac{d\psi}{dx} + K_{ic}c_i J. \tag{5.40}$$

The relationship is very complex and depends on the hindered diffusivity of a species, D_{ip}, valence of the ion, z_i, the hinderance factor for convection, K_{ic}, and the electrical potential, ψ. Many models use the above equation as a starting point, particularly for NF. Modifications have been made to account for electrostatic effects,[27] steric hindrance,[28] and the log mean pore size distribution of the membrane.[29] The resulting sets of equations must be solved numerically.

A commonly used model based on irreversible thermodynamics with a greater number of parameters than solution diffusion is the Spiegler–Kedem model.[30] The working equations of this model are:

$$J = L_p(\text{TMP} - \sigma\Delta\pi), \tag{5.41}$$

$$R_0 = 1 - \frac{C_P}{C_w} = \frac{\sigma(1-F)}{(1-\sigma F)}, \tag{5.42}$$

where

$$F = \exp\left(\frac{-J(1-\sigma)}{P_m}\right). \tag{5.43}$$

R is the true rejection of the membrane (−), σ is the reflection coefficient (−), and P_m is the overall permeability coefficient (s m^{-1}). A reflection coefficient of 1 indicates total solute rejection, whereas a coefficient of zero indicates zero rejection. Again, numerous modifications have been made to this equation to improve the accuracy of prediction under specific operating conditions.

Both the solution-diffusion and Spiegler–Kedem models are often combined with the film mass-transfer model to allow estimation of model parameters from observed rejection and solvent flux data:[31]

$$\text{Solution-diffusion}: \quad \frac{R_0}{1-R_0} = \left(\frac{J}{B}\right)\exp\left(\frac{-J}{k}\right); \tag{5.44}$$

$$\text{Spiegler–Kedem}: \quad \frac{R_0}{1-R_0} = \left(\frac{\sigma(1-F)}{1-\sigma}\right)\exp\left(\frac{-J}{k}\right). \tag{5.45}$$

5.2.5.1 Food and Bioprocess Examples

NF has the potential to simultaneously concentrate and demineralize whey products in the dairy industry. Van der Horst et al.[32] described the NF of whey UF permeate, a complex multicomponent mixture using a three-transport-parameter extended Nernst–Plank equation. To accomplish this, the

Nernst–Plank equation was combined with the film mass-transfer equation, in which the mass-transfer coefficient was calculated from an empirical correlation.

Alvarez et al.[33] used a solution-diffusion model combined with the film mass-transfer equation to determine the flux of apple juice through polyamide RO membranes under different operating conditions. Apple juice was considered to behave as a mixture of sucrose, glucose, and malic acid. The physical properties of apple juice were rigorously modelled. Total rejection was assumed. The equations to be solved are:

$$J = L_p(\text{TMP} - \Delta\pi), \tag{5.46}$$

$$C_{mi} = C_{bi}\exp\left(\frac{J}{k_i}\right). \tag{5.47}$$

The membrane permeability, L_p, and the variation of osmotic pressure, $\Delta\pi$, with the concentration of the species in the apple juice were experimentally determined. The mass-transfer coefficient, k, was determined using a correlation similar to Dittus–Boelter for turbulent flow in a pipe. The model was in good agreement with experimental data.

5.3   CASE STUDY

5.3.1   Introduction

This case study will address the detailed modelling of the hydrodynamics of a membrane process enhanced by two-phase gas/liquid flow using computational fluid dynamics. The volume of fluid (VOF) method is used to calculate bubble shape, the velocity and the pressure fields, as well as the wall shear rate. This hydrodynamic information is then combined with a mass-transfer model to predict permeate flux. The osmotic-pressure model is then used to predict the permeate flux in UF of dextran. The modelling of the system hydrodynamics is performed rigorously, with the only major constraint on this section of the model being that the permeate flux is small compared to the bulk crossflow. Therefore, the hydrodynamics section of the model is suitable for use with most membrane process.

In this case study, the particularly complex problem of two-phase flow modelling, often termed *gas sparging* in membrane applications, is addressed. However the modelling principles are equally applicable to other complex module geometries or flow scenarios. A major possible application of the model is prediction of the critical flux—the point at which fouling first occurs in applications featuring with complex geometries or flows.

When used species such as dextran, the model is particularly useful for maximizing the hydro-dynamic benefits of enhancement techniques such as gas sparging or spiral inserts. The steady-state flux in the UF of dextran, when operated under constant TMP, is controlled by concentration polarization. To improve performance in such a system, the mass-transfer coefficient must be increased. Conventionally, this is done by increasing the crossflow velocity in the module, which increases wall shear stress and therefore increases mass-transfer coefficient. However, this approach can be energy intensive because of the high pumping costs incurred. One method of increasing the mass-transfer coefficient without using high crossflow velocities is gas sparging. A gas, usually air, is injected into the feed stream just before the module and forms bubbles. As the bubbles move thorough the module, they cause high wall shear stresses that change rapidly in both magnitude and direction. This results in a higher overall wall shear, excellent mixing, and therefore an improved mass-transfer coefficient. However, the modelling of the wall shear stress in this process is challenging, requiring the use of computational fluid dynamics due to the unsteady nature of the flow. The Leveque mass-transfer equation is then used in conjunction with the osmotic-pressure model to predict the steady-state permeate flux. As an example, the predicted and experimental permeate fluxes are compared for the gas-sparged UF of 238 kDa dextran solutions.

5.3.2   Ultrafiltration and Enhancement

Any technique that thins or disturbs the mass-transfer layer on the transfer surface, and which promotes cross-stream mixing, is likely to combat concentration polarization and enhance the mass-transfer coefficient, thereby enhancing the permeate flux. Belfort[34] described a number of improved hydrodynamic enhancement techniques to ameliorate concentration polarization and to augment the permeate flux, (e.g., inserts in the flow channel, corrugated membranes, rotating blade, Taylor vortices, Dean vortices, and pulsatile flows). The introduction of gas/liquid two-phase flow has been shown to be an effective technique to ameliorate the concentration polarization by thinning or disturbing the mass-transfer boundary layer.[35]

Gas sparging, i.e., injecting air bubbles into the liquid feed to generate a two-phase flow stream, has been investigated for nearly a decade.[36] Pioneering work dates back to the first implementation of such a technique for UF by Cui in 1993. When gas and liquid flow together in a tube, a *slug* flow pattern often exists. Such a flow pattern is characterised by a quasiperiodic passage of long round-nosed bubbles—usually referred to as "Taylor bubbles," or "slugs"—separated by liquid plugs. In particular, for UF, it was found that the gas flow rate required to effect substantial improvements in permeate flux is very small. Furthermore, the liquid crossflow velocity has little effect on the permeate flux in gas-sparged UF. These two aspects of the technique mean there exists the possibility of significant savings on energy costs.[37] Therefore, it not surprising that numerous works have been dedicated to studying the development of the concentration polarization and techniques to lessen such a phenomenon. In their succinct paper, Kleinstreuer and Belford[38] reviewed early works on approximate one- and two-dimensional models. In the 1990s, computational fluid dynamics (CFD) became an attractive tool for researchers to simulate pressure driven membrane processes. The adoption of CFD revealed a detailed picture of the process and proved helpful in optimizing filtration processes.

5.3.3   Model Development

For a full computational solution in UF problems, momentum and concentration equations must be simultaneously solved to give the velocity and concentration fields within the module. However, in all but the simplest geometries,[39,40] this requires prohibitive computing power. Therefore, approximations are made. To avoid solution of the momentum equation, a velocity profile may be assumed and only the concentration equations must be solved. Both finite element[41] and finite difference methods[42] have been used to perform this solution. Alternatively, the momentum equations may be solved to find the wall shear stress for a nonporous module; then, the flux is then predicted using a standard membrane model, align as if the simulation had been carried out for a porous module. This is the approach taken here.

The first part of the model uses the computational fluid dynamics package *FLUENT* (release 5.4.8, 1998) to simulate the motion of a single bubble rising in a flowing liquid. In *FLUENT*, the control volume method—sometimes referred to as the *finite volume method*—is used to discretize the transport equations. The movement of the gas–liquid interface is tracked based on the distribution of α_G, the volume fraction of gas in a computational cell, where $\alpha_G = 0$ in the liquid phase and $\alpha_G = 1$ in the gas phase. Therefore, the gas–liquid interface exists in the cell where α_G lies between 0 and 1. A geometric reconstruction scheme that is based on the piece linear interface calculation (PLIC) method[43] is applied to reconstruct the bubble-free surface. The surface tension is approximated by the continuum-surface-force model of Brackbill, Kothe, and Zemach.[44] Turbulence is introduced by the renormalization-group-based k-epsilon zonal model. The bubble shape, velocity fields, pressure fields, and the wall shear rate around a single bubble in a membrane module with nonporous walls can be calculated.

The governing transport and volume fraction equations to be solved are:

1. *The continuity equation*:

$$\frac{\partial}{\partial t}(\rho) + \nabla(\rho\vec{v}) = 0.$$ (5.48)

2. *The momentum equation*: A single momentum equation is solved throughout the domain, and the resulting velocity field is shared among the liquid and gas phases. The momentum equation, shown below, is dependent upon the volume fractions of all phases through the properties ρ and μ:

$$\frac{\partial}{\partial t}(\rho\vec{v}) + \nabla(\rho\vec{v}\vec{v}) = -\nabla p + \nabla[\mu(\nabla\vec{v} + \nabla\vec{v}^{\mathrm{T}})] + \rho\vec{g} + \vec{F}.$$ (5.49)

3. *The volume fraction equation*: The tracking of the interface between the gas and liquid is accomplished by the solution of a continuity equation for the volume fraction of gas:[45]

$$\frac{\partial}{\partial t}(F_{\mathrm{G}}) + \vec{v}\cdot\nabla F_{\mathrm{G}} = 0.$$ (5.50)

One could solve a similar equation for the volume fraction of liquid, F_{L}, but instead F_{L} is given from Equation 5.48 by the constraint:

$$F_{\mathrm{G}} + F_{\mathrm{L}} = 1.$$ (5.51)

The physical properties of either liquid or gas are used in the transport equations when the computational cell contains a single phase. When there is an interface between the gas and liquid phases, the volume-fraction-weighted average is used to estimate the physical properties.

$$\rho = F_{\mathrm{G}}\rho_{\mathrm{G}} + (1 - F_{\mathrm{G}})\rho_{\mathrm{L}},$$ (5.52)

$$\mu = F_{\mathrm{G}}\mu_{\mathrm{G}} + (1 - F_{\mathrm{G}})\mu_{\mathrm{L}}.$$ (5.53)

To simplify the model geometry, the bubble is considered axisymmetric; thus, a two-dimensional coordinate system about the centreline of the pipe is used. The length of the domain is $11D_{\mathrm{t}}$, where D_{t} is the tube diameter. The grids used to generate the numerical results throughout this chapter are either uniform grids containing quadrilateral control elements/volumes, or uniform grids with extra refinement near the walls.

In Figure 5.1, the boundary conditions and the initial bubble shape used in the simulation are displayed. For a large tube simulation, the initial bubble shape consists of one hemisphere connected to a cylinder of the same radius. If other shapes were used (e.g., only a cylinder), the final shape of the bubble was similar, except that the convergence is slower. The initial guess for the film thickness and the bubble rise velocity are calculated using simple mass balance. No difference in the final results is discerned if a bad initial guess is adopted. The no-slip wall condition is applied to the walls. The fluid mass flux at the inlet is specified using a profile for a fully developed flow through a pipe. The governing equations are solved for a domain surrounding a Taylor bubble in a frame of reference attached to the rising Taylor bubble. With these coordinates, the bubble becomes stationary and the pipe wall moves with a velocity U_{wall}, equal to that of the Taylor bubble rise velocity, U_{TB}. The liquid is fed at the inlet with a velocity U_{inlet}, which is equal to $U_{\mathrm{TB}} - U_{\mathrm{SL}}$. A fully developed velocity profile is imposed at the inlet and the relative movement between the liquid and the wall generates a velocity profile shown in Figure 5.1. The value of U_{TB} is adjusted after the initial guess until the nose of the bubble ceases to move in the axial direction.

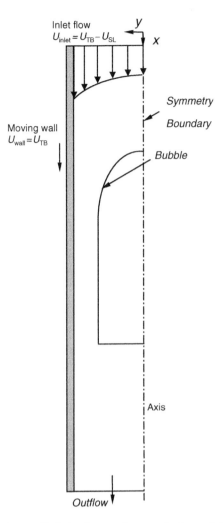

Figure 5.1   Taylor bubble rising in a vertical pipe in a moving coordinate moving with the bubble.

The local shear stress is evaluated from the CFD simulation and its absolute value is then averaged over the length of the membrane module. The average mass-transfer coefficient can then be estimated from the Leveque solution:

$$k = 1.62 \left(\frac{\gamma_w D^2}{L} \right)^{0.33}.$$
(5.54)

It should be pointed out that this equation was developed under steady shear rate[46] and is valid for laminar flow (main restriction $Re < 2000$).

The equations of the osmotic-pressure model (Equation 5.8 and Equation 5.9)

$$J = \frac{TMP - \Delta\pi}{\mu R_m},$$

$$J = k \ln\left(\frac{C_{\mathrm{w}}}{C_{\mathrm{B}}}\right),$$

can then be solved in combination with the appropriate osmotic-pressure function to give the permeate flux and the concentration at the membrane. For 283 kDa dextran:

$$\log \Delta\pi = 0.1872 + 3.343 C_{\mathrm{w}}^{0.3048}. \tag{5.55}$$

The above correlations are based on steady-state UF with negligible fouling. In UF of a macromolecule such as dextran, the value of R_{f} is expected to be negligible in comparison to R_{m} and may therefore be neglected.[47] The nature of the above model does not restrict itself to dilute solutions. In some cases, however, R_{f} is negligibly low if the solutions are dilute, but not so if they are concentrated.

5.3.4 Hydrodynamics of Gas-Sparged UF

In vertical pipes, Taylor bubbles are axisymmetric and have round noses, whereas the tail is generally assumed to be nearly flat (Figure 5.2). The Taylor bubble occupies most of the cross-sectional area of the tube. When the bubble rises through a moving liquid, the liquid that is flowing ahead of the nose of the bubble is picked up and displaced as a liquid film; it begins to flow downwards in the annular space between the tube wall and the bubble surface. Alongside the bubble, the liquid film accelerates until it reaches its terminal velocity under the condition of a long enough bubble. At the rear of the bubble, the liquid film plunges into the liquid plug behind the

Liquid film

Taylor bubble

Wake

Figure 5.2 (See color insert following page 178.) Numerical simulation of a Taylor bubble rising through glycerine in a vertical tube.

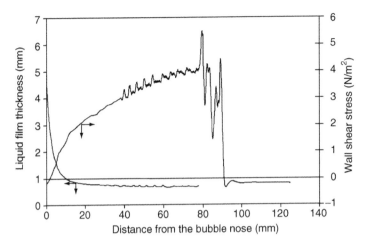

Figure 5.3 Wall shear stress distribution around a slug unit and the liquid film thickness: membrane length$=$ 1.18 m; $D_t=12.7$ mm; dextran-air system; $C_b=10$ g L^{-1} dextran (100 kDa MW); $Q_L=1.0$ L min^{-1} ($U_{SL}=0.131$ m s^{-1}); $V_b=8.3$ mL; TMP$=1.0$ bar.

bubble as a circular wall jet and produces a highly agitated mixing zone in the bubble wake. This highly agitated zone is believed to be responsible for disturbing the mass-transfer boundary layer. This mixing zone sometimes contains a dense cloud of dispersed bubbles that have been sheared off the tail of the Taylor bubble by the annular jet.

Figure 5.3 shows the wall shear stress around a slug unit (Taylor bubble + liquid plug) together with the liquid film thickness. The wall shear stress sign changes twice in a slug unit. The first change occurs near the nose of the Taylor bubble and the second occurs near the top of the liquid plug. The negative shear stress, indicating upflow, exists over the liquid plug ahead of the bubble and persists beyond the nose of the Taylor bubble, before becoming positive as the downflow is established in the liquid film around the bubble. The inverse transition from the downward film to an upward one in the liquid plug is of a burst-like type. The brief fluctuations of the wall shear stress in the film region correspond to the wavy nature of the bubble surface. Near the slug tail, the wall shear stress starts to fluctuate. The above features were observed experimentally.[48]

5.3.5 Effect of Operating Conditions on Permeate Flux

Any model must be able to correctly predict the effects of changing any of the main operating parameters. In general, for membrane processes, the applied TMP and crossflow velocity are important factors. Given the complex hydrodynamics of gas-sparged UF, the usual single important factor of crossflow velocity is replaced by a family of factors, including bubble volume, bubbling frequency, and liquid flow rate.

Sur[49] investigated the effects of TMP. The variation of permeate flux with the TMP is shown in Figure 5.4. Experiments were performed for UF of industrial-grade dextran exhibiting an average molecular weight of 283 kDa using a tubular PVDF membrane having a molecular cutoff of 100 kDa. The length of the tubular membrane was 1.18 m and the diameter was 12.7 mm. The slug frequency was controlled using a solenoid valve and set to 1.0 Hz. It can be seen that at a fixed liquid flow rate the permeate flux increases with TMP. The CFD predicted values clearly capture the same trend. The predicted values underestimate the experimental values as the model does not consider the transient nature of the shear stress around the bubble. The model has also been successfully applied to predict the effects of liquid velocity, bubble volume, and bubbling frequency.[50]

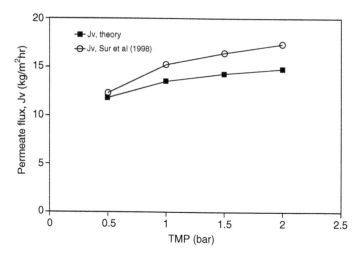

Figure 5.4 Effect of TMP on permeate flux: membrane length$=1.18$ m; $D_t=12.7$ mm; dextran-air system; $C_b=10$ g L^{-1} dextran (283 kDa MW); $Q_L=0.6$ L min^{-1} ($U_{SL}=0.079$ m s^{-1}); $Q_G=0.6$ L min^{-1} ($U_{SG}=0.079$ m s^{-1}); slug frequency$=1.0$ L s^{-1}.

5.3.6 Case-Study Conclusions

The concept of bubbling, or gas sparging, to enhance surface mass-transfer can be very effectively applied to UF membrane processes. The bubble effect is typically achieved at relatively low gas rates. The flux enhancement is most significant when nonsparged operation is dominated by concentration polarization (or, in other words, where high boundary-layer resistance occurs), such as at low liquid flow rates, higher solute/particle concentrations, and higher operating pressures.

Computational fluid dynamics has successfully been used to model gas-sparged UF. To validate the model, experimental data reported in the literature over a wide range of gas and liquid velocities, slug frequencies, transmembrane pressures, and flow directions, are compared with the CFD predictions. Good agreement was obtained between theory and experiment.

Gas sparging is a common enhancement technique for membrane processes in the water industry and is frequently used with membrane bioreactors. In this application, there exists the added benefit that the bioreactor must be aerated anyway to provide oxygen for the microbial species. Therefore, sparging of the membranes can be achieved without incurring significant additional capital or operating costs. The detailed model presented here maybe used as a tool to optimise the sparging parameters or predict critical fluxes in this important application.

5.4 CONCLUDING REMARKS

Pressure-driven membrane processes have achieved penetration into the food and bioprocess industries. The commonly encountered processes are microfiltration for the separation of micron-sized species, ultrafiltration for the separation of macromolecules, and nanofiltration and reverse osmosis for the separation of low-molecular-weight solutes. Accurate modelling of these processes is essential to allow reliable and robust process designed. The aim of a membrane process model is to predict either the steady-state permeate flux or the rejection of one or more species. The appropriate modelling strategy depends upon the process.

Microfiltration and ultrafiltration operate on a size-exclusion principle. Fouling and concentration polarization in these processes can be severe and these phenomena have resulted in the development of many models. Although concentration polarization is not a significant problem in MF applications, the film mass-transfer equation has still been applied. The replacement of the Brownian diffusion coefficient with the shear-induced diffusion coefficient has allowed modelling of MF processes. In UF, the film mass-transfer equation is principally used as part of either the gel model or the osmotic-pressure model.

If fouling is the major factor determining the permeate flux and transmission, then Hermia's unified filtration law or one of its derivatives is usually chosen. Hermia's law incorporates the four major fouling mechanisms in MF and UF: complete pore blocking, standard blocking, intermediate fouling, and cake layer fouling. If a fouling model is used, it is normally necessary to conduct preliminary experiments to determine parameters in the model.

In nanofiltration and reverse osmosis, the intrinsic properties of the membrane play a major role in determining the separation. For these processes, modeling of the membrane is often combined with the film mass-transfer equation to predict rejection, given that the solvent flux is known. The major models in this area are based on solution-diffusion, irreversible thermodynamics, or the extended Nerst–Plank equation.

Concentration polarization models, in particular the film mass-transfer equation, are widely used. For prediction of the permeate flux, a mass-transfer coefficient must be determined. This is usually achieved through the use of empirical correlations derived for heat transfer in nonmembrane systems. A range of empirical correlations for both turbulent and laminar conditions are available. Alternatively, the mass-transfer coefficient may be experimentally measured. For MF and UF processes, this is achieved by performing experiments at a range of bulk concentrations and taking the gradient from flux vs. natural log of concentration data. In NF and RO, observed rejection vs. flux data is often used to estimate the mass-transfer coefficient. Table 5.2 summarizes a number of food and bioprocess applications of mass-transfer modeling.

The models presented in this chapter may generally be considered to be of the first order of complexity. Many models are presented in the literature that use the models presented here and build upon them further. For example, the film mass-transfer equation is only a one-dimensional solution for the concentration-polarization layer; the model consequently ignores any effects of axial diffusion. More complex two-dimensional models which account for axial effects have been reported in literature (ref 51), however the simple one-dimensional film mass-transfer model is still widely used.[51].

A particular problem that arises in the food and bioprocess industries is the complexity of the feed solution. When a fouling model is used, the model parameters determined in preliminary experiments are very specific to one process and one feed stream. Consequently, great caution must be taken if parameters developed for one feed are applied to a different feed, even if the second feed appears similar in nature. For example, a cake-layer model is often used to describe the MF of yeast suspensions where the pores of the membrane are much smaller than the size of the yeast cells. The important parameter is the specific cake resistance of the cake of yeast cells. However, depending upon operating conditions and the exact properties of the yeast suspensions, this may vary by almost two orders of magnitude. Also, the majority of models for membrane processes are developed in the academic field. Thus, the literature contains many examples of typical values for model parameters obtained using small-scale equipment and model feeds. However, far fewer values are published for actual industrial plants that operate on real feed streams.

Membrane processes have been used industrially for half a century and the majority of physical phenomena connected with the processes are understood. This has lead to a wide range of models for all types of membrane processes. With large computational power now available, models will become increasingly complex as the previous restrictions on what equations can readily be solved are removed.

Table 5.2 Summary of Modeling of Pressure-Driven Membrane Processes in the Food and Bioprocess Industries

Author	Feed	Membrane	Model	Operating Conditions	Key Assumptions	Key Findings
Constenia and Lozano	Apple juice	UF Hollow fibre polysulphone MWCO 50 kDa	Gel model	Concentration Crossflow	Total rejection Pseudoplastic nature of apple juice incorporated	Solution similar to Leveque but with a smaller lead co-efficient
De Bruijn and Borquez	Apple juice	UF Tubular ceramic MWCO 50 kDa	Fouling	Constant concentration Crossflow	Total rejection	Fouling due to both internal pore blocking and cake formation
Alvarez et al.	Apple juice	RO Tubular polyamide 99% NaCl rejection	Solution diffusion and concentration polarization	Concentration Constant concentration Crossflow	Total rejection Variation of density, viscosity and diffusivity with temperature and concentration incorporated	Good agreement between model and experimental data
Clarke and Heath[52]	Skimmed milk	UF Flat sheet and spiral-wound polysulphone MWCO 5 kDa	Modified concentration polarization	Constant concentration Crossflow	Total rejection Variation of diffusivity with concentration incorporated	Using data from the flat sheet system, the flux in a spiral-wound module could be well predicted
Van der Horst et al.	UF whey permeate	NF tubular polyamide and cellulose acetate	Extended Nernst–Plank and concentration polarization	Constant concentration Crossflow	Feed can be modelled as a 3 component system	Model well described experimental results
Bhattacharya et al.	Limed sugar cane juice	UF Flat sheet cellulosic MWCO 10, 15, 20 kDa	Osmotic pressure and fouling	Stirred cell	Empirical relationship for polarization layer	Model maybe of use with full scale plant

GLOSSARY

Concentration polarization   The increase in the concentration of a species at the membrane due to rejection of that species by the membrane.

Diffusivity   Mobility of a solute in a solvent due to diffusion.

Direct osmotic concentration (DOC)   Membrane process using chemical potential as a driving force.

Flux   The amount of liquid permeating through the membrane per unit area per unit time.

Fouling   Phenomena in which the membrane interacts with or adsorbs feed components resulting in a temporary or permanent decline in the flux.

Mass-transfer coefficient   A measure of the mobility of a solute due to either natural or forced convection in a system. In the concentration-polarization equation is the ratio of solute diffusivity to polarization layer thickness.

Membrane   Semipermeable barrier.

Microfiltration (MF)   Membranes with pore sizes ranging from 0.1 to 10 µm. Used for size-exclusion-based separation of micron-sized particles.

Molecular-weight cutoff (MWCO)   Defined as the molecular weight of species for which the transmission is below 5%; frequently used to classify ultrafiltration membranes.

Nanofiltration (NF)   Membranes with high rejection of bivalent ions, but low rejection of mono-valent ions. Used for separation of low-molecular-weight solutes.

Osmotic pressure   Pressure arising from the difference in chemical potential across a semipermeable barrier.

Permeate   That which has passed through the membrane.

Reverse osmosis (RO)   Membranes with high rejection of both monovalent and bivalent ions. Used for separation of low-molecular-weight solutes.

Transmembrane pressure (TMP)   The pressure applied across the membrane between the feed and permeate sides, this is the driving force for permeation.

Ultrafiltration (UF)   Membranes with molecular weight cutoff of between 1 and 300 kDa. Used for size-exclusion-based separation of macromolecules.

NOMENCLATURE

a	Constants for osmotic pressure, Equation 5.10 (variable units)
A	Membrane area (m^2)
B	Constant in the solution-diffusion equation (m^4 kg^{-1} s^{-1})
C_B	Bulk concentration of solute (kg m^{-3})
C_g	Gel concentration of solute (kg m^{-3})
C_P	Permeate concentration of solute (kg m^{-3})
C_w	Wall concentration of solute (kg m^{-3})
c_w	Solvent concentration (mol L^{-1})
$\cot \theta$	Constant describing surface morphology, Equation 5.35
d	Diameter of tubular membrane (m)
d_p	Particle diameter (m)
D	Diffusion coefficient (m^2 s^{-1})
D_{ip}	Hindered diffusion coefficient (m^2 s^{-1})
D_s	Solute diffusivity in the membrane (m^2 s^{-1})
D_w	Solvent diffusivity in the membrane (m^2 s^{-1})
F	Farday constant, Equation 5.40 (Cmol^{-1})
F	Constant in Equation 5.42 and Equation 5.43

$\vec{F}$	External body forces (N)
F_{G}	Volume fraction of the gas phase in the computational cell
F_{L}	Volume fraction of the liquid phase in the computational cell
J	Permeate flux ($\mathrm{m^3\,m^{-2}\,s^{-1}}$)
J_0	Permeate flux at $t=0$ ($\mathrm{m^3\,m^{-2}\,s^{-1}}$)
J_{s}	Solute permeate flux ($\mathrm{m^3\,m^{-2}\,s^{-1}}$)
J^*	Critical flux ($\mathrm{m^3\,m^{-2}\,s^{-1}}$)
k	Mass-transfer coefficient ($\mathrm{ms^{-1}}$)
K	Constant in Hermia's unified fouling law (variable)
K_0	Kozney coefficient
K_{ic}	Hinderance factor for convection
K_{s}	Distribution coefficient of solute in membrane
L	Length of the tubular membrane (m)
L_{p}	Water permeability ($\mathrm{m\,s^{-1}\,Pa^{-1}}$)
M	Mass of cake deposited per unit area ($\mathrm{kg\,m^{-3}}$)
n	Constant in Hermia's unified fouling law or constant power law
P_{m}	Overall permeability coefficient ($\mathrm{m\,s^{-1}}$)
R	Gas constant ($\mathrm{Jmol^{-1}\,K^{-1}}$)
R_{o}	Observed rejection
R_{c}	Cake layer resistance ($\mathrm{m^{-1}}$)
$\hat{R}_c$	Specific cake resistance per unit depth ($\mathrm{m^{-2}}$)
R_{m}	Membrane resistance ($\mathrm{m^{-1}}$)
R_{mo}	Membrane resistance at $t=0$ ($\mathrm{m^{-1}}$)
Re	Reynolds number
s	Cake compressibility
S_{a}	Sieving coefficient
Sc	Schmidt number
Sh	Sherwood number
T	Temperature (K)
t	Time (s)
U	Crossflow velocity ($\mathrm{m\,s^{-1}}$)
V	Permeate volume collected at time t ($\mathrm{m^3}$)
V_{w}	Partial molar volume of solvent ($\mathrm{mol\,L^{-1}}$)
$\vec{v}$	Velocity vector ($\mathrm{m\,s^{-1}}$)
x	Axial coordinate (m)
y	Perpendicular coordinate (m)
z_{i}	Valence of ion
TMP	Transmembrane pressure (Pa)
$\Delta\pi$	Osmotic pressure difference (Pa)
Δx	Thickness of the active membrane layer (m)

Greek symbols

α	Specific cake resistance ($\mathrm{m\,kg^{-1}}$)
α_0	Specific cake resistance under zero applied pressure ($\mathrm{m\,kg^{-1}}$)
ε	Porosity
γ_{w}	Wall shear rate ($\mathrm{s^{-1}}$)
δ	Thickness of concentration-polarization layer (m)
μ	Solution viscosity (Pas)
μ_{o}	Consistency coefficient (Pas)
ρ	Solution density ($\mathrm{kg\,m^{-3}}$)
ρ_{c}	Density of species deposited in the cake ($\mathrm{kg\,m^{-3}}$)

τ	Tortuosity, Equation 5.29
σ	Reflection coefficient
φ	Constant $(m^3 g^{-1})$
ϕ_B	Particle volume fraction in the bulk
ϕ_w	Particle volume fraction at the wall
ψ	Electric potential (V)

REFERENCES

1. Mulder, M., *Basic Principles of Membrane Technology*, London: Kluwer Academic, 1996.
2. Cheryan, M., *Ultrafiltration and Microfiltration Handbook*, Lancaster, PA: Technomic Publishing Company, 1998.
3. Field, R. W. and Howell, J. A., *Process Engineering in the Food Industry*, Barking, UK: Elsevier Science Publishers, 1989.
4. Scott, K. and Hughes, R., Introduction to industrial membrane processes, In *Industrial Membrane Separation Technology*, K. Scott and R. Hughes, Eds., Glasgow: Chapman and Hall, 1996.
5. Fell, C. J. D., Reverse osmosis, In *Membrane Separations Technology Principles and Applications*, R.D. Noble and S.A. Stern, Eds., Amsterdam: Elsevier Science, 1995.
6. Field, R. W., Wu, D., Howell, J. A., and Gupta, G., Critical flux concept for microfiltration fouling, *Journal of Membrane Science*, 100, 259–272, 1995.
7. Belfort, G., Davis, R. H., and Zydney, A. L., The behavior of suspensions and macromolecular solutions in crossflow microfiltration, *Journal of Membrane Science*, 28, 1–58, 1994.
8. Green, G. and Belfort, G., Fouling of ultrafiltration membranes: lateral migration and the particle trajectory model, *Desalination*, 35, 129–147, 1980.
9. Zydney, A. L. and Colton, C. K., A concentration polarization model for the filtrate flux in cross-flow microfiltration of particulate suspensions, *Chemical Engineering Communications*, 47, 1–21, 1986.
10. Li, H., Fane, A. G., Coster, H. G. L., and Vigneswaran, S., An assessment of depolarization models of crossflow microfiltration by direct observation through the membrane, *Journal of Membrane Science*, 172, 135–147, 2000.
11. Porter, M. C., Concentration polarization with membrane ultrafiltration, *Industrial and Engineering Product Research and Development*, 11, 234, 1972.
12. Eykamp, W., Microfiltration and ultrafiltration, In *Membrane Separations Technology Principles and Applications*, R.D. Noble and A. Stern, Eds., Amsterdam: Elsevier Science, 1995.
13. Schock, G. and Miquel, A., Mass transfer and pressure loss in spiral wound modules, *Desalination*, 64, 339–352, 1987.
14. Constenla, D. T. and Lozano, J. E., Predicting stationary permeate flux in the ultrafiltration of apple juice, *Lebensmittel-Wissenschaft und-Technologie*, 29, 587–592, 1996.
15. Tracey, E. and Davis, R. H., Protein fouling of track-etched polycarbonate microfiltration membranes, *Journal of Colloid and Interface Science*, 167, 104–116, 1994.
16. Mota, M., Teixeira, J., and Yelshin, A., Influence of cell-shape on the cake resistance in dead-end and crossflow filtration, *Separation and Purification Technology*, 27, 137–144, 2002.
17. Chandler, M. and Zydney, A., A high throughput screening for membrane process development, *Journal of Membrane Science*, 237, 181–188, 2004.
18. Hermia, J., Constant pressure blocking laws. Application to power-law non-newtonian fluids, *Transactions of the Institute of Chemical Engineers*, 60, 183–187, 1982.
19. Palacio, L., Ho, C.-C., Prádanos, P., Hernández, A., and Zydney, A. L., Fouling with protein mixtures in microfiltration: BSA–lysozyme and BSA–pepsin, *Journal of Membrane Science*, 222, 41–51, 2003.
20. Chiang, B. H. and Cheryan, M., Modelling of hollow-fibre ultrafiltration of skimmilk under mass-transfer limiting conditions, *Journal of Food Engineering*, 6, 241–255, 1987.
21. de Bruijn, J. and Bórquez, R., Analysis of the fouling mechanisms during cross-flow ultrafiltration of apple juice LWT. *Food Science and Technology*, 8, 861–871, 2006.
22. Bhattacharya, P. K., Agarwal, S., De, S., and Gopal, R., Ultrafiltration of sugar cane juice for recovery of sugar: analysis of flux and retention, *Separation and Purification Technology*, 21, 247–259, 2001.

23. Drew, D. A., Schonberg, J. A., and Belfort, G., Lateral inertial migration of a small sphere in fast laminar flow through a membrane duct, *Chemical Engineering Science*, 46, 3219–3224, 1991.
24. Altmann, J. and Ripperger, S., Particle deposition and layer formation at the crossflow microfiltration, *Journal of Membrane Science*, 124, 119–128, 1997.
25. Sherwood, J. D., The force on a sphere pulled away from a permeable half space, *Physiochemical Hydrodynamics*, 10, 3–12, 1988.
26. Hilal, N., Al-Zoubi, H., Darwish, N. A., Mohammad, A. W., and Abu Arabi, M., A comprehensive review of nanofiltration membranes: treatment, pre-treatment, modelling and atomic force microscopy, *Desalination*, 170, 281–308, 2004.
27. Bowen, W. R., Mohammad, A. W., and Hilal, N., Characterisation of nanofiltration membranes for predictive purposes—use of salts, uncharged solutes and atomic force microscopy, *Journal of Membrane Science*, 126, 91–105, 1997.
28. Wang, X. L., Tsuru, T., Nakao, S., and Kimura, S., The electrostatic and steric-hindrance model for the transport of charged solutes through nanofiltration membranes, *Journal of Membrane Science*, 135, 19–32, 1997.
29. Bowen, W. R. and Welfoot, J. S., Modelling of membrane nanofiltration—pore size distribution effects, *Chemical Engineering Science*, 57, 1393–1407, 2002.
30. Spiegler, K. S. and Kedem, O., Thermodynamics of hyperfiltration (reverse osmosis): criteria for efficient membranes, *Desalination*, 1, 311–326, 1966.
31. Murthy, Z. V. P. and Gupta, S. K., Estimation of mass transfer coefficient using a combined nonlinear membrane transport and film theory model, *Desalination*, 109, 39–49, 1997.
32. van der Horst, H. C., Timmer, J. M. K., Robbertsen, T., and Leenders, J., Use of nanofiltration for concentration and demineralization in the dairy industry: model for mass transport, *Journal of Membrane Science*, 104, 205–218, 1995.
33. Alvarez, V., Alvarez, S., Riera, F. A., and Alvarez, R., Permeate flux prediction in apple juice concentration by reverse osmosis, *Journal of Membrane Science*, 127, 25–34, 1997.
34. Belfort, G., Membrane modules: comparison of different configurations using fluid mechanics, *Journal of Membrane Science*, 35, 245–270, 1988.
35. Cui, Z. F., Experimental investigation on enhancement of crossflow ultrafiltration with air sparging in effective membrane process—new perspectives, In *Effective Membrane Processes*, R. Patterson, Ed., London: Mechanical Engineering Publication, 1993.
36. Cui, Z. F., Chang, S., and Fane, A. G., The use of gas bubbling to enhance membrane processes, *Journal of Membrane Science*, 221, 1–35, 2003.
37. Cui, Z. F. and Wright, K. I. T., Gas–liquid two-phase cross-flow ultrafiltration of BSA and dextran solutions, *Journal of Membrane Science*, 90, 183–189, 1994.
38. Kleinstreuer, C. and Belford, G., Mathematical modelling of fluid flow and solute distribution in pressure-driven membrane modules, In *Synthetic Membrane Processes—Fundamentals and Water Applications*, G. Belford, Ed., New York: Academic Press, p. 131, 1994.
39. Henriksen, P. and Hassager, O., Simulation of transport phenomena in ultrafiltration, *Chemical Engineering Science*, 48, 2983–2999, 1993.
40. Wiley, D. E. and Fletcher, D. F., Computational fluid dynamics modelling of flow and permeation for pressure-driven membrane processes, *Desalination*, 145, 183–186, 2002.
41. Huang, L. and Morrissey, M. T., Finite element analysis as a tool for crossflow membrane filter simulation, *Journal of Membrane Science*, 155, 19–30, 1999.
42. Lee, Y. and Clark, M. M., A numerical model of steady-state permeate flux during cross-flow ultrafiltration, *Desalination*, 109, 241–251, 1997.
43. Youngs, D. L., Time-dependent multi-material flow with large fluid distortion, In *Numerical Methods for Fluid Dynamics*, K.W. Morton and M.J. Baibnes, Eds., New York: Academic Press, p. 273, 1982.
44. Brackbill, J. U., Kothe, D. B., and Zemach, C., A continuum method for modelling surface tension, *Journal of Computational Physics*, 100, 335–354, 1992.
45. Hirt, C. W. and Nichols, B. D., Volume of fluid (VOF) method for the dynamics of free boundaries, *Journal of Computational Physics*, 39, 201–225, 1981.
46. Wang, Y., Howell, J. A., Field, R. W., and Wu, D., Simulation of cross-flow filtration for baffled tubular channels and pulsatile flow, *Journal of Membrane Science*, 95, 243–258, 1994.

47. Ghosh, R. and Cui, Z. F., Mass transfer in gas-sparged ultrafiltration: upward slug–flow in tubular membranes, *Journal of Membrane Science*, 162, 91–102, 1999.

48. Nakoryacov, V. E., Kashinsky, O. N., Petukhov, A. V., and Gorelik, R. S., In *Study of Local Hydrodynamic Characteristics of Upward Slug Flow*, 7, Oxford: University of Oxford, pp. 560–566, 1989.

49. Sur, H. W. and Cui, Z. F., Enhancement of microfiltration of yeast suspension using gas sparging — the effect of feed conditions. *Separation and Purification Technology*, 15, 313–319, 2005.

50. Taha, T. and Cui, Z. F., CFD modelling of gas-sparged ultrafiltration in tubular membranes, *Journal of Membrane Science*, 210, 13–27, 2002.

51. Paris, J., Guichardon, P., and Charbit, F., Transport phenomena in ultrafiltration: a new two-dimensional model compared with classical models, *Journal of Membrane Science*, 207, 43–58, 2002.

52. Clarke, T. E. and Heath, C. A., Ultrafiltration of skim milk in flat-plate and spiral-wound modules, *Journal of Food Engineering*, 33, 373–383, 1997.

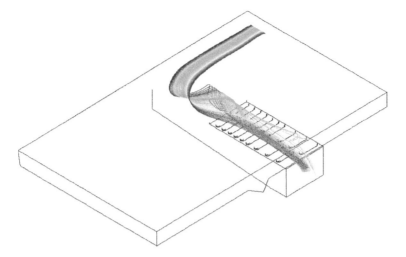

Figure 3.16 Unwanted vortex generated by improper geometry over fryer pan return.

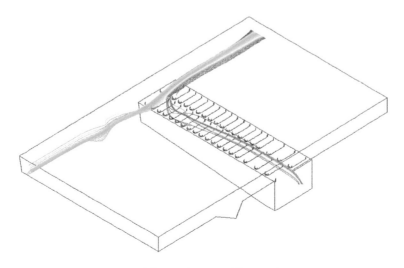

Figure 3.17 Vortex removed using FiDAP™ to analyze new 3D geometry.

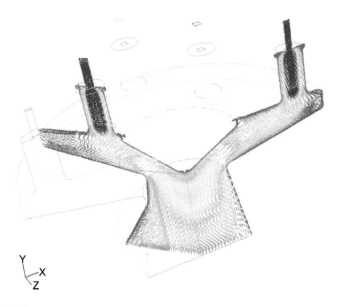

Figure 3.20 Plot of velocity vectors in the die.

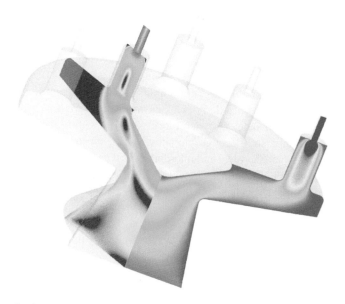

Figure 3.21 Plots of velocity magnitude (right side) and non-Newtonian viscosity (left side).

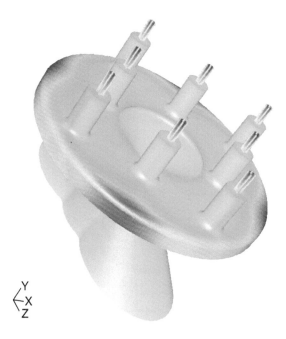

Y
X
Z

Figure 3.22 Temperature profiles on the walls of die.

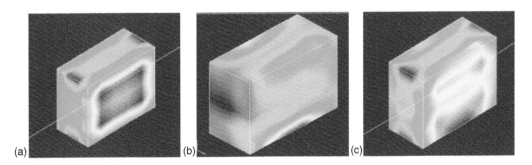

(a)   (b)   (c)

Figure 4.33 Computed temperature contours showing the food surface for (a) infrared only, (b) microwave only, and (c) combined microwave and infrared heating.

Liquid film ◄

Taylor bubble ◄

Wake ◄

Figure 5.2 Numerical simulation of a Taylor bubble rising through glycerine in a vertical tube.

CHAPTER 6

Simultaneous Heat and Mass Transfer

Xiao Dong Chen

CONTENTS

6.1 INTRODUCTION

Transporting energy and mass across interfaces is of crucial importance in food processing. In many ways, food processing is about preserving, creating, or manipulating food structures. Food microstructures may be created through crystallisation, emulsification, gelatinisation, foaming, and mixing. Additionally, sheeting, extrusion, and moulding are typical ways of making micro-structures as well as macrostructures. These microstructures can affect how water is transported within the structures. As a result, drying also may create a food structure. The nature of the molecules, e.g., polymer or not, or sheared or not, also impacts on the physical changes such as local shrinkage during drying, which in turn impacts on drying rate. Drying and product interactions are very complicated to model mathematically.

Food processing involves transfer of species and chemical reactions, which is therefore temperature dependent. Sometimes heat is required to be distributed evenly throughout the products to make their quality uniform. Other times, heat may need to be more intense inside a product (for a thorough cook of the material encapsulated) or to be more focused at the exterior surface (for a crust to form for instance). Heating has a profound influence on drying rate.

Species exchanges at pore or cell levels occur, thus controlling mass transfer process. The mass transport properties are strongly temperature dependent, which are affected more acutely at the mol-ecular level. Across the structure or composition boundaries, mass transfer rate is altered either due to the microscale alignments of the structure component or the macroscale fluid flow conditions. Fluid flow definitely affects heat transfer to the extent termed as the *convective heat transfer.*

Manipulating heat transfer operation is often easier than manipulating mass transfer operations. When evaporation or condensation occurs, mass transfer, to a large degree, is coupled with heat transfer. Simple analogies based on the mathematical models of heat transfer are sometimes not accurate for analyzing evaporative mass transfer such as air-drying.

In the chapter by Datta,[1] the physics-based mathematical models of heat transfer have been described. In that chapter, the detailed partial differential Default (PDEs) governing the heat and mass transport processes also have been provided. These fundamental equations can be resolved to cover the wide range of heat transfer problems, e.g., heat conduction (wherever there is a tempera-ture gradient), heat convection (cooling with air or water), and heat radiation (which do not involve phase change). With the exception of freezing (which does involve phase changes), the following processes have not been included: condensation (e.g., retorting using steam condensation), boiling (e.g., frying and other boiling style cooking), solidification including crystallisation (e.g., freezing ice bars and moulding chocolate), melting (e.g., thawing of meat), evaporation (e.g., falling film evaporation of milk), and steaming (e.g., softening wheat grains or making the steam buns).

The more volumetric ways of imposing heat into food materials are ohmic (or Joule) heating, radio-frequency (RF) heating, and microwave heating. These processes can "penetrate" the products better, and depending on the energy sorption properties of the food constituents, hot spots can arise, which may not be advantageous. In the same chapter by Datta,[1] the relevant equations for the heat transfer aspects of the above problems have been detailed. These processes involve molecular diffusion and convective mass transfer as well. The processes that have not been detailed are: infrared heating, membrane separation, absorption and ion exchange, etc.

In food processing, simultaneous heat and mass transfer operations and, sometimes, chemical reactions are involved. In the current chapter, the mass transfer, simultaneous heat and mass transfer problems in particular, have been addressed. Here the processes are described in an

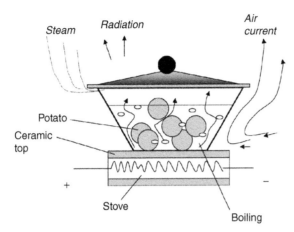

Figure 6.1 An illustration of heat transfer and mass transfer processes when cooking in a domestic pot.

intuitive way to provide the "feel" of these processes. This kind of discussion is important to allow the often semi-empirical models to be introduced with greater confidence.

The conditions of mass transfer limiting, heat transfer limiting or coupled heat, and mass transfer processes are discussed using air-drying as an example. The limiting conditions are very useful for simplifying the model equations so that the full partial differential equations, such as those given in the above-mentioned chapter by Datta[1] may not need to be solved. Furthermore, the intended simplifications may draw greater attention to the key fundamentals of the underlined physics, while the mathematics involved are not demanding.

The details of air-drying of food materials are complex to model mathematically. In the current chapter, modeling air-drying is used to show the fundamental principles of coupled heat and mass transfer phenomena. With some innovative yet careful considerations, similar arguments could be applied to baking and frying operations. Later in this chapter, a series of worked examples from the literature is presented with comments.

To begin the intuitive description, it is shown what a person may do in everyday life, i.e., cooking by boiling a food mixture in a pot in Figure 6.1. This example shows some of the most important modes of heat or mass transfer such as: conduction of heat through the stove top into the pot, convective heat loss from the pot and the sides of the stove into the air (natural convection), boiling heat transfer inside the pot, convective heat transfer into the potato, vapor transfer into the air, etc.

6.2 MODELING AIR-DRYING

6.2.1 Background

Drying of foods is a cost-effective and traditional way to preserve food quality. It is known that intermediate to low moisture contents are beneficial for quality keeping (minimized bacterial growth, minimized lipid oxidation, and less stickiness). Ancient practice has demonstrated that lowering the water content of natural food materials such as fish, meat, etc., can make the shelf life of the biological material longer. Drying has enhanced immensely the human existence and the human capability to survive. Today drying is used extensively on a large scale in order to produce more desirable foods that then can be distributed and enjoyed by people globally.

The drying of liquid foods forms powdered products, which can be used extensively as food ingredients for making new variety of foods. Many economies such as New Zealand and Australia rely on such liquid-to-solid conversions.

The principles behind drying have been investigated for many years. Due to the complications of such processes not being directly visualized, progress developing an effective yet sufficient basic understanding is still slow. When water content is so high to the extent that the solid material surface is saturated with liquid water, evaporation from the surface appears to be the same as evaporation from a free liquid water surface. This stage, after an initial cooling or a warming period is often called the "constant drying rate" stage, where the evaporation (drying) rate is simply expressed as follows:

$$\dot{m}_v = h_m \cdot (\rho_{v,s} - \rho_{v,\infty}),\tag{6.1}$$

where $\dot{m}_v$ is the mass flux of water vapor away from the liquid–gas interface (kg m^{-2} s^{-1}) and h_m is the mass transfer coefficient (m s^{-1}) which can be estimated using a correlation given in Appendix 1. $\rho_{v,s}$ is the saturated vapor concentration at the interface ($\rho_{v,sat}$) at the surface temperature (T_s), and $\rho_{v,\infty}$ is the vapor concentration in the bulk flow (kg m^{-3}).

The complexity comes when the above equation overpredicts the rate of water removal as the solid structure effect starts to take effect.

There have been a number of continuum type mechanisms proposed and the associated mathematical models established. These include the liquid diffusion concept proposed by Lewis[2] in 1921, the capillary flow concept by Buckingham[3] in 1907, the evaporation–condensation concept by Henry[4] in 1939, dual (temperature, water content gradient) and triple (temperature, water content, and pressure gradient) driving force mechanisms by Luikov[5] in 1986, another dual driving force mechanism by Philip and De Vries[6] in 1957, and De Vries[7] in 1958, dual phase (liquid and vapor) transfer mechanism proposed by Krischer as summarized by Fortes and Okos[8] in 1980. Whitaker, in 1977[9] and 1999,[10] respectively, proposed the detailed transport equations to account for the macro- and microscale structures in biological materials. Three phase (solid, vapor, and liquid) conservations and their local volume-averaged behaviors are considered. The mechanisms for moisture transfer are largely the same as those proposed by Luikov[11] in 1975 as well as Philip and De Vries,[6] except that the small scale phenomena (local pores, pore channels, shells, voids, etc.,) have been taken into account. This theory is based on a known distribution of the macroscale and microscale unit structures which allow local volume-averaging to be carried out.

The more modern approach is the pore-network model, which can include a number of scale levels but, however, is still in the development phase and has not been rigorously validated experimentally.[12,13] In addition, the detailed pore structure and network (geometries and distribution) is actually very difficult to establish quantitatively for a food material, which may hinder the applicability of the model approach to food drying. Besides the liquid diffusion mechanism considered as the single driving force for drying, the above approaches all involve complicated mathematics (partial differential equation sets and numerical schemes for gaining a stable solution) and some model coefficients which cannot be individually determined. Because the numerous coefficients need to be considered, modeling the single trend of the overall water loss during drying accurately is generally possible with these models. However, the practical significance of the detailed physics associated with these models diminishes if only such simple trends are of the practical interest.

Nonetheless, the above-mentioned models are helpful when spatial distribution of both the water content and temperature is required. As more information is required about the product surface and core during drying so that the differential changes of color, bioactivity and stress, etc., between the surface and the core can be determined, these models are useful. Conversely, as a compromise, liquid water diffusion is the model that has been used most extensively because it has had a mixture of successes in both the overall trend modeling and local profile predictions. It is the simplest approach among the models mentioned earlier.

Based on the liquid diffusion theory, it has been possible to devise laboratory protocols to measure the liquid diffusivity (for instance, the regular regime method by Coumans[14] in 1987 and Yamamoto[15] in 2001). The variations among the reported measured "effective diffusion coefficients" are large, which are due perhaps to the diverse microstructures (pore sizes, pore channels, shrinkage considered or not, etc.,) presented by different natural or processed materials as well as their sorption characteristics affected by chemical composition. This effective diffusivity is therefore often considered to be a fitting parameter in order for the model prediction to match the experimental weight loss data.

Air-drying has been modeled as an effective liquid diffusion process using Crank's basic solution of diffusion,[16] and an effective diffusion coefficient is obtained most frequently through fitting a model solution to the weight loss data. Often the effective diffusivity is correlated against the drying air temperature, making the diffusivity only applicable to constant drying temperature, sample size, and geometry. This is not fundamental enough, and a better result may be found. Many forms of effective liquid water diffusivity have been generated that vary in trends considerably at times, adding to the confusion of the underlined drying mechanisms. Even with the modern techniques for establishing the (spatial) liquid water content profiles at different timings of drying, one is tempted to fit the data only to generate the water content dependent liquid water (effective) diffusivity. Apparently the more fundamental this approach may have appeared, the isothermal condition was assumed. This is, again, an arguable proposition.

A closer look at the Crank's solution of the system with surface evaporation reveals that the boundary condition imposed by Crank in 1975[16] is not the same as Equation 6.1. The following equation was used instead:

$$-D_{\text{eff},1} \cdot \frac{\partial C_1}{\partial x}\bigg|_s = \beta \cdot (C_{1,s} - C_{1,\infty}), \tag{6.2}$$

where β is the corresponding mass-transfer coefficient (m s^{-1}), $C_{1,s}$ is the concentration of liquid water at the interface, and $C_{1,\infty}$ corresponds to the liquid concentration in the material after infinitely long time (i.e., the water content in equilibrium with the drying air humidity) (kg m^{-3}). $D_{\text{eff},1}$ is a constant liquid water diffusivity in the porous material. This is one way to represent the evaporation driving force.

However, during the air-drying process, liquid water does not mean to be able to "jump" out of the material being dried without converting to vapor first. Equation 6.2 may be viewed as nonphysics-based model. The only case when the Crank's boundary condition yields the same solution as Equation 6.1 is when the convection is very strong, making the mass transfer coefficients very large (i.e., the infinite Biot number for mass transfer (Bi_m)). When $Bi_m \to \infty$, the boundary liquid water content may be assumed to be constant throughout the drying process. The following partial difference Equation (for the slab geometry as an example) governing the mass transfer process and the long-time scale solution of this PDE, are given below, respectively, which are often used to yield the effective liquid water diffusivity:[16,17]

$$\frac{\partial C_1}{\partial t} = D_{\text{eff},1} \frac{\partial}{\partial x}\left(\frac{\partial C_1}{\partial x}\right), \tag{6.3}$$

and its solution for negligible external mass-transfer resistance:

$$\frac{\bar{X} - X_\infty}{X_0 - X_\infty} = \frac{8}{\pi^2} \sum_{n=0}^{\infty} \frac{1}{(2n+1)^2} \exp\left(-(2n+1)^2 \frac{\pi^2}{4L^2} D_{\text{eff},1} \cdot t\right), \tag{6.4a}$$

where $\bar{X}$ is the remaining average water content on dry basis (kg water kg solids^{-1}), X_0 is the initial water content (kg kg^{-1}), X_∞ is the final water content in equilibrium with the drying air conditions. The liquid concentration (C_1) is related to the dry basis water content (X) by $C_1 = X \rho_s \rho_s$ is the solid concentration $(kg\ m^{-1})$. Note that here, a constant (effective diffusivity) is assumed. L is the total thickness of the material being dried $(=2b)$ (in meters), b is the half-thickness (in meters), and the material is dried symmetrically. This model also assumes zero volume change, negligible external mass-transfer resistance and an isothermal process. The uniform initial water content is also necessary.

For a long time after drying starts, only the first term of Equation 6.4a is sufficient for a good approximation to the full solution (i.e., Equation 6.4a):

$$\ln\left(\frac{\bar{X}-X_\infty}{X_0-X_\infty}\right) \approx \ln\left(\frac{8}{\pi^2}\right) - \left(\frac{\pi^2}{L^2}D_{eff,1}\right)t. \tag{6.4b}$$

This approach, however, is likely to induce significant error and correlates the weight loss data well only towards the end of the drying period as discussed by Sikiatden and Roberts.[18]

The effective diffusivities vary greatly between materials and between individual studies, to the extent of the order of 10,000 according to Sablani et al.[19] Applying the values reported in literature is almost certainly involving some trial and error. In other words, the effective diffusivities may be viewed as empirical fitting parameters. Because the mathematical modeling of drying these days involves heat transfer as well so by nature the model is different from the simple Equation 6.3, where isothermal drying is considered. Thus, the nature of strong temperature dependence of the effective diffusivity cannot be ignored. More recently, efforts have been put into establishing the isothermal drying condition in an elaborate way[18,20,21] to conform the lab conditions to the required assumptions by the Crank model. Other methods, such as the regular regime method, have the same issue of assuming the same constant temperature inside the material and in the bulk.[14,15] The diffusivities obtained in the methods other than isothermal method would need to be derived when the drying rate becomes low in the late stage of drying where the material's temperature gets close to that of the bulk drying air.

For simplicity, air-drying also has been modeled by assuming that the process is heat transfer limited (i.e., no mass transfer resistance). Opposite to this, though more widely accepted in literature, is that air-drying of porous material can be treated as being mass transfer limited (uniform temperature condition). The evaporation is frequently assumed to occur at a sharp moving liquid–vapor interface, which then recedes into the porous solid structure as drying proceeds.

On observing the drying related literature, it may be said that a good model should predict accurately both the moisture loss and temperature-time history. Because the water loss versus time is a simple trend to model in most cases, many simple explicit time functions (power, exponential functions, etc.,) can be used for drying under constant conditions. Where possible, it is thus desirable to be able to predict well the spatial distribution of temperature. Often, a model that fits well with the weight loss versus time does not generate accurate temperature-time predictions. In most occasions, only the weight loss is modeled. As mentioned earlier, isothermal drying models are reported often yet the justification of these is sketchy. Furthermore, surface properties like stickiness and color, etc., have become more important in practical operations, and therefore the spatial distributions of temperatures and water contents (and local mechanical stress, pore size, etc.,) will be very useful. The interactions between the microstructure, composition and drying process will be the key to understanding of all these distributions. In the following discussion, diffusion of species is considered to be the only mechanism for water transport during air-drying.

6.2.2 One-Dimensional Modeling of Air Drying

6.2.2.1 Basic Formula

The visualization drying mechanisms in two- or three-dimensional drying problems is not straightforward. To visualize the underlined physics more easily, only one-dimensional conditions

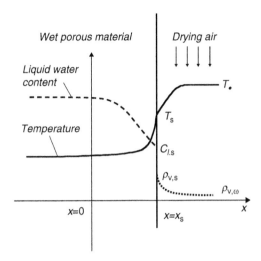

Figure 6.2 A simple schematic diagram showing the system considered in this study (the shaded background represents the complex nature of the microstructures inside the solid materials being dried).

are considered in detail in this section. The perfect flat and smooth surface is at the start a simplification, especially for drying natural food materials. For liquid foods such as milk, a "perfectly smooth surface" is, in most situations, a reasonable assumption.

Using liquid diffusion alone is not enough to be able to explain the process of air drying. Sooner or later, a vapor concentration profile has to be established within the porous material being dried. As an improvement from the pure liquid diffusion model with an effective diffusivity (or diffusivity function against temperature and water content), the mechanisms of drying are the liquid water diffusion and water vapor diffusion. The models that take into account the spatial distributions of temperature and water content (liquid water and water vapor) are summarized as the following two types (here, one-dimensional conditions are represented for easier understanding; see Figure 6.2).

Without Source Term.

$$\frac{\partial C_1}{\partial t} = \frac{\partial}{\partial x}\left(D_1 \cdot \frac{\partial C_1}{\partial x}\right) \tag{6.5}$$

$$\frac{\partial C_v}{\partial t} = \frac{\partial}{\partial x}\left(D_v \cdot \frac{\partial C_v}{\partial x}\right) \tag{6.6}$$

$$\frac{\partial T}{\partial t} = \frac{1}{\rho C_p}\frac{\partial}{\partial x}\left(k \cdot \frac{\partial T}{\partial x}\right) \tag{6.7}$$

In the above model, there is no obvious interaction between the mechanisms described in Equation 6.5 and Equation 6.6, except that the vapor diffusivity should be treated as a function of porosity, which is a function of the liquid (water) content of the porous material. The thermal conductivity k should also be a function of water content, which changes as drying proceeds. At the boundary of porous structure and air flow, i.e., the convective boundary condition at $x=x_s$ (see Figure 6.2), for vapor transfer, diffusive transport of vapor is balanced by the convective transport of vapor into the air stream at the boundary, respectively.

If the drying is symmetrical, $x=0$ would be chosen as the symmetry and $x=x_s$ ($x=+b$ or $x=-b$). At $x=0$, then, the boundary condition is the "impermeable and adiabatic" one:

$$\left.\frac{\partial C_1}{\partial x}\right|_s = 0 \tag{6.8}$$

$$\left.\frac{\partial C_v}{\partial x}\right|_s = 0 \tag{6.9}$$

$$\left.\frac{\partial T}{\partial x}\right|_s = 0. \tag{6.10}$$

The vapor concentration at the boundary can be determined by balancing the diffusive water-vapor transfer (in the porous structure side at the boundary) with the convective-vapor transfer. It is, therefore, not necessary to assume the equilibrium relationship between the water vapor at the boundary and the liquid water content at the boundary. In summary, the boundary conditions at $x=x_s$ may be written as

$$-D_1 \cdot \left.\frac{\partial C_1}{\partial x}\right|_s = 0 \tag{6.11}$$

$$-D_v \cdot \left.\frac{\partial C_v}{\partial x}\right|_s = h_m \cdot (\rho_{v,s} - \rho_{v,\infty}) \tag{6.12}$$

$$k \cdot \left.\frac{\partial T}{\partial x}\right|_s = h \cdot (T_\infty - T_s) - (\Delta H_L + C_{pv} \cdot (T_\infty - T_s)) \cdot h_m \cdot (\rho_{v,s} - \rho_{v,\infty}) \tag{6.13a}$$

or

$$k \cdot \left.\frac{\partial T}{\partial x}\right|_s \approx h \cdot (T_\infty - T_s) - \Delta H_L \cdot h_m \cdot (\rho_{v,s} - \rho_{v,\infty}). \tag{6.13b}$$

Equation 6.13 is based on the assumption that evaporation only occurs at the boundary as far as the energy transport is concerned. Equation 6.12 can be expressed in pressure difference by taking the ideal gas law into account. Equation 6.13b is different from Equation 6.13a as the enthalpy term $C_{pv}(T_\infty - T_s)$ is usually significantly smaller than the latent heat of evaporation, especially when the temperature difference is small.

The relationship between the vapor concentration at the interface but on the side of the solid ($x=x_s-$) (i.e., $C_{v,s}$) and the vapor concentration at the interface but in the gas ($x=x_s+$) (i.e., $\rho_{v,s}$) is approximately $C_{v,s} \approx \varepsilon_s \rho_{v,s}$, where ε_s is the porosity at the surface of the porous material, which may be approximated to be the same as that of the bulk material).

The heat and mass transfer coefficients (h and h_m, respectively) are usually considered in food engineering literature as "standard" parameters that can be determined using a classical Nusselt number ($Nu=h\delta/k$) or ($Nu=hl/k$) correlation (see Appendix A). The Sherwood number ($Sh=h_m\delta/D$) or ($Sh=h_mL/D$) correlation is simply a substitution of the Nusselt number correlation using the Schmidt number ($Sc=v/D$). When evaporation flux is high at the boundary, the boundary layer of air is pushed towards the bulk flow (the boundary layer becomes thicker). This is the *blowing effect*. The mass transfer coefficient is said to be in the regime called high mass flux boundary described in

Bird et al.[22] At the boundary, the vapor moves perpendicular to the airflow direction, expanding the boundary layer thus lowering the mass transfer coefficient compared with the one predicted using the conventional heat and mass transfer analogy. The Sherwood number would need to be corrected (reduced in value usually), whilst the extent of reduction in Nusselt number is not as significant and sometimes the reduction ignored (for examples, those shown by Lin and Chen,[23] Kar and Chen,[24] and Chen.[25]

In the literature, frequently only one mass transfer equation is adopted, i.e., Equation 6.5 for liquid transfer. In this case, the liquid water transfer is treated as the effective water transport. The effective liquid diffusivity is expected to "encapsulate" the effects of both liquid and vapor transport. After such a simplification, the system described by Equation 6.5 through Equation 6.7 is reduced to:

$$\frac{\partial C_1}{\partial t} = \frac{\partial}{\partial x}\left(D_{\text{eff,l}} \cdot \frac{\partial C_1}{\partial x}\right) \tag{6.14}$$

$$\frac{\partial T}{\partial t} = \frac{1}{\rho C_p}\frac{\partial}{\partial x}\cdot\left(k\cdot\frac{\partial T}{\partial x}\right). \tag{6.15}$$

The thermal conductivity may also be considered as the effective value, i.e., $k = k_{\text{eff}}$, which would be a function of water content, various components such as protein, carbohydrate, fat and minerals, and temperature. The densities and specific capacities can be estimated for the first approximation using the formulas given in the chapter by Datta.[1] Under this simplified framework, it is often assumed that the vapor concentration at the boundary is the one that is in equilibrium with the liquid water content at the same interface. The boundary condition for mass transfer can then be expressed as:

$$-D_{\text{eff,l}}\frac{\partial C_1}{\partial x}\bigg|_{s} = h_{\text{m}}\cdot(\rho_{\text{v,s}} - \rho_{\text{v,}\infty}), \tag{6.16}$$

and $C_{\text{v,s}}$ is a function of boundary temperature and boundary liquid water content, through the equilibrium sorption isotherms.

$$RH_{s} = \frac{\rho_{\text{v,s}}}{\rho_{\text{v,sat}}(T_{s})} = f(T_{s}, X_{s}), \tag{6.17}$$

where $\rho_{\text{v,sat}}(T_{s})$ is the saturated vapor concentration at interface temperature T_{s} as mentioned earlier. Most often, Equation 6.17 can be explicitly written as a GAB model equation.[26] If the surface (direct) liquid-to-vapor conversion still occurs and is significant, Equation 6.17 would not be correct as the surface liquid water content in the bulk sense is not in equilibrium with the vapor phase. When the drying proceeds into the material, Equation 6.17 may be more valid for a location some distance away from the interface.

Equation 6.17 has not been justified in a rigorous scientific manner, and it is rather a necessity for modeling. There is no strict reason why Equation 6.16 can be applied to the boundary instead of, for instance, some location within the porous material.

Furthermore, many previous studies have measured the effective liquid diffusivity under constant drying air (and said well mixed) conditions and taking no consideration of the temperature of the actual sample being dried.

More advanced understanding leads to the incorporation of the local evaporation term rather than leaving the evaporation seemingly all occur at the boundary.[27] In this case, the following system is considered:

With Source Term.

$$\frac{\partial C_1}{\partial t} = \frac{\partial}{\partial x}\left(D_{\text{eff},1} \cdot \frac{\partial C_1}{\partial x}\right) - \dot{E}_{\text{v}} \tag{6.18}$$

$$\frac{\partial C_{\text{v}}}{\partial t} = \frac{\partial}{\partial x}\left(D_{\text{eff},\text{v}} \cdot \frac{\partial C_{\text{v}}}{\partial x}\right) + \dot{E}_{\text{v}} \tag{6.19}$$

$$\frac{\partial T}{\partial t} = \frac{1}{\rho C_{\text{p}}} \cdot \frac{\partial}{\partial x}\left(k_{\text{eff}} \cdot \frac{\partial T}{\partial x}\right) - \frac{\Delta H_{\text{L}}}{\rho C_{\text{p}}} \cdot \dot{E}_{\text{v}}. \tag{6.20}$$

Note from now on, the diffusivities are all called the *effective diffusivities*. The source term $\dot{E}_{\text{v}}$ describes the local moisture evaporation/condensation. This term is very important in the physics of drying as it signifies explicitly the moisture exchange between the local solid structures or porous entities that hold liquid water, and the local voids (e.g., intercellular spaces in plant materials such as apple, carrot, etc.,) or the pore-channels into which the moisture can be evaporated. When evaporation occurs, $\dot{E}_{\text{v}}$ (kg m^{-1} s^{-1}) is positive. If there is any heat generation due to chemical reactions or volumetric heating such as ohmic or microwave heating, the above equations will be added with additional source term(s) as described in Datta.[1] The exact format of this source term is not known. If the porous system is the packed moist porous solid particles, a local evaporation (and condensation) term can be approximated around each particle (for similar pressure and uniform particle size).[28–32]

$$-\rho_{\text{sp}} \cdot \frac{dX}{dt} = h_{\text{eff},\text{m}} \cdot n_{\text{p}} \cdot \frac{A_{\text{p}}}{V_{\text{p}}} \cdot (\rho_{\text{v,s}} - \rho_{\text{v,voids}}), \tag{6.21}$$

where ρ_{sp} is the concentration of the solid mass of the particle (kg m^{-3}) and X is the water content of the particle phase on dry basis (kg kg^{-1}), n_{p} is the number concentration of the particles (L m^{-3}), A_{p} is the surface area of each particle (m^2) and V_{p} is the particle volume (m^3). $h_{\text{eff},\text{m}}$ is the effective mass transfer coefficient (m s^{-1}). The surface vapor concentration can be expressed as a function of the mean temperature, mean water contents of the small packed particles, which can be established experimentally. It is also possible to further employ the diffusion PDEs such as the ones described earlier but for spherical coordinates to describe the moisture transfer within each particle. This, however, makes the modeling rather complicated. It is noted that the use of Equation 6.21 does not mean that the internal moisture gradient has been ignored. Equation 6.18 through Equation 6.20 was solved with a slightly different equation for local rate of drying by Zhang and Datta in 2004.[27]

6.2.2.2 One-Dimensional Water Removal from a Temperature-Controlled Column: A Hypothetical Experimentation to Evaluate the Mechanisms Described in Section 6.2.2.1

In the above section, one can see that the equations are not "water-tight" fundamental equations, which can be free from empirical assumptions. Understanding the key physics involved in the simultaneous heat and mass transfer processes in foods is highly challenging. Here several intuitive analyses (Section 6.2.2.2, Section 6.2.2.3, and Section 6.2.2.4) are given. This kind of approach would also be useful in modeling other simultaneous heat and mass transfer processes.

Constructing an ideal isothermal drying device without volumetric heating such as microwave or radio frequency heating, may not be possible unless the mass transfer resistance is not high inside

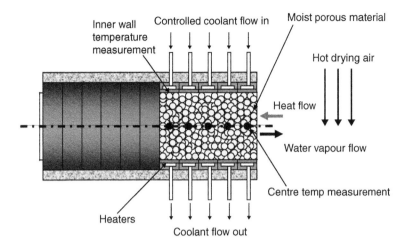

Figure 6.3   A schematic of the one-dimensional calorimeter for the "thinking" experiments.

the porous material being dried. Despite this difficulty, as mentioned earlier, many have considered isothermal drying in their model analysis even though their experiments were not isothermal.

It is more likely for one to establish a reasonable one-dimensional drying device. This concept can be seen in Figure 6.3, which may be called an *experimental finite-difference approach* (EFDA). This idea is not new, and it was practiced in establishing a one-dimensional spontaneous heating column for coal.[33] In a device such as this, for simplicity, a nonshrinkable porous particle bed is placed in the column (for example, having a diameter of 10 cm). Therefore, there is no bed movement during the drying process. Moisture is extracted from the column from the exit plane on the right hand side.

Supposing that this column is made of thin walled steel, wrapped with the finite-elements of discretely arranged (metallic) thin-walled heater bands (heater films), controlled separately to maintain the temperature at the center line in the corresponding finite-element section, cooling is then possible. Cooling is also possible using cooling coils wrapped around the thin heater films again attempting to maintain the temperatures in every section as mentioned earlier. The heat input or the cooling power of each acted on each section is controllable and measurable. These values along the column would be a good indicator of the water removal mechanisms in a way that has not been measured yet.

Assuming that these heating or cooling elements are sufficiently narrow to give reasonable resolution of the heating/cooling power (W m^{-2}) profiles, the heating/cooling power profile along the column could be plotted for a specific operation mode. When the drying apparatus is operated in order that the temperature (assumed to be of little variation along the radial direction) along the column is the same as the drying air, and the drying air has almost zero humidity to provide the maximum driving force, the expected spatial profiles of heating/cooling after a period of drying are illustrated in Figure 6.4.

Three different cases may be considered: (1) only boundary has evaporation as that governed by Equation 6.5 through Equation 6.7 with the boundary conditions in Equation 6.11 through Equation 6.13, or those that do not have source term in both the heat and mass transfer PDEs; (2) evaporation only occurs at a sharp moving front (which has been instigated in many studies whether it is assumed to be mass or heat transfer limiting situations)—the sharp peak would be present after some time because drying proceeds from the boundary (x_s); and (3) evaporation occurs in a region starting from the boundary (which would correspond to the conditions where the source terms are considered in both heat- and mass-transfer PDEs).

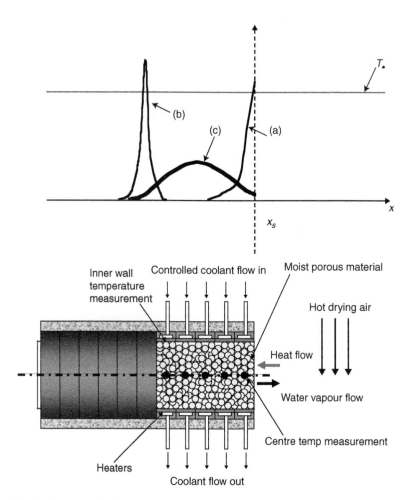

Figure 6.4 The heating power distribution for three different model approaches to maintain an isothermal drying condition: (a) no source term in energy and mass conservation equations; (b) the existence of a sharp evaporation front; (c) with source term in energy and mass conservation equations where the source term is a spatially distributed function.

In all these cases shown in Figure 6.4, the system (controlled by the devices) attempts to preserve the uniform temperature status. Evaporation would induce temperature reduction in the region of evaporation, thus the heater(s) corresponding to that section (and the adjacent sections) would have to be switched on to compensate.

As such, different assumptions and corresponding partial differential equations can lead to different heating/cooling profiles, indicating that models 1, 2, and 3 are conceptually (or physically) different, even though these models can be made to fit the experimental water-loss data. When the sample is sufficient small, the spatial distribution of temperature within the material can be negligible.

6.2.2.3 Microstructural Interpretation of Drying Profiles That Support the Model Analysis in Section 6.2.2.1

Assuming the initial temperature of the material is only slightly lower than the wet-bulb temperature of the drying medium for simplicity an ideal capillary system is shown in Figure 6.5, where the heating and water vapor transfer are opposite to each other. The capillaries here have identical diameters at the microlevel, and the walls are hydrophilic with no interexchange

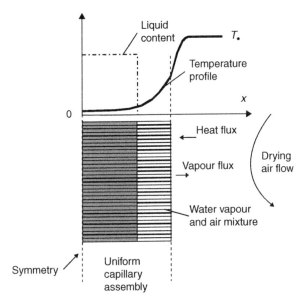

Figure 6.5 Water removal in air-drying at a direction perpendicular to the capillary assembly (bundle) with identical capillary characteristics (diameter and wall material).

of heat and mass across the capillary walls (impermeable), assuming they were initially filled with water completely and evaporation starts to happen. The evaporation occurs uniformly for all the tubes for the same convection condition at the exits of all the tubes. There would be an obvious receding front of the liquid–gas interface moving inwards as drying proceeds. The thickness of the moving evaporation front would reflect the meniscus of the liquid–gas interface.

Considering the system of the capillaries such that different diameters and permeable walls (e.g., a kind of membrane) are involved, and interexchange of heat is also possible across the capillary walls, the evaporation rates among the tubes under the same drying condition applied at the exits would be different. There would be nonuniform receding liquid–gas interfaces, giving the distribution of the averaged liquid water content along the x-direction broader than that shown in Figure 6.6. Furthermore, if the walls are made of materials that are hygroscopic, liquid diffusion or spreading along the wall surfaces is also possible.

Though the interexchange of moisture and heat in-between the tubes may attempt to even the evaporation rates and liquid water contents, a broader distribution of the liquid water content is still expected. Furthermore, due to the extended liquid–gas interfaces, evaporation would not just occur at the meniscus only. Evaporation will happen in a region of finite dimension, i.e., the occurrence of an evaporation zone at the macrolevel.

Nevertheless, in the above two systems, for the straight tubes considered and under only one-dimensional condition as shown here, there is a likelihood of receding front (curved ones included) of some description. However, this kind of system is not common.

When the system is "homogenized" at the microlevel (a more realistic situation)—an example is shown in Figure 6.7—capillaries would be oriented at many directions and interlinked or networked. Even locally, the capillary diameter sizes can be uneven. The heat and mass transfer would be multidirectional, but of course following the law of physics, i.e., following the directions of the driving forces. Locally and microscopically, the receding front(s) would be a fuzzy one, depending on the local microstructure. Liquid movement may be diffusive or driven by capillary forces and traveling in relatively easier passages. Similarly for air and vapor transfer, certain difficult (yet wet) patches may be bypassed by a main receding front (if there was one) and left to be dried more gradually. Here, the sorption/desorption characteristics of the materials distributed

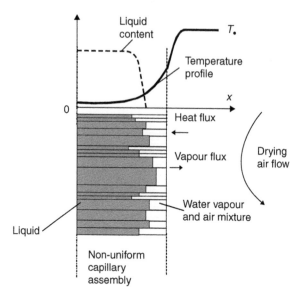

Figure 6.6   Water removal in air-drying at a direction perpendicular to the capillary assembly (bundle) with varying capillary characteristics (diameters and wall materials).

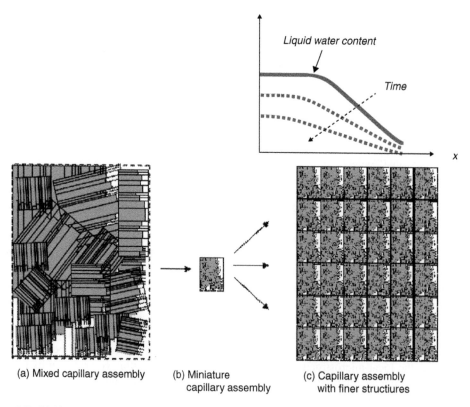

Figure 6.7   Making a porous material with complex structures at various scales.

enter to play key roles. All these chararcterics would make the liquid water content (averaged over these microscopic regions) to be distributed over a region between the really wet core and the boundary of the moist material. This shows a spatial transition rather than sharp receding liquid water front (see Figure 6.7).

Compounding this phenomena, the capillary wall's thickness (the apparently solid structure) and the walls' own porous microstructures (yet another smaller level of pore networks or systems), and their unevenness in spatial distribution would add to the overall picture making the transition more uniformly. The materials that create the walls of the microstructures are also important as they can have quite different affinities towards water molecules (these are reflected by their equilibrium isotherms or liquid water holding capacity at the same relative humidity and temperature).

The liquid water content distribution, where water vapor is in coexistence is shown Figure 6.7. What may be stressed is that the rather gradual liquid water content profile which is not the kind possessing a sharp moving front, thus indicating the effect of mass transfer and local exchange of water between the moist solids and gas in the voids or channels. Evaporation may occur in most of the transitional region where the rate is dependent on the local driving force for vapor transfer.

The above arguments readily can be generalized to packed particulate systems where the individual particles can have their own macrostructure and sorption characteristics (see Figure 6.8a), while the main voids (where easier vapor paths can be found) would be the voids in-between the packed particles. In fruits and vegetables, the cellular structure plays a very important role as the cell walls present major water transfer resistance (see Figure 6.8b).

The perception of a moving (liquid) front or the (sharp) evaporation front can lead to different approaches to drying modeling. Mass transfer from the sharp moving front and the vapor exit surface is often modeled using a simple effective diffusion concept (with an expanding resistance layer). This mass transfer is regarded sometimes as the rate limiting process form air-drying, i.e., the mass-transfer limiting process (for example, Nesic and Vodnik[34]). In this case, the vapor concentration at the evaporation front is taken to be the highest (i.e., the saturated vapor concentration inside the pores or pore-network channels).

Most recently, this relatively dry region has been taken as being the very dry crust (with only bound water) so that a heat transfer limiting process is furnished. In this case, all heat that is transferred to this evaporation front is consumed by water evaporation.[35–38] It is interesting to note that Farid[37] and Nesic and Vodnik[34] actually modeled the same set of data for skim milk droplet drying in air, and they have obtained similar level of accuracy in predictions! This means that even in recent times, the concepts of mass-transfer limiting and heat-transfer limiting are not clear and there is a need to have better evaluation methods.

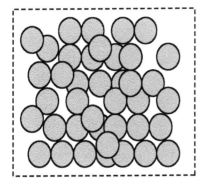

(a) Packed particle bed (milk powders etc)

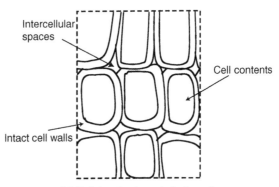

Intercellular spaces

Cell contents

Intact cell walls

(b) Cellular structures in fruits and vegetables

Figure 6.8 Packed particulate (or unit) system.

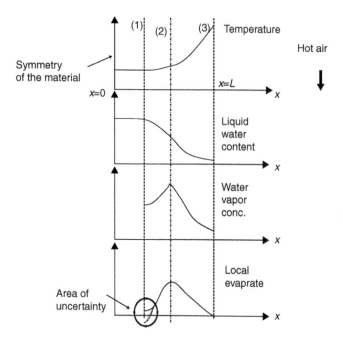

Figure 6.9 A visualization of the drying process of a hygroscopic porous material (line 3 denotes the convection heat and mass transfer boundary).

Based on the discussion earlier about the spatially distributed evaporation profile, one can arrived at Figure 6.9, which shows more realistic perception of the drying processes.

It is also interesting to note from the process shown in Figure 6.9, that due to the temperature distribution into the moist material being dried in the normal air-drying situation (where the air temperature is higher than the porous material being dried), it is not necessary to have the highest water vapor content at the innermost boundary where liquid water content starts departing from the initial value. The vapor concentration should be higher than the boundary value or else there would be no or little drying. It is possible to intuitively reason that there is a hump that can exist somewhere in the transitional region of the liquid water content (between dashed lines 1 to 3).

Zhang and Datta,[27] in fact, predicted this kind of distribution without much significance being discussed. Because there is a hump in the vapor concentration, there must be a vapor flow inwards (towards the left on Figure 6.8). A condensation mechanism also may exist in the region marked in the same figure as being the area of uncertainty.

As such, the process of having a high temperature in the air and low humidity in the same air would induce an inward transport of vapor as well as one that goes outwards (thus drying is evident). Furthermore, a part of the water evaporated in the lower part of the transitional liquid water content region is transported into the structure and condensed at the lower temperature location as long as there is porosity (spaces for vapor to go into). This is an interesting phenomena, as it clearly a more effective heat transfer mechanism than just heat conduction.

This phenomena helps increase more rapidly the temperature of the core wet region, which by itself has a higher heat conductivity due to the high water content (hence lower or no porosity). This mechanism has an impact on the preservation of active ingredients such as probiotic bacteria encapsulated inside a wet porous matrix subjected to drying. This result means that the fuller picture of the drying would involve the heat transfer and moisture transfer characteristics in at least two regions: one that is between the dashed line 1 and 2 and one that is between 2 and 3. Line (2) is chosen qualitatively to mark the inwards and outgoing vapor flows. The microstructure in region 2 and 3 must influence (or restrict) water vapor transfer (porosity and tortuosity), thus leading to reduced vapor diffusivity. Together with the effect of chemical composition, the microstructure also impacts on the liquid water transfer towards the boundary 3. The structure and indeed the porosity have a large impact on the thermal conductivity of this relatively dry layer (region between line 2 and 3). Conversely, the porosity and structure inside the region between line 1 and 2 are also important affecting vapor transport in this area. One would expect that this region to have a lower porosity (thus lower vapor transfer coefficient—the vapor diffusivity). However, due to the high water content, the heat conductivity would be higher here as mentioned earlier. In addition to this, the vapor transfer and condensation mechanism as mentioned earlier would make this heat conductivity is effectively even higher. As such, the main heat transfer retardation by the presence of the solid materials would be in the region of line 2 and 3, and the main retardation of water vapor going outwards would also be in the region 2–3. Note here that there should generally be two lines denoted by line 2 dividing the heat transfer regions and the mass transfer regions separately. The lowest liquid water content in region 2–3 clearly is determined by the nature of the material and the drying air conditions through the equilibrium water content concept (equilibrium isotherms).

Increasing temperature to a large extent may make the trend of liquid water content steeper; and a "water fall" like behavior, where the vapor wave is apparently moving inwards, and a more tidal like liquid water content versus distance profile emerges. For food and biomaterials, it likely is to have a sharp liquid water-front if the temperature can rise in the drier region quickly. In any case, how the material swells and shrinks locally would have an impact on the dried product quality.

The above arguments, in particular the profiles shown in Figure 6.9, have been supported by the microscale transient observations using Magnetic resonance imaging (MRI). MRI has become in recent years a useful tool in investigating the relationship between the material microstructure and moisture transfer mechanisms. A number of studies have been directly targeted at moisture transfer.[39–45] It has been illustrated that the simple receding evaporation-front assumption may not be a realistic proposition,[42,43] though some earlier studies suggested it was the case for some hygroscopic materials.[46] There was no sharp front of evaporation observed in the latest MRI studies. The spread of the lowering liquid water content as drying proceeds relies on capillary diffusion of liquid water.[43] The moisture transfer or transport devices or units such as capillaries, intercell spaces, void or channel networks between packed particles (which themselves may also be porous presenting another (perhaps finer) level of transfer devices or units), all naturally possess nonuniformity. The spread of the evaporation zone or a transitional or "mushy" zone from the still very wet core and the already dried surface region therefore is expected. Referring to Figure 6.9, in the region across regions 2 and 3, capillary effect is expected to be important. The pre-treatment (soaking) using surface active reagent solutions may help accelerating the water transfer process. Thicker material may be affected more by this kind of treatment.[47] This realization may open a door to a more scientific understanding of the effect of surface tension lowering agents (such as surfactants). This may also be related to explaining the results on the phenomena associated with the air-drying of osmotic-dehydrated fruit materials.

Furthermore, the spreading of the liquid water (the shape of the spatial distribution of the liquid water content) would be dependent on the hygroscopic nature of the porous material being dried. A hydrophilic porous solid system may have a smaller possibility to show a relatively "sharp moving evaporation front" compared with a hydrophobic system. This understanding has been particularly helpful in supporting the ideas of revising the conventional Biot and Lewis number calculations when air drying is of interest so that the conditions for model simplifications can be made more realistically, and the model concepts may be discerned with more quantitative backings[48,49] (see later sections on Biot and Lewis number analysis). The effective diffusivity functions published in literature can be compared and discussed based on their relationships with the scientific insights already discovered with MRI or other insightful tools. It is now recognized that the material microstructure and its nature (composition, pore sizes, etc.,) are interactive with the transfer phenomena[50,51] and the formation of microstructure has much to do with the speed and schedule of the transfer processes.[52–54]

6.2.2.4 Solving the Equation Sets that Govern the Moisture-Transfer Process

The above mentioned partial differential equation sets for resolving the temperature and moisture profiles in moist food materials during drying have to be solved using numerical methods. This in general involves the use of finite difference and finite element methods. Since the nonlinear nature of the models described above, in particular, when the models are highly coupled between heat and mass transfer, instability and sometimes numerical diffusion (an artifact) can be introduced.

There are two ways of validating the numerical solutions: an exact solution with some simplifications of the model employed may be devised so that the numerical one can be compared against it (such as that demonstrated by Wu and Irudayaraj in 1996[55]); or a detailed experimentation is conducted (note that only measuring the weight loss and a mean temperature is insufficient for validating a detailed model such as the ones discussed in Section 6.2.2.1). Where possible, the spatial distribution of temperature (or at least, a surface and a core temperature) should be measured.

There are standard texts available in literature which can be referred to when the PDE systems that govern transport processes need to be solved numerically, such as Patankar[56], Rice and Do[57] and Ozilgen.[58] There are also numerous papers published in the area such as Irudayaraj et al.[59,60]

6.2.3 Mass- or Heat-Transfer Limiting

There is not a straightforward concept as far as drying modeling is concerned. When mass transfer is the limiting process, which may usually be the case when the temperature difference between the material to be dried and that of the drying gas is not very large, one may not need to consider the spatial distribution of temperature and can therefore simplify the modeling process. If heat transfer is limiting, then the temperature gradient within the medium being dried must be considered. Thus, the mass transfer PDE may not be required to work out water loss. The reason is that in this case the water loss is primarily determined by how much and how quickly the heat can be supplied to the evaporation surfaces. This is a reasonable approach to modeling frying of a thin potato chip in hot oil[61].

In air-drying, however, there is the expectation that the spatial distribution of water content inside a material is significant, i.e., the boundary could be dry, but the core could still be very wet. The question is whether or not the temperature within the material being dried can be considered to be uniform. At first glance, all these should correlate with two classical numbers, i.e., the Biot number for examining the temperature uniformity and the Lewis number for examining heat or mass transfer limiting.

6.2.3.1 Biot Number Analysis

The Biot number criterion is used to investigate the temperature uniformity of a material being heated or cooled. Conventionally, the Biot number is introduced through steady state heat conduction in a slab with one side cooled by convection (see Figure 6.10a).[62] The conductive heat flux through the wall is set to be equal to the heat flux due to convection:

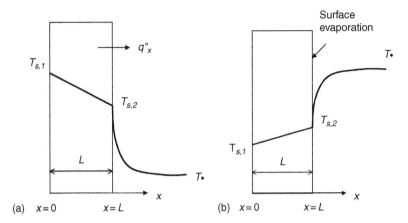

Figure 6.10 (a) The definition of Biot number; (b) schematic diagram of the system being heated with surface evaporation.

$$q''_x = k \cdot \frac{T_{s,1} - T_{s,2}}{L} = h \cdot (T_{s,2} - T_\infty). \tag{6.22}$$

The ratio of the temperature differences can then be expressed as

$$\frac{T_{s,1} - T_{s,2}}{T_{s,2} - T_\infty} = \frac{h \cdot L}{k}. \tag{6.23}$$

The Biot number is defined by

$$Bi = \frac{h \cdot L}{k}. \tag{6.24}$$

When this ratio is less than 0.1, i.e., the internal temperature difference is smaller than 10% of the external temperature difference then the internal temperature distribution may be neglected for simplicity in modeling. For a spherical object, the characteristic length, L, may be set to be the radius of the sphere.

One can rewrite Equation 6.23 in the form of the temperature ratio:

$$\frac{T_{s,1} - T_{s,2}}{T_{s,2} - T_\infty} = \frac{h \cdot L}{k} = \frac{L/k}{1/h} = \frac{R_{cond}}{R_{conv}}, \tag{6.25}$$

where R represents the thermal resistances, and the subscripts "cond" and "conv" represent the conduction and convection, respectively. Therefore, the Biot number also can be considered to be the ratio of the internal resistance to the external resistance. It appears that a small temperature ratio and a small resistance ratio essentially suggest the same thing. It can be illustrated that they can yield different results when the problem is no longer a pure heat conduction problem.[48] More fundamentally, the resistance ratio is a better argument. The thermal conductivity of the particle would be affected by water content and porosity (when filled with air).

Chen and Peng[48] have recently introduced a new formula, which accounts for evaporation from the heat exchange surface. Similar to the conventional analysis (Equation 6.22), considering the addition of evaporative loss, the following heat balance can be obtained (here the temperature of the environment is greater than the material being dried; see Figure 6.10b):

$$h \cdot (T_\infty - T_{s,2}) - \Delta H_L \cdot \dot{S}_v = h^* \cdot (T_\infty - T_{s,2}) = k \cdot \frac{T_{s,2} - T_{s,1}}{L}. \tag{6.26}$$

Here, $h*$ is an equivalent convection heat transfer coefficient and $\dot{S}_v$ is the surface based evaporation or drying rate (kg m^{-2} s^{-1}).

$$h^* = h - \frac{\Delta H_L \cdot \dot{S}_v}{(T_\infty - T_{s,2})}. \tag{6.27}$$

This leads to a more appropriate Biot number for a surface evaporative system:

$$Bi^* = Bi - \frac{\Delta H_L \cdot \dot{S}_v}{(T_\infty - T_{s,2})} \cdot \frac{L}{k} \tag{6.28}$$

When evaporation occurs from the heat exchanger surface, the modified Biot number (Bi^*) represents the temperature uniformity more precisely than Bi. Whereas a conventional and large Bi indicates a nonuniform temperature distribution does not mean the temperature is not reasonably uniform. Bi^* is a smaller value than Bi indicating a more uniform temperature distribution. The significance of this equation can be demonstrated in the following for evaporation from a water droplet (which is relevant to drying of a coal particle with high moisture content at the beginning of the drying process).

The following calculations are based on the laboratory data obtained by Lin and Chen[23,63] on drying of a single milk droplet (one-dimensional but spherical condition). The laboratory conditions and the techniques employed are given:

Droplet diameter $2L = 1.43$ mm
Heat transfer coefficient measured $h = 99.1$ W·m^{-2} K^{-1}
Thermal conductivity of water $k = 0.63$ W·m^{-1} K^{-1}
Evaporation rate $\dot{S}_v = 1.73 \times 10^{-3}$ kg·s^{-1}·m^{-2}
Interfacial temperature $T_s = 23.4°C$
Drying air temperature $T_\infty = 67.5°C$
Latent heat of evaporation $\Delta H_L = 2445 \times 10^3$ J·kg^{-1}

The conventional analysis, based on Equation 6.24, yields $Bi = 0.11$ for the above example, which is even slightly greater than the critical value of 0.1 mentioned earlier. Based on Equation 6.28, however, one can find that the Bi required for uniform temperature assumption is 0.21 to maintain the Bi^* being 0.1, i.e.,

$$Bi_{cri} = Bi^*_{cri} + \frac{2445 \times 10^3 \times 1.73 \times 10^{-3}}{(67.5 - 23.4)} \frac{1.43 \times 10^{-3}}{2 \times 0.63} \approx 0.21 \tag{6.29}$$

due to the evaporation (cooling) effect, which consumes much of the temperature diving force from the outside of the material being dried.

Equation 6.28 can be extended to account for the cases when internal mass transfer resistance do play a role. Here one can express the surface based evaporation rate using the overall mass transfer coefficient concept:

$$Bi^* = Bi - \frac{\Delta H_L U_m (\rho_{v,c} - \rho_{v,\infty})}{\frac{k}{L}(T_\infty - T_{s,2})}, \tag{6.30}$$

where L may be used as the characteristic length generically of slab, cylinder, sphere, etc. (in meters). $\rho_{v,c}$ is the vapor concentration (based in the bulk) at the location marked by the characteristic dimension (δ_c) inside the material, which may be taken as the saturated vapor concentration at the material (mean) temperature (a high bound estimate). The overall mass transfer coefficient U_m may be expressed approximately as the following by Chen and Nelson (unpublished 2005):

$$U_m \approx \frac{1}{\frac{1}{h_m} + \frac{\delta_c}{D_{eff,v}}}.$$ (6.31a)

The characteristic dimension (δ_c) signifies the effect of the vapor concentration profile and the solid matrix resistance to vapor transfer (see Figure 6.9). For a symmetric material being dried, this dimension for mass transfer should be smaller than the corresponding characteristic dimension for heat conduction (for example, $\delta_c <$ the 0.5 half-thickness for a slab being heated).[64] The effective vapor diffusivity may be estimated using the porosity (ε) and the tortuosity (τ) correction:[65]

$$D_{eff,v} \approx \frac{\varepsilon}{\tau} \cdot D_{v,air},$$ (6.31b)

where $D_{v,air}$ is the vapor diffusivity in air (m^2 s^{-1}). The tortuosity (τ) is not usually a known paramete, so an estimate between 2 up to 20 may be used.[65] Equation 6.31b does not seem to have accounted for the sticking of the molecules in transfer onto the material internal surfaces and could still give a high estimate of the real value.

The mass transfer coefficient (h_m) could also be estimated using one of the heat and mass transfer analogies, such as the one for a flat plate:[62]

$$\frac{h}{h_m} \approx \frac{k_{air}}{D_{v,air}} \cdot \left(\frac{D_{v,air}}{\alpha_{air}}\right)^{0.3},$$ (6.32)

which is based on the heat transfer coefficient h (which could be found using the correlations in the Appendix).

One can see the new number, called in the author's name to be differentiated from the classical Biot number (Bi), is as follows for air drying, i.e., a Chen–Biot number.

For surface evaporation case shown in Figure 6.10b,

$$Ch\_Bi = Bi - \frac{\Delta H_L \cdot \dot{S}_v}{(T_\infty - T_{s,2})} \cdot \frac{L}{k}.$$ (6.33)

It is difficult to evaluate the Chen–Biot number for the case when drying is occurring inside the material. As mentioned above, the following formula has been proposed (though not yet applied to a food system) as follows:[66]

$$Ch\_Bi = Bi - \frac{\Delta H_L}{\frac{k}{L} \cdot \left(\frac{T_\infty - T_{s,2}}{\rho_{v,c} - \rho_{v,\infty}}\right) \left(\frac{1}{h_m} + \frac{\delta_c}{D_{eff,v}}\right)}.$$ (6.34)

Alternatively, the term of the mass transfer resistance due to diffusion in the above equation is replaced by that using the effective liquid water diffusivity $D_{eff,l}$ but perhaps maintaining the same characteristic dimension for heat conduction.

Equation 6.34 reduces to the surface evaporation case when the characteristic thickness δ_c approaches to zero.

6.2.3.2 Lewis-Number Analysis

The Lewis number (Le) is defined as the ratio of the thermal diffusivity to the mass diffusivity (for water vapor transfer). For drying of moist porous materials, this would mean that if $Le < 1$, then the heat penetration (through conduction) into the particle is slower than the penetration of the water vapor front. If $Le \ll 1$, then the heat transfer is limiting the drying process. When $Le \approx 1$, the heat input and moisture removal are highly coupled. When $Le \gg 1$, the process is mass-transfer limited. This becomes particularly informative and important for drying after the initial moisture rich condition. This process can be visualized as follows, by considering the heating of a semi-infinite porous media in a constant temperature environment. It is known that the thermal penetration depth (δ_T) can be approximated as:[67]

$$\delta_T \approx \sqrt{12 \cdot \alpha \cdot t_T}, \tag{6.35}$$

where α is the thermal diffusivity ($= k/\rho C_p$) (m^2 s^{-1}) and t_T is thermal penetration time (s). Correspondingly, the vapor "penetration" (mass penetration denoted by the subscript M) may be expressed as:

$$\delta_M \approx \sqrt{12 D_{\text{eff},v} \cdot t_M}, \tag{6.36}$$

where t_M is the mass penetration time (in s). For reaching the same distance into the material, i.e., $\delta_T = \delta_M$, the ratio of the time required for mass to penetrate to that for heat to penetrate is essentially the Lewis number:

$$Le = \frac{t_M}{t_T}. \tag{6.37}$$

The physics is apparent: if $Le \ll 1$, mass transfer occurs much faster, thus heat transfer is limiting.

A conventional Lewis number analysis using the effective vapor diffusivity calculated with Equation 6.31b would, for a skim milk droplet drying for instance, yield a value smaller than 0.1.[37]

If one considers the source terms in describing drying, Equation 6.18 through Equation 6.20 can be used.

For simplicity, the source term similar to that in Equation 6.18 may be approximated as

$$\dot{E}_v \approx -\rho_s \cdot \frac{dX}{dt}. \tag{6.38}$$

This represents the local rate of evaporation. Assuming that C_v is the vapor concentration that is in equilibrium with the liquid water content inside the solid structure, the following relationship exists:

$$RH = \phi = \frac{C_v}{C_{v,\text{sat}}(T)} = f(X,T) \text{ or } X = F(\phi,T), \tag{6.39}$$

which is the equilibrium isotherm function. Therefore, one may write the following:

$$\dot{E}_v \approx -\rho_s \left(\frac{\partial X}{\partial \phi} \frac{\partial \phi}{\partial t} + \frac{\partial X}{\partial T} \frac{\partial T}{\partial t} \right) = \rho_s \left[\frac{\partial X}{\partial \phi} \cdot \frac{\partial \left(\frac{C_v}{C_{v,\text{sat}}(T)} \right)}{\partial t} + \frac{\partial X}{\partial T} \cdot \frac{\partial T}{\partial t} \right]. \tag{6.40}$$

If the temperature may be taken as an average value especially when the equilibrium isotherm functions are insensitive to temperature in the range considered, the following simplified Equation 6.40 can be obtained.

$$\dot{E}_v \approx -\rho_s \cdot \left(\frac{\partial X}{\partial \phi} \cdot \frac{\partial \phi}{\partial t} \right) \approx -\rho_s \cdot \frac{\partial X}{\partial \phi} \cdot \left(\frac{1}{C_{v,\text{sat}}(\bar{T})} \cdot \frac{\partial C_v}{\partial t} \right). \tag{6.41}$$

Thus vapor conservation Equation 6.19 can be rewritten into:

$$\frac{\partial C_v}{\partial t} \approx \frac{\bar{D}_{\text{eff},v}}{1 + \rho_s \cdot \left(\frac{\partial X}{\partial \phi} \cdot \frac{1}{C_{v,\text{sat}}(\bar{T})} \right)} \cdot \frac{1}{r^2} \frac{\partial}{\partial r} \left(r^2 \cdot \frac{\partial C_v}{\partial r} \right), \tag{6.42}$$

by taking a mean, $\bar{D}_{\text{eff},v}$, for simplification.

The equivalent mean effective diffusivity ($\bar{D}_{\text{eff,v}}$) is then

$$\bar{D}_{\text{eff,v}}^{\text{new}} \approx \frac{\bar{D}_{\text{eff,v}}}{1 + \rho_s \cdot \left(\frac{\partial X}{\partial \phi} \cdot \frac{1}{C_{\text{v,sat}}(\bar{T})} \right)}. \tag{6.43}$$

The effective Lewis number (i.e., the Chen–Lewis number) can then be written as

$$Ch\_Le = \frac{\bar{\alpha}_{\text{eff}}}{\bar{D}_{\text{eff}}^{\text{new}}} = \left(\frac{\bar{k}_{\text{eff}}}{\bar{\rho} \cdot \bar{C}_p} \right) \Bigg/ \frac{\bar{D}_{\text{eff,v}}}{1 + \rho_s \cdot \left(\frac{\partial X}{\partial \phi} \frac{1}{C_{\text{v,sat}}(\bar{T})} \right)}. \tag{6.44}$$

Equation 6.44 may be used to estimate a more likely high bound of Lewis number (here the Chen–Lewis number). For the same skim milk drying case as mentioned earlier (i.e., the case for Equation 6.29), Equation 6.44 can yield an estimate of the Chen–Lewis number (Ch–Le) on the order of 100, indicting mass transfer limiting.

6.2.3.3 Combination of Biot and Lewis Numbers

There is no strict scientific proof regarding why one can use Bi and Le number together in some fashion to justify isothermal, uniform temperature or else. Nevertheless, these are the two most relevant dimensionless parameters in literature that seem most relevant. Sun and Meunier, in 1987,[68] conducted a comprehensive numerical analysis on nonisothermal sorption in adsorbents, which showed that the following rule exists:

- The isothermal model would be valid if $LeBi > 100$ and the uniform temperature profile model would be a good model if $Le > 10$.

Note here that the Le and Bi are all based on the conventional definitions. It is expected that the above rules are conservative. In the desorption process the temperature profile inside the porous material would tend to be more gradual thus the criteria can be relaxed. The driving force for heat transfer, i.e., the difference between the drying air (or drying medium in general) and the porous material, i.e., ($T_\infty - T_s$), would also affect the temperature uniformity when drying proceeds.

6.2.4 Drying of Shrinkable Materials

When shrinking is considered, an apparent or effective velocity, u_{ap}, may be considered. Equation 6.14 and Equation 6.15 can be modified to

$$\frac{\partial C_1}{\partial t} + u_{\text{ap}} \cdot \frac{\partial C_1}{\partial x} = \frac{\partial}{\partial x} \left(D_{\text{eff,1}} \cdot \frac{\partial C_1}{\partial x} \right); \tag{6.14a}$$

$$\frac{\partial T}{\partial t} - u_{\text{ap}} \cdot \frac{\partial T}{\partial x} = \frac{1}{\rho C_p} \cdot \frac{\partial}{\partial x} \left(k_{\text{eff}} \cdot \frac{\partial T}{\partial x} \right). \tag{6.15a}$$

For a symmetric drying situation, $L = 2b$, the shrinking velocity, u_p, (taken as positive value when shrinking occurs) can be expressed as

$$u_{\text{ap}} = u_{\text{ap}}(x = +b) \cdot \frac{x}{b}, \tag{6.45}$$

as an approximation (Karim and Hawlader 2005). This means that the effect of boundary movement is the most influential at the boundary and has no effect at the symmetry. The velocity at the shrinking boundary corresponds to the movement of b over time.

$$u_{\mathrm{ap}}(x = +b) = -\frac{db(t)}{dt}. \tag{6.46}$$

To ensure the smooth transition of u_{ap} at $x=0$, which should be the case at the symmetry, u_{ap} may be better replaced by a nonlinear function instead so that $\frac{du_{\mathrm{ap}}}{dx} = 0$ there. Of course, $\frac{d^2 u_{\mathrm{ap}}}{dx^2} > 0$ is necessary to present a minimum shrinkage at the center.

This velocity effect should also be reflected in the boundary conditions. For example, Equation 6.12 and Equation 6.13 need to be changed to:

$$\left(-D_{\mathrm{eff},1} \cdot \frac{\partial C_1}{\partial x} - u_{\mathrm{ap}} \cdot C_1\right)_s = h_{\mathrm{m}} \cdot (\rho_{v,s} - \rho_{v,\infty}), \tag{6.12a}$$

$$\left(k_{\mathrm{eff}} \cdot \frac{\partial T}{\partial x} + \rho C_p u_{\mathrm{ap}} T\right)_s = h \cdot (T_\infty - T_s) - \Delta H_L \cdot h_{\mathrm{m}} \cdot (\rho_{v,s} - \rho_{v,\infty}). \tag{6.13a}$$

It is again noted that u_{ap} is taken to be positive when shrinking occurs. The effective diffusivity should also be corrected for shrinkage effect as it was often measured and interpreted by assuming no-shrinkage (for instance, using fixed L with Equation 6.4a). This can be found readily in the work by Crank:[16]

$$\frac{D_{\mathrm{eff},1\,\mathrm{corr}}}{D_{\mathrm{eff},1}} = \left(\frac{b}{b_0}\right)^2, \tag{6.47}$$

where "corr" represents the corrected value and b_0 is the initial thickness (in meters). The change in b is usually measured and correlated to water content.

6.3 WORKED EXAMPLES

6.3.1 Spray-Drying of Skim Milk Droplet in a Plug-Flow Dryer

Here, a simplified spray dryer, i.e., a plug-flow dryer, is simulated, which does not involve the complications of the particle trajectories as in a full-scale computational fluid dynamics modeling. In this case, the droplet drying kinetics and its influence upon the overall dryer wide performance can be readily demonstrated by Patel.[69] In previous literature, the partial differential equation for mass diffusion with shrinkage was used for simulating the drying process of a single droplet.[70] It is, however, not yet practical to trace a large number of particles in computational fluid dynamics (CFD) simulations of spray drying with this kind of detailed approach. Therefore, simplified models are useful. In the following section, the simple kinetics approaches are described.

6.3.1.1 Mass-Transfer Model

The mass-transfer model calculates the moisture concentration profiles for both droplet and hot drying medium phases. It also allows an engineer to understand the effect of various parameters such as moisture content of the droplet, air humidity of the drying medium and temperatures on the drying rate. As the first stage of spray drying, the atomization process transforms the bulk liquid feed into a large number of individual droplets. Here a uniform distributed uniform size droplets passes through the dryer length co-currently with the drying air.

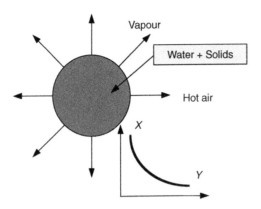

Figure 6.11   Single skim milk droplet in the spray drying chamber.

The modeling for drying of droplet using fluids such as skim milk is quite complicated because skim milk is a mixture of several constituents such as fats, proteins, carbohydrates, starches (lactose), minerals, vitamins, enzymes and water. The single skim milk droplet that is made of water and skim milk solids is shown in Figure 6.11. The droplet has moisture concentration X on dry basis (kg kg^{-1}) at time t (s). When the droplet comes into contact with the hot drying air, the moisture starts escaping from the droplet surface at a specific rate. As the droplet falls down in the spray chamber, the moisture concentration obviously decreases up to the exit moisture concentration of the bulk drying air, Y kg water per kg dry air. At the exit conditions, the droplet and moist air are nearly in dynamic and thermal equilibrium with each other and the moisture concentrations of both phases are nearly the same.

Mass Balance around the Droplet Phase. Figure 6.12 shows the step-wise mass balance around a single droplet in the spray-drying chamber. At any time t, the droplet is at a distance h meter from the atomizer. The droplet moisture content is X kg kg^{-1} with the m_s (kg) of milk solids at that specific point. After time increment dt, the droplet travels an infinitesimal distance dh (m) with losing an infinitesimal amount of moisture dX to the bulk air at specific drying rate N_v. The amount of milk solids in the droplet remains constant during drying as only water vapor escapes from the droplet surface during evaporation. The mass balance around the single droplet, P, for time step of

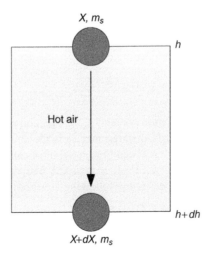

Figure 6.12   Mass balance of a single droplet.

dt can be written in the following mathematical form:

$$m_s \cdot X = m_s \cdot (X + dX) + N_v \cdot A_P dt. \tag{6.48}$$

On rearrangement,

$$\frac{dX}{dt} = -\frac{N_v \cdot A_P}{m_s}, \tag{6.49}$$

Or, written in another form,

$$\frac{dm_w}{dt} = -N_v \cdot A_P, \tag{6.50}$$

where m_w is the mass of water. A_P is the effective surface area of the particle and that can be calculated as

$$A_P = \pi \cdot d_P^2. \tag{6.51}$$

Here, d_P is the particle diameter. Equation 6.50 represents the rate of water evaporation during drying of droplets. Equation 6.49 calculates the moisture content profile for the single droplets. At this point one needs to calculate the drying rate N_v. Several approaches are available in the literature to estimate the drying rate, including the diffusion model which consists of the second-order transient partial differential Equation (like that of Sano and Keey[70]), but the present work considers the following two approaches, which are semiempirical approaches as the diffusion model requires many parameters to be pre-known.

Characteristic Drying Rate Curve Approach (CDRCA). In 1958, Van Meel[71] suggested that a unique characteristic drying curve can be drawn for the material being dried at specific drying conditions for convective drying. This approach is based on the assumption that at each volume-averaged free moisture content, there is a corresponding specific drying rate relative to the unhindered rate in the first drying period that is independent of the external drying conditions.[72] This approach takes into account the commonly acknowledged drying rate characteristics at different water contents.

Here, the relative drying rate (the drying rate relative to the maximum drying rate in the first drying period) is defined as:

$$f = \frac{N_v}{\hat{N}_v} \tag{6.52}$$

where N_v is the specific drying rate and $\hat{N}_v$ is the unhindered drying rate that is the drying rate in the first drying period. The characteristic moisture content, ϕ, can be defined as:

$$\phi = \frac{\bar{X} - X_e}{X_{cr} - X_e}, \tag{6.53}$$

where $\bar{X}$ is the volume-averaged moisture content, X_{cr} is the moisture content at the critical point (the point at which the falling rate period starts) and X_e is the equilibrium moisture content (the moisture content of the droplet when it is at equilibrium with the partial pressure of the surrounding air). This representation is attractive as it leads to a simple lumped-parameter expression for the specific drying rate, which is illustrated by the following equation:[72,73]

$$N_v = f \cdot \hat{N}_v = f \cdot [\beta_1(p_{vs} - p_{vb})], \tag{6.54}$$

where β_1 is an external mass-transfer coefficient (s m^{-1}), p_{vs} is the partial pressure on the droplet surface, and p_{vb} is the partial pressure of the bulk drying air. Here, the critical water content is considered to be a unique value, beyond that the evaporation is the same as that of the pure water droplet. Equation 6.54 has been used extensively as the basis for understanding the behavior of industrial drying plants owing to its simplicity and the separation of the parameters that influence the drying process: the material itself f, the design of the dryer β_1 and the process conditions ($p_{vs} - p_{vb}$). The characteristic drying rate curve approach, however, is clearly an approximation.[72]

For a material with a linear falling-rate curve and having no constant rate period (initial moisture content X_i becomes the critical moisture content X_{cr}), and the dimensionless drying rate f is proportional to $(X - X_e)/(X_i - X_e)$, Langrish and Kockel[73] have rewritten the lumped parameter expression in the following form:

$$N_v = f \cdot [\beta_1 \cdot (p_{vs} - p_{vb})] = \left(\frac{X - X_e}{X_i - X_e}\right) \cdot [\beta_1 (p_{vs} - p_{vb})]$$

$$= (X - X_e) \cdot [\beta_2 \cdot (p_{vs} - p_{vb})], \tag{6.55}$$

where β_2 is another coefficient (m s^{-1} kg^{-1}) as a result. Here they considered the critical water content is the same as the initial water content (i.e., no constant drying-rate period). At dynamic equilibrium of the droplet-air system, the rate of heat transfer is equal to the latent heat of vaporization times the rate of mass transfer. Hence,

$$\frac{dQ}{dt} = \alpha \cdot A_P \cdot (T_b - T_{WB}) = -N_v \cdot A_P \cdot \Delta H_L, \tag{6.56}$$

where ΔH_L is the latent heat of vaporization (J kg^{-1}). From Equation 6.55 and Equation 6.56,

$$\alpha \cdot (T_b - T_{WB}) = -(X_i - X_e)[\beta_2 \cdot (p_{vs} - p_{vb})] \cdot \Delta H_L. \tag{6.57}$$

Equation 6.57 shows that the vapor-pressure driving force could be made proportional to the wet-bulb depression. The specific drying rate expression from Equation 6.55 can be rewritten in terms of the wet-bulb depression as

$$N_v = (X - X_e)[\beta_3 \cdot (T_b - T_{WB})] \cdot \frac{m_s X}{A_P}, \tag{6.58}$$

where, T_b is the bulk-air temperature and T_{WB} is the wet-bulb temperature. Equation 6.58 predicts the drying rate profile, and in conjunction with Equation 6.49 it estimates the moisture content profile of the droplets. The drying rate calculations provide a means of computing the amount of liquid that will be lost due to evaporation and the amount of liquid left in the exit particle. The coefficient β_3 (s^{-1}K^{-1}) in Equation 6.58 can be calculated by

$$\beta_3 = \frac{\alpha}{\frac{m_s}{A_P} \cdot X \cdot \Delta H_L}. \tag{6.59}$$

In Equation 6.56, Equation 6.57, and Equation 6.59, α is the heat transfer coefficient and can be calculated from the well-known Ranz–Marshall correlation[69] as

$$Nu = \frac{\alpha \cdot d_P}{k_b} = 2 + 0.6 \cdot Re^{1/2} \cdot Pr^{1/3}. \tag{6.60}$$

Here, Re is the Reynolds number and Pr is the Prandtl number, which can be described in terms of the individual characteristics of the droplet and bulk air as:[65]

$$Re = \frac{d_P v_P \rho_b}{\mu_b}, \tag{6.61}$$

$$Pr = \left(\frac{c_p \mu}{k}\right)_b, \tag{6.62}$$

where v_P is the velocity of particle. In the above expressions, the subscript "P" denotes the particle and "b" denotes the bulk air phase. The physical properties of bulk air such as specific heat (c_p), density (ρ), viscosity (μ) and thermal conductivity (k) can be calculated using correlations. All these gas-phase properties were calculated at an average film temperature (T_f) that is the arithmetic average of the particle temperature and the drying air temperature.

The mass transfer driving force ($X - X_e$) in the drying rate equation requires the determination of equilibrium moisture content data at drying conditions. The equilibrium moisture content can be calculated using desorption isotherms for skim milk. Several isotherms are available in the literature, for example SPS, Keey, Henderson, and Luikov models,[74] which represent the direct dependency of temperature on equilibrium moisture content and BET and GAB models don't show the direct dependency of temperature. Most of the published models in the literature have been found for ambient (lower) temperatures and for lower humidity values up to 45%. Here, the modified GAB model[75], which has been fitted at elevated temperatures (up to 90°C) and at higher relative humidity levels (up to 100%), was used to estimate the equilibrium moisture content data. The modified GAB equation used for calculating equilibrium moisture content (X_e) has the following form:

$$X_e = \frac{C \cdot K \cdot m_o \cdot a_w}{(1 - K \cdot a_w)(1 - K \cdot a_w + C \cdot K \cdot a_w)}, \tag{6.63}$$

where C and K are equation coefficients and m_o is the monolayer moisture content. The coefficients C and K are theoretically related to the sorption enthalpy (energy of interaction between the first and distant adsorbed molecules at individual water monolayer) and expressed as:

$$C = C_o \cdot \exp\left(\frac{\Delta H_1}{RT}\right), \tag{6.64}$$

and

$$K = K_o \exp\left(\frac{\Delta H_2}{RT}\right), \tag{6.65}$$

where C_o and K_o are fitting parameters, ΔH_1 and ΔH_2 are enthalpies of sorption of water, R is universal gas constant, and T is absolute drying air temperature (T_b). The fitting parameters C_o, K_o, ΔH_1, and ΔH_2 were obtained by Lin.[63] The above equations clearly show the dependency of temperature on the equilibrium moisture content.

The parameter a_w in the GAB equation is the water activity of the food sample. It is assumed that the skim milk droplet during drying is in continuous contact with the hot drying air, and they are in thermal equilibrium with each other. Hence, according to the definition of water activity, water activity of the droplet is equal to the fractional relative humidity (ψ) of the hot drying medium:

$$aw = \frac{\% \cdot \psi}{100}. \tag{6.66}$$

The relative humidity is defined as the partial pressure of water vapor (p_v) for the drying air divided by the pure component vapor pressure (saturated vapor pressure, p_{sat}) of water at the same temperature as the droplet.[76] Thus, water activity

$$aw = \frac{p_v}{p_{sat}}. \tag{6.67}$$

The partial vapor pressure of water vapor can be found by using the ideal gas law equation and can be represented in the following form:

$$p_v = \frac{P \cdot Y}{Y + (M_w/M_{dry\ air})},$$ (6.68)

where P is the total pressure of the system. Usually, the spray drying operations are carried out at atmospheric pressures; therefore, 1 atm total pressure was used during modeling of droplet drying process. The parameter Y is the absolute humidity of air (moisture concentration of drying medium, measured in kg water per kg dry air). The moisture concentration profile of the drying medium is obtained from the mass balance around the drying medium. In Equation 6.68, M_w and $M_{dry\ air}$ are molecular weights of water and dry air, respectively. The saturated vapor pressure (the maximum value of partial vapor pressure at the same temperature) of water was estimated using the three-coefficient Antoine equation as

$$\log p_{sat} = A - \frac{B}{T + C}.$$ (6.69)

In the above Antoine equation, T is the absolute drying medium temperature (T_b) and A, B, and C are coefficients for water vapor.

Reaction Engineering Approach (REA). Another possible model to estimate the drying profile of the droplets is the reaction engineering approach,[28] which considers the drying process as a competitive process between evaporation reaction and condensation reaction. If the droplet surface is not completely covered up with the water (i.e., no constant-rate drying period), the drying rate is well correlated by the following equation where the driving force is the vapor concentration depression:[62]

$$N_v = -\beta_1 \cdot (\rho_{v,s} - \rho_{v,b}),$$ (6.70)

where $\rho_{v,s}$ is the vapor concentration at solid–gas interface, $\rho_{v,b}$ is the bulk vapor concentration and β_1 is the mass transfer coefficient. The bulk vapor concentration $\rho_{v,b}$ can be estimated from the ideal gas law as

$$\rho_{v,b} = \frac{p_v \cdot M_w}{R \cdot T_b}.$$ (6.71)

The mass transfer coefficient β_1 can be calculated with the aid of Ranz–Marshall correlation, which has the following form:

$$Sh = \frac{\beta_1 \cdot d_P}{D_v} = 2 + 0.6 \cdot Re^{1/2} Sc^{1/3}.$$ (6.72)

In Equation 6.72, Sh is the dimensionless Sherwood number, D_v is the vapor–air diffusion coefficient and Sc is the dimensionless Schmidt number. The Reynolds number, Re, is given by the Equation 6.61 and the Schmidt number, Sc, can be given by

$$Sc = \frac{\mu_b}{\rho_b \cdot D_v}.$$ (6.73)

However, in Equation 6.3.23, the vapor concentration at the gas–solid interface $\rho_{v,s}$ is the unknown parameter, but it is familiar that $\rho_{v,s}$ should be less than saturated vapor concentration $\rho_{v,sat}$ for falling-rate drying period. Chen and Xie[28] attempted to express $\rho_{v,s}$ as a function of temperature in the following form:

$$\rho_{v,s} = \psi \cdot \rho_{v,sat}(T_s).$$ (6.74)

Here, T_s is the interface temperature, which can be considered as the particle temperature (T_p) because it is assumed that there is no temperature gradient within the particle itself. The parameter ψ is the fractionality coefficient relative to the saturation condition. The coefficient ψ will become unity if the liquid water fully covers up the droplet surface (vapor concentration at gas–solid surface is equal to the saturated vapor concentration) and will become smaller than unity when the drying begins to be hindered by the presence of solid components. Hence, the fractionality coefficient ψ becomes the equilibrium relative humidity according to the theoretical definition of the relative humidity. The fundamental assumption of the reaction engineering approach states that evaporation is an "activation reaction" that has to overcome an "energy barrier," but condensation does not have to. Based on this assumption, Chen and Xie[28] expressed the equilibrium relative humidity $\psi(X,T)$ as

$$\psi = \exp\left(-\frac{\Delta E_v}{R \cdot T}\right),\tag{6.75}$$

where ΔE_v is a "correction factor" in apparent activation energy for drying due to the increasing difficulty of removing water from the particle at low moisture content levels. The correction factor ΔE_v is expected to be zero when liquid water fully covers up the droplet surface ($\psi = 1$) and should gradually increase to a large value when the moisture content of the particle decreases to a small value. If an absolutely dry environment ($\psi = 0$) is applied to the droplets, the activation energy should approach infinity. Chen and Lin[31] expressed the apparent activation energy in terms of temperature and vapor concentration as individual characteristics of the droplets by combining Equation 6.74 and Equation 6.75 in the following form:

$$\Delta E_v = -RT \cdot \exp\left(\frac{\rho_{v,s}}{\rho_{v,sat}}\right).\tag{6.76}$$

By combining the Equation 6.74 and Equation 6.75 and substituting them into the drying rate, Equation 6.70, one may have the following expression for the drying rate profile:

$$N_v = -\beta_1 \cdot \left(\rho_{v,sat} \cdot \exp\left(-\frac{\Delta E_v}{RT}\right) - \rho_{v,b}\right).\tag{6.77}$$

Alternatively, one can express the above equation in terms of the rate of water evaporation as following:

$$\frac{dm_w}{dt} = -\beta_1 \cdot A_P \rho_{v,sat} \cdot \exp\left(\frac{\Delta E_v}{R \cdot T}\right) + \beta_1 \cdot A_P \cdot \rho_{v,b}.\tag{6.78}$$

The first term on the right hand side is a zero order drying reaction and that is the only activation process. The second term on the right hand side is a first order wetting reaction, but it is not activation process. Hence, Equation 6.78 represents that the drying is a competitive (or changing) process between evaporation and condensation (wetting) reactions and this competitive process would lead to the possible shrinkage of the droplets.

However, $\rho_{v,s}$ is still unknown in Equation 6.76 and the apparent activation energy need to be estimated to figure out drying profile of the droplets. Chen and Xie[28] and Chen and Lin[31] have introduced the normalized activation energy, which is the apparent activation energy relative to the equilibrium activation energy to estimate apparent activation energy (ΔE_v). The equilibrium activation energy is the possible lowest value of apparent activation energy. They have correlated the normalized activation energy as a function of free moisture content ($X - X_e$). The correlation was given by the following equation:

$$\frac{\Delta E_v}{\Delta E_{v,e}} = f(X - X_e) = a \cdot \exp\left[b \cdot (X - X_e)^c\right].\tag{6.79}$$

Here, the activation energy ratio $\Delta E_v/\Delta E_{v,e}$ is the normalized activation energy, $\Delta E_{v,e}$ is the "equilibrium" activation energy, and a, b, and c are constants. The values of constants a, b, and c for skim milk are taken from Chen and Lin (2004). The equilibrium moisture content $X_e(T_b)$ can be calculated by the GAB model, Equation 6.63. The equilibrium activation energy at drying air conditions can be given by the following equation:

$$\Delta E_{v,e} = -R \cdot T_b \cdot \ln\left(\frac{\rho_{v,b}}{\rho_{v,sat}}\right). \tag{6.80}$$

It can be seen from Equation 6.79 that the normalized activation approaches zero when the free moisture content of the droplet is high, meaning that the drying process becomes the evaporation of pure water. When the free moisture content decreases to a lower value $(X \rightarrow X_e)$, the normalized activation energy approaches to unity, i.e., the apparent activation energy ΔE_v becomes the equilibrium activation energy $\Delta E_{v,e}$, and the droplet would reach equilibrium with the drying air. This way, the relation between the activation energy and the free moisture content of the material can be used to predict the drying rate and water evaporation rate profile during modeling of a drying process.

Mass Balance in the Drying Air. A similar mass balance can be written for the drying medium by considering Figure 6.13. The hot drying medium is in dynamic equilibrium with a single vertical stream of droplets. As the hot drying air passes through the dryer, it escapes vapor from the droplet surface, and the air becomes more humid. The spray dryer geometry is divided into equal sized mesh of well-defined area to simulate the governed system. In Figure 6.13, one single mesh is shown which contains θ number of droplets within the specified area. The hot drying air enters in the mesh at distance h and it exits at distance $h+dh$.

The inlet humidity of drying air is Y kg water per kg dry air, and at the exit the humidity is $(Y+dY)$ kg kg^{-1}. As mentioned before, it is assumed here that only vapor is being transferred during a droplet drying processing. The mass balance can be formulated (by balancing amount of vapor) in the form of a following mathematical equation:

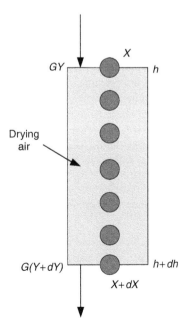

Figure 6.13 Mass balance of the drying air.

$$\theta \cdot m_s \cdot X - G \cdot Y = \theta \cdot m_s \cdot (X + dX) - G(Y + dY), \tag{6.81}$$

where, G is the mass flow rate of drying air in kg s^{-1}.
Simplifying Equation 6.71, one may obtain

$$\theta \cdot m_s \cdot dX = G \cdot dY. \tag{6.82}$$

Because changes of various physical properties and drying parameters along the dryer length are of interest, the velocity of the particle should be defined relative to the dryer length to get the humidity profile along the dryer length:

$$v_P = \frac{dh}{dt}. \tag{6.83}$$

By introducing velocity parameter in the Equation 6.82,

$$\theta \cdot m_s \cdot \frac{dX}{dt} = v_P \cdot G \frac{dY}{dh}. \tag{6.84}$$

On rearrangement of Equation 6.84,

$$\frac{dY}{dh} = \frac{\theta}{G \cdot v_P} \cdot \frac{dm_w}{dt}. \tag{6.85}$$

Equation 6.85 can be used to determine the moisture concentration profile of the drying medium. The water evaporation rate dm_w/dt can be determined by combining Equation 6.50 and any of the drying rate equations formulated before.

6.3.1.2 Heat-Transfer Model

The heat balance around the droplet phase is written based on the assumption that there is no temperature gradient within the droplet as long as the Biot number is less than 0.1. Another possible argument for ignoring the temperature gradient inside the droplet is that the characteristic time for internal heat transfer is much smaller than for internal mass transfer. Moreover, a part of the heat transferred to the surface is used immediately for the evaporation of water. Therefore, the droplet surface temperature can be considered as the average droplet temperature.

The heat balance around the evaporating droplet, moving in the hot air stream, is shown in Figure 6.14. The droplet has the initial temperature T_P (K). When the droplet comes into contact with the hot drying air, the droplet acquires the sensible heat (that raises the droplet temperature) and the latent heat (that changes the feed phase from liquid to vapor). As the droplet-hot air system is in dynamic and thermal equilibrium with each other, the droplet temperature should approach the bulk air temperature and both temperatures should be same at the exit. The energy balance around a droplet being dried may be given by the following heat-transfer model:[73]

$$m_P c_P \frac{dT_P}{dt} = \alpha \cdot A_P \cdot (T_b - T_P) + \Delta H_L \cdot \frac{dm_w}{dt}. \tag{6.86}$$

The left-hand side of the Equation 6.86 expresses the heat change of the droplet during convective drying, where c_p is the heat capacity of the particle; α is the heat-transfer coefficient due to convection (W m^{-2} K^{-1}). The first term on the right-hand side represents the convective heat being transferred from the hot drying air to a droplet because of the temperature difference between two phases. The parameter α is the heat-transfer coefficient and can be calculated using Ranz–Marshall correlation that was described by Equation 6.60. The second term on the right-hand side represents the amount of heat supplied to the evaporating liquid for change of phase from liquid

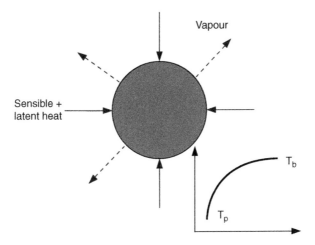

Figure 6.14 Energy balance around a single droplet.

to vapor. Equation 6.86 can be used to predict the droplet temperature profile during spray drying operation.

A heat balance can be formulated for the bulk drying medium using the step-wise heat balance by considering an infinitesimal section as shown in Figure 6.15. Here it is assumed that the heat losses through the dryer walls are negligible. The enthalpy of the bulk air at the entrance of the illustrated mesh is E_b (J kg^{-1}). The hot air supplies the convective heat to the droplets and in this way the enthalpy of air is being changed. By considering Figure 6.15, the energy balance over drying air can be written in the form of mathematical equation:

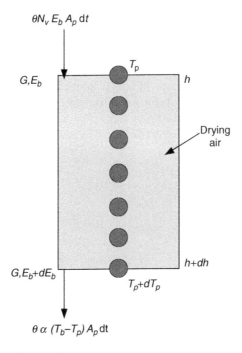

Figure 6.15 Energy balance in the gas phase.

$$G \cdot E_b - G \cdot (E_b + dE_b) = \alpha \cdot (T_b - T_P) \cdot \theta \cdot A_P \cdot dt - N_v \cdot E_b \cdot \theta \cdot A_P \cdot dt. \tag{6.87}$$

On reduction and rearrangement of the above equation,

$$G \cdot dE_b = \left[N_v \cdot E_b - \alpha \cdot (T_b - T_P) \right] \cdot \theta \cdot A_P \cdot dt. \tag{6.88}$$

The enthalpy of the gas (drying air) can be given by the following equation:

$$E_b = c_p \cdot T_b + \Delta H_v \cdot Y, \tag{6.89}$$

where, c_p is the specific heat of the bulk drying air. The bulk drying air is the mixture of vapor and dry air. The specific heat of bulk mixture can be calculated as:

$$c_{pb} = c_{p \text{ dry air}} + Y \cdot c_{p \text{ vapor}}. \tag{6.90}$$

From Equation 6.88 through Equation 6.90,

$$G \cdot c_{p \text{ dry air}} \cdot dT_b + G \cdot (\Delta H_v + c_{p \text{ vapor}} \cdot T_b) dY$$

$$= \theta A_P \left[N_v \cdot \left\{ c_{p \text{ dry air}} \cdot T_b + (\Delta H_v + c_{p \text{ vapor}} \cdot T_b) Y \right\} - \alpha \cdot (T_b - T_P) \right]. \tag{6.91}$$

On introducing the velocity of particle in above equation, one may have

$$G \cdot c_{p \text{ dry air}} \cdot \frac{dT_b}{dh} + G \cdot (\Delta H_v + c_{p \text{ vapor}} \cdot T_b) \cdot \frac{dY}{dh}$$

$$= \frac{\theta \cdot A_P}{v_P} \cdot \left[N_v \cdot \left\{ c_{p \text{ dry air}} \cdot T_b + (\Delta H_v + c_{p \text{ vapor}} \cdot T_b) Y \right\} - \alpha \cdot (T_b - T_P) \right]. \tag{6.92}$$

Equation 6.92 predicts the temperature profile of the drying medium if the moisture concentration profile for the drying medium is known.

6.3.1.3 Momentum Transfer Model

The change in the droplet velocity is calculated by using momentum balance around a single droplet. This requires the consideration of all forces acting along the falling droplet. The force balance around a single droplet for co-current flow is shown in Figure 6.16. Basically, three forces act onto the falling droplet during one-dimensional analysis: the downward gravity force (F_g), the upward drag force (F_D) and the upward buoyant force (F_B). The resulting net force (F_A) is balanced with the acceleration (or deceleration) of the droplet. The individual forces can be calculated using the following equations:

$$F_g = \frac{\pi}{6} \cdot d_p^3 \cdot \rho_p \cdot g, \tag{6.93}$$

where g is the gravitational constant.

$$F_D = \frac{\pi}{4} \cdot d_p^2 \cdot C_D \cdot \rho_b (v_b - v_p)^2 \tag{6.94}$$

$$F_B = \frac{\pi}{6} \cdot d_p^3 \cdot \rho_b \cdot g \tag{6.95}$$

$$F_A = \frac{\pi}{6} \cdot d_p^3 \cdot \rho_p \cdot \frac{dv_p}{dt}. \tag{6.96}$$

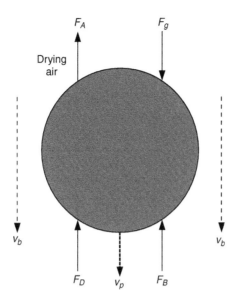

Figure 6.16 Moment balance of a single droplet traveling in drying air.

The drag force represented in Equation 6.94 dominates the droplet motion and consists of the friction force and form drag. The parameter C_D is the drag coefficient and depends mainly on the droplet's Reynolds number. Here, the Reynolds number should be calculated using the relative velocity $(v_b - v_P)$. For a spherical evaporating droplet, the drag coefficient can be defined as:[77]

$$C_D = \frac{24}{Re} \cdot \frac{1 + 0.2 \cdot Re^{0.63}}{(1 + B)^{0.2}}, \tag{6.97}$$

where B is the Spalding number (transfer coefficient) that can be estimated by

$$B = \frac{Cp_v \cdot (T_b - T_p)}{\Delta H_v}. \tag{6.98}$$

In Equation 6.98, Cp_v is the specific heat of water vapor at the film temperature. The force balance around a single droplet can be expressed as

$$F_A = F_g - F_D - F_B. \tag{6.99}$$

By using Equation 6.93 to Equation 6.96, one can derive the following equation that represents the momentum balance around a falling droplet:

$$\frac{dv_P}{dt} = \left[\left(\frac{\rho_P - \rho_b}{\rho_P} \right) \cdot g - \left\{ \frac{1.5 C_D \cdot \rho_b}{d_p \cdot \rho_P} (v_b - v_P)^2 \right\} \right]. \tag{6.100}$$

Equation 6.100 represents the velocity profile of the evaporating droplet. The velocity of the bulk drying medium can be calculated using the mass flow rate and dryer cross-sectional area as

$$v_b = \frac{G}{\rho_b \times \text{Cross–sectional Area}}. \tag{6.101}$$

6.3.1.4 Prediction of Physical Properties of the Products

It is important to predict the physical properties of the final product before it is manufactured. In the present study, models are formulated to estimate the physical properties in terms of particle density and particle size. The particle density of milk powders is an important property in terms of the physical quality and functionality. The particle density also determines the bulk density that is an important property from the point of view of the economy and market requirements. In determining the particle density, the shrinkage of the droplet is accounted for by using a shrinking balloon approach.[78] The shrinking balloon approach assumes that the volume of the particle is decreased by only the volume of the moisture leaving the particle that means the ideal shrinkage. Then, the rate of moisture evaporation is balanced with the rate of moisture transfer from the center to the droplet surface. During evaporation, the particle density will increase since the moisture is escaping due to the rapid evaporation. The moisture being transferred at any specific time t can be written as

$$dm_w = d(\rho_w \cdot V_w) = \rho_w \cdot d(V_w). \tag{6.102}$$

But it is assumed that the volume change of the particle is equal to the volume of moisture being escaped; therefore Equation 6.102 can be written as

$$dm_w = \rho_w \cdot d(V_p) = \rho_w \cdot d\left(\frac{m_p}{\rho_p}\right). \tag{6.103}$$

According to the definition of moisture content, the mass of water is described as:

$$m_w = m_s \cdot X \cdot \tag{6.104}$$

By substituting Equation 6.104 into Equation 6.103, one can obtain

$$m_s \cdot dX = \rho_w \cdot d\left(\frac{m_p}{\rho_p}\right) = \rho_w \cdot d\left(\frac{m_s + m_w}{\rho_p}\right). \tag{6.105}$$

On rearrangement,

$$\frac{dX}{\rho_w} = d\left(\frac{m_s + m_w}{m_s} \cdot \frac{1}{\rho_p}\right) = d\left(\frac{1 + X}{\rho_p}\right). \tag{6.106a}$$

Equation 6.106a can be expressed in the following integration form:

$$\frac{1}{\rho_w} \cdot \int dX = \int d\left(\frac{1 + X}{\rho_p}\right) \tag{6.106b}$$

The above equation can be solved using the following boundary conditions:

$$\text{At } X = 0, \quad \rho_p = \rho_s$$
$$\text{When } X \neq 0, \quad \rho_p = \rho_p. \tag{6.107}$$

Using the boundary conditions, Equation 6.106bcan be integrated and rearranged in the following form:

$$\rho_p = \rho_s \cdot \frac{1 + X}{1 + \frac{\rho_s}{\rho_w} \cdot X}. \tag{6.108}$$

The particle density can be predicted using Equation 6.108. The equation clearly shows the dependence of particle density on moisture content of the material being dried. The moisture content profile for the material can be calculated by using mass transfer model.

The particle size model may be estimated based on the assumption that the particle remains completely spherical during the entire processing. The mass of water balance can then be written as:

$$\rho_w \cdot dV_w = \rho_w \cdot dV_p = d(V_p \cdot \rho_p). \qquad (6.109)$$

Integration of the above equation with following boundary conditions,

$$\text{when } V_p = V_{p_0}, \quad \rho_p = \rho_{p_0};$$
$$\text{when } V_p \neq 0, \quad \rho_p = \rho_p,$$

leads to

$$\ln \frac{V_p}{V_{p_0}} = \ln \frac{\rho_{p_0} - \rho_w}{\rho_p - \rho_w}. \qquad (6.110)$$

By defining the relations between volume of particle and diameter of particle, one can write that

$$d_p = d_{p_0} \cdot \sqrt[3]{\frac{\rho_{p_0} - \rho_w}{\rho_p - \rho_w}}. \qquad (6.111)$$

The size of the dried particles can be predicted using Equation 6.111. The formulated model shows that the particle size depends on initial droplet diameter and initial droplet density. This requires suitable experimental techniques to measure droplet diameter and droplet density. The final density of the particle can be evaluated using Equation 6.108. It can be seen that a continuous change in particle diameter takes place during drying, which has a significant effect on heat and mass transfer coefficients and on drag coefficient.

The predicted results using the two approaches were similar in trend in many occasions. Figure 6.17 and Figure 6.18 show the REA model predictions of temperature, water content and

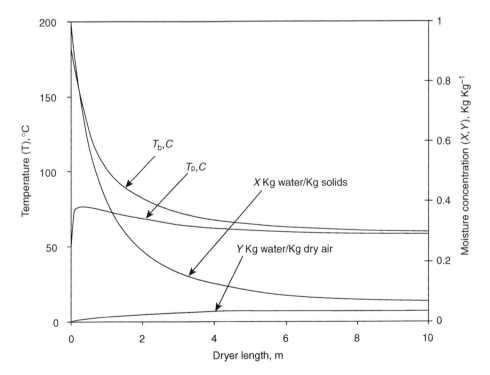

Figure 6.17 Predicted trends of temperatures, humidity and moisture content.

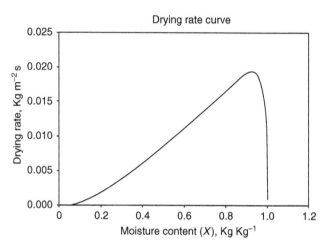

Figure 6.18   Predicted drying rate characteristics.

rate of drying at different stages of drying in the plug flow dryer. The simulation conditions are given in Table 6.1.

The models discussed thus far assumed uniform temperature within the droplet/particle being dried. It is expected, however, that the surface temperature would be higher than the core temperature. This surface temperature plays a role in determining the surface properties like stickiness. Thus far it has been impossible to measure or to realistically estimate this nonuniformity for such small droplet/particles. Future work is expected to investigate exactly how high the surface temperature in comparison with the core temperature.

6.3.2   Baking Bread

Bread has been one of the major cereal based products in West and is found to be increasingly popular in various forms in the East. Bread baking is an ancient art. Only until recently, due to the huge quantity demanded by the consumers and the high consistency of the products needed, the process of bread baking has become an act of engineering. The engineering studies of this topic have been reported by Therdthai et al.[79–82] and Zhou.[83]

Table 6.1   Summary of the Modeling Conditions for Figure 6.17 and Figure 6.18

Parameter	Value
Droplet diameter	0.0002 m
Droplet temperature (T_p)	50°C
Total solids content	50%
Droplet moisture content (X)	1 kg water/kg solids
Mass flow rate of skim milk	10,000 droplets/s
Mass flow rate of skim milk	0.18 kg/h
Air temperature (T_b)	180°C
Air humidity (Y)	0.0065 kg water/kg dry air
Mass flow rate of air	0.0009 kg/s
Volumetric flow rate of air	3.72 m³/h

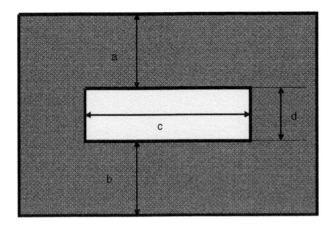

Figure 6.19 Schematic diagram of bread baking (the thick lines denote the walls of the baking oven; the central block is the bread; a, b, c, and d are, respectively, the important dimensions used in the modeling). (Modified from a diagram kindly provided by W. B. Zhou at National University of Singapore, 2005.)

At the very least, baking bread involves radiation, convection and water movement. The heat and the relatively dry conditions in the baking oven drive out moisture from the bread dough to some extent. The dough also expands due to the CO_2 generated as part of the pre-fermentation. The bubble growth in the dough while baking places a key role in the texture and final acceptance of the bread quality. Here only the heat and moisture transfer aspects are looked at.

As a simple example to show the principle, a large bread slab is considered (see Figure 6.19). Assuming no expansion occurs, the following equations govern the heat and mass transfer processes.

6.3.2.1 Inside the Bread

- Heat transfer:

$$\frac{\partial T}{\partial t} = \frac{1}{\rho c_{\mathrm{p}}} \frac{\partial}{\partial x}\left(k\frac{\partial T}{\partial x}\right) + \frac{\lambda}{c_{\mathrm{p}}}\frac{\partial W}{\partial t}, \qquad 0 < x < x_{L}/2, \qquad t > 0. \tag{6.112}$$

- Vapor transfer:

$$\frac{\partial V}{\partial t} = \frac{\partial}{\partial x}\left(D_{\mathrm{V}}\frac{\partial V}{\partial x}\right), \qquad 0 < x < x_{L}/2, \qquad t > 0. \tag{6.113}$$

- Liquid water transfer

$$\frac{\partial W}{\partial t} = \frac{\partial}{\partial x}\left(D_{\mathrm{W}}\frac{\partial W}{\partial x}\right), \qquad 0 < x < x_{L}/2, \qquad t > 0. \tag{6.114}$$

• Boundary and initial conditions:

$$-k\left[\frac{\partial T}{\partial x}\right]_{x=0} = h_r(T_r - T_s) + h_c(T_{air} - T_s) - \lambda \rho D_w \left[\frac{\partial W}{\partial x}\right]_{x=0},$$

$$\left[\frac{\partial T}{\partial x}\right]_{x=x_L/2} = 0, \quad t > 0. \tag{6.115}$$

$$T(x,0) = T_0(x), \quad 0 \leq x \leq x_L/2.$$

$$\left[\frac{\partial V}{\partial x}\right]_{x=0} = h_V(V(0,t) - V_{air}), \quad \left[\frac{\partial V}{\partial x}\right]_{x=x_L/2} = 0, \quad t > 0.$$

$$V(x,0) = V_0(x), \quad 0 \leq x \leq x_L/2. \tag{6.116}$$

$$\left[\frac{\partial W}{\partial x}\right]_{x=0} = h_W(W(0,t) - W_{air}), \quad \left[\frac{\partial W}{\partial x}\right]_{x=x_L/2} = 0, \quad t > 0.$$

$$W(x,0) = W_0(x), \quad 0 \leq x \leq x_L/2. \tag{6.117}$$

where $T(x,t)$ is temperature in K, x is space coordinate in m, and t is time in s. $V(x,t)$ is water vapor content in (kg water)/(kg product), D_V is water vapor diffusivity in $m^2 s^{-2}$. ρ is apparent density in $Kg\,m^{-3}$, c_p is specific heat in $J\,kg$, k is thermal conductivity in $J\,Kg^{-1}K^{-1}$, λ is latent heat of evaporation of water in $W\,m^{-2}K^{-1}$, $W(x,t)$ is liquid water content in (kg water)/(kg product), x_L is the thickness of the bread slab. h_r (simplified from Equation 6.30) and h_c are heat transfer coefficients due to radiation and convection respectively, in $W\,m^{-1}K^{-1}$. T_r and T_{air} are radiation source temperature and surrounding oven air temperature, respectively, in K. $T_s = T(0,t)$ is bread surface temperature in K. D_W is liquid water diffusivity in m^2 s. h_V and h_W, in $L\,m^{-1}$, are mass transfer coefficients of water vapor and liquid water at the bread surface, respectively. V_{air} and W_{air} are water vapor content and liquid water content of the oven air, respectively. T_0, V_0, and W_0 are initial temperature, initial water vapor content and initial liquid water content of the bread, respectively.

It is interesting to note that there are two separate transfer coefficients used for vapor and liquid water transfer at the outer boundary of the bread, which are not conventional.

The transfer coefficients were estimated by using natural convection theory for calculating Nu, Pr, and Gr etc. They were correlated to temperature and water content. The values used were: $h_V = 3.2 \times 10^9/T^3$, $h_W = 1.4 \times 10^{-3}\,T + 0.27W - 4.0 \times 10^{-4}TW - 0.77W^2$. Note here that these equations can only be solved numerically. Here, the Implicit Euler method was used to carry out computations (Time step $= 30$ s, space step $= 0.125$ cm, initial temperature $T_o = 25°C$, oven temperature $= 210°C$, humidity $= 0\%$).

The simulated results are shown qualitatively Figure 6.20. Obviously, the liquid water content at the surface of the beard is rapidly removed, and the surface temperature rises quickly. The inside region dries more slowly, and the temperature rise is delayed as expected. The half-way temperature is quite close to the center temperature, which may be due to the large evaporation effect in the surface region, which consumes the largest portion of the temperature gradient.

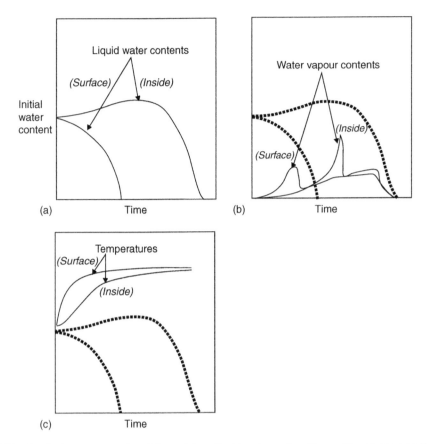

Figure 6.20 Qualitative representation of the moisture (liquid and vapor) profiles and temperature profiles of surface and inside of a bread sample. (Modified from a diagram kindly provided by W. B. Zhou at National University of Singapore, 2005.)

The complex geometry of the industrial baking devices requires the full-three-dimensional modeling of the oven spaces surround the baking line (where the breads are carried in and out of the line). Natural convection must also be taken into account. Here, the Navier–Stokes equations for fluid flow need to be solved together with the heat and mass conservation equations.

6.3.3 Further Examples on Drying, Frying, and Baking

To complete the chapter, an added list of useful references, which cover the modeling of food drying, baking and frying, has been provided with broad classifications under drying,[83–179] frying,[180–200] and baking.[201–213] It may be useful for the readers to examine these references for their specific interest in the topic of simultaneous heat and mass transfer in food processing. Furthermore, a review of the food related studies on transport phenomena has been given by Welti-Chanes et al.[214] which may also be of interest to the readers.

6.4 CONCLUDING REMARKS

Transport phenomena have been one of the most well studied engineering topics in science and engineering. Applications of the known principles in food processing area take many forms, which can be innovative and satisfying. Foods are complex materials to work with and most processes are of time-dependent by nature. The variability in food sources, structure changes during processing, post-process handling, etc. all have impact on the reliability of the type of mathematical model established. Food properties are of prime interest in any model development. The combination of the transport phenomena modeling and the models of food quality changes (chemical, physical, or microstructural) will likely be the most topical approach in the near future. In this chapter, though never intended to be comprehensive, only selected examples are given to show indeed the diversity involved in modeling heat and mass transfer processes which all have achieved certain degree of success. It is hoped that the materials provided are reasonably thought provoking. There is clearly a great room for further validations and improvements in the area.

NOMENCLATURE

$\bar{X}$	Average or mean water content on dry basis, $kg\ kg^{-1}$
$\hat{N}v$	Drying rate for constant drying-rate period, $kg\ m^{-2}\ s^{-1}$
$\dot{m}_v$	Mass flux of water vapor, $kg\ s^{-1}$
$\dot{E}_v$	Internal local evaporation rate, $kg\ m^{-}\ s^{-1}$
$\dot{S}_v$	Rate of surface evaporation, $kgm^{-2}\ s^{-1}$
ΔE_v	Apparent activation energy, $J\ mol^{-1}$
$\Delta E_{v,e}$	Equilibrium activation energy, $J\ mol^{-1}$
ΔH_1	Heat of sorption of water in GAB model, $J\ mol^{-1}$
ΔH_2	Heat of sorption of water in GAB model, $J\ mol^{-1}$
ΔH_L	Latent heat of water vaporization, $J\ kg^{-1}$
ΔH_v	Latent heat of vaporisation, $J\ mol^{-1}$
ΔH_w	Heat of wetting, $J\ kg^{-1}$
A	Antoine equation coefficient or surface area, m^2
A_h	Cross section area of air disperser hole, m^2
A_n	Cross section area of the single droplet nozzle, m^2
A_{ori}	Cross section area of the nozzle's orifice, m^2
A_p	Surface area of single droplet or particle, m^2
a_w	Water activity
b	Half-thickness, m
B	Spalding number, Antoine equation coefficient
Bi	Biot number (for heat transfer $Bi=hL/k$ or for mass transfer $Bi_m=h_mL/D$)
$Bi*$	Modified Biot number
C	GAB equation parameter, Antoine equation coefficient or concentration, $kg\ m^{-3}$
C_0	GAB equation coefficient
C_D	Drag coefficient
Ch_Bi	Chen–Biot number as defined in the text
Ch_Le	Chen–Lewis number as defined in the text
C_p or c_p	Specific heat capacity, $J\ kg^{-1}\ K^{-1}$
D	Diffusivity, $m^2\ s^{-1}$
d_D	Diameter of spray dryer, m
d_h	Diameter of air disperser hole, m
d_n	Outer diameter of single droplet nozzle, m

d_{ori}	Diameter of the nozzle's orifice, m
d_p	Diameter of droplet or particle, m
D_v	Vapor-in-air diffusivity, $m^2 s^{-1}$
E	Enthalpy, $J kg^{-1}$
F	Frequency of Droplets, s^{-1}
f	Relative drying rate
F_A	Acceleration force, N
F_B	Buoyancy force, N
F_D	Drag force, N
F_g	Gravitational force, N
g	Gravitational acceleration, $m s^{-1}$
G	Mass flow rate of the bulk-drying medium, $kg s^{-1}$
h	Heat transfer coefficient, $W m^{-2} K^{-1}$, or distance from the tip of the atomiser (dryer height), m
$h*$	Equivalent heat transfer coefficient, $W m^{-2} K^{-1}$
h_m	Mass transfer coefficient, $m s^{-1}$
k	Thermal conductivity, $W m^{-1} K^{-1}$
K	GAB equation parameter
K_0	GAB equation coefficient
L	Thickness of the material, m
Le	Lewis number ($Le = \alpha/D$)
M	Molecular weight, $g mol^{-1}$
m_o	Monolayer moisture content in GAB model, $kg kg^{-1}$
M_f	Mass flow rate of feed, $kg s^{-1}$
m_s	Mass of solids, kg
m_w	Mass of water, kg
n	Number concentration of the particles, $L m^{-3}$
Nu	*Nusselt* number ($Nu = hl/k$) (k here is that of the fluid thus differing from that for Biot number calculation)
N_v	Drying rate, $kg m^{-2} s^{-1}$
P	Product rate, $kg s^{-1}$, or pressure, $N m^{-2}$
Pr	*Prandtl* number ($Pr = v/\alpha$, where v is the kinematic viscosity, $m^2 s^{-1}$)
p_{sat}	Saturated vapor pressure, atm
p_v	Partial pressure of vapor, atm
Q	Convective heat flow, J
q	Heat flux, $W m^{-2}$
R	Particle radius (m)
r	Radial coordinate (m)
R	Ideal gas law constant
Re	Reynolds number ($Re = \rho u L/\mu$)
RH	Relative humidity
S	Spacing between two subsequent droplets, m
Sc	Schmidt number ($Sc = v/D$)
Sh	*Sherwood* number ($Sh = h_m L/D$) (D here is that in the fluid thus differing from that for *Biot* number for mass transfer calculation)
T	Temperature, K
t	Time, s
t_D	Drying time, s
t_R	Residence time, s
T_{sat}	Adiabatic saturation temperature of drying air, K

T_{wb}	Wet-bulb temperature of drying air, K
U_m	Overall mass transfer coefficient, m s^{-1}
v	Velocity, m s^{-1}
V	Volume, m^3, or, water vapor content, kg kg^{-1}
V_b	Volumetric flow rate of bulk-air, m^3 s^{-1}
V_f	Volumetric flow rate of feed, m^3 s^{-1}
v_{jet}	Velocity of liquid jet or column, m s^{-1}
v_r	Relative velocity, m s^{-1}
v_t	Terminal settling velocity, m s^{-1}
x	Distance or x-coordinate, m
X_{cr}	Critical moisture content on dry basis, kg kg^{-1}
X_e	Equilibrium moisture content on dry basis, kg kg^{-1}
Y	Air humidity, kg kg^{-1}

Greek Symbols

ρ	Density, kg m^{-3}
λ	Latent heat of water vaporization, J kg^{-1}
σ	Stefan–Boltzmann constant
ϕ	Characteristic moisture content
ρ	Density, kg m^{-3}
μ	Dynamic viscosity, Pa s
η	Efficiency
α	Heat transfer coefficient, W m^{-2} K^{-1}, or thermal diffusivity ($\alpha = k/\rho c_p$), m^{2}s^{-1}
β	Mass transfer coefficient, m s^{-1}
ε	Porosity
ψ	Relative humidity ($=RH$)
δ	Thickness, m
τ	Tortuosity
μ	Viscosity, Pa s
β_1	External mass-transfer coefficient, s m^{-1}
β_2	Coefficient, s(m kg)$^{-1}$
β_3	Coefficient, s^{-1} K^{-1}
ρ_v	Vapor concentration, kg m^{-3}

Subscripts

∞	Bulk, surrounding or equilibrium
1,2	Surface 1 and surface 2
A	Ambient condition
air	Air properties
av	Average
b	Bulk drying medium phase
c	Characteristic parameters
cond	Conduction
conv	Convection
cr	Critical conditions
D	Dryer
e	Equilibrium or ambient conditions
E	Exit condition
eff	Effective parameters

f	Feed
h	Air disperser hole
I	Inlet condition
l	Liquid
L	Thickness
M	Mass
m	Mass transfer
n	Nozzle
o	Center temperature or initial values
p	Particle parameters
s	Solid, surface
sat	Saturated conditions
sp	Porous solid particle
T	Thermal
V, v	Water vapor
W, w	Liquid water
x	Space coordinate
y	Space coordinate
z	Space coordinate

APPENDIX A: TYPICAL CONVECTIVE MASS-TRANSFER CORRELATIONS

Heat convection is a well-studied subject and is well documented in many texts.[62] The mass transfer calculations can be done, as the first approximation, based on the heat transfer correlations. The mass transfer coefficient (h_m) is obtained from established correlations for Sherwood number ($h_m L)/D$ originally for Nusselt number hl/kL is the characteristic length of the object (m) and k is the thermal conductivity of the bulk fluid (W m^{-1} K^{-1}). The two principal numbers are correlated to the Reynolds number ($(\rho u L)/\mu$), the Prandtl number (v/α) or Schmidt number (v/D), respectively. u is the bulk fluid velocity (m s^{-1}). ρ is the fluid density (kg m^{-3}), μ is the viscosity (Pa s) and α is the thermal diffusivity (m^2 s^{-1}), and D is the mass diffusivity (m^2 s^{-1}).

All the physical properties used in the calculations usually are determined at the film temperature ($T_f = (T_s + T_\infty)/2$) and film concentration ($C_f = (C_s + C_\infty)/2$). Table 6.A1 shows the typical correlations, which may be used as the first approximation for evaluating mass transfer coefficients.

ACKNOWLEDGMENTS

The author is very grateful for the kind assistance provided by Kamlesh Patel during the manuscript preparation. In particular his permission to quote elaborately his thesis work is much appreciated. Thanks also to Professor Weibiao Zhou for providing his material on CFD of baking processes to the author. Finally the author thanks his wife, Lishun Meng for her patience when some parts of the manuscript had to be completed during overtime.

Table 6.A1 Commonly Used Mass-Transfer Correlations for Average Sherwood Number Calculations

Configuration	Geometry	Conditions	Correlation
External flow (*forced convection*)	Flat plate	Laminar, $0.6 \le Pr \le 50$, L: length of the plate (m)	$\overline{Sh}_L = 0.664 Re_L^{1/2} Sc^{1/3}$
		Mixed, $Re_{L,c} = 5\times10^5$, $Re_L \le 10^8$, $0.6 \le Pr \le 50$, L: length of the plate (m)	$\overline{Sh}_L = (0.037 Re_L^{4/5} - 871) Sc^{1/3}$
	Cylinder (pipe or tube)	$Re_L Pr > 0.2$, L: diameter (L=d)	$\overline{Sh}_L = 0.3 + \dfrac{\left[0.62 Re_L^{1/2} Sc^{1/3} \times \left[1 + (0.4/Sc)^{2/3}\right]^{-1/4}\right]}{\left[1 + (Re_L/282,000)^{5/8}\right]^{4/5}}$
	Sphere	L: diameter (L=d)	$\overline{Sh}_L = 2 + 0.6 Re_L^{1/2} Sc^{1/3}$
External flow (*natural convection*)	Vertical plate	L: length of the plate	$\overline{Sh}_L = \left\{ 0.825 + \dfrac{0.387 Ra_L^{1/6}}{\left[1 + \left(\frac{0.492}{Sc}\right)^{9/16}\right]^{8/27}} \right\}^2$
	Upper surface of heated plate	$L = A_s/P$, A_s: plate surface area, P: plate perimeter, $Ra_L = g\beta (T_{wall} - T_\infty) L^3 / \alpha\nu$, $10^5 \le Ra_L \le 10^{10}$	$\overline{Sh}_L = 0.54 Ra_L^{1/4}$, $10^4 \le Ra_L \le 10^7$ $\overline{Sh}_L = 0.15 Ra_L^{1/3}$, $10^7 \le Ra_L \le 10^{11}$
		$L = A_s/P$, A_s: plate surface area, P: plate perimeter, $10^5 \le Ra_L \le 10^{10}$	$\overline{Sh}_L = 0.27 Ra_L^{1/4}$
	Cylinder (pipe or tube)	L = d (diameter), $Ra_L \le 10^{12}$	$\overline{Sh}_L = \left\{ 0.60 + \dfrac{0.387 Ra_L^{1/6}}{\left[1 + \left(\frac{0.559}{Sc}\right)^{9/16}\right]^{8/27}} \right\}^2$
	Sphere	L = d (diameter), $Pr \ge 0.7$, $Ra_L \le 10^{11}$	$\overline{Sh}_L = 2 + \dfrac{0.589 Ra_L^{1/6}}{\left[1 + \left(\frac{0.469}{Sc}\right)^{9/16}\right]^{4/9}}$
Internal flow (*forced convection*)	Laminar	Fully developed, uniform wall heat flux, $Pr \ge 0.6$, $L = 4A_c/P$ (hydraulic diameter), A_c: flow cross-sectional area, P: wetted perimeter	$\overline{Sh}_L = 4.36$
		Fully developed, uniform wall temperature, $Pr \ge 0.6$	$\overline{Sh}_L = 3.66$
		Turbulent, fully developed, $0.6 \le Pr \le 16,700$, Length/diameter ≥ 10, $Re_L \ge 10,000$, μ_s: viscosity at wall	$\overline{Sh}_L = 0.027 Re_L^{4/5} Sc^{1/3} \left(\dfrac{\mu}{\mu_s}\right)^{0.14}$
Internal flow (*natural convection*)	Horizontal cavity heated from below	$Ra_L = g\beta (T_1 - T_2)L^3/\alpha\nu > 1708$, T_1: the bottom surface temperature, T_2: the upper surface temperature, β: thermal expansion coefficient ($\approx 1/T_1$), L: distance between two horizontal walls, $3\times10^5 \le Ra_L \le 7\times10^9$	$\overline{Sh}_L = 0.069 Ra_L^{1/3} Sc^{0.074}$
	Vertical cavity heated from one side and cooled at the other	$2 < H/L < 10$, $Pr < 10$, $10^3 \le Ra_L \le 10^{10}$, H: height of the cavity, L: distance between two vertical walls	$\overline{Sh}_L = 0.22 \left(\dfrac{Sc}{0.2+Sc} Ra_L\right)^{0.28} \left(\dfrac{H}{L}\right)^{-0.25}$
Combined natural and forced convection		n is often quoted to be 3 though 7/2 and 4 are better for transverse flows (i.e., two effects are "opposite to each other" and the sign in the correction is "−")	$\overline{Sh}_L^n = \overline{Sh}_{forced}^n \pm \overline{Sh}_{natural}^n$

Source: Modified from Incropera, F. P. and DeWitt, D. P., *Fundamentals of Heat and Mass Transfer*, 4th and 5th ed., New York: Wiley, 1990 and 2002.

REFERENCES

1. Datta, A. K., Physics based models of heat and mass transfer, Heat Transfer, In *Handbook of Food and Bioprocess Modeling Techniques*, New York: CRC Press, 2005.

2. Lewis, W. K., The rate of drying of solid materials, *Industrial and Engineering Chemistry*, 13, 427–432, 1921.

3. Buckingham, E. A., *Studies on the Movement of Soil Moisture*. Bulletin 38, Washington, DC: U.S. Department of Agriculture, 1970.

4. Henry, P. S. H., Diffusion in absorbing media, *Proceedings of the Royal Society of London*, 171A, 215–241, 1939.

5. Luikov, A. V., *Drying Theory*, Moscow: Energia, 1986.

6. Philip, J. R. and De Vries, D. A., Moisture movement in porous materials under temperature gradients, *Transactions of the American Geophysical Union*, 38(5), 222–232, 1957.

7. De Vries, D. A., Simultaneous transfer of heat and moisture transfer in porous media, *Transactions of the American Geophysical Union*, 39(5), 909–916, 1958.

8. Fortes, M. and Okos, R., Drying theories: Their bases and limitations applied to food and grain, *Advances in Drying*, 1, 119–154, 1980.

9. Whitaker, S., Simultaneous heat, mass and momentum transfer in porous media. A theory of drying. In *Advances in Heat Transfer*, Vol. 13, pp. 119–203, New York: Academic Press, 1977.

10. Whitaker, S., *The Method of Volume Averaging*, Boston: Kluwer Academic, 1999.

11. Luikov, A. V., *Heat and Mass Transfer in Capillary-Porous Bodies*, Oxford: Pergamon Press, 1975.

12. Nowicki, S. C., Davis, H. T., and Scriven, L. E., Microscopic determination of transport parameters in drying porous media, *Drying Technology*, 10(4), 925–946, 1992.

13. Prat, M., Percolation model of drying under isothermal conditions in porous media, *International Journal of Multi-Phase Flow*, 19, 691–704, 1993.

14. Coumans, W. J., Power law diffusion in drying processes. Ph.D. diss., Technical University of Eindohoven, The Netherlands, 1987.

15. Yamamoto, S. A., Short-cut method for determining concentration dependent diffusivity in liquid foods and polymer solutions for regular regime drying curves, *Drying Technology*, 19(7), 1479–1490, 2001.

16. Crank, J., *The Mathematics of Diffusion*, 2nd ed., Oxford: Clarendon Press, 1975.

17. Karim, M. A. and Hawlader, M. N. A., Drying characteristics of banana: Theoretical modelling and experimental validation, *Journal of Food Engineering*, 70, 35–45, 2005.

18. Sikiatden, J. and Roberts, J. S., Measuring moisture diffusivity of potato and carrot (cor and cortex) during convective hot air and isothermal drying, *Journal of Food Engineering*, 74, 143–152, 2006.

19. Sablani, S., Rahman, S., and Al-Habsi, N., Moisture diffusivity in foods an overview, In *Drying Technology in Agriculture and Food Sciences*, A. S. Mujumdar, Ed., Enfield, NH: Science Publishers, pp. 35–59, 2000.

20. Roberts, J. S. and Tong, C. H., The development of an isothermal drying apparatus and the evaluation of the diffusion model on hygroscopic porous material, *International Journal of Food Properties*, 6(1), 165–180, 2003.

21. Roberts, J. S. and Tong, C. H., Drying kinetics of hygroscopic porous materials under isothermal conditions and the use of a first order reaction kinetics model for predicting drying, *International Journal of Food Properties*, 6(3), 355–367, 2003.

22. Bird, R. B., Stewart, W. E., and Lightfoot, E. N., 1960 and 2002. *Transport Phenomena*, 1st and 2nd ed., New York: Wiley.

23. Lin, S. X. Q. and Chen, X. D., Improving the glass-filament method for accurate measurement of drying kinetics of liquid droplets, *Trans. IChemE Part A: Chemical Engineering Research and Design*, 80, 401–410, 2002.

24. Kar, S., Chen, X. D., and Lin, S. X. Q., Effect of high mass flux on heat and mass transfer through a flat plate, *Journal of Process Mechanical Engineering*, 218(E), 213–220, 2004.

25. Chen, X. D., Lower bound estimates of the mass transfer coefficient from an evaporating liquid droplet-the effect of high interfacial vapor velocity, *Drying Technology*, 23(1–2), 59–69, 2005.

26. Rahman, S., *Food Properties Handbook*, New York: CRC Press, 1995.

27. Zhang, J. and Datta, A. K., Some considerations in modelling of moisture transport in heating of hygroscopic materials, *Drying Technology*, 22(8), 1983–2008, 2004.
28. Chen, X. D. and Xie, G. Z., Fingerprints of the drying of particulate or thin layer food materials established using a simple reaction engineering model, *Trans. IChemE Part C: Food and Bio-product Processing*, 75C, 213–222, 1997.
29. Chong, L. V. and Chen, X. D., A mathematical model of self-heating of spray dried food powders containing fat, protein, sugar and moisture, *Chemical Engineering Science*, 54, 4165–4178, 1999.
30. Chen, X. D., Pirini, W., and Ozilgen, M., The reaction engineering approach to modelling drying of thin layer of pulped kiwifruit flesh under conditions of small *Biot* numbers, *Chemical Engineering and Processing*, 40, 311–320, 2001.
31. Chen, X. D. and Lin, S. X. Q., Air drying of milk droplet under constant and time-dependent conditions, *AIChE Journal*, 51(6), 1790–1799, 2005.
32. Lin, S. X. Q. and Chen, X. D., Prediction of air-drying of milk droplet under relatively high humidity using the reaction engineering approach, *Drying Technology*, 23(7), 1395–1406, 2005.
33. Chen, X. D., Spontaneous combustion of moist coal in 2-m long column, Ph.D. diss., New Zealand: The University of Canterbury, 1991.
34. Nesic, S. and Vodnik, J., Kinetics of droplet evaporation, *Chemical Engineering Science*, 46(2), 527–537, 1991.
35. Farid, M. M., A unified approach to the heat and mass transfer in melting, solidification, frying and different drying processes, *Chemical Engineering Science*, 56, 5419–5427, 2001.
36. Farid, M. M., The moving boundary problems from melting and freezing to drying and frying of food, *Chemical Engineering and Processing*, 41, 1–10, 2002.
37. Farid, M. M., A new approach to modelling of single droplet drying, *Chemical Engineering Science*, 58(13), 2985–2993, 2003.
38. Farid, M. M. and Butcher, S., A generalised correlation for heat and mass transfer in freezing, drying, frying, and freeze drying, *Drying Technology*, 21(2), 231–247, 2003.
39. Guillot, G., Trokiner, A., Darrasse, L., Dupas, A., Ferdossi, F., Kassab, G., Hulin, J. P., Rigord, P., and Saint-Jalmes, H., NMR imaging applied to various studies of porous media, *Magnetic Resonance Imaging*, 9, 821–825, 1991.
40. Hills, B. P., Wright, K. M., Wright, J. J., Carpenter, T. A., and Hall, L. D., An MRI study of drying in granular beds of nonporous particles, *Magnetic Resonance Imaging*, 12(7), 1053–1063, 1994.
41. Bennett, G., Gorce, J. P., Keddie, J. L., McDonald, P. J., and Berglind, H., Magnetic resonance profiling studies of the drying of film-forming aqueous dispersions and glue layers, *Magnetic Resonance Imaging*, 21, 235–241, 2003.
42. Mantle, M. D., Reis, N. C., Griffith, R. F., and Gladden, L. F., MRI studies of the evaporation of single liquid droplet from porous surfaces, *Magnetic Resonance Imaging*, 21, 293–297, 2003.
43. Reis, N. C., Griffiths, R. F., Mantle, M. D., and Gladden, L. F., Investigation of the evaporation of embedded liquid droplets from porous surfaces using magnetic resonance imaging, *International Journal of Heat and Mass Transfer*, 46, 1279–1292, 2003.
44. Ruiz-Cabrera, M. A., Foucat, L., Bonny, J. M., Renou, J. P., and Daudin, J. D., Assessment of water diffusivity in gelatine gel from moisture profiles. I. Non-destructive measurement of 1D moisture profiles during drying from 2D nuclear resonance images, *Journal of Food Engineering*, 68, 209–219, 2005.
45. Ruiz-Cabrera, M. A., Foucat, L., Bonny, J. M., Renou, J. P., and Daudin, J. D., Assessment of water diffusivity in gelatine gel from moisture profiles. II. Data processing adapted to material shrinkage, *Journal of Food Engineering*, 68, 221–231, 2005.
46. Chou, S. K., Hawlader, M. N. A., and Chua, M. J., Identification of the receding evaporation front in convective food drying, *Drying Technology*, 15(5), 1353–1367, 1997.
47. Zhang, Y., Study of the effect of surfactant treatment on apple and potato drying, ME thesis, New Zealand: The University of Auckland, 2006.
48. Chen, X. D. and Peng, X. F., Modified *Biot* number in the context of air-drying of small moist porous objects, *Drying Technology*, 23(1–2), 83–103, 2005.
49. Chen, X. D., *Lewis* number in the context of drying of hygroscopic materials, *Separation and Purification Technology*, 48(2), 121–132, 2006.

50. Gladden, L. F., Structure–transport relationships in porous media, *Magnetic Resonance Imaging*, 14(7/8), 719–726, 1996.
51. Aguilera, J. M. and Stanley, D. W., *Microstructural Principles of Food Processing and Engineering*, 2nd ed., Gaithersburg, MD: Aspen, 1999.
52. Kim, E. H. J., Chen, X. D., and Pearce, D., Surface characterisation of four industrial spray–dried dairy powders in relation to chemical composition, structure and wetting property, *Colloids and Surfaces, B*, 26, 197–212, 2002.
53. Kim, E. H. J., Chen, X. D., and Pearce, D., On the mechanisms of surface formation and the surface compositions of industrial milk powders, *Drying Technology*, 21(2), 265–278, 2003.
54. Chen, X. D., Heat-mass transfer and structure formation during drying of food droplets, *Drying Technology*, 22(1–2), 179–190, 2003.
55. Wu, Y. and Irudayaraj, J., Analysis of heat, mass and pressure transfer in starch based food systems, *Journal of Food Engineering*, 29, 339–414, 1996.
56. Patankar, S. V., *Numerical Heat Transfer and Fluid Flow*, New York: McGraw-Hill, 1980.
57. Rice, R. G. and Do, D. D., *Applied Mathematical and Modeling for Chemical Engineers*, New York: Wiley, 1995.
58. Ozilgen, M., *Food Process Modeling and Control*, Australia: Gordon and Breach Sci Publishers, 1998.
59. Irudayaraj, J., Haghighi, K., and Stroshine, R. L., Nonlinear finite element analysis of coupled heat and mass transfer problems with an application to timber drying, *Drying Technology*, 8(4), 731–749, 1990.
60. Irudayaraj, J., Haghighi, K., and Stroshine, R. L., Finite element analysis of drying with application to cereal grains, *Journal of Agricultural Engineering Research*, 53(4), 209–229, 1992.
61. Farid, M. M. and Chen, X. D., The analysis of heat and mass transfer during frying of food using a moving boundary solution procedure, *Heat and Mass Transfer*, 34, 69–77, 1996.
62. Incropera, F. P. and DeWitt, D. P., *Fundamentals of Heat and Mass Transfer*, 4th and 5th ed., New York: Wiley, 1990.
63. Lin, S. X. Q., Drying of single milk droplets, Ph.D. diss., New Zealand, The University of Auckland, 2004.
64. van der Sman, R. G. M., Simple model for estimating heat and mass transfer in regular-shaped foods, *Journal of Food Engineering*, 60, 383–390, 2003.
65. McCabe, W. L., Smith, J. C., and Harriott, P., *Unit Operations of Chemical Engineering*, 6th ed., Boston: McGraw-Hill, 2001.
66. Chen, X. D., Critical *Biot* number for uniform temperature assumption in transient heat and mass transfer calculations, *International Journal of Food Engineering*, 1(3), 1–8, 2005.
67. Yang, S. M. and Tao, W. Q., *Heat Transfer*, 3rd ed., Beijing: Higher Education Press, 1999.
68. Sun, L. M. and Meunier, F. A., A detailed model for non-isothermal sorption in porous adsorbents, *Chemical Engineering Science*, 42(7), 1585–1593, 1987.
69. Patel, K., Modelling of a plug-flow spray dryer, ME thesis, New Zealand: The University of Auckland, 2004.
70. Sano, Y. and Keey, R. B., The drying of a spherical particle containing colloidal material into a hollow sphere, *Chemical Engineering Science*, 37(6), 881–889, 1983.
71. van Meel, D. A., Adiabatic convection batch drying with recirculation of air, *Chemical Engineering Science*, 9, 36–44, 1958.
72. Keey, R. B., *Drying of Loose and Particulate Materials*, New York: Hemisphere Publishing, 1992.
73. Langrish, T. A. G. and Kockel, T. K., The assessment of a characteristic drying curve for milk powder for use in computational fluid dynamics modelling, *Chemical Engineering Journal*, 84, 69–74, 2001.
74. Kockel, T. K., Allen, S., Hennings, C., and Langrish, T. A. G., An experimental study of the equilibrium for skim milk powder at elevated temperatures, *Journal of Food Engineering*, 51, 291–297, 2002.
75. Lin, S. X. Q., Chen, X. D., and Pearce, D. L., Desorption isotherm of milk powders at elevated temperatures and over a wide relative humidity range, *Journal of Food Engineering*, 68, 257–264, 2005.
76. Smith, P. G., *Introduction to Food Process Engineering*, New York: Kluwer Academic/Plenum, 2003.

77. Dyshlovenko, S., Pateyron, B., Pawlowski, L., and Murano, D., Numerical simulation of hydroxya-patite powder behavior in plasma jet, *Surface and Coating Technology*, 179, 110–117, 2004.

78. Parti, M. and Palancz, B., Mathematical model for spray drying, *Chemical Engineering Science*, 29(2), 355–362, 1974.

79. Therdthai, N., Zhou, W., and Adamczak, T., Optimisation of the temperature profile in bread baking, *Journal of Food Engineering*, 55, 41–48, 2002.

80. Therdthai, N., Zhou, W., and Adamczak, T., Two-dimensional CFD modelling and simulation of an industrial continuous bread baking oven, *Journal of Food Engineering*, 60, 211–217, 2003.

81. Therdthai, N., Zhou, W., and Adamczak, T., Three-dimensional CFD modelling and simulation of the temperature profiles and airflow patterns during continuous industrial baking oven, *Journal of Food Engineering*, 65, 599–608, 2004.

82. Therdthai, N., Zhou, W., and Adamczak, T., Simulation of starch gelatinisation during baking in a travelling-tray oven by integrating a three-dimensional CFD model with a kinetic model, *Journal of Food Engineering*, 65, 543–550, 2004.

83. Afzal, T. M. and Abe, T., Diffusion in potato during far infrared radiation drying, *Journal of Food Engineering*, 37(4), 353–365, 1998.

84. Akanbi, C. T., Adeyemi, R. S., and Ojo, A., Drying characteristics and sorption isotherm of tomato slices, *Journal of Food Engineering*, 73(2), 157–163, 2006.

85. Akgun, N. A. and Doymaz, I., Modelling of olive cake thin-layer drying process, *Journal of Food Engineering*, 68(4), 455–461, 2005.

86. Akpinar, E. K., Mathematical modelling of thin layer drying process under open sun of some aromatic plants, *Journal of Food Engineering*, 77(4), 864–870, 2006.

87. Akpinar, E. K., Determination of suitable thin layer drying curve model for some vegetables and fruits, *Journal of Food Engineering*, 73(1), 75–84, 2006.

88. Akpinar, E. K., Midilli, A., and Bicer, Y., The first and second law analyses of thermodynamic of pumpkin drying process, *Journal of Food Engineering*, 72(4), 320–331, 2006.

89. Aregba, A. W., Sebastian, P., Nadeau, J. P., Stationary deep-bed drying: A comparative study between a logarithmic model and a non-equilibrium model, *Journal of Food Engineering*, 77(1), 27–40, 2006.

90. Arévalo-Pinedo, A. and Murr, F. E. X., Kinetics of vacuum drying of pumpkin (Cucurbita maxima): Modeling with shrinkage, *Journal of Food Engineering*, 76(4), 562–567, 2006.

91. Azzouz, S., Guizani, A., Jomaa, W., and Belghith, A., Moisture diffusivity and drying kinetic equation of convective drying of grapes, *Journal of Food Engineering*, 55(4), 323–330, 2002.

92. Bayrock, D. and Ingledew, W. M., Fluidized bed drying of baker's yeast: Moisture levels, drying rates, and viability changes during drying, *Food Research International*, 30(6), 407–415, 1997.

93. Bayrock, D. and Ingledew, W. M., Mechanism of viability loss during fluidized bed drying of baker's yeast, *Food Research International*, 30(6), 417–425, 1997.

94. Bennamoun, L. and Belhamri, A., Numerical simulation of drying under variable external conditions: Application to solar drying of seedless grapes, *Journal of Food Engineering*, 76(2), 179–187, 2006.

95. Bilbao-Sáinz, C., Andrés, A., Chiralt, A., and Fito, P., Microwaves phenomena during drying of apple cylinders, *Journal of Food Engineering*, 74(1), 160–167, 2006.

96. Boeh-Ocansey, O., Some factors influencing the freeze drying of carrot discs in vacuo and at atmospheric pressure, *Journal of Food Engineering*, 4(3), 229–243, 1985.

97. Boeh-Ocansey, O., Freeze–drying in a fluidized-bed atmospheric dryer and in a vacuum dryer: evaluation of external transfer coefficients, *Journal of Food Engineering*, 7(2), 127–146, 1988.

98. Braud, L. M., Moreira, R. G., and Castell, M. E., Mathematical modeling of impingement drying of corn tortillas, *Journal of Food Engineering*, 50(3), 121–128, 2001.

99. Çakaloz, T., Akbaba, H., Yesügey, E. T., and Periz, A., Drying model for α-amylase in a horizontal spray dryer, *Journal of Food Engineering*, 31(4), 499–510, 1997.

100. Charan, R. and Prasad, S., Energy conservation in milk spray–drying plant, *Journal of Food Engineering*, 18(3), 247–258, 1993.

101. Chua, K. J., Mujumdar, A. S., Hawlader, M. N. A., Chou, S. K., and Ho, J. C., Batch drying of banana pieces—effect of stepwise change in drying air temperature on drying kinetics and product color, *Food Research International*, 34(8), 721–731, 2001.

102. Cihan, A. and Ece, M. C., Liquid diffusion model for intermittent drying of rough rice, *Journal of Food Engineering*, 49(4), 327–331, 2001.
103. Datta, A. K. and Ni, H., Infrared and hot air additions to microwave heating of foods for control of surface moisture, *Journal of Food Engineering*, 51(4), 355–364, 2002.
104. Di Matteo, M., Cinquanta, L., Galiero, G., and Crescitelli, S., A mathematical model of mass transfer in spherical geometry: Plum (Prunus domestica) drying, *Journal of Food Engineering*, 58(2), 183–192, 2003.
105. Di Matteo, P., Donsì, G., and Ferrari, G., The role of heat and mass transfer phenomena in atmospheric freeze-drying of foods in a fluidised bed, *Journal of Food Engineering*, 59(2–3), 267–275, 2003.
106. Fabiano A. N., Fernandes, S. R., Gaspareto, O. C. P., and Oliveira, E. L., Optimization of osmotic dehydration of papaya followed by air-drying, *Food Research International*, 39(4), 492–498, 2006.
107. Ferrari, G., Meerdink, G., and Walstra, P., Drying kinetics for a single droplet of skim-milk, *Journal of Food Engineering*, 10(3), 215–230, 1989.
108. Gastón, A. L., Abalone, R. M., and Giner, S. A., Wheat drying kinetics. Diffusivities for sphere and ellipsoid by finite elements, *Journal of Food Engineering*, 52(4), 313–322, 2002.
109. Gekas, V. and Lamberg, I., Determination of diffusion coefficients in volume-changing systems—application in the case of potato drying, *Journal of Food Engineering*, 14(4), 317–326, 1991.
110. George, J. P. and Datta, A. K., Development and validation of heat and mass transfer models for freeze–drying of vegetable slices, *Journal of Food Engineering*, 52(1), 89–93, 2002.
111. Hamdami, N., Monteau, J.-Y., and Le Bail, A., Transport properties of a high porosity model food at above and sub-freezing temperatures. Part 2: Evaluation of the effective moisture diffusivity from drying data, *Journal of Food Engineering*, 62(4), 385–392, 2004.
112. Hatamipour, M. S. and Mowla, D., Correlations for shrinkage, density and diffusivity for drying of maize and green peas in a fluidized bed with energy carrier, *Journal of Food Engineering*, 59(2–3), 221–227, 2003.
113. Hawlader, M. N. A., Uddin, M. S., Ho, J. C., and Teng, A. B. W., Drying characteristics of tomatoes, *Journal of Food Engineering*, 14(4), 259–268, 1991.
114. Hayaloglu, A. A., Karabulut, I., Alpaslan, M., and Kelbaliyev, G., Mathematical modeling of drying characteristics of strained yoghurt in a convective type tray–dryer, *Journal of Food Engineering*, 78(1), 109–117, 2007.
115. Herman-Lara, E., Salgado-Cervantes, M. A., and García-Alvarado, M. A., Mathematical simulation of convection food batch drying with assumptions of plug flow and complete mixing of air, *Journal of Food Engineering*, 68(3), 321–327, 2005.
116. Hernández, J. A., Pavón, G., and García, M. A., Analytical solution of mass transfer equation considering shrinkage for modeling food-drying kinetics, *Journal of Food Engineering*, 45(1), 1–10, 2000.
117. Ho, J. C., Chou, S. K., Chua, K. J., Mujumdar, A. S., and Hawlader, M. N. A., Analytical study of cyclic temperature drying: Effect on drying kinetics and product quality, *Journal of Food Engineering*, 51(1), 65–75, 2002.
118. Hussain, M. A., Rahman, M. S., and Ng, C. W., Prediction of pores formation (porosity) in foods during drying: Generic models by the use of hybrid neural network, *Journal of Food Engineering*, 51(3), 239–248, 2002.
119. Iguaz, A., Esnoz, A., Martínez, G., López, A., and Vírseda, P., Mathematical modelling and simulation for the drying process of vegetable wholesale by-products in a rotary dryer, *Journal of Food Engineering*, 59(2–3), 151–160, 2003.
120. Iguaz, A., San Martín, M. B., Maté, J. I., Fernández, T., and Vírseda, P., Modelling effective moisture diffusivity of rough rice (Lido cultivar) at low drying temperatures, *Journal of Food Engineering*, 59(2–3), 253–258, 2003.
121. Jain, D., Modeling the performance of greenhouse with packed bed thermal storage on crop drying application, *Journal of Food Engineering*, 71(2), 170–178, 2005.
122. Jain, D., Modeling the system performance of multi-tray crop drying using an inclined multi-pass solar air heater with in-built thermal storage, *Journal of Food Engineering*, 71(1), 44–54, 2005.
123. Karapantsios, T.D., Conductive drying kinetics of pregelatinized starch thin films, *Journal of Food Engineering*, 76(4), 477–489, 2006.

124. Kim, S. S. and Bhowmik, S. R., Effective moisture diffusivity of plain yogurt undergoing microwave vacuum drying, *Journal of Food Engineering*, 24(1), 137–148, 1995.
125. Krokida, M. K., Foundoukidis, E., and Maroulis, Z., Drying constant: literature data compilation for foodstuffs, *Journal of Food Engineering*, 61(3), 321–330, 2004.
126. Liu, Q. and Bakker-Arkema, F. W., A model-predictive controller for grain drying, *Journal of Food Engineering*, 49(4), 321–326, 2001.
127. Lombraña, J. I. and Villarán, M. C., The influence of pressure and temperature on freeze-drying in an adsorbent medium and establishment of drying strategies, *Food Research International*, 30(3–4), 213–222, 1997.
128. Luna-Solano, G., Salgado-Cervantes, M. A., Rodríguez-Jimenes, G. C., and García-Alvarado, M. A., Optimization of brewer's yeast spray drying process, *Journal of Food Engineering*, 68(1), 9–18, 2005.
129. Maroulis, Z. B., Kiranoudis, C. T., and Marinos-Kouris, D., Simultaneous estimation of heat and mass transfer coefficients in externally controlled drying, *Journal of Food Engineering*, 14(3), 241–255, 1991.
130. Maroulis, Z. B., Kiranoudis, C. T., and Marinos-Kouris, D., Heat and mass transfer modeling in air drying of foods, *Journal of Food Engineering*, 26(1), 113–130, 1995.
131. Maroulis, Z. B., Krokida, M. K., and Rahman, M. S., A structural generic model to predict the effective thermal conductivity of fruits and vegetables during drying, *Journal of Food Engineering*, 52(1), 47–52, 2002.
132. May, B. K. and Perré, P., The importance of considering exchange surface area reduction to exhibit a constant drying flux period in foodstuffs, *Journal of Food Engineering*, 54(4), 271–282, 2002.
133. Mayor, L. and Sereno, A. M., Modelling shrinkage during convective drying of food materials: A review, *Journal of Food Engineering*, 61(3), 373–386, 2004.
134. McMinn, W. A. M., Khraisheh, M. A. M., and Magee, T. R. A., Modeling the mass transfer during convective, microwave and combined microwave-convective drying of solid slabs and cylinders, *Food Research International*, 36(9–10), 977–983, 2003.
135. McMinn, W. A. M., Thin-layer modelling of the convective, microwave, microwave-convective and microwave-vacuum drying of lactose powder, *Journal of Food Engineering*, 72(2), 113–123, 2006.
136. Medeiros, G. L. and Sereno, A. M., Physical and transport properties of peas during warm air drying, *Journal of Food Engineering*, 21(3), 355–363, 1994.
137. Migliori, M., Gabriele, D., de Cindio, B., and Pollini, C. M., Modelling of high quality pasta drying: Quality indices and industrial application, *Journal of Food Engineering*, 71(3), 242–251, 2005.
138. Migliori, M., Gabriele, D., de Cindio, B., and Pollini, C. M., Modelling of high quality pasta drying: Mathematical model and validation, *Journal of Food Engineering*, 69(4), 387–397, 2005.
139. Mohapatra, D. and Rao, P. S., A thin layer drying model of parboiled wheat, *Journal of Food Engineering*, 66(4), 513–518, 2005.
140. Mulet, A., Drying modelling and water diffusivity in carrots and potatoes, *Journal of Food Engineering*, 22(1–4), 329–348, 1994.
141. Mwithiga, G. and Olwal, J. O., The drying kinetics of kale (Brassica oleracea) in a convective hot air dryer, *Journal of Food Engineering*, 71(4), 373–378, 2005.
142. Ni, H., Datta, A. K., and Torrance, K. E., Moisture transport in intensive microwave heating of biomaterials: A multiphase porous media model, *International Journal of Heat and Mass Transfer*, 42(8), 1501–1512, 1999.
143. Niamnuy, C. and Devahastin, S., Drying kinetics and quality of coconut dried in a fluidized bed dryer, *Journal of Food Engineering*, 66(2), 267–271, 2005.
144. Nieto, A., Castro, M. A., and Alzamora, S. M., Kinetics of moisture transfer during air drying of blanched and/or osmotically dehydrated mango, *Journal of Food Engineering*, 50(3), 175–185, 2001.
145. Pavón-Melendez, G., Hernández, J. A., Salgado, M. A., and García, M. A., Dimensionless analysis of the simultaneous heat and mass transfer in food drying, *Journal of Food Engineering*, 51(4), 347–353, 2002.
146. Pelegrina, A. H. and Crapiste, G. H., Modelling the pneumatic drying of food particles, *Journal of Food Engineering*, 48(4), 301–310, 2001.
147. Ponsart, G., Vasseur, J., Frias, J. M., Duquenoy, A., and Méot, J. M., Modelling of stress due to shrinkage during drying of spaghetti, *Journal of Food Engineering*, 57(3), 277–285, 2003.

148. Rastikian, K., Capart, R., and Benchimol, J., Modelling of sugar drying in a countercurrent cascading rotary dryer from stationary profiles of temperature and moisture, *Journal of Food Engineering*, 41(3–4), 193–201, 1999.

149. Ratti, C., Shrinkage during drying of foodstuffs, *Journal of Food Engineering*, 23(1), 91–105, 1994.

150. Ratti, C. and Mujumdar, A. S., Simulation of packed bed drying of foodstuffs with airflow reversal, *Journal of Food Engineering*, 26(3), 259–271, 1995.

151. Rordprapat, W., Nathakaranakule, A., Tia, W., and Soponronnarit, S., Comparative study of fluidized bed paddy drying using hot air and superheated steam, *Journal of Food Engineering*, 71(1), 28–36, 2005.

152. Rovedo, C. O., Suarez, C., and Viollaz, P., Analysis of moisture profiles, mass Biot number and driving forces during drying of potato slabs, *Journal of Food Engineering*, 36(2), 211–231, 1998.

153. Ruiz-López, I. I., Córdova, A. V., Rodríguez-Jimenes, G. C., and García-Alvarado, M. A., Moisture and temperature evolution during food drying: effect of variable properties, *Journal of Food Engineering*, 63(1), 117–124, 2004.

154. Simal, S., Rosselló, C., Berna, A., and Mulet, A., Drying of shrinking cylinder-shaped bodies, *Journal of Food Engineering*, 37(4), 423–435, 1998.

155. Sacilik, K. and Elicin, A. K., The thin layer drying characteristics of organic apple slices, *Journal of Food Engineering*, 73(3), 281–289, 2006.

156. Sacilik, K., Keskin, R., and Elicin, A. K., Mathematical modelling of solar tunnel drying of thin layer organic tomato, *Journal of Food Engineering*, 73(3), 231–238, 2006.

157. Schmalko, M. E. and Alzamora, S. M., Modelling the drying of a twig of "yerba maté" considering as a composite material: Part II: mathematical model, *Journal of Food Engineering*, 67(3), 267–272, 2005.

158. Sebastian, P., Bruneau, D., Collignan, A., and Rivier, M., Drying and smoking of meat: heat and mass transfer modeling and experimental analysis, *Journal of Food Engineering*, 70(2), 227–243, 2005.

159. Sereno, A. M. and Medeiros, G. L., A simplified model for the prediction of drying rates for foods, *Journal of Food Engineering*, 12(1), 1–11, 1991.

160. Sharma, G. P. and Prasad, S., Effective moisture diffusivity of garlic cloves undergoing microwave-convective drying, *Journal of Food Engineering*, 65(4), 609–617, 2004.

161. Sharma, G. P., Verma, R. C., and Pathare, P., Mathematical modeling of infrared radiation thin layer drying of onion slices, *Journal of Food Engineering*, 71(3), 282–286, 2005.

162. Simal, S., Femenia, A., Garau, M. C., and Rosselló, C., Use of exponential, page's and diffusional models to simulate the drying kinetics of kiwi fruit, *Journal of Food Engineering*, 66(3), 323–328, 2005.

163. Simal, S., Femenia, A., Garcia-Pascual, P., and Rosselló, C., Simulation of the drying curves of a meat-based product: effect of the external resistance to mass transfer, *Journal of Food Engineering*, 58(2), 193–199, 2003.

164. Straatsma, J., Van Houwelingen, G., Steenbergen, A. E., and De Jong, P., Spray drying of food products: 1. simulation model, *Journal of Food Engineering*, 42(2), 67–72, 1999.

165. Techasena, O., Lebert, A., and Bimbenet, J. J., Simulation of deep bed drying of carrots, *Journal of Food Engineering*, 16(4), 267–281, 1992.

166. Teixeira, M. and Tobinaga, S., A diffusion model for describing water transport in round squid mantle during drying with a moisture-dependent effective diffusivity, *Journal of Food Engineering*, 36(2), 169–181, 1998.

167. Tirawanichakul, Y., Prachayawarakorn, S., Varanyanond, W., and Soponronnarit, S., Simulation and grain quality for in-store drying of paddy, *Journal of Food Engineering*, 64(4), 405–415, 2004.

168. Tiwari, G. N., Kumar, S., and Prakash, O., Evaluation of convective mass transfer coefficient during drying of jaggery, *Journal of Food Engineering*, 63(2), 219–227, 2004.

169. Toğrul, H., Simple modeling of infrared drying of fresh apple slices, *Journal of Food Engineering*, 71(3), 311–323, 2005.

170. Toğrul, H., Suitable drying model for infrared drying of carrot, *Journal of Food Engineering*, 77(3), 610–619, 2006.

171. Tolaba, M. P., Viollaz, P. E., and Suárez, C., A mathematical model to predict the temperature of maize kernels during drying, *Journal of Food Engineering*, 8(1), 1–16, 1988.

172. Trujillo, F. J., Wiangkaew, C., and Pham, Q. T., Drying modeling and water diffusivity in beef meat, *Journal of Food Engineering*, 78(1), 74–85, 2007.

173. Truong, V., Bhandari, B. R., and Howes, T., Optimization of co-current spray drying process of sugar-rich foods. Part I—Moisture and glass transition temperature profile during drying, *Journal of Food Engineering*, 71(1), 55–65, 2005.

174. Truong, V., Bhandari, B. R., and Howes, T., Optimization of cocurrent spray drying process for sugar-rich foods. Part II—Optimization of spray drying process based on glass transition concept, *Journal of Food Engineering*, 71(1), 66–72, 2005.

175. Veli, D., Planini, M., Tomas, S., and Bili, M., Influence of airflow velocity on kinetics of convection apple drying, *Journal of Food Engineering*, 64(1), 97–102, 2004.

176. Verboven, P., Datta, A. K., Anh, N. T., Scheerlinck, N., and Nicolai, B. M., Computation of airflows effects on heat and mass transfer in a microwave oven, *Journal of Food Engineering*, 59, 181–190, 2003.

177. Viollaz, P. E. and Rovedo, C. O., A drying model for three-dimensional shrinking bodies, *Journal of Food Engineering*, 52(2), 149–153, 2002.

178. Waananen, K. M. and Okos, M. R., Effect of porosity on moisture diffusion during drying of pasta, *Journal of Food Engineering*, 28(2), 121–137, 1996.

179. Wang, N. and Brennan, J. G., A mathematical model of simultaneous heat and moisture transfer during drying of potato, *Journal of Food Engineering*, 24(1), 47–60, 1995.

180. Baik, O.-D. and Mittal, G. S., Heat and moisture transfer and shrinkage simulation of deep-fat tofu frying, *Food Research International*, 38(2), 183–191, 2005.

181. Costa, R. M. and Oliveira, F. A. R., Modelling the kinetics of water loss during potato frying with a compartmental dynamic model, *Journal of Food Engineering*, 41(3–4), 177–185, 1999.

182. Dincer, I. and Yildiz, M., Modelling of thermal and moisture diffusions in cylindrically shaped sausages during frying, *Journal of Food Engineering*, 28(1), 35–44, 1996.

183. Farkas, B. E., Singh, R. P., and Rumsey, T. R., Modeling heat and mass transfer in immersion frying. I, model development, *Journal of Food Engineering*, 29(2), 211–226, 1996.

184. Farkas, B. E., Singh, R. P., and Rumsey, T. R., Modeling heat and mass transfer in immersion frying. II, model solution and verification, *Journal of Food Engineering*, 29(2), 227–248, 1996.

185. Krokida, M. K., Oreopoulou, V., and Maroulis, Z. B., Water loss and oil uptake as a function of frying time, *Journal of Food Engineering*, 44(1), 39–46, 2000.

186. Moreira, R. G. and Barrufet, M. A., Spatial distribution of oil after deep-fat frying of tortilla chips from a stochastic model, *Journal of Food Engineering*, 27(3), 279–290, 1996.

187. Ngadi, M. O., Watts, K. C., and Correia, L. R., Finite element method modelling of moisture transfer in chicken drum during deep-fat frying, *Journal of Food Engineering*, 32(1), 11–20, 1997.

188. Ni, H. and Datta, A. K., Moisture, oil, and energy transport during deep-fat frying of food materials, *Transactions of the Institution of Chemical Engineers: Food and Bioproducts Processing*, 77C(9), 1–11, 1999.

189. Oroszvári, B. K., Bayod, E., Sjöholm, I., and Tornberg, E., The mechanisms controlling heat and mass transfer on frying of beefburgers. Part 2: The influence of the pan temperature and patty diameter, *Journal of Food Engineering*, 71(1), 18–27, 2005.

190. Oroszvári, B. K., Bayod, E., Sjöholm, I., and Tornberg, E., The mechanisms controlling heat and mass transfer on frying of beefburgers. III. Mass transfer evolution during frying, *Journal of Food Engineering*, 76(2), 169–178, 2006.

191. Oroszvári, B. K., Rocha, C. S., Sjöholm, I., and Tornberg, E., Permeability and mass transfer as a function of the cooking temperature during the frying of beefburgers, *Journal of Food Engineering*, 74(1), 1–12, 2006.

192. Oroszvári, B. K., Sjöholm, I., and Tornberg, E., The mechanisms controlling heat and mass transfer on frying of beefburgers. I. The influence of the composition and comminution of meat raw material, *Journal of Food Engineering*, 67(4), 499–506, 2005.

193. Ou, D. and Mittal, G. S., Double-sided pan-frying of unfrozen/frozen hamburgers for microbial safety using modelling and simulation, *Food Research International*, 39(2), 133–144, 2006.

194. Rajkumar, V., Moreira, R., and Barrufet, M., Modeling the structural changes of tortilla chips during frying, *Journal of Food Engineering*, 60(2), 167–175, 2003.

195. Rywotycki, R., The effect of fat temperature on heat energy consumption during frying of food, *Journal of Food Engineering*, 54(3), 257–261, 2002.
196. Rywotycki, R., A model of heat energy consumption during frying of food, *Journal of Food Engineering*, 59(4), 343–347, 2003.
197. Vijayan, J. and Singh, R. P., Heat transfer during immersion frying of frozen foods, *Journal of Food Engineering*, 34(3), 293–314, 1997.
198. Vitrac, O., Dufour, D., Trystram, G., and Raoult-Wack, A. L., Characterization of heat and mass transfer during deep-fat frying and its effect on cassava chip quality, *Journal of Food Engineering*, 53(2), 161–176, 2002.
199. Yamsaengsung, R. and Moreira, R. G., Modeling the transport phenomena and structural changes during deep fat frying: Part I: Model development, *Journal of Food Engineering*, 53(1), 1–10, 2002.
200. Yamsaengsung, R. and Moreira, R. G., Modeling the transport phenomena and structural changes during deep fat frying: Part II: Model solution & validation, *Journal of Food Engineering*, 53(1), 11–25, 2002.
201. Baik, O.-D. and Marcotte, M., Modeling the moisture diffusivity in a baking cake, *Journal of Food Engineering*, 56(1), 27–36, 2003.
202. Broyart, B. and Trystram, G., Modelling heat and mass transfer during the continuous baking of biscuits, *Journal of Food Engineering*, 51(1), 47–57, 2002.
203. Demirkol, E., Erdoğdu, F., and Palazoğlu, T. K., Analysis of mass transfer parameters (changes in mass flux, diffusion coefficient and mass transfer coefficient) during baking of cookies, *Journal of Food Engineering*, 72(4), 364–371, 2006.
204. Fan, J., Mitchell, J. R., and Blanshard, J. M. V., A model for the oven rise of dough during baking, *Journal of Food Engineering*, 41(2), 69–77, 1999.
205. Lostie, M., Peczalski, R., and Andrieu, J., Lumped model for sponge cake baking during the "crust and crumb" period, *Journal of Food Engineering*, 65(2), 281–286, 2004.
206. Lostie, M., Peczalski, R., Andrieu, J., and Laurent, M., Study of sponge cake batter baking process. II. Modeling and parameter estimation, *Journal of Food Engineering*, 55(4), 349–357, 2002.
207. Lostie, M., Peczalski, R., Andrieu, J., and Laurent, M., Study of sponge cake batter baking process. Part I: Experimental data, *Journal of Food Engineering*, 51(2), 131–137, 2002.
208. Ni, H. and Datta, A. K., Heat and moisture transfer in baking of potato slabs, *Drying Technology*, 17(10), 2069–2092, 1999.
209. Savoye, I., Trystram, G., Duquenoy, A., Brunet, P., and Marchin, F., Heat and mass transfer dynamic modelling of an indirect biscuit baking tunnel-oven. Part I: Modelling principles, *Journal of Food Engineering*, 16(3), 173–196, 1992.
210. Zanoni, B., Peri, C., and Pierucci, S., A study of the bread-baking process. I: A phenomenological model, *Journal of Food Engineering*, 19(4), 389–398, 1993.
211. Zanoni, B., Pierucci, S., and Peri, C., Study of the bread baking process—II.Mathematical modelling, *Journal of Food Engineering*, 23(3), 321–336, 1994.
212. Zhang, J. and Datta, A. K., Mathematical modeling of bread baking process, *Journal of Food Engineering*, 75(1), 78–89, 2006.
213. Zhang, J., Datta, A. K., and Mukherjee, S., Transport processes and large deformation during baking of bread, *American Institute of Chemical Engineers Journal*, 51(9), 2569–2580, 2005.
214. Welti-Chanes, J., Vergara-Balderas, F., and Bermudez-Aguirre, D., Transport phenomena in food engineering: basic concepts and advances, *Journal of Food Engineering*, 67(1–2), 113–128, 2005.

CHAPTER 7

Reaction Kinetics

Maria C. Giannakourou and Petros S. Taoukis

CONTENTS

7.1 INTRODUCTION

A number of different reactions occur in foods during processing and subsequent storage. The objective of processing is the optimization of specific food quality characteristics, such as: food flavor, color, texture, and nutritional profile; inactivation of undesirable microorganisms, enzymes, or toxicants; and extension of food shelf life. The specific conditions and methodology applied are often determined by the target quality parameters of the food because processes are designed to cause an action and selectively improve food features. Therefore, to successfully design and implement a process, it is necessary to both quantify the rates at which the attributes of interest change with time,[1] and estimate the effect of environmental factors, such as temperature, the system pH, the moisture content, catalysts, etc. on these rates.

After food processing, final food products are subject to numerous reactions during subsequent storage, transport, and handling. For each particular food there is a finite length of time after production during which it will retain a required level of quality, organoleptically and safety-wise, under stated conditions of storage. This period, which can be generally defined as the *shelf life* of the food product, is directly related to durability open dating on food labels. Such "expiration" labeling is of prime interest and influences consumer decisions and stock rotation systems.

Despite their multiparametric nature, food quality and food spoilage of a particular food can, in most cases, be well described by specific indices. Since characteristic properties are overlaid for every product, a decision has to be made at what level the decrease in a certain characteristic or the development of an undesirable one can be detected by the consumer. Additionally, it is necessary to be able to describe, in mathematical terms, this gradual deterioration as a function of time, and other important parameters that may affect food spoilage. After developing a robust and well-validated kinetic model for food degradation, it is possible to design alternative procedures and practices that would lead to significant quality retention.

Food quality deterioration can be the outcome of multiple actions, roughly grouped in three main categories: physical phenomena, chemical reactions, and microbiological spoilage. These actions often occur simultaneously throughout a food's life cycle, leading to its deterioration. Additionally, one type of action can affect and accelerate other actions. The exhaustive in depth study of the chemical and biological reactions and physical changes that occur in the food during and after processing allows the selection of the parameters that are important for its safety and which will, to a great extent, determine overall quality. These parameters are used as indices to quantitatively assess quality loss and the corresponding degradation rates.

After recognizing and kinetically studying the prevailing food deterioration indices, another major issue of food reaction kinetic studies involves the quantification of the factors influencing the reaction rates of these indices. Several intrinsic and extrinsic factors significantly influence indices of deterioration. Water activity, pH, and food constituents are endogenous factors affecting food spoilage. Environmental factors—specifically temperature, exposure to oxygen or other packaging gases, and light—have an important effect on food degradation rates. The scope of a full kinetic study is to establish the quantitative effect of each parameter to the quality loss of a specific food. The purpose of this chapter is to (1) provide an overall view of food reaction kinetics, (2) describe a methodology for modeling reaction rates, and (3) quantify the effect of important environmental parameters on these rates, using well-known examples of frequent reactions occurring in foods, mostly based on chemical mechanisms.

7.2 CATEGORIES OF PHENOMENA AFFECTING FOOD QUALITY

7.2.1 Introduction

In recent literature, major spoilage phenomena are attributed to physical, chemical, and microbial mechanisms that often act simultaneously—or even synergistically—to deteriorate product's quality. Table 7.1 summarizes some common deteriorative mechanisms in different food matrices, stating important influencing factors. Each of the main phenomena categories that lead to food degradation are reviewed in this section.

7.2.2 Physical Changes

The first type of food deterioration is attributed to physical change or instability.[2] Such mechanisms may involve moisture or mass transfer of different substances in food matrices, leading to a change of their moisture content, and, consequently, to a physical collapse, or in some cases, to a detrimental microbiological or chemical degradation. Popular examples of this category include staling of bakery product, rapid wilting of leafy vegetables, etc.

A number of physical changes are mostly influenced by temperature and humidity of the environment. For example, dry foods and cereals are expected to have a certain crispy texture, a requirement that frequently fails to be met in high humidity environments due to water absorption and loss of textural properties. This phenomenon is attributed to a change in food matrix glass transition temperature (T_g) that is closely related to food molecular mobility and, therefore, to stability and quality. Glass transition theory has recently received much attention in food science due to its various applications in carbohydrate containing and frozen foods.[4] This theory is based on the physical, second-order change of the product from a "glassy" rigid and solid phase to a "rubbery"? amorphous and softened state. The interesting aspect of this phenomenon is that it is often considered to be of kinetic nature, despite the discontinuity in heat capacity without change of phase that indicates a second-order thermodynamic phenomenon. Another effect of the change of glass transition temperature is the "freezer burn" phenomena in deep-frozen foods, following the significant sublimation of water from food surface when storage temperature is not constant. Caking of dry powders occurs when powder gains moisture; this is another effect of glass transition.

Other well-known physical phenomena that could also be considered as physical changes of polymers are the gelatinization and retrogradation of starch.

Other physical changes frequently occurring in frozen foods or carbohydrate-containing foods include crystal growth and undesirable crystal size and location change.[5] Fat bloom, the whitish, greasy haze found in cocoa butter of chocolate is an example of the migration and recrystallization of fat and appropriate chocolate tempering is necessary to minimize its detrimental effect on chocolate quality.[2]

7.2.3 Chemical Changes

Food spoilage involves chemical reactions, including the reaction or breakdown of intrinsic food components such as proteins, lipids, carbohydrates, vitamins, etc. Most of these reactions directly affect major quality attributes, such as color, flavor, texture, taste, and overall appearance of perishable food products.

Table 7.1 Common Deterioration Paths and Major Influencing Parameters in Various Food Matrices

Food Matrix	Deterioration Path	Critical Parameter	Remarks	References
Bakery products	Staling, starch retrogradation, microbial growth	Humidity, temperature, oxygen	Significant mass and moisture transfer	2
Dry foods/cereals	Moisture migration, loss of crispy texture, oxidation	Humidity, temperature	Change of T_g temperature, influencing food molecular mobility	2–4
Frozen foods	Moisture migration, ice crystal formation/recrystallization, oxidation, freezer burn, vitamin loss, color changes	Humidity, temperature, oxygen, inappropriate handling	Change of T_g temperature	2, 3, 5, 55, 56
Chocolate	Sugar and fat bloom (crystallization), oxidation	Humidity, temperature, oxygen	Appropriate tempering is necessary	2
Fruit and vegetable tissues	Enzymatic activity, browning, collapse, microbial growth	Humidity, temperature, oxygen, light		2,6
Milk powders, fruits	Maillard browning	Water activity, temperature, oxygen, light, pH, metal ions		2, 7, 39–46
Dairy products, animal tissues	Rancidity, microbial growth	Oxygen, temperature, pH, water activity		2, 8

Regarding proteins, the role of enzymes in food processing and subsequent storage is of major interest. Acting as biological catalysts, enzymes enable and often accelerate chemical reactions, causing, for example, browning, softening, or collapse of fruit and vegetable tissues. A number of enzymes have been reported to influence both vegetative and animal tissues; enzyme study and technological application is an area of great advance.[6]

Nonenzymatic browning, also known as the *Maillard browning reaction*, is an important case of complex, chemical reactions between proteins and reducing sugars, leading to color darkening and possible textural changes. Maillard browning has been widely studied in several tissues, with a special focus on the role of environmental factors—such as temperature and water activity—on the rate of brown pigment formation. Its relation to the change of the glass transition temperature, a closely linked parameter to physical changes in amorphous or frozen matrices, has been also investigated.[7] Sugars may also be caramelized or undergo retrogradation reactions.

Lipids are usually involved in oxidation reactions for fats and oils that lead to undesired rancidity of foods, such as milk and other dairy products, meat and fish tissues, cereals, coffee, and chocolate. Lipolytic enzymes may also act to accelerate lipid degradation; a lot of effort is focused on their inactivation by thermal processing or moisture decrease.[8]

Certain vitamins are susceptible to heat (C, thiamin, folate, B_6), to oxidation (C, D, E, A), or to photodegradation (riboflavin).

7.2.4 Microbiological Changes

Microbiological action is a common pathway of food spoilage, which may not be limited to loss of food quality, but may also jeopardize food safety by causing food borne illnesses. Food predictive microbiology is an area of increased interest that is focused on modeling food pathogen or spoilage microorganism growth as a function of multiple influencing parameters, so as to predict a certain microorganism behavior when a specific environmental factor is modified.[9] Microorganisms include bacteria, molds, and yeasts; each of these categories influence in a different way and up to a different level of food quality and safety.

Most foods are susceptible to microbiological spoilage. Common cases include mold actions or vegetable and fruit rots, growth of spoilage bacteria such as pseudomonads in refrigerated meats, poultry, dairy, and eggs, lactic acid bacteria, and *Brochothrix thermosphacta* in vacuum- or modified-atmosphere-packed meat products, etc.

Regarding food pathogens known to cause serious illnesses, *Clostridium botulinum* was recognized early as producing a lethal neurotoxin in canned products. *Staphylococcus aureus* often contaminates meat products or confectionaries through its enterotoxin, when appropriate sanitary conditions are not observed in food industries or retail points. Salmonella is a ubiquitous species, frequently causing major problems in ready-to-eat meals, meat products, poultry, and dairy products. *Listeria monocytogenes*, a resistant psychrotroph bacterium, has recently been a major problem in the dairy and ready-to-eat meat products industries.

In this chapter, modeling of microbiological events, such as microbial growth or inactivation, will not be separately covered and studied.[10] The models developed in predictive microbiology can be used to calculate microbial numbers given a specific food composition, a food manufacturing line, or a food distribution chain. Consequently, predictive microbiology gives improved, quantitative insight into the food properties and processes that are of importance to the safety and quality of foods.[11–13]

For the prediction of food quality, usually lag and exponential growth phases play the most important roles. Foods usually become spoiled if the number of microorganisms becomes larger than 10^7 cells per g of product. For pathogenic microorganisms, safe numbers are generally far lower. There are numerous empirical models of bacterial curves (Gompertz, Logistic, etc) of significant practical use, which cannot be extrapolated in different cases.[10] Mechanistic or dynamic models can be alternatively used. Specific growth rate and lag time of microorganisms depend on environmental conditions (mainly temperature, pH, water activity, and amount of oxygen); mathematical models have been developed to quantitatively describe this relationship.[11,12,14]

It is worth mentioning that there are some general predictive microbiology software packages available: Pathogen Modelling Program (PMP, USDA Eastern Regional Research Centre, Wyndmoor, Pennsylvania, U.S.A.) and the Food MicroModel (FMM, Leatherhead Food research Association, U.K.) are user-friendly software programs that predict bacterial growth at user-defined sets of values of temperature, pH, and NaCl concentration. The recently developed COMBASE (Institute of Food Research, Norwich, U.K., http://www.combase.cc) includes a significant database of published material on developed models in different food products under different environmental conditions, a tool of great practical importance.

7.3   BASIC PRINCIPLES OF REACTION KINETICS

7.3.1   Introduction

As briefly mentioned, different kinds of reactions continue to occur post processing, at a rate determined by the inherent properties of the food, the type of packaging, and conditions of storage and distribution.[15] These factors determine the shelf life of the food. The fundamentals of describing food kinetics, assessing food quality, and designing shelf life tests are the core of the methodology used to model reactions in foods.

In understanding the progress of reactions, knowledge of thermodynamics is useful for assessing the direction of a reaction, and thus, its driving force.[16] Thermodynamics provides information about the possibility of a certain reaction taking place in a specific system, but does not tell anything about its rate; the latter is the object of the reaction kinetics field.

The first step in kinetically studying food spoilage is to determine the most important parameters that reflect the overall quality loss, while at the same time, identify all controlling factors of the degradation process—either intrinsic (e.g. pH of the matrix) or extrinsic factors (e.g, a_w, temperature, oxygen).[1] Identifying and quantitatively estimating the influence of prevailing parameters is of crucial importance in the design of food processing so that appropriate modifications are made and specific processing conditions are applied.

The approach is based on the identification of the biological and chemical reactions that are most likely to have a critical effect on quality loss. Assuming constant environmental conditions for simplification purposes, a reaction scheme that expresses the effect of the concentration of the reactants can be developed. The ultimate objective is to model the change of the concentrations of constituents connected to food quality as functions of time.

7.3.1.1   Simple Kinetics

Molecular, irreversible reactions are typically expressed as[3]

$$\nu_1 A_1 + \nu_2 A_2 + \nu_3 A_3 + \nu_4 A_4 + \cdots + \nu_m A_m \xrightarrow{k_f} P, \tag{7.1}$$

where A_i are the reactant species, ν_j are the respective stoichiometric coefficients ($j = 1,2,\ldots,m$), P is the products, and k_f is the forward reaction rate constant. For such a scheme, the reaction rate, r, is given by

$$r = -\frac{1}{\nu_j}\frac{d[A_j]}{dt} = k_f[A_1]^{n_1}[A_2]^n \ldots [A_m]^{n_m}, \tag{7.2}$$

where n_j is the order of the reaction with respect to constituents A_j. A true molecular reaction holds that $n_j = \nu_j$.

7.3.1.2 Complex Kinetics

A common case of increased complexity is when consecutive or parallel reactions occur in the food matrix.[16] These cases include intermediate products that then react further. One simple example of this category is described by the following scheme:

$$A \xrightarrow{k_1} B \xrightarrow{k_2} C \tag{7.3}$$

The differential rate equations are then:

$$\frac{d[A]}{dt} = -k_1[A],$$

$$\frac{d[B]}{dt} = k_1[A] - k_2[B], \tag{7.4}$$

$$\frac{d[C]}{dt} = k_2[B].$$

The analytical solutions when external conditions are assumed constant are:

$$[A] = [A]_0 e^{-k_1 t},$$

$$[B] = [B]_0 e^{-k_2 t} + k_1[A]_0 \frac{e^{-k_1 t} - e^{-k_2 t}}{k_2 - k_1}, \tag{7.5}$$

$$[C] = [C]_0 + [B]_0(1 - e^{-k_2 t}) + [A]_0\left(1 + \frac{k_1 e^{-k_2 t} - k_2 e^{-k_1 t}}{k_2 - k_1}\right).$$

On the other hand, parallel equations happen when a reactant is simultaneously implicated in two or more different reactions:

$$A + B \underset{k_2}{\overset{k_1}{\rightleftarrows}} P \tag{7.6}$$

The corresponding differential equation that describes the reaction is

$$-\frac{d[A]}{dt} = k_1[A][B] + k_2[A], \tag{7.7}$$

which is difficult to solve analytically; instead, numerical solutions are preferred.

There are numerous common examples of consecutive and parallel equations in foods, such as non-enzymatic browning,[7] or chlorophyll degradation during heating.[17] In the literature, the so-called *steady state* that significantly simplifies the above-mentioned complex equations is frequently assumed. Supposing a case where in real, complex food systems, the degradation of crucial quality parameters is described by reversible, multiple-step reactions, the following form provides description:

$$aA + bB \underset{k_b}{\overset{k_f}{\rightleftarrows}} cC + dD. \tag{7.8}$$

In this case, A reacts with B to form products C and D that can back react with a rate constant of k_b. The reaction rate in this case would be:

$$r = \frac{-d[A]}{adt} = \frac{-d[B]}{bdt} = \frac{+d[C]}{cdt} = \frac{+d[D]}{ddt} = k_f[A]^a[B]^b - k_b[C]^c[D]^d. \tag{7.9}$$

At a certain stage, the rates for the forward and the reverse reaction become equal, equilibrium is reached, and the equilibrium constant K_{eq} is described by

$$K_{eq} = \frac{k_f}{k_b} = \frac{[C][D]}{[A][B]}. \tag{7.10}$$

However, for most cases of food degradation systems, either $k_b \ll k_f$, or for the time period of practical interest they are distant from equilibrium, i.e., $[C]$ and $[D]$ are very small, allowing the reaction to be treated as irreversible. The most common case is that one of the reactant species is the limiting factor and affects reaction kinetics, with the other constituents being in large excess. That allows the quality loss rate equation to be expressed in terms of specific reactants:

$$r = \frac{-d[A]}{dt} = k_{f'}[A]^\alpha. \tag{7.11}$$

where α is an apparent or pseudo order of the reaction of component A and k_f' is the apparent rate constant. Assuming that the aforementioned simplifications are used in complex food matrices, food-quality loss and shelf-life loss can be, in practice, represented by the loss of desirable quality factors A (e.g., nutrients, characteristic flavors) or the formation of undesirable factors B (e.g., off flavors, discoloration). The rates of loss for A and of formation for B are expressed as:

$$r_A = \frac{-d[A]}{dt} = k[A]^n,$$
$$r_B = \frac{d[B]}{dt} = k'[B]^{n'}. \tag{7.12}$$

Both k and k' are the apparent reaction rate constants, and n and n' are the reaction orders. It should be again stressed that Equation 7.11 and Equation 7.12 do not represent true reaction mechanisms, and n and n' are not necessarily true reaction orders with respect to the species A and B, but rather apparent or pseudo orders. The apparent reaction orders and constants are determined by fitting the change with time of the experimentally measured values of $[A]$ or $[B]$ to Equation 7.11 or Equation 7.12, by differential or integral methods.[18]

7.3.2 Reaction Order

The term *reaction order* is defined as the number of molecules participating in a reaction as reactants. In other words, reaction order is the sum of the exponents of the reactant species. In food science, it is quite common to describe a reaction as a first-order equation, even though the real mechanism is not first order. The reaction order is then referred to as *pseudo first-order* or, equivalently, as *apparent first-order*.

7.3.2.1 Common Types of Reactions

Zero-order reactions. If A is the characteristic quality parameter selected to describe the degradation of a food product, then the equations used are:

$$r = \frac{dA}{dt} = k,$$

$$A = A_0 + kt.$$

(7.13)

Examples of common zero-order reactions in foods include frozen food shelf life, enzymatic reactions in fresh foods, nonenzymatic browning, etc.

First-order reactions. The respective equations in the case of first-order reactions are:

$$r = -\frac{dA}{dt} = kA,$$

$$\ln\left(\frac{A}{A_0}\right) = kt \Rightarrow A = A_0 e^{-kt}.$$

(7.14)

Examples of common reactions in foods of first order include fresh and dry food shelf life, microbial growth in fresh foods, microbial death, oxidative color loss, vitamin loss in frozen, canned, and dry food, and thermal denaturation of proteins.[19]

Second-order reactions. The respective equations in the case of second-order reactions are:

$$r = -\frac{dA}{dt} = kA^2,$$

$$\frac{1}{A} - \frac{1}{A_0} = kt.$$

(7.15)

Second-order unimolecular reactions are characterized by a hyperbolic relationship between the concentration of the reactant or product and time.[20] A linear plot is obtained when the term $(1/A)$ is plotted vs. time.

Similarly, when studying a bimolecular reaction, Equation 7.15 becomes

$$r = -\frac{dA}{dt} = kAB.$$

(7.16)

where A and B are the reactants.

If B is assumed to be constant, then integration of Equation 7.16 gives

$$\ln\left(\frac{A}{A_0}\right) = -k't,$$

(7.17)

where k' is a pseudo-order rate constant, $k' = kB$.

A common case of such reaction is the aerobic degradation of ascorbic acid, where oxygen level plays the role of reactant B. When studying this reaction at different levels of oxygen availability, a family of pseudo-first order plots is discovered.

nth-order reactions. The respective equations in the case of *nth*-order reactions are:

$$r = -\frac{dA}{dt} = kA^n, \quad n > 1,$$

$$A^{1-n} - A_0^{1-n} = -(1-n)kt.$$

(7.18)

To determine the apparent reaction order, a trial-and-error process is often applied that is based on assuming various values of n and evaluating, through iterative methods, which value yields the best fit. There are two alternative methods to perform these assays, namely the graphical or the least-square linear fit to the corresponding equations listed above (Equation 7.13 through Equation 7.18). Alternative methods are briefly described in a following section. As stated by Labuza 1984,[21] a prerequisite for safely deciding the appropriate apparent order is to carry out the experiment to at least 50%, or preferably up to 75% conversion. Additionally, another observation is that the worse the precision of the method of measuring the quality factor A, the larger the extent of change to which the experiment should be carried out to obtain an acceptably accurate estimate of the reaction rate constant. Otherwise, any extrapolation to longer times may yield significant errors. Other possible pitfalls stated by Taoukis et al. (1997) include the existence of lag phase and the misleading use of the R^2 criterion, in case of scattered data in zero- and first-order reactions.[3]

After the apparent order of the quality deterioration reaction has been established, the next step involves the application of statistics for parameter k to estimate the error introduced. Student t-distribution for the 95% confidence limits is frequently used if a linear regression method was selected for the estimation of parameter k. Whichever method is applied to determine the apparent reaction order, n, of Equation 7.18, there are many cases stated in literature that none of the equations (Equation 7.13 through Equation 7.18) can adequately model the measured quality loss. One approach in this case is to develop a semi-empirical kinetic/mathematical model that effectively represents the experimental data. Saguy and Karel described the steps for building such a model.[22] Computer-aided multiple linear, polynomial, or nonlinear regressions can test multivariable linear models, polynomial equations; nonlinear models can be defined and their fit to the data can be tested. Empirical equations modeling the effect of different composition or process parameters can be derived from statistical experimental designs, such as surface-response methods,[23] where the coefficients determined lack any physical meaning.

It should be pointed out, however, that because the model often does not correspond to the true mechanism of the reaction, any extrapolation of kinetic results to similar systems should be done very cautiously. In certain cases, an in-depth kinetic study of specific reactions important to food quality is desirable, so that the effect of compositional changes can be studied. In these cases, the identity of the actual mechanism of the reactions is sought, if it is possible to find. Such studies are usually done in model systems, rather than in actual foods, so that the composition and the relative concentrations of the components are closely controlled and monitored.

7.3.3 Determining the Order of Reaction

After the parameter A that should describe a food system degradation is decided, it is necessary to generate data of the change of A over time. Analysis of this information to estimate the most appropriate reaction order can be completed using differential or integral methods.[1]

7.3.3.1 Differential Methods

The equation of unknown order used to describe the system change has the following form:

$$\frac{dA}{dt} = rate(r) = -k[A]^n.$$

(7.19)

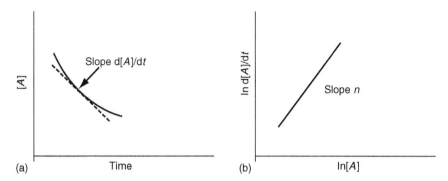

Figure 7.1 (a) Plot of concentration [A] vs. time and (b) plot of ln(d[A]/dt) vs. ln[A] to determine the slope, n.

1. [A] is plotted vs. time, and a smooth curve is drawn as shown in Figure 7.1a.
2. The slope is d[A]/dt and can be determined at various points in time by:
    a. Drawing a tangent to curve
    b. Taking d[A] for each dt (requires many data points)
3. ln dA/dt is plotted vs. ln A as in Figure 7.1b:

$$\ln \frac{d[A]}{dt} = \ln k + n \ln[A], \tag{7.20}$$

where the slope, n, is the order of the reaction, estimated from algebraic substitution.

7.3.3.2 Integral Methods

The graphical procedure is a trial-and-error process where a reaction functional form is assumed:

$$\frac{d[A]}{dt} = -kF[A] \Rightarrow \int_{A_i}^{A_f} \frac{d[A]}{F[A]} = \int_0^t -k\, dt = -kt, \tag{7.21}$$

and the term of $\int_{A_i}^{A_f} d[A]/F[A]$ is calculated from the data measurements available. If the plot or the regression of this term vs. time yields a linear function passing through the origin (0,0) with slope $-k$, then the hypothesis was correct; otherwise, a new assumption is made and the procedure is repeated.

7.3.3.3 Method of Half Lives

The half life ($t_{1/2}$) is the time required for the parameter A to lose half of its initial concentration. Table 7.2 summarizes the mathematical form of $t_{1/2}$ for the different reaction rate orders.

Table 7.2 Forms of Half-Life for the Different Reaction Orders

Order of Reaction	$t_{1/2}$ (Half-Life)
First (1)	$\dfrac{k}{0.693}$
Second (2)	$\dfrac{1}{kA_0}$
nth order ($n \neq 1$)	$\left(\dfrac{1}{(k[n-1])(2^{n-1}-1/A_0^{n-1})}\right)$

Alternatively, life can be also determined for any decided fractional decrease. If f is the decimal fraction of (A/A_0) and n is 1, then

$$\ln\frac{A}{A_0} = -kt \Rightarrow \ln\frac{fA_0}{A_0} = \ln[f] = -kt \Rightarrow k = -\frac{\ln[f]}{t_f},\qquad(7.22)$$

where t_f is the fraction life in time units. A well-known example of this methodology refers to the decrease by one log cycle of microorganisms in microbial death. In that case, $f=0,1$; t_f is called *D-value* and Equation 7.22 becomes

$$k = -\frac{\ln 0,1}{D} \Rightarrow D = \frac{2,3}{k}.\qquad(7.23)$$

7.3.4 Effect of Temperature on Reaction Rates

The prevailing effect of temperature on the rate of food related reaction rates has long been the subject of research and a significant number of kinetic studies of important indices have been published—of physical, chemical, microbiological, or sensory deterioration of foods. In general, temperature is regarded as the most important of all external factors affecting food spoilage.

7.3.4.1 The Arrhenius Equation

Of the mathematical equations that have been proposed to describe the temperature dependence of the quality loss rate, the Arrhenius relation, which was derived from thermodynamic laws and statistical mechanics principles, is the most widely used.[24] The Arrhenius relation developed theoretically for reversible molecular chemical reactions, but has been used to describe the effect of temperature on the rate of several reactions of quality loss, as follows:

$$k = k_A \exp\left(\frac{-E_A}{RT}\right),\qquad(7.24)$$

with k_A representing the Arrhenius equation constant and E_A defining the activation energy, i.e., the excess energy barrier that quality parameter A must overcome to proceed to degradation products. R is the universal gas constant (1.9872 cal/(mol K) or 8.3144 J/(mol K)).

Another form, frequently used, uses a constant reference temperature, T_{ref}, and then Equation 7.24 assumes the following form:

$$\frac{k}{k_{ref}} = e^{-\frac{E_A}{R}\left[\frac{1}{T} - \frac{1}{T_{ref}}\right]},\qquad(7.25)$$

where k_{ref} is the reaction rate constant at T_{ref}.

To estimate the temperature effect on the reaction rate of a specific quality deterioration mode, values of k are estimated at different temperatures, in the range of interest, and $(\ln k)$ is plotted vs. $(1/T)$ or, similarly $((1/T)-(1/T_{ref}))$. The case of Equation 7.25 produces a semilog graph. A straight line is obtained with a slope of $-E_A/R$, from which the activation energy is calculated.

7.3.4.2   The Q_{10} Value

The Q_{10} value of a reaction, a tool of practical importance to the food industry, is often used when the temperature dependence of biological reactions is studied. Its definition is the number of times a reaction rate changes with a 10°C change in temperature, or, equivalently, it shows the reduction of shelf life, θ_s, when the food is stored at a temperature higher by 10°C:

$$Q_{10} = \frac{k_{(T+10)}}{k_{(T)}} = \frac{\theta_s(T)}{\theta_s(T+10)}. \tag{7.26}$$

In essence, the Q_{10} approach introduces a temperature-dependence equation in the form of the following:

$$k(T) = k_0 e^{bT} \Rightarrow \ln k = \ln k_0 + bT, \tag{7.27}$$

which implies that if $(\ln k)$ is plotted vs. temperature, a straight line is obtained. Based on Equation 7.27, in an equivalent way, shelf life can be plotted vs. temperature, yielding the so-called *shelf-life plots*:

$$\theta_s(T) = \theta_{so} e^{-bT} \Rightarrow \ln \theta_s = \ln \theta_{so} - bT. \tag{7.28}$$

These plots are true straight lines only for narrow temperature ranges of 10–20°C. For such narrow intervals, the Arrhenius and the Q_{10} concept are related as follows:

$$\ln Q_{10} = 10b = \frac{E_A}{R} \frac{10}{T(T+10)}. \tag{7.29}$$

7.3.4.3   The z-Value

This term is frequently used to describe temperature dependence in microbiological studies and it is defined as the temperature change needed to change the microbial inactivation rate by a factor of 10.[20] The z-value expressed in terms of the reaction rate constant is as follows:

$$k_2 = k_1 [10]^{\left(\frac{T_2-T_1}{z}\right)} \Rightarrow \ln \frac{k_2}{k_1} = \frac{T_2-T_1}{z} \ln(10) \Rightarrow \frac{\ln(k_2/k_1)}{T_2-T_1} = \frac{\ln(10)}{z}. \tag{7.30}$$

Assuming an Arrhenius relationship for k_1 ($k_1 = k_A \exp[-E_A/(RT_1)]$) and k_2 ($k_2 = k_A \exp[-E_A/(RT_2)]$), the following ratio is estimated:

$$\ln \frac{k_2}{k_1} = \frac{-E_A}{R} \frac{T_2-T_1}{T_2 T_1} \Rightarrow \frac{\ln(k_2/k_1)}{T_2-T_1} = \frac{-E_A}{R} \frac{1}{T_2 T_1}. \tag{7.31}$$

Equating the right-hand side of the two previous equations, the relation between z-value and E_A is

$$z = \frac{\ln 10}{-E_A/R} T_1 T_2. \tag{7.32}$$

7.3.4.4 The WLF Equation

As has been often noted in the food literature, there are cases where the temperature dependence on the rate of food quality loss can deviate from the Arrhenius equation.[25] Phase change phenomena are often involved in such deviations. Glass transition is related to dramatic changes of food mechanical properties and molecular mobility, and may occur in carbohydrate-containing foods when storage conditions are suddenly modified, such as during rapid cooling or solvent removal. Examples of common cases in foods that deviates from the Arrhenius law, include: frozen carbohydrate-containing solutions or food products;[26–29] whey powder and dehydrated vegetables;[30] and osmotically dehydrofrozen fruits and vegetables.[31,32]

In such systems, there is a drastic acceleration of the diffusion-controlled reactions above T_g—a temperature zone where the system is in a rubbery, unstable phase and activation energy is not constant, but is rather a function of temperature. This behavior has been often described by an alternative equation, the Williams–Landel–Ferry (WLF) expression (Equation 7.33) that empirically models the temperature dependence of mechanical and dielectric relaxations in the range $T_g < T < T_g + 100$

$$\log \frac{k_{ref}}{k} = \frac{C_1(T - T_{ref})}{C_2 + (T - T_{ref})},$$

(7.33)

where k_{ref} is the rate constant at the reference temperature T_{ref} ($T_{ref} > T_g$) and C_1, C_2 are system-dependent coefficients. Williams, Landel, and Ferry[33], assuming $T_{ref} = T_g$ and applying WLF equation for data available for various polymers, estimated mean values of the coefficients $C_1 = -17.44$ and $C_2 = 51.6$. However, the uniform application of these constants is often problematic,[34,35] and the calculation of system-specific values is necessary. These coefficients were found to depend not only on the type of matrix in question, but also on the water content.[36] In the same study, it was observed in several cases that in the rubbery zone of the matrices, the Arrhenius equation described the temperature dependence of the loss of ascorbic acid more adequately than the WLF model.

7.3.4.5 Effects of Other Environmental Factors

Besides temperature, moisture content and water activity (a_w) are the most important factors affecting the rate of deteriorative reactions at above-freezing temperatures.[37] Water activity describes the degree of boundness of the water contained in the food and its availability to act as a solvent and participate in chemical reactions. Moisture content and water activity can influence the kinetic parameters (k_A, E_A), the concentrations of reactants, and, in some cases, even the apparent reaction order, n.

Critical levels of a_w can be recognized above which undesirable deterioration of food occurs. Microbiological decay and textural degradation are examples of quality-loss reactions that are greatly affected by levels of water activity. Regarding microbial growth, many workers concluded that the water activity is mainly the determinant of growth because it governs the osmotic stress, and the ability to grow is controlled by both the degree of that stress and the osmoregulatory capacity of a particular microbial cell.[38,39]

Besides the specific critical a_w limits, water activity has a pronounced effect on chemical reactions. This effect plays a very important role in the preservation of intermediate-moisture foods (IMFs) and dry foods. Generally, the ability of water to act as a solvent, reaction medium, and a reactant increases with increasing a_w. As a result, many deteriorative reactions increase exponentially in rate with increasing a_w above the value corresponding to the monolayer moisture. This can be represented schematically in a global food stability map, as illustrated in Figure 7.2.[3]

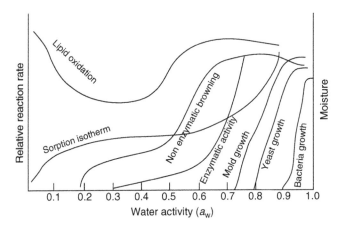

Figure 7.2 Indicative food stability map, showing the effect of a_w and moisture content on the rates of important deterioration reactions. (Adapted from Labuza, T. P., *Food Technology*, 34, 36–41, See also p. 59, 1980.)

Additionally, moisture content and a_w directly affect the glass transition temperature of the system.[4] With increasing a_w, T_g decreases. As was discussed in the previous section, transverse of T_g and change into the rubbery state has pronounced effects, especially in texture- and viscosity-dependent phenomena, but also in reaction rates and their temperature dependence. Nonenzymatic browning,[40–43] aspartame degradation,[44] and enzymatic activities in frozen-model systems[26,35] are well-studied examples of reactions significantly influenced by the glass transition temperature.

Mathematical models that incorporate the effect of a_w as an additional parameter can be used for shelf-life predictions of moisture sensitive foods.[45–47] Such predictions can be applied to packaged foods in conjunction with moisture transfer models which were developed based on the properties of the food and the packaging materials.[48]

Other factors that affect the reaction rates in foods are pH, reactant concentrations, ratio between reactants, gas composition, partial pressures, and total pressure.[15] The effect of the pH has been studied for several food reactions and food systems. Enzymatic and microbial activity are strongly affected by pH, each having a pH range for optimum activity and limits above and below which activity ceases. Most protein behavior also depends on pH, with solubility usually having a minimum near the isoelectric point, directly affecting their behavior in reactions. Acid–base catalyzed reactions important to foods, such as nonenzymatic browning and aspartame decomposition, are strongly pH dependent. With regard to the influence of pH on the Maillard reaction—the popular reaction between carbonyls and amines that strongly affects food quality—Labuza and Baisier[7] observed that the substrate loss increased with increasing pH, up to a pH level of 10, with little if any browning occurring below pH 6. Many studies are focused on assessing pH influence[49,50] and even develop kinetic models that include pH effect in Maillard reaction rate constants.[51,52]

Gas composition is an additional factor that can play a significant role in some quality loss reactions. Oxygen availability is very important for oxidative reactions and can affect both the rate and apparent reaction order depending upon whether it is limited or in excess.[15] Further, the presence and relative amount of other gases, especially carbon dioxide, strongly affect biological and microbial reactions in fresh meat, fish, fruits, and vegetables. Different food matrices have different optimum O_2–CO_2–N_2 gas composition requirements for maximum shelf life. This knowledge is the basis of successfully controlled and modified atmosphere packaging application. Total pressure is usually not an important factor except in the case of hypobaric storage, which is an alternative technology to CAP/MAP.

Recently very high pressure technology (1,000–10,000 MPa) has been used experimentally to achieve inactivation of microorganisms, modification of biopolymers (protein denaturation, enzyme inactivation or activation, degradation), increased product functionality, and quality retention.[53,54] Kinetic studies of changes occurring during high pressure processing and their effects on shelf life of the foods is an area of current research.

7.4 APPLICATION OF FOOD KINETICS IN SHELF-LIFE PREDICTION AND CONTROL

7.4.1 Case Study of Quality Degradation of Frozen Green Peas and Mushrooms

This practical example is based on experimental data generated and published by Giannakourou and Taoukis.[55,56] The purpose of these studies was to evaluate the quality loss in popular frozen vegetables (green peas and mushrooms), in a wide range of temperatures in the subfreezing zone, including the detrimental range between -3 and $-10°C$, that frequently occurs in the often problematic frozen chain (e.g., European Union survey EE1080/94/00069[57]). Additionally, these studies tested the predictive value of the developed models under dynamically fluctuating temperatures occurring during the life cycle of processed agricultural products.

The methodology utilized included three main steps, as described in the previous section (see Section 7.3.2 through Section 7.3.4) to obtain a well-established and validated kinetic model of food quality degradation:

7.4.1.1 First Step

In this phase, representative quality indices must be chosen that adequately describe the quality loss of the product. Subsequently, experimental data of these indices are obtained under isothermal conditions, and the reaction order of the corresponding deterioration is decided.

In this example, packaged frozen green peas and mushrooms were stored in controlled temperature cabinets (Sanyo MIR 153, Sanyo Electric Co, Ora-Gun, Gunma, Japan) at constant temperatures (from -1 to $-20°C$) or programmed variable temperature profiles, constantly monitored by type-T thermocouples and a multichannel datalogger (CR10X, Campbell Scientific, Leicestershire, U.K.).

1. *Green peas.* For this widely consumed frozen vegetable, the color change and vitamin C loss were effective shelf life loss indices. The chroma change, expressed by

$$DC = \sqrt{(a - a_0)^2 + (b - b_0)^2},$$
(7.34)

   was modeled by a pseudo zero-order reaction (Figure 7.3a), whereas vitamin C degradation showed an apparent first-order reaction (Figure 7.3b).
2. *White mushroom.* The main index of white mushroom deterioration is the gradual intense color change, expressed mainly by parameter L (of the CIE Lab scale) related to the lightness of the samples. The major enzyme responsible for the browning reaction is tyrosinase and the path followed is the enzymatic oxidation of natural phenolic compounds to the corresponding o-quinones that subsequently polymerize nonenzymatically to brown pigments.[58] The measured L shows an apparent first-order change (Figure 7.4).

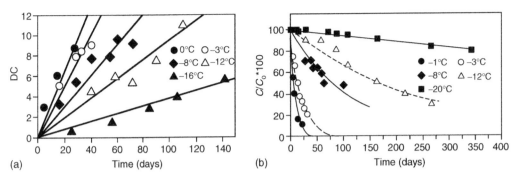

Figure 7.3 (a) Chroma change of frozen green peas at five storage temperatures vs. time. Experimental points are shown as black and white triangles, black rhombs, black and white circles, and solid lines are least square fits to the measurements. (b) Results for vitamin C loss (C is vitamin C concentration at time t and C_0 at time zero) vs. time at five storage temperatures. Experimental points are shown as black and white circles, black rhombs, black squares and white triangles, and all curves (continuous and dashed lines) represent the predicted values from the first order kinetic model.

7.4.1.2 Second Step

After determining the main quality index, the order of the corresponding quality loss reaction is decided, based on isothermal experimental data; then, the temperature dependence of the reaction is evaluated. Arrhenius kinetics are most frequently used, although there are numerous cases where other models (such as the WLF equation) were preferred, especially in the case of frozen foods where the glass transition phenomenon may significantly affect the rate of diffusion-based reactions[59–61] as stated in Section 7.3.4.

In this case study, both the Arrhenius and the WLF equation were applied, and both showed an equally adequate performance. Figure 7.5a and b show the Arrhenius plots for color loss and vitamin C degradation in frozen peas. The estimated activation energies, E_A, and 95% confidence range were 79.2 ± 19.2 kJ/mol (18.9 ± 4.6 kcal/mol, with $R^2 = 0.983$) and 136.8 ± 20.5 kJ/mol (32.7 ± 4.9 kcal/mol with $R^2 = 0.993$) for chroma and vitamin C loss, respectively, expressing the different temperature sensitivities of the two modes of deterioration. Figure 7.6 is the Arrhenius plot for enzymatic browning of white mushroom, with E_A estimated to be 155.1 ± 60.3 kJ/mol

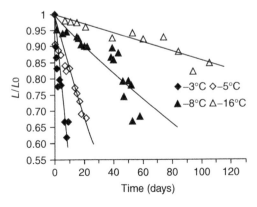

Figure 7.4 Results for enzymatic browning of white pieces of mushroom (represented by L-value loss compared to the L-value at zero time, L_0) vs. time at four storage temperatures. Experimental points are shown as black and white rhombs, black and white circles and solid lines represent the least square fits to the measurements.

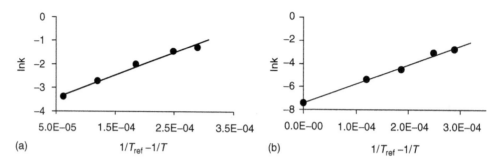

(a)   (b)

Figure 7.5   Arrhenius plot of (a) the chroma change rate and (b) the vitamin C loss rate for frozen green peas (with $T_{ref} = -20°C$).

(37.1 ± 14.4 kcal/mol with $R^2 = 0.957$), showing the extremely temperature-sensitive nature of white mushroom.

7.4.1.3   Third Step

In this phase, mathematical models already developed for isothermal studies are tested under dynamic storage conditions in programmable freezer-incubators. The agreement between experimentally measured index values and predictions from the established kinetics was satisfactory. In Figure 7.7 measurements of vitamin loss at dynamic conditions are compared to predictions at the corresponding T_{eff}. Predicted rate of loss is in good agreement with the experimental rate of loss, as shown by $k_{pred} = 0.0268 \pm 0.0038$ (1/days) vs. $k_{exp} = 0.0252$ (1/days) ($R^2 = 0.947$).

Similarly, shelf-life models for color change in green peas and mushrooms were validated under nonisothermal conditions.

The main and practical purpose of developing a robust and validated mathematical kinetic model of the behavior of principal quality indices of a food, is to expand its use at time–temperature conditions that differ from the experimental ones. The importance of this purpose is evidenced by the diversity of time–temperature history of a product during its actual distribution in the post-processing chain, to the consumer end, as confirmed by recent temperature surveys.[56] Food products are therefore found to be exposed to a variable temperature environment that frequently

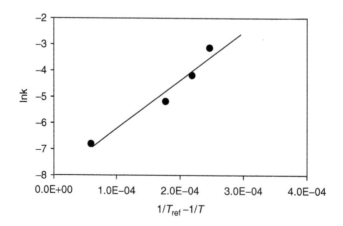

Figure 7.6   Arrhenius plot of browning rates of white mushroom with $T_{ref} = -20°C$.

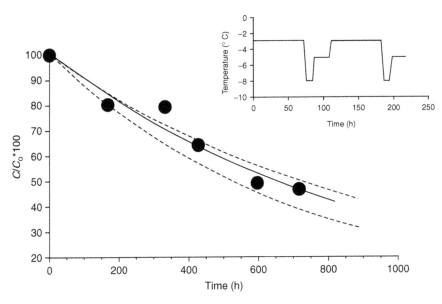

Figure 7.7 Comparison of experimental (closed circles) and predicted results of vitamin C loss for exposure at the shown variable temperature profile with $T_{eff} = -4.4°C$. The solid line represents the exponential fit of the quality measurements and dotted lines depict the upper and lower 99% confidence range of quality predicted for T_{eff}.

includes stages of abusive storage or handling conditions. Different methodologies are proposed in order to assess the kinetic parameters of certain reactions in cases of dynamic temperature conditions, including integral and differential methods,[61] the "equivalent point" theory,[62] or the "effective temperature" approach (the time/temperature/tolerance (TTT) approach).[63–65] According to this last approach, the fraction of shelf life consumed at the end of each stage, f_{con}, is calculated.

The f_{con} is estimated as the sum of the times at each constant temperature segment t_i, divided by the shelf life at that particular temperature θ_i:

$$f_{con} = \sum_i \frac{t_i}{\theta_i}, \tag{7.35}$$

where index i represents the different time–temperature steps within the particular stage of study. The remaining shelf life of products can be calculated at a reference temperature, representative of their storage conditions, after each stage as $(1 - \Sigma f_{con})\theta$, where θ is the shelf life at that reference temperature.

To understand fully this approach in assessing the effect of temperature conditions during transport and storage on food quality, an example in the real distribution chain is considered. Based on the aforementioned shelf life models, color and vitamin C loss were measured for frozen green peas, assuming a realistic scenario of 120 days in the distribution chain. Figure 7.8 shows the temperature profile and the stages assumed in this hypothetic scenario. This realistic distribution scenario includes 30 days storage in the factory warehouse, intermediate transport, 40 days stocking in a distribution center (where vegetables are distributed to different supermarkets at the retail level), and exposure at closed vertical freezers or open horizontal freezers for 20 days before purchase. The final stage of this "time–temperature history simulation" is 30 days of domestic storage, before final cooking and consumption. Temperature conditions for the initial stages were obtained in Jul 1983,[66] and for the final stages of retail and domestic storage from the data shown in Figure 7.8.

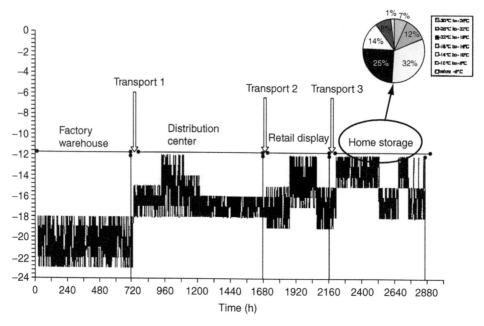

Figure 7.8 Indicative distribution scenario of frozen green peas, including four successive stages of stocking and possible intermediate transports. (Temperature data of domestic storage from Giannakourou, M. C. and Taoukis, P. S. *Journal of Food Science*, 68(1), 201–209, 2003.)

Applying the main principles of TTT approach, the shelf life of frozen green peas was estimated at the end of each stage, taking as main quality criterion either color loss ($DC = 10$ was decided by sensory results to signal the endpoint) or vitamin C loss (a 50% loss, compared to the initial concentration). A significant remark out of this comparative study is that the quality level and the remaining shelf life of a food, at any point of its distribution in the real chain, is strongly dependent on the criterion chosen to signal the acceptability limit (Figure 7.9a). In the case study of frozen green peas, at the end of the four-month cycle illustrated in Figure 7.8, the product has just expired, according to nutritional criteria (50% vitamin C loss, Figure 7.9b). However, if sensory criteria of color loss are assumed, the frozen product is still of acceptable quality for more than 200 days (221 days), if isothermally handled at $-18°C$.

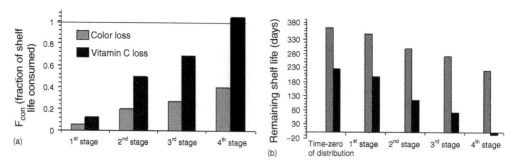

Figure 7.9 (a) Fraction of total shelf life consumed and (b) remaining shelf life estimated at $-18°C$, assuming two different acceptability criteria (color and vitamin C loss) at the end of each stage of the distribution path of frozen green peas, presented in Figure 7.8.

7.4.2 Modeling the Effect of Temperature and Relative Humidity Storage Conditions (Nonenzymatic Browning of a Cheese Powder)

Nonenzymatic browning (NEB) through the Maillard reaction is a major deteriorative factor in the storage of dehydrated dairy food products,[43,61,67,68] as already mentioned in Section 7.2.3. NEB may cause unacceptable nutritional and sensory effects in some stored food products and may be a limiting factor in the shelf life of some products. In most cases, NEB results in deterioration of flavor, darkening of color, and significant loss of nutrients in particular protein quality. Additionally, it may result in the formation of substances of questionable toxicity.[48,69] NEB as a model of a possible diffusion-controlled binary reaction[70] between an amino acid and a reducing sugar has probably been given the most attention in studies of relationships between reaction rates and the physical state controlled by the glass transition.[40–42,71–73] The kinetics of NEB have been studied as a function of many parameters (water activity, temperature, reactant concentration pH, and so forth) and even related to physical aspects such as crystallization[71,74] and viscosity.[75] The glass transition together with material composition, water activity, temperature, and other factors affect the rate of the reaction.[60,72,73,76]

Taoukis and Skiadi[69] studied the combined effect of temperature and a_w on the rate k ($k = f[T,a_w]$) of NEB of cheese powder. Because the value of water activity varies during distribution and storage of dehydrated foods due to the water vapor permeability of packaging used, it is important to be able to predict the change of moisture and a_w under different storage conditions, and thus design a packaged food of the desired quality and stability. A quantitative knowledge of dependence by the a_w of shelf life allows the determination of a_w critical limits. The combination of the shelf life quantitative dependence of a_w level and the prediction of a_w change under different storage conditions can be used (1) for the selection of the optimum food-package systems, (2) for the estimation of the effect of changes in the package or the environment parameters, and (3) for the prediction of the shelf life of the product under variable storage conditions.

In the aforementioned study, the first part focused on developing an empirical model of NEB of cheese powder, rather than exactly studying the complex mechanism of reactions occurring in the matrix. Kinetic data was obtained for brown pigment formation and an apparent zero order reaction was confirmed:

$$B - B_0 = k_{NEB}t, \tag{7.36}$$

with B being the absorbance measured spectrophotometrically and k_{NEB} being estimated by a least-square regression.

Table 7.3 summarizes the time needed for a ten-fold increase of brown pigments in cheese powder.

Subsequently, the logarithm of k_{NEB} is plotted vs. $1/T$ and the Arrhenius kinetic parameters are estimated for each level of a_w studied, and in parallel, iso-E_A lines are obtained. While the activation energy values were statistically the same, the Arrhenius constants k_A were found to be a_w-dependent, and the selected model was found to conform to the expected pattern of a maximum at the range of 0.6–0.8:

$$\ln k_A = \ln k_0 + \gamma a_w^2 \ln a_w. \tag{7.37}$$

Combining Equation 7.36 and Equation 7.37, the complete quality function for browning of cheese powder is described by the following mathematical expression:

$$B - B_0 = \exp\left(\ln k_0 + \gamma a_w^2 \ln a_w - \frac{E_A}{RT}\right)t. \tag{7.38}$$

Table 7.3 Rate of Browning of Cheese Powder and Time for a Ten-Fold Increase, t_{10}, as Affected by Water Activity and Storage Temperature

T(°C)	$a_w = 0.318$		$a_w = 0.492$		$a_w = 0.748$	
	k	t_{10} (days)	K	t_{10} (days)	K	t_{10} (days)
30	0.00242	338	0.0072	114	0.0094	87
34	0.0037	221	0.0136	60	0.0142	58
40	0.0068	120	0.0191	43	0.0289	28
45	0.0141	58	0.0472	17	0.053	16

Constants k_0, γ, and E_A were calculated using multiple nonlinear regression for all kinetic data available and the values obtained were $\ln k_0 = 30.05 \text{ d}^{-1}$, $\gamma = -23.29$, and $E_A = 97.81 \text{ kJ/mol}$.

In the second part of this study, the purpose was to model the effect of the packaging on the change of a_w of the packaged dehydrated food, and thus on the rate of browning of the cheese powder. Therefore, the sorption isotherm of the cheese powder was experimentally determined by the saturated solution equilibrated method, at all the temperatures of the kinetic experiments. As detailed by Taoukis and Skiadi,[69] samples of powder were packaged in pouches of a Cellophane (30 g/m²) + LPDE (70 g/m²) film and stored at $T = 40°C$ or controlled variable conditions and relative humidity conditions created by saturated NaCl (average $a_w = 0.748$). The quality index studied (browning) was measured at predetermined time intervals and values were compared with those predicted by the model. Film permeability was obtained by packaging desiccant in the same pouches used for the powder and measuring the rate of weight change when stored at the above isothermal and a_w-conditions.

To describe mathematically the water sorption isotherms, a linear equation has been used in a_w range of 0.2–0.6 and equation GAB in a_w range of 0–0.9:

$$\text{Linear} : m = ba_w + c$$

$$\text{GAB} : m = \frac{m_0 CKa_w}{(1 - Ka_w)(1 - Ka_w + CKa_w)}, \tag{7.39}$$

where slope b and intercept c are empirical parameters, m_0 the moisture monolayer value, and C and K parameters relating to interaction energies between sorbed water and food, and between the multiple layers of sorbed water, respectively. C and K are temperature-dependent according to the following mathematical expressions:

$$C(T) = C_0 \exp\left[(H_1 - H_m)/RT\right],$$
$$K(T) = K_0 \exp\left[(H_L - H_m)/RT\right], \tag{7.40}$$

where H represents sorption enthalpies.

Using the linear or GAB isotherm to describe the isotherms of cheese powder and the moisture transport equations through packaging (based on Fick and Henry equations), we take the following packaging equations:

$$\frac{a_{we} - a_{wi}}{a_{we} - a_w} = \frac{k}{x} \frac{A}{W_s} \frac{p_0}{b} t,$$

$$\int_{a_{wi}}^{a_w} \frac{1 + K^2[1 - C]a_w^2}{[a_{we} - a_w][1 - Ka_w^2][1 - Ka_w + CKa_w]^2} \, da_w = \frac{1}{m_0 CK} \frac{k}{x} \frac{A}{W_s} p_0 t, \tag{7.41}$$

where a_w is the water activity of the food at time t, a_i is the initial water activity of the food, a_{we} is the equilibrium water activity corresponding to the external %RH and temperature, k/x (g/day m^2 mm Hg) is the moisture permeation of the packaging material, A(m^2) is the exchange surface o the package, W_s is the dry weight of the packaged food, and p_0 (mm Hg) is the partial water vapor pressure of the storage environment.

Consequently, assuming that external conditions $T(t)$ and RH(t) are known, then functions $T(t)$ and $a_w(t)$ can be estimated, and the rate of the formation of brown pigment B can be estimated:

$$B - B_0 = \int \exp\left(\ln k_0 + (\gamma a_w(t)^2)\ln(a_w(t)) - \frac{E_A}{RT(t)} \right) dt. \tag{7.42}$$

The former equation allows for the prediction of the effect of different storage conditions, alternative packaging systems, and food-package design on shelf-life loss, giving the ability to define the optimal conditions to obtain a target quality level.

Figure 7.10a shows the effect of different storage conditions and packaging materials on the shelf life of cheese powders. From Figure 7.10b, it can be observed that when cheese powder is stored at high humidity conditions (RH=75%) and a moderate temperature of 30°C, its shelf life (designated at the point where B=0.5) is almost reduced by half (50 days) compared to the shelf life at a moderate humidity of RH=45% and the same temperature (93 days), thus confirming the significant effect of relative humidity on the durability of dehydrated food products. Equivalently, studying Figure 7.10a, it can be concluded that temperature is an important environmental parameter, significantly influencing cheese powder shelf life; an increase of 10°C reduces nearly four times the shelf life of cheese powder, if available experimental data at 20°C are arbitrarily projected until B=0.5.

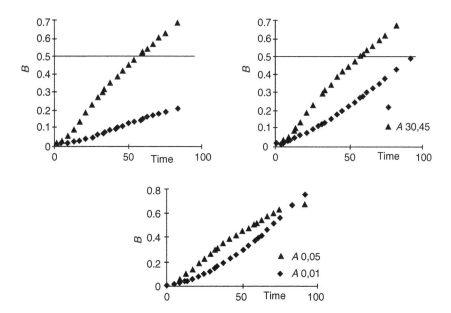

Figure 7.10 (a) Nonenzymatic browning prediction of cheese powder for different temperature conditions (upper line: 30°C, lower line: 20°C) and the same package and RH=75% conditions. Browning, B, in y-axis is measured spectrophotometrically. (b) Nonenzymatic browning prediction of cheese powder for different RH(45%, 75%) conditions and the same package and temperature conditions. (c) Nonenzymatic browning prediction of cheese powder for different packages k/x: 0.05, 0.01 and the same temperature and RH=75% conditions. Absorbance at 0.5 has been set as the end of shelf life.

7.4.3   Use of Time–Temperature Indicators as Shelf-Life Monitors

Determination of the quality function of a food product and its kinetic parameters not only allows estimation of shelf life, but also permits the effect on quality of any actual or assumed variable-temperature exposure to be calculated. Thus, the remaining shelf life at any point of a monitored distribution can be estimated, as was shown by the case study in Section 7.4.1. In the real marketing path of perishable food products, due to be stored and handled under refrigerated or frozen conditions, the actual distribution conditions are not controlled by the manufacturer and are difficult to monitor. Giannakourou and Taoukis[56] studied the effects of the realistic conditions that a packaged food can possibly face during the frozen distribution chain. Supposing average storage times at each location of the assumed distribution route of a frozen vegetable, the remaining shelf life after 160 days in the distribution ranged from -250 days to about $+300$ days for best and worst temperature scenarios, respectively, assuming that color loss is the selected acceptability criterion.

Accepting both the great deviation of the distribution conditions from what is ideal and the detrimental effects of abusive handling, it would certainly be desirable to obtain a cost-effective way to individually monitor the conditions of the products during distribution as well as a means to signal remaining shelf life. This would lead to effective control of distribution, optimized stock rotation, reduction of waste, and the meaningful information on product "freshness" that is demanded by the consumer. Time–temperature indicators (TTIs) are a potential solution.[15]

TTI are devices with an easily measurable response that reflect the accumulated time–temperature history of the product they are attached to. Their operation is based on irreversible reactions initiated at the time of their activation and proceeding with an increasing rate, as temperature is elevated, in a manner that resembles the temperature dependence of most quality loss reactions of foods.[77,78]

In essence, the usefulness of these tags in monitoring quality deterioration and estimating the remaining shelf life at any point of the distribution chain of a product depends on the successful simulation of food quality loss kinetics. Thus, the TTI will reflect the quality status of the food only if the activation energy of the reaction that describes shelf life loss is close to that of the TTI response, estimated by a thorough kinetic study of the behavior of the particular tag.[79,80]

7.5   FUTURE TRENDS

Making models more complex, in order to describe the exact mechanism of food spoilage, is an alternative to simple, empirical modeling. However, the development and validation of simple models allows for the mathematical prediction of food quality and shelf life, permitting for optimal design of process parameters. More complex approaches, such as multivariate analysis, fuzzy logic, genetic algorithms, or hybrid models, and other modeling methods that aim at describing quality changes in a more generic fashion[81] have been recently proposed in order to expand the model's use in a much wider context of processing conditions. These approaches will be extensively covered in the following chapters of this book.

The use of empirical, but well-defined and validated models can serve as the practical basis for developing a robust stochastic approach, where important environmental or product parameters are taken into account in a statistical way. Increasing attention is focused on the role and logistics of transport, storage, and handling; the benefits of taking a supply chain perspective are being appreciated and pursued.[82–85] In this context, the Monte Carlo simulation technique is frequently used in order to incorporate the realistic variabilities, or possible sources of error of important parameters into food quality and allow for process or logistics optimization.

The same iteration technique of Monte Carlo, based on the generation of alternative distribution scenarios, was applied in assessing the performance of TTIs in the real distribution chain of frozen,[55,56] or chilled foods,[77,86] in a product management system coded least shelf life first out

(LSFO). Its structure is based on validated shelf life modeling of the controlled food product, specification of the initial and the final, marginally accepted value of the selected quality parameter A_0 and A_s respectively, as well as careful temperature monitoring in the distribution chain with the appropriate TTI. LSFO aims at reducing the rejected products at the consumer end, by promoting, at selected decision making points of the product life cycle, those product units with the shorter shelf life, according to the response of the attached TTI.[55,87] A further improvement of the LSFO approach, is a chill chain management system coded shelf-life decision system (SLDS).[88,89] Compared to LSFO, SLDS policy accounts for the realistic variability of the initial quality state A_0 of the product.

The state of the TTI technology and of the scientific approach with regards to the quantitative safety risk assessment in foods will allow the undertaking of the next important step: the study and development of a TTI-based management system that will assure both safety and quality in the food chill chain. The development and application of such a system coded with the acronym SMAS is the target of a multipartner research project funded by the European Commission titled "Development and Modeling of a TTI based Safety Monitoring and Assurance System (SMAS) for chilled Meat Products" (project QLK1-CT2002-02545, 2003–2006; http://smas.chemeng.ntua.gr).

NOMENCLATURE

T	Temperature (K)
a_w	Water activity
k	Reaction rate constant
E_A	Activation energy (J/mol)
R	Universal gas constant ($=8.3144$ J/(mol K))
F	Fraction
L,a,b	Color parameters of the *Lab* scale
DC	Chroma change
m_0	Moisture monolayer value
k/x	Moisture permeation of the packaging material (g/day m^2 mmHg)
A	Exchange surface of the package (m^2)
W_s	Dry weight of the packaged food
p_0	Partial water vapor pressure of the storage environment (mmHg)

Subscripts

0	Initial value
eq	Value at equilibrium
eff	Effective
pred	Predicted
con	consumed
exp	Experimental

REFERENCES

1. Karel, M. and Lund, D. B., Reaction kinetics, In *Physical Principles of Food Preservation*, Karel, M. and Lund, D. B., Eds., New York: Marcel Dekker, pp. 28–49, 2003.
2. Singh, R. P. and Anderson, B. A., The major types of food spoilage: an overview, In *Understanding and Measuring the Shelf Life of Food*, Steele, R., Ed., Cambridge, MA: Woodhead Publishing, pp. 2–23, 2000.

51. Loudon, G. M., Mechanistic interpretation of pH-rate profiles, *Journal of Chemical Education*, 68, 973–984, 1991.

52. Martins, S. I. F. S. and Van Boekel, M. A. J. S., Kinetics of the glucose–glycine Maillard reaction pathways: influences of pH and reactant initial concentrations, *Food Chemistry*, 92, 437–448, 2005.

53. Hoover, D. G., Pressure effects on biological systems, *Food Technology*, 47(6), 150–155, 1993.

54. Polydera, A. C., Stoforos, N. G., and Taoukis, P. S., Comparative shelf life study and vitamin C loss kinetics in pasteurized and high pressure reconstituted orange juice, *Journal of Food Engineering*, 60, 21–29, 2003.

55. Giannakourou, M. C. and Taoukis, P. S., Systematic application of time temperature integrators as tools for control of frozen vegetable quality, *Journal of Food Science*, 67(6), 2221–2228, 2002.

56. Giannakourou, M. C. and Taoukis, P. S., Application of a TTI-based distribution management system for quality optimisation of frozen vegetables at the consumer end, *Journal of Food Science*, 68(1), 201–209, 2003.

57. Deep freezing temperatures in shops-temperature control in freezers and foodstuffs, Final report of European Commission sponsored Research project in four EU member states (Greece, Italy, Spain and Portugal), 24, Contract number: EC 1080/94/000069, 1995.

58. Weemaes, C., Rubens, P., De Cordt, S., Ludikhuyzem, L., Van Den Broeck, I., Hendrickx, M., Heremans, K., and Tobback, P., Temperature sensitivity and pressure resistance of mushroom poly-phenoloxidase, *Journal of Food Science*, 62(2), 261–266, 1997.

59. Kerr, W. L., Lim, M. H., Reid, D. S., and Chen, H., Chemical reaction kinetics in relation to glass transition temperatures in frozen food polymer solutions, *Journal of the Science and Food of Agri-culture*, 61, 51–56, 1993.

60. Roos, Y. H., Karel, M., and Kokini, J. L., Glass transition in low moisture and frozen foods: effects on shelf life and quality, *Food Technology*, 50(11), 95–108, 1996.

61. Dolan, K. D., Estimation of Kinetic Parameters for nonisothermal food processes, *Journal of Food Science*, 68(3), 728–741, 2003.

62. Swartzel, K. R., Equivalent-point method for thermal evaluation of continuous-flow systems, *Journal of Agricultural and Food Chemistry*, 34(3), 396–403, 1986.

63. Fu, B. and Labuza, T. P., Shelf-life testing: procedures and prediction methods, In *Quality in Frozen Food*, Erickson, M. C. and Hung, Y. H., Eds., London, UK: Chapman & Hall, pp. 29–51, 1997.

64. Van Arsdel, W. B., Coply, M. J., and Olson, R. L., *Quality and Stability of Frozen Foods*, New York: Wiley-Interscience, 1969.

65. Giannakourou, M. C. and Taoukis, P. S., Kinetic modelling of vitamin C loss in frozen green vegetables under variable storage conditions, *Food Chemistry*, 83(1), 33–41, 2003.

66. Jul, M., *The Quality of Frozen Foods*, London: Academic Press, 1983.

67. Labuza, T. P. and Saltmarch, M., The nonenzymatic browning reaction as affected by water in foods, In *Water Activity: Influences on Food Quality*, Rockland, L. B. and Stewart, G. F., Eds., New York: Academic Press, pp. 605–650, 1981.

68. Friedman, M., Food browning and its prevention: an overview, *Journal of Agricultural and Food Chemistry*, 44, 631–653, 1996.

69. Taoukis, P. S. and Skiadi, O., Shelf life prediction modeling of packaged dehydrated foods for dynamic temperature and relative humidity storage conditions, In *Proceedings of the COST 915-Copernicus CIPA-CT94-0120 Workshop on Food Quality Modelling, 4–6 June 1997, Leuven, Belgium*, Nicolai, B. M. and De Baerdemaeker, J., Eds., Luxembourg: Office for Offical Publications of the European Communities, pp. 31–38, 1998.

70. Buera, M. P. and Karel, M., Effect of physical changes on the rates of nonenzymatic browning and related reaction, *Food Chemistry*, 52, 167–173, 1995.

71. Karmas, R. and Karel, M., The effect of glass transition on Maillard browning in food models, In *Maillard Reactions in Chemistry, Food and Health*, Labuza, T. P., Reineccius, G. A., Monnier, V. M., O'Brien, J., and Baynes, J. W., Eds., Cambridge, UK: Royal Society of Chemistry, pp. 182–187, 1994.

72. Lievonen, S. M., Laaksonen, T. J., and Roos, Y. H., Glass transition and reaction rates: nonenzymatic browning in glass and liquid systems, *Journal of Agricultural and Food Chemistry*, 46, 2778–2784, 1998.

73. Lievonen, S. M. and Roos, Y. H., Nonenzymatic browning in amorphous food models: effects of glass transition and water, *Journal of Food Science*, 67(6), 2100–2106, 2002.

74. Saltmarch, M., Vagnini-Ferrari, M., and Labuza, T. P., Theoretical basis and application of kinetics to browning in spray-dried whey food systems, *Progress in Food and Nutrition Science*, 5, 331–344, 1981.

75. Eichner, K., Antioxidative effect of Maillard reaction intermediates, *Progress in Food and Nutrition Science*, 5, 441–451, 1981.

76. Miao, S. and Roos, Y. H., Comparison of nonenzymatic browning kinetics in spray-dried and freeze-dried carbohydrate-based food model systems, *Journal of Food Science*, 69, 322–331, 2004.

77. Taoukis, P. S., Modelling the use of time–temperature indicators, In *Food Process Modeling*, Tijskens, L. M. M., Hertog, M. L. A. T. M., and Nicolaï, B. M., Eds., New York: CRC Press, pp. 402–431, 2001.

78. Taoukis, P. S. and Labuza, T. P., Time–Temperature Indicators (TTIs), In *Novel Food Packaging Techniques*, Ahvenainen, R., Ed., Cambridge, UK: Woodhead Publishing, pp. 103–126, 2003.

79. Taoukis, P. S. and Labuza, T. P., Applicability of time–temperature indicators as shelf life monitors of food products, *Journal of Food Science*, 54(4), 783–789, 1989.

80. Taoukis, P. S. and Labuza, T. P., Reliability of time–temperature indicators as food quality monitors under nonisothermal conditions, *Journal of Food Science*, 54(4), 789–792, 1989.

81. Tijskens, L. M. M., Hertog, M. L. A. T. M., and Van Dijk, C., Generic modelling and practical applications, In *Proceedings of the COST 915-Copernicus CIPA-CT94-0120 Workshop on Food Quality Modelling, 4–6 June 1997, Leuven, Belgium*, Nicolai, B. M. and De Baerdemaeker, J., Eds., Luxembourg: Office for Offical Publications European Communities, pp. 145–151, 1998.

82. Browne, M. and Allen, J., Logistics of food transport, In *Food Transportation*, Heap, R., Kierstan, M., and Ford, G., Eds., London: Blackie Academic and Professional, pp. 22–50, 1998.

83. Ross, D. F., *Distribution Planning and Control*, 1st ed., New York: Chapman and Hall, 1996.

84. Broekmeulen, R. A. C. M., Modelling the management of distribution centers, In *Food Process Modelling*, Tijkskens, L. M. M., Hertog, M. L. A. T. M., and Nicolaï, B. M., Eds., 3rd ed., Washington, DC: CRC Press, pp. 432–448, 2001.

85. Tijkens, L. M. M., Koster, A. C., and Jonker, J. M. E., Concepts of chain management and chain optimisation, In *Food Process Modelling*, 3rd ed., Tijkens, L. M. M., Hertog, M. L. A. T. M., and Nicolaï, B. M., Eds., Cambridge, England: CRC Press, Woodhead Publishing Limited, pp. 448–469, 2001.

86. Giannakourou, M. C., Koutsoumanis, K., Nychas, G. J. E., and Taoukis, P. S., Development and assessment of an intelligent shelf life decision system for quality optimisation of the food chill chain, *Journal of Food Protection*, 64(7), 1051–1057, 2001.

87. Taoukis, P. S., Bili, M., and Giannakourou, M., Application of shelf life modeling of chilled salad products to a TTI based distribution and stock rotation system, In *Proceedings of the International Symposium on Applications of Modeling as an Innovative Technology in the Agri-Food Chain*, Tijkens, L. M. M. and Hertog, M. L. A. T. M., Eds., *Acta Horticulturae*, 476:131–140, 1998.

88. Koutsoumanis, K., Giannakourou, M. C., Taoukis, P. S., and Nychas, G. J. E., Application of shelf life decision system (SLDS) to marine cultured fish quality, *International Journal of Food Microbiology*, 73, 375–382, 2002.

89. Giannakourou, M. C., Koutsoumanis, K., Dermesonlouoglou, E., and Taoukis, P. S., Applicability of the intelligent shelf life decision system for control of nutritional quality of frozen vegetables, *Acta Horticulturae*, 566, 275–280, 2001.

CHAPTER 8

Probabilistic Modeling

Bart M. Nicolaï, Nico Scheerlinck, and Maarten L. A. T. M. Hertog

CONTENTS

8.1   INTRODUCTION

Most food engineering models to date are deterministic in the sense that it is assumed the product and process parameters are known in advance. In reality, many food processes are subjected to a considerable amount of uncertainty. This can be largely attributed to the variability inherent to biological materials. The variability introduced by poorly designed or controlled equipment may be important as well. Experimental measurement errors in the estimation of the parameters as well as model structure errors further increases the global process uncertainty. Consequently, the process variables, such as the core temperature, are also subject to uncertainty.

In food process design these uncertainties are often taken into account by the use of safety factors. These are often implicitly introduced by basing the design calculations on a "worst case" (fail-safe) process, in which the values leading to the most conservative process are assigned to the model parameters. However, when the number of parameters is large, the associated safety factor may be beyond reasonable magnitude, causing severe over-processing. Consequently, it would be more appropriate to base the process design on statistical considerations, taking into account all uncertainties associated with the process. Such an analysis obviously requires a complete stochastic specification of the process variables of interest, including probability density functions, mean values, and variances, as a function of the stochastic properties of the random parameters and their introduction into the governing algebraic, ordinary, or partial differential equations. Appropriate numerical techniques need to be used to solve the resulting stochastic equations.

The outline of this chapter is as follows. First, some models will be summarized that describe uncertainty, including random variables, fields, processes, and waves. Then, the Monte Carlo method will be briefly discussed, although the details are left to Chapter 16. Very generally, the Monte Carlo method may lead to excessive computing times, especially when the solution of the governing model equations is computationally intensive by itself. Therefore, alternative techniques will be evaluated that are based on a probabilistic approach rather than on numerical experiments. First, some general methods to compute the probability density function of functions of unknown stochastic variables based on transformation theory will be introduced. These methods may be used when the governing equations can be solved analytically to yield a closed solution. As an alternative, the stochastic perturbation method will be described. This method allows the calculation of the mean value and the (co)variance of the solution of the stochastic differential equation. Such information is often sufficient for engineering purposes. Then, the Fokker–Planck equation, which describes how the probability function of the solution of a stochastic differential equation evolves as a function of time, will be introduced. Although the Fokker–Planck equation is generally applicable to a broad class of stochastic differential equations, it is difficult to solve numerically except in some simple cases. It will therefore be shown how simplified expressions can be used to calculate the mean value and covariance of the solution.

All algorithms will be illustrated by some simple examples from thermal food process engineering, as they represent an important unit operation in food processing.

8.2   DESCRIPTION OF UNCERTAINTY

In this section, some models that describe randomness will be introduced for future reference. The reader who is not familiar with these concepts is referred to any textbook on probability theory or stochastic processes, such as those by Gray and Davisson[1] and Melsa and Sage.[2]

8.2.1   Random Variables

The most simple uncertainty model is that of a random variable. A random variable X is a real-numbered variable whose value is associated with a random experiment. For example, the heat capacity of a potato is a random variable that can vary between different potatoes. A random variable X can be characterized by its probability density function (pdf), $F_X(x)$, which is defined as

$$P(a \leq X \leq b) \triangleq \int_a^b f_X(x)dx, \qquad (8.1)$$

where $P(a \leq X \leq b)$ represents the probability that X takes on a value between a and b. The most common pdf is the Gaussian (normal) pdf, given by

$$f_X(x) = (2\pi\sigma^2)^{-1/2}\exp[-(x-\bar{X})^2/2\sigma^2]. \qquad (8.2)$$

Examples of other pdf's are given in Chapter 16.

The mean value, $\bar{X}$, and the variance, σ^2, of X are defined as

$$\bar{X} \triangleq E(X) \triangleq \int_{-\infty}^{\infty} xf_X(x)dx, \qquad (8.3)$$

$$\sigma^2 \triangleq E(X-\bar{X})^2 \triangleq \int_{-\infty}^{\infty} (x-\bar{x})^2 f_X(x)dx, \qquad (8.4)$$

with E the expectation operator.

Sometimes an experiment will yield values for two or more physical parameters. Assume, for example, that both the thermal conductivity as well as the volumetric heat capacity of a food material are measured simultaneously. In this case, the outcome of the experiment is a bivariate (two random variables) or multivariate (more than two random variables). The random variables can then be stacked conveniently in a random vector $\mathbf{X}$. Similar to the univariate case, the probability density function $f_\mathbf{X}(\mathbf{x})$, the mean value $\bar{\mathbf{X}}$, and covariance matrix $\mathbf{V}$ of the random vector can be defined as

$$\bar{\mathbf{X}} \triangleq E(\mathbf{X}) = \int_{-\infty}^{\infty}\int_{-\infty}^{\infty}\cdots\int_{-\infty}^{\infty} \mathbf{x}f_X(\mathbf{x})d\mathbf{x}, \qquad (8.5)$$

$$\mathbf{V} \triangleq E[(\mathbf{X}-\bar{\mathbf{X}})(\mathbf{X}-\bar{\mathbf{X}})^T]. \qquad (8.6)$$

Note that in the remainder of this chapter, vector and matrix quantities will be denoted by bold symbols. The ith diagonal entry of $\mathbf{V}$ is the variance $\sigma_{X_i}^2$ of random variable X_i; the (i,j)th entry of $\mathbf{V}$ is the covariance σ_{X_i,X_j} of random variables X_i and X_j. The correlation coefficient R_{X_i,X_j} is defined as

$$R_{X_i,X_j} = \frac{\sigma_{X_i,X_j}}{\sigma_{X_i}\sigma_{X_j}}, \qquad (8.7)$$

and $-1 \leq R_{X_i,X_j} \leq 1$.

As an example of a multivariate probability density function of dimension n, the Gaussian (normal) multivariate pdf is given by

$$f_\mathbf{X}(\mathbf{x}) = (2\pi)^{-n/2}\{\det(\mathbf{V})\}^{-1/2}\exp\left\{-\frac{1}{2}(\mathbf{x}-\bar{\mathbf{X}})\mathbf{V}^{-1}(\mathbf{x}-\bar{\mathbf{X}})^T\right\}. \qquad (8.8)$$

8.2.2 Random Processes

If a parameter changes in an unpredictable way as a function of the time coordinate, it can be described conveniently by means of a random process. The mean $\bar{X}$ and covariance V of a stationary process X with probability density function $f_X(x,t)$ are defined by

$$\bar{X} = E(X) \cong \int_{-\infty}^{\infty} x f_X(x,t) dx, \tag{8.9}$$

$$V(\tau) = E\{[X(t) - \bar{X}][X(t + \tau) - \bar{X})]\}. \tag{8.10}$$

As this process is stationary, by definition the mean and covariance do not depend on time. A Gaussian stationary white noise process W with covariance

$$V_W(\tau) = \sigma_W^2 \delta(\tau), \tag{8.11}$$

where δ is the Dirac delta, can be used to describe rapid, unpredictable fluctuations. Sample values of W are uncorrelated, irrespective of how close together in time they are. However, white noise does not exist in reality, as it has an infinite energy content. Autoregressive processes provide a tool to incorporate fluctuations which change more smoothly as a function of time. An autoregressive random process of order m is defined by the following stochastic differential equation

$$\frac{d^m}{dt^m}X(t) + a_1\frac{d^{m-1}}{dt^{m-1}}X(t) + \cdots + a_mX(t) = W(t), \tag{8.12}$$

where $a_1, a_2, \ldots, a_m$ are constants, $m \geq 1$, and $W(t)$ is a zero-mean, stationary, Gaussian white-noise process. The time scale of the fluctuations is dependent on the coefficients $a_1, \ldots, a_m$, and their high frequency content decreases with increasing order m. The (Gaussian) random variable initial condition corresponding to the stochastic differential equation, Equation 8.12, is defined as

$$E[X(t_0)] = 0, \tag{8.13}$$

$$E[X(t_0) - \bar{X}]^2 = \sigma^2. \tag{8.14}$$

Note that a *random variable* parameter X can be modeled as a trivial case of an AR(1) process:

$$\frac{d}{dt}X = 0. \tag{8.15}$$

AR(m) processes are a special case of the class of *physically realizable* stochastic processes, which comprise most of the random processes seen in practice.[1] The smoothness of AR processes depends on the order and coefficients of the governing differential equation.

In food engineering applications, the fluctuations of process conditions are often partially deterministic and partially stochastic; stochastic disturbances of, for example, the temperature in the production facility, may lead to fluctuations in the oven temperature that are partially compensated by the control system of the oven. A stochastic model of the oven temperature would ideally include both a model of the form of Equation 8.12 to describe purely random fluctuations and a model to describe the dynamics of the oven and its control system with respect to such random fluctuations. Although the development of such models is theoretically straightforward, this has not yet been achieved in practice.

8.2.3 Random Fields and Random Waves

Often, a physical quantity varies randomly as a function of the time and/or space coordinates. Examples include the temperature in an oven, the thermophysical properties of heterogeneous materials, such as foods, hydraulic properties of soils, elastic properties of construction materials, etc. The random field concept provides a convenient mathematical framework to describe such phenomena.[3]

A parameter that fluctuates in both space and time can be described by means of random waves. The random wave model is a straightforward extension of the random field model combined with the random process model. A full account of random fields and random waves is beyond the scope of this chapter, and the reader is referred to the literature.[3]

8.3 THE MONTE CARLO METHOD

A common method to solve models involving uncertainty is the Monte Carlo method. In this method, samples of random parameters are generated by means of a random number generator. For every parameter set, the model is solved by analytical or numerical means, and the solution is stored for future use. This process is repeated a large number of times, and in the end the statistical characteristics can be estimated. The Monte Carlo method is outlined in detail in Chapter 16 and will not be addressed further here. Although the Monte Carlo method is highly versatile, it relies on a large number of simulation runs, even when advanced sampling strategies are applied. Depending on the complexity of the problem that needs to be computed in every run, it may lead to excessive computing time.

8.4 TRANSFORMATION THEORY

8.4.1 Principle

For simple food engineering problems, an explicit relation between a dependent variable and a random variable is often available. If this is the case, transformation theory can be used.

Assume that the (in general, multivariate) food process variable of interest $Y=[Y_1 Y_2....Y_m]$ is related to a random vector $X=[X_1 X_2...X_n]$ according to $Y=g(X)$, where $g(X)$ a vector-valued function. Furthermore, assume that the probability density function $f_X(x)$ is known. A function of a random variable is also a random variable; therefore, Y is a random variable. It can be shown that if $n=m$, the probability density function $f_Y(y)$ can be found from[2]

$$f_Y(y) = f_X[g^{-1}(y)]\det\left[\frac{\partial}{\partial y}g^{-1}(y)\right],$$ (8.16)

where $\det(.)$ denotes the determinant, and $\det\left[\frac{\partial}{\partial y}g^{-1}(y)\right]$ is the Jacobian of the inverse transformation.

Often, m is less than n. In this case, some auxiliary dependent variables can be introduced. For example, if $m=1$, then the dependent variable of interest is a scalar Y, and Y may be defined as

$$Y = [Y,X_2,K,X_n]$$ (8.17)

and Equation 8.16 may be solved. The marginal probability distribution of Y can then be calculated by integrating $f_Y(y)$ over the range of $x_2,....,x_n$:

$$f_Y(y) = \int_{-\infty}^{\infty}\int_{-\infty}^{\infty}\cdots\int_{-\infty}^{\infty} f_Y(y)dx_2\cdots dx_n.$$ (8.18)

If only the mean $\bar{Y}$ and variance σ_Y^2 of Y are required, the following equations can be used:

$$\bar{Y} = \int_{-\infty}^{\infty} \int_{-\infty}^{\infty} \cdots \int_{-\infty}^{\infty} g(\mathbf{y}) f_{\mathbf{X}}(\mathbf{x}) dx_1 dx_2 \cdots dx_n; \tag{8.19}$$

$$\sigma_Y^2 = \int_{-\infty}^{\infty} \int_{-\infty}^{\infty} \cdots \int_{-\infty}^{\infty} [g(\mathbf{y}) - \bar{Y}]^2 f_{\mathbf{X}}(\mathbf{x}) dx_1 dx_2 \cdots dx_n. \tag{8.20}$$

In general, Equation 8.18 through Equation 8.20 must be evaluated numerically. For this purpose, the integrals are approximated by a finite sum, e.g.,

$$\int_{-\infty}^{\infty} g(x) f_X(x) dx \cong \sum_{i=1}^{q} H_i g(x_i) f_X(x_i) \tag{8.21}$$

for the univariate case, where H_i and q are the weights and the order of the quadrature formula, respectively. The q points x_i, where the integrand is to be evaluated, are called the *integration points*. Most modern integration routines utilize a Gauss variant of Equation 8.21 in which the integration points are chosen to achieve maximal accuracy for a certain class of integrands.[4] Integration orders between two and ten are commonly used, depending on the smoothness of the integrand. If the integration boundaries are infinite, as is the case if $\mathbf{X}$ is normally distributed, and if g is a relatively smooth function they can usually be approximated well by finite boundaries because the *tail* probabilities, e.g., $P(X > x)$ for the upper tail, decrease rapidly with increasing distance of x from $\bar{X}$. A convenient choice of integration interval is $[\bar{X} - 5\sigma_X, \bar{X} + 5\sigma_X]$.

For simple problems, an analytical expression for $g(X)$ may be known. If not, $g(X)$ may be evaluated numerically in the integration points.[5] For a heat transfer problem with stochastic parameters, this means that the temperature must be solved in the integration points using an appropriate numerical technique such as the finite element method.

The approximation in Equation 8.21 can be extended to multiple integrals corresponding to a problem with more than one random variable parameter. However, the number of integration points increases geometrically with the number of random variables. Obviously, if $\mathbf{g}(\mathbf{X})$ must be solved numerically, the necessary computation time may become unacceptable. The applicability of this method is, therefore, limited. However, the computed mean values and variances can be of arbitrary accuracy when appropriate integration routines and integration boundaries are chosen, so they can serve as a reference solution to which solutions obtained by other methods can be compared.

Some applications of transformation methods to food/bioprocesses are summarized in Table 8.1. Hertog[6] applied transformation methods to study shriveling of "Braeburn" apples, the color change of "Hass" avocados and stress crack development in corn grains. Schouten et al.[7] followed a similar approach to describe the variability of pigment precursor concentrations in cucumber fruit as a function of the variability in light conditions during the pre-harvest period. These authors have shown that the pdf of the variable of interest may become highly skewed as a function of time.

8.4.2 Application to Lumped Capacitance Heat Transfer

The transformation method will be illustrated based on the following simple *lumped capacitance* heat transfer problem.[8] Consider a sphere of radius r_0 with heat capacity c and density ρ. The sphere is initially at a uniform temperature, T_0. At time $t = 0$, the sphere is immersed in a water bath at temperature T_∞. The temperature of the sphere will approach T_∞ with a rate that depends on the surface heat-transfer coefficient h at the solid–liquid interface. In the lumped capacitance method, it

Table 8.1 Applications of Stochastic Methods in Food/Bioprocessing

Process	Deterministic Model	Random Parameters	Algorithm	References
Batch food sterilisation	Heat conduction	Initial and boundary conditions, thermophysical properties	Transformation, perturbation, variance propagation, Monte Carlo	5,19,20,27,30
Roasting	Heat conduction	Initial and boundary conditions, thermophysical properties	Perturbation	22
Postharvest quality changes	Stress cracking, colour change, shriveling	Material properties, initial condition, size	Transformation	6,7
Cooling of fruit	Heat conduction	Initial and boundary conditions	Variance propagation, Monte Carlo	29
Drying	Coupled heat and moisture transfer	Initial and boundary conditions, thermophysical properties	Perturbation, variance propagation, Monte Carlo	21,31

Note: For more applications of the Monte Carlo method the reader is referred to Chapter 16.

is assumed that, because of the high thermal conductivity of the solid medium, the temperature inside the solid is uniform at any time instant during the transient heat-transfer process. This hypothesis holds if the Biot number, Bi, satisfies the following constraint:

$$Bi = \frac{hL}{k} < 0.1, \tag{8.22}$$

where L is the characteristic length of the solid which is usually defined as $L=3r_0$.[8] Few food engineering problems in practice are actually lumped, but this example serves as a simple starting point to introduce the different algorithms before the more general Fourier heat conduction equation is addressed.

It is easy to show that applying an overall energy balance leads to the following differential equation:

$$\rho c \frac{d}{dt} T = \frac{3h}{r_0} (T_\infty - T). \tag{8.23}$$

After integration, the following formula for the temperature course is found:

$$T(t) = T_\infty + (T_0 - T_\infty) \exp\left(-\frac{3h}{\rho c r_0} t\right) = g(t, T_\infty). \tag{8.24}$$

Suppose now that T_∞ is not a deterministic parameter but rather a Gaussian random variable with mean $\bar{T}_\infty$ and variance $\sigma_{T_\infty}^2$. Obviously, the solution $T(t)$ of Equation 8.23 at an arbitrary time-instance is then also random. The derivation of the probability density function of $T(t)$ is relatively simple, as the function $T(t)=g(t,T_\infty)$ is a scalar function of a scalar random variable, T_∞. The inverse is given by:

$$g^{-1}(T) = T_\infty = \frac{T - T_0 \exp\left(-\frac{3h}{\rho c r_0} t\right)}{1 - \exp\left(-\frac{3h}{\rho c r_0} t\right)}, \tag{8.25}$$

such that

$$\frac{\partial}{\partial T} g^{-1}(T) = \left[1 - \exp\left(-\frac{3h}{\rho c r_0} t\right)\right]^{-1}.$$

(8.26)

Substituting Equation 8.25 and Equation 8.26 into Equation 8.16 yields

$$f(T) = \left\{2\pi\sigma_{T_\infty}^2 \left[1 - \exp\left(-\frac{3h}{\rho c r_0} t\right)\right]^2\right\}^{-1/2} \exp\left\{-\frac{\left\{T - \left[(T_0 - \bar{T}_\infty)\exp\left(-\frac{3h}{\rho c r_0} t\right) + \bar{T}_\infty\right]\right\}^2}{2\left[1 - \exp\left(-\frac{3h}{\rho c r_0} t\right)\right]^2 \sigma_{T_\infty}^2}\right\},$$

where $f(T)$ is the pdf of T. It is easy to see that $f(T)$ is Gaussian with mean

$$\bar{T}(t) = \bar{T}_\infty + (T_0 - \bar{T}_\infty)\exp\left(-\frac{3h}{\rho c r_0} t\right)$$

(8.27)

and variance

$$\sigma_T^2(t) = \sigma_{T_\infty}^2 \left[1 - \exp\left(-\frac{3h}{\rho c r_0} t\right)\right]^2.$$

(8.28)

Note that any linear function of a Gaussian random variable is also Gaussian.[2] By virtue of Equation 8.24, the temperature, T, at an arbitrary time, t, is a linear function of the Gaussian random variable, T_∞; therefore, it must follow that T is a Gaussian random variable as well. One could, therefore, just as easily have used Equation 8.19 and Equation 8.20 directly to calculate $\bar{T}$ and σ_T^2 and, hence, fully characterize $f(T)$. However, if another parameter of Equation 8.24 is random, e.g., h, then the pdf of T will, in general, not be Gaussian.

In Figure 8.1 it is shown how the probability density function of the sphere temperature evolves as a function of time. At $t=0$, the temperature was considered to be deterministic and equal to 20°C. As time progresses, the probability density function tends to a normal distribution with a standard deviation of 5°C. In Figure 8.2, the time course of $\bar{T}$ is compared with the values obtained by means of the Monte Carlo method with 100, 1000, and 5000 runs. The parameter values were as follows: $\rho = 1000$ kg/m^3, $c = 4180$ J/kg °C, $r_0 = 0.01$ m, $T_0 = 20$°C, $h = 10$ W/m^2 °C, $\bar{T}_\infty = 80$°C, and $\sigma_{T_\infty} = 5$°C. It is clear that the results of the Monte Carlo method are almost indistinguishable from those obtained with the analytical solution. In Figure 8.3 the values of σ_T^2, as calculated by the different methods, are shown. The correspondence clearly improves with increasing number of Monte Carlo runs.

8.5 STOCHASTIC PERTURBATION ANALYSIS

8.5.1 Principle

In the perturbation method, the effect of a small (stochastic) parameter deviation on the solution of an ordinary or partial differential equation is investigated. Some early analytical methods were proposed by Chen and Tien,[9] and by Beran and McCoy.[10] Tzou[11] applied similar algorithms to transient heat conduction problems with random thermal conductivity and established analytical expressions for the mean temperature and the variance of the temperature.

Assume that the random variable $Y(t)$ obeys the following differential equation:

$$\frac{d}{dt} Y(X,t) = g(Y(X,t),X) \quad \text{and} \quad Y(t = t_0) = Y_0,$$

(8.29)

where the mean value and covariance matrix of Y_0 is known and $g(Y(X,t),X)$ is a nonlinear function of $Y(X,t)$ and X; X is a vector of random variables with mean $\bar{X}$ and covariance $V_{X,X}$. $Y(X,t)$ and

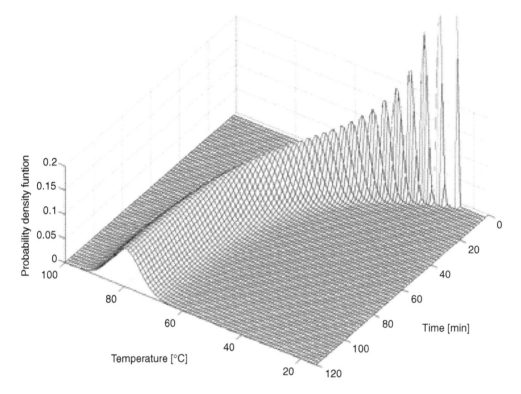

Figure 8.1 Probability density function of the temperature of a sphere as a function of the ambient temperature and time. At $t=0$, the temperature of the sphere was considered to be deterministic and equal to 20°C. As time progresses, the probability density function tends to a normal distribution with a standard deviation of 5°C.

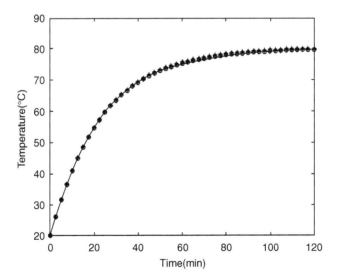

Figure 8.2 Mean temperature in sphere subjected to random variable ambient temperature. —, exact; $+$, Monte Carlo (100 runs); $*$, Monte Carlo (1000 runs); $\circ$, Monte Carlo (5000 runs).

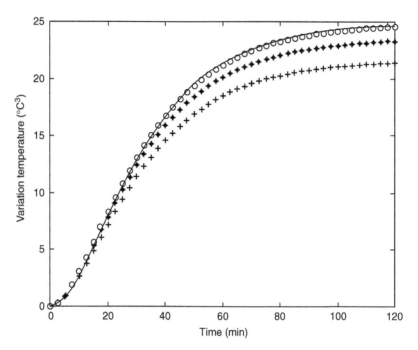

Figure 8.3. Temperature variance in sphere subjected to random variable ambient temperature. —, exact; +, Monte Carlo (100 runs); *, Monte Carlo (1000 runs); ∘, Monte Carlo (5000 runs).

$g(Y,X)$ may be expanded in a Taylor expansion to yield

$$Y(X,t) \cong \bar{Y}(t) + \frac{\partial Y(t)}{\partial X} \Delta X, \tag{8.30}$$

$$g(Y(X,t),X) \cong \bar{g} + \frac{\partial g}{\partial Y} \frac{\partial Y}{\partial X} \Delta X + \frac{\partial g}{\partial X} \Delta X, \tag{8.31}$$

where the partial derivatives are to be evaluated in the mean parameters $\bar{X}$. Substitution of Equation 8.30 and Equation 8.31 into Equation 8.29 yields:

$$\frac{d}{dt} \bar{Y}(t) = \bar{g}; \tag{8.32}$$

$$\frac{d}{dt} \left(\frac{\partial Y(t)}{\partial X} \right) = \frac{\partial g}{\partial X} \frac{\partial Y(t)}{\partial X} + \frac{\partial Y(t)}{\partial X}. \tag{8.33}$$

The covariance matrix $V_{Y,Y}(t)$ can then be calculated from:

$$V_{Y,Y}(t) = E\{[Y(t) - \bar{Y}(t)][Y(t) - \bar{Y}(t)]^T\} = E\left\{ \frac{\partial Y(t)}{\partial X} \Delta X \left[\frac{\partial Y(t)}{\partial X} \Delta X \right]^T \right\}$$

$$= \frac{\partial Y(t)}{\partial X} E[\Delta X (\Delta X)^T] \left(\frac{\partial Y(t)}{\partial X} \right)^T = \frac{\partial Y(t)}{\partial X} V_{X,X} \left(\frac{\partial Y(t)}{\partial X} \right)^T \tag{8.34}$$

The initial conditions corresponding to Equation 8.32 and Equation 8.33 are given by

$$\bar{Y}(t) = \bar{Y}_0 \tag{8.35}$$

$$\frac{\partial Y(t)}{\partial X} = \frac{\partial Y}{\partial X} \bigg|_{t=t_0} \tag{8.36}$$

at $t = t_0$. Equation 8.32 indicates that a first order approximation of the mean value of $\bar{Y}(t)$ can be found be solving the original equation, Equation 8.29, using the mean parameter vector $\bar{X}$. Equation

8.33 describes how the sensitivity of the solution with respect to the random parameter $\mathbf{X}$ changes as a function of time.

8.5.2 Application to Lumped Capacitance Heat Transfer

To illustrate the algorithm, the lumped capacitance heat-transfer problem previously introduced will again be considered. Equation 8.23 may be rewritten as

$$\frac{d}{dt}T = \frac{3h}{\rho c r_0}(T_\infty - T) = g(T, T_\infty). \tag{8.37}$$

Expanding T and T_∞ in a Taylor series yields:

$$T(T_\infty) = \bar{T} + \frac{\partial T}{\partial T_\infty}\bigg|_{\bar{T}_\infty} \Delta T_\infty; \tag{8.38}$$

$$g(T, T_\infty) = \bar{g} + \frac{\partial g}{\partial T_\infty}\bigg|_{\bar{T}_\infty} \Delta T_\infty + \frac{\partial g}{\partial T}\bigg|_{\bar{T}} \frac{\partial T}{\partial T_\infty}\bigg|_{\bar{T}_\infty} \Delta T_\infty$$

$$= \frac{3h}{\rho c r_0}(\bar{T}_\infty - \bar{T}) + \frac{3h}{\rho c r_0}\Delta T_\infty - \frac{3h}{\rho c r_0}\frac{\partial T}{\partial T_\infty}\Delta T_\infty. \tag{8.39}$$

Note that the equality holds because g is a linear function of T_∞. Substituting Equation 8.38 and Equation 8.39 into Equation 8.37 and rearranging terms yields:

$$\frac{d}{dt}\bar{T} = \frac{3h}{\rho c r_0}(\bar{T}_\infty - \bar{T}); \tag{8.40}$$

$$\frac{d}{dt}\left(\frac{\partial T}{\partial T_\infty}\right) = \frac{3h}{\rho c r_0} - \frac{3h}{\rho c r_0}\left(\frac{\partial T}{\partial T_\infty}\right). \tag{8.41}$$

Observe that both Equation 8.40 and Equation 8.41 have the same structure as the original equation, Equation 8.37. The variance σ_T^2 can then be found from

$$\sigma_T^2 = E(T - \bar{T})^2 = E\left(\frac{\partial T}{\partial T_\infty}\Delta T_\infty\right)^2 = \left(\frac{\partial T}{\partial T_\infty}\right)^2 E(\Delta T_\infty^2) = \left(\frac{\partial T}{\partial T_\infty}\right)^2 \sigma_{T_\infty}^2. \tag{8.42}$$

Direct integration of Equation 8.40 and Equation 8.41 and subsequent substitution of $\partial T / \partial T_\infty$ into Equation 8.42 again yields Equation 8.27 and Equation 8.28.

8.5.3 Application to Heat Conduction

The perturbation algorithm will now be applied to a stochastic heat conduction problem for which the lumped capacitance method is no longer valid. This is the case if the resistance to conduction within the solid is large in comparison to the resistance to convection across the fluid boundary layer, or, in other words, if $Bi > 0.1$. Key to the method is the spatial discretization of the governing partial differential equation, which leads to a system of ordinary differential equations of the form of Equation 8.29.

8.5.3.1 Finite Element Solution of Heat Conduction Problems

Transient linear heat transfer in solid foods subjected to convection boundary conditions is governed by the Fourier equation[8]

$$k\nabla^2 T = \rho c \frac{\partial T}{\partial t}, \tag{8.43}$$

$$k \frac{\partial}{\partial n_\perp} T = h(T_\infty - T) \quad \text{on } \Gamma, \tag{8.44}$$

$$T = T_0 \quad \text{at } t = t_0, \tag{8.45}$$

where T is the temperature (°C), k is the thermal conductivity (W/m °C), ρc is the volumetric heat capacity (J/m^3 °C), T_∞ is the (known) process temperature (°C), n is the outward normal to the surface, h is the convection coefficient (W/m^2 °C), and Γ is the boundary surface. The Fourier equation describes many important thermal food processes, such as thermal pasteurization and sterilization of solid foods and cooling of horticultural products. Note that only linear heat conduction problems will be considered here; heat-transfer problems involving temperature-dependent thermophysical properties such as phase-change problems (e.g., freezing and thawing of foods) or involving nonlinear boundary conditions such as radiation (e.g., oven heating of bakery products), will not be addressed. The algorithms developed below can be readily extended to such problems after linearization of the governing equations.[12]

For many realistic heat conduction problems, no analytical solutions of Equation 8.43 subjected to Equation 8.44 and Equation 8.45 are known. In this case, numerical discretization techniques, such as the finite difference or finite element method, can be used to obtain an approximate solution. The finite element method, in particular, is a flexible and accurate method for solving partial differential equations such as the Fourier equation. In the framework of the finite element method, the continuum is subdivided in elements of variable size and shape that are interconnected in a finite number, n, of nodal points. In every element, the unknown temperature is approximated by a low-order interpolating polynomial. The application of a suitable spatial discretization technique, such as the Galerkin method, to Equation 8.43 through Equation 8.45 results in the following differential system:[13]

$$\mathbf{C} \frac{d}{dt} \mathbf{u} + \mathbf{K} \mathbf{u} = \mathbf{f}; \tag{8.46}$$

$$\mathbf{u}(t = 0) = \mathbf{u}_0, \tag{8.47}$$

where $\mathbf{u} = [u_1 \ u_2 \cdots u_n]^T$ the overall nodal temperature vector, $\mathbf{C}$ is the capacitance matrix and $\mathbf{K}$ is the stiffness matrix (both $n \times n$ matrices), and $\mathbf{f}$ is an $n \times 1$ vector. For more details about the construction of these matrices, the reader is referred to the literature (see Reference 13). The system, Equation 8.46, can be solved by finite differences in the time domain.

The finite element method has been successfully used in a number of thermal food processing applications, such as sterilization of baby food jars,[14] cooling of broccoli stalks,[15] and heating of a lasagna dish.[16] For a more elaborate account, the reader is referred to Wang and Sun.[17]

8.5.3.2 Perturbation Analysis

Here, the analysis is started with the system of ordinary differential equations, Equation 8.46, that was obtained by applying the finite element method to the heat conduction equation. Note that any discretization (finite difference, finite volume) of the Fourier equation into a system of ordinary differential equations will work. For simplicity, assume that T_∞ is a random variable parameter and that all other parameters are deterministic.

It can be shown[13] that T_∞ only appears in the construction of $\mathbf{f}$, not in the stiffness or capacitance matrix. Obviously, $\mathbf{u}$ is then a function of T_∞ as well. $\mathbf{u}$ may be expanded into a first-order Taylor series around its (so far unknown) mean solution,

$$\mathbf{u} = \bar{\mathbf{u}} + \frac{\partial \mathbf{u}}{\partial T_\infty} \Delta T_\infty, \tag{8.48}$$

in which the partial derivatives must be evaluated using $\bar{T}_\infty$. Similarly, $\mathbf{f}$ is a function of T_∞ and is expanded around its mean value $\bar{\mathbf{f}}$, which can be assembled as usual but using $\bar{T}_\infty$,

$$\mathbf{f} = \bar{\mathbf{f}} + \frac{\partial \mathbf{f}}{\partial T_\infty} \Delta T_\infty. \tag{8.49}$$

The partial derivative of $\mathbf{f}$ with respect to T_∞ can be calculated by direct differentiation. Substitution of Equation 8.49 into Equation 8.46 and combining appropriate terms in ΔT_∞ yields the following system:

$$\mathbf{C} \frac{d}{dt} \bar{\mathbf{u}} + \mathbf{K}\bar{\mathbf{u}} = \bar{\mathbf{f}}; \tag{8.50}$$

$$\mathbf{C} \frac{d}{dt} \left(\frac{\partial \mathbf{u}}{\partial T_\infty} \right) + \mathbf{K} \left(\frac{\partial \mathbf{u}}{\partial T_\infty} \right) = \frac{\partial \mathbf{f}}{\partial T_\infty}. \tag{8.51}$$

Equation 8.50 expresses that the mean value of the temperature, $\bar{\mathbf{u}}$, may be found by solving the original heat-transfer problem using mean values of the parameters. From Equation 8.51, it can be derived that the temperature covariance matrix $\mathbf{V}_{\mathbf{u},\mathbf{u}}(t)$ can now be computed from

$$\mathbf{V}_{\mathbf{u},\mathbf{u}}(t) = E\{[\mathbf{u}(t) - \bar{\mathbf{u}}(t)][\mathbf{u}(t) - \bar{\mathbf{u}}(t)]^T\} = \frac{\partial \mathbf{u}(t)}{\partial T_\infty} \frac{\partial \mathbf{u}^T(t)}{\partial T_\infty} \sigma_{T_\infty}^2. \tag{8.52}$$

The variance of the temperature $u_i(t)$ at node i is the ith diagonal entry of $\mathbf{V}_{\mathbf{u},\mathbf{u}}$. It follows from Equation 8.52 that initial conditions corresponding to the system, Equation 8.50 and Equation 8.51, are given by

$$\bar{\mathbf{u}}(t = 0) = \bar{\mathbf{u}}_0 \tag{8.53}$$

$$\frac{\partial \mathbf{u}}{\partial T_\infty} (t = 0) = \mathbf{0} \tag{8.54}$$

where $\mathbf{0}$ is a null vector of appropriate dimension. As a consequence, an additional differential system must be solved.

To illustrate the above algorithm, a typical thermal food process with a random variable ambient temperature will now be analyzed. The problem consists of a cylindrical container (radius $r_0 = 3.41$ cm, height $L = 10.02$ cm) filled with a 30% solids-content tomato concentrate with $k = 0.542$ W/m °C, $\rho c = 3.89 \times 10^6$ J/m³ °C. The following process conditions were applied: $T_0 = 65$°C, $\bar{T}_\infty = 125$°C, and $h = 100$ W/m² °C. The standard deviation of the ambient temperature is set equal to 1°C, as this seems a reasonable value that can be obtained by proper control of the equipment. For the finite element analysis the region, $[0, r_0] \times [0, L/2]$ is subdivided into 100 four-node axisymmetric quadrilateral elements. The time step is set equal to 36 s.

A semi-analytical solution for the mean value and the variance at selected positions inside the can was calculated by evaluating Equation 8.19 and Equation 8.20 by numerical quadrature. The analytical solution $T = g(t, T_\infty)$ is known[18] and involves the solution of a transcedental equation and two infinite series, which were truncated after six terms. For more details the reader is referred to Nicolaï et al.[19,20]

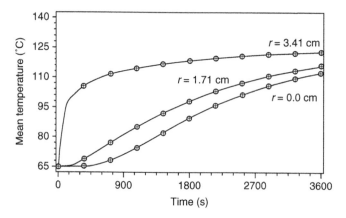

Figure 8.4 Mean temperature at various positions in an A1 can with random variable ambient temperature. —, semi-analytical solution; *, Monte Carlo (100 runs); +, Monte Carlo (1000 runs); ∘, first-order perturbation.

In Figure 8.4 and Figure 8.5, the mean temperatures and variances are shown. The results obtained with the different methods are similar for both the mean values as well as the variances of the temperature. For $t \rightarrow \infty$ the variances of the temperatures inside the container approach the ambient temperature variance.

The relative CPU time (actual CPU time divided by the CPU time required to solve a deterministic problem) for the perturbation method, the Monte Carlo method with 100 runs and the Monte Carlo method with 1000 runs was equal to 1.4, 26, and 254, respectively. The required CPU time for the perturbation method was less than twice that for a deterministic problem.

In Table 8.1, some applications of the perturbation method are given. The application of stochastic finite element perturbation methods to heat conduction problems with random variable thermophysical parameters has been described.[19] An extension to heat conduction with random field parameters has also been presented.[12,20] The authors found that variation on the initial temperature of the food tends to be smoothed during the process. Also, random variation on the thermophysical properties and surface heat transfer coefficient initially increases the

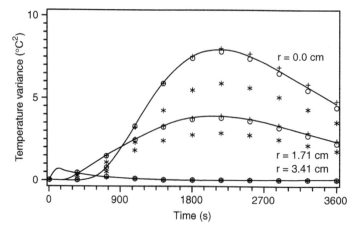

Figure 8.5 Temperature variance at various positions in an A1 can with random variable ambient temperature. —, semi-analytical solution; *, Monte Carlo (100 runs); +, Monte Carlo (1000 runs); ∘, first-order perturbation.

variability of the temperature inside the food but, again, fades out toward the end of the heating process. Random variation on the ambient temperature, on the other hand, causes the variation of the temperature inside the food to increase monotonically as a function of time. The duration of the thermal process will thus determine which properties or process parameters the food engineer has to focus on to reduce the variability of the process. Further, the authors found that random variable properties cause more variability of the temperature inside the food than random field properties. The random variable problem can, hence, serve as a worst case scenario.

Scheerlinck et al.[21] applied the perturbation algorithm to coupled heat and mass transfer problems described by Luikov's equations. Aguirre et al.[22] applied the perturbation algorithm to roasting of coffee beans. They assumed that the initial condition, thermal properties, and boundary conditions were random variables and showed that there might be a considerable uncertainty about the actual effect of the roasting treatment.

8.6 THE FOKKER–PLANCK EQUATION AND THE VARIANCE PROPAGATION ALGORITHM

8.6.1 Principle

Although the perturbation method is very powerful, it can only be used to solve problems involving parameters that may be random but do not change unpredictably in time. Also, it does not provide the full probabilistic information of the solution in terms of joint probability density functions. In this case, some results from stochastic systems theory can be used to calculate the mean value and variance of the solution of a stochastic heat-transfer problem.

The study of stochastic ordinary differential equations originated in the beginning of the twentieth century with the work of Einstein and Langevin on Brownian motion.[23] The formal mathematical theory of stochastic differential calculus was developed only forty years later by Ito[24] and involves a new definition of the integral-concept in a stochastic setting. Stochastic systems theory concerns the behavior of systems subjected to stochastic input signals and is based on Ito-calculus. Assume that the evolution of a random process $\mathbf{X}(t)$ is governed by the following differential equation:

$$\frac{d}{dt}\mathbf{X}(t) = \mathbf{g}[\mathbf{X}(t),t] + \mathbf{h}[\mathbf{X}(t),t]\mathbf{W}(t) \quad \text{and} \quad \mathbf{X}(t = t_0) = \mathbf{X}_0, \tag{8.55}$$

where the pdf of $\mathbf{X}_0$ is known, $\mathbf{g}[\mathbf{X}(t),t]$ and $\mathbf{h}[\mathbf{X}(t),t]$ are nonlinear functions of $\mathbf{X}(t)$, and $\mathbf{W}(t)$ is a zero-mean white-noise process with covariance matrix

$$V_{\mathbf{W},\mathbf{W}}(\tau) = \mathbf{\Psi}_{\mathbf{W}}\delta(\tau). \tag{8.56}$$

It can be shown[2] that the probability density function $f_{\mathbf{X}(t)}(\mathbf{x})$ is described by the following Fokker–Planck equation:

$$\frac{\partial}{\partial t}f_{\mathbf{X}(t)}(\mathbf{x}) = -\mathrm{tr}\left\{\left[\frac{\partial}{\partial \mathbf{x}}\mathbf{g}(\mathbf{x},t)\right]f_{\mathbf{X}(t)}(\mathbf{x}) + \mathbf{g}(\mathbf{x},t)\frac{\partial}{\partial \mathbf{x}}f_{\mathbf{X}(t)}(\mathbf{x})\right\}$$

$$+ \frac{1}{2}\mathrm{tr}\left\{\frac{\partial}{\partial \mathbf{x}}\left(\frac{\partial}{\partial \mathbf{x}}\right)^{\mathrm{T}}\mathbf{h}(\mathbf{x}(t),t)\mathbf{\Psi}_{\mathbf{W}}\mathbf{h}^{\mathrm{T}}(\mathbf{x}(t),t)f_{\mathbf{X}(t)}(\mathbf{x})\right\}, \tag{8.57}$$

where tr$\{\mathbf{A}\}$ denotes the trace of a matrix $\mathbf{A}$, i.e., the sum of the principal diagonal elements of $\mathbf{A}$. The boundary condition for Equation 8.57 is given by

$$f_{\mathbf{X}(t)}(\mathbf{x}) = f_{\mathbf{X}_0}(\mathbf{x}) \quad \text{at } t = 0. \tag{8.58}$$

Only for trivial problems can an analytical solution to Equation 8.57 and Equation 8.58 be found. However, it can be shown[2] that first-order approximate expressions for the mean vector $\bar{\mathbf{X}}(t)$ and the covariance matrix $\mathbf{V}_{\mathbf{X},\mathbf{X}}(t)$ can be derived after some manipulations from the Fokker–Planck equation:

$$\frac{d}{dt}\bar{\mathbf{X}}(t) = \mathbf{g}(\bar{\mathbf{X}}(t)); \tag{8.59}$$

$$\frac{d}{dt}\mathbf{V}_{\mathbf{X},\mathbf{X}}(t) = \frac{\partial \mathbf{g}[\bar{\mathbf{X}}(t),t]}{\partial \mathbf{X}(t)}\mathbf{V}_{\mathbf{X},\mathbf{X}}(t) + \mathbf{V}_{\mathbf{X},\mathbf{X}}(t)\left(\frac{\partial \mathbf{g}[\bar{\mathbf{X}}(t),t]}{\partial \mathbf{X}(t)}\right)^{\mathrm{T}} + \mathbf{h}[\bar{\mathbf{X}}(t),t]\Psi_{\mathbf{W},\mathbf{W}}(t)\mathbf{h}^{\mathrm{T}}[\bar{\mathbf{X}}(t),t] \tag{8.60}$$

Equation 8.59 and Equation 8.60 are called the *variance propagation algorithm*. Equation (8.59) is a matrix differential equation of the Lyapunov type.

8.6.2 Application to Lumped Capacitance Heat Transfer

The procedure based on the same simple lumped capacitance problem described by Equation 8.23 will be outlined. For simplicity, assume that $T_\infty(t)$ can be written as

$$T_\infty(t) = \bar{T}_\infty + \tilde{T}_\infty(t), \tag{8.61}$$

where $\bar{T}_\infty$ is a (constant) mean ambient temperature and $\tilde{T}_\infty(t)$ is a time fluctuation that will be modeled as an AR(1) random process:

$$\frac{d}{dt}\tilde{T}_\infty(t) + a_1\tilde{T}_\infty(t) = W(t), \tag{8.62}$$

where W is a zero-mean, white-noise process with variance $\sigma_{\tilde{T}_\infty}^2 = \sigma_{\tilde{T}_\infty}^2$.

Equation 8.23 and Equation 8.62 may be combined into the following global system:

$$\frac{d}{dt}\mathbf{X} = \mathbf{g}(\mathbf{X}) + \mathbf{h}W(t), \tag{8.63}$$

where

$$\mathbf{X} = \begin{bmatrix} T \\ \tilde{T}_\infty \end{bmatrix}, \tag{8.64}$$

$$\mathbf{g} = \begin{bmatrix} \dfrac{3h}{\rho c r_0}(\bar{T}_\infty + \tilde{T}_\infty) - \dfrac{3h}{\rho c r_0}T \\ -a_1\tilde{T}_\infty \end{bmatrix}, \tag{8.65}$$

$$\mathbf{h} = \begin{bmatrix} 0 \\ 1 \end{bmatrix}. \tag{8.66}$$

When Equations 8.63 through Equation 8.66 are substituted into the Fokker–Planck equation

(Equation 8.57),

$$\frac{\partial}{\partial t} f_{\mathbf{X}(t)}(\mathbf{x}) = -\mathrm{tr}\left\{ \left[\begin{array}{cc} -\frac{3h}{\rho c r_0} & \frac{3h}{\rho c r_0} \\ 0 & -a_1 \end{array} \right] f_{\mathbf{X}(t)}(\mathbf{x}) + \left[\begin{array}{cc} \frac{3h}{\rho c r_0}(\tilde{T}_\infty + \tilde{T}_\infty) - \frac{3h}{\rho c r_0} T \\ -a_1 \tilde{T}_\infty \end{array} \right] \frac{\partial}{\partial \mathbf{x}} f_{\mathbf{X}(t)}(\mathbf{x}) \right\}$$

$$+ \frac{1}{2} \sigma_W^2 \mathrm{tr}\left\{ \frac{\partial}{\partial \mathbf{x}} \left(\frac{\partial}{\partial \mathbf{x}} \right)^{\mathrm{T}} \left[\begin{array}{cc} 0 & 0 \\ 0 & 1 \end{array} \right] f_{\mathbf{X}(t)}(\mathbf{x}) \right\} \tag{8.67}$$

If, for simplicity of notation, the joint probability density function of T and $\tilde{T}_\infty$ is defined as $f(T,\tilde{T}_\infty) \triangleq f_{\mathbf{X}(t)}(\mathbf{x})$, then Equation 8.67 may be further elaborated into

$$\frac{\partial}{\partial t} f(T,\tilde{T}_\infty) = \left(\frac{3h}{\rho c r_0} + a_1 \right) f(T,\tilde{T}_\infty) - \left[\frac{3h}{\rho c r_0}(\tilde{T}_\infty + \tilde{T}_\infty) - \frac{3h}{\rho c r_0} T \right] \frac{\partial}{\partial T} f(T,\tilde{T}_\infty)$$

$$+ a_1 \tilde{T}_\infty \frac{\partial}{\partial \tilde{T}_\infty} f(T,\tilde{T}_\infty) + \frac{1}{2} \sigma_W^2 \frac{\partial^2}{\partial \tilde{T}_\infty^2} f(T,\tilde{T}_\infty) \tag{8.68}$$

This equation is a parabolic equation in two dimensions (T and $\tilde{T}_\infty$). Because Equation 8.23 is linear in T_∞ and because T_∞ is a Gaussian process, it can be shown that the solution of Equation 8.68 must be bivariate Gaussian. Therefore, Equation 8.8 (with $n=2$) may be substituted into Equation 8.68 and, after algebraic manipulations of considerable length, it is possible to derive equations for $\bar{T}$ and $\mathbf{V}_{T,\tilde{T}_\infty}$ that can be solved. Another approach is to solve Equation 8.68 numerically to describe the evolution of the probability density function of the soluble solids content in apple fruit.[25] These routes will not be explored here. Instead, the variance propagation algorithm will be directly applied to calculate $\bar{T}$ and $\mathbf{V}_{T,\tilde{T}_\infty}$.

The following system is then obtained:

$$\frac{d}{dt} \bar{T} = \frac{3h}{\rho c r_0}(\bar{T}_\infty - \bar{T}); \tag{8.69}$$

$$\frac{d}{dt} \bar{T}_\infty = 0; \tag{8.70}$$

$$\frac{d}{dt} \mathbf{V}_{\mathbf{x},\mathbf{x}} = \left[\begin{array}{cc} -\frac{3h}{\rho c r_0} & \frac{3h}{\rho c r_0} \\ 0 & -a_1 \end{array} \right] \mathbf{V}_{\mathbf{x},\mathbf{x}} + \mathbf{V}_{\mathbf{x},\mathbf{x}} \left[\begin{array}{cc} -\frac{3h}{\rho c r_0} & 0 \\ \frac{3h}{\rho c r_0} & -a_1 \end{array} \right] + \left[\begin{array}{cc} 0 & 0 \\ 0 & \sigma_W^2 \end{array} \right], \tag{8.71}$$

where

$$\mathbf{V}_{\mathbf{x},\mathbf{x}} = \left[\begin{array}{cc} \sigma_T^2 & \sigma_{T,\tilde{T}_\infty} \\ \sigma_{T,\tilde{T}_\infty} & \sigma_{\tilde{T}_\infty}^2 \end{array} \right], \tag{8.72}$$

and $\sigma_{T,\tilde{T}_\infty}$ is the covariance of T and $\tilde{T}_\infty$. The initial conditions are given by

$$\bar{T}(t = 0) = T_0 \tag{8.73}$$
$$\bar{T}_\infty(t = 0) = \bar{T}_\infty \tag{8.74}$$

$$\mathbf{V}_{\mathbf{x},\mathbf{x}}(t = 0) = \left[\begin{array}{cc} 0 & 0 \\ 0 & \sigma_{\tilde{T}_\infty}^2 \end{array} \right]. \tag{8.75}$$

Equation 8.69 expresses that the mean solution can be found by solving the original differential equation for the mean value of the random parameter. Equation 8.70 confirms that the mean value of the random parameter is constant (which was expected because an autoregressive process is stationary).

Equation 8.71 can be elaborated further to yield

$$\frac{d}{dt}\sigma_T^2 = -\frac{6h}{\rho c r_0}\sigma_T^2 + \frac{6h}{\rho c r_0}\sigma_{T,\tilde{T}_\infty}, \tag{8.76}$$

$$\frac{d}{dt}\sigma_{T,\tilde{T}_\infty} = \frac{3h}{\rho c r_0}\sigma_{\tilde{T}_\infty}^2 - \left(\frac{3h}{\rho c r_0} + a_1\right)\sigma_{T,\tilde{T}_\infty}, \tag{8.77}$$

$$\frac{d}{dt}\sigma_{\tilde{T}_\infty}^2 = -2a_1\sigma_{\tilde{T}_\infty}^2 + \sigma_W^2. \tag{8.78}$$

Because T_∞ is stationary, $\sigma_{\tilde{T}_\infty}^2$ is not a function of time; therefore,

$$\sigma_{\tilde{T}_\infty}^2 = \sigma_{T_\infty}^2 = \sigma_W^2/2a_1. \tag{8.79}$$

The solution of Equation 8.77 can be readily found through direct integration:

$$\sigma_{T,\tilde{T}_\infty} = \frac{3h/\rho c r_0}{3h/\rho c r_0 + a_1}\sigma_{\tilde{T}_\infty}^2\left\{1 - \exp\left[-\left(\frac{3h}{\rho c r_0} + a_1\right)t\right]\right\}. \tag{8.80}$$

After substitution of Equation 8.80 into Equation 8.78 and subsequent integration, the following expression for σ_T^2 may be derived:

$$\sigma_T^2 = \frac{3h/\rho c r_0}{3h/\rho c r_0 + a_1}\sigma_{T_\infty}^2 + \frac{3h/\rho c r_0}{3h/\rho c r_0 - a_1}\sigma_{T_\infty}^2\exp\left(-\frac{6h}{\rho c r_0}t\right)$$

$$-\frac{18h^2/\rho^2c^2r_0^2}{(3h/\rho c r_0 + a_1)(3h/\rho c r_0 - a_1)}\sigma_{T_\infty}^2\exp\left[-\left(\frac{3h}{\rho c r_0} + a_1\right)t\right]. \tag{8.81}$$

In the special case of a random variable, T_∞, the above expression may be simplified by putting $a_1 = 0$, so that

$$\sigma_T^2 = \sigma_{T_\infty}^2 + \sigma_{T_\infty}^2\exp\left(-\frac{6h}{\rho c r_0}t\right) - 2\sigma_{T_\infty}^2\exp\left[-\left(\frac{3h}{\rho c r_0}\right)t\right], \tag{8.82}$$

which is identical to Equation 8.28. Observe that the corresponding Fokker–Planck equation, Equation 8.68, in this case reduces to a hyperbolic equation.

A sample of the random-process ambient temperature and the corresponding temperature course in the sphere are shown in Figure 8.6. The parameter values were the same as before. The high-frequency fluctuations are smoothed because of the thermal inertia of the sphere. There was a very good agreement between the mean temperature of the sphere calculated by means of the Monte Carlo and the variance propagation algorithm (not shown). In Figure 8.7 the time course of the variance of the temperature of the sphere is shown. The results obtained by means of the variance propagation algorithm and the Monte Carlo method with 1000 or 5000 runs were comparable. However, the variances obtained by means of the Monte Carlo method with 100 runs are scattered.

8.6.3 Application to Heat Conduction

As was the case for the perturbation algorithm, for the extension of the variance propagation to conduction-limited problems, the discussion will begin from the spatially discretized system, Equation 8.46. As with the lumped capacitance problem, a first step is to write the stochastic

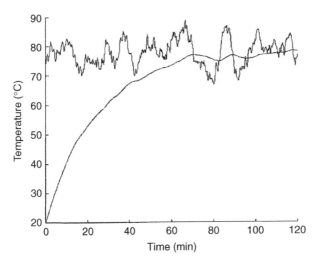

Figure 8.6 Realisation of AR(1) ambient temperature and corresponding temperature in sphere.

heat conduction in the form of Equation 8.55. It is easy to see that this can be accomplished through the following choice of $\mathbf{x}$, $\mathbf{g}$, and $\mathbf{h}$:

$$\mathbf{x} = \begin{bmatrix} \mathbf{u} \\ \tilde{T}_{\infty} \end{bmatrix}; \tag{8.83}$$

$$\mathbf{g} = \begin{bmatrix} \mathbf{C}^{-1}(-\mathbf{K}\mathbf{u} + \mathbf{f}) \\ -a_1\tilde{T}_{\infty} \end{bmatrix}; \tag{8.84}$$

$$\mathbf{h} = \begin{bmatrix} \mathbf{0} \\ 1 \end{bmatrix}, \tag{8.85}$$

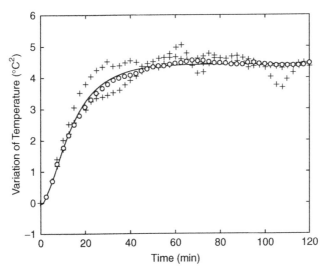

Figure 8.7 Temperature variance in sphere subjected to random process ambient temperature. —, perturbation; +, Monte Carlo (1000 runs); +, Monte Carlo (100 runs); ∘, Monte Carlo (5000 runs).

with $\mathbf{0}$ a null vector of dimension n_{nod}. The variance propagation algorithm can now be applied, and after some manipulations the following system is obtained:

$$\frac{d}{dt}\bar{\mathbf{u}} = \mathbf{C}^{-1}(-\mathbf{K}\bar{\mathbf{u}} + \bar{\mathbf{f}});$$ (8.86)

$$\frac{d}{dt}\mathbf{V}_{\mathbf{u},\mathbf{u}} = \mathbf{C}^{-1}\left[-\mathbf{K}\mathbf{V}_{\mathbf{u},\mathbf{u}} + \frac{\partial \mathbf{f}}{\partial \tilde{T}_\infty}\mathbf{V}_{\mathbf{u},\tilde{T}_\infty}^T\right] + \left[-\mathbf{K}\mathbf{V}_{\mathbf{u},\mathbf{u}} + \frac{\partial \mathbf{f}}{\partial \tilde{T}_\infty}\mathbf{V}_{\mathbf{u},\tilde{T}_\infty}^T\right]^T\mathbf{C}^{-T};$$ (8.87)

$$\frac{d}{dt}\mathbf{V}_{\mathbf{u},\tilde{T}_\infty} = \mathbf{C}^{-1}\left(-\mathbf{K}\mathbf{V}_{\mathbf{u},\tilde{T}_\infty} + \frac{\partial \mathbf{f}}{\partial \tilde{T}_\infty}\sigma_{\tilde{T}_\infty}^2\right) - \mathbf{V}_{\mathbf{u},\tilde{T}_\infty}a_1,$$ (8.88)

where the notation $\mathbf{C}^{-T}$ denotes the transpose of the inverse of $\mathbf{C}$. $\bar{\mathbf{f}}$ is assembled using $\bar{T}_\infty$. The initial condition for Equation 8.86 is given by

$$\bar{\mathbf{u}}(t = 0) = \bar{\mathbf{u}}_0,$$
$$\mathbf{V}_{\mathbf{u},\mathbf{u}} = \mathbf{0},$$ (8.89)
$$\mathbf{V}_{\mathbf{u},T_\infty} = \mathbf{0},$$ (8.90)

with $\mathbf{0}$ null matrices of appropriate dimension. Equation 8.86 through Equation 8.90 constitute the variance propagation algorithm for heat conduction problems with random process ambient temperature.

Equation 8.87 is of the general form,

$$\frac{d}{dt}\mathbf{V}(t) = \mathbf{A}\mathbf{V}(t) + \mathbf{V}(t)\mathbf{A}^T + \mathbf{B}(t),$$

where $\mathbf{V}$, $\mathbf{A}$, and $\mathbf{B}$ are square matrices of equal dimension, and is called a *Lyapunov* matrix differential equation. This equation can be numerically solved after time discretization. In the case an implicit discretization technique is applied, an algebraic Lyapunov equation is obtained that can be solved readily.[26] Equation 8.88 is of the same form as Equation 8.86, and this fact can be exploited to reduce the computational effort.

The algorithm will now be illustrated by the tomato concentrate example that was introduced in the previous section. The parameters are the same as given before, but the ambient temperature is now described by means of an AR(1) process with $\sigma_{T_\infty} = 1°C$ and $a_1 = 0.00277/s$. An implicit Euler finite difference method in the time domain was used to integrate the differential systems.

In Figure 8.8 the temperature variance at three different positions in the can are shown as calculated by means of the Monte Carlo method with 100 and 1000 runs, and the variance propagation algorithm. The agreement between the Monte Carlo method with 1000 runs and the variance propagation algorithm is good, but the Monte Carlo method with 100 runs is not very accurate. For the mean value an excellent agreement between the different method was observed (figure not shown). The relative CPU time was equal to 74, 242, and 2426, for the variance propagation, Monte Carlo with 100 runs and Monte Carlo with 1000 runs, respectively.

Applications of the variance propagation algorithm are given in Table 8.1. More extended variance propagation algorithms for heat conduction problems involving stochastic parameters have been developed.[27–29] These authors applied the variance propagation algorithm with random process and random wave parameters to evaluate the effect of random fluctuations in storage temperature on the temperature inside pineapple and cucumber during cooling.[30] They found that

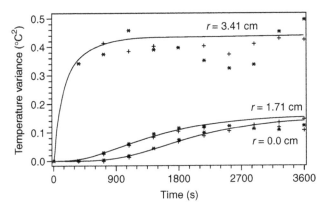

Figure 8.8   Temperature variance at various positions in an A1 can with random process ambient temperature. —, variance propagation algorithm; *, Monte Carlo (100 runs); +, Monte Carlo (1000 runs).

high frequency fluctuations of the storage temperatures are attenuated because of the thermal inertia of the fruit. The attenuation was larger near the center of the product. This is an important result because it indicates that phenomena such as brown heart in pineapple—a chilling disorder characterized by a brown discoloration of the core of the pineapple—are likely not a consequence of rapid temperature fluctuations but rather of wrong storage temperature set-points or temperature abuse. For the same reason, other products, such as cucumber, which develop chilling injuries at their surface, are expected to be more susceptible to temperature fluctuations. As in the case of random field parameters, random variable properties cause more variability of the temperature inside the food than random process properties.

The algorithm was further extended to coupled heat and mass transfer problems by Scheerlinck et al.[31]

8.7   CLOSING REMARKS/FUTURE TRENDS

In this chapter, some algorithms for stochastic heat transfer analysis have been outlined.

In the Monte Carlo method, a large number of process samples are obtained by solving the heat transfer model for artificially generated random parameter samples. Straightforward statistical analysis of the simulation results yields the mean values and variances of the temperature. Transformation methods can be used to calculate the probability density function and moments of a function of random variables but are only applicable to problems for which a (simple) analytical solution is available. The perturbation method is based on the Taylor expansion of the governing ordinary or partial differential equations. It is only applicable for random parameters of the random variable and field type. The Fokker–Planck equation and the variance propagation algorithm are based on stochastic systems theory and were originally developed for systems of ordinary differential equations with parameters that fluctuate randomly in time (random process parameters). The Fokker–Planck equation is difficult to solve by numerical means but can be used as a reference technique, as it yields the exact expression (or a numerical approximation of arbitrary accuracy) of the probability density function of the solution. It is limited to problems with a relatively small number of random variables. Both the perturbation and the variance propagation algorithm can readily be extended to spatially discretized partial differential equations, such as the Fourier equation for heat conduction problems. The variance propagation algorithm then yields a system of matrix differential equations that can be solved numerically.

For random variable problems, the perturbation algorithm is the method of choice because of its speed. For random process problems the perturbation algorithm is not applicable. In this case, the variance propagation algorithm is to be preferred above the Monte Carlo method. The Monte Carlo method, in general, requires a large amount of computing time to obtain results with an acceptable accuracy. Also, it requires a complete stochastic specification of the random parameters, while for the perturbation algorithm only the mean values of the parameters and their covariance matrix must be known. The variance propagation algorithm is restricted to problems involving autoregressive parameters with known mean value and covariance function. However, the latter two algorithms can provide only limited statistical information, such as the mean value and the variance, whereas the Monte Carlo method can also be applied to derive other statistical characteristics of the solution, such as its probability density function. Also, as both the perturbation and variance propagation algorithm are essentially based on a linearization of the governing equations around their mean solution, they are only applicable if the variability is relatively small (coefficient of variation smaller than 0.2).

ACKNOWLEDGMENTS

Author Nico Scheerlinck is a Postdoctoral Fellow with the Flanders Fund for Scientific Research (FWO Vlaanderen). The authors wish to thank the IWT for financial support (project S-6087).

GLOSSARY

Autoregressive random process A special type of random process in which there is some correlation between successive values; formally, an autoregressive random process is defined through a differential equation with a white noise input.

Finite element method Numerical discretisation method to calculate approximate solutions of partial differential equations.

Fokker–Planck equation Partial differential equation that describes the evolution of the probability density function of the solution of a stochastic differential equation.

Gauss quadrature Numerical solution of surface integrals where the integration points are chosen such to achieve maximal accuracy for a certain class of integrands.

Lumped capacitance heat transfer Heat transfer in objects with an internal heat transfer resistance that is much smaller than the external resistance.

Lyapunov equation A special type of matrix differential equation.

Monte Carlo method A method to solve a probabilistic model by random numerical sampling.

Normal (Gaussian) distribution The most important type of probability distribution characterized by a bell-shaped function.

Perturbation method A method to approximately calculate the mean value and variance of the solution of a probabilistic model.

Probability density function A function that allows the calculation of the probability of a variable to have a value in an arbitrary range.

Random field A random variable that may vary in an unpredictable way as a function of position.

Random process A random variable that may vary in an unpredictable way as a function of time.

Random variable A physical variable that may vary from experiment to experiment in an unpredictable way.

Random wave A random field that may also vary in an unpredictable way as a function of time.

Stationary process A process of which the probabilistic characteristics do not change in time.

Transformation method A method to calculate the probabilistic characteristics of a function of random variables.

Variance propagation method   A method to approximately calculate the mean value and
variance of the solution of a stochastic differential equation.

White noise   A special type of random process; successive values are completely uncorrelated no
matter how close in time they are.

NOMENCLATURE

All vector and matrix quantities are denoted with bold symbols.

a_i	Coefficient of autoregressive process
Bi	Biot number
c	Heat capacity
$\mathbf{C}$	Finite element capacity matrix
det	Determinant
E	Mean value operator
$f_X(x)$	Probability density function
$\mathbf{f}$	Finite element thermal load vector
g	Function
h	Surface heat transfer coefficient (W/m °C)
$\mathbf{h}$	Vector valued function
H_i	Weight in Gauss quadrature formula
k	Thermal conductivity (W/m °C)
$\mathbf{K}$	Finite element stiffness matrix
L	Half-height of can
$n_\perp$	Outward normal
n	Number of nodes, order
$\mathbf{0}$	Zero matrix
P	Probability
q	Order of quadrature formula
r_0	Radius (m)
R	Correlation coefficient
t	Time (s)
T	Temperature (°C)
T_0	Initial temperature (°C)
T_∞	Ambient temperature (°C)
$\tilde{T}_\infty$	Stochastic fluctuation of ambient temperature (°C)
tr	Trace of a matrix (sum of diagonal entries)
$\mathbf{u}$	Nodal temperature vector
V	Covariance function
W	White noise process
X	Random variable or process
$\bar{X}$	Mean value of X
Y	Random variable or process

Greek Symbols

$\delta(\tau)$	Dirac delta
Γ	Convection surface
ρ	Density (kg/m)

σ Standard deviation
τ Separation time

REFERENCES

1. Gray, R. M. and Davisson, L. D., *Random Processes: A Mathematical Approach for Engineers*, Prenctice-Hall Information and System Sciences Series, Englewood Cliffs, NJ: Prentice-Hall, 1986.
2. Melsa, J. L. and Sage, A. P., *An Introduction to Probability and Stochastic Processes*, Englewood Cliffs, NJ: Prentice-Hall, 1973.
3. Vanmarcke, E., *Random Fields: Analysis and Synthesis*, Cambridge, MA: MIT Press, 1983.
4. Kincaid, D. and Cheney, W., *Numerical Analysis: Mathematics of Scientific Computing*, 3rd ed., Pacific Grove, CA: Brooks/Cole, 2002.
5. Johns, W. R., Simulation of food processes with uncertain data, In *Food Engineering in a Computer Climate*, New York, Philadelphia, London: Institution of Chemical Engineers, Hemisphere Publishing Company, pp. 1–24, 1992.
6. Hertog, M. L. A. T. M., The impact of biological variation on postharvest population dynamics, *Postharvest Biology and Technology*, 26(3), 253–263, 2002.
7. Schouten, R. E., Van Kooten, O., Tijskens, L. M. M., and Jongbloed, G., Batch variability and cultivar keeping quality of cucumber, *Postharvest Biology and Technology*, 26(3), 253–263, 2002.
8. Incropera, F. P. and De Witt, D., *Fundamentals of Heat and Mass Transfer*, 4th ed., New York, Chichester, Brisbane: Wiley, 2001.
9. Chen, Y. M. and Tien, C. L., Penetration of temperature waves in a random medium, *Journal of Mathematics and Physics*, 46, 188–194, 1967.
10. Beran, M. J. and McCoy, J. J., Mean field variation in random media, *Quarterly of Applied Mathematics*, 28(2), 245–258, 1970.
11. Tzou, D. Y., Stochastic analysis of temperature distribution in a solid with random heat conductivity, *Journal of Heat Transfer*, 110, 23–29, 1988.
12. Nicolaï, B. M. and De Baerdemaeker, J., Finite element perturbation analysis of nonlinear heat conduction problems with random field parameters, *International Journal of Numerical Methods for Heat and Fluid Flow*, 5, 525–544, 1997.
13. Segerlind, L., *Applied Finite Element Analysis*, 2nd ed., New York: Wiley, 1984.
14. Naveh, D., Kopelman, I. J., and Pflug, I. J., The finite element method in thermal processing of foods, *Journal of Food Science*, 48, 1086–1093, 1983.
15. Jiang, H., Thompson, D. R., and Morey, R. V., Finite element model of temperature distribution in broccoli stalks during forced-air precooling, *Transactions of the ASAE*, 30(5), 1473–1477, 1987.
16. Nicolaï, B. M., Van den Broeck, P., Schellekens, M., De Roeck, G., Martens, T., and De Baerdemaeker, J., Finite element analysis of heat conduction in lasagna during thermal processing, *International Journal of Food Science and Technology*, 30(3), 347–364, 1995.
17. Wang, L. J. and Sun, D. W., Recent developments in numerical modelling of heating and cooling processes in the food industry: a review, *Trends in Food Science and Technology*, 14(10), 408–423, 2003.
18. Carslaw, H. S. and Jaeger, J. C., *Conduction of Heat in Solids*, Oxford: Clarendon, 1959.
19. Nicolaï, B. M. and De Baerdemaeker, J., Computation of heat conduction in materials with random variable thermophysical properties, *International Journal for Numerical Methods in Engineering*, 36, 523–536, 1993.
20. Nicolaï, B. M., Scheerlinck, N., Verboven, P., and De Baerdemaeker, J., Stochastic perturbation analysis of thermal food processes with random field parameters, *Transactions of the American Society of Agricultural Engineers*, 43(1), 131–138, 2000.
21. Scheerlinck, N., Verboven, P., Stigter, J. D., De Baerdemaeker, J., Van Impe, J., and Nicolaï, B. M., Stochastic finite element analysis of coupled heat and mass transfer problems with random field parameters, *Numerical Heat Transfer, Part B: Fundamentals*, 37(3), 309–330, 2000.
22. Aguirre C. G., Haghighi, K., Corvalan, C., and Campanella, O., A stochastic finite element methodology for analysis of coffee beans during roasting, ASAE meeting paper number 013023, St. Joseph, Michigan, USA: ASAE, 2001.

23. Gardiner, C. W., *Handbook of Stochastic Methods for Physics, Chemistry and the Natural Sciences*, Springer Series in Synergetics 13, 2nd ed., Berlin: Springer, 1997.

24. Ito, K., *On Stochastic Differential Equations*, New York: American Mathematical Society, 1951.

25. Scheerlinck, N., Peirs, A., Desmet, M., Schenk, A., and Nicolaï, B. M., Modelling fruit characteristics during apple maturation: a stochastic approach, *Journal Mathematical and Computer Modeling of Dynamical Systems*, 10(2), 149–168, 2004.

26. Bartels, R. H. and Steward, G. W., Solution of the matrix equation, *Communications of the ACM*, 15(9), 820–826, 1972.

27. Nicolaï, B. M. and De Baerdemaeker, J., Simulation of heat transfer in foods with stochastic initial and boundary conditions, *Transactions of the IChemE, Part C*, 70, 78–82, 1992.

28. Nicolaï, B. M. and De Baerdemaeker, J., A variance propagation algorithm for the computation of heat conduction under stochastic conditions, *International Journal for Heat and Mass Transfer*, 42, 1513–1520, 1998.

29. Nicolaï, B. M., Verboven, P., Scheerlinck, N., and De Baerdemaeker, J., Numerical analysis of the propagation of random parameter fluctuations in time and space during thermal food processes, *Journal of Food Engineering*, 38, 259–278, 1998.

30. Nicolaï, B. M., Verlinden, B., Beuselinck, A., Jancsók, P., Quenon, V., Scheerlinck, N., Verboven, P., and De Baerdemaeker, J., Propagation of stochastic temperature fluctuations in refrigerated fruits, *International Journal of Refrigeration*, 22(2), 81–90, 1999.

31. Scheerlinck, N., Verboven, P., Stigter, J. D., De Baerdemaeker, J., Van Impe, J. F., and Nicolaï, B. M., A variance propagation algorithm for stochastic heat and mass transfer problems in food processes, *International Journal for Numerical Methods in Engineering*, 51(8), 961–983, 2001.

PART II

Observation-Based Models

CHAPTER 9

Experimental Design and Response-Surface Methodology

Shuryo Nakai, Eunice C. Y. Li-Chan, and Jinglie Dou

CONTENTS

9.1   INTRODUCTION

The basic principle of experimental design is to plan experiments so as to be able to study the effects of certain factors on some specific results, and to identify the important influential factors with a given level of confidence. The effects of the factors and their interactions are evaluated by computing F-values, which are the ratio of effect-mean-square vs. error-mean-square. Randomization and minimization of error are two important mathematical principles for efficient selection of truly influential factors. Randomization is useful in objectively detecting uncontrollable errors, thereby avoiding the inclusion of errors arising from man-made causes in the true errors. Sometimes "blocking" is required during randomization when there is an unavoidable need for handling many factors in an experimental plan. Blocking should be made based on a controllable condition (factor), such as date, so that the influence of that factor can be determined later if required.

Response-surface methodology (RSM) is a modeling technique to illustrate the effects of factors on response surfaces for the purpose of locating the optimum. Therefore, the RSM consists of the techniques for experimental design, regression analysis, and computation of the optimum. There are two popular books for RSM: (1) *Response Surface Methodology: Process and Product Optimization Using Designed Experiments* by Myers and Montgomery[1] and (2) *Response Surfaces: Designs and Analysis* by Khuri and Cornell.[2] The former has a broader coverage, including the category of evolutionary operation such as sequential simplex optimization (SSO), which is not covered by the latter.

To avoid conflict with other chapters in this handbook on linear programming and optimization, the discussion in this chapter will be restricted to the response-surface modeling, including some optimization techniques based on response surfaces. This is indispensable because the RSM itself is an optimization technology, as discussed in the above textbook by Myers and Montgomery.[1]

Another point to be stressed is that this chapter does not give any detailed or deep elaboration on the methods, theories, algorithms, and principles. The authors have followed the approach used in the excellent chapter entitled "Optimization and Experimental Designs" by Otto,[3] which was written for a book for analytical chemists. The readers are advised to refer to his chapter for more information, as we have avoided excessive duplication and instead have focused this chapter to discuss the specific nature of techniques useful in food- and bio-processing.

9.2   BASIC PRINCIPLES

9.2.1   Factor Screening

There are two conflicting considerations in selecting factors for designing experiments. (1) Theoretically, all potential factors should be entered into an experimental design to avoid missing important factors that may appear minor but could in fact be critical in the mechanism of the reaction in question. (2) However, it is generally accepted that the greater the number of factors, the more the number of iterations that are required to reach the optimum. This increase in the number of experiments—usually exponential—would immediately increase labor and expenses of carrying out experiments, especially in the case of biological exercises.

To circumvent these problems, a preliminary selection of factors is customarily made using a factorial design, especially fractional factorial design. The $L_{16}(2^{15})$ design of Taguchi[4] has been one of the most useful designs for this purpose due to easy identification of factor-interaction columns in the design, thus being efficient in selecting influential factors or interactions without losing important information. The delicate nature and complexity of biological phenomena can be exemplified in cytokine chemistry and genetic engineering. For instance, interleukins 4, 5, and 6 may simultaneously exert their functions either antagonistically or symbiotically in the probiotic activity of orally administered lactobacilli.[5] The function of single residues in a protein sequence cannot be accurately defined by replacement or deletion of the site in the sequence.[6] In the case of multifunctional phenomena, it is possible that the scree-plot of principal components[3] does not show a sharp decline of eigenvalues, thereby implying that many factors are playing almost equally important roles in some biological phenomena. In this case, ignoring higher-order interactions for the sake of dimensionality reduction may not always be justified as in the case of Taguchi designs.[4] Decision making for the priority on whether full information should be kept so as not to miss the chance of important discovery or the best efficacy in experiments by sacrificing some rarely significant high level interactions is extremely critical.

9.2.2 Errors

Error is defined as the cumulative effects of uncontrollable factors that are the effects of undefined factors as well as any other factors that have not been selected during the designing of experiments. However, the most important error is the one that would critically affect the reliability of analysis. It is highly recommended to maintain factors as constant as possible, because the repeatability of analysis is extremely critical. Training of analysts prior to commencing experiments is a minimum prerequisite, as increasing the cycles of replication is not as efficient as intensive training in terms of reducing the size of error, which directly affects the reliability of selecting the truly influential factors.

9.2.3 Replication

Replication is a popular way to decrease the size of error. However, because the error is only reduced by a factor of $> 1/\sqrt{n}$, where n is the order of replication such that for duplication $n=2$ or for triplication $n=3$, replication is not a very efficient way to reduce error. Therefore, for the sake of the best efficiency in biological projects, it is recommended to include replicates only for the purpose of computing the standard error value, instead of replicating the entire design.

9.3 FACTORIAL DESIGNS

9.3.1 Elements of Factorial Analysis

In contrast to one-factor-at-a-time approaches, where only one factor is varied while other variables are held constant, designed experiments involving factorial analysis allow the analyst to investigate the effects of the individual (main) factors as well as interactions between the factors.

9.3.1.1 *Effects of Main Factors*

Significance is assessed using F-values (mean square of treatment/mean square of error), as previously discussed.

9.3.1.2   Effects of Factor Interactions

Significant interaction between factors is defined as the departure from parallelism. If the main effects are parallel, they are unilateral. Existence of significant interaction between factors A and B means that a specific combination of factors A and B would result in an effect that is different from the unilateral effects of factors A and B considered separately. For example, if the response in question is processing yield, a significant factor interaction would mean that a specific combination of A and B could result in unexpectedly high increase or decrease in the yield, which cannot be anticipated from knowing the main effects of A and B independently.

9.3.1.3   Confounding

Confounding, in which effects are inseparable, can occur between factors and/or interactions during the course of factorial analysis.

9.3.1.4   Symmetry

Experiments should be designed to distribute evenly within the search space of individual factors as much as possible to avoid undesirable confounding. This rule was utilized in the regulated random design of the random-centroid optimization (RCO) as described later.

9.3.1.5   Number of Experiments

In full factorial designs, all possible interactions between factors are assessed, and all combinations of levels are taken into consideration. Consequently, the required number of experiments increases exponentially as the number of factors (n) increases, and it also increases as the number of levels of factors increases. Two- and three-level full factorial designs require 2^n and 3^n experiments for n-factor analysis. For instance, the full factorial designs for two-level experiments with 5, 6, and 7 factors would require 32, 64, and 128 experiments, respectively, whereas a three-level, 5-factor design requires 243 experiments.

Because it is usually not only expensive but also time consuming to carry out each experiment, especially in the life sciences, it is of paramount importance to obtain the desired information by most efficiently conducting the minimum number of experiments. There are such techniques available, in that only a portion of the full factorial design is chosen, yet satisfactorily adequate information can be extracted. These methods are called *fractional factorial designs* and are frequently used as a preliminary step during an optimization project by selecting the truly influential factors prior to performing the optimization experiments.

9.3.2   Fractional Factorial Designs

It is reasonable to believe that the higher the order of factor-factor interactions, the less the statistical significance. By ignoring one such interaction, the total number of required experiments in the design is reduced by half. This is the basis of fractional factorial designs, which gain their efficiency by disregarding higher-order interactions, thereby analyzing main effects and lower- (e.g., second-) order interactions only. In other words, it is assumed that usually two factor interactions are adequate for judging the general trend in the response surface, and that unexpectedly high or low response values at specific combination of levels of more than two factors are not as important as the main effects or second-order interactions. Note that these assumptions lead to confounding of the higher-order interactions with some main effects or lower-order interactions.

Since the early Yates' algorithm using $+$ and $-$ signs that was employed by Box et al.,[7] many designs have been published, including one of the most popular, the Plackett–Burman design.[3] In this chapter, Taguchi's designs[4] that are based on orthogonal arrays will be discussed due to ease of explanation of the principle and ANOVA (analysis of variance) computation. Among the many of Taguchi's designs for factors with two to five levels, the most useful $L_{16}(2^{15})$ design will be explained in this chapter to illustrate the basic principles. This fractional factorial design is based on the four-factor, two-level full-factorial design; thus, the number of experiments required is $2^4 = 16$ (L_{16}). The other digits of 2^{15} in the $L_{16}(2^{15})$ designation show that there are 15 columns of two-level limit values (upper and lower limits represented by digits 1 and 2, respectively) in the design table, as shown in Table 9.1a. Columns 1–15 describe the experimental conditions as combinations of the two levels of factors for each of the 16 experiments in rows 1–16. The bottom of the table shows the list of component rows expressed as alphabetical letters, or their combinations, their grouping, and the interaction schemes coined as *linear graphs* are depicted underneath the table.

The components are used to find the columns where "column×column" interactions appear. To use these components, the simple rules to be applied are: (1) multiply components of a combination of any two columns to which factors are assigned; (2) replace with unity when squares of letters appear in the component products; and (3) find the column that matches the computed component products. The interaction effect of two factors assigned in the above rule 1 will appear in this new column. If a new factor is assigned to this column, its effect will be confounded with the interaction effect of the previously assigned two factors. Therefore, if this interaction should not be ignored, no new factor should be assigned to the same column. For example, when pH and temperature are assigned to columns 1 (component a) and 2 (component b), the pH×temperature interaction will appear in column 3 (ab). Similarly, the interaction of columns 3 (ab) and 4 (c), which is abc, will be found in column 7, whereas the interaction of columns 3 (ab) and 6 (bc), which is $ab^2c \rightarrow ac$, will appear in column 5. In theory, the most logical half-design is to eliminate the experimental condition appearing in column 15 with highest order of interaction, i.e., the four-factor interaction of abcd. If all interactions are to be ignored, this table can accommodate a maximum of 14 factors assigned to each of 14 columns, with one left-over column 15 assigned to the error term.

An easier way of assigning factors is to use the interaction schemes shown underneath Table 9.1a. The schemes are composed of lines (interaction) connecting pairs of dots or "nodes" (for factor assignment), with interactions designated as digits (interaction columns) on the lines. Scheme 1, for example, may be used to assign five factors (in columns 1, 2, 4, 8 and 15) along with interactions between each pair of factors, whereas scheme 4 is the design that may be used when all of the two-factor interactions of factor 1 with other factors are assumed to be important.

9.3.2.1 Two-Level Design Example

A hypothetical example is shown in Table 9.1b. Nine factors (WHard: water hardness, pH, PasTemp: pasteurization temperature, StTemp: storage temperature, StTime: storage time, BotSize: bottle size, BotShape: bottle shape, VarLoc: fruit variety or locality, and Sugar: sugar content) are considered to be affecting the sediment formation in a fruit drink. If a full two-level factorial design is planned, $2^9 = 512$ experiments would have to be carried out. By assuming that some higher-order interactions are not important, a fractional factorial design with only 16 experiments may be used. To decide which interactions should be computed for significance, all of the available information derived from past experience, literature data, and any other sources should be taken into consideration. Four two-factor interactions (columns 3, 4, 14, and 15) are designed to be assessed for their statistical significance. Several two-factor interactions were regarded to be nonsignificant: columns 5, 6, 9, and 10, which were used to assign factors (BotSize, PasTemp, StTime and BotShape, respectively). To avoid overlooking important interactions, it is

Table 9.1a L₁₆(2¹⁵)

Experiment No.	1	2	3	4	5	6	7	8	9	10	11	12	13	14	15
1	1	1	1	1	1	1	1	1	1	1	1	1	1	1	1
2	1	1	1	1	1	1	1	2	2	2	2	2	2	2	2
3	1	1	1	2	2	2	2	1	1	1	1	2	2	2	2
4	1	1	1	2	2	2	2	2	2	2	2	1	1	1	1
5	1	2	2	1	1	2	2	1	1	2	2	1	1	2	2
6	1	2	2	1	1	2	2	2	2	1	1	2	2	1	1
7	1	2	2	2	2	1	1	1	1	2	2	2	2	1	1
8	1	2	2	2	2	1	1	2	2	1	1	1	1	2	2
9	2	1	2	1	2	1	2	1	2	1	2	1	2	1	2
10	2	1	2	1	2	1	2	2	1	2	1	2	1	2	1
11	2	1	2	2	1	2	1	1	2	1	2	2	1	2	1
12	2	1	2	2	1	2	1	2	1	2	1	1	2	1	2
13	2	2	1	1	2	2	1	1	2	2	1	1	2	2	1
14	2	2	1	1	2	2	1	2	1	1	2	2	1	1	2
15	2	2	1	2	1	1	2	1	2	2	1	2	1	1	2
16	2	2	1	2	1	1	2	2	1	1	2	1	2	2	1
Component	a	b	a b	c	a c	b c	a b c	d	a d	b d	a b d	c d	a c d	b c d	a b c d
Group	(a)	(b)		(c)				(d)							

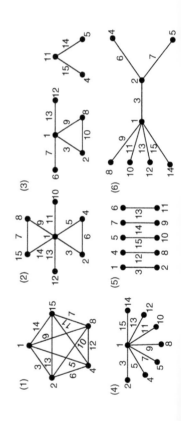

Table 9.1b Sediment Formation in the Fruit Drink

	Column	Interaction
WHard	1 (a)	1×2=ab→3
pH	2 (b)	
PasTemp	6 (bc)	2×6=c→4
StTemp	7 (abc)	7×9=bcd→14
StTime	9 (ad)	
BotSize	5 (ac)	5×10=abcd→15
BotShape	10 (bd)	
VarLoc	8 (d)	
Sugar	11(abd)	
Unused [a]	12 (cd), 13 (acd)	

[a] Other factors, e.g., food colors and preservatives, can be assigned.

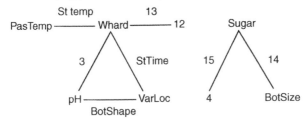

Source: Nakai's lecture note.

recommended to assign new factors, possible interactions or an error term to columns for higher-level interactions, e.g., 7, 11, 13, and 14 (all are three-factor components) in Table 9.1a when other factors are already assigned to columns 1, 2, and 8. At least one column should be left unassigned because it is required to serve for the error sum of square computation. In the present example, columns 12 and 13 are left unused for this purpose.

In total, 16 experiments under the conditions shown in Table 9.1c are carried out in random order to avoid carryover error, one experiment after another. The amounts of sediments measured as a result of each of the 16 experiments are also shown in Table 9.1c.

Table 9.1c The Amounts of Sediments Measured After Storage of Bottled Drink

Exp	W Hard	pH	3	4	Bot Size	Pas Temp	St Temp	Var Loc	St Time	Bot Shape	Sugar	12	13	14	15	Sediment -gr-
1	1	1			1	1	1	1	1	1	1					1.25
2	1	1			1	1	1	2	2	2	2					1.00
3	1	1			2	2	2	1	1	1	1					0.50
4	1	1			2	2	2	2	2	2	2					0.20
5	1	2			1	2	2	1	1	2	2					0.10
6	1	2			1	2	2	2	2	1	1					0.05
7	1	2			2	1	1	1	1	2	2					0.07
8	1	2			2	1	1	2	2	1	1					0.15
9	2	1			2	1	2	1	2	1	2					0.20
10	2	1			2	1	2	2	1	2	1					1.05
11	2	1			1	2	1	1	2	1	2					0.85
12	2	1			1	2	1	2	1	2	1					0.60
13	2	2			2	2	1	1	2	2	1					0.50
14	2	2			2	2	1	2	1	1	2					0.40
15	2	2			1	1	2	1	2	2	1					0.75
16	2	2			1	1	2	2	1	1	2					0.55

Then, ANOVA computation was performed in ordinary fashion to obtain sum of squares for treatment S_t, sum of squares for total S_T, and correction term CT:

$$S_t = \left(\sum_k \text{treatment 1} \right)^2 + \left(\sum \text{treatment 2} \right)^2 - \text{CT}, \tag{9.1}$$

$$S_T = \sum_{1}^{k} (\text{data})^2 - \text{CT}, \tag{9.2}$$

$$\text{CT} = \left(\sum_{1}^{k} \text{data} \right)^2 / k, \tag{9.3}$$

where k is the number of experiments.

An example of S_t computation is:

S_t for bottle size in column 5 is $> (1.25 + 1.00 + 0.10 + 0.05 + 0.85 + 0.60 + 0.75 + 0.55)^2 + (0.50 + 0.20 + 0.07 + 0.15 + 0.20 + 1.05 + 0.50 + 0.40)^2 - (1.25 + 1.00 + \ldots + 0.75 + 0.55)^2 / 16$.

All computations can be readily carried out using Excel™ functions SUM (total) and SUMSQ (sum of squares).[9]

The ANOVA table thus calculated is shown in Table 9.1d. None of the nine factors is significant in affecting the amount of sediment, even after the pH×PasTemp interaction has been pooled into the error (panel A of Table 9.1d). To extract useful information, SS with F values lower than unity, i.e., StTime, BotShape, VarLoc, and BotSize×BotShape, are pooled into the Error SS, thereby increasing its degrees of freedom to 7 and resulting in an increase in the power of the analysis to detect significant factors. In the ANOVA thus repeated (panel B of Table 9.1d), it is evident that pH is highly significant ($P<0.01$), and BotSize and WHard×pH interactions are less significant ($P<0.05$).

Any phenomenon is the consequence of its causes, and if all of the causes are correctly defined, there should be no error introduced into the resultant data analysis. However, in reality, it is humanly impossible to eliminate all errors throughout the entire experiment. Therefore, errors would always exist, despite the fact that highly trained skill, appropriate knowledge, and use of instruments with high accuracy may diminish the size of error to the minimum.

The response curves illustrated in Figure 9.1a demonstrate a decrease in sediment as pH decreases and bottle size is increased. The increase in sediment by increasing WHard is nonsignificant (Table 9.1d:B). Significant interaction between pH and WHard is apparent in Figure 9.1b as the departure from parallelism of the response curves. The specific combination of these two factors has an effect on the response, which in this case is the amount of sediment. The formation of sediment cannot be simply stated as "the lower the pH and WHard, the smaller the sediment," which would be true in the case of no interaction. It is interesting to note that the WHard×pH interaction is significant although WHard itself is nonsignificant (Table 9.1d:B). This phenomenon is characterized by a great reduction of sediment at low pH of 3 when water is soft (W1). At pH 6, the sediment has increased even when the water is soft (Figure 9.1b).

There are many other fractional factorial methods aside from the Taguchi designs, such as those of Plackett and Burman;[3] many Latin square designs[3] are also similar as far as undetectability of interactions is concerned. Because of the importance of detecting factor–factor interactions, these ineffective designs are not discussed further in this chapter.

9.3.2.2 Three-Level Design Example

The two-level designs are useful for the purposes of screening influential factors, but they are not adequate to represent response surfaces because they cannot illustrate concave or convex

Table 9.1d ANOVA for Sediment Formation of a Fruit Drink Before (A) and After (B) Pooling Factors or Interactions Into the Error Term on the Basis of Mean Squares

Factor/ Interaction	Column	SS	Df	MS	F
(A)					
WHard	1	0.16	1	0.16	2.18
pH	2	0.59	1	0.59	8.05
PasTemp	6	0.21	1	0.21	2.86
StTemp	7	0.13	1	0.13	1.77
StTime	9	0.04	1	0.04	0.55
BotSize	5	0.27	1	0.27	3.68
BotShape	10	0.01	1	0.01	0.14
VarLoc	8	0.00	1	0.00	0.00
Sugar	11	0.14	1	0.14	1.91
WHard×pH	3	0.27	1	0.27	3.68
StTemp×StTime	14	0.09	1	0.09	1.23
BotSize×BotShape	15	0.04	1	0.04	0.55
[pH×PasTemp	4	0.05	1	0.05]	
Error	4, 12, 13	0.22	3	0.073	
Total		2.17	15		

$F(0.05)_{1,3}=10.1$.
Source: Nakai's lecture note.

Factor/Interaction	SS	Df	MS	F
(B)				
WHard	0.16	1	0.16	3.61
pH	0.59	1	0.59	13.32[a]
PasTemp	0.21	1	0.21	4.74
StTemp	0.13	1	0.13	2.94
BotSize	0.27	1	0.27	6.10[b]
Sugar	0.14	1	0.14	3.16
WHard×pH	0.27	1	0.27	6.10[b]
StTemp×StTime	0.09	1	0.09	2.03
Error	0.31	7	0.044	
Total	2.17	15		

$F(0.05)_{1,7}=5.59$, $F(0.01)_{1,7}=12.25$.
[a] $F>F(0.01)$, $[P<0.01]$.
[b] $F(0.01)>F>F(0.05)$, $[P<0.05]$.
Source: Nakai's lecture note.

surfaces for which at least three-level designs are required. Taguchi's $L_{27}(3^{13})$ design shown in Table 9.2a is most suitable for this purpose.

The general strategies and procedures previously described for two-level designs also apply to three-level designs, with two exceptions. First, the component calculation to find interaction columns is different from that used for two-level designs; i.e., "replace cubes (instead of squares) with 1." Second, an interaction appears in two columns rather than one column. For instance, to find the interaction of columns 10 and 12, the relation of column 10 $(ab^2c^2)\times$ column 12 $(ab^2c)\rightarrow a^2b^4c^3$ is used. By the "replace cubes with 1" rule, this becomes a^2b because $b^3c^3\rightarrow 1$; when squared, this becomes a^4b^2, which turns to ab^2 located at column 4. Meanwhile, the relation of $(ab^2c^2)\times(ab^2c)^2\rightarrow a^3b^6c^4$ becomes c (column 5). In summary, the interaction of column 10 and column 12 appears in columns 4 and 5.

The example used here involves an experimental design to study the effects of four factors and three two-factor interactions on the solubilization of wheat flour by heating with HCl (Table 9.2b). The solubility of the hydrolysates from each of the 27 experiments is measured as the responses. The experimental conditions and the solubility obtained are shown in Table 9.2c.

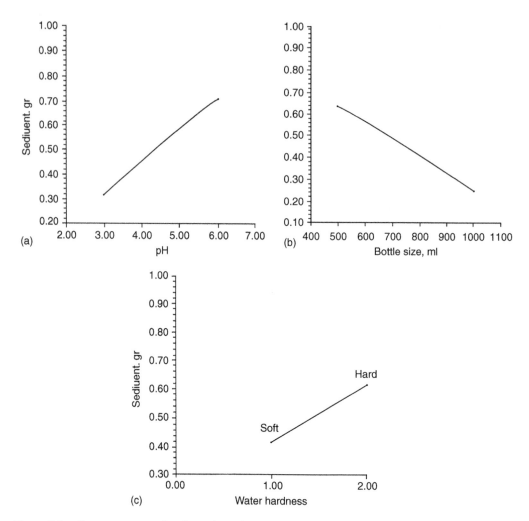

Figure 9.1a Response curves of sediment formation in the fruit drink as a function of (a) pH, (b) bottle size, and (c) water hardness.

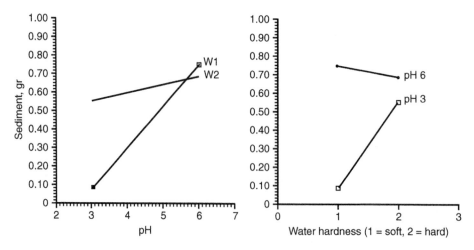

Figure 9.1b Interaction effects on sediment formation.

For ANOVA computation, the same equations (Equation 9.1 through Equation 9.3) described for the two-level design example can again be used; the only difference is that the treatment levels are 1–3 instead of 1–2. The ANOVA table is shown in Table 9.2d.

The response curves for the main effects and interactions are shown in Figure 9.2a and Figure 9.2b, respectively. The confidence limit (CL) is calculated using the following formula:

$$CL = \left[t(0.05) \text{ at } df_e / \sqrt{k} \right] \sqrt{S_e / df_e} \tag{9.4}$$

where $t(0.05)$ is the Student's t value at $P=0.05$, k is the number of data, and S_e is SS for error.

Table 9.2a $L_{27}(3^{13})$

Experiment No.	1	2	3	4	5	6	7	8	9	10	11	12	13
1	1	1	1	1	1	1	1	1	1	1	1	1	1
2	1	1	1	1	2	2	2	2	2	2	2	2	2
3	1	1	1	1	3	3	3	3	3	3	3	3	3
4	1	2	2	2	1	1	1	2	2	2	3	3	3
5	1	2	2	2	2	2	2	3	3	3	1	1	1
6	1	2	2	2	3	3	3	1	1	1	2	2	2
7	1	3	3	3	1	1	1	3	3	3	2	2	2
8	1	3	3	3	2	2	2	1	1	1	3	3	3
9	1	3	3	3	3	3	3	2	2	2	1	1	1
10	2	1	2	3	1	2	3	1	2	3	1	2	3
11	2	1	2	3	2	3	1	2	3	1	2	3	1
12	2	1	2	3	3	1	2	3	1	2	3	1	2
13	2	2	3	1	1	2	3	2	3	1	3	1	2
14	2	2	3	1	2	3	1	3	1	2	1	2	3
15	2	2	3	1	3	1	2	1	2	3	2	3	1
16	2	3	1	2	1	2	3	3	1	2	2	3	1
17	2	3	1	2	2	3	1	1	2	3	3	1	2
18	2	3	1	2	3	1	2	2	3	1	1	2	3
19	3	1	3	2	1	3	2	1	3	2	1	3	2
20	3	1	3	2	2	1	3	2	1	3	2	1	3
21	3	1	3	2	3	2	1	3	2	1	3	2	1
22	3	2	1	3	1	3	2	2	1	3	3	2	1
23	3	2	1	3	2	1	3	3	2	1	1	3	2
24	3	2	1	3	3	2	1	1	3	2	2	1	3
25	3	3	2	1	1	3	2	3	2	1	2	1	3
26	3	3	2	1	2	1	3	1	3	2	3	2	1
27	3	3	2	1	3	2	1	2	1	3	1	3	2
Component	a	b	a b	a b²	c	a	a c	b c²	a	a b² c²	b c²	a b² c	a b c²
Group	(a)		(b)						(c)				

Source: From Taguchi, G, *Design of Experiments*, Vol. 1, Maruzen, Tokyo, 1957.

Table 9.2b Wheat Flour Solubilization

Factor	Column	Interaction
Flour (F)	1 (a)	1×2=ab→3+4
HCl	2 (b)	2×5=bc→8+11
Heat treatment (H)	5 (c)	1×5=ac→6+7
Reductant (R)	9 (abc)	
Unused	10 (ab²c²), 12 (ab²c), 13 (abc²)	

The parallel curves for flour-HCl (Figure 9.2b) indicate no significant interaction between these two factors, and the effects of these factors can simply be stated as "the higher the HCl concentration, the higher the obtained solubility at all flour concentrations." However, the HCl× temperature-time interaction is significant. When autoclaved (120 °C for 15 min), almost the same high solubility was obtained at all HCl concentrations studied, but yielding solubility responses that are not parallel to those at other HCl×temperature-time combinations.

Table 9.2c Solubility of Acid-Treated Wheat Flour

Experiment	F	HCl	H	R	Solubility (%)
1	1	1	1	1	95.6
2	1	1	2	2	80.2
3	1	1	3	3	85.2
4	1	2	1	2	95.0
5	1	2	2	3	85.8
6	1	2	3	1	92.4
7	1	3	1	3	95.8
8	1	3	2	1	87.4
9	1	3	3	2	92.4
10	2	1	1	2	94.4
11	2	1	2	3	71.0
12	2	1	3	1	86.0
13	2	2	1	3	94.7
14	2	2	2	1	80.9
15	2	2	3	2	88.9
16	2	3	1	1	94.7
17	2	3	2	2	84.5
18	2	3	3	3	89.1
19	3	1	1	3	91.0
20	3	1	2	1	68.0
21	3	1	3	2	83.9
22	3	2	1	1	94.3
23	3	2	2	2	76.5
24	3	2	3	3	85.1
25	3	3	1	2	94.4
26	3	3	2	3	81.9
27	3	3	3	1	85.3

Table 9.2d   ANOVA of Acid Hydrolysates of Wheat Flour

Factor/interact ion	SS	Df	MS	F
Wheat flour	135.64	2	67.83	16.58[a]
HC1	152.91	2	76.46	18.69[a]
Temperature–Time	995.20	2	497.60	121.66[a]
Reductants	6.25	2	3. 13	0.77
Flour×HCl	1.60	4	0.40	0.10
Flour×T–T	38.04	4	9.51	2.33
HCl×T–T	88.79	4	22.20	5.43 [b]
Error	24.53	6	4.09	
Total	1442.96	26		

$F(0.01)_{2,6} = 10.43$, $F(0.05)_{2,6} = 5.14$, $F(0.01)_{4,6} = 9.15$, $F(0.05)_{4,6} = 4.53$.
[a] $P < 0.01$.
[b] $P < 0.05$.
Source: Nakai's lecture note.

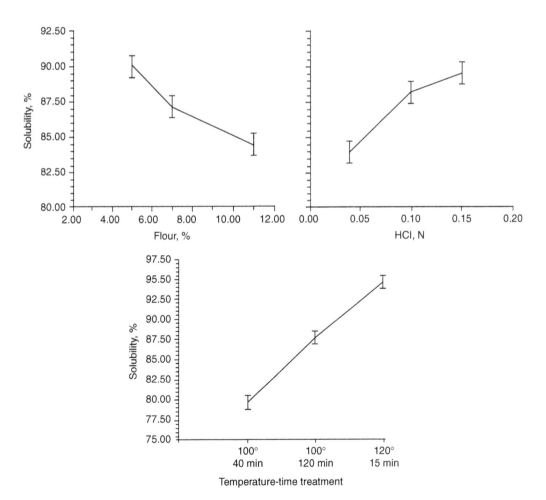

Figure 9.2a   Response curves of acid-solubilized wheat flour as a function of percent flour, HCl concentration and temperature-time treatment.

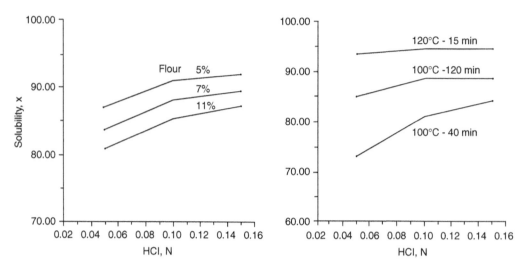

Figure 9.2b Response curves of acid-solubilized wheat showing significant (right side) and nonsignificant (left side) interaction between factors.

9.3.2.3 How to Select Designs

The orthogonal array of Taguchi[4] was constructed on the basis of a probabilistic rule that two factors, which are difficult to change the experimental conditions, should be assigned to two columns belonging to different groups in $L_{16}(2^{15})$ design. Assigning these two factors to columns in the same group is not recommended due to less chance of useful interaction effects compared to the main effects of the already assigned factors. Also, the pooling-up process of the resultant ANOVA, which is similar to the algorithm of backward stepwise multiple regression, may improve the screening efficiency of factors.

We have found that Taguchi's $L_{16}(2^{15})$ design is the most efficient and useful for general purposes, especially for factor screening. The three-level designs such as $L_{27}(3^{13})$ are valuable for approximating the response surfaces. More efficient designs than the $L_{27}(3^{13})$ design for the purpose of response-surface modeling will be discussed in the following section. Whether higher-order interactions, especially higher than second order, can be ignored as in the case of the Taguchi method[10] is debatable. In the multivariate era, especially in biotechnology, high-degree interactions may not be always ignored.

It was reported that the linear graphs of Taguchi had two disadvantages.[11] First, they do not identify unspecified interaction effects that are confounded with the main effects and the specified interaction effects. Second, they do not provide information on confounding relationships. The "interaction graphs" of Kacker[11] may circumvent these problems. According to the interaction graph for $L_{16}(2^{15})$, eight columns (1, 2, 4, 7, 8, 11, 13, and 14) can be used for assigning factors, whereas the rest (3, 5, 6, 9, 10, 12, and 15) are for assigning two-factor interactions or to be left blank for error SS computation. In the case of the example of fruit drink sediment in Table 9.1b, therefore, assigning factors to columns 5 (ac), 6 (bc), 9 (ad) and 10 (bd) may be a problem because the 2 (b)×7 (abc), 1 (a)×7 (abc), 1 (a)×8 (d), and 2 (b)×11 (abd) interactions, as well as 1 (a)× 11 (abd) and 2 (b)×8 (d) interactions, respectively, may confound with those four factors.

Our result using the Taguchi method identified significance of the 1×2 interaction appearing on column 3 (Table 9.1d:B and Figure 9.1b). In practice, columns 1 and 2 belong to different groups, i.e., group "a" and group "b," respectively, (Table 9.1a); and column 3 was left blank to calculate the 1×2 interaction effect. However, according to the interaction graph of Kacker, column 3 consisted of four interaction effects of 1×2, 4×7, 8×11, and 13×14. Because factors were not assigned to columns 4 and 13, the above-mentioned 4×7 and 13×14 do not eventually exist in this example, whereas the 8×11 interaction can also be considered nonexistent because of the near-zero sum of squares for VarLoc assigned to column 8. Therefore, significance in column 3 representing interactions should be mostly due to contributions of the interaction of factors assigned to columns 1 and 2, which belong to different groups. According to the interaction graphs of Kacker,[11] interactions consist of at least four interactions, thereby making it difficult to separate individual interactions. On the other hand, the Taguchi's linear graphs were made for detecting the most probable interaction effects, and the impact of the two disadvantages mentioned above may therefore often be negligible with no practical relevance, as illustrated in the fruit drink sediment example.

Recently, Ross[10] proposed new interaction tables by thoroughly covering possible interactions; for column 3, seven interactions may be confounded. He recommended a simpler digital series of 1, 2, 4, 7, 8, 11, 13, 14, (3, 5, 6, 9, 10, 12) for assigning more than eight factors. As the number of factors to be assigned increases, it is recommended to follow this order of selecting columns. For instance, if one has 6–8 factors to be assigned, a shorter digital series of 1, 2, 4, 7, 8, (11, 13, 14) is recommended. Column numbers in parentheses may be assigned in any order, whereas column numbers outside of the parentheses must be used for assigning factors first. Other unassigned columns are left blank to use in computation of interactions or error. Columns 1, 2, 4, 7, 8, 11, 13, and 14 all bear one or three letters as components: a, b, c, abc, d, abd, acd, and bcd, respectively. If we accept low probability of significance of three-factor interactions, assigning factors to these columns is reasonable. Columns 3, 5, 6, 9, 10, and 12, corresponding to ab, ac, bc, ad, bd, and cd, respectively, should be left blank for computing two-factor interactions. At least column 15 should be left blank for error computation. According to the two-level interaction table shown in Table 9.3, even factors assigned to one-letter columns 1(a), 2(b), 4(c), and 8(d), may each be confounded with seven two-level interactions each. Whether the two-level interactions adversely affect the main effects or not should be tested as was done in the ANOVA computation shown above.

Theoretically, a fractional factorial design cannot avoid the confounding of multiple interaction effects. However, it is worth noting that the great saving in the number of experiments that can be gained by sacrificing some information on the interactions using Taguchi's orthogonal array method is still extremely valuable. This is especially true in the case of factor screening as a prerequisite of the subsequent optimization. The important fact is that the controlling effects of main factors are always vulnerable to the influence of any two-factor interaction. No matter which design is used, the factors computed to be significant even after confounding with interaction effects should be truly significant. This is important for factor screening.

Table 9.3 Two-Level Interaction Table

Column no.	2	3	4	5	6	7	8	9	10	11	12	13	14	15
1	3	2	5	4	7	6	9	8	11	10	13	12	15	14
2	—	1	6	7	4	5	10	11	8	9	14	15	12	13
3	—	—	7	6	5	4	11	10	9	8	15	14	13	12
4	—	—	—	1	2	3	12	13	14	15	8	9	10	11
5	—	—	—	—	3	2	13	12	15	14	9	8	11	10
6	—	—	—	—	—	1	14	15	12	13	10	11	8	9
7	—	—	—	—	—	—	15	14	13	12	11	10	9	8
8	—	—	—	—	—	—	—	1	2	3	4	5	6	7
9	—	—	—	—	—	—	—	—	3	2	5	4	7	6
10	—	—	—	—	—	—	—	—	—	1	6	7	4	5
11	—	—	—	—	—	—	—	—	—	—	7	6	5	4
12	—	—	—	—	—	—	—	—	—	—	—	1	2	3
13	—	—	—	—	—	—	—	—	—	—	—	—	3	2
14	—	—	—	—	—	—	—	—	—	—	—	—	—	1

Source: Adapted from Ross, P. J., *Taguchi Techniques for Quality Engineering*, 2nd ed., McGraw-Hill, New York, 1996.

Table 9.4   Recent Applications of Experimental Designs in Food/Bio Processing

Method	Food/Bio Processes	Finding	Reference
Full factorial	MAP[a] of sponge cake	Minimum K sorbate to prevent fungal spoilage	12
Fractional factorial			
Plackett–Burman	Triglyceride GC of cocoa butter equivalents	Routine control of chocolate bars	13
Taguchi–Orthogonal	Red wine aroma	Macerating enzymes from transgenic wine yeast	14

[a] Modified atmosphere packaging.

Recent examples of applications of factorial and fractional factorial experimental designs in food/bio processes are shown in Table 9.4.

9.4   RESPONSE-SURFACE METHODOLOGY

9.4.1   Introduction

The basic procedure of the experimental designs is to compute ANOVA using factorial analysis. In the life sciences, fractional factorial designs are essential for efficient analysis in terms of costs and labor for biological experiments. After finding significant factors by these experimental designs, the subsequent aim becomes to find the optimum that appears on the response surfaces. Accordingly, the visualization of the curvature of response surfaces requires more response values per factor than provided by three-level experiments.

There have been two schools of thought with regard to approaches for approximating the response surface and finding the optimum, namely (1) extension and/or continuation of experimental designs followed by curve-fitting to illustrate the response surfaces, and (2) "evolutionary operation" and its derivatives to rotate the search toward the optimum during iterative search. The incipient design of methods in school 2 can be rather simple, and these methods are usually simpler in experimentation than those in school 1.

The earliest technique of RSM requires central composite designs to draw 2D or 3D response surfaces, modeled using quadratic and/or polynomial factorial equations. At the same time, mixture designs were developed to meet the constraint requiring that "the sum of components constituting a food formula is 1.0."

Subsequently, a stepwise approach to find the optimum was proposed, i.e., SSO (sequential simplex optimization), which belongs to school 2. However, to search for the global optimum, it was recommended to repeat the SSO process by initiating the search from different starting locations within the global search spaces. Under this circumstance, despite its speculative property and being basically inefficient in nature, a random search is the only powerful approach in terms of overall efficiency to target the global optimum.

We have proposed "RCO (random centroid optimization)"—a randomized version of SSO with an additional, characteristic 2D mapping—to approximate the true response surfaces. A 2D map is used to visualize the response surface for each factor because it is not feasible to visualize multivariate response surfaces with greater than three dimensions. Furthermore, the curve-fitting technology currently available is limited in ability to depict rugged, nonlinear global response surfaces.

Since our presentation of RCO, a variety of RCO versions have been developed, such as for the purposes of mixture designs, food formulation, container designs, site-directed mutagenesis, etc., with potential applications in many other areas of food and biological processes. These extended applications will be discussed later in this chapter.

9.4.2 Central Composite Designs

The minimum number of experiments necessary to estimate the parameters in a response-surface model depends on both the number of factors (k) and whether the model is expected to be first-order, second-order, or higher-order polynomial. For a model to account for a qth order effect in a factor, the experimental design must have $q+1$ different levels in that factor. For example, the experimental design for a first-order model with k factors could be either a 2^k full factorial design or a two-level fractional factorial design. However, for many life science experiments, the response is not first order, usually showing some curvature and factor interactions or second-order relationships. For such cases, although a 3^k full factorial design could be satisfactory to generate the necessary data to fit a second-order polynomial model, the number of experiments, n, becomes unrealistically large. For example, for $k=4$, then $n=81$. For this reason, central composite designs are recommended for RSM fitted to second-order polynomial models.

Central composite designs are a "composite" of a star design (with axial points at $\pm\alpha$), a 2^k factorial design (with design points at ±1), and a common center point of the two designs (at zero). The general form of α-values is:

$$\alpha = 2^{k/4}, \tag{9.5}$$

where k is the number of variables. The combination of the two designs leads to five levels of each factor, with $[-\alpha, -1, 0, +1,$ and $+\alpha]$ coordinates expressed as coded values. The minimum number of design points in a central composite design is 2^k+2k+1, where "1" is the center point that is usually conducted in replicate to provide an estimate of error. Thus, for example, the usual number of experiments for $k=2$, 3, and 4, with the recommended 5, 6, and 7 replicates of the center point, are 13, 20, and 31 experiments, respectively.

The example used here is the optimization of product yield by changing three factors, i.e., pH, heating temperature, and time, with α-value of 1.682 (Table 9.5 and Figure 9.3).

Table 9.5 Product Yield Experiments Using Three-Factor Central Composite Design

Experiment	PH	Temperature	Time	Yield (g)
1	−1	−1	−1	16.44
2	1	−1	−1	12.50
3	−1	1	−1	16.10
4	1	1	−1	6.92
5	−1	−1	1	14.90
6	1	−1	1	7.83
7	−1	1	1	19.90
8	1	1	1	4.68
9	−1.682	0	0	17.65
10	1.682	0	0	0.20
11	0	−1.682	0	25.39
12	0	1.682	0	18.16
13	0	0	−1.682	7.37
14	0	0	1.682	11.99
15	0	0	0	24.03[a]

[a] Average of hexaplicate: 22.22, 19.49, 22.76, 24.27, 27.88, and 27.53 to estimate error variability.

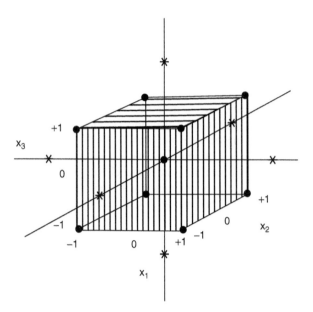

Figure 9.3 Central composite design for three factors. *α=1.682. (Adapted from Otto, M., *Analytical Chemistry*, Kellner, R., Mermet, J. -M., Otto, M., and Widmer, J. M., Ed., Wiley-VCH, Weinheim, pp 759–773, 1998.)

The limit level values corresponding to the coded values of -1.682 and $+1.682$ were assigned as follows: 4 and 9 (pH), 60° and 100 °C (Temp) and 1 and 10 min (Time). The prediction equation obtained by curve-fitting using multiple linear regression analysis was:

$$y = 24.04 - 4.72x_1 - 1.189x_2 + 0.229x_3 - 5.426x_1^2 - 0.884x_2^2 - 5.159x_3^2 - 1.674x_1x_2$$

$$- 1.146x_1x_3 - 0.971x_2x_3, \tag{9.6}$$

with $R^2=0.923$. By solving the simultaneous equations resulting from derivatization then equalizing to zero, the optimal scale or coded values were computed to be -0.392, -0.355, and 0.099 for x_1, x_2, and x_3, respectively. These values are equivalent to the factor conditions of pH 5.53 and heating at 75.8 °C for 5.76 min, with the estimated best yield of 25.19 g.

9.4.3 Mixture Designs

Mixture designs are a technique to find the combination of the food ingredients that yields the best quality of a food product. In the optimization of mixtures or food formulations, the sum of all the ingredients must always be one, or 100%. Mixture designs must therefore consider this constraint that changing the proportion of one component automatically leads to a change in the proportion of the other components. The most well-known designs for ingredient mixing are simplex-lattice designs.[15] To accommodate a polynomial equation, points are spread evenly over the whole simplex factor space and lattice; an ordered arrangement of points may have a special correspondence to a specific polynomial equation.

In this chapter, the simplest mixture design method, i.e., extreme vertices designs using limit or bound values (0 and 1.0) only, without selecting intermediate values (fractions such as 0.3), will be discussed. The limit values (bounds) of each ingredient are initially selected. Then, all possible combinations of the bounds for q ingredients are computed by selecting bound values of $q-1$ ingredients at a time. The value subtracted from 1.0 (100%) is assigned to the remaining qth ingredient, provided it is within its bounds; if not, this combination is withdrawn from the design. In theory, the maximum number of design points for a q-component design will be $q \cdot 2^{q-1}$. In practice, many points will be eliminated since the remaining qth component may exceed its bounds.

An example is shown in Table 9.6 for maximizing gel strength of fish cake (kamaboko) measured as the peak force of Instron trace (N). Surimi base, potato starch, water, and salt were the ingredients used, and their bounds were set as follows:

$$0.70 \leq \text{Base} \leq 0.85$$
$$0.10 \leq \text{Starch} \leq 0.20$$
$$0 \leq \text{Water} \leq 0.10 \tag{9.7}$$
$$0.02 \leq \text{Salt} \leq 0.03$$

For instance, a combination of vertex values of 0.70, 0.20, 0.1, and 0.02 for the four ingredients was eliminated from Table 9.6, as the total was $1.02 > 1.0$.

A quadratic model fitted to the data was:

$$y = 9.783x_1 - 508.93x_2 - 378.34x_3 - 3987.0x_4 + 629.55x_1x_2 + 265.92x_1x_3 + 3827.28x_1x_4$$
$$+ 1318.19x_2x_3 + 7177.21x_2x_4 + 4977.35x_3x_4, \tag{9.8}$$

with $R^2 = 0.981$, where y is peak force (N), x_1, x_2, x_3, and x_4 are base, starch, water, and salt, respectively. Note that the substitution using the composition constraint $x_1 + x_2 + x_3 + x_4 = 1.0$ leads to Equation 9.8 by eliminating the intercept and square terms from the full factorial equation exemplified in Equation 9.6.

Table 9.6 Extreme Vertices of Formulation of Kamaboko

	Base	Starch	Water	Salt	Gel strength (N) Measured	Predicted
1	0.78	0.10	0.10	0.02	7	6.2
2	0.77	0.10	0.10	0.03	6	6.2
3	0.78	0.20	0	0.02	15	12.7
4	0.77	0.20	0.10	0.03	14	14.6
5	0.70	0.18	0.10	0.02	13	8.7
6	0.70	0.17	0	0.03	10	10.7
7	0.85	0.13	0	0.02	18	15.7
8	0.85	0.12	0	0.03	20	15.3
9	0.70	0.20	0.08	0.02	10	9.4
10	0.70	0.20	0.07	0.03	12	12.5
11	0.85	0.10	0.03	0.02	16	13.0
12	0.85	0.10	0.02	0.03	17	13.0
Center	0.775	0.15	0.05	0.025	13	13.3

The simultaneous equations to be solved were:

$$x_1 + x_2 + x_3 + x_4 = 1.0,$$

$$0x_1 + 629.55x_2 + 265.92x_3 + 38.28x_4 + \lambda = -9.783,$$

$$629.55x_1 + 0x_2 + 1318.19x_3 + 7177.21x_4 + \lambda = 508.93, \tag{9.9}$$

$$265.92x_1 + 1318.19x_2 + 0x_3 + 4977.35x_4 + \lambda = 378.34,$$

$$3827.28x_1 + 7177.21x_2 + 4977.35x_3 + 0x_4 + \lambda = 3987.0.$$

where λ is a Lagrangian multiplier to match the number of independent variables (i.e., 5) to that of the above simultaneous equations, including the top equation for composition constraint. The greatest gel strength that can be expected was calculated to be 15.5 N using 74.2% surimi base, 19.8% potato starch, 1% water, and 4.9% salt. This predicted value is in fact lower than the actual value of 20 N measured for experiment 8 in Table 9.6. Furthermore, multiple linear regression analysis such as used here could not prevent the boundary violation for salt constraint of 2–3%. The well-known critical rule of polynomial surface derived from curve fitting is that the accuracy of approximation is restricted to values within the range of the search spaces. The salt content of 4.9% is obviously a violation of the search-space boundary, thus accurate prediction cannot be expected. These problems could be a drawback of curve-fitting optimization in comparison to an evolutionary operation approach.

9.4.4 Sequential Simplex Optimization

Sequential simplex optimization is an empirical feedback strategy of optimization, that basically searches for the optimum (either a maximum or a minimum point) by sequentially moving away from the worst response or point in a small search area, termed a *simplex*.

Simplex evolutionary operation owes most of its success to its ability to handle multiple variables including mutual interactions.[16] Thus, the probability of reaching the true optimum is diminished when the response surface contains a ridge (interaction), as illustrated by R in Figure 9.4. When the search starts from point A on the x-axis by moving parallel to the y-axis, point C appears to be the location of highest point on section profile S_a. Perpendicular to the x-axis, the section profile S_b also shows that point C on the xy plane is the highest point (so-called *ridge problem*). This figure shows that the iterative or evolutionary move of the triangular simplex finally has reached the summit, P. However, if the search instead reaches a local hilltop D (a local optimum), there is no chance to move away from it as long as only one-search-at-a time processes parallel to x and y axes are alternately used.

Spendley et al.[17] recommended the use of the following $(k+1) \times k$ matrix for initiation of SSO with k factors:

$$\begin{vmatrix} 0 & 0 & 0 & \cdots & 0 \\ p & q & q & \cdots & q \\ q & p & q & \cdots & q \\ \cdots & \cdots & \cdots & \cdots & \cdots \\ q & q & q & \cdots & p \end{vmatrix} \tag{9.10}$$

where $> p = 1/(k\sqrt{2})\{(k-1) + \sqrt{k+1}\}$ and $q = 1/(k\sqrt{2})\{\sqrt{k+1} - 1\}$. The p and q values for

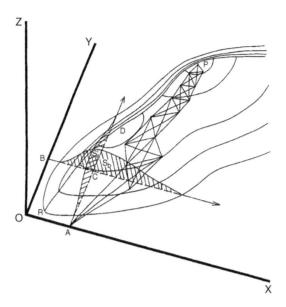

Figure 9.4 Hypothetical illustration of ridge problem, local optima, and simplex search for the global optimum.

$k = 2, 3, 4$, and 5 are $(0.966, 0.259)$, $(0.943, 0.236)$, $(0.926, 0.219)$ and $(0.912, 0.205)$, respectively, when the lower and upper limits of each factor are designated as 0 and 1, respectively.

After carrying out the experiments as designed above, the reflection vertex R is computed using $> R = \bar{P} + (\bar{P} - W)$ where W is the worst vertex and $> \bar{P}$ is the centroid vertex that is the average of all vertices within the simplex except W as shown in Figure 9.5 (right half). The response value of R vertex (Y_r) is compared as shown in the flow chart (left half of Figure 9.5). By replacing W with the best vertex selected from the flow chart, the second simplex is formulated. This cycle is iterated by moving to oblique directions (not vertical direction) until reaching a summit. An example of computing the second vertex in simplex search is shown in Figure 9.6.

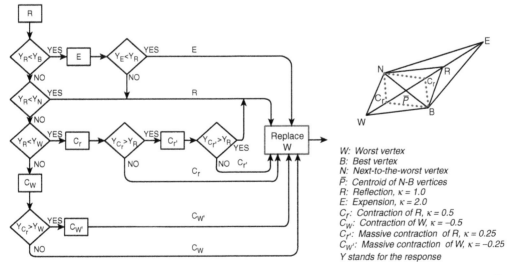

Figure 9.5 Flow chart of the simplex minimization (left side). Calculation of the new vertex using $> V = \bar{P} + (\bar{P} - W)$ (rightside).

Problem: The recovery of product A is optimized by varying the amount of additive B and heating temperature and time. The ranges set for search are 0–10% and heating at 70-80˚C for 2–12 min.

Initial simplex and recovery obtained:

Vertex	B (%)	Temp (˚C)	Time (min)	Recovery (%)	Inferiority order
1	0	70	2	60 W	W 1
2	9.4	72.4	4.4	72	W 3
3	2.4	79.4	4.4	75 B	W 4
4	2.4	72.4	11.4	68	W 2

Centrold $\bar{P}$= (9.4+2.4+2.4)/3

$\qquad$ = 4.73 $\qquad$ 74.73 $\qquad$ 6.73

Reflection R = 74.73+(74.73–70)

$\qquad$ = 9.5 $\qquad$ 79.46 $\qquad$ 11.5 $\qquad$ 80 (Vertex 5)

As response $Y_R > Y_B$, expansion is tried

Expansion E = $\qquad\qquad\qquad$ 6.73+2x(6.73–2)

$\qquad$ = 18.9 $\qquad$ 84.2 $\qquad$ 16.2 $\qquad$ 76 (Vertex 6)

As response $Y_E < Y_R$ R replaces W

Simplex 2 constructed:

Vertex	B	Temp	Time
5	9.5	79.5	11.5
2	9.4	72.4	4.4
3	2.4	79.4	4.4
4	2.4	72.4	11.4

Figure 9.6 Model computation of vertices in simplex search.

This computer program written in Quick Basic is available on request.[*]

9.4.5 Random-Centroid Optimization

Because SSO cannot guarantee finding the global optimum unless the simplex search cycle is repeated several times after shifting the search space, which will ruin the optimization efficiency, a new approach of RCO was proposed, with a random search replacing the simplex search.[18] Each search cycle in the RCO consists of a regulated random search, a centroid search and mapping, and requires much less number of experiments than that of the SSO. The number of experiments required is $2k \geq 9$ for the random search and generally less than four for the centroid search, compared to $(k + 1) \times n$ in the case of SSO, where n is the number of replications of the SSO cycle until an optimum is reached. Therefore, for optimization involving four factors, a maximum of 13 experiments is required for the first cycle of RCO, with a slightly reduced number of experiments for every succeeding cycle, whereas the SSO requires $5 \times n$ experiments, where n can frequently be greater than three for four factors, without any warranty of homing in on the global optimum.

[*] shuryo.nakai@ubc.ca.

The random design in each cycle of the RCO is regulated so that at least one experiment is located in each quartile within the search space to assure near-even distribution of the search. This requirement is in a good agreement with the above-mentioned symmetry rule for factorial designs. The maps drawn after every cycle are essential in deciding the direction of move for searching toward the global optimum. To extract the maximum available information for deciding on the search spaces of the subsequent cycles, a single- or double-factor ignoring process was introduced. An example of the effects of factor-ignoring on the mapping is shown in Figure 9.7,[18] which illustrates the results obtained with (right half) and without (left half) the factor-ignoring process. Clearer trends towards the potential location of the global optimum are observed after using the factor-ignoring processes. New search spaces different from the current spaces and with usually narrower search spaces are determined based on these maps and used as the new search spaces in the subsequent cycle.

The routine procedure for cycle 1 of RCO is as follows: (1) The factors and their search limit values are entered into the RCO software; (2) the random design of experiments is printed out, and (3) the experiments are carried out; (4) the response values of the experiments are recorded, whereupon

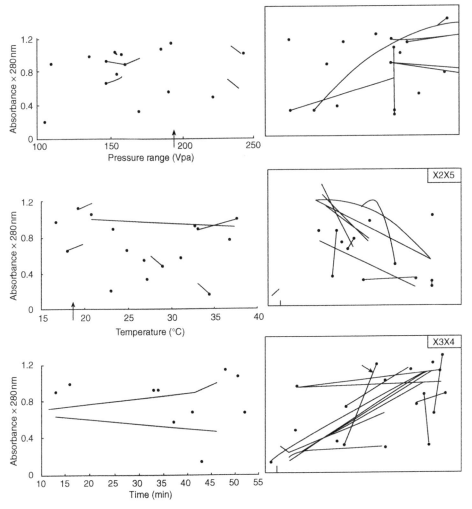

Figure 9.7   Effects of factor ignoring on maps of enzymatic activity (A_{280}). x_1: reactant concentration, x_2: NaCl, x_3: pressure, x_4: temperature, x_5: time. For instance, in the middle row for the response surface of temperature, NaCl and time were not ignored (left half) and ignored (right half). (From Nakai, S., Ogawa, M., Nakamura, S., Dou, J., and Funane, K., *International Journal of Food Property*, 6, 25–47, 2003.)

the centroid design is printed out; (5) after conducting the centroid experiments, the response values are reported and the summary data are printed; (6) the computer program immediately moves to the mapping process from which new search spaces are selected. This cycle is continued using the new search spaces in the subsequent cycles, i.e., cycles 2, 3, etc. The simultaneous shift included in the RCO package was useful in the model optimization trials for fine tuning of the search, but rarely used in general optimization purposes in food and bioprocess applications.

Based on countless trials for many model computations, the RCO was demonstrated to be successful in finding the global optimum; an example was shown in Nakai et al.[18] When the SSO was applied to a Fletcher-Powell arctangent model containing two local minima in addition to the deeper global optimum in between, the search was stalled at local optima five times during 20 optimization runs. Even if the SSO search did not stall, 60–180 iterations were needed to reach the global optimum, compared to less than 50 iterations without being stalled when the RCO was used. Furthermore, the RCO is advantageous with a possibility of making a new discovery due to its speculative nature. An example is shown later in the application of RCO to genetic modification studies of human cystatin C, when an unexpected amyloidosis problem resulting in insolubilization during isolation from yeast cells reduced the papain inhibitory activity of the isolated cystatin. RCG (RCO for genetics) found an explanation of the potential mechanism of this phenomenon.

RCO and RCG can be downloaded from ftp://ftp.agsci.ubc.ca/foodsci/. The package includes instructions on how to use the programs and also optimization model equations for practice training.

9.4.6 Extended Application of RCO

Extension of the applications of RCO for diverse research problems in food science and related disciplines are illustrated in the following examples.

9.4.6.1 Mixture Designs

The RCO approach was applied for a mixture design: the analysis of components of Raman spectra to best fit to the surface hydrophobicity of proteins determined by different fluorescent probes.[20] The component constraint of $\Sigma C_i = 1.0$, where i is the number of components C, was imposed. Using this analysis, the best fluorescent-probe method for protein surface hydrophobicity was, for the first time, selected based on quantitative assessment. In comparison to the conventional mixture-design technology, in which experimental designs become more complicated along with enhanced unreliability as the number of components increases, RCO could readily overcome this problem due to its evolutionary operation nature.

9.4.6.2 Food Formulation

Factorial analysis, the Taguchi method, the original RSM, mixture designs, and constrained simplex optimization were compared for food formulation purposes.[21] Also, formulation of a food with component constraints was successfully performed using RCO by incorporating a penalty function accommodating component constraints.[22]

9.4.6.3 Shape Design and Market Survey

Glass shape was optimized by changing elemental measures, specifically bottom width, height, top diameter of the hand-holding portion, and diameter of the round top of the glass shape,[23] which was then used as an example of market survey. The optimal preference of glass shape for various consumer groups was determined using this RCO shape optimization.

Broader application of RCO to art work or commercial designs may be possible. Multifactor optimization required for those applications, such as shape or color combinations, is feasible. Theoretically, there is no limit for the number of factors manipulated by the RCO program, which is suitable for multifactor optimization, although the RCO program currently available has a set limit to restrain optimization computation within the memory capacity of commercially available PC computers.

9.4.6.4 Site-Directed Mutagenesis

Optimization of glycosylation of human cystatin C was successfully conducted by application of RCG to obtain the best papain inhibitory activity as shown in Table 9.7, and to simultaneously

Table 9.7 Thermostability and Activity of Human Cystatin C Variants ($n=3$)

Variants (two sites)	$T_{1/2}{}^a$(°C)	Relative Activity[b]	Variants (one site)	$T_{1/2}{}^a$(°C)	Relative Activity[b]
WT[c]	68.2	1.00±0.11			
Cycle 1, random search					
L9F/C83H	67.5	0.49±0.12	L9F	ND[d]	ND
P6C/Q107L	72.4	1.19±0.13	P6C	72.8	1.15±0.03
G32Y/A95D	69.7	0.97±0.13	G32Y	71.4	0.97±0.12
D15L/G69I	63.3	1.03±0.11	D15L	68.3	0.27±0.03[e]
G4L/D40I	58.2	2.11±0.29[e]	G4L	59.4	0.13±0.01[e]
F29K/S44R	67.1	1.19±0.22	F29K.	ND	ND
E19R/H86F	68.0	0.76±0.08	E19R	68.1	0.47±0.08
L27S/A120R	62.0	0.55±0.12	L27S	60.8	0.73±0.13
S2K/V57F	68.0	1.68±0.06[e]	S2K	64.8	0.51±0.20
Cycle 1, centroid search					
D15P/H86I	71.5	2.65±0.30[e]	D15P	71.3	1.24±0.09
G12W/H86V	70.3	4.98±0.09[e]	G12W	66.4	0.51±0.08
A16W/R93V	58.4	0.28±0.09[e]	A16W	65.4	1.85±0.16[e]
L9V/C83W	ND	ND	L9V	70.5	1.00±0.02
Cycle 2, random search					
E20K/S115D	59.2	0.31±0.07[e]	E20K	59.0	2.39±0.43[e]
R8N/T71G	ND	ND	R8N	72.0	2.30±0.14[e]
D15S/C83K	ND	ND	D15S	68.3	2.01±0.18[e]
P13F/G1O8S	70.8	2.37±0.22[e]	PI3F	73.4	1.56±0.13
L9F/R93L	67.4	3.75±0.16[e]	L9F	ND	ND
E21S/L64G	ND	ND	E2IS	66.4	1.97±0.12[e]
R25Y/H90N	65.5	2.60±0.20[e]	R25Y	66.6	2.57±0.14[e]
Cycle 2, centroid search					
R8A/S98G	63.0	2.17±0.43[e]	R8A	73.1	3.17±0.07[e]
V10S/Y102G	59.7	3.60±0.35[e]	V10S	72.1	2.96±0.06[e]
V10S/R93G	67.9	4.50±0.07[e]			

Left half and right half are data for double mutants and single mutants, respectively.
[a] $T_{1/2}$(°C), half-life temperature of papain-inhibitory activity.
[b] Mean±S.D.
[c] WT, wild-type.
[d] ND, no data because of absence of detectable papain-inhibitory activity after purification process.
[e] Significant differences from WT using LSD analysis at $P<0.05$.
Source: From Ogawa, M., Nakamura, S., Scaman, C. H., Jing, H., Kitts, D. D., Dou, J., and Nakai, S., *Biochimica et Biophysica Acta*, 1599, 115–124, 2002.

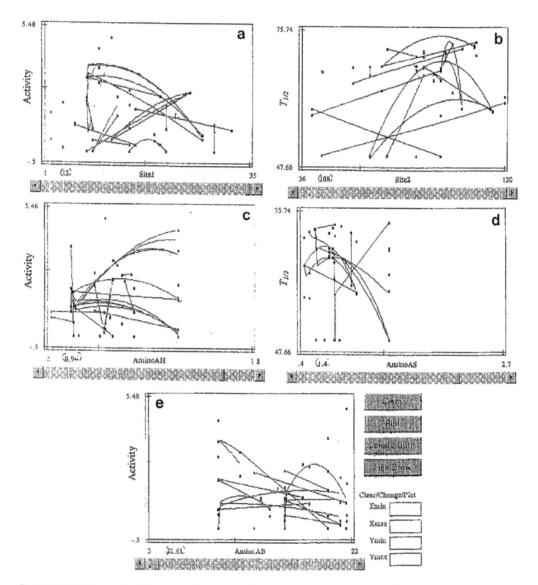

Figure 9.8 RCG maps of site-directed mutagenesis of human cystatin C. (a) and (b): different domains in the sequence. (c): α-helix propensity. (d): β-strand propensity. (e): bulkiness. (From Ogawa, M., Nakamura, S., Scaman, C. H., Jing, H., Kitts, D. D., Dou, J., and Nakai, S., *Biochimica et Biophysica Acta*, 1599, 115–124, 2002.)

avoid amyloidosis (insolubilization) during mutation and purification from *Pichia pastoris* cells.[24] An approximate five-fold enhancement of activity was obtained by mutant G12W/H86V compared to the recombinant control of wild-type enzyme.

Figure 9.8a and Figure 9.8b match the active site and binding site, respectively, with the substrate papain. Figure 9.8c and Figure 9.8d demonstrate a decrease in helix (increase in the scale) and strand, which may lead to a decrease in amyloidosis. Figure 9.8e shows that a decrease in bulkiness is favorable for the activity. The reaction mechanism of unexpected amyloidosis as suggested by maps is an advantage of RCG, which was performed without prior information on the amyloidosis.

Table 9.8   Recent Applications of Response-Surface Methodology in Food/Bio Processing

Method	Food/Bio Processes	Finding	Reference
Central composite designs	High pressure milk coagulation	Best gel strength	30
	Pork-batter gel	Best texture	31
Mixture designs	Gelled dairy deserts	Best texture	32
	High-melting milk fat	Higher melting characteristics	33
Simplex optimization	Wine blending	Best blend based on GC profiles	27
Random-centroid optimization	Food formulation	Better formulation than mixture designs	22
	Cooked Indica rice	Best taste	34
	Dietary-fiber bread	Best loaf volume	35

9.4.7   Multivariate Response Surfaces

For depicting response surface, multiple regression analysis (MRS) has been widely used as in the case of a variety of RSM. However, multicolinearity may hinder accurate prediction of objective functions by MRS.[25] This problem is best circumvented by relating the objective response variables to the principal components of the predictor variables. To analyze structure-activity relationships, principal components thus obtained were related to objective functions using artificial neural networks (ANNs).[26] The ANN packages, such as Statistica™, usually include a sophisticated graphic subroutine to depict predicted response surfaces.

9.4.8   SSO vs. RCO

Despite the advantages of RCO over SSO as discussed above, the SSO could be advantageous, especially in the case of automated optimization using computers. By using SSO as a subroutine, automated optimization of many analyses or processes could be performed. Optimization of GC patterns to obtain the best wine blending[27] and blending of GC fractions with concentrated fruit juice to best simulate fresh juice were realized. Use of RCO to do the same is difficult due to required human intervention in the response-surface mapping, but it may nonetheless be impossible. There have been many attempts at applying the computerized optimization, such as simulated annealing and genetic algorithms.[28] However, a user-friendly interface is not available to inexperienced users.[29]

As was previously mentioned, the manual simultaneous shift in the RCO program was used as a fine adjustment of search direction toward the global optimum. This process can be replaced by the SSO after adequately narrowing down the search spaces by bringing them to the neighborhood of the global optimum by first using RCO. This approach may be a good strategy in enhancing both the objectivity and the reliability for the global optimization.

Recent applications of RSM in food/bio processing are shown in Table 9.8.

GLOSSARY

Evolutionary operation (EVOP)   A group of algorithms to shift or rotate the initial design pattern to move toward the optimum. Typical examples are Box EVOP, rotating square EVOP, random EVOP, and simplex EVOP.[36]

Fractional factorial designs   In addition to computing effects of main factors, only selected interactions between factors are considered to reduce the total number of experiments to be carried out.

Full factorial designs   Effects of not only main factors but also interactions between all possible combinations of factors, including two-factor, three-factor, or even higher-level interactions are investigated.

Central composite designs   Typical designs for the original response surface methodology as shown in Figure 9.3.

Centroid   The original simplex optimization technique defines centroid as the average point of $n-1$ vertices after excluding the worst vertex (Figure 9.6). A similar definition is used in RCO.

Lagrangian multiplier   Supplemental/hypothetical independent variables when the number of experiments is less than the number of independent variables in linear regression analysis.

Mixture designs   In food formulation, ingredients are used as the independent variables in a similar form to the full factorial designs. However, the total contents of ingredients should be 100%; therefore, this equation should be included in multiple regression equations as a constraint. This is a typical case requiring a Lagrangian multiplier.

Plackett–Burman design   This fractional factorial design aims to define the effects of main factors only[3] without computing the effects of interactions.

Random-centroid optimization (RCO)   A regulated random design assisted by surface-response maps as human intervention effective for searching the global optimum.

RCO for genetic engineering (RCG)   RCO to apply for site-directed mutagenesis of peptides and proteins.

Response-surface methodology (RSM)   The original RSM use central composite designs for experiments, and then factorial (quadratic) multiple linear regression analysis to compute the correlation of factors with responses.

Sequential simplex optimization (SSO)   Spendley's matrix is used as a starting simplex that is then iteratively moved in different directions depending on larger (maximization) or smaller (minimization) response values.

Simplex   Pinnacles of vertices in SSO.

Spendley's matrix   Design of the starting simplex vertices as suggested by Spendley.[17]

Taguchi's orthogonal designs   One of the most efficient fractional factorial designs proposed by Genichi Taguchi.

REFERENCES

1. Myers, R. H. and Montgomery, D. C., *Response Surface Methodology: Process and Product Optimization Using Designed Experiments*, 2nd ed., New York: Wiley, 2002.
2. Khuri, A. I. and Cornell, J. A., *Response Surfaces: Designs and Analyses*, 2nd ed., New York: Marcel Dekker, 1996.
3. Otto, M., Optimization and experimental design, In *Analytical Chemistry*, R. Kellner, J.-M. Mermet, M. Otto, and J.M. Widmer, Eds., Weinheim: Wiley-VCH, pp. 759–773, 1998.
4. Taguchi, G., *System of Experimental Design: Engineering Methods to Optimize Quality and Minimize Costs*, New York: Unipub, 1987.
5. Cross, M. L., Stevenson, L. M., and Gill, H. S., Anti-allergic properties of fermented foods: an important immuno-regulatory mechanism of lactic acid bacteria?, *International Immunopharmacology*, 1, 891–901, 2001.
6. Giuliani, A., Benigni, R., Zbilut, J. P., Webber, C. L., Sirabella, P., and Colosimo, A., Nonlinear signal analysis methods in the elucidation of protein sequence-structure relationships, *Chemical Reviews*, 102, 1471–1491, 2002.

7. Box, G. E. P., Hunter, W. G., and Hunter, J. S., *Statistics for Experimenters: An Introduction to Design, Data Analysis, and Model Building*, New York: Wiley, 1978.

8. Taguchi, G., *Design of Experiments*, Vol. 1, Tokyo: Maruzen, 1957.

9. Blattner, P., *Excel 2000 Functions in Practice*, Indianapolis IN: Que, 2000.

10. Ross, P. J., *Taguchi Techniques for Quality Engineering*, 2nd ed., New York: McGraw-Hill, 1996.

11. Charteris, W., Taguchi's system of experimental design and data analysis: a quality engineering technology for the food industry, *Journal of the Society of Dairy Technology*, 45, 33–49, 1992.

12. Guyno, M. E., Marin, S., Sanchis, V., and Ramos, A., An attempt to minimize potassium sorbate concentration in sponge cake by modified atmosphere packaging combination to prevent fungal spoilage, *Food Microbiology*, 21, 449–457, 2004.

13. Gueon, F., Destouesse, S., Moustirats, J., Esclapez, M., Salagoity, M.-H., and Medina, B., Alternative method for the quantification by gas chromatography triacylglycerol class analysis of cocoa butter equivalent added to chocolate bars, *Journal of Agricultural and Food Chemistry*, 52, 2770–2775, 2004.

14. Gil, J. V. and Vallés, S., Effect of macerating enzymes on red wine aroma at laboratory scale: Exogenous addition or expression by transgenic wine yeasts, *Journal of Agricultural and Food Chemistry*, 49, 5515–5523, 2001.

15. Cornell, J. A., *Experiments with Mixtures: Designs, Models, and the Analysis of Mixture Data*, New York: Wiley, 1990.

16. Walters, F. H., Parker, L. R., Morgan, S. L., and Deming, S. N., *Sequential Simplex Optimization*, Boca Raton FL: CRC Press, 1991.

17. Spendley, W., Hext, C. R., and Himsworth, F. R., Sequential application of simplex designs in optimization and evolutionary operation, *Technometrics*, 4, 441–461, 1962.

18. Nakai, S., Dou, J., Victor, K., and Scaman, C. H., Optimization of site-directed mutagenesis. 1. New random-centroid optimization program for windows useful in research and development, *Journal of Agricultural and Food Chemistry*, 46, 1642–1654, 1998.

19. Nakai, S., Ogawa, M., Nakamura, S., Dou, J., and Funane, K., A computer-aided strategy for structure-function relation study of food proteins using unsupervised data mining, *International Journal of Food Property*, 6, 25–47, 2003.

20. Alizadeh-Pasdar, N., Li-Chan, E. C. Y., and Nakai, S, FT-Raman spectroscopy, fluorescent probe and solvent accessibility study of egg and milk proteins, *Journal of Agricultural and Food Chemistry*, 52, 5277–5283, 2004.

21. Arteaga, G. E., Li-Chan, E., Vazquez-Arteaga, M. C., and Nakai, S., Systematic experimental designs for product formula optimization, *Trends in Food Science & Technology*, 5, 243–254, 1994.

22. Dou, J., Toma, S., and Nakai, S., Random-centroid optimization for food formulation, *Food Research International*, 26, 27–37, 1993.

23. Nakai, S., Amantea, G., Nakai, H., Ogawa, M., and Kanagawa, S., Definition of outliers using unsupervised principal component similarity analysis for sensory evaluation of foods, *International Journal of Food Property*, 5, 289–306, 2002.

24. Ogawa, M., Nakamura, S., Scaman, C. H., Jing, H., Kitts, D. D., Dou, J., and Nakai, S., Enhancement of proteinase inhibitor activity of recombinant human cystatin C using random-centroid optimization, *Biochimica et Biophysica Acta*, 1599, 115–124, 2002.

25. Nakai, S. and Li-Chan, E., *Hydrophobic Interactions in Food Systems*, Boca Raton, FL: CRC Press, 1988.

26. Nakai, S., Chan, J. C. K., Li-Chan, E. C. Y., Dou, J., and Ogawa, M., Homology similarity analysis of sequences of lactoferricin and its derivatives, *Journal of Agricultural and Food Chemistry*, 51, 1215–1223, 2003.

27. Datta, S. and Nakai, S., Computer-aided optimization of wine blending, *Journal of Food Science*, 57, 178–182, 2005.

28. Pongcharoen, P., Hicks, C., Braiden, P. M., and Stewardson, D. J., Determining optimum genetic algorithm parameters for scheduling the manufacturing and assembly of complex products, *International Journal of Production Economy*, 78, 311–322, 2002.

29. Mardle, S. J., Pascoe, S., and Tamiz, M., An investigation of genetic algorithms for the optimization of multi-objective fisheries bioeconomic models, *International Transaction and Operation Research*, 7, 33–49, 2000.

30. Pandey, P. K., Ramaswamy, H. S., and St-Gelais, D., Effect of high pressure processing on rennet coagulation properties of milk, *Innovative Food and Emerging Technologies*, 4, 245–256, 2003.
31. Pietrasik, Z. and Li-Chan, E. C. Y., Response surface methodology study on the effects of salt, microbial transglutaminase and heating temperature on pork batter gel properties, *Food Research International*, 35, 387–396, 2002.
32. Verbeken, D., Thas, O., and Dewettinck, K., Textural properties of gelled dairy desserts containing κ-carrageenan and starch, *Food Hydrocolloids*, 18, 817–823, 2004.
33. Nor Hayati, I., Aminah, A., Mamot, S., Nor Aini, I., and Noor Lida, H. M., Physical characteristics of modified milk fat in high-melting fat preparation, *International Journal of Food Science and Nutrition*, 53, 43–54, 2002.
34. Nishimura, K., Goto, M., Imazuya, N., and Nakai, S., Optimum cooking conditions for Indica type rice by using random-centroid optimization, *Journal of Cookery Science Society of Japan*, 30, 9–16, 1997.
35. Kobayashi, Y., Nakai, S., Horimoto, Y., and Chiji, H., Optimum composition in ingredients for high-fiber bread containing wheat bran by random-centroid optimization, Bulletin of Fuji Womens' College Number 35, Series II, 31–37, 1997.
36. Low, C. W., Some techniques of evolutionary operation, *Transaction of the Institute of Chemical Engineering*, 42, T334–T344, 1964; Refer to 16.

CHAPTER 10

Multivariate Analysis

Ioannis S. Arvanitoyannis

CONTENTS

10.1   INTRODUCTION AND ELEMENTARY CONCEPTS

Over the last decade, multivariate data analysis techniques have seen wider acceptance and use in almost all fields of scientific inquiry. Although many reasons can be advocated, the following two represent the most important ones:

- The realization that in many scientific inquiries it is necessary to analyze the simultaneous relationships among three or more variables.
- The advent of the high speed computer with large storage facility and the development of readily available and easy to use software packages for implementing multivariate analysis (MVA).[1]

A modern PC can be assigned to exclusively accommodate most of the MVA packages. MVA includes all statistical methods that simultaneously analyze multiple measurements on each individual or object under investigation. Because MVA is defined as simultaneous analysis of more than two variables, many multivariate techniques are extensions of univariate analysis and bivariate analysis, i.e., simple regression is extended to the multivariate case to include several predictor variables. A confusion issue for the definition of MVA resides in its inconsistent use either for examining relationships between or among more than two variables, or only for problems in which all the multiple variables are assumed to have a multivariate normal distribution.[2]

Most commonly, multivariate statistics are applied for:[3]

- Developing taxonomies or systems of classification
- Investigating promising approaches to conceptualize or group items
- Generating hypotheses
- Testing hypotheses

Application of MVA instead of multiple univariate analyses is based on the following reasons:

- If there are numerous variables (for example, hundreds of species) multiple univariate analyses are tedious, and the problem of multiple comparisons may emerge.
- Multivariate methods take advantage of intercorrelations among variables.
- Multivariate methods can provide statistical tests of all response variables simultaneously.

In a case of few variables (e.g., 1–5), it is not advisable to employ MVA for data processing because there are minimum variables required for the proper functioning of MVA methods. However, the standard procedure is to use the simplest analysis possible to answer the question posed.

The key notion underlying the classification of MVA is the data matrix (Table 10.1). The table consists of a set of objects (p rows) and a set of measurements on the objects (q columns). The variables are characteristics of the objects and define the latter in any specific study. There are many descriptors by which associative data analyses methods can be classified:[4]

Table 10.1   A Typical Data Matrix

	Variables					
Objects	1	2	3	4	...	q
1	X_{11}	X_{12}	X_{13}	X_{14}	...	X_{1q}
2	X_{21}	X_{22}	X_{23}	X_{24}	...	X_{2q}
3	X_{31}	X_{32}	X_{33}	X_{34}	...	X_{3q}
4	X_{41}	X_{42}	X_{43}	X_{44}	...	X_{4q}
.	.	.	.	.	...	.
.	.	.	.	.	...	.
.	.	.	.	.	...	.
p	X_{p1}	X_{p2}	X_{p3}	X_{p4}	...	X_{pq}

- Purpose of study and types of assertions
- Research focus (i.e., whole bundle, specific variables)
- Specific partitioning of data matrix with regard to the type and number of variables subsets
- Number of variables in the partitioned subsets
- Type of association under study: linear, transformable to linear, or inherently nonlinear
- Measuring scales of variables: nominal, ordinal, interval, ratio, and mixed

Most decisions about associative data analysis are strongly related to the researcher's approach to variables and his particular interests: (1) nature and degree of association between two or more variables, (2) predicting the values of one or more criterion/predictor variables, and (3) assessment of statistical reliability of an association between two or more variables.

However, for classification purposes, MVA techniques can be broadly divided into two categories: dependence and inter-dependence methods. A dependence technique (DT) is defined as one in which a variable or set of variables is identified as the dependent variable to be predicted or explained by other variables known as *independent* variables. Multiple regression (MR) and analysis of variance (ANOVA) are representative DTs. The interdependence method aims to establish interrelationships among variables and includes factor analysis (FA), multidimensional scaling (MDS), and cluster analysis (CA). FA, MDS, and CA are not used for prediction purposes; they are intended to interpret the analysis output to opt for the best and most representative model. The DTs are classified on the basis of two characteristics: (1) the number of dependent variables (single-dependent and several-dependent, respectively), and (2) type of measurement scale (MS) employed by the variables. The MS can be either metric (quantitative/numerical) or nonmetric (quantitative/categorical) dependent variables.[2,3]

Dependence structures are usually analyzed by one of the following DTs; MR, discriminant analysis (DA), logit analysis, multivariate analysis of variance (MANOVA) and canonical correlation analysis (CCA). Provided that the variables to be analyzed have at least scale properties the following MA techniques are available; principal component analysis (PCA), FA, metric multidimensional scaling (MMS), CA, nonmetric multidimensional scaling (NMDS) and log-linear methods.[1]

The main assumptions for ensuring the proper function of MVA are:

- Normality (normal distribution of data)
- Homoscedasticity (dependent variables should exhibit equal levels of variance across the range of predictor variables)
- Linearity (linear associations between variables)

Table 10.2   Assumptions for Ensuring the Best Function of MVA in Conjunction with Testing Methodology and Corrective Actions

Assumption	Testing Method	Corrective Action
Normality	(1) Graphical (normal probability plot) (2) Statistical testing (rule of thumb based on skewness and kyrtosis values)	Data Transformation (inverse for flat distribution and square root/logarithm/inverse for skewed distribution)
Homoscedasticity	(1) Graphical test of multiple regression (2) Levene test for single metric variant (3) Box's *M*-test for many metric variables	
Linearity	(1) Examination of scatterplots of the variables (2) Running a simple regression analysis and to examine the residuals	(1) Transformation of one or both variables (squared, square root, inverse) (2) Creation of new variables (termed *polynomials*) to represent the nonlinear part of the relationship

Source: From Dillon, W. R. and M. Goldstein, *Multivariate Analysis*, New York: Wiley, 1984; Hair, J. F., Anderson, R. E., Tatham, R. L., and Black, W. C., *Multivariate Data Analysis*, Upper Saddle River, NJ: Prentice Hall, 1998; http://trochim.human.cornell.edu/tutorial/flynn/multivar.htm (accessed January, 2004).

All of the above-mentioned assumptions must be tested and corrective actions must be undertaken if deviations occur. Table 10.2 summarizes the current assumptions, testing methods, and suggested corrective actions for the appropriate functioning of MVA.

Multivariate data can be presented in two major ways: visualization and representation. The former includes univariate (histogram, boxplot), biavariate (scatter plot), trivariate (scatter plot [matrix], bubble plot, coplot), and multivariate (scatter plot matrix, Chernoff faces, stars); the latter comprises data matrix (*n* rows, *p* columns), mean vector (*p* rows, *l* columns), (co)variance matrix (*p* rows, *p* columns), and correlation matrix (*p* rows, *p* columns).[5]

10.2   PRINCIPAL COMPONENT ANALYSIS

The technique of PCA was first described by Pearson[6] who proposed this method as the correct solution to some of the biometricians' problems. However, the practical computing method for more than two or three variables was put forward only after three decennies (1933) by Hotelling.[7] The PCA or Karhunen–Loeve transformation is a mathematical expression of determining that linear transformation of a sample of points in *N*-dimensional space which exhibits the properties of the sample most clearly along the coordinate axe. Along the new axes, the sample variances are extremes (maxima, minima), and uncorrelated. The name comes from the principal axes of an ellipsoid that are the coordinate axes in question. The principal axes will include those along which the point sample has little or no spread (minima of variance) thus often showing linear interdependence in data.[8]

The main applications of PCA are to reduce the number of variables and to detect structure in the relationships between variables (classification of variables).[9] The assumptions, testing methodology and eventually required corrective actions are summarized in Table 10.3.

PCA transforms the original set of variables into a smaller set of linear combinations that account for most of the variance of the original set. The principal components are extracted so that the first principal component, denoted with $PC_{(1)}$ accounts for the largest amount of the total variation:

$$PC_{(1)} = w_{(1)1}X_1 + w_{(1)2}X_2 + \cdots + w_{(1)p}X_p. \qquad (10.1)$$

Table 10.3   Assumptions for Ensuring the Best Function of PCA in Conjunction with Testing Methodology and Corrective Actions

Assumption	Testing Method	Corrective Action
Multicollinearity	Visual inspection	Anti-image correlation matrix
Appropriateness	Bartlett test of sphericity	Examination of entire correlation
	Measure of sampling adequacy	matrix
Presence of underlying structure in the set of selected variables	No objective method is available	Up to the researcher
Mixing dependent and independent variables		Separate analyses
Homogeneous sample		

The second principal component $PC_{(2)}$ is an expression of the weighted linear combination of the observed variables, uncorrelated with the first linear combination, and accounting for the maximum amount of the remaining total variation (apart from $PC_{(1)}$). The zth principal component written as linear combination of the X's is as follows:

$$PC_{(z)} = w_{(z)1}X_1 + w_{(z)2}X_2 + \cdots + w_{(z)p}X_p. \tag{10.2}$$

One should bear in mind that $PC_{(z)}$ size is heavily influenced by the constraint shown in Equation 10.3:

$$w_{(z)1}^2 + w_{(z)2}^2 + \cdots + w_{(z)p}^2 = 1. \tag{10.3}$$

In a typical PCA, the following steps should be taken:[10]

- Coding the variables $X_1, X_2,...,X_p$ to have zero means and unit variances
- Calculation of covariance/correlation matrix
- Determining the eigenvalues $\lambda_1, \lambda_2,..., \lambda_p$ and their corresponding eigenvectors $a_1, a_2,...,a_p$
- Omission of the components accounting for a small proportion of the variation in the data

The implementation of PCA is mainly focused on the number of components to be retained that will account for most of the variability of the original data. Although there are several approaches, such as formal significance testing and graphical, none of them seems to enjoy universal approval. For instance, one could claim that only the components with associated eigenvalues higher than zero should be retained. Another option is to require the cumulative percentage of the variance extracted by successive components to be higher than a certain value (often 80%). A criterion originally put forward by Kaiser[11] consists of retaining components with eigenvalues greater than one. Cattell[12] introduced the scree test, in which the eigenvalues are plotted in successive order of their extraction. The number of components retained is given by the point at which the components curve above the straight line formed by the smaller eigenvalues. Cattell and Jaspers[13] suggested the number of factors to be taken as the number immediately before the straight line begins. However, this apparently simplistic approach becomes complicated when two or more breaks occur in the first half of the eigenvalues.[1]

10.3   FACTOR ANALYSIS

Although FA has a similar object to that of PCA—to describe a set of p variables $X_1, X_2, X_3,...,X_p$ in terms of a smaller number of indices or factors and clarify the relationship between these

Table 10.4 Objectives, Steps to be Undertaken and Main Underlying Errors in Implementation of FA

Objectives	Steps	Underlying Errors
Identification of a smaller set of uncorrelated variables for explaining the relationships among the original variables	Finding provisional factor loadings with PCA	Original variables are uncorrelated Infinite number of alternative solutions for the FA (which one to choose)
Determination of the number of underlying variables	Factor rotation (orthogonal or oblique) to get new factors more easily to interpret	How many factors to extract (subjective approach)
Interpretation of these new variables	Calculation of factor scores	Subjective interpretation
Evaluation of individuals and/or experimental units of the data set versus the new variables		Subjective evaluation
Employment of these new variables in other statistical analyses		

Source: From Manly, B. F. J., *Multivariate Statistical Methods*, Boca Raton, FL: CRC, 2000; Johnson, D. E., *Applied Multivariate Methods for Data Analysis*, Pacific Grove, CA: Duxbury Press, 1998; Caroll, I. D., Green, P. E., and Chaturvedi, A. *Mathematical Tools for Applied Multivariate Analysis*, San Diego, CA: Academic Press, 1997.

variables—there is one basic difference: PCA is not based on any statistical model, whereas FA employs a specific model.[10] Both objectives and underlying errors of FA are summarized in Table 10.4. Spearman[15] was the first to analyze the correlations between test scores of various types and suggested that most observed correlations could be described by the following simple model of the scores.

$$X_i = a_i F + e_i, \tag{10.4}$$

where X_i is the ith standardized score with a mean of zero and a standard deviation of one, a_i is a constant, and F is a factor value. Apart from the constant correlation ratios, the variance of X_i is given by

$$\text{var}(X_i) = \text{var}(a_i F + e_i) = a_i^2 + \text{var}(e_j). \tag{10.5}$$

A proper FA presupposes a PCA and employment of the first few principal components as unrotated factors. This approach is by no means accepted as correct, because e_i is the part of X_i that is specific exclusively to the ith test. Both F and e_i are independent and the variance of F is assumed to be unity:

$$a_i^2 + \text{var}(e_i) = 1, \tag{10.6}$$

where a_i is the factor loading. The resulting final equation for general FA model is

$$X_i = a_{i1} F_1 + a_{i2} F_2 + \cdots + a_{im} F_m + e_i, \tag{10.7}$$

where X_i is the ith score with mean zero and unit variance; $a_{i1}, a_{i2},...,a_{im}$ are the factor loadings for the ith test; $F_1, F_2,...,F_m$ are m uncorrelated common factors and e_i is factor specific and uncorrelated with any of the common factors. A correlation of test scores is only possible if they have high loading on the same factors.

A typical FA is usually carried out in three stages:

- Generation of a correlation coefficients matrix for all the variable combinations
- Extraction of factors from the correlation matrix (principal factors)
- Axes rotation, preferably with varimax that ensures the orthogonality of axes (independence among factors)

FA is considered to be less objective than the other statistical methods, especially if the sample is small and there are no replicates,[16] because it is a descriptive (qualitative) tool and not a quantitative one. Another handicap according to Seber[17] is that "even if the postulated factor model is correct then the chance of recovering with employing available methods is not high."

Generally speaking, FA is similar to PCA, except that instead of trying to account for as much of the total variance as possible, only correlations between variables are of interest as reflecting putative underlying causes or factors.

10.4 DISCRIMINANT ANALYSIS

Discriminant analysis (DA) involves deriving linear combinations of the independent variables that will discriminate between the a priori defined groups such that the misclassification error rates are minimized. The latter is possible through maximization of the between-group variance relative to the within-group variance. DA is based on a scoring system that assigns a score to each individual or object in the sample that is essentially a weighed average of the individual's or object's values on the set of independent variables.[1] The basic purpose of DA is to estimate the relationship between a single nonmetric (categorical) dependent variable and a set of metric independent variables, in the general form:[2]

$$Y(\text{nonmetric}) = X_1 + X_2 + \cdots + X_n(\text{metric}). \tag{10.8}$$

The linear combination for DA, also known as the *discriminant function*, is derived from an equation of the following form

$$Z_{mn} = a + w_1 X_{1n} + w_2 X_{2n} + \cdots + w_p X_{pn}, \tag{10.9}$$

where Z_{mn} is the discriminant z score of discriminant function m for object n, a is the intercept, w_i is the discriminant weight for independent variable i and X_{pn} stands for the independent variable p for object n.

As with rest of multivariate analyses there are several assumptions, one should bear in mind, closely related with the objectives and testing methods (of the assumptions) of DA (Table 10.5).

There are three types of DA: direct, hierarchical, and stepwise. In direct DA, all the variables enter the equations at once; in hierarchical DA, they enter according to a schedule set by the researcher; and in stepwise DA, statistical criteria alone determine the order of entry. In most analyses, the researcher has no reason for giving some predictors higher priority than others. The third (stepwise) method, therefore, is the most generally applicable.[18]

A frequently employed corrective action towards improving the interpretability of DA solutions is rotation. The latter leads to the following results:[19]

- Retention of total discriminatory power
- Groups relative position maintenance
- Rotated standardized discriminant coefficients and loadings coming closer to zero
- More even distribution of discriminating power across functions

Assigning of ungrouped individuals to groups is another crucial issue of DA. Several computer programs allow the input of data values for a number of individuals for which the true group has not been clarified. Thus, it is possible to assign these individuals to the group they are closest to, in terms of Mahalanobis distance, assuming they come from one of the m sampled groups. However, it

Table 10.5   Objectives, Assumptions, and Testing Methods of DA

Objectives	Assumptions	Testing Methods
Identification of statistically significant differences between average score profiles for a priori defined groups	Normality of independent variables	Calculation of discriminant Z scores (metric variables) Simultaneous estimation Stepwise estimation (large sample of variables)
Finding the independent variables most accounting for the differences	Linearity of relationships	Evaluation of group differences by means of centroids (Mahalanobis D^2 measure)
Opting for procedures for proper object classification based on the scores (independent variables)	Lack of multicollinearity (data set problem)	Group membership prediction accuracy assessment with graphical display
Determining the number and composition of discrimination dimensions	Equal dispersion matrices (variance and covariance matrices)	Box's M Profiling group differences on the independent variables Split-sample/cross validation techniques

Source: From Dillon, W. R. and Goldstein, M., *Multivariate Analysis*, New York: Wiley, 1984; Hair, J. F., Anderson, R. E., Tatham, R. L., and Black, W. C., *Multivariate Data Analysis*, Upper Saddle River, NJ: Prentice Hall, 1998; Hawkins, D. M., *Technometrics*, 23, 105–110, 1981; Durbin, J., *Journal of the American Statistical Association*, 1, 279–290, 1973.

is not known whether the assignment was correct. The errors that occur in the allocation of individuals from known groups indicate the assignment-process accuracy.[10]

10.5   CANONICAL CORRELATION ANALYSIS

Canonical correlation analysis (CCA) is a useful and powerful technique for exploring the relationships among multiple dependent and independent variables. The technique is primarily descriptive, although it may also be used for predictive purposes.[2] Its main advantage resides in its placing the fewest restrictions on working data contrary to other methods considered as leading to better quality results due to the many restrictions imposed.[20] CCA stands for the most generalized member of the family of multivariate statistical techniques and is strongly related to other dependence methods such as DA and FA. The general equation describing the CCA is:

$$Y_1 + Y_2 + \cdots + Y_n \text{(metric, nonmetric)} = X_1 + X_2 + \cdots + X_n \text{(metric, nonmetric)} \qquad (10.10)$$

One should bear in mind that there are several limitations that might significantly impact the CCA results:[2]

- CCA expresses the variance shared by the linear composites and not extracted from the variables.
- Canonical weights are subject to instability.
- Canonical weights are derived to maximize the correlation between linear composites, not the variance extracted.
- Canonical variates interpretation is difficult since no further aids are available such as rotation.
- Meaningful relationships are difficult to be established between dependent and independent variables due to lack of appropriate statistical methodology.

Table 10.6 Objectives, Assumptions, and Testing Methods of CCA

Objectives	Assumptions	Testing Methods
Determining whether two sets of variables are independent of one another	Linearity of correlations	Creation of two samples and performing CCA on both
Determining magnitude of relationships between two sets	Linearity of relationship	Comparison for similarity of canonical functions
Deriving a set of weights both per dependent and independent variable to correlate linear combinations per set	Multivariate normality	Sensitivity assessment of results by removing dependent or independent variable
Interpretation of dependent and independent variables relationships	Sample size. It has been recommended at least 20/40–60 times as many cases as variables for estimating one or two canonical roots, respectively	Estimation of numerous canonical correlations by removing a different dependent or independent variable
Measurement the relative contribution of each variable to the canonical function	Outliers can strongly affect the magnitudes of correlation coefficients	Application of Jackknife statistic in CCA to minimise errors

Source: From Dillon, W. R. and Goldstein, M., *Multivariate Analysis*, New York: Wiley, 1984; Hair, J. F., Anderson, R. E., Tatham, R. L., and Black, W. C., *Multivariate Data Analysis*, Upper Saddle River, NJ: Prentice Hall, 1998; Gray, H. L. and Schucany, W. R., *The Generalized Jackknife Statistic*, New York: Marcel Dekker, 1972; Mosteller, F. A. and Tukey, J., *Handbook of Social Psychology*, Vol. 2, 2nd ed., Lindzey, G. and Aaronson, R., Eds., Reading, MA: Addison-Wesley, 1968.

An approach to restrict the chance of misinterpretation is outlined in the following procedure:[1]

- A considerable amount of variance in the criterion set is shared with the predictor set
- Inspection of the among set correlations
- Comparison of canonical weights and loadings per pair of canonical variates with regard to algebraic sign and magnitude
- Implementation of cross-validation in case instability is detected

A synopsis of the major objectives and assumptions in conjunction with the validity testing methods of CCA are summarized in Table 10.6. As a concluding remark about CCA it can be said that its main drawback resides in interpretation of how strongly the two sets of variables are practically related.

10.6 CLUSTER ANALYSIS

Cluster analysis (CA) is the name for a group of multivariate techniques whose primary purpose is to group objects based on the characteristics they possess. CA classifies objects so that each object is very similar to others in the cluster based on some predetermined selection criteria. The resulting clusters are expected to show high internal (within-cluster) homogeneity and high external (between-clusters) heterogeneity. Because CA is a data reduction technique it is important to visualize its relationship with another similar technique of the same family already presented: DA. The main difference is that, in DA, there is an a priori assumption that the groups are known; i.e., all the observations are assumed to be correctly classified. Therefore, in DA, one starts with well-defined groups and attempts to discover how these groups differ; in CA, in contrast, one begins with groups that are initially undifferentiated and attempts to find whether a given group can be partitioned into clearly differing subgroups. Many disciplines have their own terminology for CA; these are topology (construction), grouping, classification (analysis), numerical taxonomy, segmentation analysis, and Q-analysis.[1,2,14]

The first step in CA is the establishment of the similarity or distance matrix. This matrix is a table in which both rows and columns are the units of analysis and the cell entries are a measure of similarity or distance for any pair of cases.[21] The clustering algorithms are classified into two general categories; hierarchical and nonhierarchical.

10.6.1 Hierarchical Cluster Analysis

Hierarchical cluster analysis (HCA) is a statistical method for identifying relatively homogeneous clusters based on measured characteristics. Hierarchical procedures are distinguished as agglomerative or divisive. In the former, each object stands for its own cluster and these clusters are further combined into aggregates. The graphical representation of the agglomerative procedure is a tree diagram or dendrogram. In the divisive method, one starts with a large cluster that gradually breaks down to smaller clusters. The most popular agglomerative algorithms and their cluster criteria are summarized in Table 10.7.

10.6.2 Hierarchical Divisive Methods

Hierarchical divisive (HD) methods begin with splitting the total number of objects into two groups, followed by further splitting. This initial split, which is of considerable importance for the

Table 10.7 Most Commonly Used Agglomerative Algorithms and Cluster Criteria for Cluster Development

Agglomerative Algorithm	Cluster Criterion	Weak Point	Strong Point
Single linkage or nearest neighbor approach	Minimum distance or the closest single pair	"Snaking" effect	(1) "Correctedness" maximisation of pair of clusters (2) Fewer clusters than other methods
Complete linkage or the furthest neighbor or diameter method	Maximum distance or the distance between their two furthest members	Refers to a single pair	(1) Intra-cluster distances minimisation (2) Compact cluster formation
Average linkage	Average distance from all individuals in one cluster to all individuals in another	Cluster production with approximately the same variance	Produced hierarchy is the same with the single or complete linkage algorihm
Centroid method	The squared Euclidian distance between the cluster means	Any difference in the distances between the centroids of pairs merged consecutively	Less affected by outliers than other hierarchical methods
Ward's method or minimum variance method or error sum of squares method	Square of the distance between the cluster means divided by the sum of the reciprocals of the number of points within each cluster	Production of clusters with approximately the same number of observations	Cluster combination with a small number of observations

Source: From Dillon, W. R. and Goldstein, M., *Multivariate Analysis*, New York: Wiley, 1984; Hair, J. F., Anderson, R. E., Tatham, R. L., and Black, W. C., *Multivariate Data Analysis*, Upper Saddle River, NJ: Prentice Hall, 1998; Johnson, D. E., *Applied Multivariate Methods for Data Analysis*, Pacific Grove, CA: Duxbury Press, 1998.

rest of the procedure, can take place according to two methods: (1) a splinter-average distance (SAD) method,[22] and (2) automatic interaction detection (AID).[23] The SAD method is based on calculating the average distance of each object to objects in a splinter group and the average distance of these same objects to other individuals in this group. The process starts by splitting out the most remote object (object of greatest distance). The splitting continues with the next object being closer to the splinter group than the main group. AID sequentially divides a total sample into mutually exclusive subgroups through a series of binary splits. Each split is determined by selecting a predictor value and its categories that maximize the reduction in the unexplained variation in the dependent variable. The final result is one group of a low criterion score and another of a high criterion score.[1]

10.6.3 Nonhierarchical Clustering Methods

Nonhierarchical clusterig (NHC) methods (k-means clustering) select an initial set of cluster seed points around which further clusters are built by assigning every point in the data set to its closest cluster seed according to the dissimilarity measures.[14] The assignment of clusters follows one of the following three approaches:[2]

1. Sequential threshold method: inclusion of all objects within a prespecified distance from an initially selected seed
2. Parallel threshold method: simultaneous selection of several cluster seeds and assignment within threshold distance to the nearest seed
3. Optimisation method: similar to the other two but more flexible due to the induced optimisation that is reassignment of objects

Table 10.8 compares the advantages and disadvantages of both hierarchical and nonhierarchical clustering methods.

Some further problems of NHC are the selection of cluster seeds and the initially required prevision of the number of clusters to exist at the end of the process. Although NHC is one of the

Table 10.8   Advantages and Disadvantages of Both Hierarchical and Nonhierarchical Clustering Methods

	Advantages	Disadvantages	Corrective Actions
Hierarchical	Fast, less computer time is required	Misleading due to persisting early combinations	—
	Effective as exploratory (first part of a joint venture with NHC)	Considerable impact of outliers	Use NHC
		Not amenable to analyse very large samples	Use NHC
		In case of random sampling, problem of "representativeness"	Avoid it
Nonhierarchical	The results are less susceptible to:		
	Outliers	Ability of researcher to select seed points	Researcher's training
	Distance measure used	Many alternative solutions	Validation is required
	Inclusion of irrelevant variables	Final result is greatly influenced by the order of data appearance	

Source: Hair, J. F., Anderson, R. E., Tatham, R. L., and Black, W. C., *Multivariate Data Analysis*, Upper Saddle River, NJ: Prentice Hall, 1998; Johnson, D. E., *Applied Multivariate Methods for Data Analysis*, Pacific Grove, CA: Duxbury Press, 1998.

continuously gaining ground methods in terms of its acceptability among scientists in view of its major and undoubtful advantages, a potential combination of both hierarchical and nonhierarchical is bound to gain the benefits of each.[24]

10.7 APPLICATIONS OF MULTIVARIATE METHODS

Multivariate analysis has been effectively used in authentication and classification of food products such as meat and dairy products. The first step of the analysis is to obtain reflectance and transmittance spectra of food products using visible or mid- or near-infrared (NIR) spectroscopy. The spectrum is usually acquired over a wide range of wavelength. These spectra are then classified using one of the interdependence methods. Sensory properties, textural images, electronic-nose performance, and physicochemical parameters (pH, color, water holding capacity (WHC) soluble protrin, electrical conductivity etc.) have been analyzed using multivariate analysis. Data obtained from different analytical instruments such as nuclear magnetic resonance spectrometers, differential scanning calorimeters, thermogravimetric analyzers, high-performance liquid chromatographs, gas chromatographs, and Fourier-transform infrared spectroscopes have also been examined using multivariate analysis for purposes of classification and quality control of food products. Table 10.9 gives a synopsis of statistical and screening methods for authenticating meat and dairy products.

10.7.1 Meat and Meat Products

The feasibility of using mid-infrared spectroscopy for addressing certain authenticity problems with selected fresh meats (chicken, turkey, and pork) has been investigated.[25] Preliminary analyses for the detection of frozen/thawed meat and semi-quantitative analysis of meat mixtures were reported. Minced chicken, pork, and turkey meat were identified by their infrared spectra. It was also possible to differentiate between fresh and frozen/thawed samples. Moreover, mid-infrared spectroscopy allowed the semi-quantitative determinations of the levels of pork and turkey mixed with chicken. The experimental parameters were optimized and, although the models used were not over-fit, no independent test was found within the frame of a full DA, thus showing that the results cannot be considered conclusive.[25]

Visible and NIR reflectance spectra (400–2498 nm) of homogenized meat samples (chicken, turkey, pork, beef, and lamb) were collected. The researchers attempted to classify the spectra into individual species using factorial discriminant analysis (FDA), soft independent modeling of class analogy (SIMCA), K-nearest neighbor analysis, and discriminant partial least squares (PLS) regression. Optimum accuracy was achieved by investigating a variety of wavelength ranges and data pre-treatments. Particular difficulty was encountered in distinguishing between chicken and turkey. Therefore, models were initially developed using five separate meat classes and again using four groups, with chicken and turkey being amalgamated into a single class. In a four-group classification, the best models produced between 85 and 100% correct identifications, whereas employment of five groups resulted in lower rates. FDA and PLS discrimination generally resulted in the best accuracy rates whereas SIMCA exhibited the poorest classification performance.[26]

Spectra of raw pork, turkey, and chicken meat were recorded in the visible, near and mid-infrared ranges. Discriminant models were initially developed separately in the mid-IR and the visible–NIR regions. A number of discrete regions of the NIR spectra were analyzed. The best predictive model achieved using mid-IR spectra correctly classified 86.5% of test samples; for visible–NIR data, the optimum classification of 91.9% was achieved in the 400–1100 nm

Table 10.9 Multivariate Analysis Methods Employed for Detecting Meat and Dairy Products Authentication

Statistical Method	Product	References
Principal component analysis	Chicken, turkey, pork	25
	Beef	32, 34, 45, 46, 55
	Pork	63, 64, 67, 68, 70, 71
	Poultry	77, 78, 84
	Lamb	85
	Dry cured hams	86, 87
	Fermented sausages	88
	Frankfurters	89, 90
	Salami	91, 92
	Cavourmas	93
	Ham	94
	Sausages	95–97
Partial least-squares regression	Chicken, turkey, pork	25, 26
	Beef	33, 37, 40–42, 46, 52
	Poultry	76, 81
	Dry cured hams	86, 87
	Dried sausages	98
	Sausages	95, 99
Factorial discriminant analysis	Chicken, turkey, pork, beef, lamb	26
	Beef	36, 37
SIMCA	Chicken, turkey, pork, beef, lamb	26
	Beef	37, 38
Discriminant analysis	Pork, turkey, chicken	27
	Beef	31, 56
	Pork	56, 57, 60, 62
	Poultry	56, 73, 83
	Dry cured ham	87
	Frankfurters	100
	Ham	101
Neural networks	Beef	29, 39
	Pork	61
	Poultry	80, 82
	Fermented sausages	88
Factor analysis	Beef	30
	Ham	101
Evolving factor analysis	Beef	30
Principal component regression	Beef	31, 42, 45
	Poultry	77, 81
	Fermented sausages	88
Multivariate discriminant analysis	Beef	32
Univariate analysis	Beef	33, 48
Generalised procrustes analysis	Beef	35
Canonical discriminant analysis	Beef	38
Stepwise multiple linear regression	Poultry	80
	Beef	38
Canonical variate analysis	Beef	41
	Beef	42
Locally weighted regression	Beef	43, 44, 47
Multiple regression analysis	Lamb	85
	Pork	71
Simple regression	Beef	44, 46, 53, 59, 66
Univariate analysis	Beef	48
Multivariate analysis	Beef	48–52
	Pork, ground meat	28, 69

continued

Table 10.9   Continued

Statistical Method	Product	References
ANOVA	Beef	54
	Pork	67
	Poultry	75
	Cecina	102
	Dried sausages	98
	Dry fermented sausages	103
	Frankfurters	103, 104
	Luncheon meat	105
	Ham	94
	Sausages	95, 106
Principal factor analysis	Beef	54
MANOVA	Pork	58
General least-squares analysis of variance	Pork	58
	Pork	65
Cluster analysis	Poultry	74, 79
Variable cluster analysis	Poultry	72
Multiple linear regression	Dried sausages	98
	Dairy Products	
Principal component analysis	Milk/cow	126–131
Discriminant factorial analysis	Milk/cow	132
Principal component analysis Hierarchical Cluster analysis	Cheddar type cheese	133
Principal component analysis	Gouda type cheese	134
ANOVA	Serra da estrella cheese	135
Principal component analysis	Danbo cheeses	136
Principal component analysis	Belgian cheeses	137
Discriminant analysis	Commercial strawberry yoghourt	119
Partial least squares Principal component analysis	Traditional yoghourt	120
Principal component analysis Generalised procrustes analysis Principal component analysis Cluster analysis	Vanilla ice cream	124

Source: From Spanier, A. M., Stangelo, A. J., and Shaffer, G. P., *Journal of Agricultural and Food Chemistry*, 40(9), 1656–1662, 1992; Djenane, D., Sanchez-Escalante, A., Beltran, J. A., and Poncales, P., *Food Chemistry*, 76, 407–415, 2002; Bertram, H. C., Engelsen, S. B., Busk, H., Karlsson, A. H., and Andersen, H. J., *Meat Science*, 66(2), 437–446, 2004; Kaneki, N., Miura, T., Shimada, K., Tanaka, H., Ito, S., Hotori, K., Akasaka, C., Ohkubo, S., and Asano, Y., *Talanta*, 62(1), 215–219, 2004; Vainionpää, J., Smolander, M., Hanna-Leena Alakomi, H.-L., Ritvanen, T., Rajamäki, T., Rokka, M., and Ahvenainen, R., *Journal of Food Engineering*, in press; Arvanitoyannis, I. S. and van Houwelingen-Koukaliaroglou, M., *Critical Reviews in Food Science and Nutrition*, 43(2), 173–218, 2003.

wavelength range. Combined visible, NIR and mid-IR data yielded the most accurate classification rate of 94.6%.[27]

Multilayer feed-forward neural networks were suggested to capture the nonlinearity between the system inputs and outputs to predict meat quality with the textural features from the ultrasonic elastograms. The efficiency of the training processes and the generalization of the networks using the gradient descent and Levenberg–Marquardt algorithms in back propagation has been investigated. In the case of difficult convergence in the gradient descent algorithm, the Levenberg–Marquardt algorithm converged effectively. The Levenberg–Marquardt algorithm modeled output variation accounting and network generalization better. Weight decay was further used in the Levenberg–Marquardt back propagation to improve the generalization of the network models. The leave-one-out procedure became part of every training process to ensure sufficient modeling on a limited number of samples.[29]

10.7.1.1 Beef

The tenderization of beef with pineapple juice was monitored by Fourier-transform infrared spectroscopy. FA and evolving factor analysis (EFA) were applied to successively collect attenuated spectra. Absorptions around the 1600–1500 and 1400 cm^{-1} regions were shown to increase with time. After subtracting the starting spectrum from each of the succeeding spectra, FA and EFA were applied to the resulting data matrix.[30]

The potential of different statistical techniques in the classification of raw beef, bovine *M. longissimus dorsi* muscles, samples in tenderness subgroups was studied. The sample was classified into three categories according to the degree of tenderness. A training set of samples was used to reveal any relationships between categories determined with NIR spectroscopic measurements. The study indicates that classical DA has advantages compared to multivariate calibration methods (i.e., principal component regression [PCR]), in this application. PCR underestimated high measurement values and overestimated low values. By employing PCR, the number of correct classifications for the intermediate subgroup amounted to 23%, whereas the use of DA increased this number up to 60%. The number of classifications in correct or neighbor subgroups for the two intermediate extreme subgroups reached 97%.[31]

The effectiveness of dual-attribute time-intensity (DATI) method for assessment of temporal changes in perceived toughness and juiciness of commercially acceptable meat cuts was examined. Usefulness of DATI in assessing temporal aspects of perception of juiciness and toughness was compared with single-attribute time-intensity (SATI) and line-scale profile. Results showed that DATI provided a good separation of attributes and was equal to or better than SATI in differentiating beef samples based on perceived juiciness and toughness. Generalised procrustes analysis (GPA) was performed on a data matrix contained the SATI and DATI parameters to visualize similarities and differences between DATI and SATI.[35]

Discrimination between fresh and frozen-then-thawed beef *M. longissimus dorsi* by combined visible–NIR reflectance spectroscopy was investigated. FDA and SIMCA were used to facilitate classification.[36] Authentication of fresh vs. frozen-then-thawed beef employing NIR reflectance spectroscopy of dried drip juice and selected chemometric techniques (PLS, FDA, SIMCA) has been examined. The best separation was obtained using FDA for spectral data from 1100 to 2498 nm and a standard freeze–thaw regime.[37]

The authentication of beef and ox kidney and liver by the use of mid-infrared spectroscopy was investigated. MIR spectra of beef, kidney, and liver were identified by PLS, canonical variate analysis, and predictive models. Using modified SIMCA, the pure beef specimens are modeled as single-class; this model identifies spectra of unadulterated beef with an acceptable error rate, while rejecting spectra of specimens containing 10–100% w/w kidney or liver. Finally, PLS regressions are performed to quantify the amount of added offal. The obtained prediction errors ($\pm4.8\%$ and $\pm4.0\%$ w/w, respectively, for the kidney and liver calibrations) are commensurate with the detection limits suggested by the SIMCA analysis.[41]

Investigations were conducted on the effectiveness of morphometric measurements (MM) of beef carcass for the prediction of weights and percentages of muscle and fat. In addition to the MM, side-weight (SW) and cross-sectional measurements (CSM) were employed for prediction with MR analysis, singly or in combination with these measurements. For the prediction of muscle and fat, the MM group was the best predictor. The CSW were effective only for the improvement of tissue-weight prediction. The CSW + MM + SM (twelfth rib) and the MM + SM (twelfth rib) combinations were best for the tissue-weight and the percentage prediction, respectively. A synergistic effect was observed between MM and SM for the improvement of tissue-percentage prediction accuracy. Regrouping the data to three breed types by morphological differences and analyzing it by a PCA improved the prediction accuracy. Carcass weight range caused differences in fat content but did not affect the prediction accuracy of muscle and fat percentages in the carcass when MM were used.[43]

10.7.1.2 Pork

The possibility of using an electronic nose (e-nose) with a conducting-polymer sensor array combined with pattern recognition routines to discriminate between varying intensities of boar taint was explored. A set of samples in a model system comprising a neutral lipid base with various combinations of androsterone and skatole were tested as if they were pork fat samples. The data set was used to develop a discriminant function for grouping pork samples into three response classes: normal, doubtful, and abnormal. Based on this, the e-nose managed to identify all the abnormal samples correctly. However, 16% of the normal samples were also classified as abnormal. It was concluded that the e-nose can satisfactorily discriminate between different levels of boar taint.[57]

The muscle pH at exsanguination and the rate of pH changes in porcine *M. longissimus dorsi* (LD) of normal, dark, firm, and dry (DFD), and pale, soft, and exudative (PSE) quality were compared. The pH was continuously measured in the LD during the first 50-min post-mortem. Calculations were attempted both on measured pH values and on pH values expressed as hydrogen-ion concentrations. A regression of pH or hydrogen-ion concentration with time was constructed for each animal. These individuals were then combined using a multivariate analysis (MANOVA in the general linear model (GLM)-procedure) to estimate regression curves for each meat quality class. The multivariate approach led to better results than general least-squares ANOVA because the variances were somewhat influenced by the time of measurement. A linear relationship was established between pH and time for normal and DFD quality and quadratic for PSE quality. The intercepts of the regression curves differed significantly between PSE and the other two quality classes only when the measured pH values were employed without transformation.[58]

The objective prediction of the ultimate quality of post-rigor pork musculatory was investigated. Randomly selected carcasses were probed at 24 h post-mortem (PM) by Danish meat-quality marbling (MQM), Hennessy grading probe (HGP), sensoptic resistance probe (SRP), and NWK pH-K21 meter (NpH). Also, filter paper wetness (FPW), lightness (L*), ultimate pH (pHu), subjective color (SC), firmness/wetness (SF), and marbling scores (SM) were recorded. Each carcass was categorized as either PSE, RSE (reddish–pink, soft, and exudative), RFN (reddish–pink, firm, and nonexudative), or DFD. When DA was used to sort carcasses into four quality groups, the highest proportion of correct classes was 65% by HGP, 60% by MQM, 52% by NpH and 32% by SRP. However, when two groups designated as PSE and non-PSE were sorted, then the proportion of the correct classification by MQM, HGP, SRP, and NpH were 87, 81, 71, and 66%, respectively. The best prediction accuracy reached 72% for a combination of MQM and NpH.[62]

Research has also been conducted on objective pork-quality on-line evaluations. "Fat-o-meat'er" carcass classification (lean content, fat, and muscle thickness), cold carcass weight, 24-h pH, and color were the measurements performed. The results showed that cold carcass weight and lean content are only slightly related to meat quality as evaluated by pH and color. CA has confirmed the importance of pH and color parameters such as L*, a*, and hue angle for the evaluation of pork quality on-line in an industrial context.[65]

The objective measurements of acidity (pH), light scattering (FOP), conductivity (PQM), light absorption (Gofo), L*, redness (a*), yellowness (b*), and WHC were evaluated for their suitability in estimating meat quality of slaughter pigs. The animals were divided in three quality categories based on pH (PSE, DFD, and normal). The predictability of the ultimate meat quality (paleness, hue, exudation) with the initial measurements in the slaughter line was rather low. Differences between light and heavy hams were identified by ANOVA. PCA of proteolytic and lipolytic enzymes was performed by the SIMCA algorithm. The number of the significant components of factors was established by the cross validation procedure. PCA showed that 56% of the total variation in the different meat quality parameters could be explained by the first three principal components. Based on the first principal component, a distinction could be made between PSE and normal meat because the intermediate slightly PSE meat class strongly overlapped both meat categories.[67]

10.7.1.3 Poultry

Poultry breasts, deboned at 2, 6, and 24 h post-mortem, provided various characteristics for establishing instrumental and sensory tenderness relationships. Sensory descriptive texture attributes were classified by variable cluster analysis into five groups representing mechanical, moisture, and chew-down characteristics. Warner–Bratzler (WB) and Allo–Kramer (AK) shear values indicated differences due to deboning time and correlated highly ($r \geq 0.90$) with mechanical and chew down sensory characteristics.[72]

The texture of cooked chicken breast is usually determined on intact pieces or strips representative of the muscle. Two sensory and two mechanical methods were evaluated for effectiveness in discerning texture differences in 1-cm^2 pieces of cooked chicken. Three post-mortem deboning times provided a texture-quality spectrum. Sensory characteristics were evaluated by descriptive analysis–trained panel (DA–TP) and by category scales–untrained panel (CS–UP). Twenty-gram breast portions of 1-cm^2 pieces were evaluated using a multiblade AK shear blade. WB shear values of intact 1.9-cm-wide strips were also recorded. DA classification based on DA–TP amounted to 0.21, whereas error rates for classification by mechanical devices were 0.3 and 0.21 for WB and AK, respectively.[73]

The effect of irradiation of refrigerated and frozen chicken on sensory properties was investigated on skinless, boneless breasts (white), and leg quarters (dark). Cluster analysis (VARCLUS) was conducted to determine panelist performance and to detect outliers. Subsequent data analysis was performed on responses from remaining panelists. ANOVA using the GLM procedure and Duncan's multiple range tests were conducted to determine significant differences among treatments. Irradiation did not affect appearance of moistness and glossiness of raw chicken (white or dark). Irradiated leg quarters while refrigerated were darker ($p \leq 0.05$) than controls (nonirradiated chicken).[74]

Commercial chickens were implanted with a microosmotic pump that released 1 IU/kg BW/day of adrenocorticotropic hormone (ACTH) and were killed after 24 and 48 h. The effect of the implantation of ACTH on liver and carcass weight, lipid, moisture and pH of livers, liver color, and fatty acid changes in liver was investigated. Data were analyzed by the ANOVA procedure of the statistical analysis system software package. The differences between treatment means were analyzed by the Duncan's multiple-range test.[75]

Artificial neural networks (ANNs) serve as alternatives to regression analysis for complex data. Based on CP or proximate analysis (PA) of ingredients, two types of ANN and linear regression (LR) were evaluated for predicting amino-acid levels in corn, wheat, soybean, meat, bone, and fishmeal. The two ANNs were a three-layer back propagation network (BP3) and a general regression neural network (GRNN). Methionine, TSAA, Lys, Thr, Tyr, Trp, and Arg were evaluated and the R^2 values were calculated per prediction method. ANN training was completed with Neuroshell 2™ using calibration to prevent over-training. Nearly 80% of the data were used as the input for the LR and the two ANNs. The remaining 20% of the randomly extracted data were used to calibrate the performance of the ANN.[82]

Characteristics of lipid fraction of two populations of ducks and two of chicken were studied. Large sensory differences in tenderness, flavor, and juiciness were noted between the two populations of each species. The best discriminant factor between the lean and fat ducks was the ratio between the percentage of linoleic acid and the total lipid content of the breast muscle. This ratio allowed the correct assignment for 98% of the animals ($p < 0.001$). For the chickens, by using only the fat score, 85% of the animals were properly classified. However, by introducing the DA, two major fatty acids (C 16:1 and C 20:4) and two minor ones (C 17:0 and C 17:1), 100% of the animals could be correctly assigned.[83]

10.7.2   Milk and Dairy Products

10.7.2.1   *Physicochemical and Sensory Characteristics of Cheese Treated with Multivariate Analysis*

Major research efforts have been made toward amelioration of production aspects and quality of cheese. The main aspects of this research include increasing cheese yield, flavour and taste improvement by lipolysis, and study of volatiles for cheese odor.[107]

Evaluations by hand and mouth were compared for texture of cheese. Seven mouth terms and five hand terms were identified and used to evaluate texture properties of 14 different types of natural and processed full-fat and reduced-fat cheeses. PCA revealed that hand and mouth evaluations differentiated the cheeses in a similar manner.[108] Descriptive sensory analysis of two cheese batches (produced from either raw or pasteurized milk) at two different ripening times (90 and 180 days) showed that: (1) cheeses made from the raw milk were firmer, with a more characteristic odor, taste, and aftertaste, (2) cheeses made from pasteurized milk were creamier with a sweeter odor and taste and a more bitter aftertaste, and (3) at 180 days, all samples were firmer, grainier, with characteristic, spicier taste and aftertaste.[109]

A new descriptive and discriminative language and the corresponding set of standard references for evaluating the main sensory characteristics of ewes' cheeses were developed. Thirteen assessors generated a very broad range of terms (attribute terms). Following discussion within the panel, the redundant, synonymous, and vague terms were discarded. From this preliminary result, objective selection of attributes was performed using PCA on the correlation matrix over assessor-averaged data and stepwise discriminant analysis (SDA). Three linear discriminant analyses (LDA) were performed to compare and validate the list of selected attributes: one with those obtained from SDA and a third one with the finally selected group.[110]

Eleven different cheeses were sensory profiled by a TP using a previously defined methodology. Using the same set of samples, a group of 300 consumers scored their preferences in a blind test; their perceptions of the samples were taken into account.[111] Table 10.9 (second part) summarizes several applications of chemometrics and multivariate analysis on representative dairy products.

10.7.2.2   *Multivariate Analysis and Technological Developments in Cheese Manufacturing*

The ability to predict sensory texture properties of feta cheese made from ultrafiltered milk (UF–feta cheese) was investigated with uniaxial compression, small shear deformation measurements, and indices of proteolysis. In PCA, some of the instrumental analyses were highly correlated. PCA of the six sensory attributes showed that mainly one type of information was present in the sensory results. Partial least-squares regression (PLS) of all results revealed that stress at fracture from uniaxial compression was the individual instrumental parameter having the highest correlation with the sensory texture attribute.[112]

A study of the effect of pasteurization in the manufacture of Idiazabal ewe's-milk cheese was conducted. Pasteurization resulted in a decrease in amino-acid release in the cheese during ripening when a specific starter culture was used, but not with another starter culture. Multivariate analysis confirmed increases in asparagine, serine, and taurine, and decreases in aspartic acid and glycine with pasteurization, irrespective of the starter used.[113]

A face-centred cubic (FCC) experimental design was used to investigate the effects of extraction time, solvent, and sample temperature on simultaneous distillation–extraction (SDE). The backward stepwise selection of the regression coefficients led to the identification of seven groups of compounds of similar behavior with respect to the three variables investigated. The same set of models was found by performing PCA on the t-values of the regression coefficients of the unrefined polynomials and cluster analysis on the scores of the 13 acids obtained with PCA.[114] Martin and Coulon[115] analyzed the role of various factors involved in milk production and its clotting ability. The considerable variations (30–48 mm and 8–22 min for curd firmness and rennet clotting time, respectively) in 271 milks were attributed to lactation stages, milk-protein genetic variants, and feeding practices depending on seasonal variations. Implementation of PCA resulted in two main axes (PC1 and PC2) describing only 35% of total variation. PC1 characterizes the milks rapidly forming gels (firm texture) and PC2 characterizes the milks of a weak pH and rapid gellation.

10.7.2.3 Genetic Effect on Cheese Properties

Rahali and Menard[116] investigated the influence of genetic variants of β-lactoglobulin (β-Lg) and κ-casein (κ-Cn) on milk composition and on 83 Camembert cheeses' capacities. Eight genetic combinations of β-Lg and κ-Cn were studied. Discriminant factor analysis showed that the effects of β-Lg and κ-Cn were additive and allowed for grading genetic combinations according to their cheese-making quality.

The influence of three casein haplotypes (BBB, BA^2A, and CA^2B) was compared with regard to the composition of milk and its cheese production capability. The physicochemical composition (fat, total casein content, elasticity, firmness, cohesion) of 89 samples (40 BBB, 30 BA^2A, and 19 CA^2B) selected on the basis of genotype and lactation stage was determined. The cheese yield coming from BBB milk was the highest because of better retention of the amount of fat in the curd. At the end of the ripening period (45 days) the BBB cheeses were endowed with superior mechanical properties (firmer, less elastic, and more breakable) than the other two haplotypes' counterparts.[117] Heil and Dumond[118] studied the sensory properties of goat cheeses of various genetic types of αs1-casein and sampled at early, middle, and final stages of lactation. The sensory properties were recorded on mold-ripened cheese (Pelardon-type) and semi-hard cheeses of Gouda-type. Nine and eight descriptors were used for Pelardon and Gouda cheese, respectively. A 15-member TP was asked to classify the above mentioned cheeses on a 5-point scale (1–5). Genetic-type instead of lactation stage emerged as the most crucial parameter for the reported texture and flavor changes.

10.7.2.4 Yogurt

Samples of strawberry yogurts were determined by country, yogurt producer, and marketed age group using PCA. From the PCA data generated, new strawberry flavors were developed along the creamy-vanilla, juicy, green-floral, and berry-ripe-banana dimensions. The new flavors and three commercial ones were analyzed by trained panelists using both PCA and PLS, and hedonically by 160 consumers between the ages of 6 and 16 years. PLS and correlation analysis identified the positive consumer-liking drivers for strawberry flavor to be *vanilla, creamy*, and *balsamic* attributes, whereas *wild* and *floral* attributes decreased liking.[119]

The sensory properties of traditional acidic and mild, less acidic yogurts were evaluated by a TP using a descriptive approach. Important flavor differences were found between two classes of yogurt. They were mainly due to differences in acidity and not to different concentrations of the three impact aroma compounds acetaldehyde, 2,3-butanedione, and 2,3-pentanedione. The flavor descriptors *astringent, bitter*, and *acid* as well as the aftertastes *persistent, astringent*, and *lemon* revealed a positive correlation with decreasing pH. To better comprehend the overall sensory data, PCA of significantly different attributes (according to ANOVA) was carried out on the correlation matrix. The resulting biplot of the first two components shows that the main variation perceived by the panel is in the attribute *acid* (77% of the total variation).[120,121]

Fifteen assessors, trained solely in visual profiling based on the quantitative descriptive analysis (QDA) method, analyzed 16 formulations of mousse desert products over ten visual texture

Table 10.10 Mean Value, Standard Deviation and Percent Coefficient of Variation of 30 Chemical, Physical, Microbiological, and Sensory Variables Evaluated on 35 Greek Traditional Sausages

Variables	Mean Value	Standard Deviation	% CV
Physical			
Water activity (a_w)			
in the core (internal)	0.94	0.02	2.06
at the surface (external)	0.93	0.03	3.30
Firmness (N)	26.19	19.31	73.72
Color at the surface			
L^*	29.78	2.66	8.92
a^*	5.68	2.30	40.59
b^*	8.84	2.17	24.62
Color at cross section			
L^*	43.50	4.12	9.47
a^*	7.89	3.24	41.01
b^*	13.07	2.13	16.30
Chemical			
Nitrites	0.77	0.90	116.23
Nitrates	38.19	71.19	186.41
TBA	1.56	1.23	78.54
PH	5.76	0.53	9.23
Moisture	43.98	9.18	20.86
Protein	19.19	3.53	18.41
Fat	33.50	9.16	27.35
Ash	3.33	0.60	17.99
Sodium chloride (NaCl)	3.31	0.64	18.26
Microbiological			
Aerobic plate count (APC)	8.21	0.87	10.61
Lactic acid bacteria (LAB)	8.08	0.84	10.43
Psychrotrophic bacteria	6.32	1.21	16.49
Micrococci/staphylococci	5.22	0.73	14.05
Enterobacteriaceae	4.30	1.33	31.06
Sensory			
Fresh sausages			
Appearance	3.54	0.64	18.20
Flavor	3.42	0.73	21.40
Firmness	3.41	0.73	21.44
Overall aceeptability	3.21	0.76	23.76
Grilled sausages			
Taste	3.18	0.87	27.39
Flavor	3.40	0.88	25.7
Overall acceptability	3.14	0.85	28.10

(L^* surface). Extreme attribute values (high or low) for firmness, moisture, and fat content, low salt content, and low taste were related to low consumer preference.

10.9 FUTURE TRENDS

Multivariate analysis is a very promising field for further development and application. Anyone reviewing the literature will likely be surprised to discover that MVA has experienced exponential

Table 10.11 Results from Principal Component Analysis (Eigenvalues and Cumulative Percentage), Carried out by SPSS, for 14 Selected Variables

Variable	Factor	Eigen Value	Pct of Var	Cum Pct
Moisture	1	4.22241	30.2	30.2
Fat	2	2.86606	20.5	50.6
Sodium chloride	3	1.54177	11.0	61.6
TBA	4	1.17378	8.4	70.0
PH	5	1.05591	7.5	77.6
a_w internal	6	0.91260	6.5	84.1
Firmness	7	0.64718	4.6	88.7
$L^*_{surface}$	8	0.50926	3.6	92.3
$a^*_{surface}$	9	0.42317	3.0	95.4
APC	10	0.29617	2.1	97.5
External appearance	11	0.17466	1.2	98.7
Overall acceptability (fresh)	12	0.12458	0.9	99.6
Odor-taste	13	0.03323	0.2	99.9
Overall acceptability (grilled)	14	0.01923	0.1	100.0

growth in terms of applications in medicine, chemistry, engineering, agriculture, and food science, among other fields. MVA has been widely recognized as a very powerful tool for handling data. In view of the continuously increasing costs of experimentation, it becomes imperative to convert the rows and columns of data into meaningful plots through which the information is clearly understood and conveyed. Some representative applications of MVA include:[140–146]

- Quality control and quality optimization (foods, drugs, cosmetics, automobiles)
- Process optimization and process control
- Development and optimization of measurement methods
- Prospecting for oil, water, minerals
- Classification of bacteria, viruses, varieties, samples of same or different origin
- Analysis of economics, market search, consumer behavior

Table 10.12 Rotated Factor Matrix (Varimax–Kaiser Normalization) Calculated by SPSS

Variable	Factor 1	Factor 2	Factor 3	Factor 4	Factor 5
Moisture	0.87892	0.22893	−0.06356	−0.28454	−0.09504
a_w internal	0.87042	0.09487	0.17694	−0.27482	−0.03815
Firmness	−0.78887	0.11750	0.07659	−0.12817	0.23346
Fat	−0.73739	−0.26211	0.06029	0.40453	0.06023
$a^*surface$	0.71998	−0.08385	−0.44076	0.15709	0.24535
Odor-taste	0.06711	0.95250	−0.15324	−0.06277	0.02350
Overall acceptability (grilled)	0.15230	0.88866	−0.28823	−0.16159	0.00441
Overall acceptability (fresh)	−0.25109	0.74628	0.18846	0.025330	0.40056
APC	0.29358	0.64644	0.29354	0.15501	−0.09652
TBA	−0.05863	−0.20001	0.73481	0.10068	0.03522
PH	0.12779	−0.24099	−0.56664	−0.54619	−0.02497
$L^*surface$	−0.18426	−0.08173	0.08307	0.83934	0.03551
Sodium chloride	−0.20091	−0.04507	−0.21745	0.18517	0.73901
External appearance	−0.01001	0.44500	0.24179	0.024101	−0.01210

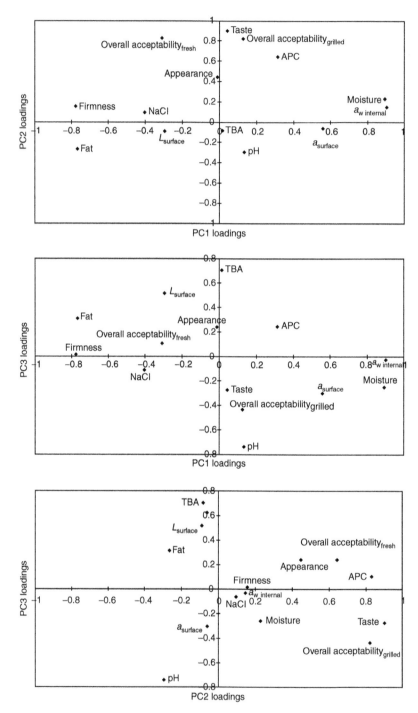

Figure 10.1   PC1 vs PC2, PC1 vs PC3 and PC2 vs PC3 for physio-chemical and sensory analysis of traditional Greek sausages.

- Design of new products
- Authentication of products (foods of different cultivar, different geographical origin, adulterated products)
- Pattern recognition
- Neural networks (food science, weather prediction, credit scoring and practically applicable in any field)
- Identification of crucial factors for ecosystem and landscape ecology

However, one should also understand that the application of MVA bears the potential pitfall of over-fitting. This discrepancy can be rather easily detected by using an appropriately sized, independent test set. This presupposes the selection of a good proportion of the observations available and not using them in the modeling process. After the model is constructed, it should be applied to the test data. Should the solution stand for a good model of genuine structure in the data, it should perform equally well when applied to the test set.[147]

GLOSSARY

Arch effect   A distortion or artifact in an ordination diagram, in which the second axis is an arched function of the first axis. It is caused by the unimodal distribution of species along gradients. The arch appears in correspondence analysis and other ordination techniques. One of the main purposes of detrended correspondence analysis is to remove the arch effect. Principal components analysis creates a more serious artifact called the *horseshoe effect*.

Bartlett test of sphericity.   Statistical test for the overall significance of all correlations within a correlation matrix.

Beta diversity   Also called *species turnover* or *differentiation diversity*, beta diversity is a measure of how different samples are from each other, and/or how far apart they are on gradients of species composition. Alternatively, it is a measure of the "length" of an ecological gradient or ordination axis in terms of species composition. An axis or gradient with low beta diversity will be similar in species composition at both ends. Some ordination techniques (e.g., PCA) behave best at low beta diversity, and others (e.g., DA, CCA) behave best at high beta diversity.

Bootstrap   A reasonably new computer-intensive method to obtain confidence intervals, to estimate parameters, or in some cases to test hypotheses, the bootstrap is considered a "resampling method," and is closely related to the jackknife and to randomization tests.

Categorical variable   A variable that is represented by several different types; for example: lake/river/stream, farm/pasture/unmanaged, pitfall trap/fence trap/direct sighting. For most multivariate analyses, categorical variables must be converted to $k-1$ dummy variables (where $k=$ the number of categories).

Centroid   The (weighted) mean of a multivariate data set which can be represented by a vector. For many ordination techniques, the centroid is a vector of zeroes (i.e., the scores are centered and standardized). In a direct gradient analysis, a categorical variable is often best represented by a centroid in the ordination diagram.

Classification   The methodology of placing things/samples into groups. Classification can be completely subjective, or it can be objective and computer-assisted. *Hierarchical classification* means that the groups are nested within other groups. There are two general kinds of hierarchical classification: divisive and agglomerative.

Correlation matrix   A square, symmetric matrix consisting of correlation coefficients. The rows and the columns represent the variables. The diagonal elements are all equal to one, for the simple reason that the correlation coefficient of a variable with itself equals one.

Correspondence analysis An eigenanalysis-based ordination method, also known as *reciprocal averaging*. Correspondence analysis has a problem: the arch effect. This effect is caused by nonlinearity of species response curves. The arch is not as serious as the horseshoe effect of PCA, because the ends of the gradient are not convoluted. Another related problem of correspondence analysis is that the ends of the gradient are compressed.

Covariance matrix A square, symmetric matrix in which the rows and columns are variables and the entries are covariances. The diagonal elements (i.e., the covariance between a variable and itself) will equal the variances.

Cross-validation Procedure of dividing the sample into two parts; the analysis sample is used in the estimation of the discriminant functions or logistic regression model, and the holdout sample is used to validate the results. Cross-validation avoids the "overfitting" of the discriminant function.

Detrended Correspondence Analysis (DCoA) An eigenanalysis-based ordination technique derived from correspondence analysis. DA performs detrending to counteract the arch effect, a defect of correspondence analysis.

Discriminant loadings Measurement of the simple linear correlation between each independent variable and the discriminant Z score for each discriminant function also known as *structure correlations*.

Distance matrix A square and (usually) symmetric matrix in which the rows and the columns represent (usually) samples. The entries represent some index of the difference between samples, i.e., Euclidean distance, Manhattan (city block) distance, Bray–Curtis dissimilarity, the Jaccard coefficient, or any of a huge number of possibilities.

Dummy variable A binary variable of ones and zeroes that is one if the observation belongs to a category and zero if it does not.

Eigenanalysis The process of finding eigenvectors and eigenvalues.

Eigenvalue A central concept in linear algebra (i.e., matrix algebra); an eigenvector of a matrix is a vector that, when multiplied by the matrix, gives the vector back again, except that it has been multiplied by a particular constant, called the *eigenvalue*. For a square matrix, there are as many eigenvectors and eigenvalues as there are rows and columns in the matrix. The eigenvalues are usually ranked from highest to lowest, and termed the *first, second, third*, etc. *eigenvalues* or *latent roots*.

Fuzzy sets and fuzzy set ordination Fuzzy sets allow various grades of membership. Classical set theory would define an arbitrary elevation or threshold, above which all plots must belong, and below which no plots belong. Fuzzy set theory would allow a plot to belong with 25% membership (for a relatively low elevation) or 75% (for a relatively high elevation).

Gaussian ordination A little-used ordination technique that arranges samples along ordination axes such that the fit of the species response curves to the Gaussian curve is maximized. The fit can be measured by r^2.

Horseshoe effect A distortion in ordination diagrams. It is more extreme than the arch effect because the ends of the first gradient are involuted. The horseshoe effect can be observed for very long gradients in PCA.

Inertia A measure of the total amount of variance in a data set directly related to the physical concept of inertia, which is the tendency for an object in motion to stay in motion, and the tendency for an object at rest to remain at rest. For weighted averaging methods such as DA and CCA, the inertia is related to the spread of species modes (or optima) in ordination space.

Jackknife A (usually) computer-intensive method to estimate parameters and/or to gauge uncertainty in these estimates. The name is derived from the method that each observation is removed (i.e., cut with the knife) one at a time (or two at a time for the second-order jackknife, and so on) to get a feeling for the spread of data.

Latent root   Another name for *eigenvalue*.

Latent value   Another name for *eigenvalue*.

Linear combination   A linear combination of a set of variables is a new variable (y_i) THAT can be expressed as follows: $y_i = \mathring{a}(b_j x_{ij})$, where b_j is the "coefficient" of variable j, and x_{ij} is value of observation i of variable j. In MR, predicted values of the dependent variable are linear combinations of the independent (or explanatory) variables. In CCA and RDA, sample scores are linear combinations of the environmental variables.

Mantel test   A method for comparing matrices to each other, also known as *matrix correlation*.

Monotonic distributions   Describes species response curves in which species only increase along environmental gradients, or only decrease along environmental gradients. A monotonic distribution can be linear or more complex. If most species have a monotonic distribution, then it is best to use PCA and RDA, but if most species have unimodal distributions, then it is best to use DA and CCA.

Monte Carlo tests   A synonym of *randomization tests*. A Monte Carlo permutation test is when the actual data values are maintained, but they are randomly permuted to obtain the distribution of the test statistic. Exactly how they are permuted depends upon the null hypothesis to be tested.

Multidimensional scaling   Currently, this is often a synonym for *nonmetric multidimensional scaling*, but it previously referred to *principal coordinates analysis*.

Multiresponse permutation procedure   Usually abbreviated MRPP. A randomization test that evaluates differences in species composition, based on some distance measure.

Noise   Refers to chance variation in nature that interferes with our ability to see pattern and infer processes. In its simplest form, noise is synonymous with *statistical error* (e.g., the error term in a regression).

Nominal variable   A variable which can be represented as a binary: yes/no, on/off, present/absent. A nominal variable is usually summarized by a dummy variable.

Nonhierarchical Clustering (NHC) methods   Selection of initial set of cluster seed points around which further clusters are built.

Nonmetric Multidimensional Scaling (NMDS)   The most widely used distance-based ordination method. The user needs to prespecify the number of dimensions, and then the method will minimize the stress (a measure of poorness of fit between the ordination and measured ecological distances). See also *distance matrix*.

Outlier   An observation that is substantially different from the other observations (extreme value).

Partial analysis   An analysis (e.g., regression, correlation, ANOVA, ordination) in which the effects of covariables are "factored out" or nullified. Examples of partial analysis include partial correlation, partial DA, partial CCA, and ANOVA, among others.

Principal Coordinates Analysis (PCoA)   A distance-based ordination method in which the distances between sites in the ordination diagram is maximally correlated with the distances.

Randomization test   The purpose of inferential statistics is to evaluate whether a number that summarizes something of interest is greater than (or less than) one would expect just due to chance (i.e., if H_0 is true). This number can be one of the well-known parametric statistics (t, F, chi-squared, r, etc.), or nonparametric statistics (Mann–Whitney U, Spearman r, etc.).

Similarity index   A measure of the similarity of species composition between two samples. Examples include the Sørensen coefficient and the Jaccard coefficient. Most similarity indices have values of zero for samples that share absolutely no species, and 1 or 100% for samples that have identical species composition.

Similarity matrix   A square and (usually) symmetric matrix in which the entries are similarities between samples. Similarity matrices are easily produced from, or converted into, distance matrices. The diagonal entries are usually 1 or 100%, meaning a sample is usually 100% similar to itself.

Singular matrix   A square matrix that cannot be inverted. In multivariate methods, a singular matrix can occur if one variable is precisely a linear combination of the other variables. This may occur if data are expressed in a percentage basis, or there are is a categorical variable expressed as a series of dummy variables.

Singular value decomposition   A way of manipulating matrices that is similar to, and ultimately equivalent to, eigenanalysis.

Standardization   A way of scaling variables so that different variables, measured in different units, can be compared. The most common forms of standardization include ranking, logarithmic transformations, placing on a 0–1 scale (according to the formula $[x-\text{min}]/[\text{max}-\text{min}]$; this is used in fuzzy set ordination), and subtracting the mean and dividing by the standard deviation. The last two kinds of standardization produce variables which are perfectly correlated ($r=1$) with the raw data. The last kind is by far the most common, and unless otherwise stated, is what should be assumed by "standardized variables."

Stepwise analysis   A MR method (including RDA and CCA, which are special cases of MR) in which explanatory (independent) variables are selected on the basis of whether they explain a "significant" amount of variation in your dependent variable(s). There are several approaches to stepwise analysis: forward selection, backwards selection, and combined analysis.

Weighted average   An average, except that different observations are given differing importances or "weights." In ordination, the weights are typically the abundances (perhaps transformed) of species. Weighted averages are intrinsic to correspondence analysis and related methods.

Z   The symbol used to represent a "regionalized variable," a spatially varying variable.

REFERENCES

1. Dillon, W. R. and Goldstein, M., *Multivariate Analysis*, New York: Wiley, 1984.
2. Hair, J. F., Anderson, R. E., Tatham, R. L., and Black, W. C., *Multivariate Data Analysis*, Upper Saddle River, NJ: Prentice Hall, 1998.
3. http://trochim.human.cornell.edu/tutorial/flynn/multivar.htm (accessed January, 2004).
4. Tatsuoka, M. M. and Tiedeman, D. V., Statistics as an aspect of scientific method in research on teaching, In *Handbook of Research on Teaching*, N.L. Gage, Ed., Chicago: Rand McNally, pp. 142–170, 1963.
5. http://www.cs.vu.nl/~geurt/Multivar/College4.pdf (accessed February, 2004).
6. Pearson, K., On lines and planes of closest fit to a system of points in space, *Philosophical Magazine*, 2, 557–572, 1901.
7. Hotelling, H., Analysis of a complex of statistical variables into principal components, *Journal of Educational Psychology*, 24, 417–441, 1933.
8. http://www.rkb.home.cern.ch/AN16pp (accessed February, 2004).
9. Thurstone, L. L., *Multiple Factor Analysis*, Chicago: University of Chicago Press, 1947.
10. Manly, B. F. J., *Multivariate Statistical Methods*, Boca Raton, FL: CRC, 2000.
11. Kaiser, H. F., The varimax criterion for analytic rotation in factor analysis, *Psychometrika*, 23, 187–200, 1958.
12. Cattell, R. B., The screen test for the number of factors, *Multivariate Behavioral Research*, 1, 140–161, 1966.
13. Cattell, R. B. and Jaspars, J., A general plasmode for factor analytic exercises and research, *Multivariate Behavioral Research Monographs*, 67(3), 1–212, 1967.
14. Johnson, D. E., *Applied Multivariate Methods for Data Analysis*, Pacific Grove, CA: Duxbury Press, 1998.

15. Spearman, C., "General intelligence" objectively determined and measured, *American Journal of Psychology*, 15, 201–293, 1904.

16. Chatfield, C. and Collins, A. J., *Introduction to Multivariate Analysis*, London: Chapman and Hall, 1980.

17. Seber, G. A. F., *Multivariate Observations*, New York: Wiley, 1984.

18. Kinnear, P. and Gray, C. D., *SPSS for Windows Made Simple*, East Sussex, UK: Psychology Press, 1997.

19. Watson, C. J., An additional approach for interpretation of multiple discriminant analysis for business research, *Journal of Business Research*, 9, 1–11, 1981.

20. Green, P. E. and Carroll, J. D., *Mathematical Tools for Applied Multivariate Analysis*, New York: Academic Press, 1978.

21. http://www2.chass.ncsu.edu/garson/pa765/cluster.htm (accessed January, 2004).

22. MacNaughton-Smith, P., Williams, W. T., Dale, M. B., and Mockett, L. G., Dissimilarity analysis: A new technique of hierarchical division, *Nature*, 201, 426–428, 1962.

23. Sonquist, J. A. and Morgan, J. N., *The Detection of Interaction Effects*, Monograph No. 35, Ann Arbor, MI: Institute for Social Research, University of Michigan, 1964.

24. Milligan, G. W., An examination of the effect of six types of error perturbation on fifteen clustering algorithms, *Psychometrika*, 45, 325–342, 1980.

25. Al-Jowder, O., Kemsley, G., and Wilson, R. H., Mid-infrared spectroscopy and authenticity problems in selected meats: A feasibility study, *Food Chemistry*, 59(2), 195–201, 1997.

26. McElhiney, J., Downey, G., and Fearn, T., Chemometric processing of visible and near infrared reflectance spectra for species identification in selected raw homogenized meats, *Journal of Near Infrared Spectroscopy*, 7(3), 145–154, 1999.

27. Rannou, H. and Downey, G., Discrimination of raw pork chicken and turkey meat by spectroscopy in the visible, near and midpinfrared ranges, *Analytical Communications*, 34(12), 401–404, 1997.

28. Skarpeid, H. J., Kvaal, K., and Hildrum, K. I., Identification of animal species in ground meat mixtures by multivariate analysis of isoelectric focusing protein profiles, *Electrophoresis*, 19(18), 3103–3109, 1998.

29. Huang, Y., Lacey, R. E., and Whittaker, A. D., Neural network prediction modeling based on elastographic textural features for meat quality evaluation, *Transactions of ASAE*, 41(4), 1173–1179, 1998.

30. Iizuka, K. and Aishima, T., Tenderisation of beef with pineapple juice monitored by Fourier transform infrared spectroscopy and chemometric analysis, *Journal of Food Science*, 64(6), 973–977, 1999.

31. Naes, T. and Hildrum, K. I., Comparison of multivariate calibration and discriminant analysis in evaluating NIR spectroscopy for determination of meat tenderness, *Applied Spectrocscopy*, 51(3), 350–357, 1997.

32. Luno, M., Beltran, J. A., Jaime, I., and Roncales, P., Textural assessment of clenbuterol treatment in beef, *Meat Science*, 51(4), 297–303, 1999.

33. Toscas, P. J., Shaw, F. D., and Beilken, S. L., Partial least squares (PLS) regression for the analysis of instrument measurements and sensory meat quality data, *Meat Science*, 52(2), 173–178, 1999.

34. Zimoch, J. and Gullett, E. A., Temporal aspects of perception of juiciness and tenderness of beef, *Food Quality and Preference*, 8(3), 203–211, 1997.

35. Zimoch, J. and Findley, C. J., Effective discrimination of meat tenderness using dual attribute time intensity, *Journal of Food Science*, 63(6), 940–944, 1998.

36. Downey, G. and Beauchēne, D., Discrimination between fresh and frozen then thawed beef m. longissimus dorsi by combined visible–near infrared reflectance spectroscopy: A feasibility study, *Meat Science*, 45(3), 353–363, 1997.

37. Downey, G. and Beauchēne, D., Authentication of fresh vs. frozen-then thawed beef by near infrared reflectance spectroscopy of dried drip juice, *Food Science and Technology-Lebensmittel-Wissenschaft & Technologie*, 30(7), 721–726, 1997.

38. Ding, H. B. and Xu, R. J., Differentiation of beef and kangaroo meat by visible/near-infrared reflectance spectroscopy, *Journal of Food Science*, 64(5), 814–817, 1999.

39. Basset, O., Dupont, F., Hernandez, A., Odet, C., Abouelkaram, S., and Culioli, J., Texture image analysis: Application to the classification of bovine muscles from meat slice images, *Optical Engineering*, 38(11), 1950–1959, 1999.

40. McElhiney, J., Downey, G., and O'Donnell, C., Quantitation of lamb content in mixtures with raw minced beef using visible, near and mid-infrared spectroscopy, *Journal of Food Science*, 64(4), 587–591, 1999.

41. Al-Lowder, O., Defernez, M., Kemsley, E. K., and Wilson, R. H., Mid-infrared spectroscopy and chemometrics for the authentication of meat products, *Journal of Agricultural and Food Chemistry*, 71, 210–3218, 1999.

42. Wang, Z. Y., Isaksson, T., and Kowalski, B. R., A new approach for distance measurement in locally weighted regression, *Analytical Chemistry*, 66(2), 249–260, 1994.

43. Zembayashi, M., Effectiveness of morphometric measurements for improvement of the prediction accuracy of beef carcass composition, *Meat Science*, 51(4), 339–347, 1999.

44. Johnson, E. R. and Baker, D. A., Use of linear measurements of M-longissimus to predict the muscle content of beef carcasses, *Meat Science*, 45(3), 321–327, 1997.

45. Byrne, C. E., Downey, G., Troy, D. J., and Buckley, D. J., Non-destructive prediction of selected quality attributes of beef by near-infrared reflectance spectroscopy between 750 and 1098 nm, *Meat Science*, 49(4), 399–409, 1998.

46. Priyanto, R., Johnson, E. R., and Taylor, D. G., Investigations into the accuracy of prediction of beef carcass composition using subcutaneous fat thickness and carcass weight 1. Identifying problems, *Meat Science*, 46(2), 147–157, 1997.

47. Johnson, E. R., Priyanto, R., and Taylor, D. G., Investigations into the accuracy of prediction of beef carcass composition using subcutaneous fat thickness and carcass weight 2. Improving the accuracy of prediction, *Meat Science*, 46(2), 159–172, 1997.

48. Gerhardy, H., Quality of beef from commercial fattening systems in Northern Germany, *Meat Science*, 40, 103–120, 1995.

49. Thyholt, K., Enersen, G., and Isaksson, T., Determination of endpoint temperatures in previously heat treated beef using reflectance spectroscopy, *Meat Science*, 48(1–2), 49–63, 1998.

50. Ellekjaer, M. R., Assessment of maximum cooking temperatures of previously heat treated beef. 2. Differential scanning calorimetry, *Journal of the Science of Food and Agriculture*, 60(2), 255–261, 1992.

51. Rossi, V., Denoyer, C., and Berdague, J. L., Effect of storage in air or cooked beef meat-analysis of desorbed volatile compounds, *Science des Aliments*, 15(4), 381–392, 1995.

52. Sanderson, R., Lister, S. J., Dhanoa, M. S., Barnes, R. J., and Thomas, C., Use of near infrared reflectance spectroscopy to predict and compare the composition of carcass samples from young steers, *Animal Science*, 65(1), 45–54, 1997.

53. Gerrard, D. E., Gao, X., and Tan, J., Beef marbling and color score determination by image processing, *Journal of Food Science*, 61(1), 145–146, 1996.

54. Spanier, A. M., Stangelo, A. J., and Shaffer, G. P., Response of beef flavor to oxygen depletion and an antioxidant chelator mixture, *Journal of Agricultural and Food Chemistry*, 40(9), 1656–1662, 1992.

55. Djenane, D., Sanchez-Escalante, A., Beltran, J. A., and Poncales, P., Ability of tocopherol, taurine, and rosemary, in combination with vitamin C, to increase the oxidative stability of beef steaks packaged in modified atmosphere, *Food Chemistry*, 76, 407–415, 2002.

56. Cozzolino, D. and Murray, I., Identifiction of animal meat muscles by visible and near infra red reflectance spectroscopy, *Lebensm.-Wiss.u.-Technol*, 37(4), 447–452, 2004.

57. Annor-Frempong, I. E., Nute, G. R., Wood, J. D., Whittington, F. W., and West, A., The measurement of the responses to different odour intensities of "Boar Taint" using panel and an electronic nose, *Meat Science*, 50(2), 139–151, 1998.

58. Enfalt, A. C., Lundstorm, K., and Engstrand, U., Early post-mortem pH decrease in porcine M. longissimus dorsi of PSE, normal and DFD quality, *Meat Science*, 34(2), 131–143, 1993.

59. Annor-Frempong, I. E., Nute, G. R., Whittington, F. W., and Wood, J. D., The problem of taint in pork. 2. The influence of skatole, androsterone and indole, presented individually and in combination in a model lipid base, on odour perception, *Meat Science*, 47(1–2), 49–61, 1995.

60. Wicke, M., Vonlengerken, G., Maak, S., and Fiedler, I., Prediction of PSE meat by biochemical and morphological traits of skeletal muscle in living pigs, *Archiv für Tierzucht*, 36(6), 631–638, 1993.

61. Muller, B., Better output prediction in hog slaughtering with neural networks? *Fleischwirtschaft*, 77(1), 80–83, 1997.

62. Joo, S. T., Kauffman, R. G., Warner, R. D., Borggaard, G., Stevenson-Barry, J. M., Rhee, M. S., Park, G. B., and Kim, B. C., Objectively predicting ultimate quality of post-rigor pork musculature: II Practical classification method on the cutting line, *Asian–Australasian Journal of Animal Sciences*, 13(1), 77–85, 2000.

63. Karlsson, A., The use of principal component analysis for evaluating results from pig meat quality measurements, *Meat Science*, 31(4), 423–433, 1992.

64. Garrido, M. D., Pedauye, J., Banon, S., and Laencina, J., Objective assessment of pork quality, *Meat Science*, 37(3), 411–420, 1994.

65. Chizolini, R., Novelli, E., Badiani, A., Delbono, G., and Rosa, P., Objective evaluation of pork quality—results of online measurements, *Meat Science*, 34(1), 79–93, 1993.

66. Annor-Frempong, I. E., Nute, G. R., Whittington, F. W., and Wood, J. D., The problem of taint in pork. 3. Odour profile of pork fat and the interrelationships between androsterone, skatole and indole concentrations, *Meat Science*, 47(1–2), 63–76, 1997.

67. VanOeckel, M. J., Casteels, M., Warnants, N., DeBoever, J. L., Boucque, C. V., and Bosschaerts, L., Instrumental evaluation of meat quality characteristics of Belgian slaughter pigs, *Fleischwirtschaft*, 77(1), 84–87, 1997.

68. Toldra, F., Flores, M., Aristoy, M. C., Virgili, R., and Parolari, G., Pattern of muscle proteolytic and lipolytic enzymes from light and heavy pigs, *Journal of the Science of Food and Agriculture*, 71(1), 124–128, 1996.

69. Swatland, H. J. and Uttaro, B., Optoelectrical assessment of pork quality using parallel hypodermic needles, *Archiv für Tierzucht*, 41(4), 379–385, 1998.

70. Bertram, H. C., Engelsen, S. B., Busk, H., Karlsson, A. H., and Andersen, H. J., Water properties during cooking of pork studied by low-field NMR relaxation: Effects of curing and the RN-gene, *Meat Science*, 66(2), 437–446, 2004.

71. Kaneki, N., Miura, T., Shimada, K., Tanaka, H., Ito, S., Hotori, K., Akasaka, C., Ohkubo, S., and Asano, Y., Measurement of pork freshness using potentiometric sensor, *Talanta*, 62(1), 215–219, 2004.

72. Lyon, B. G. and Lyon, C. E., Sensory descriptive profile relationships to shear values of deboned poultry, *Journal of Food Science*, 62(4), 885–888, 1997.

73. Lyon, B. G. and Lyon, C. E., Texture evaluations of cooked, diced broiler breast samples by sensory and methods, *Poultry Science*, 75(6), 813–819, 1996.

74. Hashim, I. B., Resureccion, A. V. A., and McWatters, K. H., Descriptive sensory analysis of irradiated frozen or refrigerated chicken, *Journal of Food Science*, 60(4), 664–666, 1995.

75. King, Y. T. and Chen, T. C., Chemical and physical characteristics of chicken livers following adrenocorticotropic hormone-induced stress, *Journal of Food Science*, 63(4), 589–591, 1998.

76. Bruun-Jensen, L., Skovgaard, I. M., and Skibsted, L. H., The antioxidative activity of RRR-α-tocopherol vs. RRR-δ-tocopherol in combination with ascorbyl palmitate in cooked, minced turkey, *Food Chemistry*, 56(4), 347–354, 1996.

77. Kent, M. and Anderson, D., Dielectric studies of added water in poultry meat and scallops, *Journal of Food Engineering*, 28(3–4), 239–259, 1996.

78. Meykier, A., Genot, C., and Gandemer, G., Oxidation of muscle phospholipids in relation to their fatty acid composition with emphasis on volatile compounds, *Journal of the Science of Food and Agriculture*, 79(6), 797–804, 1999.

79. Siegmund, B. and Pfannhauser, W., Changes of the volatile fraction of cooked chicken meat during chill storing: Results obtained by the electronic nose in comparison to GC–MS and GC olfactometry, *Zeitschrift für Lebensmittel Unterschung und Forschung-A Food Research and Technology*, 208(5–6), 336–341, 1999.

80. Santé, V. S., Lebert, A., Le Pottier, G., and Ouali, A., Comparison between two statistical models for prediction of turkey breast meat color, *Meat Science*, 43(3–4), 283–290, 1996.

81. Ding, H. B., Xu, R. J., and Chan, D. K. O., Identification of broiler chicken meat using a visible/near infrared spectroscopic technique, *Journal of the Science of Food and Agriculture*, 79(11), 1382–1388, 1999.

82. Roush, W. B. and Cravener, T. L., Artificial neural network prediction of amino acid levels in feed ingredients, *Poultry Science*, 76(5), 721–727, 1997.
83. Girad, J. P., Culioli, J., Denoyer, C., Berdague, J. L., and Touraille, C., Comparison between two populations of two poultry species according to their fat composition, *Archiv für Geflugelkunde*, 57(1), 9–15, 1993.
84. Vainionpää, J., Smolander, M., Hanna-Leena Alakomi, H. -L., Ritvanen, T., Rajamäki, T., Rokka, M. and Ahvenainen, R., Comparison of different analytical methods in the monitoring of the quality of MAD broiler chicken using PCA, *Journal of Food Engineering*, 65(2), 273–280, 2004.
85. Rousset-Akrim, S., Young, O. A., and Berdagué, J.-L., Diet and growth effects in panel assessment of sheepmeat odour and flavor, *Meat Science*, 45(2), 169–181, 1997.
86. Hinrichsen, L. L. and Pedersen, S. B., Relationship among flavor, volatile compounds, chemical changes and microflora in Italian-type dry cured ham during processing, *Journal of the Science of Food and Agriculture*, 43, 2932–2940, 1995.
87. Guerrero, L., Gou, P., Alonso, P., and Arnau, J., Study of the physicochemical and sensorial characteristics of dry cured hams in three pig genetic types, *Journal of the Science of Food and Agriculture*, 70, 526–530, 1996.
88. Eklöv, T., Johansson, G., Winquist, F., and Lundström, I., Monitoring Sausage Fermentation using an electronic nose, *Journal of the Science of Food and Agriculture*, 76, 525–532, 1998.
89. Gerbanowski, A., Rutledge, D. N., Feinberg, M. H., and Ducauze, C. J., Multivariate regression applied to time domain-nuclear magnetic resonance signals: Determination of moisture in meat products, *Science des Aliments*, 17, 309–323, 1997.
90. Mittal, G. S. and Barbut, S., Effects of fat reduction on frankfurters' physical and sensory characteristics, *Food Research International*, 27, 425–431, 1994.
91. Casiraghi, E., Pompei, C., Dellaglio, S., Parolari, G., and Virgili, R., Quality attributes of milano Salami, an Italian dry cured sausage, *Journal of Agriculture and Food Chemistry*, 44, 1248–1252, 1996.
92. Dellaglio, S., Casiraghi, E., and Pompei, C., Chemical, physical and sensory attributes for the characterization of an Italian dry cured sausage, *Meat Science*, 42(1), 25–35, 1996.
93. Arvanitoyannis, I. S., Bloukas, J. G., Pappa, I., and Psomiadou, E., Multivariate data analysis of Cavourmas-a Greek cooked meat product, *Meat Science*, 54(1), 71–75, 2000.
94. Solange, B., Berdagué, J. L., Bousset, J., Cornet, M., Gandemer, G., Touraille, J., and Monin, G., Relations between compositional traits and sensory qualities of French dry cured ham, *Meat Science*, 37, 229–243, 1994.
95. Ellekjaer, M. R., Isaksson, T., and Solheim, R., Assessment of sensory quality of meat sausages using near infrared spectroscopy, *Journal of Food Science*, 59(3), 456–464, 1994.
96. Ellekjaer, M. R., Naes, T., and Baardseth, P., Milk proteins affect yield and sensory quality of cooked sausages, *Journal of Food Science*, 61(3), 660–666, 1996.
97. Papadima, S. N., Arvanitoyannis, I. S., Bloukas, J. G. and Fournitzis. G. C., Chemometric model for describing Greek traditional sausages, *Meat Science*, 51(3), 271–277.
98. Waade, C. and Stahnke, L. H., Dried sausages fermented with *Staphylococcus xylosus* at different temperatures and with different ingredient levels. Part IV. Amino acid profile, *Meat Science*, 46(1), 101–114, 1997.
99. Chunyu, S. and Reinhard, O., Rapid determination of the value-determining constituents in sausage products by near-infrared transmission spectroscopy, *Zeitschrift für Lebensmittel Unterschung und Forschung-A Food Research and Technology*, 201(3), 226–229, 1995.
100. Bloukas, J. G., Arvanitoyannis, I. S., and Siopi, A. A., Effect of natural colorants and nitrites on color attributes of frankfurters, *Meat Science*, 52(3), 257–265, 1999.
101. DePena, M. P., Cid, M. C., and Bello, J., A method for identification of frozen meat used for production of cooked ham, *Meat Science*, 48(3–4), 257–264, 1998.
102. Reyes-Cano, R., Dorantes-Alvarez, L., Hernandez-Sanche, H., and Gutierrez-Lopez, G. F., A traditional intermediate moisture meat: Beef Cecina, *Meat Science*, 36, 365–370, 1994.
103. Zalacain, I., Zapelena, M. J., De Pena, M. P., Astiasarán, I., and Bello, J., Use of lipase from Rhizomucor miehei in dry fermented sausages elaboration: Microbial, chemical and sensory analysis, *Meat Science*, 45(1), 99–105, 1997.
104. Desmond, E. M. and Kenny, T. A., Preparation of surimi-like extract from beef hearts and its utilisation in frankfurters, *Meat Science*, 50(1), 81–89, 1998.

105. Madsen, L. M., Stapelfeldt, H., Bertelsen, G., and Skibsted, L. H., Cochineal as a colorant in processed pork meat. Color matching and oxidative stability, *Food Chemistry*, 46, 265–271, 1993.

106. Lyons, P. H., Kerry, J. F., Morrissey, P. A., and Buckley, D. J., The influence of added whey protein carrageenan gels and tapioca starch on the textural properties of low fat pork sausages, *Meat Science*, 51(1), 43–52, 1999.

107. Boyazoglou, J. and Morand-Fehr, P., Mediterranean dairy sheep and goat products and their quality. A critical review, *Small Ruminant Research*, 40, 1–11, 2001.

108. Drake, M. A. and Gerard, P. D., Ability of hand evaluation versus mouth evaluation to differentiate texture of cheese, *Journal of Texture Studies*, 30, 425–441, 1998.

109. Mendia, C., Ibanez, F. C., Torre, P., and Barcina, Y., Effect of pasteurization on the sensory characteristics of a ewe's-milk cheese, *Journal of Sensory Studies*, 14(4), 415–424, 1999.

110. Barcenas, P., Elortondo, F. J. P., Salmeron, J., and Albisu, M., Development of a preliminary sensory lexicon and standard references of ewes' milk cheeses aided by multivariate statistical procedures, *Journal of Sensory Studies*, 14, 161–179, 1999.

111. Barcenas, P., Perez de San Roman, R., Perez-Elortondo, F. J., and Albisu, M., Consumer preference structures for traditional Spanish cheeses and their relationship with sensory properties, *Food Quality Preference*, 12, 269–272, 2001.

112. Wium, H. and Qvist, K. B., Prediction of sensory texture of Feta cheese made from ultrafiltered milk by uniaxial compression and shear testing, *Journal of Texture Studies*, 29, 215–232, 1998.

113. Ordonez, A. I., Ibanez, F. C., Torre, P., and Barcina, Y., Effect of ewe's milk pasteurization on the free amino acids in Idiazabal cheese, *International Dairy Journal*, 9, 135–141, 1999.

114. Careri, M., Magnia, A., Mori, G., and Musci, M., A new multivariate approach for the optimisation of the simultaneous distillation–extraction technique for free fatty acids using a face centred cube experimental design: Application to Parmigiano-Regiano cheese, *Analytica Chimica Acta*, 386, 169–180, 1999.

115. Martin, B. and Coulon, J. B., Facteurs de production du lait et characteristiques des fromages. I. Influence des facteurs de production sur l'aptitude a la coagulation des laits de troupeaux, *Lait*, 75, 61–80, 1995.

116. Rahali, V. and Menard, J. L., Influence des variants genetiques de la β-lactoglobuline et de la κ-caseine sur la composition du lait et son aptitude fromagere', *Lait*, 71, 275–297, 1991.

117. Nuits-Petit, V., Delacroix-Buchet, A., and Vassal, L., Influence de trois haplotypes des caseines αs1, β-et κ-frequents en race bovine Normande sur la composition du lait et l'aptitude a la fabrication fromagere', *Lait*, 77, 625–639, 1997.

118. Heil, F. and Dumont, J. P., Caracteristiques organoleptiques de fromages de chevre fabriques a partir de laits contenant des variants genetiques differents de la caseine αs1, *Lait*, 73, 559–565, 1993.

119. Ward, C. D. W., Koeferli, C. S., Schwegler, P. P., Schaeppi, D., and Plemmons, L. E., European strawberry yogurt market analysis with a case study on acceptance drivers for children in Spain using principal component analysis and partial least squares regression, *Food Quality and Preference*, 10, 387–400, 1999.

120. Ott, A., Hugi, A., Baumgartner, M., and Chaintreau, A., Sensory investigation of yogurt flavor perception: Mutual influence of volatiles and acidity, *Journal of Agricultural and Food Chemistry*, 48, 441–450, 2000.

121. Ramos, L., Eljarrat, E., Hernandez, L. M., Rivera, J., and Gonzalez, M. J., Levels of polychlorinated biphenyls, polychlorinated dibenzo-*p*-dioxins and dibenzofurans in commercial yogurt samples in Spain. Comparison with different dairy products, *Analytica Chimica Acta*, 402, 241–252, 1999.

122. Imram, N., Color and appearance of chilled mousse: A factorial study of ingredient and processing effects, *Nutrition and Food Science*, 1, 19–23, 1999.

123. Koeferli, C. R. S., Piccinali, P., and Sigrist, S., The influence of fat, sugar and non-fat milk solids on selected taste, flavor and texture parameters of a vanilla ice-cream, *Food Quality and Preference*, 7(2), 69–79, 1996.

124. King, B. M., Arents, P., and Moreau, N., Cost/efficiency evaluation of descriptive analysis panels—I. Panel size, *Food Quality and Preference*, 6, 245–261, 1995.

125. Koeferli, C. S., Schwengler, P. P., and Hong-Chen, D., Application of classical and novel sensory techniques in product optimization, *Lebensmittel-Wissenschaft und Technologie*, 31, 407–417, 1998.

126. Horimoto, Y. and Nakai, S., Classification of pasteurized milk using principal component similarity analysis of off-flavors, *Food Research International*, 31(4), 279–287, 1998.

127. Dufour, E., Subirade, M., Loupil, F., and Riaublanc, A., Whey proteins modify the phase transition of milk fat globule phospholipids, *Lait*, 79, 217–228, 1999.

128. Lopez, C. and Dufour, E., The composition of the milk fat globule surface alters the structural characteristics of the coagulum, *Journal of Colloid and Interface Science*, 233, 241–249, 2001.

129. Winquist, F., Krantz-Rulcker, C., Wide, P., and Lundstrom, I., Monitoring of freshness of milk by an electronic tongue on the basis of voltammetry, *Measurement Science and Technology*, 9, 1937–1946, 1998.

130. Horimoto, Y., Lee, K., and Nakai, S., Classification of microbial defects in milk using a dynamic headspace gas chromatograph and computer-aided data processing. 1. Principal component similarity analysis, *Journal of Agricultural and Food Chemistry*, 45, 733–742, 1997.

131. Saputra, D., Payne, F. A., Lodder, R. A., and Shearer, S. A., Selection of near-infrared wavelengths for monitoring milk coagulation using principal component analysis, *Transactions of ASAE*, 35(5), 1597–1605, 1992.

132. Demarigny, Y., Beuvier, E., Buchin, S., Pochet, S., and Grappin, R., Influence of raw milk microflora on the characteristics of Swiss-type cheeses: Biochemical and sensory characteristics, *Lait*, 77, 151–167, 1997.

133. Pripp, A. H., Shakeel, U.-R., McSweeney, P. L. H., and Fox, P. F., Multivariate statistical analysis of peptide profiles and free amino acids to evaluate effects of single-strain starters on proteolysis in miniature Cheddar-type cheeses, *International Dairy Journal*, 9, 473–479, 1999.

134. Califano, A. N., A, A. N., and Bevilacqua, E., Multivariate analysis of the organic acids content of Gouda type cheese during ripening, *Journal of Food Composition and Analysis*, 13, 949–960, 2000.

135. Dahl, S., Tavaria, F. K., and Malcata, F. X., Relationships between flavor and microbiological profiles in Serra da Estrela cheese throughout ripening, *International Dairy Journal*, 10, 255–262, 2000.

136. Sorensen, J. and Benfeldt, C., Comparison of ripening characteristics of Danbo cheeses from two dairies', *International Dairy Journal*, 11, 355–362, 2001.

137. Dewettinck, K., Dierckx, S., Eichwalder, P., and Huyghebaert, A., Comparison of SDS-PAGE profiles of four Belgian cheeses by multivariate statistics, *Lait*, 77, 77–89, 1997.

138. Pompei, C. and Spagnolello, A., Chemometric model for cooked ham, *Italian Journal of Food Science*, 9, 3–11, 1997.

139. Massart, D. L., Vandenginste, B. G. M., Deming, S. N., Michotte, Y., and Kaufman, L., *Chemometrics: A Textbook*, Amsterdam, The Netherlands: Elsevier, 1988.

140. Schalkoff, R., *Pattern Recognition*, New York: Wiley, 1992.

141. http://www.env.duke.edu/landscape/classes/env358 (accessed January, 2005).

142. http://www.foofproductdesign.com/archive (accessed January, 2005).

143. http://www.umetrics.com/methodtech_mva.asp (accessed January, 2005).

144. http://www.umetrics.com/download/KB/Multivariate/casestudies (accessed January, 2005).

145. Arvanitoyannis, I. S., Wine authenticity, In *Food Authenticity and Traceability*, M. Lees, Ed., Cambridge, UK: Woodhead Publishing, pp. 426–456, 2003.

146. Le Gall, G. and Colquhoun, I. J., NMR spectroscopy in food authentication, In *Food Authenticity and Traceability*, M. Lees, Ed., Cambridge, UK: Woodhead Publishing, pp. 131–155, 2003.

147. Kemsley, E. K. Lies, damned lies and chemometrics, Presented at Food Authenticity '96, University of East Anglia, Norwich, http://www.ifr.bbsrc.ac.uk/materials/sensors/publications.html (accessed January, 2005). 1996.

CHAPTER 11

Data Mining

Geoffrey Holmes

CONTENTS

Table 11.1 A Dataset in the Correct Format for WEKA

@relation heart-disease-simplified
@attribute age numeric
@attribute sex { female, male}
@attribute chest_pain_type { typ_angina, asympt, non_anginal, atyp_angina}
@attribute cholesterol numeric
@attribute exercise_induced_angina { no, yes}
@attribute class { present, not_present}
@data
63,male,typ_angina,233,no,not_present
67,male,asympt,286,yes,present
67,male,asympt,229,yes,present
38,female,non_anginal,?,no,not_present

who presents with typical angina and a cholesterol reading of 233; the angina was not exercise induced, and heart disease was not present.

Most data mining software acknowledges that data can come from a variety of sources, and it will provide ways of importing data from other formats such as binary, comma-separated values (CSV), universal resource locators (URLs), and databases.

Once the data has been transformed, the best way to get to know it is to use simple visualization tools. For nominal attributes, this usually means histograms across the set of values. These will usually conform to the person performing the visualization's background or domain knowledge of the data's distribution. Graphs of the values of numeric attributes are useful for discovering outliers. Both two and three-dimensional visualizations will uncover dependencies between attributes (and possibly the class attribute), and these can be very useful. Figure 11.1 shows a sample visualization

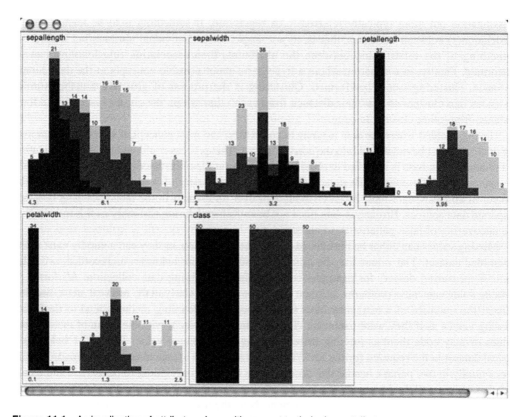

Figure 11.1 A visualization of attribute values with respect to their class attribute.

of some data using WEKA. This dataset has four numeric attributes and a nominal class attribute that can take on one of three values. As can be seen, the histogram for the class attribute shows that there are 50 instances of each class (a total of 150 instances in the dataset). The other histograms show how the numeric ranges of each attribute is distributed with respect to the class. This can show the attributes that best separate the classes. Here, it can be seen that sepalwidth and sepallength do not discriminate the classes as well as the other two attributes.

Other visualizations in two and even three dimensions are possible and can be helpful in showing dependencies, but in all cases, domain experts—people who truly understand the data—need to be consulted so that erroneous decisions are not made early in the process. Given that data mining involves large amounts of data, it is sensible to ask what to do if there is an excessive amount of data. The answer is to sample, and most packages make it straightforward to do this.

Of course, looking at the data is not the only operation that should be performed; others include discretization, normalization, resampling, attribute selection, transforming and combining attributes, and so on. These operations are needed to massage the data into more understandable forms. For example, suppose data on someone's preference for a chocolate bar was collected. Questions about certain features (attributes) of the bar such as its texture may have been asked, and other attributes such as its sweetness may have been measured. Further, a scale of one to ten for the individual's overall rating of the bar may have been used. Options for data mining at this point are to perform regression using the overall rating as the class or to perform classification by enumerating each of the ten grades into a set. Neither option will be terribly satisfactory. The regression will provide a real-valued number (somewhere between two possible grades in all likelihood), and classification will be weak because there probably will not be a sufficient number of instances of each of the ten grades to build a decent model.

Discretization could be applied to the overall rating to generate a smaller number of classes. Perhaps all one to three instances could be replaced with poor; four to six replaced with average; and the rest with good. There is a better chance now of constructing a decent model for each of these new grades. The chance element could be removed by discretizing into a known number of classes on the basis of ensuring an equal number of instances in each new class. Many options for discretization are available, and experimentation is needed to find the best for a given dataset. When discretizing attributes that are not the class attribute, options are available that either take into account the class labels (supervised) or not (unsupervised). In practice, supervised methods are preferred as they have typically led to superior results on classification tasks.

Many of the other techniques deal with the fact that a dataset can be large both in its number of instances and in its number of attributes. In the former, sampling can occur, and in the latter, the best attributes can be sought. In reality, it is not only the best attributes that need to be found, but it is also the worst. Adding a random (irrelevant) attribute can significantly degrade the performance of a data mining algorithm. In data mining, this is referred to as attribute selection, and it contains two parts

- A search method (for example, best-first, forward selection, backward elimination, random, exhaustive, genetic algorithm, and ranking)
- An evaluation method (correlation-based, cross-validation, information gain, chi-squared, and so on)

The first part searches through attribute space looking for the most useful (predictive) attributes. It measures utility by employing the second part. Two architectures for putting these two parts together have emerged. The first is called the filter approach that bases its assessment of attributes on general characteristics of the data. As such, it is purely a preprocessing step and is usually relatively fast. The alternate approach uses a learning method to select attributes. This method is called the wrapper approach and is typically much slower because of the running time of the learning method during cross-validation.

Some techniques exist for automatically cleansing data. For example, some methods attempt to relearn models with previously misclassified instances removed. Naturally, this can be a dangerous thing to do, and it is always better to let a domain expert check misclassified instances for possible data entry errors.

Noise is another source of difficulty in data. This can arise as class noise where two identical instances are given different class labels or as attribute noise. Attribute noise can arise in a variety of settings, for example, through faulty equipment giving noisy measurements. If the noise is systematic, then it is better to leave it in the data as the pattern of this noise may be uncovered. Unsystematic noise, however, needs to be removed as it will go undetected and will lead to spurious model construction farther down the data mining process.

11.3   DATA MINING METHODS

There are a number of criteria that have to be met for a data mining method to be considered practical. Minimally, an algorithm must have the ability to deal with

- Numeric attributes, missing values, and noise
- Very large datasets

Very large is not normally quantified. Rather, algorithms are sought with a complexity that scales with the number of instances. There are not many algorithms that pass the criteria, but there are ways of making use of more complex methods by periodically applying them to parts of the data (more on this later). In this field, some terms are overloaded, so discovering patterns and finding a model are equivalent. The models can be represented in many ways but typically come in the form of a tree or set of rules of the form, if Condition(s), then Consequence.

11.3.1   Classification

The WEKA system contains over 50 methods for performing data mining. The majority of methods perform classification. It is not possible to cover all of these; instead, focus will be on a few of the major paradigms. The first are decision trees. A decision tree attempts to build a tree in a top down (root first) recursive fashion that employs a divide-and-conquer methodology. The tree is made up of interior nodes that perform a simple test on an attribute and leaf nodes that contain a class label. The procedure first selects the attribute at the root node of the tree, and it creates a branch for each of the possible outcomes for the test. Next, the instances are split into subsets, one for each branch leading from the test node. Finally, the procedure is recursively applied to each branch using only the instances that reach the branch. The process stops if all the instances of a branch have the same class label.

At each iteration, an attribute to split the data should be chosen. The choice of best attribute is problematic. It would be best to choose the one that will ultimately lead to the smallest tree, but all possibilities cannot be exhaustively searched at all nodes. As a result, some form of educated guess or heuristic is needed to make the choice. Most heuristics attempt to choose the attribute test that leads to the purest separation of the classes. A popular purity criterion is called information gain. This measure increases with the average purity of the subsets that an attribute test produces, so the strategy is to choose the attribute test that results in the greatest information gain.

Consider the following problem. An optometrist has recorded data on the use of contact lenses in a range of patients and wants to know if there is a simple way of recommending contact lenses to new patients. The attributes of this problem are the age of the patient, the type of eye problem he or she has, if they have astigmatism, and the patient's tear production rate. The class attribute is the type of lens that was recommended. A typical decision tree for this problem is shown in Figure 11.2. The tree has the tear production rate at its root. This was considered the most important attribute when splitting up the data. The next most important attribute is the astigmatism attribute and so on. As can be seen, patients presenting with reduced tear production rates are not recommended for contact lenses. However, if their rate is normal, and they have no astigmatism, then soft lenses are preferred. Patients are recommended hard lenses if they have normal tear production, astigmatism, and are short-sighted (myopic).

Decision trees explain the relative importance of attributes in terms of their ability to divide the data by the class attribute, and they give a model of a problem that can be used for advice (prediction) in cases where the attribute values are known but the class attribute is unknown.

There are many decision tree algorithms. The best known, and probably most widely used, methods are C4.5[4] and its commercial successor C5.0.[5] The algorithms generally follow the top down procedure outlined above but differ in their attribute selection criteria. The result of this variation, however, almost always produces no difference in accuracy. Decision tree algorithms have a complexity of approximately $O(mn \log n) + O(n(\log n)^2)^3$ for a dataset of n instances and m attributes. This is a far cry from the linear or $O(n)$ criteria previously mentioned.

A quite different, but commonly used, approach that has its roots in Bayes' theorem is an algorithm called *NaiveBayes*.[6] Decision trees make greedy choices of attributes in growing a tree whereas NaiveBayes makes no assumptions about the relative importance of attributes. The only assumption it does make is that each attribute is statistically independent. Therefore, for a given class value, the values of each attribute for that class value are independent of each other. In practice, of course, this is highly unlikely, but equally improbably, it does not seem to lead to poor performance. The method can be easily misled by irrelevant attributes, so it is recommended that this method is used in combination with attribute selection.

An alternative to the greedy approach taken by decision trees that is also quite different from NaiveBayes is a method based on rote learning. This method has a long history in the pattern recognition literature and is known as nearest-neighbor or instance-based learning. Instead of building a tree or counting attribute values as above, the instances themselves are used as the underlying knowledge. If an instance has no class label then simply search through all the stored instances and find the closest match. The class label of this instance is the one predicted for the new instance. This approach is termed *lazy* because no decisions are made during learning that cannot be undone.

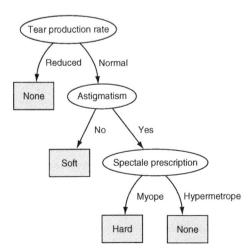

Figure 11.2 Decision tree for optometrist contact recommendation system.

For this scheme to operate, a distance function must be defined so that "closest match" makes sense. Typical distance functions are the Euclidean distance and the Manhattan or city block metric. Nominal attribute values have to be handled (for example, by setting the distance to one if values are equal and zero otherwise), and a method for dealing with missing values is also required. As with NaiveBayes, attributes are equally treated although weighting attributes can be beneficial. This method is both time and space consuming but can be made manageable.[7]

One of the most recent developments in data mining that is receiving much attention are algorithms for learning linear classifiers called support vector machines (or SVMs).[8] Many data mining methods produce models that are in some sense too close or too specific to the data from which they were learned. This phenomenon is called *overfitting*. The great strength of SVMs is their resilience to this phenomenon. The method attempts to learn something called the *maximum margin hyperplane*. The instances closest to the hyperplane are called support vectors, and they define the hyperplane. Unlike the instance-based methods that require many instances to be stored in memory, in SVMs, all other instances can be deleted without effect on the hyperplane. If a linear hyperplane is not sufficient, then other non-linear boundaries can be attempted.

SVMs have been found to be most applicable in domains with many attributes where each attribute incrementally adds to the concept. Example applications include human face identification, handwritten digit recognition, protein secondary structure prediction, and text classification.

11.3.2 Regression

Numeric prediction is also a requisite part of a data miner's toolkit. Simple methods exist for linear regression, and when applicable, they work very well. In circumstances where the data is non-linear, more sophisticated methods must be used. One method of doing this is to try to break up the data into parts that are approximately linear and then apply linear regression to those parts. It is a method that combines decision trees and linear regression. The most recent incarnation of this philosophy is the model tree[9] that has attribute tests on interior nodes and linear regression models at the leaves. Other systems such as CART[10] and CUBIST[11] work in a similar fashion.

11.3.3 Association Rules

The aim in association rule learning is to find a correlation between arbitrary attribute-value pairs in data. The class attribute is treated in the same way as all other attributes in this method. The most often quoted application is called market basket analysis where supermarket basket goods are analyzed to answer questions such as when someone buys X, what else do they tend to buy? Obvious correlations such as bread and milk are uncovered alongside some less obvious combinations, classics being beer and diapers.

The single biggest issue in association learning is managing the large number of possible associations. Two parameters called support and confidence are normally used to restrict the number of associations. Support is measured by the number of instances that are correctly predicted by a rule, and confidence is the number of correct predictions it makes relative to all the instances it applies to. For example, if six shopping baskets were viewed and four of them had bread and milk together and all had bread, then for the rule, if bakery item = bread, then fridge item = milk, support is the number of baskets that have both bread and milk (4), and confidence is the proportion of baskets with bread that also have milk (4 from 6 or 66%). In other words, people seek frequent items with strong correlations with other items.

Support and confidence are actually specified before rules are generated with support over 4 and confidence over 80%. An efficient algorithm called *Apriori* is used to generate the rules.[12] Some issues remain as confidence is not that useful for items that are very common such as milk, and although Apriori is efficient, it can still be overwhelmed by large datasets. Research is continuing into alternative algorithms, for example [13].

11.3.4 Clustering

The target concept in most data mining tasks is usually well-defined. For association rule learning, it is not as important but is at least present. In clustering, there is typically no such concept, and the algorithm must attempt to discover collections of instances that are similar to each other but distinct from other collections. There are many clustering algorithms and also some differences in the clustering goals. For example, some methods attempt to place instances in exclusive clusters whereas others accept that clusters overlap; therefore, an instance may belong to several clusters.

All clustering methods must deal with the notion of similarity, and this usually means defining a distance function between two instances. Various distance metrics have been proposed (see the section on instance-based learning). Clustering methods vary tremendously in approach. Some are probabilistic and others deterministic; some are hierarchical and others flat; and some work incrementally and others on batches of instances. The classic k-means[14] algorithm has the longest history, is simple, and works well in practice. Other systems that work quite differently include the expectation maximization (EM) algorithm,[15] COBWEB,[16] AUTOCLASS,[17] and fractal clustering.[18]

11.3.5 Meta Learning

Meta learning is one of the most significant recent developments in data mining. The basic idea is to try to get more out of an existing classifier by building several models of the training data rather than one. This is achieved by presenting several versions of the data to an algorithm and storing the resulting models in a committee structure. At prediction time, each model votes, and the majority vote is used to determine the outcome. Generating the versions of the dataset gives rise to different approaches. Taking samples of the data gives rise to a technique known as bagging.[19] An alternative approach is to use weights on each instance then re-weighting according to performance on the last model. Instances that are incorrectly classified are given a boost in weight so that they may be the focus of the next model. This technique is known as boosting.[20]

In both techniques, it is possible to enhance the basic performance of a classification algorithm. The maximum performance on some datasets is reached early whereas, for others, many models have to be constructed. The automatic discovery of the correct number of models to build for a given dataset is still an open research question.

11.3.6 Online Learning

There are application areas where it is possible to imagine a data source that continuously provides data (for example, a manufacturing line). In these circumstances, the data cannot be processed all together, partly because it is arriving all the time and partly because it could not possibly be stored and processed quickly enough. It would be useful, however, to be able to provide some form of monitoring capability to keep an eye on the data to ensure that it does not deviate from known limits.

One solution to this problem is to learn a model from a finite source and then make continuous predictions on the basis of that source. New models could be re-trained offline and brought into operation at set times. Alternatively, models could be constantly updated at the same time that predictions are being made.[21] There is strong interest for more research in this area.

11.4 EVALUATION

Thus far, this chapter has not addressed how models are assessed. The situation is somewhat different for each of the methods previously outlined. Clustering, for example, is the most problematic as it can be subjectively and objectively judged, and there appears to be no universally accepted standard. In regression, there is a choice of statistics that can be used to assess performance. Correlation coefficients, mean absolute error, root mean squared error, relative absolute error, and root relative absolute error can all be computed.

In classification, the picture is much simpler. The natural performance measure is error rate that is the proportion of errors made by a classifier over a set of instances. An error occurs if the class label predicted by a model differs from the class label of the instance under consideration. One error rate that is easy to compute is the error rate a model attains on the data from which it learned. This is known as the resubstitution error. This error rate is notoriously optimistic and is not a good measure of performance.

More typically, a separate test set of data, independent of the training data, is used to test performance. The test data is drawn from the same distribution as the training data, but it is not used in the construction of the model. Many data mining methods have parameters that need to be tuned to a particular application. In these circumstances, a third distinct set of data, called validation data, is required. This data will be used to tune parameters.

When there is a large amount of data, then a single training, test and possibly validation dataset may be all that is required. For smaller datasets, it is problematical because subtle variations in the training data can lead to different outcomes in performance. In such circumstances, a technique known as k-fold cross-validation is typically used. This technique divides the data into k subsets of equal size. Then, each subset, in turn, is used for testing and the remainder used for training. In practice, the subsets are stratified before cross-validation to ensure that the distribution of the classes is approximately equal in each subset.

The overall error is computed as an average of the errors at each of the k-folds. The standard approach is to choose k equal to 10. This value has been shown to be a good choice experimentally. It represents something of a trade between size and computation. Researchers often go a step farther and repeat the 10-fold cross-validation ten times. This step further reduces the error estimate's variance, further ruling out chance effects.

Given that the performance of a classifier can be measured, it seems natural to then ask how to compare two classifiers on the same data. If the repeated 10-fold cross-validation procedure for two classifiers is performed, then the variance is reduced, but the reliability of the results is questionable. To establish confidence that there is a real difference in the results, significance tests are performed (typically a t-test).

There are many other techniques for evaluation such as the bootstrap and ROC (Received Operating Characteristic) curves. The topic of evaluation is a topic of research. Surprisingly, perhaps, there is no agreed standard for comparing the performance of two methods.[22]

11.5 APPLICATIONS

The number of application areas for data mining has grown enormously in the last few years. In broad terms, the areas cover medicine, commerce, manufacturing, bioinformatics, and so on. There are many commercial and open-source data mining tools, and these have helped to increase application development. In this section, attention is focused on some of the applications that have been developed with the WEKA system.

WEKA was originally developed to process agricultural data because of the importance of this application area in New Zealand. However, because of significant growth in methods and data engineering capability, it is now commonly used in all forms of data mining application.

The major conferences regularly release competition datasets and give prizes to the best (they release training and validation data and hold back test data). These datasets are usually from a commercial source and represent real-world applications. Research papers are often written on the outcomes of these competitions to further understanding of best practice.

In New Zealand, there are many research centers dedicated to agriculture and horticulture. These centers provided many of the early applications of the WEKA system. Some examples include predicting the internal bruising sustained by several varieties of apple as they make their way through a pack-house on a conveyor belt,[23] predicting the quality of a mushroom from a photograph in real-time in order to provide automatic grading,[24] and classifying kiwifruit vines into twelve classes in order to determine which one of twelve pre-harvest fruit management treatments has been applied to the vines. The data for this project was visible-NIR spectra.[25] The applicability of WEKA in this domain was also the subject of further research[26] that showed a high level of satisfaction with WEKA and some advice on improvements. Other work directly connected to the food industry can be found in[27-31], a synopsis for which is provided in Table 11.2.

In the field of bioinformatics, WEKA has been extensively used. Applications include automated protein annotation in[32,33]; probe selection for gene expression arrays[34]; experiments with automatic cancer diagnosis[35]; development of a computational model for frame-shifting sites[36]; and several others.

Another application area that has received widespread attention is text classification. The idea is to automatically classify pieces of text so that they can be appropriately grouped together. The Reuters news agency, for example, would be able to group news items without having to employ people to do it. The problem is extremely difficult because of subject overlap. Detecting the main subject area of an article is harder than it seems. Other applications revolve around the provision of digital libraries and search engines. In these systems, it is important that articles provide accurate keywords so that query results are meaningful. Research on the automatic extraction of keywords has shown a great deal of promise.[37]

Many of the future applications involving WEKA will be developed in an online setting. The recent work on data streams[38] enables WEKA to be used in situations where a possibly infinite source of data is available. As previously mentioned, such situations are common in manufacturing industries with 24/7 processing. The challenge is to develop models that constantly monitor data in order to detect changes from steady state. Such changes may be indicative of failure in the process and may provide operators with early warning signals that equipment needs re-calibration or replacement. Both case studies overleaf were analyzed with the WEKA software.

Table 11.2 Synoptic Table of Applications

Reference Number	Key Findings
24	Drop height and location of impact determine likelihood of bruising
25	It is possible to perform automatic mushroom grading for quality
26	Agricultural users satisfied with machine learning as a data analysis technique
27	Multivariate techniques can be used to tackle problems in food science such as sugar quality prediction and final product sensory quality
28	Time series analysis of point-of-sale data. General business solutions incl. seasonal, performance, promotional analysis
30	AI techniques are useful in food processes because they automatically identify non-linear behavior, explain themselves and find the major influences affecting performance
31	Found context-dependent variables in a cheesemaking process evaluation
40	Successful application to the process of tulip bulb-forcing

11.5.1 Case Study 1: Baking Industry

Automated plant bakeries are now commonplace, and although a large amount of operational data is collected, very little of it is analyzed. The following study[39] looked for relationships between the process data and the product quality. The process data was collected from two dough mixers, a divider, check weigher, rounder, and intermediate prover.

Data acquisition issues surrounded the timing of collection, for example, whether to record every second, minute, or longer time unit. Not all the data was recorded by instruments connected to the process; for example, photo images of the finished product were taken to arrive at quality measures. Some data was not recorded because of computer constraints. In any data mining application, it is highly likely that extra data is required, data that is thought to have an impact on the process. The ease of collecting this information should not be underestimated.

11.5.1.1 Data Preparation

In most applications, visualization is a useful tool for detecting outlers or errors in the data. Some automatic tools can also be tried, for example, by computing the leverage of an example. In this study, missing data needed to be added and links made between data collected from different parts of the process (typically by different recording instruments).

Because this application is time dependent, averaging was necessary to arrive at meaningful instances. Additional attributes were also generated such as the time a loaf traveled between stages of the process. The data was clustered so that average values could be incorporated to form cluster instances so that a cluster of dough could be tracked through the entire process.

11.5.1.2 Analysis

Loaves are made up of five pieces of dough dropped into a tin by a divider. By tracking individual pieces of dough, it was determined that the divider regularly produced heavier pieces of dough in three positions (generally the same positions) than the other two. It was established that the position of the dough in the tin has an impact on the final quality of the loaf.

Many applications focus on the problems within a process, in this case, the issue of top collapse (the loaf has a convex shape on top). The cause of such collapse is generally not well-understood. A decision tree was used on data labeled high and low (top collapse), and a relationship was found as follows:

If Attribute X1 > Value 1 then top collapse is high
Else   If Attribute X2 > Value 2 then top collapse is low
          Else   If Attribute X3 > Value 3 then top collapse is low
                    Else top collapse is high

The details of the attributes and values can be found in [39]. The important aspect of this study is that it is possible to determine from the above that attribute X1 is the most important attribute in determining top collapse and that, altogether, only three variables (from those recorded) are important.

11.5.2 Case Study 2: Horticulture

This study attempted to apply data mining to three related datasets connected with the kiwifruit industry. The question addressed was to determine if data mining could be used at maturity clearance time to assess the risk of physiological pit occurring within individual *Actinidia chinensis* cv. Hort16A orchard lines that are stored for more than 6–8 weeks.

Hort16A fruit that are prone to physiological pit have been shown to differ in fruit composition and maturity when compared to unaffected fruit. Although the compositional and maturity differences may not be directly causal for physiological pit, the combination of these affected fruit properties may allow at-risk lines to be identified. Such an approach is akin to metabolic profiling where changes in gas chromatograph/mass spectroscopy profiles of plants can be used as a diagnostic technique.

11.5.2.1 Methods

Three industry databases provided the raw data for this study. The first database, orchard identity, included data on orchard geographic location. The second database, maturity clearance, contained the 2003 season data on clearance date, fruit maturity characteristics, and fruit age. The third database, storage library, held data on defects recorded in 2003 on a representative sample of fruit from each orchard line.

These relational database tables needed to be transformed into a single flat table before the data mining algorithms could be applied. In this study, an instance was taken as the information unique to a specific maturity area. Disorder data was discretized into groups of equal frequency, resulting in an instance having a low or high potential to express a disorder such as physiological pit. The final table consisted of 769 instances (each representing a maturity area) and 17 attributes.

Many learning algorithms were applied to the data, and the percentage of instances that were correctly predicted was used to assess the predictive accuracy of the models they produced. The evaluation method used was a 10×10 cross-validation (see above). The algorithms' performance on instances containing information about the presence of other defects was compared against the performance of these same algorithms against instances where the presence of the other defects was excluded.

11.5.2.2 Results

The physiological pit risk was correctly predicted 75–82% of the time, depending on the prediction algorithm used and the presence or absence of other defects in the instance data (Table 11.3). These results were considerably higher than the default accuracy (50%), and three algorithms were significantly better than NaiveBayes (the least computational method). This gives evidence that the learning algorithms were able to find patterns in the data.

Table 11.3 Percent Correct on Predicting Physiological Pit Risk Under a 10×10 Cross-Validation

Algorithm	Without Extra Defect Counts	With Extra Defect Counts
NaiveBayes	78.24	75.17
J48	77.56	76.20
ADTree	78.62	78.00
RandomForest	80.40	**81.31**
PART	76.85	75.37
JR	78.53	76.37
Logistic	81.04	**80.27**
SMO	**81.85**	**80.40**
Average	79.14	77.89

Values in bold are significantly better than NaiveBayes according to a 5% corrected *t*-test.

Adding the extra defect information caused predictions of algorithms such as NaiveBayes to suffer and others like RandomForest to improve. This behavior could be explained if some of the attributes' being introduced are highly correlated with other attributes. Correlated attributes are known to hinder schemes like NaiveBayes, whereas schemes like RandomForest are more resistant to this effect.

11.5.2.3 Conclusion

These finding indicate the potential to predict storage disorders such as physiological pit in Hort16A from data already collected within a kiwifruit supply chain. This work needs to be validated over several seasons. Where possible, the weighting that the various data mining learning algorithms place on specific attributes should be determined, allowing the physiological basis of the models to be assessed.

11.6 CONCLUSION

Although only in its infancy, the field of data mining has enormous potential in the quest to extract knowledge from data. Many exciting applications are under development in many different fields of endeavor. The provision of open-source toolkits such as the WEKA system enables users with application level experience to analyze data. Although young, the field inherits much of its solid foundation from the fields of statistics, machine learning, and databases.

In building applications of this technology, it is important to learn from the past, particularly the experience that has been formed in the field of applied statistics where data pre-processing, algorithm application, and result evaluation have featured as standard methodology for some time. The great leap forward, of course, is the processing power that is now available. The inclusion of databases as a subfield of data mining has seen new research emerge that attempts to place data mining more explicitly in database systems, for example, extending SQL to perform data mining queries. In fact, many commercial database systems already have data mining features.

Data mining does not replace any of the other data analysis techniques that have been developed and used for decades. For many processes, where background knowledge and underlying physical principles are well-understood, mechanistic models are likely to prove more successful than models induced from data. One of the great challenges in data mining is the incorporation of background knowledge. Very few systems even permit it. More research is needed to find ways to use this information in existing methods and to develop hybrid systems that combine the best features of the mechanistic and inductive models.

ACKNOWLEDGMENTS

The author would like to thank the WEKA team for providing daily inspiration.

GLOSSARY

Attribute A variable normally associated with input as an observation used to characterize an object or a process. For example, if characteristics of mushrooms were recorded, then the length of the stalk might be an attribute. Also referred to as independent variable or feature.
Attribute selection Process by which irrelevant attributes are removed from consideration. In the

statistics literature, this is known as dimensionality reduction. As the above, can be called feature selection.

Classification   Process by which a learning model is developed to describe a finite set of target values. Often referred to as supervised learning.

Clustering   Process where groups of data are discovered that are similar to each other yet are distinct from other groups. Often referred to as unsupervised learning.

Concept   A description of an object in terms of its characteristics (attributes), including a special target attribute that is dependent on those characteristics.

Data mining   An emerging field related to machine learning and statistics that attempts to uncover previously unknown information from large amounts of data.

Decision tree   A data structure comprising interior nodes that represent tests on attributes and leaves that hold classification decisions.

Generalization   Extending the definition of a concept beyond its description as a collection of instances.

Heuristics   Rules of thumb that are used to guide decision making.

Incremental learning   Learning in stages by updating a model when new information becomes available.

Inductive learning   Learning by employing facts and observations obtained from a teacher or an environment.

Instance   An example observation normally arranged as a set of attribute values. Can be referred to as an example.

Machine learning   A field of artificial intelligence dedicated to the development of theories and practical systems for learning from data.

Model   A description of data, usually in the form of a data structure, such as a tree or set of rules. Models are the embodiment of knowledge in learning systems.

Prediction   Process of using an instance of values and a model to produce a target value.

Regression   Process by which a learning model is developed to describe a continuous set of target values. Statisticians refer to this as classification with a real-valued target variable.

Supervised learning   Branch of learning where observations are labeled by a teacher.

Training data   Data used to build a model from instances with pre-specified labels in the case of classification.

Test data   Data used to test a model built from training data.

Unsupervised learning   Branch of learning where hypotheses are developed, concerning a collection of facts and observations without any a priori knowledge of where the data came from.

REFERENCES

1. Samuel, A. L., Some studies in machine learning using the game of checkers, In *Computers and Thought*, Feigenbaum, E. A. and Feldman, J., Eds., New York: McGraw-Hill, pp. 71–105, 1963.
2. Soloway, E. M., Learning = interpretation + generalization: A case study in knowledge-directedlearning, PhD diss., Amherst: University of Massachusetts, 1967.
3. Witten, I. H. and Frank, E., *Data Mining: Practical Machine Learning Tools and Techniques With Java Implementations*, San Francisco, CA: Morgan Kaufmann, 2000.
4. Quinlan, J. R., *C4.5: Programs for Machine Learning*, San Matgeo, CA: Morgan Kaufmann, 1993.
5. C5.0 is available from Rulequest Research. http://www.rulequest.com..
6. Langley, P., Iba, W., and Thompson, K., An analysis of Bayesian classifiers, In *Proceedings of Tenth National Conference on Uncertainty in Artificial Intelligence*, Swartout, W., Ed., Menlo Park, CA: AAAI Press, pp. 223–228, 1992.
7. Aha, D., Tolerating noisy, irrelevant, and novel attributes in instance-based learning algorithms, *International Journal of Man-Machine Studies*, 36(2), 267–287, 1992.

8. Platt, J., Fast training of support vector machines using sequential minimal optimization, In *Advances in Kernel Methods—Support Vector Learning*, Scholkopf, B., Burges, C., and Smola, A., Eds., Cambridge, MA: MIT Press, pp. 185–208, 1999.

9. Wang, Y. and Witten, I. H., Induction of model trees for predicting continuous classes, In *Proceedings of the Poster Papers of the European Conference on Machine Learning*, van Someren, M. and Widmer, G., Eds., Prague, Czech Republic: University of Economics, Faculty of Informatics and Statistics, pp. 128–137, 1997.

10. Breiman, L., Friedmann, J. H., Olshen, R. A., and Stone, C. J., *Classification and Regression Trees*, Monterey, CA: Wadsworth, 1984.

11. CUBIST is available from Rulequest Research. http://www.rulequest.com.

12. Agrawal, R., Imielinski, T., and Swami, A., Mining association rules between sets of items in large databases, In *Proceedings of the ACM SIGMOD International Conference on Management of Data*, Buneman, P. and Jajodia, S., Eds., Washington, DC: ACM press, pp. 207–216, 1993.

13. Han, J., Pei, J., Yin, Y., Mao, R., Mining frequent patterns without candidate generation: a frequent-pattern tree approach, *Data Mining and Knowledge Discovery*, 8, 53–87, 2004.

14. Hartigan, J. A., *Clustering Algorithms*, New York: Wiley, 1975.

15. McLachlan, G. and Krishnan, T., *The EM algorithm and extensions*, Wiley Series in Probability and Statistics, New York: Wiley, 1997.

16. Fisher, D., Knowledge acquisition via incremental conceptual clustering, *Machine Learning*, 2(2), 139–172, 1987.

17. Cheeseman, P. and Stutz, J., *Bayesian classification (AutoClass): Theory and results Advances in Knowledge Discovery and Data Mining*, Menlo Park, CA: AAAI Press, 1995 pp. 153–180

18. Barbara, D. and Chen, P., Using the fractal dimension to cluster datasets, *International Conference on Knowledge Discovery and Data Mining*, Boston, MA: ACM press, pp. 260–264, 2000.

19. Breiman, L., Bagging predictors, *Machine Learning*, 24(2), 123–140, 1996.

20. Freund, Y. and Schapire, R. E., Experiments with a new boosting algorithm, In *Proceedings of the Thirteenth International Conference on Machine Learning*, Saitta, L. and Bari, L., Eds., San Francisco, CA: Morgan Kaufmann, pp. 148–156, 1996.

21. Frank, E., Holmes, G. R., Kirkby, and Hall M., Racing committees for large datasets, In *Proceedings of the International Conference on Discovery Science*, Berlin: Springer LNCS 2534, pp. 153–164, 2002.

22. Bouckaert, R., Choosing between two learning algorithms based on calibrated tests, In *Proceedings of the International Conference on Machine Learning*, San Francisco, CA: Morgan Kaufmann, pp. 51–58, 2003.

23. Holmes, G., Cunningham, S. J., Dela Rue, B., and Bollen, F., Predicting apple bruising using machine learning, *Acta Horticulturae*, 476, 289–296, 1998.

24. Kusabs, N., Bollen, F., Trigg, L., Holmes, G., and Inglis, S., Objective measurment of mushroom quality, In *Proceedings of the New Zealand Institute of Agricultural Science and the New Zealand Society for Horticultural Science Annual Convention*, 51, 1998.

25. Holmes, G. and Hall, M., A development environment for predictive modelling in foods, *International Journal of Food Microbiology*, 73, 351–362, 2002.

26. McQueen, R. J., Holmes, G., and Hunt, L., User satisfaction with machine learning as a data analysis method in agricultural research, *New Zealand Journal of Agricultural Research*, 41(4), 577–584, 1998.

27. Bro, R., van den Berg, F., Thybo, A., Andersen, C. M., Jørgensen, B. M., and Andersen, H., Multivariate data analysis as a tool in advanced quality monitoring in the food production chain, *Trends in Food Science & Technology*, 13(6-7), 235–244, 2002.

28. Liu, L.-M., Bhattacharyya, S., Sclove, S. L., Chen, R., and Lattyak, W. J., Data mining on time series: an illustration using fast-food restaurant franchise data, *Computational Statistics & Data Analysis*, 37(4), 455–476, 2001.

29. O'Neill, R. T. and Szarfman, A., Some US Food and Drug Administration perspectives on data mining for pediatric safety assessment, *Current Therapeutic Research*, 62(9), 650–663, 2001.

30. Goyache, F., Bahamonde, A., Alonso, J., Lopez, J., del Coz, J. J., Quevedo, J. R., Ranilla, J., Luaces, O., Alvarez, I., Royo, L. J., and Diez, J., The usefulness of artificial intelligence techniques to assess subjective quality of products in the food industry, *Trends in Food Science & Technology*, 12(10), 370–381, 2001.

31. Guillaume, S. and Charnomordic, B., Knowledge discovery for control purposes in food industry databases, *Fuzzy Sets and Systems*, 122(3), 487–497, 2001.

32. Kretschmann, E., Fleischmann, W., and Apweiler, R., Automatic rule generation for protein annotation with the C4.5 data mining algorithm applied on SWISS-PROT, *Bioinformatics*, 17, 920–926, 2001.

33. Bazzan, A. L., Engel, P. M., Schroeder, L. F., and da Silva, S. C., Automated annotation of keywords for proteins related to mycoplasmataceae using machine learning techniques, *Bioinformatics*, 18, 35S–43S, 2002.

34. Tobler, J. B., Molla, M. N., Nuwaysir, E. F., Green, R. D., and Shavlik, J. W., Evaluating machine learning approaches for aiding probe selection for gene-expression arrays, *Bioinformatics*, 18, 164S–171S, 2002.

35. Li, J., Liu, H., Ng, S. K., and Wong, L., Discovery of significant rules for classifying cancer diagnosis data, *Bioinformatics*, 19, 93ii–102ii, 2003.

36. Bekaert, M., Bidou, L., Denise, A., Duchateau-Nguyen, G., Forest, J. P., Froidevaux, C., Hatin, I., Rousset, J. P., and Termier, M., Towards a computational model for -1 eukaryotic frameshifting sites, *Bioinformatics*, 19, 327–335, 2003.

37. Frank E., Paynter, G. W., Witten, I. H., Gutwin, C, and Nevill-Manning, C. G., Domain-specific keyphrase extraction, In *Proceedings of the International Joint Conference on Artificial Intelligence*, San Francisco, CA: Morgan Kaufmann, pp. 668–673, 1999.

38. Holmes, G., Pfahringer, B., and Kirkby, R., Mining data streams with option trees, Working Paper 08/03, Computer Science Department, University of Waikato, 2003.

39. Wilson, A. J., Morgenstern, M. P., Pfahringer, B., and Leschi, C., Data mining bread quality and process data in a plant bakery, In *Proceedings of the 12th ICC Cereal and Bread Congress*, Woodhead Publishing Limited, 2004.

40. Verdenius, F., Methodological aspects of designing induction-based applications, *SIKS Dissertation Series*, University of Amsterdam, 2004.

FURTHER READING

As previously mentioned, an excellent starting point is the book by Witten and Frank.[3] Each chapter of this book has a further reading section where more specific material can be sourced. Verdenius[40] provides an excellent overview of the area with examples of agricultural applications. The KDD Nuggets website at http://www.kdnuggets.com is another excellent source of information. Some of the major conferences and journals in this area are the International Conference on Machine Learning, European Conference on Machine Learning, Knowledge Discovery and Data Mining, IEEE International Conference on Data Mining, Machine Learning Journal, Journal of Machine Learning Research, IEEE Transactions on Knowledge and Data Engineering, and Data Mining and Knowledge Discovery Journal.Many software suites integrate various aspects of the data mining process. Some of the best known commercial software packages are SAS Enterprise Miner (http://www.sas.com), SPSS (http://www.spss.com), and DBMiner (http://www.dbminer.com). A full overview of software, including surveys, can be obtained from the KDNuggets site previously mentioned.

CHAPTER 12

Artificial Neural Network Modeling

Shyam S. Sablani

CONTENTS

12.1 INTRODUCTION

Artificial neural networks are computational structures inspired by biological neural systems. Conventional computational models are particularly well suited to executing sequences of instructions that have been precisely formulated for them. On the other hand, biological neural systems are well suited for tasks/operations such as speech, vision, information retrieval, generalization, and complex spatial and temporal pattern recognition in the presence of noisy distorted data, all of which are extremely difficult to accomplish by conventional computing methods. Therefore, the motivation for artificial neural networks (ANNs) is to achieve many of those desirable abilities of the biological neural systems.

An artificial neural network is a massively parallel-distributed processor made up of simple processing units, which has a natural tendency for strong empirical knowledge and making it available for use [1]. ANNs have the ability to learn from examples (i.e., data) through iteration without requiring a prior knowledge of the relationship of the parameters and therefore generalize (i.e., establish relationship between input and output data). The neural networks are also capable of dealing with uncertainties, noisy data, and nonlinear relationships.

Increasingly, artificial neural networks are being used as effective general-purpose, nonlinear regression tools and for developing models governed by complex relationships. ANNs are especially useful for classification and function approximation/mapping problems which have lots of data available but to which hard and fast rules (such as those that might be used in an expert system) cannot easily be applied. Most neural networks that can learn to generalize effectively from complex data are similar to classical statistical methods. One of advantages of ANNs over the conventional statistical methods is that they do not require assumptions about the distribution of the data to analyze it. ANNs are more tolerant of imperfect or incomplete data. They perform better when there are complex nonlinear relationship in the data, especially with regard to classification, pattern recognition and forecasting.

There are several problems in the food and bioprocessing area that cannot be characterized and solved using a physics-based modeling approach. In those situations artificial neural network modeling can be used as potential alternatives to physics-based models in food and bioprocessing. The detail on physics base modeling is presented in Chapter 2 through Chapter 6. It is practically impossible to cover all aspects of artificial neural networks and their application in a single book chapter. Several books [1–4] and review articles [4–7] have been published on artificial neural networks. Most of these books highlight the fundamental aspects of artificial neural networks such as components of ANN and how they work. Most of the review articles have focused on providing some background and basic principles related to ANN and their possible applications in food processing without emphasis on how to formulate and model a particular problem using ANN modeling methods. The primary difficulty for a beginner in using any new modeling technique is to formulate a problem and choose the right analysis. The scope of this chapter is to introduce a limited amount of fundamentals on artificial neural networks to a beginner to get him/her started on using this modeling technique. The major emphasis is on how to develop artificial neural network-based models using selected examples. Each application example is presented with step-by-step solution procedure that includes major results.

12.2 PRINCIPLES OF NEURAL NETWORKS

12.2.1 Biological Neural Networks

A biological neural system consists of several neurons. The idea of neurons as structural constituents of the brain was introduced by Ramon y Cajal [8] in his pioneering work in understanding the biological nervous system. Typically, neurons are five to six orders of magnitude slower than silicon logic gates; events in a silicon chip happen in the nanosecond (10^{-9} s) range, whereas neural events happen in the millisecond (10^{-3} s) range. However, the brain makes up for the relatively slow rate of operation of a neuron by having a truly staggering number of neurons with massive interconnections between them. It is estimated that there are approximately 10 billion neurons in the human cortex, and 60 trillion synapses or connections. The net result is that the brain is an enormously efficient structure. The energetic efficiency of the brain is approximately 10–16 Joules per operation per second, whereas the corresponding value for the best computers in use today is about 1906 Joules per operation per second [1,9,10].

Neurons come in a wide variety of shapes and sizes in different parts of the brain. A typical neuron contains three major parts: soma (cell body), axon, and dendrites. The axon (the transmission lines), and dendrites (the receptive zones), constitute two types of cell filaments that are distinguished on morphological grounds. An axon has a smoother surface, fewer branches, and greater length, whereas a dendrite has an irregular surface and more branches. At the end of the axon, it contacts dendrites of neighboring neurons at a special contact organ, called the synapse, where the signals are passed between neurons (Figure 12.1). The synapses are considered as elementary structural and functional units that mediate the interactions between neurons. The signals are transmitted electrically and affected by chemical transmitters released at the synapse. The chemical transmitters also affect the response of the neuron that receives the signals.

The incoming signals from neighboring neurons are in an excitatory state if they cause firing, or an inhibitory state if they hinder the firing response. The condition for firing or not firing is decided by the state of aggregation of impulses. If it is in excitatory mode and exceeds a certain level (called the threshold value), then the neuron will generate a pulse response and transmit it through its axon. Thus, the activation depends on the number of signals received, the strength of the incoming signals, and the synaptic strength of the connections. The magnitude of signals is not significantly different among biological neurons. Therefore, we can treat the neurons as passing information by means of binary signals.

Therefore a neuron can be considered as a simple signal processing unit with multiple inputs from other neurons and only a single output that is distributed to other neurons. The neuron aggregates the incoming signals, and when the signal exceeds a certain threshold level, it will produce an output signal and transmit it to other neurons. Each neuron performs only a simple function, but when neurons are massively connected together, they can perform complicated tasks.

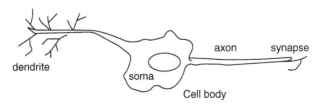

Figure 12.1 The biological neuron model.

12.2.2 Artificial Neural Networks

An artificial neural network (ANN) is a computational structure and is characterized by net topology, node characteristics, and learning rules. The basic processing unit is called a neuron (node), which performs the simple functions of summing inputs and nonlinear mapping. Each connection (synapse) comes with a numerical value, called a weight that expresses the strength of connection. The potential advantages and capabilities of ANN are:

1. Nonlinearity. An artificial neuron can be linear or nonlinear. A neural network, made up of an interconnection of neurons, is itself nonlinear. The nonlinearity is distributed throughout the network.
2. Input–output mapping. ANN is capable of learning and mapping input–out data. The learning involves modification of the synaptic weights of a neural network by applying a set of labeled learning/training data.
3. Adaptivity. Neural networks have a built-in capability to adapt their synaptic weights to changes in the surrounding environment. In particular, a neural network trained to operate in a specific environment can be easily retrained to deal with minor changes in the operating environmental conditions.
4. Fault tolerance. The ANN is tolerant to noisy and incomplete data, because the information is distributed in the massive processing nodes and connections. Minor damage to parameters in the network will not degrade overall performance significantly.
5. High computational speed. Potentially, the massive parallel nature of a neural network makes it fast for the computation of certain tasks.

12.2.2.1 Model of an Artificial Neuron

McCulloch and Pitts [11] were the first to propose a formal model of an artificial neuron (Figure 12.2a). The model was based on the highly simplified considerations of the biological model. The elementary computing neuron functions as an arithmetic logic-computing element. The inputs x_i, for $i=1,2,\ldots,n$ are 0 or 1, depending on the absence or presence of the input impulse. The weights of connections between the ith input x_i and the neuron are represented by w_i. When $w_i>0$, the input is excitatory, and when $w_i<0$, it is inhibitory.

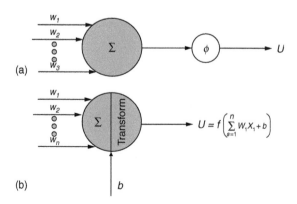

Figure 12.2 Artificial neuron models: (a) McCulloch and Pitts model and (b) general neuron model.

The neurons output can be described as

$$u = \sum_{i=1}^{n} w_i x_i. \tag{12.1}$$

The general neuron model includes an externally applied bias denoted by b. The bias has the effect of increasing or lowering the net input of the activation function $f(.)$, depending on whether it is positive or negative. The activation function limits the amplitude of the output of a neuron. The general model (output) of neuron can be written as:

$$y = f\left(\sum_{i=1}^{n} w_i x_i + b\right). \tag{12.2}$$

The bias, b, is an external parameter of an artificial neuron that can also be combined in Equation 12.1 as follows:

$$y = f\left(\sum_{i=0}^{n} w_i x_i\right). \tag{12.3}$$

In Equation 12.3, a new synapse is added and its input is

$$x_0 = +1, \tag{12.4a}$$

and its weight is

$$w_0 = b. \tag{12.4b}$$

The input and output signals are not limited to binary form, and the activation function can be a continuous function other than the threshold function used in the earlier model (Figure 12.2b). The activation function is typically a monotonic, nondecreasing, nonlinear function. Some of the most commonly used activation functions are [5]:

- Threshold function:

$$f(x) = \begin{pmatrix} 1, & x > \phi \\ 0, & x < \phi \end{pmatrix} \tag{12.5}$$

- Sigmoid function:

$$f(x) = \frac{1}{1 + e^{-\alpha x}} \tag{12.6}$$

- Hyperbolic function:

$$f(x) = \tanh(\alpha x) = \frac{e^{\alpha x} - e^{-\alpha x}}{e^{\alpha x} + e^{-\alpha x}} \tag{12.7}$$

- Linear function:

$$f(x) = \begin{pmatrix} 1, & x \geq \phi \\ x/\phi, & 0 < x < \phi \\ 0, & x \leq 0 \end{pmatrix} \tag{12.8}$$

• Gaussian function:

$$f(x) = e^{-\alpha x^2}.$$ \hfill (12.9)

12.2.2.2 Neural Network Learning Processes

The ability of a network to learn from its environment and to improve its performance is a significant property of a neural network. Learning is a process of forcing a network to yield a particular response to a specific input. A particular response may or may not be specified to provide external correction. The learning of a network can be achieved by a supervised or an unsupervised mode. In supervised learning, a set of input and desired output data, called training set, is supplied. At each instant of time when the input is applied, the desired response of the system is provided. The difference between the actual and the desired response serves as an error measure and is used to correct network parameters externally. Error signals are then used to update weights and threshold of networks. Examples of supervised learning are: perceptron learning, Hebbian learning, Widrow–Hoff learning, delta learning, and back-propagation learning. In unsupervised learning, only input data is fed into the network, because the output is unknown, and thus no explicit error information is given. In this learning process the network is tuned to the statistical regularities of the input data and then it develops the ability to form internal representations for coding features of the input, and thereby creates new outputs automatically. Examples of this type of learning are winner-take-all learning, Hamming net and MAXNET learning, and adaptive resonance theory learning [3].

Back-Propagation Learning. This is one of the most popular and extensively used learning algorithms for network training. The back propagation uses the supervised training technique where the network weights and biases are initialized randomly at the beginning of the training phase. For a given set of inputs to the network, the response to each neuron in the output layer is calculated and compared with the corresponding desired output response. The errors associated with the desired output response are adjusted in such a way that it reduces these errors in each neuron from the output to the input layers. This procedure is repeated over the entire training dataset for a specified number of times (training/learning runs), usually several thousands times, to be chosen by the user.

12.3 NETWORK ARCHITECTURE

The common structure of a neural network is formed of layers of neurons interconnected in different ways (Figure 12.3). Neural networks can have various structures depending on the way in which the neurons are interconnected and on the flow of signals through the network. Haykin [3] classified neural network structures into three classes of architectures.

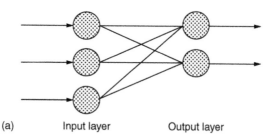

(a)   Input layer   Output layer

Figure 12.3.   Structures of artificial neural networks: (a) single layer feedforward network, (b) multi-layer feedforward network, and (c) recurrent network.

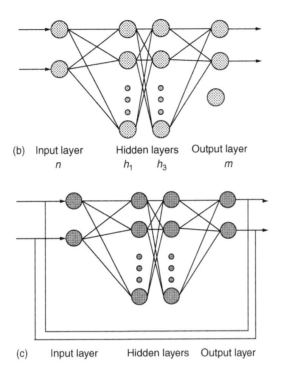

(b) Input layer   Hidden layers Output layer
         n h_1 h_3 m

(c)   Input layer   Hidden layers   Output layer

Figure 12.3 (*Continued*)

12.3.1 Single-Layer Feedforward Networks

This is the simplest form of layered networks. In this structure, neural network has an input layer consisting of input data that directly projects onto an output layer of neurons, but not vice versa (Figure 12.3a). This is strictly a feedforward or acyclic type. The single layer refers to the output layer of neurons (computational nodes). The input layer of source nodes/data is not counted since no computation is performed in this layer.

12.3.2 Multilayer Feedforward Networks

This class of feedforward neural networks has one or more hidden layers of neurons. The neurons present in these layers are known as hidden neurons. The function of hidden neurons is to intervene between the external input and the network output in some useful manner. By adding one or more hidden layers, the network is enabled to extract higher order statistics due to increased connections. The hidden layers are particularly useful when the size of the input layer is large. This network has an input, an output and one or more hidden layers (Figure 12.3b). The neurons in the input layer receive input signals from the user. These signals are carried to the first hidden layer through the connections. The output signal from the first hidden layer becomes the input to the second hidden layer and so on. The signals are transmitted this way to the output layer which produces the network output. The number of neurons in the input and output layers correspond to the number of input and output signals or variables. The number of hidden layers and the number of neurons in each hidden layer can be varied. The network shown in Figure 12.3b can be referred as an n–h_1–h_2–m network with n input neurons, h_1 neurons in the first hidden layer, h_2 neurons in the second hidden layer, and m neurons in the output layer. Such a neural network is said to be *fully connected* in the sense that every neuron in each layer of the network is connected to every other neuron in the adjacent forward layer. If, however, some of the connections are missing from the network, then the network is referred to as *partially connected*.

12.3.3 Recurrent Networks

This network structure differs from a feedforward neural network because it has at least one feedback loop. For example, a recurrent network may consist of a single layer of neurons with each neuron feeding its output signals to the inputs of all the other neurons (Figure 12.3c). This type of network may or may not have hidden neurons (Figure 12.3c). The presence of feedback loops has a profound impact on the learning capability of the network and on its performance.

12.4 DEVELOPMENT OF AN ANN-BASED MODEL

In principle, artificial neural networks can compute any mathematical function. In food related areas, ANNs have most often been employed as flexible, nonlinear regression, and classification models. Because it is a data-driven modeling technique, it requires (training) data that include both the inputs (independent variables) and the desired results (dependent variables). For example, consider the developments of an ANN model to determine the effect of process variables on heat/mass transfer kinetics and quality factors. In this case, heat/mass transfer rates, temperature/-mass concentration profiles, quality factors such as color, texture, nutrient retention/degradation, and shrinkage/collapse can be taken as dependent variables, whereas thermophysical and mass-transfer properties of food material, temperature, air velocity, air relative humidity, and pressure can be considered as independent variables. Multiple dependent variables can be simultaneously correlated with multiple independent variables in ANN modeling, which is unique to this modeling technique.

The development of an ANN model involves three basic steps. These include the generation of (or compilation of available) data required for training, the training of ANN networks, and the selection and validation of the optimal configuration of the ANN model.

12.4.1 Data Generation

The data for training can be obtained by experiments, mathematical simulations, or compiled from the literature. For example, neural networks can be applied to datasets obtained from many food-processing operations without going through the rigor of extensive training, or understanding the mathematical background of a problem [6]. Although the same data can be used for both training and testing/validation, it is preferable to have two different datasets.

12.4.2 Training of ANN Networks

To train a model, several parameters (including the number of hidden layers and the number of neurons in each hidden layer, the learning rule, the transfer/activation function, the random number seed for initialization of connection weights, the error minimization algorithm, and the number of training/learning runs) have to be specified. These parameters could be varied based on the complexity of the problem. Given the lack of clear guidance in the literature concerning the selection of the above parameters, a trial-and-error procedure must be followed.

12.4.3 Selection of Optimal Network

The performances of the various ANN configurations can be compared using several statistical parameters such as the mean relative error (MRE), the mean absolute error (MAE), and the standard deviations in the relative (STD$_R$) and absolute (STD$_A$) errors:

$$\text{MAE} = \frac{1}{N} \sum_{i=1}^{N} \Delta f_A, \tag{12.10}$$

$$\text{STD}_A = \sqrt{\frac{\sum_{i=1}^{N} \left(\Delta f_A - \overline{\Delta f_A} \right)^2}{N-1}}, \tag{12.11}$$

$$\text{MRE} = \frac{1}{N} \sum_{i=1}^{N} \Delta f_R, \tag{12.12}$$

$$\text{STD}_R = \sqrt{\frac{\sum_{i=1}^{N} \left(\Delta f_R - \overline{\Delta f_R} \right)^2}{N-1}}, \tag{12.13}$$

where $\Delta f_A = |f_P - f_D|$ and $\Delta f_R = |(f_P - f_D)/f_D|$.

The parameter f_P represents the predicted output from the neural network model for a given input whereas f_D is the desired output from the same input. The coefficient of determination, R^2, of the linear regression line between the predicted values from the neural network model and the desired output is also used as a measure of performance. The best, or optimal, configuration of the neural network is decided based on the minimization of error parameters. After the optimal network configuration is selected, its performance is evaluated with the testing data set that is not used in the training of the network.

12.4.4 Validation of the Optimal ANN Model

The performance of the optimal neural network is validated using a smaller data set not used in the training procedure. As such, the network's prediction ability is tested. Thereafter, the network weights and coefficients associated with the optimal ANN model are presented in the form of simple algebraic equations, so that these can be used for further predictions without the need of the neural network software program that was used for modeling.

12.5 APPLICATIONS

The applications of ANN can be classified into four categories, such as product grading and classification, food quality assessment, food process/property modeling, and process control. Only the applications of ANN related to the modeling of food processes and properties are highlighted in this chapter. In most of these cases, ANN has been used as a modeling tool similar to that of nonlinear regression analysis. Specific neural network applications in food and bioprocess modeling include thermal processing, freezing, drying, frying, reaction kinetics, food quality, and the prediction of thermal and physical properties of foods (Table 12.1). The neural network modeling, combined with fuzzy logic and genetic algorithm, has been used in the processes of

Table 12.1 Application of Neural Network in Food and Bioprocessing

Application	Problem	Data	ANN Models	Reference
Product quality during baking	(a) Danish pastry: prediction of height, % open folding and hardness of dough; (b) Rye bread: pH, specific volume and hardness of dough	Experimental industrial process data	Feed forward, back propagation, sigmoid function	12
Thermal properties during thawing	Prediction of thermal conductivity and specific heat as a function of temperature	Thawing experiments with 10% gelatin gel	Feed forward, back propagation, Sigmoid function	13
Fluid flow in pipes	Explicit calculation of friction factor (f) in pipeline flow: Bingham plastic	Numerical method (Regula–Falsi method) was used as implicit procedure to estimate f	Feed forward, back propagation, hyperbolic tangent function	14
Thermal processing	Prediction of optimal variable retort temperature for conduction heated foods	Finite difference computer simulation for data generation	Feed forward, back propagation, hyperbolic tangent function	15
Mechanical properties during drying	Prediction of porosity as a function of temperature, moisture content, initial porosity and product type	Experimental data for fruits and vegetables from literature	Feed forward, back propagation, linear, sigmoid, and tangent hyperbolic function	16
Electrical property	Prediction of electrical conductivity as function of fat, protein, lactose and temperature	Experimental data for milk from literature	Feed forward, back propagation, sigmoid, tangent hyperbolic	17
Predictive microbiology	Prediction of thermal inactivation of bacteria as a function of temperature, pH and water activity	Experimental data of destruction of E. coli from literature	Feed forward, back-propagation, transfer function not clear	18
Frying	Prediction of temperature, moisture and fat content as a function of process variables and properties of meat balls	Physical model was used to generate data	Feed forward, back-propagation, transfer function not clear	19
Fluidized bed drying	Estimation of heat transfer coefficient from dimensionless number Re, Ar, H/d	Experimental data from drying experiments with silica gel particles	Feed forward, back-propagation, transfer function not given	20
Fluid flow in pipes	Estimation of pressure drop as a function of fluid parameters (m, n, density), diameter of pipe and mass flow rate	Experimental data for five fluid fluids were generated using Brookfield and rotational viscometers	Back-propagation, generalized regression networks and quick-propagation	21
Freezing time	Prediction of freezing time of food of any shape	Used Pham model to generate freezing time data	Feed forward, back-propagation, tangent hyperbolic function	22
Kinetics of food quality parameters	Modeled kinetics of eight quality attributes of dry peas during cooking	Experimental data	Feed forward, back-propagation, sigmoid transfer function	23
Food quality	Prediction of quality parameters (i.e., rheological properties) as a function of cheese composition and processing conditions	Experimental data for 48 types of cheeses from literature	Feed forward, back propagation, Sigmoid transfer function	24

control and optimization, respectively. In addition, several researchers have presented neural network-based simple algebraic equations for prediction purposes. The details on genetic algorithm and fuzzy logic have been presented in the Chapter 13 and Chapter 15 of this handbook. Here, three examples are given to detail how the artificial neural network modeling approach is used in the different areas of food and bioprocessing.

12.5.1 Modeling of Thermal Conductivity

12.5.1.1 Background

Thermal conductivity, k, of food materials is one of the important properties used to estimate the rate of conductive heat transfer during processes, such as freezing, sterilization, drying, cooking, and frying. Three factors affect the thermal conductivity of foods: composition, structure, and processing conditions. Water content plays a significant role due to the relative magnitude of conductivities of water in food. The nonaqueous part of food such as fats and oils also influences thermal conductivity of fatty foods. The structural factors are porosity, pore size, shape and distribution, arrangement, or distribution of different phases such as air, water, ice, and solids. The processing factors are temperature, pressure, and mode of heat or energy transfer.

There has been consistent effort spent in developing generalized correlations to predict thermal conductivity of food materials for the use in process design and optimization [25]. Sweat [26] proposed a linear correlation for predicting the thermal conductivity of fruits and vegetables giving predictive results within $\pm 15\%$ for most experimental values. This model, however, is valid for situations where moisture content (wet basis) greater than 0.60 and does not account for temperature and apparent porosity effects. According to Sweat, there was a strong relation between water content and thermal conductivity of all fruits and vegetables tested except for apples, which were highly porous. Therefore, it was suggested that a general correlation should include a porosity (i.e., apparent porosity) term. Considering the wide variations in materials and processing conditions, it is difficult to develop an analytical model for the prediction of thermal conductivity.

12.5.1.2 Problem Definition

The chosen problem is to develop an ANN based model to predict thermal conductivity (k) as a function of water content (X_w), temperature (T), and apparent porosity (ε_a) of the food material. Data needed for development of an ANN model were obtained from the literature. A total of 676 data points collected from 12 journal papers for 10 different materials was used in the training and validation of ANN models.

12.5.1.3 Solution Procedure

Collection of Thermal Conductivity Data. Several researchers have measured thermal conductivity of different fruits and vegetables. Most data on thermal conductivity have been modeled and reported as a function of limited temperature and moisture range. However, as described above, thermal conductivity is also influenced by the amount of air/void fraction present in the food material. A thermal conductivity data set was prepared from literature that included the thermal conductivity for a given moisture content, temperature, and apparent porosity. The thermal conductivity data of several food products for the range of processing conditions are presented in Sablani and Rahman [27].

Training the Neural Networks. A multi-layer feed forward network structure with input, output, and hidden layer(s) was used in this study, as shown in Figure 12.4. Several ANN models were trained using the thermal conductivity data. A back-propagation algorithm was utilized in training

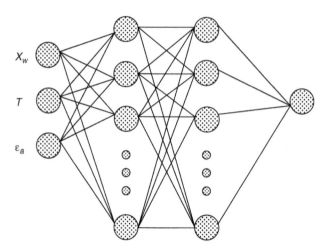

Input layer   Hidden layers   Output layer

Figure 12.4 Schematic of multilayer neural network (X_w=moisture content, T=Temperature, ε_a =apparent density and k=thermal conductivity).

of ANN models and a hyperbolic-tangent transfer function was used in all cases. The input layer consisted of three neurons which corresponded to product moisture content, temperature, and apparent porosity. The output layer had one neuron representing thermal conductivity (Figure 12.4). The number of hidden layers varied from 1 to 2. The number of neurons within each of these layers varied from 2 to 10, with increments of two. This resulted in a total of 10 networks.

The commercial software package Neural Works Professional II/Plus (Neural Ware, Pittsburgh, PA), was employed in this study. To train a model, several parameters including the learning rule, the transfer function, the learning coefficient ratio, the random number seed, the error minimization algorithm, and the number of learning cycles, had to be specified [28]. These parameters could be varied based on the complexity of the problem. While some of the parameters were kept constant during our study, others were varied to develop the optimum ANN configuration. The parameters that were kept constant included the transfer function (the hyperbolic-tangent transfer function), the learning rule (the normalized-cumulative delta rule), the random number seed (257), and the learning rate (0.9), momentum (0.6). The error-minimization process was achieved using the gradient-descent rule [28] while the number of training cycles was set at 200,000. All of the remaining model parameters (as specified above) were kept constant throughout the training processes. The performance of various ANN configurations was compared using appropriate statistical parameters (Equation 12.10 through Equation 12.13).

Result. Thermal conductivity data for different food products plotted as a function of moisture content shows a strong dependence of k on temperature and relatively less influence of apparent porosity (Figure 12.5). The thermal conductivity data set of the 676 cases was divided into two groups. In the first group, 540 cases were taken for training/testing and in the second group 136 cases for validation, chosen randomly from the set of 676 cases. The error measures associated with different ANN configurations for prediction of thermal conductivity with different data sets are presented in Table 12.2. The network model with two hidden layers and four neurons in each hidden layer resulted in the best prediction. The MRE and MAE for this configuration were 12.6% and 0.081 W/m K, respectively. The prediction performance for predicted values of k is compared in Figure 12.6. The results demonstrated good agreement between the predicted and the desired values of thermal conductivity (R^2=0.957). Even though the coefficient of determination was very good

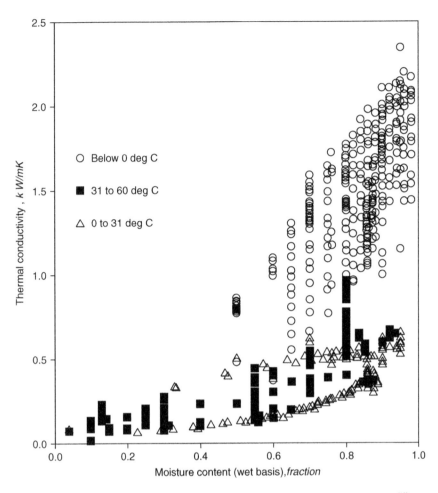

Figure 12.5 Thermal conductivity values for different foods as a function of moisture content at different temperatures. (Adapted from Sablani, S. S. and Rahman, M. S., *Food Res. Int.*, 36, 617–623, 2003.)

Table 12.2 Prediction Errors in the Thermal Conductivity with Different Neural Network Configurations (Randomly Selected Data Set [~80%] Training and Testing 540 Cases)

Number of Hidden Layers	Number of Neurons in Each Hidden Layer	MRE (%)	STD$_R$ (%)	MAE (W/m K)	STD$_A$ (W/m K)	R^2
1	2	18.63	21.11	0.118	0.138	0.922
1	4	18.34	22.43	0.112	0.119	0.938
1	6	18.31	23.33	0.104	0.115	0.943
1	8	14.84	16.61	0.093	0.111	0.949
1	10	19.84	22.73	0.119	0.130	0.926
2	2	25.08	32.05	0.128	0.119	0.932
2	4	12.63	14.73	0.081	0.103	0.957
2	6	14.05	16.17	0.095	0.111	0.957
2	8	13.11	15.41	0.082	0.106	0.956
2	10	13.96	15.24	0.092	0.110	0.955

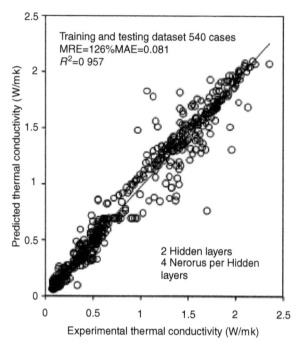

Figure 12.6   Experimental versus neural network predicted values of thermal conductivity with training data set using the optimal network. (Adapted from Sablani, S. S. and Rahman, M. S., *Food Res. Int.*, 36, 617–623, 2003.)

($p > 0.95$), the standard deviations in relative errors were high 16.6%. This was due to a small prediction error (in terms of absolute) at lower values of thermal conductivity which translated into very high relative errors. The magnitude of errors reported for experimental measurement of thermal conductivity was generally in the same order of magnitude as found earlier for narrow experimental ranges and materials [25,26,29].

Depending upon the material, the ANN model either underpredicted (apple, raisin, potato, rice, corn starch) or overpredicted (pear, carrot) thermal conductivity values, although for some materials (i.e., starch, sucrose, ovalbumin) the predicted values of thermal conductivity closely matched those of the experimental values. The ANN model was able to predict thermal conductivity values with acceptable accuracy both above and below freezing temperatures. Sablani and Rahman [27] also developed a multiple regression model, with a polynomial of degree four, using literature data. The R^2 and mean relative error in the model were 0.91 and 81.6%, respectively. The influence of moisture content, temperature, and porosity on thermal conductivity is highly nonlinear in nature and a simple multiple regression equation was not able to capture this complexity.

The performance of the optimal neural network (two hidden layers and four neurons in each hidden layer) was validated using a smaller data set consisting 136 cases. This network predicted thermal conductivity values with an MRE of 16.2% and MAE 0.088 W/m K. The standard deviations in relative and absolute errors were 21.2% and 0.095 W/m K, respectively. The network weights and coefficients associated with this ANN model are presented in the form of simple algebraic equations in Appendix A. These equations can be programmed in MS EXCEL and used to predict thermal conductivity values for known moisture content, temperature, and apparent porosity of food product. They can also be programmed using any computer language (FORTRAN, C, BASIC, etc.) and can be incorporated as a simple subroutine in the numerical analysis of heat transfer during processing where moisture, temperature, and apparent porosity-dependent thermal conductivity values are needed. In literature, such equations and coefficients based on ANN are missing, and therefore, it limits the usefulness of neural network based models. The coefficients, in the form of simple algebraic equations, make the ANN model user friendly.

12.5.2 Estimation of Friction Factor in Pipe Flow

12.5.2.1 Background

Calculating the drop of pressure in pipeline flow due to friction is important during the design of the pipeline and the selection of the pump. The mechanical energy balance equation (commonly known as the "engineering Bernoulli equation" for an incompressible fluid in a pipe) is used for the hydraulic analysis of flow conditions of viscous nonNewtonian fluids [30,31]. The equation is derived from the principle of conservation of momentum. It includes internal, potential, pressure, and kinetic energy terms for a fluid in motion. Numerous assumptions are made in developing this equation. These include a constant fluid density, the absence of thermal energy effects, single phase, uniform material properties, and uniform equivalent pressure [32]. The mechanical energy balance for an incompressible fluid in a pipe may be written as:

$$\left[\frac{(\overline{u_2})^2}{\alpha_2} - \frac{(\overline{u_1})^2}{\alpha_1}\right] + g(z_2 - z_1) + \frac{P_2 - P_1}{\rho} + \sum F + W = 0, \tag{12.14}$$

where the subscripts 1 and 2 refer to two specific locations in the system and ΣF is the summation of all friction losses. The summation of all friction losses can be determined from the following equation:

$$\sum F = \sum \frac{2f(\overline{u_1})^2 L}{D} + \sum \frac{k_f(\overline{u})^2}{2}. \tag{12.15}$$

These losses include those from pipes of different diameters and a contribution from each individual valve and fitting. Pressure losses in other types of in-line equipment (such as strainers) should also be included in ΣF term.

The power-law fluid model ($\sigma = K\gamma^n$) is one of the most useful models in pipeline design work for nonNewtonian fluids. It has been studied extensively and found to accurately express the behavior of many fluid foods, which commonly exhibit a shear-thinning ($0 < n < 1$) behavior. Dodge and Metzner [33] correlated experimental results for the Fanning friction factor for turbulent flow of shear thinning fluids in smooth pipes as a generalized form of the von Karman equation:

$$\frac{1}{\sqrt{f}} = \left(\frac{4}{n^{0.75}}\right) \log_{10}\left[Re'f^{(1-(n/2))}\right] - \left(\frac{0.4}{n^{1.2}}\right). \tag{12.16}$$

This correlation is shown in Figure 12.7. The figure also includes some lines representing extrapolation of Equation 12.16 for values of Re' and n beyond the measurements made by Dodge and Metzner [33]. Recent studies tend to confirm the findings of Dodge and Metzner [33], but do not significantly extend the range of their applicability. Having determined the value of the friction factor, f, for a specified flow rate and having therefore determined Re', the pressure gradient can be calculated by

$$\Delta P_f = \frac{2fL\rho\overline{u}}{D}. \tag{12.17}$$

The hydraulic analysis of pipe networks often involves the implementation of a tedious and time-consuming iterative procedure that requires extensive use of computers [34]. Numerical procedures such as the finite difference or finite element formulations allow for the hydraulic

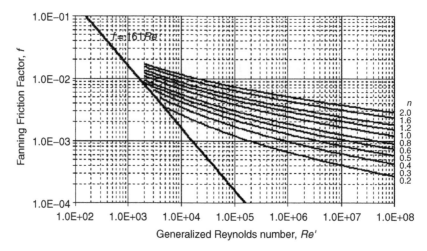

Figure 12.7 Friction factor chart for power law fluids. (Adapted from Sablani, S. S. and Shayya, W. H., *J. Food Eng.*, 57, 327–335, 2003.)

analyses of these systems, but with excessively large systems of algebraic equations to be solved iteratively [35]. The number of equations varies with the number of pipe elements and flow outlets within the system and may reach the order of thousands. The system of nonlinear algebraic equations has to be solved iteratively, making it impractical to solve f iteratively within each pipe section and during any given iteration.

12.5.2.2 Problem Definition

To devise a noniterative procedure using ANN for estimating the friction factor, f, for turbulent power law fluids flow in closed pipes to avoid the need for a time-consuming, iterative solution. This was accomplished using a numerical method to generate values of the friction factor, f, for a range of Reynolds numbers, Re', and n values in tube flow. These values were then used in the development of an ANN model.

12.5.2.3 Solution Procedure

Data Generation. The friction factor, f, in Equation 12.3 must be handled either by trial-and-error or after implementing an implicit solution procedure such as the Newton–Raphson or Regula–Falsi method. To implement these implicit solution procedures, Equation 12.3 is first written as:

$$F(f) = \frac{1}{\sqrt{f}} - \left(\frac{4}{n^{0.75}}\right) \log_{10}\left[Re'f^{(1-(n/2))}\right] + \left(\frac{0.4}{n^{1.2}}\right), \tag{12.18}$$

where f is the unknown and $F(f)$ is the function to be reduced to zero once the implicit solution procedure converges to the solution. The Regula–Falsi method (RFM) was chosen to solve above equation [36]. A total of 546 f values were computed for 39 Re' and 14n values. The data was then used in the training of artificial neural network models.

Training of Artificial Neural Networks. Following the definition of network structure and components, several ANN models were trained and tested using the developed training data set. The input layer consisted of two neurons that corresponded to Re' and n, whereas the output layer had one neuron representing the friction factor, f (Figure 12.8). The number of hidden layers was varied from 1 to 2, while the neurons within each of these layers were varied from 2 to 16, in

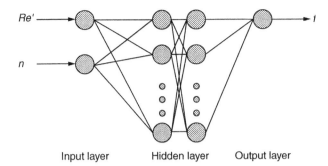

Figure 12.8 Schematic of feedforward multilayer neural network (Re=Reynolds number, n=flow behavior index, f=friction factor).

increments of two. This resulted in a total of 16 networks. The optimal configuration was based upon minimizing the difference between the neural network predicted values and the desired outputs. The data for training the ANN model were generated using the numerical procedure described in the previous section. A data set consisting of 7240 points (40 values of Re' ranging from 2000 to 10^8 and 181 values of n ranging from 0.2 to 2.0) resulting from the combination of Re' and n as inputs and f as output was used for training the ANN model. The other details about the computer program and parameters are presented in previous example. The performances of the various ANN configurations were compared using error parameters (Equation 12.10 through Equation 12.13).

Results. The RFM was used in this study for the purpose of generating training and validation data sets. The training data set consisted of 7240 cases, with Re' ranging from 2000 to 10^8 and n ranging from 0.2 to 2.0. After a given ANN configuration was trained using the training data set, its performance was evaluated using the same data set. The ANN configuration (out of 16) that minimized the five error measures described in the previous section, and optimized R^2, was selected as the optimum. The whole analysis was repeated thrice (i.e., without transformation of data, logarithmic transformation of only one input parameter: Re', and logarithmic transformation of both input parameters: Re' and n).

Input data without transformation. In the initial attempt, the original data set obtained from the numerical procedure was used to develop the ANN models. Its configuration was varied, as discussed above, and an optimal configuration was chosen. However, the performances of the 16 ANN configurations were found to be inadequate because the MRE was always in excess of 22.3% (which is unacceptably high). The error measures associated with the various ANN configurations of this analysis are presented in Table 12.3. The mean relative error for the optimal ANN configuration, which included two layers and 8 neurons within each layer, was 12.1%, while the maximum and the standard deviation of the relative error were 64.6 and 17.7%, respectively. The coefficient of determination of the optimal network was 0.684.

Transformation of input data. As mentioned earlier, the development of ANN models does not require any prior knowledge of the relationships among model inputs and outputs. However, having some idea about such relationships may provide for the fine-tuning of the ANN model. In some instances, these ideas have shown exceptional improvements in ANN model performance [37]. It is clear from Equation 12.16 and Figure 12.7 that the friction factor, f, is a logarithmic function of input parameters. For this reason, an attempt was made to improve the performance of the ANN model by transforming the input parameters Re' and n (one at a time) using a logarithmic function to the base ten before supplying the data to the ANN model. As a first step only Re' was transformed on the logarithmic scale (n was kept without transformation). The result was a remarkable improvement in the prediction performance of the ANN model since the MRE of all 16 configurations was less than 1.6%. The best ANN configuration included two hidden layers with 6 neurons in each

HANDBOOK OF FOOD AND BIOPROCESS MODELING TECHNIQUES

Table 12.3   ANN Analysis Using the Back-Propagation Algorithm without Transformations of the Input Parameters Re' and n or the Output Parameter f

Number of Hidden Layers	Number of Neurons in Each Hidden Layer	MRE	STD_R	MAE	STD_A	R^2
1	2	12.5	15.5	0.00138	0.00161	0.645
1	4	13.1	16.7	0.00143	0.00169	0.641
1	6	13.0	16.8	0.00142	0.00170	0.644
1	8	14.0	17.9	0.00151	0.00179	0.640
1	10	13.3	13.3	0.00157	0.00144	0.631
1	12	13.5	17.5	0.00146	0.00174	0.638
1	14	13.1	17.5	0.00139	0.00176	0.638
1	16	13.5	15.5	0.00147	0.00145	0.634
2	2	15.3	16.2	0.00165	0.00134	0.627
2	4	13.4	18.6	0.00143	0.00188	0.677
2	6	13.3	14.9	0.00143	0.00140	0.656
2	8	12.1	17.7	0.00129	0.00190	0.684
2	10	12.3	18.0	0.00128	0.00184	0.678
2	12	17.6	15.8	0.00206	0.00127	0.633
2	14	17.8	22.8	0.00186	0.00176	0.635
2	16	11.9	14.1	0.00134	0.00154	0.672

layer. The mean relative error for this optimal configuration was 0.27%, with a 0.36% standard deviation.

The prediction performance of ANN models improved further when both input parameters Re' and n were transformed on a logarithmic scale. The error measures associated with the different ANN configurations for this case are presented in Table 12.4. The best ANN configuration included two hidden layers with 12 neurons in each layer. The mean relative error for this optimal configuration was 0.19%, with a standard deviation of 0.27%, and a coefficient of determination of 0.999. However, considering the explicit nature of the neural network based approach, the simplest ANN configuration (i.e., one hidden layer with two neurons) can be considered a good predictor since its mean relative error is 0.91% with a standard deviation in relative error of 0.92% and a coefficient of determination of 0.999. For this reason, the simplest ANN model is recommended to users since it is very easy to implement. The network weights and coefficients associated with this ANN model are presented in Appendix B. This simple algebraic set of equations may be used for the noniterative estimation of the friction factor. The equations can simply be programmed and incorporated in any numerical-based hydraulic analysis program.

Verification of the optimal ANN model. The performance of the optimal ANN model (2 hidden layers and 12 neurons in each hidden layer) was validated using a larger data set (72,400 data points generated in the normal range of Re' and n using RFM) not previously used in the training of the ANN model. The optimal ANN predicted f with a mean relative error of 0.17%, a standard deviation in relative error of 0.20%, and a coefficient of determination of 1.000. The large data set was also used with the simplest ANN configuration of one hidden layer consisting of two neurons. This ANN configuration predicted the friction factor with a mean relative error of 0.50%, a standard deviation of relative error of 0.60%, and a coefficient of determination of 0.999.

Given its simplified form, the simpler model was recommended for the noniterative calculation of the friction factor for viscous nonNewtonian fluids. The ANN model allows for the explicit solution of f without the need to employ a time-consuming iterative or trial-and-error solution scheme. Such a model will be useful for flow problems that involve repetitive calculations of the friction factor, such as those encountered in the hydraulic analysis of flow conditions in pipe network problems with viscous nonNewtonian fluids.

Table 12.4 ANN Analysis Using the Back-Propagation Algorithm and Logarithmic Transformations of the Input Parameters Re' and n (No Transformation of the Output Parameter f)

Number of Hidden Layers	Number of Neurons in Each Hidden Layer	MRE	STD_R	MAE	STD_A	R^2
1	2	0.913	0.922	0.00010	0.00008	0.999
1	4	0.664	0.666	0.00008	0.00007	0.999
1	6	0.474	0.636	0.00005	0.00007	1.000
1	8	0.490	0.627	0.00006	0.00007	0.999
1	10	0.655	0.531	0.00008	0.00006	1.000
1	12	1.000	0.777	0.00012	0.00008	0.999
1	14	0.785	0.772	0.00009	0.00008	0.999
1	16	1.194	0.864	0.00014	0.00008	0.999
2	2	0.619	0.716	0.00007	0.00007	1.000
2	4	0.635	0.477	0.00007	0.00004	1.000
2	6	0.350	0.426	0.00004	0.00004	1.000
2	8	0.211	0.225	0.00003	0.00003	1.000
2	10	0.280	0.335	0.00003	0.00004	1.000
2	12	0.188	0.268	0.00002	0.00003	1.000
2	14	0.415	0.348	0.00005	0.00003	1.000
2	16	0.168	0.243	0.00002	0.00003	1.000

12.5.3 Estimation of Heat-Transfer Coefficients

12.5.3.1 Background

The determination of surface temperatures, heat source rates, and thermophysical properties by utilizing measured temperatures inside solid bodies is classified as inverse heat conduction problems (IHCPs). Such problems are encountered in a multitude of food and process engineering applications. Examples include: sterilization of particulate liquids in continuous systems (aseptic processing), cooling of fresh produce, frying and freezing of food, and biological materials [38–40]. The estimation of the heat transfer coefficient also falls under the category of an IHCP. This approach requires experimental measurement of the transient temperatures inside a body of known geometry at a specified location, usually at the center, and estimation of transient temperatures at the same location by solving the governing heat conduction equations with an assumed convective boundary condition (i.e., the Biot number, Bi). In doing so, Bi is varied systematically to produce computed temperature/time histories closely matching the experimentally measured temperature histories. The procedure involved is iterative in nature and needs a long computation time. Several algorithms based on finite difference and finite element methods have been developed for solving the IHCP. An excellent discussion of the difficulties encountered in solving the IHCP, and several solution methods used, can be found in Beck et al. [39] and Beck and Arnold [40].

12.5.3.2 Problem Definition

Use an artificial neural network approach to develop a single and direct procedure for estimating the heat transfer coefficient to avoid the use of a time-consuming, iterative solution. This has relevance in food processing operations such as transient heat transfer analysis during drying, frying and freezing of small fruit and vegetable cubes, and sterilization of particulate liquids in continuous systems (aseptic processing). All of these require knowledge of heat transfer coefficients.

12.5.3.3 Solution Procedure

Data Generation in the DHCP. Consider the problem of transient heat conduction in an isotropic cube exposed to a forced flow of a viscous fluid. The thermophysical properties of the fluid and solid, as well as the heat transfer coefficient at all faces of the cube were assumed to be constant. The governing tri-dimensional heat conduction equation in nondimensional form is:

$$\frac{\partial^2 \theta}{\partial X^2} + \frac{\partial^2 \theta}{\partial Y^2} + \frac{\partial^2 \theta}{\partial Z^2} = \frac{\partial \theta}{\partial Fo}. \tag{12.19}$$

The initial and boundary conditions that are imposed on Equation 12.19 are:

$$\text{For } Fo = 0; \quad \theta = 1 \quad \text{for all } X, Y, \text{ and } Z \tag{12.20a}$$

$$\frac{\partial \theta}{\partial X} = 0 \quad \text{at} \quad X = 0; \quad \text{for all } Y \text{ and } Z, \quad Fo \geq 0 \tag{12.20b}$$

$$\frac{\partial \theta}{\partial Y} = 0 \quad \text{at} \quad Y = 0; \quad \text{for all } X \text{ and } Z, \quad Fo \geq 0 \tag{12.20c}$$

$$\frac{\partial \theta}{\partial Z} = 0 \quad \text{at} \quad Z = 0; \quad \text{for all } X \text{ and } Y, \quad Fo \geq 0 \tag{12.20d}$$

$$\frac{\partial \theta}{\partial X} = -Bi\theta \quad \text{at} \quad X = 1; \quad \text{for all } Y \text{ and } Z, \quad Fo > 0 \tag{12.20e}$$

$$\frac{\partial \theta}{\partial Y} = Bi\theta \quad \text{at} \quad Y = 1; \quad \text{for all } X \text{ and } Z, \quad Fo > 0 \tag{12.20f}$$

$$\frac{\partial \theta}{\partial Z} = -Bi\theta \quad \text{at} \quad Z = 1; \quad \text{for all } X \text{ and } Y, \quad Fo > 0, \tag{12.20g}$$

where θ is the nondimensional temperature, X, Y, and Z are the nondimensional coordinates, Fo is the Fourier number or dimensionless time and Bi is the Biot number. The finite-element-based computer software FIDAP (Fluent, Inc., NH) was used to solve this conduction problem with a convective boundary condition. The details of the solution procedure are presented in Sablani et al. [38]. The Biot number varied from 0.01 to 10. The increment of Bi increased with increasing Bi. The FIDAP program was run several times with different values of Bi, thus obtaining the temperature history at $X=0$. Because the nondimensional center temperature varied linearly with the Fourier number when plotted on a semi-log scale, the temperature profile could be characterized using the slope, S, of this curve. The slope was obtained from calculated temperature histories at the center for the 65 Bi values. Thus, 65 cases were used in the development of the ANN models. The data set of 65 cases consisting of a Biot number and the corresponding slope was divided into two groups. The first group consisted of 51 cases for training/testing of ANN models, whereas the second group had 14 cases for validation of the ANN model, chosen randomly from the set of 65 cases.

Training of the Artificial Neural Network Model. The feedforward network structure [41] was used in this example, as shown in Figure 12.9. Several ANN models were trained and tested using the training data set. In the case of cube, for the inverse problem, the input layer consisted of a neuron corresponded to the input parameter, i.e., slope, S, whereas the output layer had one neuron representing the Biot number, Bi. The number of hidden layers and the neurons within each hidden layer can be varied based on the complexity of the problem and the data set. In order to reduce the chances of memorization of the behavior among the data set (rather than generalization), the

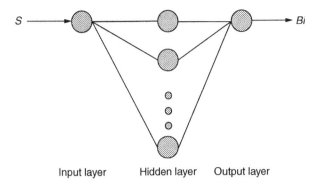

Figure 12.9 Schematic of multilayer neural network used for the inverse conduction problem (S=slope and Bi=Biot number).

number of hidden layers and neurons in these layer ought to be minimized [14]. Only one hidden layer was chosen while the neurons in that layer were varied from 2 to 10, in increments of 2. This resulted in a total of five networks. The optimal configuration was based upon minimizing the difference between the ANN predicted values and the desired outputs. The other details about the computer program and parameters are presented in example A. The performance of various ANN configurations was compared using the statistical parameters (Equation 12.12 and Equation 12.13).

The Iterative Parameter Estimation Approach. The transient temperatures at the center of the cube with known physical and thermal properties were estimated by solving the governing heat conduction Equation 12.19 and Equation 12.20) with an assumed Biot number /convective heat transfer coefficient using an FIDAP program. The Biot number/heat transfer coefficient was then varied to produce several time-temperature profiles and then slopes were estimated from these temperature profiles. The Bi for transient temperatures obtained from the experiment was then estimated by minimizing the following function, called the cost function:

$$E = \sum_{i=1}^{N} [S_{exp} - S_{num,i}]^2. \tag{12.21}$$

The slope, S_{exp}, was obtained from experimental time/temperature data as described in the previous section. The transient temperatures at the center of the cube were obtained for a range of Bi. Using transient temperature data, the $S_{num,i}$ were computed. The cost function, E, was determined using Equation 12.21 for various Bi values and at the minimum value of E, the corresponding Bi was taken as experimental Bi. In this iterative procedure, the stopping criteria used was set as $E \leq 10^{-6}$ at lower range ($1 < Bi < 1.6$) and $E \leq 10^{-4}$ at higher range ($5 < Bi < 8$) of Biot number. Minimizing E with respect to parameter Bi leads to:

$$\frac{\partial E}{\partial Bi} = 0 \Rightarrow \sum_{i=1}^{M} \frac{\partial S_{num,i}}{\partial Bi_i} (S_{exp} - S_{num,i}). \tag{12.22}$$

Statistical Considerations in Parameter Estimation. Experimental measurements of temperatures are not always exact. The measurement errors in temperature produce error on the estimation of slope that may be amplified by an ill-posed character of the inverse problem. A typical approach to verify the robustness of the inverse algorithm is to introduce Gaussian noise with an average zero mean and a constant variance σ^2 (or standard deviation, σ) to measured temperature [40,42]. The statistical properties of estimated parameters, with and without noise, are then correlated. The confidence limits in the estimated parameters (i.e., Bi) with a confidence interval at 99% was estimated as $+2.576\sigma$ [40,42].

Results. A data set of 51 conditions was used for training the ANN models. Different ANN configurations were trained using the original as well as the transformed variables. In each analysis, the ANN configuration (out of five) that minimized the four error measuring parameters and optimized R^2 was ultimately selected as the optimum. In the first attempt, the ANN models were trained using an original data set without applying any transformation to the Biot number or to the slope, S. The configuration of the ANN model was varied, as discussed above. However, the performance of many ANN configurations was not very satisfactory (Table 12.5), since the MRE and STD_R in the prediction of Bi always exceeded 9% and 23, respectively. This was particularly true for the prediction Bi in the lower range ($Bi = 0.01–1.0$).

In principle, ANN models do not require any prior knowledge of the relationships between dependent and independent variables. However, as shown in the previous example, transformation of the independent or/and dependent variables can improve their predictive performance [36,37]. For examples, friction factor in pipe flow problems was correlated with Reynolds number on logarithmic scale [36], heat transfer coefficient in tubes was correlated with thermal, physical and flow properties in terms of dimensionless numbers such as Nusselt, Reynolds, Prandtl, and Eckert numbers [43]. A plot of Bi versus S indicated that arctangent relationship between Bi and S [44]. Therefore, both Bi and S were transformed using the inverse tangent functions $\tan^{-1} Bi$ and $\tan^{-1} S$ before feeding to the ANN model. This transformation led to a significant improvement in the prediction performance of all ANN models. The optimal ANN configuration included two neurons in the hidden layer (Table 12.6). The MRE for this optimal configuration was 1.7%, with a standard deviation of 4.9%. Other trigonometric functions such as exponential transformation were also used but they did not improve prediction performance. The prediction error (i.e., relative error) of optimal network in the higher range of Bi ($7.0 < Bi < 10.0$) was between 4 and 7%. Consequently, the standard deviation in relative error was slightly higher than mean relative error.

The simplest ANN models with two neurons can be considered a very good predictor. This particular model shows excellent accuracy (MRE of 1.4%) for the prediction of Bi in the range of Bi between 0.04 and 10.0. The network weights and coefficients associated with this ANN model are presented in the form of simple algebraic equations in Appendix C. These equations can be used to predict Bi from the slope of experimental measured transient temperatures.

Verification of the ANN models. The predictive performance of ANN model was validated using a data set of 14 cases, which were not used in the initial training of the ANN models. The simple ANN model (2 hidden neurons) predicted Bi with a mean relative error of 2.3%, a standard deviation in relative error of 5.5%, and a coefficient of determination of 1.000. Once again, the standard deviations are higher than the mean relative errors because less than 0.04 of the errors in the predictive performance for Bi were rather high ($\sim 20\%$). Otherwise, the mean relative errors in the prediction of Bi using the ANN model was less than 1% in the Biot number range between 0.04 and 10.0.

Uncertainty analysis. Artificial neural networks are capable of handling uncertainties [41,45]. To test the generalization capability of artificial neural networks, a random noise was introduced to

Table 12.5　**Associated Prediction Errors of the Biot Number, *Bi*, for Cube/Fluid Assembly with Different ANN Configurations before Transformations of Data**

Number of Neurons in Hidden Layer	MRE (%)	STD_R (%)	R^2
2	12.6	33.7	0.999
4	14.4	39.3	0.999
6	15.9	44.5	0.999
8	9.88	23.3	0.999
10	43.1	97.9	0.999

Table 12.6 Associated Prediction Errors of the Biot Number, *Bi*, for Cube/Fluid Assembly with Different ANN Configurations after Transformations of Data

Number of Neurons in Hidden Layer	MRE (%)	STD$_R$ (%)	R^2
2	1.71	4.88	1.000
4	2.56	8.60	1.000
6	3.24	10.9	1.000
8	2.39	7.50	1.000
10	3.34	11.6	1.000

the training dataset. The Gaussian distribution, with a zero mean and a standard deviation of 5% in slope, was introduced in each input data point (i.e., in slope, *S*). This is the worst-case error in estimated slope. The error distribution is chosen such that, with a 99% probability, the error in measured temperature is less than or equal to the worst-case error. The sensitivity of the optimal network was examined using the full dataset (51 cases) with the noise. A set of 100 different files with noisy data based on Gaussian distribution was created (with a total of 5100 cases). The prediction accuracy of the optimal network with uncertain data was close to that of original data set without noise. The results of the 100 data set are collected in a graph (Figure 12.10). The prediction accuracy of neural network with noisy data, at higher range of Biot number was in the same range (i.e., 10%) as observed with the original data set. The established neural network exhibited small uncertainty in random errors; in *Bi*, there were less than eight. This demonstrated the capability of ANN in dealing with uncertainties and noise.

The ANN model developed was noniterative, which yield results within 2.0% of those obtained by iterative solution of the governing conduction equation. Though an analytical solution is available to determine the temperature in an arbitrary rectangular parallelepiped subjected to convective heat transfer, estimation of the heat transfer coefficient/*Bi* from known time temperature data still remains iterative in nature. The ANN model presented here can easily be used without any

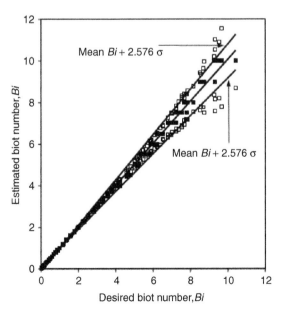

Figure 12.10 Biot number estimation results for Gaussian distribution with zero mean a standard deviation of 5% in slope. (Adapted from Sablani, S. S., Kacimov, A., Perret, J., Mujumdar, A. S., and Campo, A., *Int. J. Heat Mass Transfer*, 48, 665–679, 2005.)

elaborate programming. The present concept of using neural networks for estimating heat transfer coefficient can easily be extended for complex shapes and temperature dependent boundary conditions. However, as the complexity of the problem increases, different transformation of input/output variables may be required.

12.6   CONCLUDING REMARKS

In this chapter, the basic principles of artificial neural networks have been described. Neural networks have the capability to generalize the behavior among data without requiring a prior knowledge of the relationship of the parameters. The dataset may be experimental, simulated, or compiled from the literature. Applications of neural network modeling in food and bioprocessing were presented through different examples of modeling thermal conductivity, estimation of friction factor in pipe flow, and heat transfer coefficients. There have been many interesting applications of ANN modeling reported in the literature including the analysis of reflectance or transmission spectra of a variety of food products by chemometric techniques. Neural network modeling is becoming a very promising tool in predictive modeling of food and bioprocesses. When combined with fuzzy logic, neural network modeling has potential applications in process control and automation.

GLOSSARY

Activation   The time-varying value that is the output of a neuron.

Artificial neural networks   Computational structure whose architecture is modeled after the brain.

Artificial neuron   A simple computational unit that performs a weighted sum on incoming signals, adds a threshold or bias term to this value to yield a net input, and maps this last value through an activation function to compute its own activation.

Back-propagation   A name given to the process by which the Perceptron neural network is "trained" to produce good responses to a set of input patterns.

Bias   The net input (or bias) is proportional to the amount that incoming neural activations must exceed in order for a neuron to fire.

Connectivity   The amount of interaction in a system, the structure of the weights in a neural network, or the relative number of edges in a graph.

Feedforward network   An artificial *neural network* in which the flow of activity is in one direction, from input *neurons* to output neurons.

Generalization   A measure of how well a network can respond to new images on which it has not been trained, but which are related in some way to the training patterns.

Input layer   Neurons whose inputs are fed from the outside world.

Learning rule   The algorithm used for modifying the connection strengths, or weights, in response to training patterns while training is being carried out.

Output neuron   A neuron within a neural network whose outputs are the result of the network.

Pattern recognition   The ability to recognize a given sub-pattern within a much larger pattern.

Perceptron   An artificial neural network capable of simple pattern recognition and classification tasks. It is composed of three layers where signals only pass forward from nodes in the input layer to nodes in the hidden layer, and finally out to the output layer. There are no connections within a layer.

Recurrent network   A neural network in which the output of some neurons feeds back via intervening connections to become input to them.

Sensitivity analysis   The process which determines the sensitivity of a predictive model to small fluctuations in predictor value.

Supervised learning   A class of data mining and machine learning applications and techniques in which the system builds a model based on the prediction of a well defined prediction field. This is in contrast to unsupervised learning where there is no particular goal aside from pattern detection.

Synapse   The space in which a signal passes from one *neuron* to another.

Threshold   A quantity added to (or subtracted from) the weighted sum of inputs into a neuron, which forms the neuron's net input. Intuitively, the net input (or bias) is proportional to the amount that the incoming neural activations must exceed in order for a neuron to fire.

Training set   A neural network is trained using a training set. A training set comprises information about the problem to be solved as input stimuli.

Weight   In a neural network, the strength of a synapse (or connection) between two neurons. Weights may be positive (excitatory) or negative (inhibitory). The thresholds of a neuron are also considered weights, since they undergo adaptation by a learning algorithm.

NOMENCLATURE

$2a$	Side of cube (m)
Bi	Biot number (ha/k)
D	Pipe diameter (m)
E	Cost function
f	Fanning friction factor, dimensionless or Correction factor
Fo	Fourier number ($\alpha t/a^2$), linear problem
g	Acceleration due to gravity, 9.81 (m/s^2)
h	Heat transfer coefficient (W/m^{2}K)
k	Thermal conductivity (W/m K)
k_f	Friction-loss coefficient, dimensionless
K	Consistency coefficient, (Pa s^n)
L	Length of pipe (m)
n	Flow-behavior index, dimensionless
N	Number of measurement
P	Pressure (Pa)
Re'	Reynolds number, $(\rho u^{2-n}D^n/8^{n-1}K)[4n/(3n+1)]^n$, dimensionless
Q	Volumetric flow rate in a pipe (m^3/s)
Q	Volumetric average velocity in the tube (m/s)
ΔP_f	Pressure drop due to friction (Pa)
R^2	Regression coefficient
S	Temperature-time slope, ($d\theta/dFo$)
t	Time (s)
T	Temperature (K)
x, y, z	Linear coordinates (m)
X	Moisture content (wet basis), fraction
X, Y, Z	Dimensionless linear coordinates (x/a, y/a, z/a)

Greek Symbols

α	Kinetic energy correction coefficient, dimensionless, thermal diffusivity (m^2/sec)
γ	Shear rate (1/s)
σ	Shear stress (Pa)
σ	Standard deviation

θ Dimensionless temperature $(T_f - T)/(T_f - T_i)$
ε Volume fraction

Subscripts
a   Air for k or apparent in ε
e   Effective
f   Fluid
i   Initial
r   Reference
w   Water
s   Solid

APPENDIX A

Neural network based equations for calculation of friction factor (f) for power law fluids for known Reynolds number (Re') and consistency index (n)

$$Y0 = \text{LOG } 10(Re)$$

$$Y1 = \text{LOG } 10(n)$$

$$X2 = Y0(0.426) + (-2.405)$$

$$X3 = Y1(2.00) + (0.398)$$

$$X4 = \tanh((-2.291) + (-1.733)X2 + (0.252)X3)$$

$$X5 = \tanh((0.790) + (0.418)X2 + (-0.555)X3)$$

$$X6 = \tanh((0.733) + (0.826)X4 + (-0.632)X5)$$

$$f = X6(0.0139) + (0.00862)$$

APPENDIX B

Neural-network-based equations for estimation of effective thermal conductivity, k (W/m K) for known moisture content (M, fraction, wet basis), temperature ratio (T, T/T_r, T in K) and apparent porosity (ε)

$$X2 = M(2.13) + (-1.09)$$

$$X3 = T(3.18) + (-3.69)$$

$$X4 = \varepsilon(2.86) + (-1)$$

$$X5 = \tanh[(-0.66) + (-0.99)X2 + (-3.28)X3 + (-0.38)X4]$$

$$X6 = \tanh[(1.42) + (-0.82)X2 + (-0.54)X3 + (2.11)X4]$$

$$X7 = \tan h[(-3.49) + (2.11)X2 + (-3.66)X3 + (1.00)X4]$$

$$X8 = \tanh[(-0.18) + (0.92)X2 + (0.36)X3 + (0.042)X4]$$

X9 = tanh[(0.21) + (−0.62)X5 + (−0.72)X6 + (−0.01)X7 + (0.73)X8]

X10 = tanh[(0.025) + (0.058)X5 + (−0.043)X6 + (0.17)X7 + (−0.189)X8]

X11 = tanh[(0.54) + (0.55)X5 + (0.69)X6 + (−1.13)X7 + (0.19)X8]

X12 = tanh[(−0.15) + (−2.76)X5 + (0.53)X6 + (−2.92)X7 + (0.021)X8]
X13 = tanh[(0.078) + (0.14)X9 + (−0.032)X10 + (−0.31)X11 + (−0.29)X12]

k = X13(1.95) + (1.18)

APPENDIX C

Direct estimation of Biot number, Bi from the slope (S) of temperature ratio (on logarithm scale) versus Fourier number for cube

Y = a tan(S)

X2 = Y(1.676) + (1.022)
X3 = tanh((−0.354) + (0.444)X2)

X4 = tanh((−1.700) + (−1.218)X2)
X5 = tanh((0.296) + (−1.185)X3 + (0.884)X4)

Bi = tan(X5 × 1.218 + 0.741)

REFERENCES

1. Aleksander, I. and Morton, H., *An Introduction to Neural Computing*, London: Chapman and Hall, 1990.
2. Baughman, D. R. and Liu, Y. A., *Neural Networks in Bioprocessing and Chemical Engineering*, London: Academic Press, 1995.
3. Haykin, S., *Neural Networks: A Comprehensive Foundation*, 2nd ed., New Jersey: Prentice Hall, 1999.
4. Sarle, W. S., Comp.ai.neural-nets FAQ, Part 1 of 7: Introduction, www.faqs.org/faq/ai-faq/neural-nets/part4/section-1.html, accessed in April 2006.
5. Singh, R. K. and Ou-Yang, F., Neuro-fuzzy technology for computerized automation, In *Computerized Control Systems in the Food Industry*, G.S. Mittal, Ed., New York: Marcel Dekker, pp. 119–178, 1997.
6. Jindal, V. K. and Chauhan, V., Neural network approach to modeling food processing operations, In *Food Processing Operation Modeling*, J. Irudayaraj, Ed., New York: Marcel Dekker, pp. 305–342, 2001.
7. Sarle, W. S., Comp.ai.neural-nets FAQ, Part 4 of 7: Books, data, etc., www.faqs.org/faq/ai-faq/neural-nets/part4/section-1.html, accessed in April 2006.
8. Ramon y Cajal, *Histologie du Systeme Nerveux de L'homme et des Vertbres*, Paris: Malonie Edition Francaise Revue: Tome I, 1952: Tome II, 1955, Madrid: Consejo Superior de Investigaciones Cientifices, 1911.
9. Karayiannis, N. B. and Venetsanopoulos, A. N., *Artificial Neural Networks: Learning Algorithm, Performance Evaluation, and Applications*, Boston: Kluwer Academic, 1993.
10. Rojas, R., *Neural Networks: A Systematic Introduction*, Berlin, Germany: Springer, 1996.
11. McCulloch, W. S. and Pitts, W., A logical calculus of the ideas imminent in nervous activity, *Bulletin of Mathematical Biophysics*, 5, 115–133, 1943.
12. Rousu, J., Flander, L., Suutarinen, M., Autio, K., Kontkanen, P., and Rantanen, A., Novel computational tools in bakery data analysis: A comparative study, *Journal of Food Engineering*, 57, 45–56, 2003.

13. Boillereaux, L., Cadet, C., and Le Bail, A., Thermal properties estimation during thawing via real-time neural network, *Journal of Food Engineering*, 57, 17–23, 2003.
14. Sablani, S. S., Shayya, W. H., and Kacimov, A., Explicit calculation of the friction factor in pipeline flow of Bingham plastic fluids: a neural network approach, *Chemical Engineering and Sciences*, 58, 99–106, 2003.
15. Chen, C. R. and Ramaswamy, H. S., Modeling and optimization of variable retort temperature (VRT) thermal processing using coupled neural networks and genetic algorithms, *Journal of Food Engineering*, 53, 209–220, 2002.
16. Hussain, M. A., Rahman, M. S., and Ng, C. W., Prediction of pores formation (porosity) in foods during drying: Generic models by the use of hybrid neural network, *Journal of Food Engineering*, 51, 239–248, 2002.
17. Therdthai, N. and Zhou, W., Artificial neural network modeling of electrical conductivity property of recombined milk, *Journal of Food Engineering*, 50, 107–111, 2001.
18. Lou, W. and Nakai, S., Application of artificial neural networks for predicting the thermal inactivation of bacteria: A combined effect of temperature, pH and water activity, *Food Research International*, 34, 573–579, 2001.
19. Mittal, G. S. and Zhang, J., Artificial neural network for the prediction of temperature, moisture and fat contents in meatballs during deep-fat frying, *International Journal of Food Science and Technology*, 36, 489–497, 2001.
20. Zbicinski, I. and Ciesielski, K., Extension of the neural networks operating range by the application of dimensionless numbers in prediction of heat transfer coefficient, *Drying Technology*, 18, 649–660, 2002.
21. Adhikari, B. and Jindal, V. K., Artificial neural networks: a new tool for prediction of pressure drop of non-Newtonian fluid foods through tubes, *Journal of Food Engineering*, 46, 43–51, 2000.
22. Mittal, G. S. and Zhang, J., Prediction of freezing time for food products using a neural network, *Food Research International*, 33, 557–562, 2000.
23. Xie, G. and Xiong, R., Use of hyperbolic and neural network models in modeling quality changes of dry peas in long time cooking, *Journal of Food Engineering*, 41, 151–162, 1999.
24. Ni, H. and Gunasekaran, S., Food quality prediction with neural networks, *Food Technology*, 52(10), 60–65, 1998.
25. Rahman, M. S., Thermal conductivity of four food materials as a single function of porosity and water content, *Journal of Food Engineering*, 15, 261–268, 1991.
26. Sweat, V. E., Experimental values of thermal conductivity of selected fruits and vegetables, *Journal of Food Science*, 39, 1080–1083, 1974.
27. Sablani, S. S. and Rahman, M. S., Using neural networks for predicting thermal conductivity of food as a function of moisture content, temperature and porosity, *Food Research International*, 36, 617–623, 2003.
28. Neural Ware, *NeuralWorks: Reference Guide, Software Reference for Professional II/Plus and NeuralWorks Explorer*, Pittsburgh: Neural Ware, 1993.
29. Rahman, M. S., Chen, X. D., and Perera, C. O., An improved thermal conductivity prediction model for fruits and vegetables as a function of temperature, water content and porosity, *Journal of Food Engineering*, 31, 163–170, 1997.
30. Denn, M. M., *Process Fluid Mechanics*, Englewood Cliffs, NJ: Prentice-Hall, 1980.
31. Brodkey, R. S. and Hershey, H. C., *Transport Phenomena*, New York: McGraw-Hill, 1988.
32. Steffe, J. F. and Singh, R. P., Pipeline design calculations for Newtonian and non-Newtonian fluids, In *Handbook of Food Engineering Practice*, K.J. Valentas, E. Rotstein, and R.P. Singh, Eds., Boca Raton, FL: CRC Press, pp. 1–35, 1997.
33. Dodge, D. W. and Metzer, A. B., Turbulent flow of non-Newtonian systems, *AIChE Journal*, 5, 189–204, 1959.
34. Mohtar, R. H., Bralts, V. F., and Shayya, W. H., A finite element model for the analysis and optimization of pipe networks, *Transaction of the American Society of Agriculture Engineers*, 34, 393–401, 1990.
35. Bralts, V. F., Kelly, S. F., Shayya, W. H., and Segerlind, L. J., Finite element analysis of microirrigation hydraulics using a virtual emitter system, *Transaction of the American Society of Agriculture Engineers*, 36, 717–725, 1993.

36. Sablani, S. S. and Shayya, W. H., Neural network based non-iterative calculation of the friction factor for power law fluids, *Journal of Food Engineering*, 57, 327–335, 2003.

37. Sablani, S. S., A neural network approach for non-iterative calculation of heat transfer coefficient in fluid-particle systems, *Chemical Engineering and Processing*, 40, 363–369, 2001.

38. Sablani, S. S., Kacimov, A., Perret, J., Mujumdar, A. S., and Campo, A., Non-iterative estimation of the heat transfer coefficients using artificial neural network models, *International Journal of Heat and Mass Transfer*, 48, 665–679, 2005.

39. Beck, J. V., Blackwell, B., and St. Clair, C. R., *Inverse Heat Conduction*, New York: Wiley-Interscience, 1985.

40. Beck, J. V. and Arnold, K. J., *Parameter Estimation in Engineering and Science*, New York: Wiley, 1977.

41. Hornik, K., Stinchombe, M., and White, H., Multilayer feed forward network are universal approximator, *Neural Network*, 2, 359–366, 1989.

42. Le Niliot, C. and Lefevre, F., A parameter estimation approach to solve the inverse problem of point heat sources identification, *International Journal of Heat and Mass Transfer*, 47, 827–841, 2004.

43. Scalabrin, G. and Piazza, L., Analysis of forced convection heat transfer to supercritical carbon dioxide inside tubes using neural networks, *International Journal of Heat and Mass Transfer*, 46, 1139–1154, 2003.

44. Sreekanth, S., Ramaswamy, H. S., Sablani, S. S., Prasher, S. O., A neural network approach for inverse heat transfer problems. In *American Society of Mechanical Engineers, National Heat Transfer Conference*, Baltimore, MD, August 10–12, 1997.

45. Fan, H. Y., Lu, W. Z., Xi, G., and Wang, S. J., An improved neural network based calibration method for aerodynamic pressure probes, *ASME Journal of Fluids Engineering*, 125, 113–120, 2003.

CHAPTER 13

Genetic Algorithms

T. Morimoto

CONTENTS

13.1 INTRODUCTION

In recent years, consumer demands for better quality of foods have increased, and it has been necessary to reduce the costs of food and bio production processes. The use of an optimization technique offers one solution for overcoming these problems. It is, however, not easy to solve such problems using conventional mathematical approaches because the food and bio production systems are complex and uncertain. There exist systematically complicated properties such as nonlinearity, time variation, and multivariability in their systems. Intelligent approaches are useful for treating and optimizing such complex systems to which conventional mathematical approaches are not easily applied. One approach is the use of genetic algorithms. Recently, genetic algorithms have been applied for optimizations of a wide variety of complex systems such as food, bio, and agricultural production processes.

In this chapter, the basic concept of genetic algorithms is first introduced and then two applications of the genetic algorithms for optimizations of model parameterization and the optimal control of the temperature during a fruit-storage process are focused on.

13.2 FUNDAMENTAL INTRODUCTION TO GENETIC ALGORITHMS

Genetic algorithms for computer simulation were mainly developed by Holland [1,2]. The research has been extended further by his students and colleagues [3,4]. Their studies were first concentrated on the mechanism of an adaptive system—how biological systems adapt to new environment after the large climate change [5].

Genetic algorithms are search techniques for an optimal value, mimicking the mechanism of biological evolution. They have a high ability to find an optimal value (global optimal value or at least near global one) of a complex objective function, without falling into local optima [3,4,6]. They can deal with both continuous and discrete optimization problems. They are especially useful for combinatorial optimization problems characterized by a set of several decision variables (e.g., several kinds of physical variables or l-step setpoints of one control variable). This is a kind of discrete type. In this case, the optimal combination of several decision variables is determined.

The search space in the combinatorial (or multimodal) optimization problem is usually very large and complex. However, genetic algorithms allow an optimal value to be easily and quickly sought from a very large solution space parallel to a multi-point search procedure, by mimicking the mechanics of biological evolution. Such genetic operators as crossover and mutation lead to a multi-point search procedure. A global optimal value can be successfully found in parallel with this multi-point search procedure [2,4,7]. This is because this genetics-based search technique allows the search point to be efficiently focused on the most promising part of the solution space by inheriting important information from parent to offspring. This most promising part is called a "building block" in the building block hypothesis [4]. The building block means "excellent descendant," as it is formed during the searching process by applying crossover and mutation. It usually proliferates if it is not destroyed by crossover and mutation and, finally, an optimal value can be created.

The search way for an optimal value is based on the principle of a Darwinian-type survival of the fittest in natural evolution, where excellent individuals can mate with other individuals and survive for the next generation [3]. The most excellent individual, which is the most evolved one, is an optimal value. Figure 13.1 shows a conceptual search way for a global optimal value using the multi-point search process based on the genetic algorithm. The horizontal axis is individual (contains decision variable) and the vertical axis is fitness value (contains the value of objective function). Here, because the search space in a real optimization problem is usually enormous and complex, we suppose an objective function having many peaks. Two parents, A and B, are first mated at random, and then two offsprings A and B are generated through genetic operations. Furthermore, superior offspring, which have higher fitness than the parents, remain for the next generation. Here, offspring A is superior one. In the next generation, furthermore, more superior offsprings are selected and remain for the next generation. A global optimal value (or at least a near-global one), which is given as the most excellent offspring (the most evolved one), can finally be obtained by repeating these procedures, without falling into local optima.

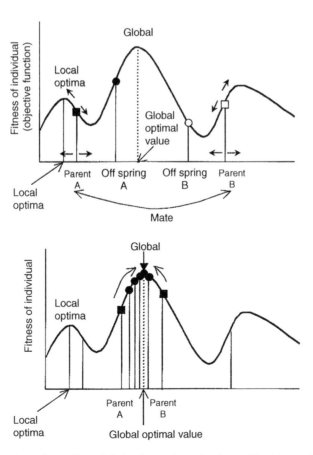

Figure 13.1 Conceptual search way for a global optimal value using the multi-point search procedure based on the genetic algorithm.

Thus, the genetic algorithms are quite different from traditional methods such as gradient search techniques (or hill-climbing methods), linear programming, dynamic programming, and direct-search algorithms, which are characterized by deterministic and numerical. Most of traditional analytical approaches, such as derivative and differential equations, require precise knowledge of the system. Any slightly wrong knowledge causes a decrease in the performance of optimization. It is therefore difficult to obtain the global optimal value of a complex objective function (a multi-modal nonlinear function including local optima) by using traditional methods. This is because the search point in the traditional method (gradient method) moves from one point to another (neighboring) point in turn in the search space based on the arbitrary rule and, consequently, it sometimes stops at the nearest peak (local optima), not the global one of the complex objective function. This is one of the major drawbacks of traditional methods. It is also clear that since most of procedures of the traditional methods are rigid or deterministic, they often suffer from low flexibilities in computation. However, the traditional methods have some merits. For example, most of them consume less time for computation, and their optimal values are analytically guaranteed. In contrast, because the procedures of the genetic algorithms are probabilistic (or stochastic) and require little knowledge, they have high flexibilities in computation [7]. These are remarkable features of the genetic algorithms. However, their procedures are time consuming, and the optimal value is not analytically guaranteed because of their probabilistic procedures.

The genetic-algorithm procedures are not analytical; they are biological. The biological optimization (or adaptation) method through the undirected mechanism of biological evolution and natural selection, which is an iteration approach through trials and errors, might be the best problem solvers for optimization (or adaptation) because living organisms on earth have acquired many excellent functions for survival and have overcome many struggles for existence using this method [2]. So, it can be seen that the genetic algorithms are practical and applicable to a wide range of complex optimization problems.

From these viewpoints, genetic algorithms have been widely applied to optimization problems of complex systems in many fields. Table 13.1 shows applications of genetic algorithms to food production processes, bioprocesses, and agricultural production processes. Many good books on the genetic algorithm have been also published [4,38,39].

13.2.1 Definition of Individual

To use genetic algorithms, an "individual" for evolution should be defined in the first step. Each individual represents a candidate for an optimal value (one possible solution) to the problem that is to be solved. If the solution consists of the combination of mth decision variables $(x_1, x_2, ..., x_m)$ then the individual can be given as follows:

$$\text{Individual} = (x_1, x_2, ..., x_m),$$

Table 13.1 Applications of Genetic Algorithms for Food Production Processes, Bioprocesses and Agricultural Production Processes

Applications	References
Optimization of bioprocesses	8–15
Optimization of crop-production processes	16–19
Optimization of fruit-storage processes	20–25
Model parameterization	26–30
System identification	31, 32
Planning	33, 34
Optimization of agricultural machinery	35, 36
Image processing (feature extraction)	37

where each variable, x_i, is called a *chromosome*. In this case, the optimal combination of $(x_1, x_2, ..., x_m)$ is obtained. It is noted that genetic algorithms are well suited for solving combinatorial optimization problems. A set of individuals is called a *population* that evolves toward better solutions. The individual number included in a population is called a *population size*. Genetic algorithms work with a population involving many individuals:

$$\text{Population} = \begin{matrix} \text{Individual } 1 = (x_{11}, x_{12}, ..., x_{1m}) \\ \text{Individual } 2 = (x_{21}, x_{22}, ..., x_{2m}) \\ .... \\ \text{Individual } N = (x_{N1}, x_{N2}, ..., x_{Nm}) \end{matrix}$$

where N is the population size. The population size increases with applying genetic operations, such as crossover and mutation, to individuals. In general, smaller population size tends to converge to a local optima [40].

13.2.2 Coding

The second step is to transform the chromosomes $(x_1, x_2, ..., x_m)$ into finite-length character strings to make a fit to the genetic operations. Binary digit strings are usually employed for the coding of the chromosomes. The "coding" means the transform of the variable from phenotype (decimal representation) to genotype (binary representation). On the other hand, the reverse transformation, which means restoration, is called a *decoding*. There are two main coding ways. One is the case of a binary-valued representation (binary coding) consisting of 0 and 1, which are most commonly used. One character (0 or 1) in the binary strings (chromosome) is called a "gene." For example, an individual in this method can be expressed as follows:

$$\text{Individual} = (x_1, x_2, ..., x_m) = (010111, 111000, ..., 101010).$$

In this case, an individual is expressed by 6-bit binary strings. The length of the binary strings depends on the range of numerical values that we want to use. The 6-bit binary strings provide numerical values between 0 ($=000000$) and 63 ($=111111$). Thus, the genetic algorithm usually works with a set of binary encoded parameters.

The other is the case of an integer-valued representation. In this case, an individual is given by integral numbers as follows [41,42].

$$\text{Individual} = (x_1, x_2, ..., x_m) = (2, 4, 1, 8, 5, ...).$$

When an individual is composed of many variables, this coding might be effective.

13.2.3 Definition of Fitness

The fitness is an indicator for measuring an individual's quality for survival. All individuals' performances are evaluated based on their fitness values at each generation (iteration step). The fitness is similar to the objective function in conventional optimization problems. Individuals having higher fitness are good ones. During the evolution process, therefore, relatively good individuals reproduce, and relatively bad individuals with lower fitness die in each generation. Finally, an individual having maximum fitness is obtained as an optimal solution. Any nonlinear and discontinuous functions can be selected as the fitness [2].

13.2.4 Genetic Operators

There are mainly three types of genetic operators: (1) crossover, (2) mutation, and (3) selection and reproduction in the genetic algorithms. There are also other operators such as copy, in addition to an island model. Concerning the copy operator, excellent individuals are more copied and reproduced. In the island model, several individuals in another population (island model) are inserted into the original population to maintain the diversity of the population.

13.2.4.1 Crossover

The most important genetic operator is crossover. The crossover combines the features of different individuals (two parents) to form two new individuals (two similar offspring). It operates by swapping corresponding components in their binary strings representing the parents. The cross-over has several types: one-point crossover, multi-points crossover, and uniform crossover.

Now, suppose an individual consisting of one chromosome (individual=chromosome) that is coded by 6-bit binary strings. Figure 13.2a and b show the operations of one-point and multi-point crossovers. In the case of one-point crossover, two individuals (e.g., 000$\overline{101}$ and 001111, which are, respectively, given by 3 and 47 in decimal) are selected at random from the population and mated with each other. The method is as follows. These binary strings are first cut at the center of the 6-bit binary strings (3-bit position) and then two new individuals (000111 and 001$\overline{101}$, which are, respect-ively, given by 7 and 43 in decimal) are generated by swapping all the characters to the right of this point (1–3-bit position). This point is called the "crossover point." In many cases, the crossover point, which is given by the center of the binary strings, is fixed. In the multipoint crossover, on the other hand, it has several crossover points, and the crossing is carried out at their positions in the binary strings, with the same method described above. Here, the case of two-point crossover is described in Figure 13.2b. In this case, the first 3-bit and the last 3-bit binary strings in two individuals are swapped with each other. The two crossover points are usually fixed because of its easy application.

On the other hand, a one-point crossover operation in the case of a real-valued parameter set is shown in Figure 13.3 [41,42].

The individual number for crossover depends on the crossover rate, P_c, which has the value of more than 50% so that new individuals (offspring), which are different from the parents, can be

<table>
<tr><td>Parents   Offsprings</td><td>Parents   Offsprings</td></tr>
<tr><td>000<u>011</u> (3) → 000<u>111</u> (7)
↓↑
101<u>111</u> (47) → 101<u>011</u> (43)</td><td>101000<u>101</u> → <u>001</u>000111
↓↑   ↓↑
<u>001</u>001<u>111</u> → 101001<u>101</u></td></tr>
<tr><td>(a) One-point crossover</td><td>(b) Multi-point crossover</td></tr>
<tr><td>Parents   : 000011 and 101111
Offsprings : 000111 and 101011</td><td>Parents   : 101000101 and 001001111
Offsprings : 001000111 and 101001101</td></tr>
</table>

Figure 13.2 Methods of one-point and multi (two)-point crossovers in the case of binary strings (the numeral in the paranthesis shows the decimal number for the binary number).

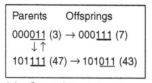

Parents   Offsprings
(3, 2, 5, <u>6, 7, 1</u>) → (3, 2, 5, <u>2, 3, 0</u>)
          ↓↑
(4, 7, 1, <u>2, 3, 0</u>) → (4, 7, 1, <u>6, 7, 1</u>)

Figure 13.3 Method of a one-point crossover in the case of a real-valued parameter set.

created sufficiently. The crossover rate is usually ranged from 0.6 to 0.8. In each generation, NP_c individuals undergo crossover (N: individual number). The higher the crossover rate, the more excellent the individuals that are created in the population. If the crossover rate is too low, the evolution (searching process) may stagnate due to the lower number of new individuals with ultimately poor performance.

13.2.4.2 Mutation

The mutation inverts one or more components, selected at random, of the binary strings from 0 to 1 or vice versa. The new individuals generated by the mutation undergo random changes. The mutation operation increases the variability of the population and helps to prevent premature convergence to local optima in the evolution process [7]. Figure 13.4 shows the operation of a two-point mutation. In the method, one individual (e.g., 000110, 6 in decimal) is first selected at random, and then a new individual (010100, 20 in decimal) is created by inverting two characters (genes), selected at random, from 0 to 1 or 1 to 0.

On the other hand, a two-point mutation in the case of a real-valued parameter set is shown in Figure 13.5 [42]. Here, two values (2 and 6) at two positions selected at random are converted into new two values (8 and 3) based on the random number.

The individual number for mutation depends on the mutation rate, P_m. The mutation rate in natural evolution is usually very small. So, mutation can be considered as a secondary operator for evolution. Therefore, the mutation rate is traditionally given by low values at the range of 0.01–0.1. In each generation, NP_m individuals undergo mutation (N: individual number). If the value is too low, however, the possibility to fall into local optima increases. The lack of mutation induces poorer performances in evolution. On the other hand, the higher mutation rate helps prevent premature convergence to local optima. Recently, high mutation rates up to 0.4 or 0.6 have been found beneficial [43,44]. It is, however, noted that a significantly high mutation rate leads to an essentially random search.

13.2.4.3 Selection and Reproduction

In the process of evolution, new individuals (offspring) are continuously created from the old ones (parents) every generation. Selection and reproduction are operators for choosing and reproducing excellent individuals. Excellent individuals with higher fitness values are fundamentally chosen and reproduced, and they have better chances for crossover and mutation than those with lower fitness values in order to create the more excellent individuals [45]. Figure 13.6 shows an example of the reproduction. Individuals with higher fitness are fundamentally reproduced and

Parent	0 0 0 1 1 0 (6)
	↓ ↓
Offspring	0 1 0 1 0 0 (20)

Figure 13.4 Method of a two-point mutation in the case of binary strings (the numeral in the paranthesis shows the decimal number for the binary number).

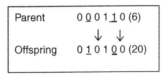

Parent	(3, 2, 5, 6, 7, 1)
	↓ ↓
Offspring	(3, 8, 5, 3, 7, 1)

Figure 13.5 Method of a two-point mutation in the case of a real-valued parameter set.

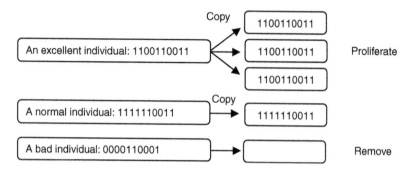

Figure 13.6 Example of reproduction. Excellent individuals with higher fitness values are chosen and reproduced and have a better chance to do crossover and mutation than their less fitness values.

proliferated by a copy operator with a generation number. In contrast, individuals with lower fitness values are removed from the population. There are several selection strategies: a roulette wheel selection, an elitist strategy and an expected-value selection.

Roulette Wheel Selection (Proportional Selection). The roulette wheel selection (proportional selection) is the most simple and fundamental selection method. In this method, individuals are selected based on their probability of selection, in proportion to their fitness values. Table 13.2 shows individuals' fitness values and their probability of selection, P_s. Figure 13.7 shows the conceptual diagram of the roulette wheel in the case of Table 13.2. In this method, each individual is allocated a slot in the roulette wheel. The width of each slot is proportional to the individual's fitness value so that the individual can get a chance to reproduce new individuals (offspring) according to its fitness value. Let f_i be the fitness in an individual, i. The probability of selection, P_s, is described as follows:

$$P_s = \frac{f_i}{\sum_{i=1}^{N_s} f_i}. \tag{13.1}$$

Individuals are selected according to the ratio P_s of an adequate individual's fitness f_i to the total fitness Σf_i obtained by summing the fitness values over all individuals. From Table 13.2, a total fitness value of 126 is obtained by summing the fitness of all individuals. Here, individual 1 has a fitness value of 53, which corresponds to 42% of the total fitness. Consequently, the 42% area of the roulette wheel is allocated to individual 1 and each spin turns up individual 1 with a probability of 0.42. This method has an undesired property: there is a danger that the best individual may disappear from the population due to the probabilistic selection method.

Elitist Strategy. In the elitist strategy, the best individual with highest fitness is compulsively copied with no operations of crossover and mutation in each generation. It can always survive and is

Table 13.2 Individuals' Fitness Values and Their Probability of Selection in the Roulette Wheel Selection

No.	Individual	Fitness	Percent of Total
1	100011	53	42
2	100111	35	28
3	011001	19	15
4	001101	15	12
5	000011	4	3
Total		126	100

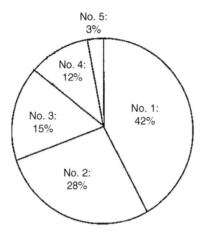

Figure 13.7 The conceptual diagram of the weighted roulette wheel in the case of Table 13.1.

guaranteed a long life from generation to generation. This method has an advantage of faster evolution because the best individual always remains in each generation and the next generation inherits its structure, without destruction. However, there is a danger of falling into local optima because the diversity of the population becomes low because of the concentration of only better individuals.

Expected-Value Selection. In the expected-value selection, individuals are selected based on their expected-values, which come from their fitness values. The expected-value (i.e., the probability of selection, P_s) of each individual is computed using Equation 13.1 The individual number to be reproduced is then determined based on its expected-value. Table 13.3 shows individuals' fitness values, their expected-values, and the individual numbers to be reproduced. In this case, the number of $P_s \times 10$ is reproduced. In this method, the best and excellent individuals always remain to some extent because it is not a probabilistic approach like a roulette wheel selection and excellent individuals are certainly selected based on their expected-values.

As mentioned above, there are various selection and reproduction methods. Many researchers have investigated a more effective method. It can especially be seen that the combinatorial use of the elitist and roulette wheel strategies is useful for selection.

13.2.5   Searching Procedure of an Optimal Value (Artificial Evolution Process)

The searching process for an optimal solution by using genetic algorithms is analogous to a natural evolution process. As generation progresses, each individual in the population evolves toward the better individual with a higher fitness value. Individuals with higher fitness dominate

Table 13.3 Individuals' Fitness Values, Their Expected-Values and Individual Numbers to Be Reproduced in the Expected-Value Selection

No.	Individual	Fitness	Expectedvalue	Individual Number Reproduced
1	100011	53	0.42	4
2	100111	35	0.28	3
3	011001	19	0.15	2
4	001101	15	0.12	1
5	000011	4	0.03	0
Total		126		

in the population with generation. The most evolved individual is an optimal value. Figure 13.8 shows the basic procedure of the genetic algorithm used for searching of an optimal value.

Step 1: *Initial population.* An initial population $P(0)$ consisting of N_i types of individuals is generated at random ($N=N_i$). Each individual (binary string) is created based on the random number. The genetic algorithm starts with the generation of the initial population.

Step 2: *Fitness computation.* The fitness values of all (N) individuals are calculated using a fitness function (or objective function), and their performances are evaluated.

Step 3: *Selection and reproduction.* N_s types of individuals, which are superior individuals having higher fitness, are selected and reproduced based on the roulette wheel strategy. The best individual with highest fitness is also remained based on the elitist strategy. After the selection, there are N ($=N_s+N_{rep}$) types of individuals (N_{rep} is the individual number reproduced).

Step 4: *Crossover.* Two individuals (two parents) are chosen at random from the population including N types of individuals and then the crossover operator is applied to those. Through one crossover operation, two new individuals (two offsprings) are created and inserted into the population. The number of the crossover N_c ($=NP_c$) depends on the crossover rate P_c.

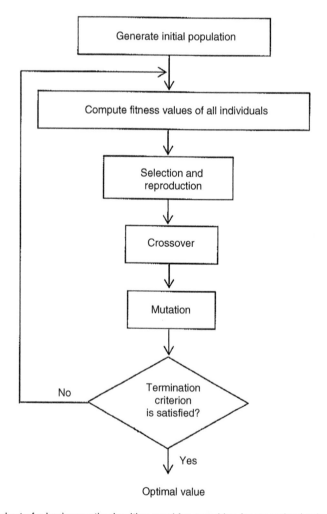

Figure 13.8 Flow chart of a basic genetic algorithm used for searching for an optimal value.

After the crossover, there are $N\,(=N_s+N_{rep}+N_c)$ types of individuals.

Step 5: *Mutation*. One individual (one parent) is chosen at random from the population and the mutation operator is applied to it. Through one mutation operation, one individual (one offspring) is generated and then it is inserted into the population. The number of the mutation $N_m\,(=NP_m)$, depends on the mutation rate, P_m. After the mutation, there are $N\,(=N_s+N_{rep}+N_c+N_m)$ types of individuals.

Step 6: Steps 2–5 are repeated until an adequate termination condition is satisfied. The iteration usually stops when the fitness value of the best individual continues to keep the same maximum value with increasing generation number. An optimal value is given by an individual with maximum fitness.

Recently, it has been noted that various searching techniques have been studied to obtain a global optimal value more quickly and efficiently. The methods improving the searching procedure are mentioned in Section 13.2.7.

13.2.6 A Simple Genetic Algorithm

The most simple and fundamental method of genetic algorithms is called "simple genetic algorithm." It is usually composed of three genetic operators: (1) 1-point crossover, (2) 1-point mutation, and (3) selection (or reproduction).

13.2.7 Improvement of Evolution Performance

The evolution process is time-consuming and its convergent speed is low. Problems such as premature local convergence and the bias by genetic drift, which are probably caused by the loss of diversity of the population, arise in the evolution process. There is also no guarantee to yield a global optimal solution in the search by the genetic algorithms. In recent years, therefore, new evolution techniques have been discussed in order to improve the evolution performance.

The diversity of the population significantly affects the speed of the evolution. The lower diversity easily falls into a local optima. In the elitist strategy, all individuals tend to become uniform with comparatively higher fitness values. This problem is especially severe when the population size is small. The higher mutation rate helps prevent premature convergence to local optima. In the roulette wheel selection, however, it should be noted that a higher mutation rate tends to destroy the good structure (or scheme) of better individuals. Morimoto et al. [20,22] added a number of individuals, generated at random, to the population in each generation to maintain the diversity of the population at higher level. Through this procedure, a global optimal solution could be successfully obtained. Tanese [46] also proposed a parallel genetic algorithm (GA) to maintain the diversity of the population at higher levels. In this method, two GAs in parallel work under two populations (original population and another one) and the immigration of some individuals is conducted between two populations in every arbitrary generation. Van Rooij et al. [40] also claimed that the use of a parallel GA is effective for increasing the speed of the evolution.

From the hints of new natural evolution theories, new evolution schemes have been developed in order to improve the evolution performance. Kubota et al. [47] applied a virus evolutionary algorithm to improve the performance of the evolution. In this method, the structures of chromosomes were effectively changed by the infection of retrovirus. Ohkura and Ueda [48] proposed a new genetic algorithm based on neutral mutations to solve deceptive function optimization problems having no search direction based on the building block hypothesis. On the other hand, Bersini and Varela [49] developed an immune system for improving a local search *performance near the best solution* by genetic algorithms. This technique, which is called a *genetic immune recruitment mechanism* (GIRM), makes up a weak point of the local search of GA by conducting a test of the similarity of each individual.

13.2.8 Problems of the Genetic Algorithms

As mentioned above, the genetic algorithms have a high ability to find a global optimal value of a complex objective function, without falling into local optima. However, the major problem is that there is no guarantee that they will always yield a global optimal value. This is because the searching (evolution) procedures, especially the methods of creating individuals by genetic operators, depend on arbitrary probabilistic roles and, consequently, the process which reaches a global optimal value becomes uncertain. However, better solutions (individuals) appear with generation. It is clear that one can at least obtain the near-global optima using the genetic algorithm. In such case, it is very important to confirm whether an optimal value obtained is a global optima or local one. Comparing some optimal values obtained under different search conditions, e.g., different crossover and mutation rates, initial population, population size, and different methods for crossover and mutation, which significantly affect the search performance, carries out this confirmation. Moreover, it is also important to perfectly confirm an optimal solution by examining values around it through trial and error. Through these procedures, a global optimal solution can be obtained. Hu et al. [50] developed a new technique, which is a combinatorial approach of genetic algorithms and gradient method, to quickly find a global optimal solution. In this method, the genetic algorithm is first used for searching for a near optimal solution and then a gradient method, which can be regarded as a local searching algorithm, is employed for finding a global optimal. Renders and Flasse [51] also proposed a hybrid method combining genetic algorithms and "hill-climbing" methods to yield a global optimal solution. These techniques are very effective for searching for the global optima quickly because genetic algorithms are time-consuming techniques.

13.3 APPLICATIONS OF GENETIC ALGORITHMS TO MODEL PARAMETERIZATION

13.3.1 Optimization Problem

Many types of mathematical models have been developed for prediction and analysis of a system in the crop, food, and bio-production processes. The aim of modeling usually focuses on determining the appropriate values of several parameters (coefficients) used in their models. Manual adjustment is often used for determining their best values. However, such tasks as a model parameterization are usually troublesome and time consuming.

Here, the example that Sequeira et al. [26] used the genetic algorithm to automatically determine the optimal parameters of the photosynthesis model (mathematical model) is introduced. The model, which is used for the photosynthesis of C_3 and C_4 species by Johnson et al. [52], is given by a direct analog of the lower root of the quadratic equation. The gross photosynthesis (P) of *Panicum maximum* to variations in incident radiation is given by

$$P = \frac{1}{2\theta}\left(\alpha I + P_m - \sqrt{(\alpha I + P_m)^2 - 4\theta\alpha I P_m}\right) \tag{13.2}$$

where α is a photochemical efficiency, I is irradiance, P_m is an asymptotic value of P at the saturating irradiance, and θ is a scaling parameter. In this study, the optimal values of three parameters (α, θ, and P_m) are determined. This model is also defined as a semiempirical function because all parameters have biological meanings. The genetic algorithm is used to search for the optimal values of three parameters (α, θ, and P_m).

The optimal parameters are determined so that the average error between the estimated and observed values of the gross photosynthesis can be minimized. Therefore, an objective function f_i which depends on three parameters (α, θ, and P_m), can be defined as follows:

$$F_1(a_i, \theta_i, P_{mi}) = \sum_{j=1}^{n}(x_{j,\text{real}} - x_{j,\text{est}})^2,\qquad(13.3)$$

where n is the number of observations, $x_{j,\text{real}}$ is the observed value, $x_{j,\text{est}}$ is the model-prediction value for observation j, and i is the individual (parameter set) being evaluated.

As for the restraints of three parameters, since there is no pre-knowledge on three parameters, their ranges (search spaces) were given as wide as possible.

Thus, the optimization problem here is to search for the optimal values (optimal combination) of three parameters (α, θ, P_m) that minimize the objective function f_i given by Equation 13.3, under the restraints of $\alpha_{\min} \le \alpha \le \alpha_{\max}$, $\theta_{\min} \le \theta \le \theta_{\max}$, and $P_{m\ \min} \le P_m \le P_{m\ \max}$.

$$\text{minimize } F_1$$
$$\text{subject to } \alpha_{\min} \le \alpha \le \alpha_{\max}, \quad \theta_{\min} \le \theta \le \theta_{\max}, \quad P_{m\cdot\ \min} \le P_m \le P_{m\cdot\ \max}\qquad(13.4)$$

It is noted that if each parameter is coded as a 9-bit binary strings, the combinatorial number (the size of the search space) of three parameters considered is $(2^9) \times (2^9) \times (2^9) = 512^3 = 13{,}421{,}772$. The genetic algorithm allows the best combination of three parameters to be efficiently found out from among this 13,421,772 combination.

13.3.2 Definition of Individual

An optimal value here is given by the optimal combination of three parameters (α, θ, and P_m). So, an individual is given by the parameter set of α, θ, and P_m. They were all coded as 6- or 9-bit binary strings. If each parameter was given by 9-bit binary strings, it has numerical values between 0 (000000000) and 512 (111111111).

$$\text{Individual } i = (\alpha_i, \theta_i, P_{mi}) = (10101010,\ 100111000,\ 111010011).$$

13.3.3 Definition of Fitness

Fitness is an indicator for measuring an individual's survival quality. All individuals are scored by the fitness function and ranked according to their fitness values. During the evolution process, individuals having higher fitness remained for the next population, and individuals with lower fitness die in each generation. Finally, an individual having the maximum fitness can be regarded as an optimal solution. Fitness is similar to the objective function in conventional optimization problems.

Here, because the optimal values of three parameters are determined so that the average error between the estimated and observed values of the gross photosynthesis is minimized, the fitness function (objective function) can be defined as follows:

$$\text{Fitness} = F_1\qquad(13.5)$$

13.3.4 Flow Chart of the Genetic Algorithm

Figure 13.9 shows the flow chart of searching for the optimal combination of three parameters using genetic algorithms. The real values used in the genetic algorithm are shown in Table 13.4. They are empirically determined [4,26]. The procedure is as follows: (1) The initial population is generated at random. (2) All individuals in the population are passed to the simulation model (Equation 13.2) and their fitness values are calculated using the Equation 13.3. (3) All individuals are ranked based on their fitness values. (4) Individuals with higher fitness are selected for reproduction and inserted into the next generation. Highly fit individuals tend to reproduce in

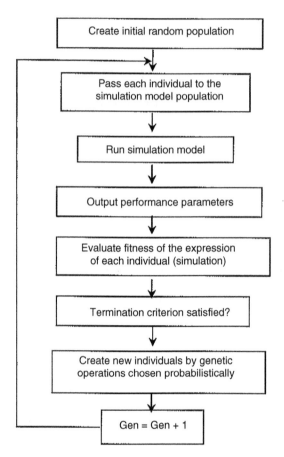

Figure 13.9　Flow chart of a genetic algorithm as applied to model parameterization (Gen: generation number). (From Sequeira, R. A., Olson, R. L., Willers, J. L., and McKinion, J. M., *Computers and Electronics in Agriculture*, 11(2,3), 265–290, 1994.)

the next generation. (5) The selected individuals are altered based on crossover and mutation. The crossover and mutation rates were empirically determined to be 0.60 and 0.03, respectively. Two-point crossover was used in this experiment. The probability that an individual $C(t)$ at the present generation t survives for the next generation $(t+1)$ is given by its fitness value to the average fitness value of all individuals in the present population:

$$P(C(t + 1)) = \text{Fitness}(C(t))/\text{Average population fitness}.$$

Table 13.4　Real Values Used in the Genetic Algorithm Procedure

GA Parameter	Tested Range
Population size	100
Crossover rate	0.60
Mutation rate	0.03
Number of gene	3–6
Number of generations	10.000
Crossover type	Two-point crossover
Selection method	Roulette + Elitist
Stopping criteria	Specified by number of trials
Fitness: forecast accuracy	Sums of squares
String length	30–70

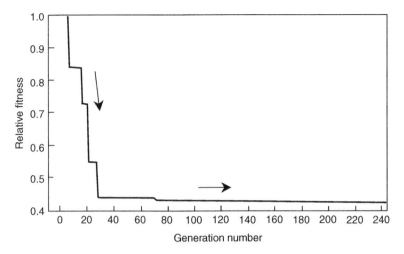

Figure 13.10   Searching process of an optimal value. (From Sequeira, R. A., Olson, R. L., Willers, J. L. and Mckinion, J. M. *Computers and Electronics in Agriculture*, 11(2,3), 265–290, 1994.)

The better individuals with higher fitness remained for the next generation to form the basis of the new population. These procedures, from (2) to (5), continue in a manner until the average fitness of all individuals in the population show slight changes (i.e., converges). This is the termination criterion.

13.3.5   Searching Process of an Optimal Value (Artificial Evolution Process)

Figure 13.10 shows the evolution curve in searching for of an optimal value. This process is analogous to a biological evolution process, where individuals usually evolve toward maximization. In this case, however, because this optimization problem is minimization, all individuals evolve toward minimization. The horizontal axis is the generation number for evolution, and the vertical axis is the fitness of the best individual in each generation. The fitness value corresponds to the degree of the individual's evolution. As shown in the figure, the fitness dramatically decreased with the generation number and then lowered down to the minimum value at about the 70th generation

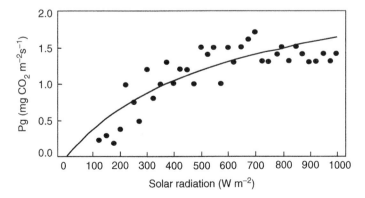

Figure 13.11   Relationship between the solar radiation and the gross photosynthesis of the plant, obtained from simulation of the model optimized by the genetic algorithm. The black circles are the observed values of the gross photosynthesis. (From Sequeira, R. A., Olson, R. L., Willers, J. L. and Mckinion, J. M. *Computers and Electronics in Agriculture*, 11(2,3), 265–290, 1994.)

number. The search was stopped when the fitness continued to keep the same minimum value. The individual that gave the minimum fitness is considered to be the optimal value.

13.3.6   Model Performance

Figure 13.11 shows the nonlinear curve, representing the relationship between the solar radiation and the gross photosynthesis of the plant, which was obtained from simulation of the model optimized. The black circles are observed values. It is found that the estimated nonlinear curve is well fitted to the observed data. Off course, traditional nonlinear curve-fitting methods are also useful for this simple optimization problem. Notice that the genetic algorithm-based curve-fitting method is useful for more complex problems.

13.4   APPLICATIONS OF GENETIC ALGORITHMS TO DYNAMIC OPTIMIZATION OF FRUIT-STORAGE PROCESSES

13.4.1   Dynamic Optimization Problem

Freshness retention of vegetables and fruits during storage is one of the most important issues in the post-harvest technology. The most fundamental manipulated variable for controlling freshness is a storage temperature.

The storage temperature is usually maintained constant at a low level. This is because the low temperature effectively reduces microbial spoilage and water loss of the fruit. In recent years, however, there has been much interest in heat treatments that reduce the quality loss of fruit during storage [53,54]. It has been reported that heat treatment is effective for inhibiting ethylene production and delaying the ripening [55–59], for controlling insect pests and reducing chilling injury of fruit [56,60]. An intermittent heat treatment is also more effective than a single treatment in delaying the ripening of fruit [22,61,62]. It has been also reported that heat treatment can improve fruit quality [63–65]. This is probably due to thermotolerance of the fruit acquired by heat stress.

It is well known that the exposure of living organisms to heat stress produces several types of heat shock proteins (HSPs) in their cells, which acquire transient thermotolerance [66,67]. Recently, the relationships between the heat treatment and HSPs have been investigated to elucidate the effects of heat treatment [68]. Acquiring thermotolerance may lead to the reduction of water loss of fruits during storage [59]. It is, therefore, important to know how to apply the heat stress to the fruit in order to minimize loss of quality. An optimal control technique will give us the solution.

Figure 13.12 shows the conceptual diagram of the dynamic optimization problem that minimizes the water loss of the fruit during storage by applying the heat stress [23,25]. For realizing optimization, the control process was divided into l steps. Hence, the optimization problem here is to find the variable temperature profile (optimal l-step set points of temperature; $T_1, T_2,...,T_l$) to minimize the rate of the water loss of the fruit. That is, an optimal value is given by the optimal combination of the l-step set points for temperature. The control input is the temperature, and the controlled output is the rate of water loss of the fruit during storage.

Let $WT(k)$ ($k=1,2,...,N$) be a time series of the rate of the water loss, as affected by temperature $T(k)$, at time k. An objective function, $F_2(T)$, is given by the average value of the rate of the water loss during the last period ($N_L \leq k \leq N$) of the control process. N_L and N are the first and last time points in the evaluation period.

$$F_2(T) = \sum_{k=N_L}^{N} \frac{W_T(k)}{N - N_L + 1},$$ (13.6)

Note that the rate of the water loss was evaluated at the last step (last process) of the control process ($N_L \leq k \leq N$) in the control process. This is because the influence of heat stress is thought to

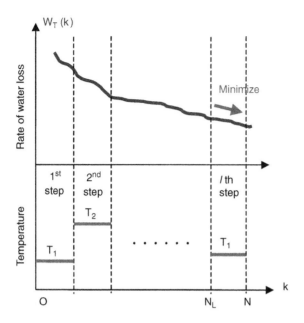

Figure 13.12 A dynamic optimization problem that reduces (minimizes) the rate of the water loss of the fruit during storage by the *l*-step set-points of the temperature including heat stresses $\{T_1, T_2,...,T_l\}$. Here, the optimal trajectory (*l*-step set-points) of the temperature is obtained.

appear at the latter half stage if heat stress was applied to the fruit during the first period of the control process.

As for the constraint of the temperature, we found 15 and 40°C as the minimum and maximum shelf-life temperatures in Japan from previous literature [53–55] and considerations for a one-day application of heat stress.

Thus, the dynamic optimization problem here is to find the optimal *l*-step set points of the temperature (or the optimal combination of *l*-step set points of the temperature) that minimizes the objective function $F_2(T)$ from among numerous output responses $W_{T(k)}$, obtained from simulation of a model (here a neural-network model is used) to any combination of the *l*-step set points of the temperature under the constraint of $15 \leq T(k) \leq 40°C$.

$$\text{minimize } F_2(T)$$
$$\text{subject to } 15 \leq T(k) \leq 40°C \tag{13.7}$$

Had the 4-step control process been used, the combination number of the 4-step set points is $(40-5+1)^4 = 456{,}976$. The use of genetic algorithms allows the best combination to be very efficiently sought from among such numerous combinations.

13.4.2 Plant Materials and Measuring Systems

Tomatoes (*Lycopersicon esculentum* Mill. cv. Momotaro) were used for the experiment. Mature green tomatoes of uniform size (about 8 cm in diameter) were stored in a storage chamber (Tabai-espec, LHU-112M), where the temperature and relative humidity were strictly controlled by a personal computer with an accuracy of $\pm 0.1°C$ and $\pm 2\%$ RH, respectively. The rate of water loss was estimated from the weight loss in the tomato, by hanging a cage containing three tomatoes using an electronic balance. In this case, the electronic balance was set outside of the chamber in order to remove the effect of the temperature change. The relative humidity was maintained constant at $60 \pm 2\%$ RH while only the temperature was flexibly changed based on a system control manner. The sampling time was 10 min.

13.4.3   Definition of Individual

Because an optimal value to be obtained here is the l-step set points of temperature, an individual can be given by the l-step set points of temperature $\{T_1, T_2,...,T_l\}$. They were all coded as 6-bit binary strings, which gave numerical values between 0 (000000) and 63 (111111).

$$\text{Individual } i = \{T_{i1}, T_{i2}, ...T_{il}\} = \{101010, \ 111000..., \ 010011\}.$$

13.4.4   Definition of Fitness

As mentioned above, because fitness is similar to the objective function in conventional optimization problems, it can be represented by Equation 13.6:

$$\text{Fitness} = F_2(\text{T}). \tag{13.8}$$

13.4.5   Genetic Operations

A single crossover and two-point mutation were used as genetic operators. The selection of individuals was carried out based on the elitist strategy by which an individual with minimum fitness compulsorily remains for the next generation. However, its searching performance can easily fall into a local optimum because only the superior individuals with higher fitness are picked in each generation. In this study, therefore, quite different individuals (100) in another population were added into the original population in order to maintain the diversity and obtain a global optimal value.

13.4.6   Searching Process of an Optimal Value

Figure 13.13 shows the block diagram of the searching procedure used here:

Step 1: An initial population $P(0)$ consisting of N_i ($=6$) types of individuals is generated at random.
Step 2: N_o ($=100$) types of new individuals in another population are added to the original population to maintain the diversity of the original population. Here, the other population is independent from the original population.
Step 3: The one-point crossover and two-point mutation operators are applied to the individuals selected at random. The crossover and mutation rates are 0.8 and 0.6, respectively. Through these operations, N ($=N_i+N_o+N_c+N_m=$ about 1500) types of individuals are obtained.
Step 4: The fitness values of all individuals are calculated using the identified neural-network model and their performances are evaluated.
Step 5: N_s ($=300$) types of superior individuals (individuals with higher fitness) are selected and retained for the next generation based on the elitist strategy.
Step 6: Steps 2–5 are repeated until the fitness continues to keep the same maximum value with increasing generation number. An optimal value is given by an individual with maximum fitness.

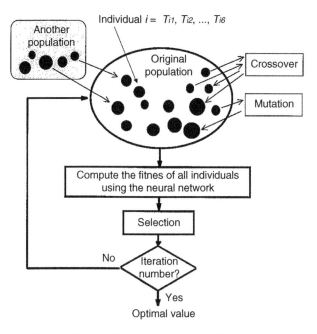

Figure 13.13 Flow chart of the genetic algorithm used for searching for an optimal value.

13.4.7 An Intelligent Control System for Dynamic Optimization

In this study, dynamic optimization of heat treatment for reducing the water loss in fruit during storage was investigated using an intelligent control technique combining with neural networks (simulation model) and genetic algorithms.

Figure 13.14 shows the block diagram of a control system for realizing the optimization (minimization) of the rate of water loss of fruit as affected by temperature during the storage process [16,18,23,25]. It consists of a decision system and a feedback control system. The decision system, consisting of neural networks and genetic algorithms, determines the optimal set point trajectory of the temperature. In the decision system, the rate of water loss, as affected by

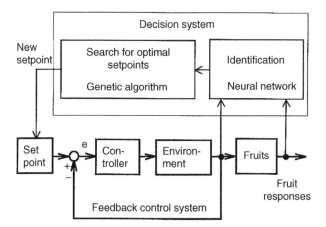

Figure 13.14 The block diagram of an intelligent control system for realizing the optimization of the water loss of fruit as affected by the temperature during the storage process.

temperature, is first identified using the neural network, and then the optimal combination of the *l*-step set points of the temperature that minimize the objective function is searched for through simulation of the identified neural-network model using the genetic algorithm. The genetic algorithm is used for searching for the best individual (optimal combination of the *l*-step set points of the temperature is equal to the input signal) from among numerous trials obtained from simulation of the neural-network model.

It was found that if these two procedures, identification and the search for an optimal value, are repeated periodically during the storage process to adapt to the time variation of the physiological status of the fruit, then both optimization and adaptation can be satisfied.

13.4.8   Dynamic Responses of the Rate of Water Loss

First, dynamic responses of the rate of water loss were observed in order to confirm the effect of heat stress. Figure 13.15 shows a typical response of the rate of water loss as affected by the temperature. The temperature was first increased from 25 to 35 to 40°C and then decreased from 40 to 35 to 25°C, respectively. Comparing to the two values of the rate of water loss at the same temperature, before increasing and after dropping the temperature, it is found that the values after dropping the temperature are lower than those before increasing the temperature at both the 25 and 35°C conditions. These results suggest that a temperature operation that first rises to the high level (35 to 40°C) and then drops to the prior level has a tendency to reduce the rate of water loss, as compared to when the temperature was maintained constant throughout the control process.

Figure 13.16 shows eight types of dynamic changes in the rate of water loss (controlled output) as affected by temperature (control input) for about 192 h. These data are applied for identification using neural networks. A black box model (dynamic model), representing the dynamic relationship

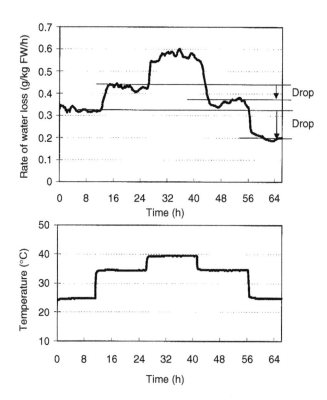

Figure 13.15   A dynamic response of the rate of the water loss, as affected by the temperature.

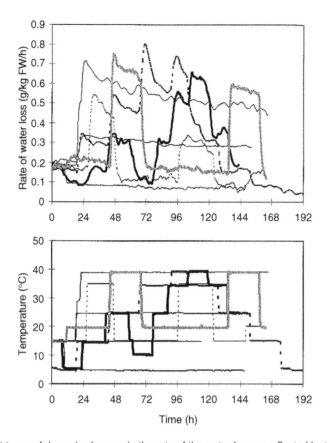

Figure 13.16 Eight types of dynamic changes in the rate of the water loss, as affected by temperature.

between the temperature and the rate of water loss, is built through identification. That is, these data are all training data sets for the neural network. The temperature was flexibly changed between 5 and 40°C to identify clearly the dynamics of the rate of water loss as affected by temperature. Short-term heat stresses of 40°C for about 24 h were included in several temperature operations. From the figure, it is found that, in all cases, the rate of water loss dynamically changes with the temperature.

13.4.9 Identification of the Rate of Water Loss to Temperature Using the Neural Network

Next, the training data in Figure 13.16 are identified using the neural network in order to make a dynamic model for simulation. The system parameter number and the hidden neuron number of the neural network were determined based on the cross-validation.

Figure 13.17 shows the comparison of the estimated response, calculated from the neural network model, and the observed response for the rate of water loss. A testing data set, which is quite different from the training data sets, shown in Figure 13.7; was used for this comparison. It was found that the estimated response was closely related to the observed response. This means that a suitable search simulator was found for obtaining an optimal value.

Figure 13.18 shows the estimated relationship between the temperature and the rate of water loss of a tomato, calculated from the simulation of the identified neural-network model. Open circles represent real observed data. The rate of water loss increases with temperature. In the range over 35°C, it has a tendency to decrease with temperature. This means that the water loss

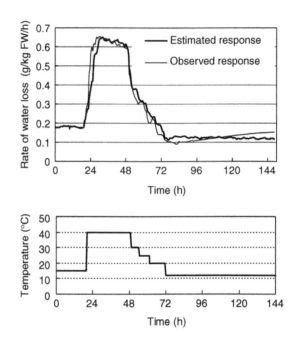

Figure 13.17 Comparison of the estimated and observed responses of the rate of the water loss.

was significantly suppressed by high temperature. Thus, it is found that the relationship between temperature and the rate of water loss is nonlinear.

13.4.10 Searching Process of an Optimal Value

As mentioned earlier, 24 h was selected as the heat–stress application time, which means one step for dynamic optimization. Here, the length of the control process for optimization was six days, so it was divided into six steps.

Next, the optimal combination of the 6-step set points for temperature that minimized the objective function was searched for through simulation of the identified neural-network model using the genetic algorithm. In this method, the optimal combination was selected from among

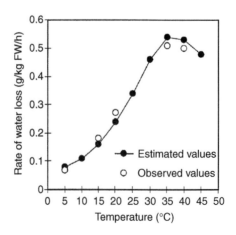

Figure 13.18 The static relationship between the temperature and the rate of the water loss, obtained from simulation of the identified model.

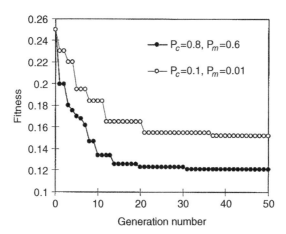

Figure 13.19 An evolution curve in searching for an optimal value when the evaluation length is the latter half stage of the control process (96–168 h).

numerous responses of the rate of water loss, as affected by any 6-step set point trajectory of the temperature, obtained from simulation.

Figure 13.19 shows the searching process for an optimal value under the different crossover and mutation rates. The horizontal axis is the generation number for evolution and the vertical axis is the fitness of the best individual in each generation. As shown in the figure, the fitness dramatically decreased with the generation number and then lowered down to the minimum value. The search was stopped when the fitness continued to keep the same minimum value and that individual was considered to give the minimum fitness as an optimal value.

It was also found that the convergence speed was larger for the higher crossover and mutation rates (P_c=0.8 and P_m=0.6) than for the lower crossover and mutation rates (P_c=0.1 and P_m= 0.01). The fitness could not decrease to the minimum value and fell into a local optimum when the crossover and mutation rates decreased to lower values. The searching performance usually depends on the diversity of the population [16,20]. A global optimal value could be obtained if the diversity in the population was always maintained at a high level in each generation. Higher crossover and mutation rates were shown to be effective in keeping a higher diversity in the population, but excessively high crossover and mutation rates are time consuming. The values of P_c=0.8 and P_m=0.6, which were determined through a trial and error, were high enough to avoid a local optimum. It is also effective to add new individuals generated in another population to maintain the diversity of the population.

As mentioned in Section 13.2.8, there is no guarantee that genetic algorithms yield a global optimal solution. It is, therefore, important to confirm whether the optimal value is global or local. Here, the confirmation was mainly carried out using a round-robin algorithm, which systematically searches for all possible solutions around the optimal solution at the proper step. This is because a near-global optimal solution can at least be obtained by genetic algorithms. An optimal solution was confirmed with a different initial population and different methods of crossover and mutation. Through these procedures, a global optimal solution was confirmed.

Through these investigations, we had two optimal values, under different evaluation lengths of the control process. For example, when the evaluation length was at the latter half stage of the control process, a single heat stress application of 40°C during the first 24 h, T_l={40, 15, 15, 15, 15, 15°C} was found to be an optimal value. The length of each step is 24 h. A double heat stress application, T_l={40, 15, 40, 15, 15, 15°C}, was also found to be an optimal value when the evaluation length was restricted to the final step (only the last two steps=2 days) of the control process. Two optimal values (single and double heat stresses) were characterized by the combination of the highest temperature (40°C) and the lowest temperature (15°C) and the rapid rise and drop operations of the temperature.

13.4.11 Optimal Control Performances in a Real System

Finally, the optimal values for single and double heat stresses obtained were applied to a real storage system. Figure 13.20 shows an optimal control performance of the rate of water loss when an optimal value (a single heat stress; $T_l = 40, 15, 15, 15, 15, 15°C$) was applied to the fruit. The bold line shows the case of optimal control, and the fine line shows the case of a constant-temperature $T_1 = \{15, 15, 15, 15, 15, 15°C\}$. The initial temperature was kept at 15°C for 24 h and then the optimal control started. The 6-day control process from 24 to 168 h was divided into six steps. In this case, the evaluation length was the latter half step of the control process ($96 - 168\,h = 3$ days), and the constraint of the temperature was $15 \leq T \leq 40°C$. It was found that, after the single heat stress application, the rate of water loss became lower in the optimal control than in the constant-value control.

Figure 13.21 shows an optimal control performance of the rate of water loss when an optimal value (double heat stresses; $T_l = 40, 15, 40, 15, 15, 15°C$) was applied to the fruit. The bold line shows the case of the optimal control, and the fine line shows the case of a constant-temperature ($T_l = \{15, 15, 15, 15, 15, 15°C\}$). In this case, the evaluation length was only the last two steps of the control process ($120 - 168\,h = 2$ days), and the constraint was $15 \leq T \leq 40°C$. The initial temperature was kept at 15°C for 24 h and then the optimal control started.

From the optimal control performance, the rate of water loss after the second heat stress (double heat stress) is lower than that after the first heat stress (single heat stress). Thus, the rate of water loss had a tendency to decrease after each application of the heat stress. After the double heat stresses, therefore, the value becomes much lower than that in the constant-value control. However, the degree of reduction caused by the heat stress decreases with the application number of the heat stress. In addition, the values of the rate of water loss during the second heat stress application were

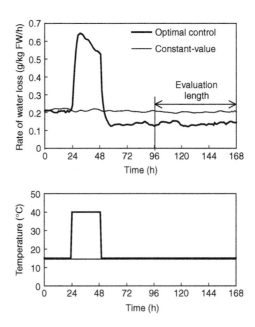

Figure 13.20 An optimal control performance of the rate of the water loss when the evaluation length is the latter half stage of the control process (96–168 h) under the temperature range ($15 \leq T(k) \leq 40°C$).

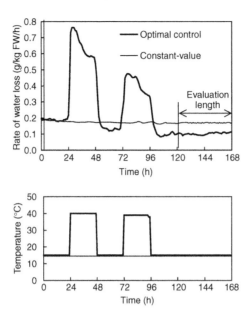

Figure 13.21 An optimal control performance of the rate of the water loss when the evaluation length is only the last two stages of the control process (120–168 h) under the temperature range ($15 \leq T(k) \leq 40°C$).

much lower than that during the first heat stress application. This is because the first heat stress significantly suppressed the water loss of the fruit. It was confirmed that this significant reduction after the second heat stress application continues for at least 3 or 4 more days from other experiments.

In this study, we focused on the rate of water loss, not the total amount of water loss, in order to apply a dynamic control for optimization. This is because the rate of water loss, against the temperature, is more sensible and controllable than the amount of the water loss. It is also clear that, since the total amount of water loss is obtained by integrating the rate of water loss, the response speed is always slow.

The reduction of the water loss caused by the heat stress suggests that the heat–stress fruits acquired a transient thermo-tolerance. Controlling temperature, so that it first rises to the highest level and then drops to the lowest level, seems to be especially effective at reducing the water loss of the fruit during storage, as compared with 15°C constant control.

These results suggest that genetic algorithms are very useful for dynamic optimization of plant production systems, and a control method that changes flexibly and optimally on the basis of fruit responses is a better way to maintain fruit quality during storage than a conventional control manner that simply maintains the temperature at the lowest level.

13.5   CONCLUSIONS

Biological systems are usually characterized by complexity and uncertainty. They have complex properties such as the strong nonlinearity, time variation, large scale, and so on in their systems. It is, therefore, very difficult to solve optimization problems of such systems. Intelligent approaches such as fuzzy reasoning, neural networks, and genetic algorithms are useful for dealing well with such complex systems.

Genetic algorithms are a class of heuristic and iterative search techniques based on the mechanics of biological evolution. Their procedures are biological, probabilistic, flexible, heuristic, iterative, and trial-and-error based.

In this section, the basic concept of the genetic algorithms was first introduced and then applied to two optimization problems in the area of bio-production, model parameterization and dynamic optimization of the fruit storage process. The former is the determination of the optimal values of several parameters used in a photosynthetic model, given by a mathematical equation, and the latter is the minimization of the water loss of the fruit during storage by the control of the storage temperature. In both cases, their global optimal values (or near-global ones) could be rapidly and successfully obtained, in parallel with a multi-point search procedure brought by the genetic operators, even in case of a very large and complex objective function, without falling into local optima.

It is, therefore, suggested that the genetic-based optimization technique such as genetic algorithm might be the best problem solver for any types of optimization problems because living organisms on earth have acquired many excellent functions for survival, adaptation and evolution, and also have overcome many struggles for existence using this method. The genetic algorithms need only the objective function (fitness value) to guide its search direction. There is no requirement for derivatives or other knowledge. These are major features of the genetic algorithms. Experimental results showed that the genetic algorithm outperforms other traditional mathematical approaches. Thus, it can be seen that the genetic algorithms can be applicable to any type of optimization problems.

However, it is also important to know the drawbacks of genetic algorithms. There is no guarantee to always yield a global optimal solution because of their probabilistic rules, not deterministic ones. It is, however, observed that a near-global optimal solution can be at least obtained using genetic algorithms. Thus, it is important to confirm a global optimal solution using other methods.

Finally, it can be concluded that the genetic algorithms are practical and effective tools for solving optimization problems in the complex systems such as the crop, food, and bio-production systems to which mathematical (or analytical) approaches are not easily applied. They are applicable to a wider class of optimization problems.

13.6   FUTURE TRENDS AND POTENTIAL DEVELOPMENT

In recent years, complexities and uncertainties of food and bio production systems have increased as consumer demands for better quality heighten and their control systems have reached large scale, aiming at more effective methods. The development of a more effective optimization technique for such complex systems is severely required.

Genetic algorithms have a high ability to find a global optimal value even in the complex objective function, without falling into local optima. They are also useful for both continuous and discrete optimization problems. Therefore, the genetic algorithms are applicable to any types of optimization problems to which traditional mathematical approaches are not easily applied.

Genetic algorithms are also able to create a new organism (new control method or new control system) through crossover and mutation. So, they are applicable to a self-organization system such as a protein synthesis based on the base sequence in a gene level.

APPENDIX

International Conferences on Genetic Algorithms

In recent years, many international conferences on genetic algorithms have been held in the world. International Conference on Genetic algorithms (ICGA) organized by the International Society on Genetic Algorithms has been held every two years since 1985. Annual Conference on Evolutionary Programming (EP) organized by the Evolutionary Programming Society has been held at Indiana police every year since 1992. IEEE Conference on Evolutionary Computation has also been held at Indiana police every year since 1994.

NOMENCLATURE AND GLOSSARY

Chromosome   Each variable (or parameter), represented by binary digit strings, in an individual.

Crossover   A genetic operator for swapping corresponding components of the binary strings in the two individuals, by which the features of two individuals (two parents) are combined and two new individuals (two offspring) are generated.

Elitist strategy   One of selection methods, where the best individual with highest fitness is compulsively copied and remained for next generation with no operations of crossover and mutation in each generation.

Fitness   An indicator for measuring individual's quality for survival, which is equal to an objective function in a conventional optimization problem.

Gene   One character (0 or 1) in a variable (chromosome) represented by binary digit strings.

Individual   One possible solution (a candidate for an optimal value) to the problem, which consists of several chromosomes.

Mutation   A genetic operator for inverting one or more components of the binary strings from 0 to 1 or vice versa.

Population   A set of individuals.

Population size   The number of individuals in a population.

Roulette wheel selection (proportional selection)   Fundamental selection method, where individuals are probabilistically selected and reproduced in proportion to their slot sizes in the roulette wheel which correspond to their fitness values.

Selection   A genetic operator for choosing and reproducing excellent individuals for crossover and mutation.

REFERENCES

1. Holland, J. H., *Adaptation in Natural and Artificial Systems*, Ann Arbor: The University of Michigan Press, 1975.
2. Holland, J. H., Genetic algorithms, *Scientific American*, July, 44–50, 1992.
3. De Jong, K., Adaptive systems design. A genetic approach, *IEEE Transactions on Systems, Mans, and Cybernetics*, 10, 556–574, 1980.
4. Goldberg, D. E., *Genetic Algorithms in Search, Optimization and Machine Learning*, Reading, MA: Addison-Wesley, 1989.
5. Holland, J. H., *Adaptation in Natural and Artificial Systems*, Cambridge, Massachusetts: MIT Press, 1992.
6. Grefenstette, J. J., Optimization of control parameters for genetic algorithms, *IEEE Transactions on Systems, Man, and Cybernetics*, 16(1), 122–128, 1986.
7. Krishnakumar, K. and Goldberg, D. E., Control system optimization using genetic algorithms, *Journal of Guidance, Control, and Dynamics*, 15(3), 735–739, 1992.
8. Matsuura, K., Shiba, H., Hirotsune, M., and Hamachi, M., Optimizing control of sensory evaluation in the Sake mashing process by decentralized learning of fuzzy inferences using a genetic algorithm, *Journal of Fermentation and Bioengineering*, 80(3), 251–258, 1995.
9. Weuster-Botz, D., Pramatarova, V., Spassov, G., and Wandrey, C., Use of a genetic algorithm in the development of a synthetic growth medium for *Arthrobacter Simplex* with hydrocortisone Δ^1-dehydrogenase activity, *Journal of Chemical Technology and Biotechnology*, 64, 386–392, 1995.

10. Hanai, T., Honda, H., Ohkusu, E., Ohki, T., Tohyama, H., Muramatsu, T., and Kobayashi, T., Application of an artificial neural network and genetic algorithm for determination of process orbits in the Koji making process, *Journal of Bioscience and Bioengineering*, 87(4), 507–512, 1999.

11. Ranganath, M., Renganathan, S., and Srinivasa Rao, C., Genetic algorithm based fuzzy logic control of a fed-bach fermentor, *Bioprocess Engineering*, 21(3), 215–218, 1999.

12. Parbhane, R. V., Unniraman, S., Tambe, S. S., Nagaraja, V., and Kulkarni. B. D., Optimum DNA curvature using a hybrid approach involving an artificial neural network and genetic algorithm, *Journal of Biomolecular Structure and Dynamics*, 17(4), 665–672, 2000.

13. Nagata, Y. and Chu, K. H., Optimization of a fermentation medium using neural networks and genetic algorithms, *Biotechnology Letters*, 25(21), 1837–1842, 2003.

14. Singh Cheema, J. J., Sankpal, N. V., Tambe, S. S., and Kulkarni, B. D., Genetic programming assisted stochastic optimization strategies for optimization of glucose to gluconic acid fermentation, *Biotechnology Process*, 18(6), 1356–1365, 2002.

15. Dutta, J. R., Dutta, P. K., and Banerjee, R., Modeling and optimization of protease production by a newly isolated *Pseudomonas* sp. using a genetic algorithm, *Process Biochemistry*, 40(2), 879–884, 2005.

16. Morimoto, T., Torii, T., and Hashimoto, T., Optimal control of physiological processes of plants in a green plant factory, *Control Engineering Practice*, 3(4), 505–511, 1995.

17. Morimoto, T., Hatou, K., and Hashimoto, Y., Intelligent control for plant production system, *Control Engineering Practice*, 4(6), 773–784, 1996.

18. Morimoto, T. and Hashimoto, Y., AI approaches to identification and control of total plant production systems, *Control Engineering Practice*, 8(5), 555–567, 2000.

19. Hashimoto, Y., Murase, H., Morimoto, T., and Torii, T., Intelligent systems for agriculture in Japan, *IEEE Control System Magazine*, 21(5), 71–85, 2001.

20. Morimoto, T., De Baerdemaeker, J., and Hashimoto, Y., An intelligent approach for optimal control of fruit-storage process using neural networks and genetic algorithms, *Computers and Electronics in Agriculture*, 18(2,3), 205–224, 1997a.

21. Morimoto, T., Suzuki, J., and Hashimoto, Y., Optimization of a fuzzy controller for fruit storage using neural networks and genetic algorithms, *Engineering Applications of Artificial Intelligence*, 10(5), 453–461, 1997b.

22. Morimoto, T., Purwanto, W., Suzuki, J., and Hashimoto, Y., Optimization of heat treatment for fruit during storage using neural networks and genetic algorithms, *Computers and Electronics in Agriculture*, 19, 87–101, 1997c.

23. Morimoto, T. and Hashimoto, Y., An intelligent control for greenhouse automation, oriented by the concepts of SPA and SFA—an application to post-harvest process, *Computers and Electronics in Agriculture*, 29(1,2), 3–20, 2000.

24. Morimoto, T. and Hashimoto, Y., A decision and control system mimicking a skilled grower's thinking process for dynamic optimization of the storage environment, *Environment Control in Biology*, 41(3), 29–42, 2003.

25. Morimoto, T., Tu, K., Hatou, K., and Hashimoto, Y., Dynamic optimization using neural networks and genetic algorithms for tomato cool storage to minimize water loss, *Transactions of the ASAE*, 46(4), 1151–1159, 2003.

26. Sequeira, R. A., Olson, R. L., Willers, J. L., and McKinion, J. M., Automating the parameterization of mathematical models using genetic algorithms, *Computers and Electronics in Agriculture*, 11(2,3), 265–290, 1994.

27. Matsuura, K., Hirotsune, M., Hamachi, M., and Nunokawa, Y., Modeling of the sensory evaluation of sake by Dempster–Shafer's measure and genetic algorithm, *Journal of Fermentation and Bioengineering*, 79, 45–53, 1995.

28. Park, L. J., Park, C. H., Park, C., and Lee, T., Application of genetic algorithms to parameter estimation of bioprocess, *Medical and Biological Engineering and Computing*, 35(1), 47–49, 1997.

29. Pabico, J. P., Hoogenboom, G., and McClendon, R. W., Determination of cultivar coefficients of crop models using a genetic algorithm: A conceptual framework, *Transactions of the ASAE*, 42(1), 223–232, 1999.

30. Mayer, D. G., Belward, J. A., and Burrage, K., Robust parameter settings of evolutionary algorithms for the optimization of agricultural systems models, *Agricultural Systems*, 69, 199–213, 2001.

31. Purwanto, W., Morimoto, T., and Hashimoto, Y., Simulative estimation for the identification of the cumulative response of a plant using neural networks, *Journal of Society of High Technology in Agriculture*, 8(2), 44–50, 1996.
32. Ranganath, M., Renganathan, S., and Gokulnath, C., Identification of bioprocesses using genetic algorithm, *Bioprocess Engineering*, 21(2), 123–127, 1999.
33. Parmar, R. S., McClendon, R. W., and Potter, W. D., Farm machinery selection using simulation and genetic algorithms, *Transactions of the ASAE*, 39(5), 1905–1909, 1996.
34. Parsons, D. J., Optimising silage harvesting plans in a grass and grazing simulation using the revised simplex method and a genetic algorithm, *Agricultural Systems*, 56(1), 29–44, 1998.
35. Hansson, P. A., Optimization of agricultural tractor cab suspension using the evolution method, *Computers and Electronics in Agriculture*, 12, 35–49, 1995.
36. Noguchi, N. and Terao, H., Path planning of an agricultural mobile robot by neural network and genetic algorithm, *Computers and Electronics in Agriculture*, 18(2–3):187–204, 1997.
37. Chtioui, Y., Bertrand, D., and Barba, D., Feature selection by a genetic algorithm. Application to seed discrimination by artificial vision, *Journal of the Science of Food and Agriculture*, 76, 77–86, 1998.
38. Davis, L., *Genetic Algorithms and Simulated Annealing*, Los Altos, CA: Morgan Kaufmann, 1987.
39. Deb, K., *Optimization for engineering design: Algorithms and examples*, London: Prentice-Hall, 1995.
40. Van Rooij, A. J. F., Jain, L. C., and Johnson, R. P., *Neural Network Training Using Genetic Algorithms*, Singapore: World Scientific Publishing, 1996.
41. Matsuura, K., Shiba, H., Nunokawa, Y., Shimizu, H., Shioya, S., and Suga, K., Calculation of optimal trajectories for fermentation processes by genetic algorithm, *Seibutsu-Kogakkaishi*, 71(3), 171–178, 1993 in Japanese
42. Cartwright, H. M., The genetic algorithm in science, *Pesticide Science*, 45, 171–178, 1995.
43. Bäck, T., *Mutation Parameters, Handbook of Evolutionary Computation*, New York: Oxford University Press, 1997.
44. Mayer, D. G., Belward, J. A., and Burrage, K., Survival of the fittest—genetic algorithms versus evolution strategies in the optimization of systems models, *Agricultural Systems*, 60, 113–122, 1999.
45. Kuo, T. and Hwang, S. Y., A genetic algorithm with disruptive selection, *IEEE Transactions on Systems, Man, and Cybernetics1*, 26(2), 299–307, 1996.
46. Tanese, K., Distributed genetic algorithms. In *Proceedings of the 3rd International Conference on Genetic Algorithm and Their Applications*, 455–460, 1989.
47. Kubota, N., Shimojima, K., and Fukuda, T., The role of virus infection in virus-evolutionary genetic algorithm, *Proceedings of IEEE International Conference on Evolutionary Computation*, 182–187, 1999.
48. Ohkura, K. and Ueda, K., A genetic algorithm with neutral mutations for deceptive function optimization, *Transactions of Japanese Society of Instrument and Control Engineers*, 32(10), 1461–1469, 1996.
49. Bersini, H. and Varela, F. J., The immune recruitment mechanism: a selective evolutionary strategy, *Proceedings of 4th International Conference on Genetic Algorithms* 1994 pp. 520–526
50. Hu, J., Kumamaru, K., and Inoue, K., A hybrid robust identification using genetic algorithm and gradient method, *Transactions of the Japanese Society of Instrument and Control Engineers*, 32(5), 714–721, 1996.
51. Renders, J. M. and Flasse, S. P., Hybrid methods using genetic algorithms for global optimization, *IEEE Transactions on Systems, Man, and Cybernetics*, 26(2), 243–258, 1996.
52. Johnson, I. R., Parsons, A. J., and Ludlow, M. M., Modelling photosynthesis is monocultures and mixtures, *Australian Journal of Plant Physiology*, 16, 501–516, 1989.
53. Lurie, S., Review postharvest heat treatments, *Postharvest Biology and Technology*, 14, 257–269, 1998.
54. Ferguson, L. B., Ben-Yehoshua, S., Mitcham, E. J., McDonald, R. E., and Lurie, S., Postharvest heat treatments: introduction and workshop summary, *Postharvest Biology and Technology*, 21, 1–6, 2000.
55. Biggs, M. S., William, R., and Handa, A., Biological basis of high-temperature inhibition of ethylene biosynthesis in ripening tomato fruit, *Physiclogia Plantarum*, 72, 572–578, 1988.
56. Lurie, S. and Klein, J. D., Acquisition of low-temperature tolerance in tomatoes by exposure to high-temperature stress, *Journal of the American Society for Horticulture Science*, 116(6), 1007–1012, 1991.

57. Lurie, S. and Klein, J. D., Ripening characteristics of tomatoes stored at 12°C and 2°C following a prestorage heat treatment, *Scientia Horticulturae*, 51, 55–64, 1992.
58. McDonald, R. E. and McCollum, T. G., Prestorage heat treatments influence free sterols and flavor volatiles of tomatoes stored at chilling temperature, *Journal of the American Society for Horticulture Science*, 121(3), 531–536, 1996.
59. Paull, R. E. and Chen, N. J., Heat treatment and fruit ripening, *Postharvest Biology and Technology*, 21, 21–37, 2000.
60. Lurie, S. and Sabehat, A., Prestorage temperature manipulations to reduce chilling injury in tomatoes, *Postharvest Biology and Technology*, 11, 57–62, 1997.
61. Artés, F., Garcia, F., Marquina, J., Cano, A., and Fernadez-Trujilo, J. P., Physiological responses of tomato fruit to cyclic intermittent temperature regimes, *Postharvest Biology and Technology*, 14, 283–296, 1998.
62. Artés, F., Sanchez, E., and Tijskens, L. M. M., Quality and shelf life of tomatoes improved by intermittent warming, *Lebensmittel Wissenschaft und Technologie*, 31, 427–431, 1998.
63. Liu, F. W., Modification of apple quality by high temperature, *Journal of the American Society for Horticulture Science*, 103, 730–732, 1978.
64. Shellie, K. C. and Mangan, R. L., Postharvest quality of "Valencia" orange after exposure to hot, moist, forced air for fruit fly disinfestation, *HortScience*, 29, 1524–1527, 1994.
65. Hofman, P. J., Tubbings, B. A., Adkins, M. F., Meiburg, G. F., and Woolf, A. B., Hot water treatments improves "Hass" avocado fruit quality after cold disinfestation, *Postharvest Biology and Technology*, 24, 183–192, 2000.
66. Chen, H. H., Shen, Z. Y., and Li, P. H., Adaptability of crop plants to high temperature stress, *Crop Science*, 22(July–August), 719–725, 1982.
67. Kimpel, J. A. and Key, J. L., Heat shock in plants, *Trends in Biochemical Sciences*, 10, 353–357, 1985.
68. Sabehat, A., Weiss, D., and Lurie, S., The correlation between heat–shock protein accumulation and persistence and chilling tolerance in tomato fruit, *Plant Physiology*, 110, 531–537, 1996.

CHAPTER 14

Fractal Analysis

Mohammad Shafiur Rahman

CONTENTS

14.1 INTRODUCTION

Fractal analysis is mainly applied when other methods fail or become tedious to solve complex or chaotic problems. Many natural patterns are either irregular or fragmented to such an extreme degree that Euclidian or classical geometry could not describe their form (Mandelbrot 1977, 1987). Any shape can be characterized by whether or not it has a characteristic length (Takayasu 1990). For example, a sphere has a characteristic length defined as the diameter. Shapes with characteristic lengths have an important common property of smoothness of surface. A shape having no characteristic length is called self-similar. Self-similarity is also known as scale-invariance, because self-similar shapes do not change their shape under a change of observational scale. This important symmetry gives a clue to understanding complicated shapes, which have no characteristic length, such as the Koch curve or clouds (Takayasu 1990). The idea of a fractal is based on the lack of characteristic length or on self-similarity. The word fractal is a new term introduced by Mandelbrot (1977) to represent shapes or phenomena having no characteristic length. The origin of this word is the Latin adjective *fractus* meaning broken. The English words "fractional" and "fracture" are derived from this Latin word *fractus*, which means the state of broken pieces being gathered together irregularly.

Most branches of science and engineering are now using fractal analysis for characterizing natural or synthetic particles, complex physical or chemical processes, and complex signatures of instruments. Peleg (1993) and Barrett and Peleg (1995) reviewed the applications of fractal analysis in food science. These included particulates characterization, non-linear kinetics, agglomeration and crystallization, mixing of viscous liquids, diffusion in non-uniform media, and characterizing jagged signatures. Fractal dimensions have been successfully used to describe the ruggedness and geometric complexities of both natural and synthetic particles (Peleg and Normand 1985; Yano and Nagai 1989; Nagai and Yano 1990; Graf 1991; Barletta and Barbosa-Canovas 1993; Peleg 1993; Rahman 1997). Peleg (1993) also applied fractal geometry in the study of the shape of broccoli. Similarly, fractal analysis has also been applied to characterize native and physically or chemically transformed food particles. Fractal analysis can predict the efficiency of the transformation process and food particle properties, such as adsorption capacity, solubility, puffing ability, chemical reactivity, and emulsifying ability to optimize food ingredient selection for product development and process design (Rahman 1997). Applications have also been made in studying textural properties of foods (Barrett et al. 1992; Rohde, Normand, and Peleg 1993; Barrett and Peleg 1995). Examples are: acoustic signature analysis of crunchy food (Peleg 1993), image analysis (Barrett et al. 1992; Peleg 1993), analysis of the cell size distribution of puffed corn extrudates (Barrett and Peleg 1995), fractal reaction kinetics (Kopelman 1988), diffusion in fractal surfaces (Nyikos and Pajkossy 1988); gel strength by rheology and fractal analysis (Bremer, van Vliet, and Walstra 1989); pore size distribution for porosimetry data (Ehrburger-Dolle, Lavanchy, and Stoeckle 1994); and moisture sorption isotherms (Suarez-Fernandez and Aguerre 2000). The fractal dimensions of solid surfaces are known not to be constant but to range from 2 (flat) to 3 (volume-filling). They depend not only on the composition of the material but also on how it was produced. This fractal property has a strong influence on the efficiency of chemical reactions since most chemical reactions take place on the surface and the higher the dimension, the greater the efficiency.

14.2   FRACTAL BASIC

The properties of fractals can be summarized as: (i) no characteristic length within the lower cutoff and upper cutoff; (ii) self similarity, which means fractal objects can be decomposed into parts, each obtainable from the whole by a similitude (fractal dimension is not same as the Euclidian dimension, i.e., $\delta \neq d$); (iii) scale invariance; (iv) power law relation is valid; and (v) δ is a global characteristic of the system. Because dealing with fractal objects requires a computer, Mandelbrot's success in popularizing the concept of fractal geometry was overwhelming as powerful microcomputers became widely available (Peleg 1993). If self-similarity is maintained over a limited but pertinent range of length scales, fractal geometry can be effectively used to describe this range without causing conceptual concerns.

14.3   FRACTAL DIMENSION

14.3.1   What Is Fractal Dimension?

In classical Euclidian geometry a point has a zero dimension, a line one, an area two, and a volume three. In physics time axis is often added and as much it is considered that space-time is four-dimensional. All these empirical dimensions take integer values. Defining fractal dimension is no problem, but a basic difficulty occurs when applying them in real physics or systems (Takayasu 1990). The mass contained in a sphere of radius r varied as:

$$m \propto r^d \tag{14.1}$$

where m is the mass (kg), and r is the radius (m), respectively. In the case of spheres one-dimension can predict the mass, but the mass of a rectangular varies in two directions. Thus rectangle mass is not self-similar only in one direction and it can be characterized by two exponents c and d. So the mass in rectangular geometry of sizes a and b in x and y directions as:

$$m \propto Aa^c + Bb^d \tag{14.2}$$

Similarly the mass of ellipsoid bodies can vary in 3 directions. A rugged line is shown in Figure 14.1. The fractal dimension can be between 1 and 2. It is obvious that the greater the tortuosity of the line, the higher its fractal dimension above one.

14.3.2 Estimation of Fractal Dimension

Takayasu (1990) identified five methods to define dimension: (i) changing coarse-graining level; (ii) using the fractal measure relations; (iii) using the correlation function; (iv) using the distribution function; and (v) using the power spectra or spectrum. The dimensions obtained by different methods may differ from each other. In some cases they agree but in other cases they differ; this is a basic and difficult problem in applying fractal analysis.

14.3.2.1 *Changing the Coarse-Graining Level*

A complicated curve is shown in Figure 14.1. Let one edge of the curve be the starting point and from there we can draw a circle of radius r. Then connect the starting point and the point where the circle intersects the curve by a straight line. The intersection can be considered as a new starting point and the above procedure can be repeated until it comes to the other edge. Then the number of segments $N(r)$ can be approximated to the curve and $N(r)$ depends on the unit length segments r. If the curve is a straight line, then the number of segments depends on:

$$N \propto \frac{1}{r} = r^{-1} \tag{14.3}$$

The exponent 1 of r in the above equation also agrees with the dimension of a straight line. When r is larger, then tiny rugged curvatures of the curve are not noticed. Thus for a rugged curve it can be written as:

$$N \propto r^{-\delta} \tag{14.4}$$

where δ is the fractal dimension which characterizes the ruggedness of the curve. An adsorption of a monolayer of identical molecules on the surface can be used to characterize the fractal surface of

Figure 14.1   A complicated curve. (From Rahman, M. S., *Handbook of Food Properties*, Boca Raton, FL: CRC Press, 1995.)

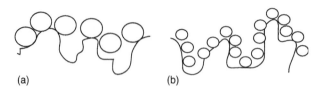

Figure 14.2 (a) Number of gas moles is lower; (b) number of gas moles is higher. (From Rahman, M. S., *Handbook of Food Properties*. Boca Raton, FL: CRC Press, 1995.)

porous structures. Usually the number is expressed as number of moles per gram of adsorbent. The number of monolayer values will vary with the size of the adsorbate molecule as shown in Figure 14.2. There is a lower cutoff and an upper cutoff molecular size, between which the number of molecules will vary. Thus the adsorption of gas molecules can be used to characterize the surface of the solid matrix.

14.3.2.1.1 Richardson's Method

Richardson (1961) proposed a structured walk procedure for characterizing the fractal dimension of a rugged boundary. A series of polygons of side x are constructed on the perimeter using a pair of compasses (Figure 14.3). The perimeter of a fractal shape can be measured many times with many different scales of measurement. The scale of measurement is called the stride length. Larger stride lengths give a lower perimeter, written as:

$$p = nx + \gamma \tag{14.5}$$

In dimensionless form, the above equation can be expressed as:

$$\xi = n\lambda + \frac{\gamma}{L} \tag{14.6}$$

A plot of ξ against λ on a log–log graph yields a straight line of slope s where $s = 1 - \delta_r$ or $\delta_r = 1 - s$. The relationship of δ_r equal to 1.0 indicates the smoothness and higher values of δ_r indicates the ruggedness or roughness of the boundary. A log–log plot of stride length versus perimeter is called a *Richardson plot*. The *fractal dimension*, δ_r is the morphometric ruggedness of a particle

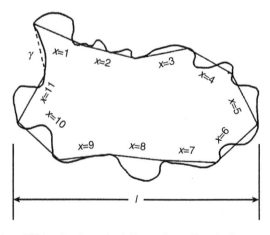

Figure 14.3 Characteristics of Richardson's method; l is maximum Feret's diameter or length of profile, γ is the length of final step. (From Rahman, M. S., *Handbook of Food Properties*. Boca Raton, FL: CRC Press, 1995.)

boundary, which can be estimated by Richardson's method. In many cases, particles may have two fractal components with two linear segments having two different slopes (Graf 1991). The linear segment corresponding to large stride lengths is called the structure, and the linear segment corresponding to small stride lengths is called the texture (Kaye 1989). The value of critical stride length can only be predicted by visualization of the Richardson plot. Photographs from Scanning Electron Microscope (SEM) analysis can be used to construct Richardson plots. The Richardson method provides a quantitative measure of the ruggedness of the particle boundary. Fractal dimensions estimated by Richardson method varied from 1.02 to 1.10 for native and modified starch (Rahman 1997), and from 1.06 to 1.10 for instant coffee particles (Peleg and Normand 1985). Higher values indicate more ruggedness or roughness of the boundary, which was corroborated by scanning electron microscopy of native and modified starch. There are objects whose morphology does not have true self-similarity on any scale, but whose Richardson plot is still linear. In such a case the slope can be used to determine a natural or apparent fractal dimension whose magnitude only serves as a measure of jaggedness or ruggedness on a scale from 1.0 (smooth) to 2.0 (the upper theoretical limit) (Rohde, Normand, and Peleg 1993).

14.3.2.1.2 Other Algorithms Similar to Richardson's Method

Brittle foods, such as puffed cereals and extrudates have very irregular and irreproducible strain-stress relationships (Figure 14.4). The fractal dimensions of digitized experimental force–displacement curves of cheese balls were determined with the Richardson, Mikowski, Kolmogirov, and Korcak algorithms (Borges and Peleg 1996). These algorithms used by the authors were described and explained in great detail by Hastings and Sugihara (1993) and Russ (1994). The Richardson method is based on calculating the jagged signature's length as the sum of linear segments, which progressively decrease in size at each iteration. The relationship between the line length at each iteration and the corresponding segment size in algorithmic coordinates is known as the Richardson plot. The apparent fractal dimension of the original line is the absolute magnitude of the slope of its Richardson plot plus one. Kolmogorov is based on counting the number boxes filled by rugged lines. The boxes on the plane where at least one point belongs to the jagged signature can be found as the box size progressively diminishes at each iteration. The Kolmogorov dimension is calculated from the slope of the number of filled boxes versus the box size (or number of boxes) in the logarithmic coordinates. The Minkowski method is akin to the Richardson algorithm and it is based on covering the signature by a chain of circles whose diameter diminishes at each iteration. The line's length at each iteration is calculated from the circles' combined area. The calculated signature length versus the corresponding circle size is plotted in logarithmic coordinates. The slope of the resulting line provides the original signature's fractal dimension. The Korcak method is based on determining the cumulative distribution of segments

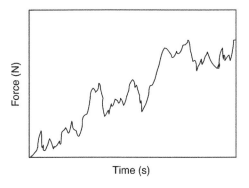

Figure 14.4 Force–time graph for a brittle food.

longer than a chosen size which are formed when the signature intersects a baseline. For a true fractal signature, the plot of the cumulative number versus the changing size in logarithmic coordinates converges to a straight line whose slope provides the fractal dimension. For all these algorithms the force column of each data file can be transformed in the "Fractal" program which is provided with Russ (1994). The files were analyzed to determine their apparent fractal dimension with the Richardson, Kolmogorov, Minkowski, and Korcak algorithms, all of which are standard options in the program. Borges and Peleg (1996) found all four methods, but especially the first three, produced a consistent dimension despite the fact that the curves themselves were not truly fractal. Similarly Damrau, Normand, and Peleg (1997) determined the fractal dimensions of jagged force–displacement signatures using these algorithms.

The fractal dimension of a jagged line can be determined in more than one way (Barnsley 1988). One of the more convenient methods is based on the blanket algorithm (Normand and Peleg 1988; Barrett et al. 1992). According to this method, the processed image is covered by a *blanket* with a thickness that increases stepwise. The surface area of the image is calculated at each step by dividing the blanket's volume by its thickness and the equivalent of a Richardson plot can be constructed from these data. The same can be done for a one-dimensional image (line). In this case, the line length is calculated by dividing the area of the formed coated image by the thickness at each step (Normand and Peleg 1988). Application of the blanket algorithm to mechanical signatures first involves fitting the force–deformation relationship to a polynomial or other function that describes its general shape and then subtracting the fitted function from the original data. The difference between the two curves exclusively represents the jaggedness of the relationship (Barrett and Peleg 1995).

Barletta and Barbosa-Canovas (1993) determined the fractal dimension of agglomerated food powders by image analysis, which provided particle contour. They used fast, hybrid, and exact procedures to obtain the profile perimeter using several stride lengths. Detailed descriptions of the above procedures are discussed by Clark (1986). Each method specifies a different algorithm to trace the contour from a file containing a string of coordinate points that represent the outline of the profile. Once the data of the perimeter and yardstick in dimensionless units are available, the next step is to find the range of scrutiny under which contour showed a self-similar curve. This was done by scanning several ranges of dimensionless units (based on the maximum feret's diameter) and performing linear regressions.

There are various techniques to determine the fractal dimension of an object's surface from its digitized image (Peleg 1993). The latter can be a micrograph or a photograph taken with or without magnification. The basis of most, if not all, of these methods is to convert the pixel gray level into a numerical value to which various algorithms can be applied (Peleg et al. 1984; Gagenpain and Rogues-Carmes 1986; Normand and Peleg 1986; Liao, Cavalieri, and Pitts 1988; Bartlett 1991). In principle, the algorithms are similar to those used to determine the fractal dimension of jagged lines, except that the image, or map, is three-dimensional; with the axes usually assigned to the gray level (usually Z) and to the pixel coordinates (X,Y) (Peleg 1993). The inherent drawback of this method is the fact that the gray level of the pixels is determined not only by the physical features of the surface itself but also by the illumination intensity and the specimen's orientation relative to the illumination source (true in electron as well as light microscopy). Thus, the distinction between the real surface morphology and induced shadows can sometimes be difficult. In principle, there are ways to resolve images that eliminate such ambiguities, one of which is to reconstruct a truly stereoscopic image using more than one photograph of the same object or site taken at different angles (Peleg 1993). Normand and Peleg (1986) measured the fractal dimension of agglomerate coffee using an image processing technique. From the digitized image, the fractal dimension can be calculated by the box counting method (Kaye 1989). The steps are as follows:

1. A square mesh of a certain size L is laid over the object on the digitized image.
2. The number of mesh boxes N that contain part of the image is counted.

3. The slope is calculated for the slope of logarithmic plot of N versus L.
4. The fractal dimension can be calculated from: $\delta = \text{slope} + 1$.

Hagiwara et al. (1997) used computer software for fractal analysis based on the box counting method (Bourke 1993). Similarly Pedreschi, Aguilera, and Brown (2000) and (2002) performed length-scale and area-scale analyses, two of the scale-sensitive fractal analyses, by the software Surfrax (www.surfract.com). The data of food surfaces were measured with a scanning laser microscope.

Rohde, Normand, and Peleg (1993) developed symmetrized-dot pattern (SDP) from force–time curves of brittle food. It was demonstrated that although the original force–time relationships were irregular and irreproducible, consistent patterns and values did emerge in the transformed data. It has long been known that humans are particularly sensitive to symmetric patterns. This has led to the development of a data presentation method known as a SDP (Pickover 1990). It was originally used in speech analysis by transforming the irregular acoustic signature into an image resembling a snowflake. Basically, the method consists of duplication of the original pattern along a selected number of symmetry axes (i.e., six in a snowflake-like image) and addition of a mirror image of each duplicate across its corresponding symmetry axis. It is roughly equivalent to a computerized kaleidoscope. This form of data is aesthetically appealing, but it does not add any information not already contained in the original signature (Rohde, Normand, and Peleg 1993).

14.3.2.2 Using a Fractal Relation

In the literature there are numbers of scaling equations available to determine the fractal dimensions, which characterize a process or material. In this case, experimental data can be used to determine the fractal dimensions. In this section selected examples are presented.

14.3.2.2.1 Scaling Equation to Characterize Gel based on Permeability

Bremer, van Vliet, and Walstra (1989) developed a scaling equation to characterize gel structure from permeability measurements through the gel. In a gel each lattice site in a cluster is occupied with either a particle or a volume element of solution. The number of lattice sites occupied by a particle is:

$$N_p = \left(\frac{R}{r}\right)^{\delta} \tag{14.7}$$

where N_p is the number of lattice sites occupied by a particle, R is the radius of the fractal cluster, and r is the radius of one lattice site, which can be equal to the radius of one primary particle if a lattice site is of the same size as the particle size. The number of lattice sites (N_a) that have been taken by an aggregate for a three-dimensional lattice are related as follows:

$$N_a = \left(\frac{R}{r}\right)^{3} \tag{14.8}$$

The above relation is limited to $R/r \gg 1$. The volume fraction of particles (ϕ_a) in an aggregate can be expressed as:

$$\phi_a = \frac{N_p}{N_a} = \left(\frac{R}{r}\right)^{\delta - 3} \tag{14.9}$$

The sum of all sites occupied by the individual fractal aggregates will be equal to the total number of lattice sites in the gel:

$$\sum_{i=1}^{n} N_{ai} = N_T \tag{14.10}$$

$$\sum_{i=1}^{n} \phi_{ai} N_{ai} = \phi_p N_T \tag{14.11}$$

where ϕ_p is the overall volume fraction of the particles and n is the total number of aggregates. From Equation 14.10 and Equation 14.11, the overall volume fraction of particles can be written as:

$$\phi_p = \frac{\sum_{i=1}^{n} (R_i)^\delta}{r^{\delta-3} \sum_{i=1}^{n} (R_i)^3} \tag{14.12}$$

$$\phi_p = \left(\frac{R_{av}}{r}\right)^{\delta-3} \tag{14.13}$$

The above equation describes a relationship between the overall volume fraction and the average size of the clusters in a gel, given a fixed value of the fractal dimension. R_{av} is the three-dimensional average cluster radius such that:

$$R_{av} = \left[\frac{\sum_{i=1}^{n} (R_i)^3}{\sum_{i=1}^{n} (R_i)^\delta}\right]^{1/(3-\delta)} \tag{14.14}$$

If a gel consists of fractal clusters, the gel will be scale invariant. The fractal dimension can be determined by measuring the permeation (Bremer, van Vliet, and Walstra 1989). The flux through the ith fractal structural element can be written from Poiseuille's law as:

$$Q_i \propto (R_i)^4 \tag{14.15}$$

The flux through the gel in the case of a constant pressure gradient can be scaled as:

$$Q \propto \sum_{i=1}^{n} Q_i \propto \sum_{i=1}^{n} (R_i)^4 \tag{14.16}$$

The above equation considers the number of pores in the gel as constant. The absolute size of pore surface area in gel scale can be expressed as:

$$S \propto (R_i)^2 \tag{14.17}$$

Permeability from Darcy's law can be written as:

$$k \propto \frac{Q}{S} \propto \frac{\sum_{i=1}^{n} (R_i)^4}{\sum_{i=1}^{n} (R_i)^2} \infty (R_{av})^2 \tag{14.18}$$

If the cluster size distribution is self-preserving during the aggregation process then $R_{42} \propto R_{3D}$, Equation 14.13, Equation 14.14 and Equation 14.18 can combine as:

$$k = \left(\frac{r^2}{\tau}\right)(\phi_p)^{2/(\delta-3)} \tag{14.19}$$

where τ is a constant, comparable with the tortuosity factor in the Kozeny–Carman equation. The above equation gives a relation between the structure of the gel (fractal dimension), volume fraction of the particles in the gel, size of these particles and permeability of the gel. Bremer, van Vliet, and Walstra (1989) plotted log k versus log ϕ_p for a casein gel and observed a straight line with a slope of -2.6. The fractal dimension of the casein gel was 2.23 before significant microsyneresis.

14.3.2.2.2 Scaling Equation to Characterize a Gel from Rheology

Bremer, van Vliet, and Walstra (1989) derived a relation between the modulus and the volume fraction of particles to determine fractal dimensions of a gel. The modulus of a gel network can be written as:

$$G = \sum_{i=1}^{n} N_i C_i \left(\frac{d^2 A}{dx^2}\right)_i \tag{14.20}$$

where G is the modulus of the network, x is the direction of applied force, N is the number of stress carrying strands per unit cross sectional area perpendicular to x, dA is the change in Gibbs energy when the particles in the strands are moved apart over a distance dx and C is the characteristic length. The term $C_i(d^2A/dx^2)_i$ can be considered constant for varying concentrations at the time at which G is measured. In this case:

$$G \propto N_i \tag{14.21}$$

In a fractal network due to scale invariance the number of stress-carrying strands per unit cross sectional area can be scaled as:

$$N \propto \frac{1}{S} \propto \frac{1}{R^2} \tag{14.22}$$

Combining Equation 14.21 and Equation 14.22 the modulus can be written as:

$$G \propto \frac{1}{R^2} \tag{14.23}$$

From the above equation which can be transformed to:

$$R = a\phi_p^{1/(\delta-3)} \tag{14.24}$$

the final equation for a modulus can be written as:

$$G \propto \phi_p^{2/(3-\delta)} \quad \text{or} \quad G = K\phi_p^{2/(3-\delta)} \tag{14.25}$$

where K is a constant independent of the volume fraction but dependent on the size of primary particles and on the interactions between these particles. From the slope of a plot log G versus log ϕ_p, the fractal dimension of a gel can be estimated. The fractal dimension estimated for Na-caseinate and skimmed milk (rennet induced) are 2.23 and 2.17, respectively (Bremer, van Vliet, and Walstra 1989). Shih et al. (1990) developed the scaling equation depending on the strength of the links between the neighboring flocs compared to that within the flocs. The links are classified into two types: strong-links and weak-links. In the strong-link regime, the links between neighboring flocs have a larger elasticity than those within the flocs. For the gel with a strong-link, the dependence of the elasticity E with concentration is described as:

$$E \propto \phi_p^{(3+x)/(3-\delta)} \tag{14.26}$$

where δ is the fractal dimension of the flocs ($\delta \leq 3$), and x is the backbone fractal dimension of the flocs, which varies between 1.0 and 1.3. On the other hand, in the weak link regime, the links in the flocs have a larger elasticity than those between the neighboring flocs: for the gel with a weak-link, the dependence of the E on particle concentration can be expressed as:

$$E \propto \phi_{\mathrm{p}}^{1/(3-\delta)} \tag{14.27}$$

Hagiwara et al. (1997) estimated the fractal dimension of protein gels from rheology and image analysis of laser microscopy photographs and found similar results. Similarly, Marangoni and Rousseau (1996) determined the fractal dimension of plastic fat crystal network from scaling equation from rheology. Other scaling equations were developed for coagulation formation with time as a function of processing time (Jullien and Botet 1987), measuring the transport of elastic waves or solute or solvent along a fractal structure (Pfeifer, Avnir, and Farin 1984; Jullien and Botet 1987), aggregation of colloidal silica by x-ray scattering techniques (Cannell and Aubert 1986), x-ray scattering of porous bodies (Bale and Schmidt 1984), turbidity data of gels (Bremer, van Vliet, and Walstra 1989), flocculation and gel formation process (Kolb et al. 1986), yeast flocculation (Davis and Hunt 1986), energy requirements in solid grinding processes (Suzuki and Yano 1989), gas adsorption processes (Pfeifer and Avnir 1983; Pfeifer, Avnir, and Farin 1984), and antigen-antibody binding kinetics on biosensors (Sadana and Ram 1994). Hongsprabhas, Barbut, and Marangoni (1999) determined the fractal dimension from the scaling equation based on rheology and image from scanning and transmission electron micrographs; they found both the methods provided similar fractal dimensions.

14.3.2.2.3 Scaling Equation for Pore-Size Distribution

Pfeifer and Avnir (1983) derived a scaling equation for pore size distribution as:

$$\frac{dV_{\mathrm{v}}}{dr_{\mathrm{v}}} \propto (r_{\mathrm{v}})^{2-\delta_{\mathrm{v}}} \tag{14.28}$$

The fractal dimension can be estimated from the slope of the plot of $\log(dV_{\mathrm{v}}/dr_{\mathrm{v}})$ versus $\log r_{\mathrm{v}}$. In terms of applied pressure the scaling equation can be written as (Ehrburger-Dolle, Lavanchy, and Stoeckle 1994):

$$\frac{dV_{\mathrm{v}}}{dP} \propto P^{\delta_{\mathrm{v}}-4} \tag{14.29}$$

The data (dV_{v}/dP versus P) for the above relationships (Equation 14.28 and Equation 14.29) could be generated from the mercury porosimetry. The fractal dimension δ_{v} is the characteristic size distribution of micropores in particles of the same size. Pore size distribution method provides the characteristics of micropore formation in the particle. Two or three linear portions can also be observed in the plot of $\log(dV_{\mathrm{v}}/dr_{\mathrm{v}})$ versus $\log r_{\mathrm{v}}$. Initial and final segments indicate the mechanical properties of the solid and the fractal dimension can be higher than 3. Ehrburger-Dolle, Lavanchy, and Stoeckle (1994) observed three linear sections in $\log(dV_{\mathrm{v}}/dP)$ versus $\log P$ or $\log(dV_{\mathrm{v}}/dr_{\mathrm{v}})$ versus $\log r_{\mathrm{v}}$ plots of different active carbon particles and concluded that only the middle linear segment will give the actual fractal dimension. The initial or final slope may lead to a fractal dimension much higher than 3 as observed by Friesen and Mikula (1988), Ehrburger-Dolle, Lavanchy, and Stoeckle (1994), and Rahman (1997) which was unrealistic. A possible explanation was proposed by Ehrburger-Dolle, Lavanchy, and Stoeckle (1994) as: below lower critical pressure (lower cutoff), mercury is filling the inter-grain voids, which is not fractal in nature and above higher critical pressure (upper cutoff); the result reflects in explaining the mechanical behavior of the sample. Above the higher critical pressure, fine pores may also be formed by cracking, which is also dependent on the chemical strength of the particles or indicates the compressibility of the particles. Fractal dimensions were 2.71, 2.96, and 3.11 for fresh and air-dried samples (20 and 3 h drying at 80°C), respectively. Dimensions were increased with the increasing drying time indicating the formation of micro-pores on the surface during air-drying (Rahman, Al-Zakwani, and Guizani 2005). Rahman (1997) determined the fractal dimension of native, gelatinized, and ethanol deformed starch and the fractal dimension values as 3.09, 3.10, and 2.45 respectively. Lower values indicate the removal of micro-pores within the starch particles by ethanol modification.

14.3.2.3 Using Correlation Function

The correlation function is a fundamental statistical quantity, and it can be used to find the fractal dimension. In theoretical models, the exponential function $[\exp(-r/r_0)]$ or the Gaussian function $[\exp(-r^2/r_0^2)]$ is used to correlate variables. However, these functions do not possess the fractal property because both of them have a characteristic length r_0 (Takayasu 1990). On the other hand, when the distribution is a fractal property, the correlation follows a power law (Takayasu 1990). In case of fractal there is no characteristic length and the rate of decrease of the correlation is always at the same rate:

$$C \propto r^{-\alpha} \tag{14.30}$$

The relation between the exponent α and the fractal dimension δ is simply

$$\delta = d - \alpha \tag{14.31}$$

where d denotes the dimension of space.

14.3.2.4 Using the Distribution Function

$P(r)$ is the probability that an arbitrarily chosen black spot has a radius greater than r. A change of scale corresponds to transferring r to λr. The above mentioned fractal property of the distribution requires the invariance:

$$P \propto P(\lambda r) \tag{14.32}$$

The only functional form that satisfies the above inequality is the power law:

$$P \propto r^{-\delta} \tag{14.33}$$

The exponent δ is the fractal dimension of the distribution. If we count the number of observable black spot, then the number should be proportional to P.

14.3.2.5 Using Power Spectrum

Many irregular patterns, for example, the acoustic signature of a crunchy food, can be treated as a form of a periodic or wave function. For N values of a sequence $X(t)$ the discrete Fourier transformation is given by

$$X \propto \sum_{f=0}^{N-1} C_f[\exp(2\pi ft)] \tag{14.34}$$

where f is the frequency. The Fourier transformation is a conversion of the original function. The power spectrum helps identifying frequencies of particular importance. The conversion itself can be done with the aid of the Fast Fourier transform (FFT) algorithm (Sandefur 1990), which is an integral part of most statistical software packages for microcomputers commonly used in engineering and physical analyses. A dense jagged pattern has a spectrum in which the high frequencies are prominent (Peleg 1993). If the fluctuation can be transformed to a power spectrum then the only spectrum $S(f)$ with these properties is given by the power law:

$$S \propto f^{-\beta} \tag{14.35}$$

When a spectrum is described by this power law, an important relationship exists between the exponent β and the fractal dimension δ of the graph of the signal (Takayasu 1990)

$$\beta = 5 - 2\delta \tag{14.36}$$

where $1 < \delta < 2$. When $2 < \delta < 3$ then:

$$\beta = 7 - 2\delta \tag{14.37}$$

The most convenient and common tool for such an analysis is "Fast Fourier Transform," or FFT (Ramirez 1985). The FFT converts, or transforms, the original curve into a power spectrum of frequencies, thus making interpretation easier or more meaningful. The transform is based on the possibility of expressing or approximating a function, $f(x)$, in a given interval to any degree of accuracy by the Fourier series (Kreyszig 1972):

$$f(x) = a_0 + \sum_{i=1}^{\infty} [a_n \cos(nx) + b_n \sin(nx)] \tag{14.38}$$

where a and b are constants, and the n's set of frequencies with the dimensions and units of $1/x$. The sum $a_n^2 + b_n^2$ of any frequency is defined as the corresponding power. Plots of power versus frequency, known as power spectra, represent a transform of the original function of the variable x to a new function of its underlying frequencies. In such plots special frequencies are characterized by a prominent power magnitude and can therefore be easily identified. The FFT incorporates a special algorithm that facilitates the conversion of the original function to its power spectrum, hence the name FFT. The procedure is very common in engineering and physical analysis and is almost standard in computer mathematical and statistical packages. One primary application of the method in engineering is to determine of whether a system vibrates at a frequency (1/time) near to its natural frequency, the result of which may be instability or mechanical damage.

Barrett et al. (1992) and Rohde, Normand, and Peleg (1993) applied power spectrum method based on FFT procedure to analyze the force–deformation data, yielding a power spectrum. The frequencies of the dimension are the reciprocal of the deformation length unit. The frequency scale (mm^{-1}) is obtained from dividing the original scale (s^{-1}) by the compression velocity. The power spectrum, which represents the relative intensity of the various frequencies, enables identification of special frequencies having particular physical or structural significance. The power spectra of puffed extrudates derived from FFT could be used to identify the length scale of structural features where fractures took place, and its shape could be more directly related to structural features and textural properties. The blanket algorithm was the most convenient measure of overall ruggedness, since it was expressed by a single number (Barrett et al. 1992) A summary of selected applications of fractal analysis in food systems are provided in the Table 14.1.

14.4 CONCLUSIONS

Fractal analysis is a powerful tool for characterizing materials and processes. The applications of fractal analysis in modeling food systems are limited. Availability of commercial programs for fractal analysis and current imaging technology could encourage the application of fractal

Table 14.1 Applications of Fractal in Food Systems

Application	Technique Used	References
Characterization of different starch particles	Richardson's method	Rahman (1997)
Characterization of pores	Scaling equation for Poremaster	Rahman (1997) and Rahman, Al-Zakwani, and Guizani (2005)
Characterization of pores	Scaling equation for x-ray scattering	Bale and Schmidt (1984)
Mechanical signature of brittle foods	Blanket algorithm based on Richardson's method	Rohde, Normand, and Peleg (1993), Damrau, Normand, and Peleg (1997), and Borges and Peleg (1996)
Mechanical signature of brittle foods	Symmetrized-dot pattern (SDP) method	Rohde, Normand, and Peleg (1993)
Mechanical signature of brittle foods	First Fourier Transformation (FFT)	Rohde, Normand, and Peleg (1993) and Barrett et al. (1992)
Surface boundary	Richardson's method	Barletta and Barbosa-Canovas (1993), Rahman (1997), and Peleg and Normand (1985)
Surface roughness	Algorithms for length-scale (Richardson's method), and area-scale	Pedreschi, Aguilera, and Brown (2000) and (2002)
Image characterization	Box counting method	Hagiwara et al. (1997) and Hongsprabhas, Barbut, and Marangoni (1999)
Fat crystal and gel network	Scaling equation from rheology	Marangoni and Rousseau (1996), Hagiwara et al. (1997), and Hongsprabhas, Barbut, and Marangoni (1999)
Yeast flocculation process	Scaling equation for light scattering	Davis and Hunt (1986)
Binding process to biosensor	Scaling equation from kinetics and transport equation	Sadana and Ram (1994)

techniques in a broader array of food-related fields. More studies are needed to explore physical meaning and industrial applications for fractal analysis in foods.

GLOSSARY

Acoustic signature   Sound signals generated during biting and chewing recorded by acoustic instruments. The signature is usually very rugged.

Blanket method   According to this method, the processed image is covered by a blanket with a thickness that increases stepwise. The surface area of the image is calculated at each step by dividing the blanket's volume by its thickness and the equivalent of a Richardson plot can be constructed from these data.

Correlation function   Correlation function is a fundamental statistical quantity used to correlate variables, and it can be used to find the fractal dimension.

Fourier transform   Fourier function is a form of a periodic or wave function which be used to relate a variables as a function of time. The Fourier transform is a conversion of the original function. The power spectrum helps identifying frequencies of particular importance. All these analysis could be done by statistical packages for microcomputers commonly used in engineering and physical analyses.

Fractal analysis   The origin of "fractal" word is the Latin adjective "fractus" meaning broken. Fractal analysis is to represent shapes or phenomena having no characteristic length. It is usually applied in complex or chaotic problems when traditional methods fail or become tedious to solve.

Fractal dimension Fractal dimension is a fractional number which characterizes a fractal object or process. In classical Euclidian geometry a point has a zero dimension, a line one, an area two, and a volume three. However fractal dimension cannot be integer number.

Image A micrograph or photograph taken with or without magnifications of an object by microscopic analysis.

Monolayer Amount of gas molecules needed to cover completely a solid porous or non-porous material. It could be affected by surface characteristics of a solid surface.

Richardson's plot In this method a series of polygons of side x are constructed on the perimeter of a rugged boundary using a pair of compasses. A log–log plot of stride length versus perimeter is called a Richardson plot and fractal dimension can be estimated from the slope.

Scaling equation Equation developed to characterize a product or process based on the physico-chemical principles which could be used to estimate fractal dimension by fitting the experimental data.

Self similarity Self similarity means fractal objects can be decomposed into parts, each obtainable from the whole by a similitude.

REFERENCES

Bale, H. D. and Schmidt, P. W., Small-angle x-ray-scattering investigation of submicroscopic porosity with fractal properties, *Physical Review Letters*, 53(6), 596–599, 1984.

Barletta, B. J. and Barbosa-Canovas, G. V., Fractal analysis to characterize ruggedness changes in trapped agglomerated food powders, *Journal of Food Science*, 58(5), 1030–1046, 1993.

Barnsley, M., *Fractals Everywhere*, Boston: Academic Press, 1988.

Barrett, A. H. and Peleg, M., Applications of fractal analysis to food structure, *Food Science and Technology*, 28, 553–563, 1995.

Barrett, A. M., Normand, M. D., Peleg, M., and Ross, E., Characterization of the jagged stress-strain relationships of puffed extrudates using the fat Fourier transform and fractal analysis, *Journal of Food Science*, 57, 227–235, 1992.

Bartlett, M. L., Comparison of methods for measuring fractal dimension, *Australian Physical & Engineering Sciences in Medicine*, 14(3), 146–152, 1991.

Borges, A. and Peleg, M., Determination of the apparent fractal dimension of the force–displacement curves of brittle snacks by four different algorithms, *Journal of Texture Studies*, 27, 243–255, 1996.

Bourke, P., *Fractal Dimension Calculator User Manual Version 1.5.*, Auckland: Auckland University, 1993.

Bremer, L. G. B., van Vliet, T., and Walstra, P., Theoretical and experimental study of the fractal nature of the structure of casein gels, *Journal of the Chemical Society, Faraday Transactions 1*, 85(10), 3359–3372, 1989.

Cannell, D. S., and Aubert, A., Aggregation of colloidal silica, *On Growth and Form: Fractal and Non-Fractal Patterns in Physics*, Dordrecht, The Netherlands: Martinus Nijhoff Publishers, 1986.

Clark, N. N., Three techniques for implementing digital fractal analysis of particle shape, *Powder Technology*, 46, 45–52, 1986.

Damrau, E., Normand, M. D., and Peleg, M., Effect of resolution on the apparent fractal dimension of jagged force–displacement relationships and other irregular signatures, *Journal of Food Engineering*, 31, 171–184, 1997.

Davis, R. H. and Hunt, T. P., Modeling and measurement of yeast flocculation, *Biotechnology Progress*, 2(2), 91–97, 1986.

Ehrburger-Dolle, F., Lavanchy, A., and Stoeckle, F., Determination of the surface fractal dimension of active carbons by mercury porosimetry, *Journal of Colloid and Interface Science*, 166, 451–461, 1994.

Frisen, W. I. and Mikula, R. J., Mercury porosimetry of coals. Pore volume distribution and compressibility, *Fuel*, 67, 1516–1520, 1988.

Gagenpain, J. J. and Rogues-Carmes, C., Fractal approach to two dimensional and three dimensional surface roughness, *Wear*, 109, 119, 1986.

Graf, J. C., The importance of resolution limits to the interpretation of fractal descriptions of fine particles, *Powder Technology*, 67, 83–85, 1991.

Hagiwara, T., Kumagai, H., Matsunaga, T., and Nakamura, K., Analysis of aggregate structure in food protein gels with the concept of fractal, *Bioscience, Biotechnology and Biochemistry*, 61(10), 1663–1667, 1997.

Hastings, H. M. and Sugihara, G., *Fractals, A User's Guide for the Natural Sciences*, Oxford: Oxford University Press, 1993.

Hongsprabhas, P., Barbut, S., and Marangoni, A. G., The structure of cold-set whey protein isolate gels prepared with Ca^{++}, *Food Science and Technology*, 32, 196–202, 1999.

Jullien, R. and Botet, R., *Aggregation and Fractal Aggregates*, Singapore: World Scientific Publishing, 1987.

Kaye, B. H., Image analysis techniques for characterizing fractal structures, In *The Fractal Approach to Heterogeneous Chemistry*, D. Avnir, Ed., Chichester: Wiley, pp. 53–63, 1989.

Kolb, M., Botet, R., Jullien, R., and Herrmann, H. J., *On Growth and Forms: Fractal and Non-Fractal Patterns in Physics*, Dordrecht, The Netherlands: Martinus Nijhoff Publishers, 1986.

Kopelman, R., Fractal reaction kinetics, *Science*, 241, 1620–1626, 1988.

Kreyszig, E., *Advanced Engineering Mathematics*, 3rd ed., New York: Wiley, 1972.

Liao, K., Cavalieri, R.P, and Pitts, M. J., Fractal and Hausdorff dimensional analysis of digitized fruit images, *American Society of Agricultural and Biological Engineers Paper*, 88–6017, 1988.

Mandelbrot, B. B., *Fractals Form, Chance, and Dimension*, San Francisco, CA: W. H. Freeman, 1977.

Mandelbrot, B. B., Fractals, In *Encyclopedia of Physical Science and Technology*, New York: Academic Press, pp. 579–593, 1987.

Marangoni, A. G. and Rousseau, D., Is plastic fat rheology governed by the fractal nature of the fat crystal network?, *Journal of the American Oil Chemists' Society*, 73(8), 991–994, 1996.

Nagai, T. and Yano, Y., Fractal structure of deformed potato starch and its sorption characteristics, *Journal of Food Science*, 55, 1334–1337, 1990.

Normand, M. D. and Peleg, M., Determination of the fractal dimension of a particle silhouette using image-processing techniques, *Powder Technology*, 5, 271–275, 1986.

Normand, M. D. and Peleg, M., Evaluation of the blanket algorithm for ruggedness assessment, *Powder Technology*, 54, 255–259, 1988.

Nyikos, L. and Pajkossy, T., Diffusion to fractal surfaces, *Electrochemica Acta*, 31(10), 1347–1350, 1988.

Pedreschi, F., Aguilera, J. M., and Brown, C. A., Characterization of food surfaces using scale-sensitive fractal analysis, *Journal of Food Process Engineering*, 23, 127–143, 2000.

Pedreschi, F., Aguilera, J. M., and Brown, C. A., Characterization of the surface properties of chocolate using scale-sensitive fractal analysis, *International Journal of Food Properties*, 5(3), 523–535, 2002.

Peleg, M., Fractals and foods, *Critical Reviews in Food Science and Nutrition*, 33(2), 149–165, 1993.

Peleg, M. and Normand, M. D., Characterization of the ruggedness of instant coffee particle shape by natural fractals, *Journal of Food Science*, 50, 829–831, 1985.

Peleg, S., Naor, J., Hartley, R., and Avnir, D., Multiple resolution texture analysis and classification, *IEET Transaction Pattern Analysis Machine Intelligence*, 6, 518, 1984.

Pfeifer, P. and Avnir, D., Chemistry in noninteger dimensions between two and three. I. Fractal theory of heterogeneous surfaces, *Journal of Chemical Physics*, 79(7), 3558–3565, 1983.

Pfeifer, P., Avnir, D., and Farin, D., Scaling behavior of surface irregularity in the molecular domain: From adsorption studies to fractal catalysts, *Journal of Statistical Physical*, 35(5/6), 699–714, 1984.

Pickover, C. A., *Computers, Patterns, Chaos and Beauty*, New York: St. Martin's Press, 1990.

Rahman, M. S., *Handbook of Food Properties*, Boca Raton, FL: CRC Press, 1995.

Rahman, M. S., Physical meaning and interpretation of fractal dimensions of fine particles measured by different methods, *Journal of Food Engineering*, 32, 447–456, 1997.

Rahman, M. S., Al-Zakwani, I., and Guizani, N., Pore formation in apple during air drying as a function of temperature and pore size distribution, *Journal of the Science of Food and Agriculture*, 85(6), 979–989, 2005.

Ramirez, R., *The FFT: Fundamentals and Concepts*, Englewood Cliffs, NJ: Prentice-Hall, 1985.

Richardson, L. F., The problem of contiguity: An appendix of statistics of deadly quarrels, *General Systems Yearbook 6*, pp. 139–187, 1961.

Rohde, F., Normand, M. D., and Peleg, M., Effect of equilibrium relative humidity on the mechanical signatures of brittle food materials, *Biotechnology Progress*, 9, 497–503, 1993.

Russ, J. C., *Fractal Surfaces*, New York: Plenum Press, 1994.

Sadana, A. and Ram, A. B., Fractal analysis of antigen-antibody binding kinetics: Biosensor applications, *Biotechnology Progress*, 10, 291–298, 1994.

Sandefur, J. T., *Discrete Dynamical Systems- Theory and Application*, Oxford: Clarendon, 1990.

Shih, W. H., Shih, W. Y., Kim, S. I., Liu, J., and Aksay, I. A., Scaling behavior of the elastic properties of colloidal gels, *Physical Review A*, 42, 4772–4779, 1990.

Suarez-Fernandez, C. and Aguerre, R. J., Multilayer sorption isotherms, In *Food Engineering 2000*, Fito, P., Ortega-Rodriguez, E., and Barbosa-Canovas, G.V., Eds., New York: Chapman & Hall, pp. 65–87, 2000.

Suzuki, T. and Yano, T., Interpretation of the relationship between energy and size reduction on the basis of the fractal concept, *Journal of Society Powder Technology Japan*, 26, 99, 1989.

Takayasu, H., *Fractals in the Physical Sciences*, Manchester: Manchester University Press, 1990.

Yano, T. and Nagai, T., Fractal surface of starchy materials transformed with hydrophilic alcohols, *Journal of Food Engineering*, 10, 123–133, 1989.

CHAPTER **15**

Fuzzy Modeling

Haitham M. S. Lababidi and Christopher G. J. Baker

CONTENTS

15.1 INTRODUCTION

Complexity in the real world generally arises from uncertainty in the form of ambiguity. Humans have been subconsciously addressing complex and ambiguous problems since they could think, whereas current computers fail to do so. This is simply because humans have the capacity to reason approximately about the behavior of complex systems, thereby maintaining only a general understanding of the problem. Fortunately, this generality and approximation is sufficient to comprehend complex systems.

Models are usually developed to study and simulate the behavior of physical systems. A mathematical model, for instance, is a set of equations that describes the desired operation of a process. The size and complexity of mathematical models increase with the increasing number of dependent and independent variables involved. As complexity decreases, the precision afforded by computational methods becomes more useful in modeling the system.[1]

The relationship between the degree of complexity of a system and the precision inherent in the model of the system is exemplified in Figure 15.1. Models represented by mathematical equations provide precise descriptions of systems with little complexity and, hence, little uncertainty. Such models range from fundamental (first-principles) models, with the highest precision, to data-driven empirical models whose precision is based on the available data describing the system's behavior. Such models are known as *white-box models*. For systems that are slightly more complex but for which significant data exist, model-free empirical methods provide a powerful and robust means to reduce some uncertainty through learning from available data. For example, in artificial neural networks, a structure is chosen for the network and the parameters (connections, weights, and

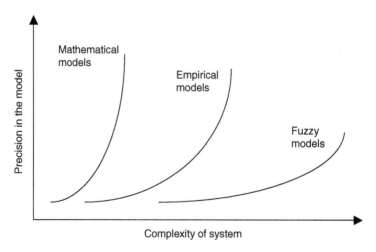

Figure 15.1 Complexity of a system versus precision in the model of the system. (From Ross, T. J., *Fuzzy Logic with Engineering Applications*, McGraw-Hill, New York, 1995.)

threshold values) are tuned to fit the observed data as well as possible. These parameters are not human-interpretable and do not offer any insight about the modeled system. Such models are known as *black-box models*.

As complexity increases, fewer numerical data may exist, and only ambiguous or imprecise information may be available. In this case, fuzzy reasoning may provide a way to describe the behavior of the system by allowing functional mapping between input and output observations. The imprecision in fuzzy models is therefore generally quite high. Fuzzy models can be viewed as *gray-box models* because they allow the modeler to extract and interpret the knowledge contained in the model as well as incorporating a priori knowledge into it.

Fuzzy systems can focus on modeling problems that are characterized by imprecise or ambiguous information and are less efficient in situations where precision is apparent. Despite the fact that all models shown in Figure 15.1 may be considered mathematical abstractions of the real physical system, the point is to match the model type with the character of the uncertainty exhibited in the modeling problem.

Fuzzy set theory provides a means for representing uncertainties. But what about probability theory that, for a long time, has been the primary tool for handling uncertainty in mathematical models? Probability theory is, in fact, ideal for problems characterized by random uncertainty. This exists in cases where the outcomes of any particular realization of the process are strictly a matter of chance and can be precisely described by statistical functions. However, not all uncertainties associated with complex systems and addressed by humans on a daily basis are random in nature. Nonrandom uncertainty is well-suited to treatment or modeling by the fuzzy set theory, which provides the tools for modeling the uncertainty associated with vagueness, with imprecision, and/or with a lack of information.

Lotfi Zadeh made the following statement in his seminal paper of 1965:[2]

> The notion of fuzzy set provides a convenient point of departure from the construction of a conceptual framework, which parallels in many respects the framework used in the case of ordinary sets, but is more general than the latter and, particularly, may prove to have a much wider scope of applicability, particularly in the fields of pattern classification and information processing. Essentially, such a framework provides a natural way of dealing with problems in which the source of imprecision is the absence of sharply defined criteria of class membership rather than the presence of random variables.

The application of fuzzy set theory and fuzzy logic has shown a tremendous payoff in areas where intuition and judgment still play major roles in the model. Fuzzy logic seems to be most successful in two situations:

1. Very complex models where understanding is strictly limited or quite judgmental
2. Processes where human reasoning, human perception, or human decision making are inextricably involved

Human knowledge is becoming increasingly important. A theory is needed to formulate this knowledge in a systematic manner and incorporate it into engineering systems together with other information such as mathematical models and sensory measurements. As a general principle, a good engineering theory should be capable of making effective use of all available information. For many practical systems, important information comes from two sources: (1) human experts who describe their knowledge about the system in natural language; and (2) sensory measurements and mathematical models that are derived according to physical laws.[3] An important task, therefore, is to combine these two types of information. The main question is how to formulate human knowledge into a similar framework used to formulate sensory measurements and mathematical models. Fortunately, a fuzzy system essentially performs this transformation.

In general, a fuzzy system is any system whose operation is mainly based on fuzzy concepts, such as reasoning, arithmetic, algebra, topology, or programming, among others.[4] However, in the frame of this chapter, the term *fuzzy system* refers to developing a rule-based system that uses (1) fuzzy variables represented as linguistic terms; (2) fuzzy sets to represent the linguistic values of such variables; (3) fuzzy rules to describe causal relationships between the variables; and (4) fuzzy inference to compute the responses of the fuzzy system to a given input. Input variables are fuzzified by associating them with linguistic terms whose values are defined by membership functions. An inference mechanism applies fuzzy logic operations on the fuzzy rule-base to derive the implication of individual rules and determine an overall conclusion represented as a fuzzy output. Finally, this output is translated into a crisp value using a defuzzification method.

In this chapter, the fundamentals of fuzzy modeling are discussed, with particular emphasis on food-processing applications. Section 15.2 reviews relevant publications in the literature. These relate principally to food quality and production-line control issues and equipment selection. Section 15.3 describes the fundamentals of fuzzy set theory. Its application in fuzzy modeling is discussed in Section 15.4, which addresses such topics as linguistic variables, membership functions, fuzzy rule bases, fuzzy inference, and fuzzification/defuzzification processes. Section 15.5 presents a typical example of a fuzzy model, describing the operation of a bread extrusion process. Finally, Section 15.6 discusses advanced techniques for developing and tuning membership functions. These include neural-fuzzy systems and genetic algorithms.

15.2 APPLICATIONS OF FUZZY MODELING IN FOOD INDUSTRY

Food processes largely rely on operators' rules of thumb and are not fully automated. A study by Iiyukhin and coworkers[5] shows that 59% of food manufacturing plants are not automated. The main reason is that it is difficult to develop a realistic model due in part to (1) the many dimensions that must be taken into account in parallel; and (2) the nonlinearity and coupling between the variables involved in the system.[6]

A number of examples of fuzzy modeling relating to the food industry have been published in the literature. Some of the more recent ones are summarized below. Broadly speaking, most of the applications are in the area of image analysis and quality control of food products. Fuzzy modeling as a precursor to fuzzy process control is finding useful applications in food processing. The following sections will focus on the image analysis and fuzzy control applications found in the literature. A brief section on equipment selection using fuzzy logic is also presented.

15.2.1 Image Analysis and Quality Control

The main problem in food process automation is that only a limited number of variables are measurable online.[7] A number of these variables, such as color, odor, taste, appearance, and texture, are subjective and are usually evaluated qualitatively as linguistic terms. For such applications, symbolic sensors are more useful than classical sensors that provide direct numeric values. The aim of a symbolic sensor is to perform a linguistic symbolization of a phenomenon from numeric measurements.[8] Therefore, the symbolic sensor acquires a numeric measurement and then performs the numeric–linguistic conversion itself, taking into account the subjectivity of the problem. Fuzzy set theory provides an optimal frame for the treatment of this numeric–linguistic conversion, reasoning with linguistic terms, and converting the linguistic conclusions back to numerical representations that may be communicated to other instruments.

In food processes, fuzzy symbolic sensing is normally employed as a classification technique used in conjunction with computer vision to evaluate food quality. Classification is an important image-processing operation in which objects are identified and allocated to one of the finite sets describing individual classes. This involves comparing the measured features of a new object with those of a known object or other known criteria, and it determines whether the new object belongs to a particular category of objects.

In detailed and comprehensive reviews, Du and Sun[9,10] described the classification techniques that are used in food processing. They stated that statistical fuzzy logic and neural networks are the main methods of classification employed in the literature. Of these, fuzzy logic has the advantage of simulating the human experience of generating complex decisions using approximate and uncertain information. The authors concluded that the performance of a fuzzy system depends on how well it is tuned. This becomes difficult in multidimensional problems.

A fuzzy classifier for sorting apples on the basis of watercore severity was presented by Shahin and coworkers.[11] The classifier was able to separate the apples into three classes with an overall accuracy of 80%. However, they found that a neural classifier performed better with an overall accuracy of 88%.

Hu et al.[12] employed both a conventional and a fuzzy classifier with a four-level hierarchy based on the "generalized K-nearest neighbor rules" to grade herring roe and indicated that the fuzzy classifier produced more accurate results (89%). Croft et al.[13] achieved results of similar accuracy (85%–95%) and reasonable repeatability for herring roe using an automated fuzzy system.

Center and Verma[14] presented an overview of the application of fuzzy modeling to biological and agricultural systems. Of the examples they quoted, only a few are relevant to food processing. These include the sorting of tomatoes on the basis of quality,[15] identification of physiological states in fermentation processes for use in distributed control,[16,17] feeding strategies on large dairy farms,[18] and grading beef quality.[19]

In a study by Verma,[15] tomato quality was described as a complex combination of attributes such as color (brightness and hue), sensory color, and fruit size, shape, and firmness. Six fuzzy models were developed and linked together to form a fuzzy decision support system (DSS) that was employed to predict the highest quality and the time required for a given tomato to attain this quality. The DSS thus enabled the tomatoes to be grouped for sale on a certain date on which their quality had peaked. On the basis of limited testing, the predictions of the model were claimed to be very accurate.[15]

Jahns and coworkers[20] combined image analysis and fuzzy logic to grade tomatoes on the basis of quality. The authors proposed a fuzzy model for mapping visual fuzzy parameters into an output parameter termed "total visual quality." The input fuzzy parameters were the *contour, shape, color,* and *color uniformity.* Contour and shape were processed by fuzzy reasoning to yield an output called *geometry,* whereas color and color distribution were combined to yield an output called *impression.* Both gave rise to the overall *appearance* fuzzy variable, which had to be checked for cracks and faults. The output fuzzy variable *cracks* was a knockout criterion. When no discontinuities in color or reflection could be detected, a further grading to *size* gave an overall *visual quality* as a result. Image analysis measurements were fuzzified into the four input variables. Each pair of input variables was used to infer the two intermediate fuzzy variables (geometry and impression), which in turn were used to infer the existence of cracks, or appearance and size that yielded the final output, termed *visual quality.* All input and output fuzzy variables were defined as linguistic variables with triangular membership functions (see Section 15.3). The authors reported that an optimization or tuning of the input membership functions by back-propagation was possible if enough consumer interviews on total quality were available.

Kavdir and Guyer[21] employed fuzzy logic as an aid in automating the grading of Golden Delicious apples. They justified their choice of a fuzzy model as opposed to techniques such as template matching and fixed-object modeling used in industrial applications because of the inherent variability of the agricultural environment (weather, soil, etc.) and the morphological diversity of the apples. The principal objectives of their study were to evaluate techniques for constructing and tuning the fuzzy membership functions and to compare the predictions of their model with those of a human expert. To construct the fuzzy model, five quality features (color, defects, shape, weight, and size) were measured for each apple using a variety of nondestructive testing techniques. These features were grouped into three fuzzy attributes: color, defects, and size. The model graded the apples into three categories: good, medium, and bad. A trial-and-error approach was employed to develop the membership functions. For color and defects, these were triangular and trapezoidal in shape. However, an exponential function was used to simulate the human expert in grading apples in terms of size. The authors[21] devised a total of 27 rules, which were of the form "If the color is *greenish* AND there is *no defect* AND the apple is *large and well formed*, THEN the quality is *Good*." The Center-of-Area Method (COAM) was used for defuzzification. The fuzzy model predicted the quality of around 89% of the 181 apples correctly. In general, misclassification errors occurred mainly between adjacent groups. The authors concluded that the specification of membership functions in terms of shape and boundary had a clear effect on the predictions. They suggested that statistics of the class populations, such as average, standard deviation, and minimum and maximum values, could help in the determination of membership functions.

As part of an automated inspection system for chocolate chip cookies, Davidson et al.[22] developed fuzzy models to predict consumer preferences based on a number of physical attributes. These included size, shape, baked dough color, and fraction of the top surface area occupied by chocolate chips. Five separate polls, each with 30 participants, were undertaken during the course of the study. The first three (so-called *calibration polls*) were conducted to formulate the fuzzy models of consumer preference; two additional polls were undertaken to validate the models. The authors concluded that such models could find additional use in the bakery for automating quality assurance decisions (acceptance or rejection of individual cookies prior to packaging) and improving the baking oven control.

Ioannou and coworkers[23] employed fuzzy modeling to reproduce an assessment by experts of the degree of crusting in sausages. This important defect is observed during manufacture and is characterized by the appearance of a watertight crust on the surface of the sausage. The degree of crusting is normally determined by experts (the plant operators) through visual observation of a sausage slice. It is classified into five groups: 0 for sausages without defect and 1 (thinnest) to 4 (thickest) for defective sausages. Three factors are considered by the experts in determining the degree of crusting: the length of black (LB) and red (LD) areas in the outer part of the sausage and the color of the center (COL) of the slice. The input parameters LB and LD were fuzzified as linguistic terms, such as nonexistent, a bit long, long, and very long, and triangular membership functions were used. The color parameter COL was used to adjust the thresholds for LB and LD, values of which were expressed in numbers of pixels. The numerical values of LB and LD were first fuzzified, and the degree of crusting was determined using appropriate rules in the rule base; for example, "IF *LB* is *long* AND *LD* is *long* THEN crusting degree (CD) is 3." In the defuzzification step, the results of all the triggered rules were combined to yield the crusting degree. The multiplication defuzzification concept was used in which the crisp value of CD was obtained by multiplying the membership values of LB and LD, and the classification degree was inferred by the rules. The authors compared the predicted values of the degree of crusting with the average assessment made by two experts. Satisfactory agreement was achieved in 68 of the 76 cases evaluated. The authors concluded that, overall, their technique appeared very promising.

Sun and Brosnan[24,25] investigated the feasibility of using computer vision for automated quality inspection of pizzas. Part one of their study[24] focused on the quality of the pizza base and the sauce spread, while toppings were considered in part two.[25] The authors analyzed 20 pizza base samples and 25 base samples with tomato paste spread over them. The bases were classified as *standard*, *poor pressing*, *poor alignment*, and *flowing base*, whereas the sauce spread quality was classified as *reject underwipe*, *acceptable underwipe*, *even spread*, *acceptable overwipe*, and *reject overwipe*. Acceptable samples should have a standard base and even spread sauce. The criteria established for an acceptable base were the minimum values of the area and the spatial ratios and a maximum value for the circularity. The overall accuracy of the system was 87% when compared with human quality inspection. On the other hand, sauce spread quality was evaluated by two indices: sauce area percentage and heavy area percentage. A fuzzy evaluation score (FES) was calculated that reflected the overall quality of the sauce spread. A comparison between the results produced by the fuzzy model and the human experts showed a misclassification rate of 52%. The authors concluded that errors in the human classifications due to subjectivity and inconsistency, as well as inaccuracies in the algorithm, contributed to the discrepancies. When two categories of classification (accept and reject) were employed, an overall accuracy of 92% was achieved.

Analysis of the images of pizza toppings is very difficult, as different toppings are similar in appearance, and overlapping is likely to occur.[25] A total of 25 pizzas were analyzed for topping quantity using three numerical indices representing ham, mushroom, and topping area percentages. The same five linguistic descriptors as employed for sauce spread were used to fuzzify these indices using triangular membership functions, as before. A FES was again calculated. The results showed that the topping area percentage exhibited the least fuzziness and the ham area percentage exhibited the most. The misclassification rate over the five levels examined was 24%. However, when only two quality levels (accept and reject) were considered, the accuracy was 100%.

Davidson and Ryks[26] investigated the use of fuzzy modeling to analyze specific microbial risks in food-processing operations. Monte Carlo simulations have been widely used in such studies. However, such simulations require extensive data to define appropriate probability distribution functions for risk factors such as microbial loadings, processing effects, consumption levels, and health effects. In the early stages of a risk assessment, however, it is frequently found that only limited data are available. The authors demonstrated that fuzzy modeling is a more satisfactory approach under these circumstances because fuzzy values can describe variability in model parameters on the basis of limited information from the literature, quality control data, or expert opinion. They suggested that their approach would allow an evolution to probability-based methods as the knowledge-base is developed without the need to completely reformulate the risk-assessment model.

15.2.2 Fuzzy Models in Fuzzy Control Applications

Food processes may be difficult to control or model by conventional methods where simplifications or linearization are often employed. This difficulty is due to tightly coupled control loops, nonlinearities, and unpredictable noise. Some food processes are combinations of continuous and batch operations. Such processes are usually controlled by experienced operators, who frequently make observations and judgments, particularly about sensory attributes, that are difficult to reproduce using online sensors. It is not easy to transfer the knowledge from human experts to mathematical models when designing a control system for food processes.[7] However, fuzzy modeling techniques make it possible to use operators' observations and their experience in operating the process in computer-based control systems. As a result, fuzzy control systems offer considerable potential for increasing automatic control applications in the food industry.[27]

Fuzzy control is a nonlinear predictive (or model-based) control system. Such systems require a process model to predict the states of the controlled variables and a controller model to drive the variables to their desired values. Fuzzy control systems can be based entirely on fuzzy techniques. In this case, both the process model and controller model are fuzzy. Alternatively, predictive fuzzy process models may be combined with conventional control techniques in hybrid systems. Furthermore, fuzzy control can be classified into *nonadaptive* fuzzy control and *adaptive* fuzzy control.[3,28] In nonadaptive fuzzy control, the structure and parameters of the fuzzy model are fixed and do not change during real-time operation. In adaptive fuzzy control, the structure and/or parameters of the fuzzy model change during real-time operation. Nonadaptive fuzzy control is simpler but requires more knowledge of the process model or heuristic rules. Adaptive fuzzy control, on the other hand, is more difficult to implement but requires less information and may perform better.

Davidson[27] outlined the basic elements of fuzzy control systems and discussed their application in food processes. The author focused on the use of fuzzy techniques in controlling dryers and fermentation processes. A number of fuzzy control applications of food-related processes are cited in the literature. These include the control of a grain dryer,[29] aseptic processing,[30] fermentation processes,[31-35] cheese ripening,[6,36] food frying,[37] and browning of food products.[38,39] The following section will highlight a number of these applications.

Perrot and coworkers[6] described the development of a fuzzy DSS to aid in the control of cheese ripening. The authors constructed their DSS in three stages: acquisition of the expert knowledge, building the fuzzy algorithm, and validating the approach. The acquisition process involved assimilating the linguistic terms employed in the factory and converting these into numerical sensory indicators in the manner described in Ioannou et al.[39] and Curt et al.[44] Formulation and preliminary evaluation of the algorithm was achieved with the aid of experiments on a pilot plant. Further validation was undertaken by testing it on-line on cheeses produced in a ripening cellar in the factory.

Murnleitner et al.[34] conducted an experimental study to develop a fuzzy control system for the two-stage anaerobic pre-treatment of food industry wastewater. They produced a fuzzy model that successfully predicted the biological state of the reactors and initiated suitable control measures to avoid overload. The system could handle large fluctuations in the concentration of the substrate and the volumetric loading.

Rywotycki[39] employed a two-stage fuzzy model as the basis for controlling a fryer. The objective was to increase the efficiency of the frying process while maintaining an acceptable product quality. In the first stage of the model, the required temperature of the fat, T_3, is inferred from two inputs: the thermal power necessary to heat the raw product, Q_2; and the makeup fat placed in the fryer, Q_3. Input variables were fuzzified and assigned linguistic values: *very small*, *small*, *medium*, *big*, and *very big*. An example of one of the rules is: "IF Q_2 is *very small* AND Q_3 is *medium* THEN T_3 is *medium*." The fat temperature, T_3, is directly related to the desired quality attributes of the fried product, which were rated in consumer surveys as *very good*, *good*, *neutral*, *poor*, and *bad*. T_3 is the input to the second stage of the fuzzy controller that determines the required travel speed of the fryer conveyor belt, v_p. The resulting travel speed was defuzzified using the centroid or center-of-mass method. The proposed fuzzy controller was tested for the frying of french fries in the temperature range 414–453 K. The author reported an increase in the frying efficiency by several percentage points while, at the same time, matching the quality of the product to consumer expectations.

Ioannou et al.[38] employed fuzzy modeling to develop a diagnostic algorithm to describe non-enzymic browning of a food product based on milk and eggs. Their model was subsequently employed as the basis for controlling the browning process on the manufacturing plant.[39] The developed fuzzy model was based on three sensory indicators—the percentage of spot areas (V1), the color of the spotted areas (V2), and the color of the nonspotted areas (V3)—each of which was described by a set of linguistic terms and a membership function. These fuzzy variables were amalgamated into a series of fuzzy rules that were used to define a browning global appearance (BGA) index (*hardly browned*, *slightly browned*, *browned*, *darker browned*, and *very browned*). For example, "IF V1 is *small* AND V2 is *dark brown* AND V3 is *beige*, THEN the BGA is *hardly browned*." The diagnostic model was validated on 40 products against the assessment of the operator; the compatibility was 92.5%.

The fuzzy model described above was incorporated by Ioannou et al.[39] into a fuzzy control system (based on the Takagi–Sugeno method) that was fitted to the continuous browning oven on the production line. The controller varied the power input to four banks of stainless steel lamps located above the product conveyor in the browning oven. The effectiveness of the controller was validated in a series of trials. In the example given in the paper, it was tested on the production line for a 70-minute period, during which the controller adequately coped with three process upsets that required corrective action to be taken. Moreover, the actions undertaken by the controller were 90% consistent with those that would have been taken by the operator.

15.2.3 Process Equipment Selection

Fuzzy modeling can also play a useful role in the selection of industrial plants because many of the parameters influencing this process are often ill-defined and uncertain. The present authors, for example, developed a fuzzy expert system to partially de-skill the selection of dryers for food materials. These devices are particularly complex, as they not only reduce the moisture content of the product but can also affect its quality-related attributes (e.g., color, flavor, etc.) as well. The selection process involves a large number of imprecise variables, many of which can only be defined in linguistic terms. Examples of the latter include *stickiness, cohesiveness, fragility, temperature sensitivity*, etc. The authors therefore employed a flexible system that featured multiple fuzzy knowledge bases that could be accessed over the Web.[40] Separate rule bases were developed for batch dryers[41] and continuous dryers.[42] A preliminary selection was first made to determine which of these was the most appropriate in the particular circumstances. The flexible structure of the algorithm also enabled, for example, design programs and help files to be accessed during the selection process. The original knowledge-base was subsequently extended to aid in the specification of detailed spray dryer configurations.[43]

As the above survey shows, fuzzy modeling has a useful role to play in food-industry applications in situations where quantitative knowledge is limited or imprecise. Grading and quality assessment of food products have perhaps received the most attention. This is not surprising because such judgments are made on the basis of human preferences rather than mathematical equations. It is surprising that fuzzy modeling has not been employed more widely in sensory analysis and profiling, which would appear to be a logical extension of some of the published studies described herein. The apparent ability of fuzzy modeling to provide an adequate description of several food-manufacturing processes is encouraging and should ultimately lead to the more widespread use of fuzzy modeling in the industry.

15.3 FUZZY SET THEORY

For the classical set theory, an element x in the universe U is either a member of some crisp set A or not. This binary membership can be expressed mathematically as:

$$\chi_A(x) = \begin{cases} 1, & x \in A \\ 0, & x \notin A \end{cases}.$$ (15.1)

Lotfi Zadeh[2] extended the notion of binary membership to accommodate various *degree of membership* in the real continuous interval [0,1], where the infinite number of values between the endpoints (0 = no and 1 = full membership) can represent various degrees of membership. Consider an element x in some set $\underset{\sim}{A}$ in the universe U. This set was termed by Zadeh as a *fuzzy set*, whose membership function is expressed as

$$\mu_{\underset{\sim}{A}}(x) \in [0,1],$$ (15.2)

where the symbol $\mu_{\underset{\sim}{A}}(x)$ is the degree of membership of element x in fuzzy set $\underset{\sim}{A}$. Therefore, $\mu_{\underset{\sim}{A}}(x)$ is a value in the unit interval that measures the degree to which element x belongs to fuzzy set $\underset{\sim}{A}$.

For example, consider two crisp sets W and H defining "warm" and "hot" temperatures, respectively. Any specific temperature, T, may be categorized as "warm" if its value is less than

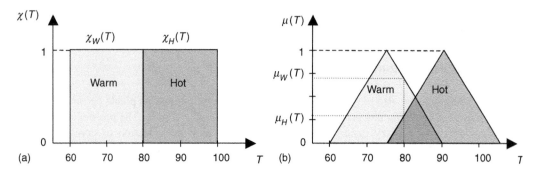

Figure 15.2 Temperature membership function in (a) crisp sets and (b) fuzzy sets.

80°C and "hot" if greater than 80°C, as shown in Figure 15.2a. In this case, 79.9°C would be classified as "warm" ($\chi_W (79.9°C) = 1$ and $\chi_H (79.9°C) = 0$) whereas 80.1°C would be classified as "hot" ($\chi_W (80.1°C) = 0$ and $\chi_H (80.1°C) = 1$). Hence, for classical set theory, transition from one set to another set is instantaneous. With fuzzy sets, however, the transition may be gradual because an element can have partial membership in multiple sets. Figure 15.2b defines the membership functions for two fuzzy sets W and H describing "warm" and "hot" temperatures, respectively. In the fuzzy formulation, a particular temperature may belong to the two sets with certain degree of membership. For instance, according to Figure 15.2b, $T = 80°C$ belongs to the "warm" set with degree of membership $\mu_W(80°C) = 0.67$ and to the "hot" set to degree $\mu_H(80°C) = 0.3$.

Zadeh's notation convention for a fuzzy set A when the universe of discourse, U, is discrete and finite, is as follows:

$$A = \left\{ \frac{\mu_A(x_1)}{x_1} + \frac{\mu_A(x_2)}{x_2} + \cdots \right\} = \left\{ \sum_i \frac{\mu_A(x_i)}{x_i} \right\}. \tag{15.3}$$

Accordingly, for the temperature universe, T, shown in Figure 15.2b, the "warm" and "hot" fuzzy sets are defined as:

$$W = \left\{ \frac{0}{60} + \frac{1}{75} + \frac{0}{90} \right\} \quad \text{and} \quad H = \left\{ \frac{0}{75} + \frac{1}{90} + \frac{0}{105} \right\}. \tag{15.4}$$

In this notation, the numerator in each term is the membership of the corresponding element in the denominator, and the plus sign (+) does not denote algebraic summation but rather the collection or aggregation of elements.

When the universe, U, is continuous and infinite, a fuzzy set A is denoted by:

$$A = \int \frac{\mu_A(x_i)}{x_i} \tag{15.5}$$

where the integral sign is not an algebraic integral in the usual sense; rather, it is a notation that represents the grade of membership of x in a continuous universe.

15.3.1 Fuzzy Logic

Fuzzy logic is a method to formalize the human capacity of imprecise reasoning or approximate reasoning. In fuzzy logic, all truths are partial or approximate. It can be considered as an interpolation process between the binary extremes of true and false.

In classical predicate logic, a simple proposition P is a linguistic statement that can be identified as strictly *true* or strictly *false*. For binary (Boolean) predicate logic, the truth of a proposition, denoted as T_P, is assigned a value of 1 (true) or 0 (false). In contrast, fuzzy logic is a multivalued logic where propositions are assigned to fuzzy sets. When a fuzzy proposition $\underset{\sim}{P}$ is assigned to a fuzzy set $\underset{\sim}{A}$, the degree of truth for the proposition $\underset{\sim}{P}$, denoted $T_{\underset{\sim}{P}}$, is equal to the membership value of x in fuzzy set $\underset{\sim}{A}$, given by:

$$T_{\underset{\sim}{P}} = \mu_{\underset{\sim}{A}}(x) \quad 0 \leq \mu_{\underset{\sim}{A}}(x) \leq 1. \tag{15.6}$$

To illustrate the fuzzy logic approach, consider the expression: "For *low* throughput, reduce air flow." The proposition here is "throughput is low." In classical set theory, the truth of this proposition may be defined as:

$$T_P = \chi_{\text{low throughput}}(x) = \begin{cases} 1, & x \leq 50 \text{ kg/h} \\ 0, & x > 50 \text{ kg/h} \end{cases}. \tag{15.7}$$

In this case, a computer program would not recognize a 50.1 kg/h as being a member of the "low throughput" set. In fact, both 50.1 and 1,000 kg/h may be equally recognized as not low, and consequently, an incorrect decision would be taken.

In contrast, the fuzzy logic approach to handling the proposition, "throughput is low," starts by associating the crisp variable, *throughput*, with linguistic terms such as *low*, *medium*, and *high*. Then the truth or confidence of the proposition is defined as:

$$T_{\underset{\sim}{P}} = T_{(\text{throughput is low})} = \mu_{\underset{\sim}{A}_{\text{low}}}(x) \quad 0 \leq \mu_{\underset{\sim}{A}_{\text{low}}}(x) \leq 1. \tag{15.8}$$

Membership functions for the linguistic terms describing the *throughput* are shown in Figure 15.3. In this case, for a throughput value of 110 kg/h, for instance, the truth of the proposition "throughput is low" is $T_{(\text{throughput is low})} = \mu_{\underset{\sim}{A}_{\text{low}}}(110) = 0.2$. Nevertheless, this crisp value of the throughput is described at the same time with another proposition, "throughput is medium," with truth value 0.8 (see Figure 15.3).

15.3.2 Fuzzy Set Operations

Consider two fuzzy sets, $\underset{\sim}{A}$ and $\underset{\sim}{B}$, in the universe U. For a given element $x \in U$, the following function operations are defined for *union*, *intersection*, and *complement* operators:

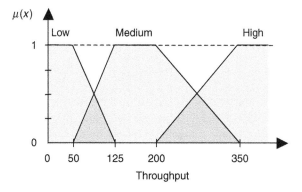

Figure 15.3 Membership function for throughput.

$$\mu_{\underline{A}\cup\underline{B}}(x) = \mu_{\underline{A}}(x) \vee \mu_{\underline{B}}(x) = \max(\mu_{\underline{A}}(x), \mu_{\underline{B}}(x)),$$ (15.9)

$$\mu_{\underline{A}\cap\underline{B}}(x) = \mu_{\underline{A}}(x) \wedge \mu_{\underline{B}}(x) = \min(\mu_{\underline{A}}(x), \mu_{\underline{B}}(x)),$$ (15.10)

$$\mu_{\bar{\underline{A}}}(x) = 1 - \mu_{\underline{A}}(x),$$ (15.11)

where $\vee$ and $\wedge$ are maximum and minimum operators, respectively. In fact, all of the operations on classical sets also hold for fuzzy sets, except for the *law of the excluded middle* and the *law of contradiction*, which deal with the union and intersection, respectively, of a set $\underline{A}$ with its compliment. These are expressed as:

$$\underline{A} \cup \bar{\underline{A}} \neq U,$$ (15.12)

$$\underline{A} \cap \bar{\underline{A}} \neq \varnothing.$$ (15.13)

To illustrate the fuzzy set operations, consider the two fuzzy sets describing the temperature that are plotted in Figure 15.2b and expressed by Equation 15.4. Applying the operations given by Equation 15.9 through Equation 15.13 results in the following expressions, which are also illustrated in Figure 15.4:

$$\underline{W} \cup \underline{H} = \left\{\frac{0}{60} + \frac{1}{75} + \frac{0.5}{87.5} + \frac{1}{90} + \frac{0}{105}\right\}, \quad \underline{W} \cap \underline{H} = \left\{\frac{0}{75} + \frac{0.5}{87.5} + \frac{0}{90}\right\},$$

$$\bar{\underline{W}} = \left\{\frac{1}{60} + \frac{0}{75} + \frac{1}{90}\right\}, \quad \underline{W} \cup \bar{\underline{W}} = \left\{\frac{1}{60} + \frac{0.5}{67.5} + \frac{1}{75} + \frac{0.5}{82.5} + \frac{1}{90}\right\},$$

$$\underline{W} \cup \bar{\underline{W}} = \left\{\frac{0}{60} + \frac{0.5}{67.5} + \frac{0}{75} + \frac{0.5}{82.5} + \frac{0}{90}\right\}.$$

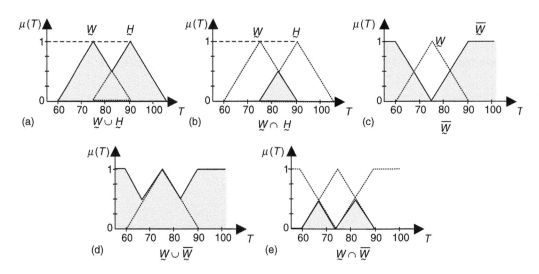

Figure 15.4 Graphical representation of (a) union, (b) intersection, (c) complement, (d) law of the excluded middle, and (e) law of contradiction.

15.4   FUZZY MODELING

The objective of any model is to find the state of a system, X, which can be implicitly expressed as:

$$X_i = f(\mathbf{X}, \mathbf{X}', \mathbf{X}'', \ldots), \quad X_i \in \mathfrak{R}, \quad i \in \{1, \ldots, n\}, \tag{15.14}$$

where $X = [X_1, X_2, \ldots, X_n]$ is the state vector consisting of n dependent variables, while X' and X'' are the first and second derivatives of the state vector, respectively. The state vector X consists of m input variables, $\mathbf{Z}$, and $n - m$ output variables, $Y(X = Z \cup Y)$.

Fuzzy modeling of an engineering system becomes advantageous when the function f (Equation 15.14) is so complex that it cannot be deduced purely by physical modeling. Even when a physical or statistical model does exist, for a relatively simple system or after many simplifying assumptions have been made, the nature of the variables in the model and the methods of measuring them may make it difficult to quantify them precisely.

Fuzzy models are known as *fuzzy systems*. They are knowledge-based (or rule-based) systems constructed from human knowledge in the form of fuzzy IF–THEN rules that are represented in terms of linguistic variables that are characterized by continuous membership functions. The starting point of constructing a fuzzy system is to extract a collection of fuzzy IF–THEN rules from human experts or based on domain knowledge. The strength of fuzzy systems is that they provide systematic procedures for transforming the knowledge-base into a nonlinear mapping. This transformation enables us to use fuzzy systems in engineering applications in the same manner that we use mathematical models. In this case, the function f in Equation 15.14 is replaced by a relation R representing the fuzzy model in the form of fuzzy rules, whereas the state vector, X, representing the crisp input and output variables, is replaced by a fuzzy state vector, $\underset{\sim}{X}$, representing the fuzzy input and output variables $\underset{\sim}{Z}$ and $\underset{\sim}{Y}$, respectively.

The structure of the fuzzy system described in this document is shown in Figure 15.5. The heart of this system is the middle block that consists of a fuzzy rule base and an inference engine. It receives fuzzy input variables and produces fuzzy output variables. Both the input and output fuzzy variables are represented by linguistic variables and membership values. Based on the values of the fuzzy input variables, the inference engine processes the fuzzy IF–THEN rules in the rule base and deduces the fuzzy values of the output variables.

Engineering problems are normally represented in terms of real-valued (crisp) input and output variables. Hence, to utilize the fuzzy rule base and the inference engine, these inputs should first be converted into the fuzzy domain. This is achieved by adding a *fuzzifier* that transforms the crisp input variables into fuzzy inputs ($Z \mapsto \underset{\sim}{Z}$), and a *defuzzifier*, that transforms the fuzzy outputs of the

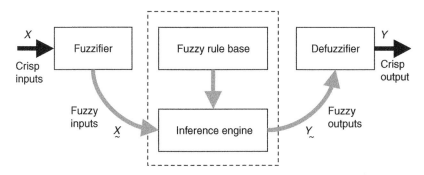

Figure 15.5   Simplified structure of a fuzzy system.

system into crisp outputs ($\underline{Y} \mapsto Y$). Both fuzzy inputs and outputs are represented in terms of linguistic variables.

The main objective of the rest of this chapter is to describe the constituents of fuzzy systems and to outline the procedure for constructing fuzzy models that can be effectively used in modeling engineering systems. The topics that will be addressed are: (1) linguistic variables, (2) membership functions, (3) fuzzy rule base, (4) fuzzy inference, and (5) fuzzification and defuzzification.

In constructing a new fuzzy system, one is usually faced with the following questions:[45]

- How to define membership functions? This involves describing a given variable by a set of linguistic terms, defining each linguistic term within its universe of discourse and membership function, and determining the best shape for each of these membership functions.
- How to obtain the fuzzy rule base? This involves extracting knowledge from someone with sufficient experience to provide a comprehensive knowledge base for a complex system that cannot be modeled physically. This task is referred to as *knowledge acquisition*.
- What are the best ways to perform fuzzy reasoning or inferencing? This involves defining the mechanism by which the fuzzy rules are processed to provide the transformation between input and output information.
- What is the best defuzzification technique for a given problem?
- How to improve and refine the computational efficiency of the fuzzy model?

15.4.1 Linguistic Variables

We often use words to describe variables in our daily life. Operators of a given process use terms such as "moderate," "high," or "extremely high" when expressing the current values of operating conditions. An expression such as "the pressure is very high" assigns the variable "pressure" a value of "very high." This value would be clearly understood by other operators; the pressure is greater than the maximum acceptable level. When a variable takes numbers as its value, there exists a well-established mathematical framework to formulate it. But when a variable takes words as its values, classical mathematical theory would be unable to formulate it. In this case, to provide a formal framework, the concept of linguistic variables should be utilized. Thus, a *linguistic variable* is a variable that can take words in natural language as its values, where the words are characterized by fuzzy sets defined in the universe of discourse in which the variable is defined. Linguistic variables are also referred to as *fuzzy variables*.

A linguistic variable $X_{\bar{i}}$ can be defined by the quadruple

$$X_{\bar{i}} = (x, U, T(x), M(x))_i \qquad (15.15)$$

where x is a text expression in natural language that expresses the name of the variable, such as *pressure*, *throughput*, and *moisture content*. U is the universe of discourse; $U = [U_L, U_U]$ defines the interval of real values that X_i can belong to, where $X_i \in U$. In other words, it is the actual physical domain in which the linguistic variable X_i takes its quantitative (crisp) values. For example, the domain of the "throughput" linguistic variable (Figure 15.3) may be defined as $U = [0,500]$. $T(x)$ is the set of linguistic terms that X_i can take. In Figure 15.3, the set of linguistic terms describing "throughput" are defined as $T = \{\text{low,medium,high}\}$. $M(x)$ are the membership functions that provide the transformation from crisp numbers to linguistic terms. For example, M relates the linguistic terms *low*, *medium*, and *high* to the membership functions shown in Figure 15.3. Membership functions are discussed further in Section 15.4.2.

Sometimes, more than one word is used to describe a linguistic variable. In this case, its value may be considered a composite of *primary terms* and *linguistic hedges*. Primary terms are the basic linguistic terms describing the linguistic variable, whereas linguistic hedges act as intensifiers of the primary terms. They are adjectives (nouns) or adverbs (verbs) like "very," "slightly," "more or less," "fairly," "almost," "barely," "mostly," "roughly," "approximately," and others. These linguistic hedges have the effect of modifying the membership function of the primary terms. As an example, consider the following definitions of linguistic hedges applied on the primary linguistic term L:

$$\mu_{\text{very } L} = (\mu_L)^2; \tag{15.16}$$

$$\mu_{\text{slightly } L} = \sqrt{\mu_L}; \tag{15.17}$$

$$\mu_{\text{not } L} = 1 - \mu_L; \tag{15.18}$$

$$\mu_{\text{not very } L} = 1 - (\mu_L)^2. \tag{15.19}$$

To illustrate how hedges are used, define the membership function of L as,

$$\mu_L = \left\{ \frac{1}{1} + \frac{0.8}{2} + \frac{0.6}{3} + \frac{0.4}{4} + \frac{0.2}{5} \right\}.$$

Then we have,

$$\mu_{\text{very } L} = \left\{ \frac{1}{1} + \frac{0.64}{2} + \frac{0.36}{3} + \frac{0.16}{4} + \frac{0.04}{5} \right\},$$

$$\mu_{\text{slighly } L} = \left\{ \frac{1}{1} + \frac{0.8944}{2} + \frac{0.7746}{3} + \frac{0.6325}{4} + \frac{0.4472}{5} \right\},$$

$$\mu_{\text{not } L} = \left\{ \frac{0}{1} + \frac{0.2}{2} + \frac{0.4}{3} + \frac{0.6}{4} + \frac{0.8}{5} \right\}, \text{ and}$$

$$\mu_{\text{not very } L} = \left\{ \frac{0}{1} + \frac{0.36}{2} + \frac{0.64}{3} + \frac{0.84}{4} + \frac{0.96}{5} \right\}.$$

15.4.2 Membership Functions

Membership functions characterize the fuzziness in a fuzzy set in a graphical form for eventual use in the mathematical formulation of fuzzy systems. There are an infinite number of ways to graphically depict the membership functions. Thus, defining membership functions is subjective and context-dependent.

Standard membership functions that are commonly used are shown in Figure 15.6. These are functions of a set of parameters (a, b, c, and d) and can describe a large class of shape variations. Figure 15.6a and b define the trapezoidal and triangular functions, which are piece-wise membership functions; smooth continuous sigmoidal functions are defined in Figure 15.6c and d.

There are many ways to assign membership values or functions to fuzzy variables. This assignment can be intuitive, or it can be based on some algorithmic or logical operations. Automated methods for developing and tuning membership functions are discussed in Section 15.6.

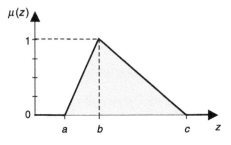

$$M_i(z) = \mu(z,a,b,c,d) = \begin{cases} 0, & z \leq a \\ \dfrac{z-a}{b-a}, & a \leq z \leq b \\ 1, & b \leq z \leq c \\ \dfrac{d-z}{d-c}, & c \leq z \leq d \\ 0, & z \geq d \end{cases}$$

(a) Trapezoidal-shaped membership function

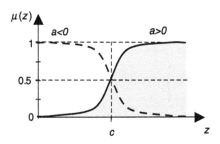

$$M_i(z) = \mu(z,a,b,c) = \begin{cases} 0, & z \leq a \\ \dfrac{z-a}{b-a}, & a \leq z \leq b \\ \dfrac{c-z}{c-b}, & b \leq z \leq c \\ 0, & z \leq d \end{cases}$$

(b) Triangular-shaped membership function

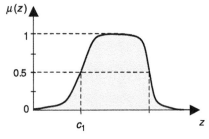

$$M_i(z) = \mu(z,a,c) = \dfrac{1}{1+e^{-a(z-c)}}$$

(c) Sigmoidally-shaped membership function

$$M_i(z) = \mu(z,a_1,c_1,a_2,c_2) = \dfrac{1}{1+e^{-a_1(z-c_1)}} - \dfrac{1}{1+e^{-a_2(z-c_2)}}$$

(d) Difference between two sigmoidal membership function

Figure 15.6 Shape and mathematical definition of standard membership functions: (a) trapezoidal, (b) triangular, (c) sigmoidal, and (d) difference of two sigmoidals.

15.4.3 Fuzzy Rule Base

In the field of artificial intelligence, there are various ways to represent human knowledge. Perhaps the most common way is the IF–THEN *rule-based* form. A *fuzzy* IF–THEN *rule* is a conditional statement expressed as:

$$\text{IF } [\textit{fuzzy proposition}] \text{ THEN } [\textit{fuzzy proposition}]. \tag{15.20}$$

The fuzzy proposition in the IF-part of the rule acts as condition (or conditions) to be satisfied, while the fuzzy proposition in the THEN-part provides the conclusion (or conclusions) of the rule. The propositions in the IF and THEN parts of the rule are known as *antecedents* and *consequents*, respectively.

There are two types of fuzzy proposition: atomic fuzzy propositions and compound fuzzy propositions. An *atomic fuzzy proposition* is a single statement of the form:

$$[\text{FP}] = \text{`}z \text{ is } L\text{'} \tag{15.21}$$

where z is a linguistic variable and L is a linguistic value of z. For example, propositions describing the throughput fuzzy variable defined in Figure 15.3 may be expressed as: "throughput is low" and "throughput is not medium." A *compound fuzzy proposition* is a composition of atomic fuzzy propositions using connectives "and" and "or" that represent fuzzy intersection and fuzzy union, respectively. Compound fuzzy propositions may be used to represent values of two or more fuzzy variables. Compound fuzzy propositions combining two fuzzy variables are expressed as:

$$[\text{FP}_{\text{and}}] = \text{`}z_1 \text{ is } L_1\text{'} \text{ and } \text{`}z_2 \text{ is } L_2\text{'}; \tag{15.22}$$

$$[\text{FP}_{\text{or}}] = \text{`}z_1 \text{ is } L_1\text{'} \text{ or } \text{`}z_2 \text{ is } L_2\text{'}. \tag{15.23}$$

Compound fuzzy propositions are understood as *fuzzy relations*. Their membership functions are obtained using the fuzzy set operations. The truth value of the propositions in Equation 15.22 and Equation 15.23 are obtained by combining the membership functions of the linguistic terms, L_1 and L_2, using the fuzzy intersection and union fuzzy operator given by Equation 15.10 and Equation 15.9, respectively. Hence,

$$T(\text{FP}_{\text{and}}) = \mu_{\text{FP}_{\text{and}}}(z_1, z_2) = \mu_{L_1 \cap L_2}(z_1, z_2) = \min(\mu_{L_1}(z_1), \mu_{L_2}(z_2)), \tag{15.24}$$

and

$$T(\text{FP}_{\text{or}}) = \mu_{\text{FP}_{\text{or}}}(z_1, z_2) = \mu_{L_1 \cup L_2}(z_1, z_2) = \max(\mu_{L_1}(z_1), \mu_{L_2}(z_2)). \tag{15.25}$$

15.4.3.1 *Aggregation of Fuzzy Rules*

The rule base of most fuzzy systems consists of more than one rule. A rule is said to be *fired* if it is satisfied. This means that the proposition in the antecedent (condition) part of the rule is true, which implies that the proposition in the consequent (conclusion) part of the rule is also true. But what if more than one rule in the rule base is fired? The process of obtaining the overall consequent from the individual consequents contributed by each rule in the rule base is known as *aggregation* of rules. Selecting an aggregation strategy depends on the type of satisfaction of a system of rules. There are two types: conjunctive and disjunctive systems of rules.[46] For the former the rules must be jointly satisfied while for the latter at least one rule is required. Consider, for instance, a system of r rules where the consequent of rule i is y^i. In the case of a conjunctive system of rules, an "and" connective is used, and the aggregated output, y, is determined by the fuzzy intersection of all individual rule consequents, y^i, as:

$$y = y^1 \cap y^2 \cap \cdots \cap y^r. \tag{15.26}$$

Using the fuzzy intersection operation (Equation 15.10), the overall membership function is thus defined as:

$$\mu_y(y) = \mu_{y^1}(y) \wedge \mu_{y^2}(y) \wedge \cdots \wedge \mu_{y^r}(y)$$
$$= \min(\mu_{y^1}(y), \mu_{y^2}(y), \ldots, \mu_{y^r}(y)) \tag{15.27}$$

For the case of a disjunctive system of rules, the rules are connected by the "or" connectives, and the aggregated output is determined by the fuzzy union of individual rule contributions:

$$y = y^1 \cup y^2 \cup \cdots \cup y^r \tag{15.28}$$

Using the fuzzy union operation (Equation 15.9), the overall membership function is defined as:

$$\mu_y(y) = \mu_{y^1}(y) \vee \mu_{y^2}(y) \vee \cdots \vee \mu_{y^r}(y)$$
$$= \max(\mu_{y^1}(y), \mu_{y^2}(x), \ldots, \mu_{y^r}(y)) \tag{15.29}$$

15.4.4 Fuzzy Inference

Due to the fact that fuzzy propositions are interpreted as fuzzy relations, the key task now is how to interpret the IF–THEN rule. To do this, let us first rewrite the statement in Equation 15.20 as:

$$\textbf{IF } [\text{FP}_1] \textbf{ THEN } [\text{FP}_2] \tag{15.30}$$

where the fuzzy propositions FP_1 and FP_2 constitute the antecedent and consequent of the fuzzy rule, respectively. In classical propositional calculus, the expression "IF p THEN q" is written as $p \rightarrow q$. The variables p and q are propositional variables whose values are either true (T) or false (F), while the implication, $\rightarrow$, is defined as a connective, such that if both p and q are true or false then $p \rightarrow q$ is true; if p is true and q is false then $p \rightarrow q$ is false; and if p is false and q is true then $p \rightarrow q$ is true.

Fuzzy implications may be considered as extensions of the classical implications and are usually referred to as *fuzzy inference*. Hence, the role of the inference engine (Figure 15.5) is to process the rules in the fuzzy rule base using a fuzzy implication procedure. Fuzzy inference is the process of formulating the mapping from a given input to an output using fuzzy logic. This mapping provides a basis from which decisions can be made. There are two types of inference systems that are commonly used in implementing fuzzy models: Mamdani-type and Sugeno-type inferences. These two types of inference systems vary somewhat in the way outputs are represented and determined.[47] For Mamdani-type inference, the rule in Equation 15.30 may be expressed as:

$$\textbf{IF } \text{'}z_1 \text{ is } L_1\text{'} \quad \text{and} \quad \text{'}z_2 \text{ is } L_2\text{'} \quad \textbf{THEN} \quad \text{'}y \text{ is } L_3\text{'} \tag{15.31}$$

where L_1, L_2, and L_3 are fuzzy sets. On the other hand, for the Sugeno-type inference, fuzzy rules are expressed as:

$$\textbf{IF } \text{'}z_1 \text{ is } L_1\text{'} \quad \text{and} \quad \text{'}z_2 \text{ is } L_2\text{'} \quad \textbf{THEN} \quad \text{'}y = g(z)\text{'} \tag{15.32}$$

It is obvious from Equation 15.31 and Equation 15.32 that the difference between both representations is in the way the consequent part of the rule (i.e., the proposition FP_2) is expressed. For a Mandani-type system, both the antecedents and consequents are defined in terms of fuzzy sets and membership functions. However, for a Sugeno-type system, the consequent part is a "crisply" defined function. This function can be constant or linear with respect to the crisp values of the input variables. Mamdani-type inference is the most widely used in fuzzy systems.[3] It will be described first, followed by the Sugeno-type inference.

15.4.4.1 Mamdani-Type Fuzzy Inference

Fuzzy systems that use Mamdani-type inference are referred to as "standard fuzzy systems." Applying this type of inference to the fuzzy rule in Equation 15.30 results in a fuzzy relation Q_M with a membership function $\mu_{Q_M}(z,y)$, which is a function of both the input and output fuzzy variables z and y. This indicates that this membership function is a combination of the membership functions representing the input and output variables. There are a number of implication techniques for obtaining the values of $\mu_{Q_M}(z,y)$. The two most widely used techniques are correlation-minimum and correlation-product implications. The first technique is also referred to as *Mamdani's implication* and defined as:

$$\mu_{Q_M}(z,y) = \min[\mu_{FP_1}(z), \mu_{FP_2}(y)]. \tag{15.33}$$

The correlation-product technique is defined as:

$$\mu_{Q_M}(z,y) = \mu_{FP_1}(z) \cdot \mu_{FP_2}(y). \tag{15.34}$$

The appropriate choice of an implication operator is typically context dependent. Although the implications defined by Equation 15.33 and Equation 15.34 are suitable for most applications, there are other implication techniques, such as:

$$\mu_{Q_M}(z, y) = \max[\mu_{FP_2}(y), 1 - \mu_{FP_1}(z)], \tag{15.35}$$

$$\mu_{Q_M}(z, y) = \min[1, [1 - \mu_{FP_1}(z) + \mu_{FP_2}(y)]], \tag{15.36}$$

and

$$\mu_{Q_M}(z, y) = \begin{cases} 1, & \mu_{FP_1}(z) \leq \mu_{FP_2}(y) \\ \mu_{FP_2}(y), & \text{otherwise} \end{cases}. \tag{15.37}$$

These other methods have been introduced as computationally effective under certain conditions of the membership values, $\mu_{FP_1}(z)$ and $\mu_{FP_2}(y)$. The implication technique defined by Equation 15.35 reduces to Mamdani's implication (Equation 15.33) when $\mu_{FP_1}(z) \geq 0.5$ and $\mu_{FP_2}(y) \geq 0.5$. The implications defined by Equation 15.36 and Equation 15.37 are known as *Lukasiewicz's implication* and *Brouwerian's implication*, respectively.[46]

To demonstrate the Mamdani-type inference, consider the following simple two-rule system where each rule comprises two conditions in the antecedent part and one conclusion in the consequent part:

$$\begin{aligned} &\textbf{IF } \text{'}z_1 \text{ is } L_{11}\text{'} \quad \text{and} \quad \text{'}z_2 \text{ is } L_{21}\text{'} \quad \textbf{THEN} \quad \text{'}y \text{ is } P_1\text{'} \\ &\textbf{IF } \text{'}z_1 \text{ is } L_{12}\text{'} \quad \text{and} \quad \text{'}z_2 \text{ is } L_{22}\text{'} \quad \textbf{THEN} \quad \text{'}y \text{ is } P_2\text{'} \end{aligned} \tag{15.38}$$

where L_{ij} is the jth linguistic term (fuzzy set) of the ith input variable and P_j is the jth linguistic term for the output variable y. The overall (aggregated) conclusion for these two rules is obtained by first combining the conditions in the antecedent part of the rules using the fuzzy intersection operator (Equation 15.24), followed by determining the conclusion of each rule using Mamdani's implication technique (Equation 15.33). Individual conclusions from each rule are then aggregated using the fuzzy union operator (Equation 15.29). This inference procedure is known as *max–min* method, which results in the following output membership function:

$$\mu_{Q_M}(z,y) = \max\left\{\begin{array}{l} \min\left[\min(\mu_{L_{11}}(z_1), \mu_{L_{21}}(z_2)), \mu_{P_1}(y)\right], \\ \min\left[\min(\mu_{L_{12}}(z_1), \mu_{L_{22}}(z_2)), \mu_{P_2}(y)\right] \end{array}\right\}. \qquad (15.39)$$

Applying the correlation-product implication defined by Equation 15.34 results in a similar inference procedure, known as a *max-product* method, which results in the following formulation for the output membership function:

$$\mu_{Q_M}(z,y) = \max\left\{\begin{array}{l} \min(\mu_{L_{11}}(z_1), \mu_{L_{21}}(z_2))\mu_{P_1}(y), \\ \min(\mu_{L_{12}}(z_1), \mu_{L_{22}}(z_2))\mu_{P_2}(y) \end{array}\right\}. \qquad (15.40)$$

A graphical representation illustrating the Mamdani-type inference of the two rules given by Equation 15.38 is shown in Figure 15.7. For each rule, the graphical inference in Figure 15.7 shows that the minimum function (Equation 15.24) is applied because the conditions in the antecedent are connected by a logical "and" connective. The minimum membership value for the antecedent propagates through to the consequent where the output membership function is truncated for the max–min method and scaled down for the max-product method. Then the resulting membership functions for each rule are aggregated using the graphical equivalence of Equation 15.29 for disjunctive rules. Hence, the resulted aggregated membership function is comprised of the outer envelopes of the individual truncated membership forms for each rule.

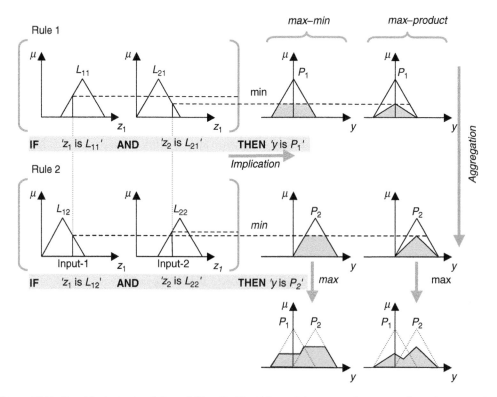

Figure 15.7 Graphical representation of Mamdani-type fuzzy inference using max–min and max-product implication methods.

15.4.4.2 Sugeno-Type Fuzzy Inference

A Takagi–Sugeno–Kang (TSK) fuzzy system is a special case of what is known as "functional fuzzy systems." The TSK approach was proposed in an effort to develop a systematic approach to generating fuzzy rules from a given input–output data set. For TSK systems, instead of a linguistic term with an associated membership function the consequent part of the fuzzy rule is a function $y = g(z)$, which is usually defined in terms of the crisp values of the input variables. Virtually any function can be used, which makes the functional fuzzy system very general.[28] The functional fuzzy system is referred to as a *first-order Sugeno model* if $g(z)$ is a linear function of input variables. For the TSK rule defined in Equation 15.32, $g(z)$ is given as:

$$y = g(z) = a_0 + a_1 z_1 + a_2 z_2, \tag{15.41}$$

where a_0, a_1, and a_2 are constants. If all coefficients are zeros except a_0, the $g(z)$ mapping is constant and the fuzzy system is called a *zero-order Sugeno model*, which is a special case of the Mamdani system in which each rule's consequent is specified as a fuzzy singleton.

To demonstrate the Sugeno-type inference, consider the following simple two-rule system, which is similar to the one in Equation 15.38 with an output y given as a zero-order Sugeno model:

$$\textbf{IF } \text{`}z_1 \text{ is } L_{11}\text{'} \quad \text{and} \quad \text{`}z_2 \text{ is } L_{21}\text{'} \quad \textbf{THEN } \text{`}y = C_1\text{'}$$
$$\textbf{IF } \text{`}z_1 \text{ is } L_{12}\text{'} \quad \text{and} \quad \text{`}z_2 \text{ is } L_{22}\text{'} \quad \textbf{THEN} \quad \text{`}y = C_2\text{'} \tag{15.42}$$

where C_1 and C_2 are constants. A graphical representation of the Sugeno-type inference for the two rules is shown in Figure 15.8. Values of the output variable are singleton spikes at C_1 and C_2. The implication of each rule is simply the multiplication of the membership values of the antecedent part of the rule with the singletons. This is, in fact, equivalent to taking the minimum between the resulting antecedent value and the singleton spike (Figure 15.8). Hence, for Sugeno-type inference, both the correlation-minimum and correlation-product implication methods are identical. Individual conclusions from each rule are aggregated using the fuzzy union operator (Equation 15.29), which results in simply including all scaled singletons. This implication procedure can be expressed mathematically as:

$$\mu_{Q_M}(z, y) = \max \left\{ \begin{array}{l} \min(\mu_{L_{11}}(z_1), \mu_{L_{21}}(z_2))C_1, \\ \min(\mu_{L_{12}}(z_1), \mu_{L_{22}}(z_2))C_2 \end{array} \right\}. \tag{15.43}$$

At this stage, the output of the fuzzy inference system has been obtained. But for both the Mamdani-type and Sugeno-type inference techniques, the fuzzy output is expressed as a fuzzy set with a defined membership function (see Equation 15.39, Equation 15.40, and Equation 15.43) as represented in Figure 15.7 and Figure 15.8. For this output to be useful, it should be converted to a crisp value so that it may be used by a human or by an instrument. This task is the subject of the next Section 15.4.5.

15.4.5 Fuzzification and Defuzzification

Most data used in engineering applications or measured by sensors are real-valued (crisp), such as temperature, flow rate, and pressure. On the other hand, for a fuzzy system to be useful, its output should also be crisp so that it can be readily used by other systems. For this reason, it is essential to construct an interface around the fuzzy inference engine. This interface is achieved through the fuzzifier and defuzzifier blocks shown in Figure 15.5. The role of the fuzzifier is to convert crisp input variables into fuzzy input variables. This action is known as *fuzzification*. Conversely, the role

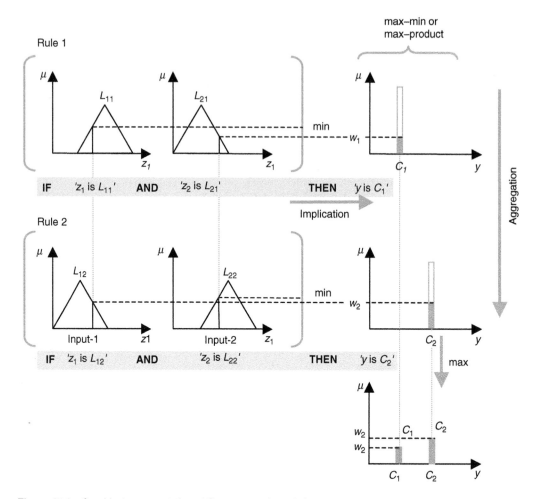

Figure 15.8 Graphical representation of Sugeno-type fuzzy inference.

of the defuzzifier is to convert the fuzzy output variables resulting from interpreting the fuzzy rules into crisp output values. This action is known as *defuzzification*.

Fuzzification is a mapping from a space of real crisp values to a space of a fuzzy set. In other words, it is a mapping between a real-valued input state vector $Z \in \mathcal{R}$ and a fuzzy input state vector $\underset{\sim}{Z} \in \underset{\sim}{A}$, where $\underset{\sim}{A}$ is a fuzzy set. Consequently, fuzzification is the process of converting numerical input variables into fuzzy linguistic variables defined by the quadruple in Equation 15.15 above. This means that each input variable should be associated with a set of linguistic terms defined as fuzzy sets with membership functions (Section 15.4.2).

On the other hand, defuzzification is the conversion of a fuzzy quantity to a precise quantity, just as fuzzification is the conversion of a precise quantity to a fuzzy quantity. It is a mapping from the space of a fuzzy set to a space of crisp values. Consequently, it is a mapping between a fuzzy output state vector $\underset{\sim}{Y} \in \underset{\sim}{A}$ and real-valued output state vector $Y \in \mathcal{R}$.

A number of defuzzification methods have been proposed in the literature.[4,28,46,48,49] Five methods will be described below. Their objective is to determine a crisp value of a defuzzified variable y* given an output fuzzy variable y that belongs to a fuzzy set $\underset{\sim}{A}$ defined in the universe U (where $\underset{\sim}{A}$ is combination of the fuzzy sets in which y is defined, i.e., $\underset{\sim}{A} = \underset{\sim}{A}_1 \cup \underset{\sim}{A}_2 \cup \cdots \cup \underset{\sim}{A}_N$). The membership function of a typical fuzzy output shown in Figure 15.9 will be used to demonstrate the computation of the different defuzzification methods described below.

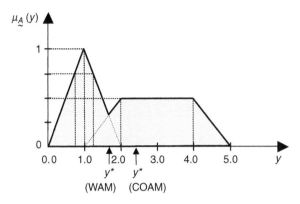

Figure 15.9 Defuzzification of the aggregated output membership function.

15.4.5.1 Center-of-Area Method

This method is also known as *center-of-gravity* or *centroid* method:

$$y^* = \frac{\int\limits_U y\mu_{\underset{\sim}{A}}(y)dy}{\int\limits_U \mu_{\underset{\sim}{A}}(y)dy}, \tag{15.44}$$

where $\int$ denotes an algebraic integration. Ross[46] noted that this method is the most prevalent and physically appealing of all the defuzzification methods. It may be considered the most accurate defuzzification method; however, it is computationally slow and difficult and poses some doubts when overlapping zones exist.[45] For the fuzzy output membership function shown in Figure 15.9, the COAM can be determined as:

$$y^* = \frac{\int\limits_0^1 y^2 \cdot dy + \int\limits_1^{4/3} y(2-y)\cdot dy + \int\limits_{4/3}^2 y\left(\frac{y-1}{2}\right)\cdot dy + \frac{1}{2}\int\limits_2^4 y\cdot dy + \int\limits_4^5 y\left(\frac{5-y}{2}\right)\cdot dy}{\int\limits_0^1 y\cdot dy + \int\limits_1^{4/3} (2-y)\cdot dy + \int\limits_{4/3}^2 \left(\frac{y-1}{2}\right)\cdot dy + \frac{1}{2}\int\limits_2^4 \cdot dy + \int\limits_4^5 \left(\frac{5-y}{2}\right)\cdot dy} = 2.275.$$

A practical and approximate way to solve Equation 15.44 is to divide the aggregated area of the output membership function into small segments. In this case, a summation operator replaces the integration operator, and Equation 15.44 becomes:

$$y^* = \frac{\sum\limits_{i=1}^N y_i\mu_{A_i}(y)}{\sum\limits_{i=1}^N \mu_{A_i}(y)}, \tag{15.45}$$

where N is the number of segments, y_i is the location of the segment, and $\mu_{A_i}(y)$ is the membership value at y_i. Approximation embedded in Equation 15.45 is fully justified when the membership function is defined as fuzzy sets comprising singletons. Otherwise, for continuous membership functions, the approximation accuracy increases with increasing number of segments.

Singleton output membership functions result mainly from the Sugeno-type inference systems (see Section 15.4.4). For such systems, the overall crisp output is obtained by using Equation 15.45 where y_i is the location of the ith singleton and $\mu_{A_i}(y)$ is its height. For instance, a crisp output may be simply obtained from the resulting singletons in Figure 15.8 by:

$$y^* = \frac{w_1 C_1 + w_2 C_2}{w_1 + w_2}. \tag{15.46}$$

15.4.5.2 Weighted-Average Method

The weighted-average method (WAM) is given by the expression:

$$y^* = \frac{\sum\limits_{i=1}^{N} \bar{y}_i \mu_{\underset{\sim}{A}i}(\bar{y}_i)}{\sum\limits_{i=1}^{N} \mu_{\underset{\sim}{A}i}(\bar{y}_i)}, \tag{15.47}$$

where Σ denotes an algebraic sum and N is the number of aggregated membership functions. This method is formed by weighting each membership function in the output by its respective maximum membership value. The WAM is normally used for symmetrical membership functions.[46] However, Ali and Zhang[45] used a method called the *weighted-section method* (WSM), which is identical to the WAM, and claimed that it is sensitive to unsymmetric membership functions. The same method is termed as the mean of maxima (MOM) method by Peña-Reyes.[4]

Defuzzification of the fuzzy output membership function in Figure 15.9 using WAM is performed as:

$$y^* = \frac{(1.0)(1.0) + (3.0)(0.5)}{1.0 + 0.5} = 1.67.$$

For a Mamdani-type fuzzy system, either COAM or WAM defuzzification techniques could be applied to the aggregated membership function, and a value such as y^*, shown in Figure 15.7 would result. On the other hand, for Sugeno-type systems, Equation 15.45 and Equation 15.47 are equivalent, and either one may be used to determine the weighted average of the singletons.

In fact, the COAM and WAM are the most commonly used defuzzification methods. Although the COAM is more computationally intensive and many consider it more accurate, others prefer the WAM due to its simplicity.

15.4.5.3 Max-Membership Principle

This defuzzification method is also known as the height method and is limited to peaked output functions. The defuzzified output y^* is given by the expression:

$$\mu_{\underset{\sim}{A}}(y^*) \geq \mu_{\underset{\sim}{A}}(y) \quad \forall \, y \in U. \tag{15.48}$$

For the fuzzy output function in Figure 15.9, y^* is simply the location of the maximum peak for which $\mu_{\underset{\sim}{A}}(y) = 1$ and, hence, $y^* = 1.0$.

15.4.5.4 Mean-Max Membership

This method is also known as *middle-of-maxima*. It is identical to the max-membership principle (Equation 15.48) except that the locations of the maximum membership can be a plateau rather than a single point. In this case, y^* would be determined by finding the mean of the locations of the maximum plateau. That is,

$$y^* = \frac{a+b}{2}, \tag{15.49}$$

where a and b are the minimum and maximum limits of the maximum plateau. For instance, if the output membership function in Figure 15.9 is chopped at $\mu(y)=0.75$ then $a=0.75$ and $b=1.25$ while $y^*=(0.75+1.25)/2=1.0$.

15.4.5.5 Center of Sums

This method involves the algebraic sum of individual output fuzzy sets, instead of their union. The main advantage of this method is that it is not restricted to symmetric membership functions. However, there are two drawbacks: the intersecting areas are added twice, and the method involves finding the centroids of the individual membership functions. The defuzzified value y^* is obtained by the following equation:

$$y^* = \frac{\sum\limits_{i=1}^{N} \bar{y}_i \int\limits_{U} \mu_{A_i}(y)\,dy}{\sum\limits_{i=1}^{N} \int\limits_{U} \mu_{A_i}(y)\,dy}, \tag{15.50}$$

where $\bar{y}_i$ is the centroid of the ith membership function, while the integral represents the area of the individual membership functions. The center of sums method is similar to the WAM, except that the weights used are the areas of the respective membership functions, whereas in the WAM the weights are individual membership values.

For the output membership function in Figure 15.9, the centroid of the two membership functions are at $\bar{y}_1 = 1$ and $\bar{y}_2 = 3$. Hence, y^* can be obtained as:

$$y^* = \frac{1\times(0.5\times2\times1) + 3\times(0.5\times(4+2)\times0.5)}{(0.5\times2\times1) + (0.5\times(4+2)\times0.5)} = 2.2.$$

After describing the defuzzification methods, a natural question to ask is "Which method should be selected?" The answer to this question is that the defuzzification method is context- or problem-dependent.[46] As with other issues in fuzzy systems, the method of defuzzification should be assessed in terms of the correctness of the answer in the context of the data available. Hellendoorn and Thomas[50] have specified the following criteria against which to assess the suitability of the defuzzification methods:

1. *Continuity*: a small change in the input of a fuzzy model should not produce large change in the output. This criterion is specifically important for fuzzy models used in control applications.
2. *Disambiguity*: the defuzzification method should always result in a unique value.
3. *Plausibility*: the defuzzified output should lie approximately in the middle of the support region and should exhibit a high degree of membership. For our example, the COAM resulted in an output close to the middle of the region, while that of the WAM lay in the first quarter of the region. However, the WAM value is very close to the region with maximum membership values compared with the COAM value. This is, in fact, a good indication that selection of the defuzzification methods is problem-dependent.

4. *Computational simplicity*: less time-consuming methods are favored. It is clear that the COAM is more computationally intensive than the WAM.

15.5 EXAMPLE: FUZZY MODELING OF A BREAD EXTRUSION PROCESS

Application of neural networks and fuzzy logic in the modeling and control of extrusion cooking was attempted by Linko and coworkers.[51] They reported input–output data for extruded flat bread produced with a twin-screw extruder. They reported 15 experimental data sets with three inputs and four outputs. The input variables were *moisture content of the dough, feed rate*, and *screw speed*. The output variables were *product expansion rate, main motor current, pressure at the die plate*, and *bulk density*.

The aim of the current example is to demonstrate the process of building a fuzzy system to model the bread extrusion process. Product expansion rate, E, has been selected as an output variable. Experimental data adopted from Linko et al.[51] is listed in Table 15.1. The fuzzy model that will be developed here has two input variables and one output variable. The selected input variables are moisture content, M, of the dough and speed of the screw, S. The third input variable will be used later in other examples.

First, the universe of discourse for each variable is defined: $M \in [17,21]$, $S \in [200,300]$, and $E \in [4.1,5.6]$. This is followed by defining the fuzzy sets and membership functions. Three linguistic terms (fuzzy sets) will be defined for each input variable: $T_M = \{$"dry,""normal," "wet"$\}$, and $T_S = \{$"slow,""normal,""fast"$\}$. The membership functions for M and S are intuitively defined, as plotted in Figure 15.10a and b, respectively.

15.5.1 Mamdani-Type System

For a Mamadani-type fuzzy system, the fuzzy sets and membership functions for the output variable must also be defined. The linguistic terms assigned to the expansion rate are $E = \{low,$

Table 15.1 Experimental Data for Bread Extrusion Example

	Inputs		Output
Moisture M (%)	Screw Speed S (rpm)	Feed F (g/min)	Expansion E (%)
17.52	213	113	5.3
20.48	213	113	4.8
17.52	213	187	5.0
20.48	213	187	4.6
17.52	287	113	5.5
20.48	287	113	4.6
17.52	287	187	5.1
20.48	287	187	4.6
17.00	250	150	5.6
21.00	250	150	4.1
19.00	250	100	5.0
19.00	250	200	4.8
19.00	200	150	4.9
19.00	300	150	4.7
19.00	250	150	5.1

Source: Linko, P., Uemura, K., and Eerikainen, T., *Food Engineering in a Computer Climate*, IChemE, Symposium Series No. 126, EPCE Event No. 452, Rugby, UK, 401, 1992.

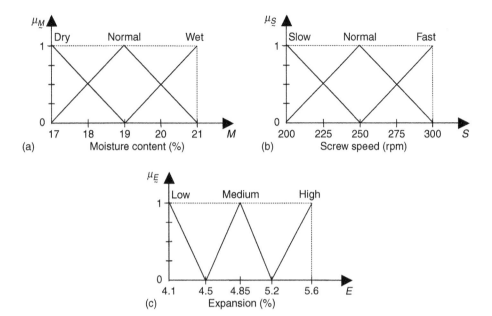

Figure 15.10 Membership functions representing the (a) moisture content, (b) screw speed, and (c) product expansion rate, for the bread extrusion example.

medium, high}, and the membership functions are plotted in Figure 15.10c. The task now is to develop the rule base.

The rule base is normally constructed in close cooperation with human experts. In this case, where the rules are extracted from experimental data, a practical way to construct the rule base is to use the input–output matrix form shown in Figure 15.11. The linguistic terms describing the output fuzzy variable, E, were assigned based on the numerical data given in Table 15.1. One fuzzy rule is developed for each combination of the input variables. Hence, the rule base would consist of nine rules. However, Figure 15.11 shows that the expansion rate, E, is "high" for all values of screw speed, S, when the moisture, M, is "dry," and E is "medium" for all values of S when M is "normal." This reduces the rule-base to the following four rules:

Rule 1:	IF	M is dry			THEN E is high	
Rule 2:	IF	M is normal			THEN E is medium	
Rule 3:	IF	M is wet	AND	S is slow	THEN E is medium	(15.51)
Rule 4:	IF	M is wet	AND	S is normal	THEN E is low	
Rule 5:	IF	M is wet	AND	S is slow	THEN E is medium	

These rules will be processed using the Mamdani-type inference for the crisp inputs $M=20.48\%$ and $S=213$ rpm. The inference mechanism is represented graphically in Figure 15.12 for both the max–min and the max-product methods. The fuzzy model predicted an expansion rate as $E=4.72\%$ for the max–min method and 4.74% for the max-product method. The inference steps can be summarized as follows:

1. Fuzzifying the crisp inputs results in M as "normal" with $\mu_{normal}(M)=0.26$ and "wet" with $\mu_{wet}(M)=0.74$, and S as "slow" with $\mu_{slow}(S)=0.74$ and "normal" with $\mu_{normal}(S)=0.26$.

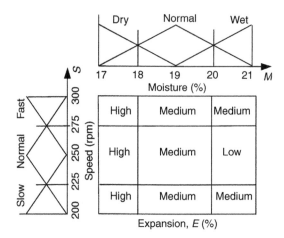

Figure 15.11 Input–output matrix for developing the fuzzy rule-base for the bread extrusion example.

2. The fuzzy values of the input variables triggered all rules except rule 1. However, rule 5 did not fire because "S is fast" is not a true proposition. Therefore, as illustrated by Figure 15.12, rules 2, 3, and 4 were fired.

3. Correlation-minimum implication was used for the max–min method and correlation-product implication was used for the max-product method. Both implications of the fired rules resulted in three fuzzy values for E, each with a defined membership function.

4. The three individual fuzzy outputs were aggregated using the fuzzy union operator (Equation 15.29).

5. COAM defuzzification technique (Equation 15.44) was applied on the aggregated membership functions to determine the crisp value of E.

Predictions of the fuzzy model are compared with experimental data in Table 15.2. The average absolute error for both the max–min and max-product is 2.8%, which is quite low even though only intuition was used in developing the membership functions and the rule-base. It is worth

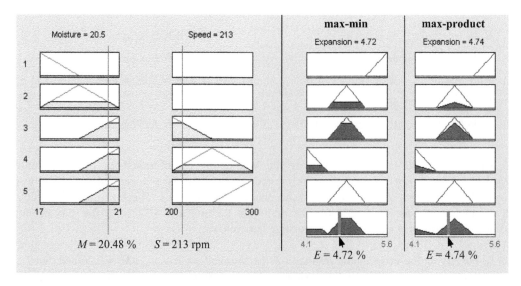

Figure 15.12 Graphical Mamdani-type inference for the bread extrusion example.

Table 15.2 Prediction Results of the Bread Extrusion Fuzzy System Using Mamdani and Sugeno Inferences

Inputs		Expansion E (%)			
			Mamdani Inference		
Moisture M (%)	Screw Speed S (rpm)	Exp.	Max–min	Max-prod	Sugeno Inference
17.52	213	5.3	5.19	5.24	5.37
20.48	213	4.8	4.72	4.74	4.61
17.52	213	5.0	5.19	5.24	5.37
20.48	213	4.6	4.72	4.74	4.61
17.52	287	5.5	5.19	5.24	5.37
20.48	287	4.6	4.72	4.74	4.66
17.52	287	5.1	5.19	5.24	5.37
20.48	287	4.6	4.72	4.74	4.66
17.00	250	5.6	5.47	5.47	5.50
21.00	250	4.1	4.23	4.23	4.10
19.00	250	5.0	4.85	4.85	5.00
19.00	250	4.8	4.85	4.85	5.00
19.00	200	4.9	4.85	4.85	5.00
19.00	300	4.7	4.85	4.85	5.00
19.00	250	5.1	4.85	4.85	5.00
Error (%)		Minimum	1.0	1.0	0.0
		Maximum	5.6	4.9	7.4
		Average	2.8	2.8	2.6

mentioning that a number of input data sets are repeated because the third input variable, F, is not considered in this example. Moreover, the model predicted the same value of the output for most of the values of M (Table 15.2). This is due to the fact that Rule 1 and Rule 2 included only M in their antecedent part.

15.5.2 Sugeno-Type System

A Sugeno-type system for the bread extrusion example is implemented in this section. The system assumes that the function in the consequent part of the fuzzy rules is a constant. The rule base consists of the following five rules:

Rule 1:	IF	M is dry			THEN $E = 5.5$	
Rule 2:	IF	M is normal			THEN $E = 5.0$	
Rule 3:	IF	M is wet	AND	S is slow	THEN $E = 4.6$	(15.52)
Rule 4:	IF	M is wet	AND	S is normal	THEN $E = 4.1$	
Rule 5:	IF	M is wet	AND	S is fast	THEN $E = 4.7$	

The constants in the consequent part of the rules are derived from the experimental data listed in Table 15.1. For the crisp inputs $M=20.48\%$ and $S=213$ rpm, the Sugeno-type model predicted the expansion rate as $E=4.61\%$. The inference mechanism is represented graphically in Figure 15.13. It starts with steps 1 and 2, described above, followed by rule implication, which is carried out using the correlation-product method. For each rule, the implication produces a singleton. The singletons

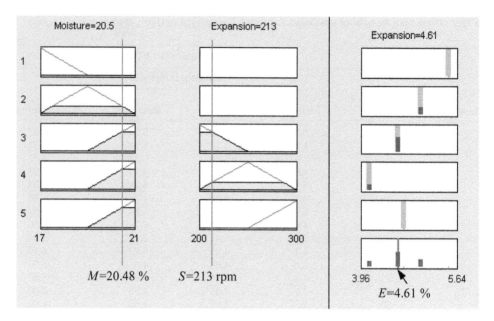

Figure 15.13   Graphical Sugeno-type inference for the bread extrusion example.

are then aggregated by simply adding them in one plot. Finally, the WAM defuzzification method (Equation 15.47) is used to determine the value of the crisp output.

As reported in Table 15.2, the Sugeno-type fuzzy model predicted the experimental data quite closely, with an average absolute error of only 2.6%.

15.5.3   Implementing the Fuzzy Model in MATLAB®

The *Fuzzy Logic Toolbox*[52] consists of a collection of functions built on the MATLAB numeric computing environment.[53] It provides useful tools for creating and editing fuzzy inference systems and a number of interactive tools for accessing a number of the functions through a graphical user interface (GUI). There are five primary GUI tools: the fuzzy inference system (FIS) editor, the membership function editor, the rule editor, the rule viewer, and the surface viewer.

In this section, we will implement a fuzzy system for the bread extrusion example discussed above using the fuzzy logic toolbox. The implementation starts with the FIS editor that displays general information about the fuzzy system. In this editor, we specify the input and output variables and the main parameters of the inference mechanism. This includes the type of inference (Mamdani or Sugeno), setting the fuzzy operations, rule implication, aggregation and defuzzification methods. The FIS editor for the bread extrusion example is shown in Figure 15.14.

The next step is to define the membership functions for the input and output variables. The shape and parameters of the transfer functions associated with each variable are constructed using an interactive GUI. An example membership function editor displaying the speed of the screw input variable is shown in Figure 15.15. The rule-base is then constructed using the rule editor, which allows the fuzzy rule statements to be constructed automatically. The rule editor for our example is shown in Figure 15.16.

At this stage, the implementation of the fuzzy system for the bread extrusion example has been completed. There are two useful tools for testing and running the developed fuzzy model. The first tool is the rule viewer, which displays a roadmap of the whole fuzzy inference process. An example rule viewer window is shown in Figure 15.17. This tool can be used to determine the value of the output variable for given values of input variables. The second tool is the surface viewer, which

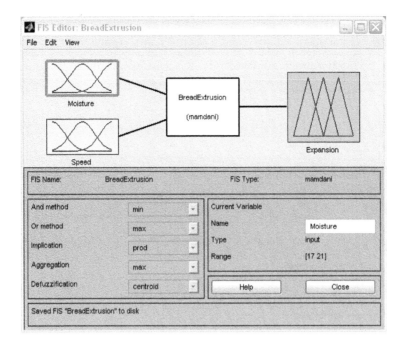

Figure 15.14   Fuzzy inference editor for the bead extrusion fuzzy system.

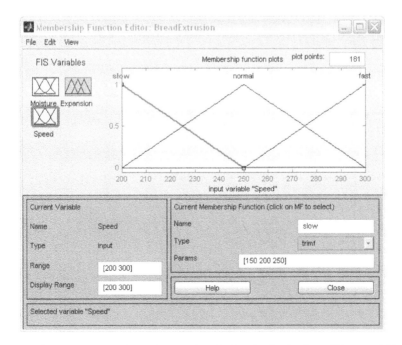

Figure 15.15   Membership editor displaying the membership function for the "speed" input variable.

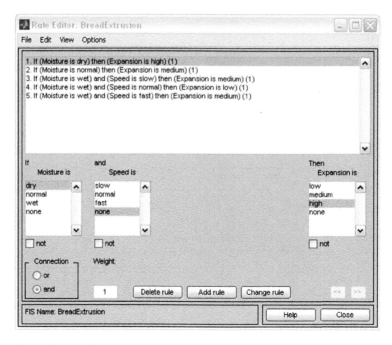

Figure 15.16 Rule editor used for constructing the rule base.

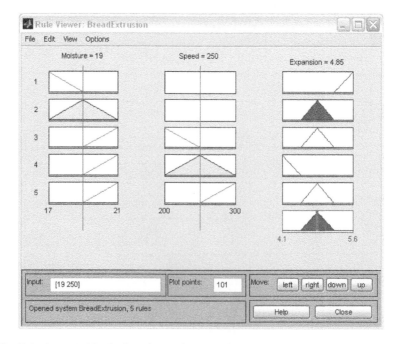

Figure 15.17 Rule viewer tool for the bread extrusion example.

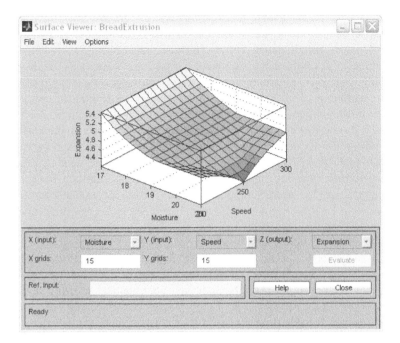

Figure 15.18 Surface viewer tool displaying the mapping from input to output variables for the bread extrusion example.

presents a three-dimensional curve that represents the mapping from the two input variables (moisture content and speed) to the output variable (expansion rate). The surface viewer for our example is shown in Figure 15.18.

15.6 DEVELOPING AND TUNING MEMBERSHIP FUNCTIONS

Because all information contained in a fuzzy set is described by its membership function that embodies all its fuzziness, its description is the essence of a fuzzy property or operation. Nevertheless, a question that is usually raised is whether there is a unique and well-defined method for specifying membership functions, similar to writing a set of differential equations for an engineering problem. From an engineering standpoint, assigning a membership function is to take into account effects of uncertainty. Hence, demanding precise specifications would contradict the whole concept of fuzzy modeling. According to Bandemer and Gottwald[54] the choice of the type and the precise values for the parameters have, in general, little influence on the results, decisions, and conclusions obtained as long as a local monotonicity is preserved. For two different membership specifications, μ_1 and μ_2, local monotonicity means:

$$\forall z_1, z_2 \in U : \mu_1(z_2) \leq \mu_1(z_1) \Leftrightarrow \mu_2(z_2) \leq \mu_2(z_1). \tag{15.53}$$

Intuition is the easiest method for defining membership functions. It is simply derived from the capacity of humans to develop membership functions through their own innate intelligence and understanding. This method has been used in developing the membership functions for the bread extrusion fuzzy system developed above. It is remarkable that good results were obtained and prediction errors were consistently small. Besides the intuition method, Ross[46] listed four methods for assigning and tuning membership functions. These include (1) inference, (2) rank ordering, (3) neural networks, (4) genetic algorithm, and (5) inductive reasoning.

In cases where experimental data or practical observations are available, there are a number of methods to develop or tune membership functions. These involve, more or less, an optimization technique that minimizes the differences between the available data and the predictions of the fuzzy system. In addition to developing the membership functions, there is an increasing interest in augmenting fuzzy systems with learning capabilities. The most successful approach is to hybridize the approximate reasoning of fuzzy systems with the learning capabilities of neural networks and the evolutionary power of genetic algorithms. Integration of neural networks and genetic algorithms with fuzzy systems cover different levels of complexity, from the simplest case of parameter optimization to the highest level of complexity of learning the fuzzy rules in the rule-base.[4,28,46]

It is important to distinguish between tuning and learning problems. Tuning is more concerned with the optimization of an existing fuzzy system, whereas learning constitutes an automated design method for fuzzy systems that starts from scratch. Tuning processes assume a predefined rule base and have the objective of finding a set of optimal parameters for the membership functions. Learning processes perform more elaborate searches and aim at developing the rule base and membership functions from input–output data.

Artificial neural networks, genetic algorithms, and fuzzy logic belong to the same family of bio-inspired methodologies.[4] In fact, they model, to different extents, natural processes such as evolution, learning, or reasoning. The dynamic and continuously growing research on these subjects have allowed for the identification of the strengths and weaknesses of each methodology and have motivated their hybridization in order to take advantage of their complementary features. This originated hybrid techniques known as *neuro-fuzzy systems* and *genetic fuzzy modeling*. Integrating neural networks and genetic algorithms with fuzzy systems is the subject of the remainder of this chapter. For each approach, a brief background will first be presented, followed by a description as to how they are integrated with fuzzy systems.

15.6.1   Hybrid Neural-Fuzzy Systems

Derived from their biological counterparts, artificial neural networks (ANNs) are based on the concept that a highly interconnected system of simple processing elements can learn complex interrelationships between independent and dependent variables. Basic concepts and background on ANNs are introduced in Chapter 12 of this handbook. In this section, an explanation is given of how neural network capabilities are utilized to determine and tune the membership functions of fuzzy systems. First, a brief introduction to neural networks is given.

15.6.1.1   Artificial Neural Networks

An ANN consists of an interconnection of basic neuron processing elements. Each element computes some function of its inputs and passes the results to connected elements (neurons) in the network. The knowledge of the system comes out of the entire network of the neurons. Figure 15.19 shows a common structure of a neural network with n input variables and m output variables. Input variables are passed to the hidden layer that contains p neurons. ANNs may have more than one hidden layer. The ANN shown in Figure 15.19 is a feed-forward network. There are other types of networks that differ in the manner in which neurons are interconnected. These include feedback, lateral, time-delayed, and recurrent networks.[55–57] Each ANN type has its own specific applications; however, feed-forward networks are the most popular and are commonly used with fuzzy systems.

ANNs solve problems by adapting to the nature of available input–output data. This is accomplished in two phases: learning and testing. In learning, a set of training data is fed into the network to determine its parameters, weights, and thresholds. It often involves optimizing some energy function. In testing, another set of data is used to test the generalization of a trained ANN.

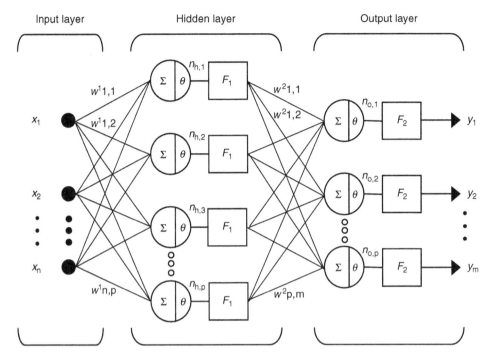

Figure 15.19 A general structure of an artificial neural network.

To train an ANN, weights $w^i_{j,k}$ are randomly assigned to the connections between neurons. Then, input data records from the training data set are passed through the neural network to compute the values of the output variables. Predicted output values are compared with the actual values, and an error is evaluated. Next, a technique called *back-propagation* is used to determine a new set of weights that can approximate the output more closely. This procedure is iterated until the error value of the output is within some prescribed limits. Finally, a testing-data set is used to verify how well the network can simulate the nonlinear model.

15.6.1.2 *Integrating Neural Networks and Fuzzy Systems*

Neural networks provide learning capability, whereas fuzzy methods provide flexible knowledge representation capability. Integrating these two methodologies can lead to better capabilities that take advantage of the strengths of each methodology and at the same time overcome some of the limitations of the individual ones. There are basically two ways that fuzzy logic and neural networks can be combined. In the first way, fuzzy logic is introduced into neural networks to enhance their knowledge representation capabilities. This leads to a *fuzzy-neural system* in which fuzzy concepts are introduced within neural networks at the levels of inputs, weights, aggregation operations, activation functions, and outputs. The second way is using neural networks in fuzzy modeling to provide fuzzy systems with learning capabilities. This leads to a *neural-fuzzy system* (also known as a *neuro-fuzzy system*), in which a fuzzy system is represented as a modified neural network, resulting in a fuzzy inference system that is enhanced by neural-network capabilities. Among the various neural-fuzzy systems reported in literature, an adaptive network fuzzy system (ANFIS) is the most affective and has spawned numerous applications.[58] ANFIS is discussed later in this section.

Although fuzzy logic can encode expert knowledge directly using rules with linguistic labels, designing and tuning the membership function is usually not an easy task. Moreover, useful applications of fuzzy modeling are restricted mainly to those fields where expert knowledge is available and the number of input variables is small. Neural-network learning techniques can automate this process and substantially reduce development time and cost while improving performance. Neural networks are also utilized to process data and to tune membership functions of fuzzy systems as well as to extract fuzzy rules from numerical data.

Fuzzy logic concepts can be incorporated into a neural-network structure at any level. For instance, some or all components of a neuron can be replaced with fuzzy operations. This results in a *fuzzy neuron* that replaces summation operator by the fuzzy aggregation operator. A *min-fuzzy neuron* applies a fuzzy intersection operator, whereas a *max-fuzzy neuron* applies a fuzzy union operator. A neural network with fuzzy neurons becomes a *multi-layer fuzzy-neural network*.

Three main areas may be identified in hybrid neural-fuzzy systems:[3]

1. Fuzzy-rule extraction from neural networks. This approach attempts to extract the fuzzy rules from the knowledge embedded in trained neural networks.[59,60]
2. Neuro-fuzzy systems. These are fuzzy inference systems implemented as neural networks, taking advantage of their structural similarity. ANFIS is a well-known neuro-fuzzy system that will be discussed in more details in Section 15.6.1.3.
3. Interpretability-oriented neuro-fuzzy systems. A recent family of neuro-fuzzy systems is constructed respecting certain interpretability-related constraints to keep permanent readability of the system during the learning process.[4] Examples of such systems include NEFCON and NEFPROX,[61] which are based on a three-layer neuro-fuzzy architecture whose weights are constraints to respect the integrity of the fuzzy linguistic variables.

15.6.1.3 Adaptive Network Fuzzy Inference Systems

Adaptive Network Fuzzy Inference Systems (ANFIS) is an adaptive neuro-fuzzy inference system proposed by Jang in 1993.[62] The basic idea behind ANFIS is to provide a method for the fuzzy modeling procedure to learn information about a data set in order to determine the parameters of the membership functions that best allow the associated fuzzy inference system to track the given input–output data with minimum error. Such hybrid systems can be optimized via powerful, well-known neural-network learning algorithms. An ANFIS consists of a six-layer generalized network with supervised learning. The structure and inference mechanism of ANFIS will first be demonstrated on two simple rules, followed by an illustration of the implementation of such systems in MATLAB.

To demonstrate the process of developing a neuro-fuzzy system, consider the two Sugeno rules represented by Equation 15.42. Recall that the implication of these rules is given by Equation 15.43, whereas the defuzzified output y^* is evaluated using Equation 15.46. The parameters, w_1 and w_2 in Equation 15.46 are the implication results of the rules. After all, when a crisp input $z = (z_1, z_2)$ is presented, the inference mechanism should produce the crisp output:

$$y^* = \frac{\min(\mu_{L_{11}}(z_1), \mu_{L_{21}}(z_2)) \cdot C_1 + \min(\mu_{L_{12}}(z_1), \mu_{L_{22}}(z_2)) \cdot C_2}{\min(\mu_{L_{11}}(z_1), \mu_{L_{21}}(z_2)) + \min(\mu_{L_{12}}(z_1), \mu_{L_{22}}(z_2))}. \tag{15.54}$$

A fuzzy-neural network for implementing this mechanism is shown in Figure 15.20. The network consists of six layers: an input layer, an output layer, and four hidden layers. The first hidden layer is the *fuzzification layer*. The input to this layer is a set of crisp input variables $z = (z_1, z_2)$, and the output is the corresponding set of fuzzy variables:

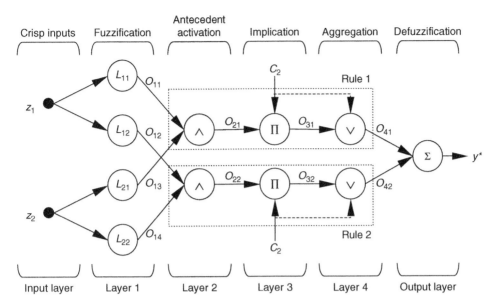

Figure 15.20 Graphical representation of an ANFIS network for two fuzzy rules.

$$(O_{11},O_{12},O_{13},O_{14}) = (\mu_{L_{11}}(z_1),\mu_{L_{12}}(z_1),\mu_{L_{21}}(z_2),\mu_{L_{22}}(z_2)). \tag{15.55}$$

The four neurons in this layer represent the membership functions L_{11}, L_{12}, L_{21}, and L_{22}, which are specified in some parametric manner from a family of membership functions, such as those shown in Figure 15.6. The outputs from layer 1 are fed to layer 2, which is an *antecedent activation* layer. This layer consists of two min-fuzzy neurons that apply a fuzzy intersection operator (Equation 15.24). The output of layer 2 is

$$(O_{21},O_{22}) = (\min(\mu_{L_{11}}(z_1),\mu_{L_{21}}(z_2)),\min(\mu_{L_{12}}(z_1),\mu_{L_{22}}(z_2))). \tag{15.56}$$

A rule is fired if its antecedent membership value is greater than zero. This value is forwarded to layer 3, which is an *implication* layer. This layer performs rule implications by applying the correlation-product operator (Equation 15.34). The outputs of layer 3 are, in fact, two singletons resulted from implication of the two rules:

$$(O_{31},O_{32}) = (\min(\mu_{L_{11}}(z_1),\mu_{L_{21}}(z_2))\cdot C_1,\min(\mu_{L_{12}}(z_1),\mu_{L_{22}}(z_2))\cdot C_2). \tag{15.57}$$

The last hidden layer is an aggregation layer. Layer 4 aggregates the resulted singletons, and its output is

$$(O_{41},O_{42}) = \left(\frac{\min(\mu_{L_{11}}(z_1),\mu_{L_{21}}(z_2))\cdot C_1}{\min(\mu_{L_{11}}(z_1),\mu_{L_{21}}(z_2)) + \min(\mu_{L_{12}}(z_1),\mu_{L_{22}}(z_2))} \atop \frac{\min(\mu_{L_{12}}(z_1),\mu_{L_{22}}(z_2))\cdot C_2}{\min(\mu_{L_{11}}(z_1),\mu_{L_{21}}(z_2)) + \min(\mu_{L_{12}}(z_1),\mu_{L_{22}}(z_2))} \right). \tag{15.58}$$

Finally, the output layer is a *defuzzification* layer that determines the crisp value of the output variable by simply summing the inputs:

$$y^* = (O_{41}+O_{42}) = \frac{\min(\mu_{L_{11}}(z_1),\mu_{L_{21}}(z_2))\cdot C_1 + \min(\mu_{L_{12}}(z_1),\mu_{L_{22}}(z_2))\cdot C_2}{\min(\mu_{L_{11}}(z_1),\mu_{L_{21}}(z_2)) + \min(\mu_{L_{12}}(z_1),\mu_{L_{22}}(z_2))}. \tag{15.59}$$

The procedure outlined above, and the graphical representation shown in Figure 15.20 can be extended to any number of rules. An efficient learning algorithm is needed for this representation to be more useful. In conventional neural networks, the back-propagation algorithm is effectively used to learn or adjust the weights. In ANFIS, the parameters of the antecedents and consequents of the fuzzy rules play the role of weights. The parameters used in specifying the membership functions in the IF-part are known as *antecedent parameters*, whereas the parameters used in defining the functions $g(z)$ in the THEN-part are known as *consequent parameters*. The objective of ANFIS learning algorithms is to find the best values of these parameters from sample data. A detailed description of the training algorithms is beyond the scope of this chapter and can be found elsewhere.[58] In Section 15.6.1.4, the development of ANFIS neuro-fuzzy system will be illustrated using the Fuzzy Logic Toolbox of MATLAB.

15.6.1.4 Developing ANFIS in MATLAB: Bread Extrusion Problem

There exist some modeling problems in which the intuitive method is not sufficient. In such cases, it is difficult to discern the shape of the membership functions and to predict their appropriate parameters. Rather than choosing the parameters associated with membership functions arbitrarily, such parameters may be chosen so that they are tailored to a set of input–output data. In this section, the use of neuro-fuzzy techniques, in particular ANFIS, will be demonstrated in both tuning as well as developing membership functions. This will be achieved by using the *ANFIS tool* that is part of the Fuzzy Logic Toolbox of MATLAB.[52] Using a given input–output data set, this tool enables the construction of a fuzzy inference system (FIS) whose membership function parameters are tuned using either a back-propagation algorithm alone or in combination with a least-squares type of method.

The fuzzy systems that were developed in Section 15.5 for modeling the bread extrusion process are based mainly on intuition. In this section, the same problem is revisited to demonstrate how neural networks may be utilized to enhance the performance and prediction accuracy of fuzzy systems. Three FIS models will be developed and compared with the Sugeno model developed in Section 15.5. The results of this model are presented in Table 15.2 and will be referred to as *model-0*. The input–output data set listed in Table 15.1 will be used in training the ANFIS networks for each model.

For the first model (*model-1*), the ANFIS tool will be used to tune the membership and rule-base parameters of *model-0*, whereas for the second model ("*model-2*") a new Sugeno FIS will be developed from scratch without prior knowledge of these parameters. So far, the three models are for two input variables, M and S, and one output variable, E. For this reason, the third model (*model-3*) will be developed for three inputs and one output. The additional input variable is the feed flow rate, F, which is included in Table 15.1.

The front end of the ANFIS tool is an "Anfis Editor" (Figure 15.21), which may be activated using the *anfisedit* command. The lower half of the editor contains means for loading the input–output data, in addition to constructing, training, and testing fuzzy systems. The upper half provides a display for several pieces of useful information. For instance, the dialog shown in Figure 15.21 displays a comparison between the training and testing data points for *model-1*.

Development of *model-1* starts by loading the input–output data set, followed by loading the FIS model that was previously developed for *model-0*. The same data set has been used as both training and testing data sets. Tuning of the parameters of *model-1* has been performed by training an automatically developed ANFIS, using 50 epochs. The resulted training error is 0.142. Predictions of *model-1* are compared with the training data set in Figure 15.21.

For *model-2*, after loading the data set, the type of the membership functions are specified using the dialog shown in Figure 15.22. For these applications, triangular membership functions are used, which are defined in MATLAB as "trimf." Training is then performed for the network structure shown in Figure 15.23, which is automatically generated. Using 50 epochs, the training error for *model-2* is 0.113. It is clear from Figure 15.23 that the rule base of the fuzzy system that has been

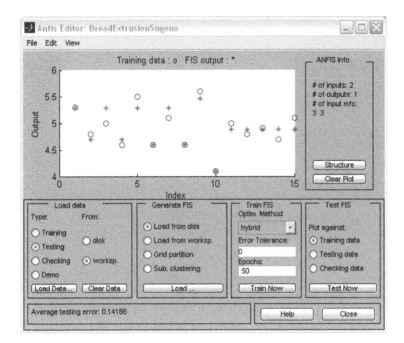

Figure 15.21 GUI for the ANFIS tool for the bread extrusion problem (*model-1*).

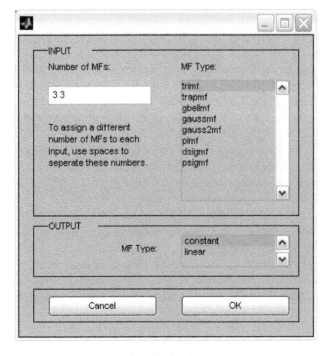

Figure 15.22 Specifying the shape of the membership functions.

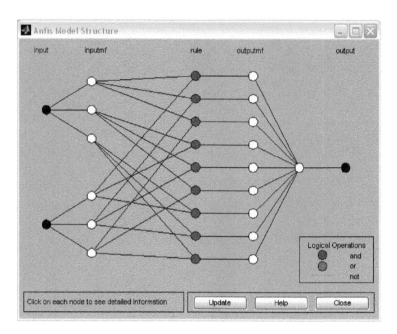

Figure 15.23 Automatically generated ANFIS structure for the bread extrusion model.

generated by ANFIS consists of nine rules, whereas only five rules are used by all models developed above.

Finally, *model-3* is developed by first loading the data set for three inputs (M, S, and F) and one output (E) and specifying three triangular membership functions for each input variable. Results of the ANFIS training are shown in Figure 15.24. After 60 epochs, the training error approaches zero, which means that *model-3* may have achieved a perfect fit.

Prediction results of different models are compared in Table 15.3 against the experimental data set.

Despite the fact that *model-0* gave acceptable prediction accuracy, the other models that were developed by ANFIS performed much better. For *model-1* and *model-2*, the prediction error went down to 2.2 and 1.8, respectively. However, *model-3* showed a perfect fit. The following comments may be derived from the analysis of the results and the structure of the resulting models:

- The shapes of the membership functions for all FIS models are comparable to those shown in Figure 15.10a and b, and small variations were noticed in the parameters of the tuned and developed membership functions.
- For *model-0* and *model-1*, the rule bases were forced to have five fuzzy rules (see the rules in Equation 15.52). However, ANFIS training for *model-2* resulted in a rule base with nine fuzzy rules. For this reason, *model-2* showed better performance when compared with the other two models.
- Discrepancies between the various developed FIS models are mainly due to differences in the rule bases. Besides the number of rules, considerable differences were found in the values of the $g(z)$ function in the consequent part of the rules (Equation 15.52).

In conclusion, utilizing the training capabilities of neural networks is quite beneficial; it improves the prediction performance of fuzzy systems. The ANFIS method is effective in determining the optimum parameters of the membership functions and the rule-bases.

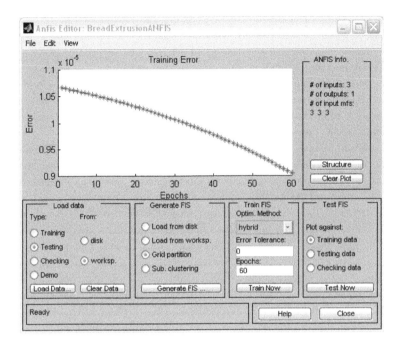

Figure 15.24 ANFIS training results for case 3.

15.6.2 Genetic Algorithms

Genetic algorithms (GAs) belong to the domain of evolutionary computing, which is based on the principles of natural evolution, and specifically utilize the concept of Darwin's theory of

Table 15.3 Prediction Results of the Bread-Extrusion Fuzzy System That Are Tuned and/or Developed Using the ANFIS Tool

Inputs			Expansion E (%)				
M (%)	S (rpm)	F (g/min)	Exp.	Model-0	Model-1	Model-2	Model-3
17.52	213	113	5.3	5.37	5.28	5.15	5.30
20.48	213	113	4.8	4.61	4.70	4.70	4.80
17.52	213	187	5.0	5.37	5.28	5.15	5.00
20.48	213	187	4.6	4.61	4.70	4.70	4.60
17.52	287	113	5.5	5.37	5.28	5.30	5.50
20.48	287	113	4.6	4.66	4.60	4.53	4.60
17.52	287	187	5.1	5.37	5.28	5.30	5.10
20.48	287	187	4.6	4.66	4.60	4.53	4.60
17.00	250	150	5.6	5.50	5.45	5.60	5.60
21.00	250	150	4.1	4.10	4.10	4.10	4.10
19.00	250	100	5.0	5.00	4.89	4.97	5.00
19.00	250	200	4.8	5.00	4.89	4.97	4.80
19.00	200	150	4.9	5.00	4.89	4.90	4.90
19.00	300	150	4.7	5.00	4.89	4.70	4.70
19.00	250	150	5.1	5.00	4.89	4.97	5.10
			Minimum	0.0	0.0	0.0	0.0
Error (%)			Maximum	7.4	5.6	3.9	0.0
			Average	2.6	2.2	1.8	0.0

evolution. Darwin's theory basically stressed the fact that the existence of all living things is based on the rule of survival of the fittest. It also postulates that new breeds or classes of living things come into existence through the processes of reproduction, crossover, and mutation among existing organisms.[63] There are several types of evolutionary algorithms, among which GAs are the best known. Other algorithms are *genetic programming*, *evolution strategies*, and *evolutionary programming*. Although different in their specifics, these algorithms are all based on the same general principle.[4] A brief introduction to genetic algorithms will be given next, followed by a demonstration of how these algorithms can be used to determine membership functions. Details on computational procedures and algorithms are readily available in a number of textbooks.[64–67]

GAs can be viewed as either a search method or an optimization technique that performs an intelligent search for a solution from a nearly infinite number of possible solutions. Although there are many possible variants of the basic GA, the fundamental underlying mechanism consists of three operations: evaluation of individual fitness, formation of a gene pool (intermediate population) through selection mechanisms, and recombination through crossover and mutation operators.

The algorithm proceeds as follows: an initial population of individuals, $P(0)$, is generated at random or heuristically. For every evolutionary step, t, known as a *generation*, the individuals in the current population, $P(t)$, are decoded and evaluated according to some predefined quality criterion, referred to as *fitness*, or a *fitness function*. Then, a subset of individuals, $P'(t)$, known as the *mating pool*, is selected to reproduce, according to their fitness. Thus, high-fitness individuals stand a better chance of reproducing whereas low-fitness ones are more likely to disappear.

New individuals enter the population and may be considered as new points in the search space. These are generated by altering the selected population $P'(t)$ via the application of crossover and mutation so as to produce a new population, $P''(t)$. Crossover enables the evolutionary process to move towards promising regions of the search space. Mutation prevents premature convergence to local optima by randomly sampling new points in the search space. Finally, the new individuals, $P''(t)$, are introduced into the next-generation population, $P(t+1)$. The termination condition may be specified as some maximal number of generations or as the attainment of an acceptable fitness level.

In a GA, individuals in a population are coded as finite strings. Each of the strings is decoded into a set of parameters that it represents and is passed though a numerical model of the problem space (a fuzzy system, in this case). The numerical model gives out a solution and, based on the quality of this solution, the string is assigned a fitness value. With this fitness value, the three genetic operators (reproduction, crossover, and mutation) are used to create a new generation of strings that is expected to perform better than the previous generations. The new set of strings is again decoded and evaluated, and the process is continued until convergence is achieved within a population.

Many fitness selection procedures are currently in use, one of the simplest being fitness-proportionate selection,[4] where individuals are selected within a probability proportional to their fitness. Crossover is performed using a "crossover rate" between two selected individuals, called *parents*, by exchanging parts of their genomes to form two new individuals, called *offsping*. The mutation operation is carried out by flipping randomly selected parts of the strings preserving the size of the population. There are several variations of GAs[66] with different selection mechanisms (e.g., ranking, tournament, and elitism), crossover operators (e.g., multipoint crossover), and mutation operators (e.g., adaptive mutation). These and other advanced topics related to genetic algorithms are found in books such as Mitchell,[64] Michalewicz,[65] and Banzhaf et al.[67]

15.6.2.1 *Genetic Tuning of Membership Functions*

As briefly described above, GAs can be used to optimize the internal parameters of fuzzy membership functions. The objective here is to find the most appropriate parameters of the membership functions so that the total error between the outputs of the fuzzy model and the desired outputs derived from the learning data is minimized.

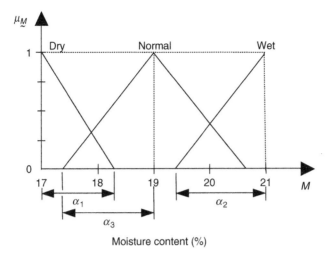

Figure 15.25 Physical representation of the membership function parameters.

To illustrate how GAs may be utilized in developing and/or tuning the membership functions of fuzzy systems, consider the membership functions defined in Figure 15.25 for the bread extrusion example. The shape of these membership functions can be simply adjusted by fixing the full membership points ($\mu_M = 1$) and allowing the zero points ($\mu_M = 0$) to vary. For each of the fuzzy variables in Figure 15.25, the membership function on the right side is constrained to have the right-angle wedge at the upper limit of the range of the fuzzy variable, and the left side function is constrained to have the right-angle wedge at the lower limit of the range. In addition, the middle membership function is assumed to be symmetrical, and its peak point is constrained at the middle of the range.

Consider, for instance, the fuzzy moisture content M, which is parameterized in Figure 15.10. The right side of the "wet" membership function is fixed at (21,1), i.e., $M = 21$ and $\mu_M = 1$, the left side of the "dry" membership function is fixed at (17,1), and the peak of the "normal" membership function is fixed at (19,1). As illustrated in Figure 15.25, the lengths of their bases that are represented by the three parameters, α_1, α_2, and α_3 are needed to describe the shape of the three membership functions.

This mechanism is used in encoding the membership functions of the three fuzzy variables. Therefore, the number of unknown variables for the bread extrusion problem is nine. These variables are encoded as a concatenated string. The GA procedure starts with generating an initial random population of individuals (strings). Each string contains a combination of membership functions. To determine the fitness of each string, the square of the errors that are produced when estimating the value of the output y of the fuzzy system are calculated, given the inputs x and parameters of the membership functions α's represented by the string. Relative fitness values are then used to determine which strings are to be eliminated and the number of copies of each remaining string that make the next generation of strings. Crossover and mutation are then applied on the strings, thus developing a new generation of strings. These strings undergo the same process of decoding and evaluation as the previous generation. The process of generating and evaluating strings is continued until we get the membership functions with the best fitness value.

15.6.2.2 *Genetic Learning of the Rule-Base*

GAs may be employed as an evolutionary learning process to automate the knowledge-base generation. This process can be considered as an optimization or search problem, where the task is to find an appropriate rule base from a set of input–output observations. Using an approach similar to the one described for tuning membership functions, the main concept is to parameterize the different components of the fuzzy system (fuzzy rules and fuzzy variables) and find those parameter values that are optimal with respect to the optimization criteria. The parameters constitute the optimization space, which is transformed into a suitable genetic representation on which the search process operates.

For fuzzy rule-base learning, there are three main approaches that have been applied in the literature:[4] the *Pittsburgh approach*, the *Michigan approach*, and the *iterative rule learning approach*. For the Pittsburgh approach, each individual string (genetic code or chromosome) represents an entire fuzzy system, whereas in the Michigan approach each individual represents a single rule and the fuzzy inference system is represented by the entire population. The main shortcoming of the Pittsburgh approach is its computational cost, because a population of the entire fuzzy system has to be evaluated each generation. The iterative rule learning approach combines the speed of the Michigan approach with the simplicity of fitness evaluation of the Pittsburgh approach. In this approach, each individual encodes a single rule, and a new rule is adapted and added to the rule set, in an iterative fashion, in every run of the GA. An evolutionary algorithm is used iteratively for the discovery of new rules until an appropriate rule-base is built.

Other hybridization techniques of GAs and fuzzy systems have been proposed in the literature. These include genetic-neuro-fuzzy systems that are the result of adding genetic or evolutionary learning capabilities to systems in which fuzzy and neural concepts are integrated. Such systems incorporate fuzzy numbers to represent the weights, perform fuzzy operations in the nodes of the network, and/or incorporate fuzzy nodes that represent membership functions. In addition, the learning process applies GAs to obtain the weights of the neural network, to adapt the transfer functions of the nodes, and/or to adapt the topology of the net.

Recently, a new evolutionary approach called *Fuzzy CoCo* has evolved and attracted numerous research activities. Fuzzy CoCo is a cooperative coevolutionary approach to fuzzy modeling wherein two coevolving species are defined.[4] Inspired by natural coevolution, *artificial coevolution* refers to the simultaneous evolution of two or more species with coupled fitness. This means that the fitness of one individual depends on the fitness of individuals of other species and/or its interaction with them. Such coupled evolution is the main difference between the Fuzzy CoCo and the noncoevolutionary GA-based fuzzy systems.

Coevolving species can either compete or cooperate. In a competitive-coevolutionary algorithm, the fitness of an individual is based on direct competition with individuals of other species. Increased fitness of one of the species implies a diminution in the fitness of the other species. On the other hand, in a cooperative-coevolutionary algorithm, the fitness of an individual depends on its ability to collaborate with individuals from other species.

REFERENCES

1. Ross, T. J., *Fuzzy Logic with Engineering Applications*, New York: McGraw-Hill, 1995.
2. Zadeh, L. A., Fuzzy sets, *Information and Control*, 8, 338, 1965.
3. Wang, L. X., *A Course in Fuzzy Systems and Control*, Upper Saddle River, NJ: Prentice-Hall International, 1997.
4. Peña Reyes, C. A., *Coevolutionary Fuzzy Modeling*, New York: Springer, 2004.
5. Iiyukhin, S. V., Haley, T. A., and Singh, R. K., A survey of automation practices in food industry, *Food Control*, 12, 285, 2001.
6. Perrot, N., Agioux, L., Ioannou, I., Mauris, G., Corrieu, G., and Trystram, G., Decision support system design using the operator skill to control cheese ripening: application of the fuzzy symbolic approach, *Journal of Food Engineering*, 64, 321, 2004.
7. Singh, R. K. and Ou-Yang, F., Neuro-fuzzy technology for computerized automation, In *Computerized Control Systems in the Food Industry*, G.S. Mittal, Ed., New York: Marcel Dekker, p. 119, 1997.

8. Mauris, G., Benoit, E., and Foulloy, L., Fuzzy symbolic sensors: from concept to application, *Measurement*, 12, 357, 1994.
9. Du, C. J. and Sun, D. W., Recent developments in the applications of image processing techniques for food quality evaluation, *Trends in Food Science and Technology*, 15, 230, 2004.
10. Du, C.-J. and Sun, D. W., Learning techniques used in computer vision for food quality evaluation: a review, *Journal of Food Engineering*, 72, 39–55, 2006.
11. Shahin, M. A., Tollner, E. W., Evans, M. D., and Arabnia, H. R., Watercore features for sorting red delicious apples: a statistical approach, *Transactions of the American Society of Agricultural Engineers*, 42, 1889, 2001.
12. Hu, B. G., Gosine, R. G., Cao, L. X., and de Silva, C. W., Application of a fuzzy classification technique in computer grading of fish products, *IEEE Transactions of Fuzzy Systems*, 6(1), 144, 1998.
13. Croft, E. A., de Silva, W., and Kurnianto, S., Sensor technology integration in an intelligent machine for herring roe grading, *IEEE/ASME Transactions on Mechatronics*, 1, 204, 1996.
14. Center, B. and Verma, B. P., Fuzzy logic for biological and agricultural systems, *Artificial Intelligence Review*, 12, 213, 1998.
15. Verma, B. P., Application of fuzzy logic in postharvest quality decisions, In *Proceedings of the National Seminar on Postharvest Technology of Fruits*, University of Agricultural Sciences (Augus 7–9) Bangalore, India, pp. 22–28, 1995.
16. Konstantinov, K. and Yoshida, T., An expert approach for control of fermentation processes as variable structure plants, *Journal of Fermentation and Bioengineering*, 70(1), 48, 1990.
17. Konstantinov, K. and Yoshida, T., On-line monitoring of representative structural variables in fed-batch cultivation of recombinant *Escherichia coli* for phenylaline production, *Journal of Fermentation and Bioengineering*, 70(6), 420, 1990.
18. Edan, Y., Grinspan, P., Maltz, E., and Kahn, H., Fuzzy logic for applications in the dairy industry. Technical Report no. 92-3600, St Joseph, MI: ASAE, 1992.
19. Nakamishi, S., Takagi, T., and Kurosania, M., Expert systems of beef grading by fuzzy inference and neural networks, In *Proceedings of the 3rd International Fuzzy Association and World Congress*, p. 360, 1993.
20. Jahns, G., Nielsen, H. M., and Paul, W., Measuring image analysis attributes and modeling fuzzy consumer aspects for tomato quality grading, *Computers and Electronics in Agriculture*, 31, 17, 2001.
21. Kavdir, I. and Guyer, D. E., Apple grading using fuzzy logic, *Turkish Journal of Agriculture and Forestry*, 27, 375, 2003.
22. Davidson, V. D., Ryks, J., and Chu, T., Fuzzy models to predict consumer ratings for biscuits based on digital image features, *IEEE Transactions on Fuzzy Systems*, 9(1), 62, 2001.
23. Ioannou, I., Perrot, N., Hossenlopp, J., Mauris, G., and Trystram, G., The fuzzy set theory: a helpful tool for the estimation of sensory properties of crusting sausage appearance by a single expert, *Food Quality and Preference*, 13(7–8), 589, 2002.
24. Sun, D. W. and Brosnan, T., Pizza quality evaluation using computer vision: part 1, pizza base and spread sauce, *Journal of Food Engineering*, 57, 81, 2003.
25. Sun, D. W. and Brosnan, T., Pizza quality evaluation using computer vision: part 2, pizza topping analysis, *Journal of Food Engineering*, 57, 91, 2003.
26. Davidson, V. J. and Ryks, J., Comparison of Monte Carlo and fuzzy math simulation methods for quantitative microbial risk assessment, *Journal of Food Protection*, 66(10), 1900, 2003.
27. Davidson, V. J., Fuzzy control for food processes, In *Computerized Control Systems in the Food Industry*, G.S. Mittal, Ed., Marcel Dekker: New York, p. 1799, 1997.
28. Passino, K. M. and Yurkovich, S., *Fuzzy Control*, Menlo Park, CA: Addison Wesley Longman, 1998.
29. Zhang, Q. and Litchfield, J. B., Knowledge representation in grain drier fuzzy logic controller, *Journal of Agricultural Engineering Research*, 57, 269, 1994.
30. Singh, R. K. and Ou-Yang, F., Knowledge-based fuzzy control of aseptic processing, *Food Technology*, 48(6), 423, 1994.
31. Whitnell, G. P., Davidson, V. J., Brown, R. B., and Hayward, G. L., Fuzzy predictor for fermentation time in a commercial brewery, *Computers and Chemical Engineering*, 17(10), 1025, 1993.
32. Von Numers, C., Nakajima, M., Siimes, T., Asama, H., Linko, P., and Endo, I. A., Knowledge-based system using fuzzy inference for supervisory control of bioprocesses, *Journal of Biotechnology*, 34, 109, 1994.

33. Zhang, X.-C., Visala, A., Halme, A., and Linko, P., Functional state modeling and fuzzy control of fed-batch aerobic baler's yeast process, *Journal of Biotechnology*, 37, 1, 1994.
34. Murnleitner, E., Becker, T. M., and Delgado, A., State detection and control of overloads in the anaerobic wastewater treatment using fuzzy logic, *Water Research*, 36, 201, 2002.
35. Georgieva, G., Wagenknecht, M., and Hampel, R., Takagi-Sugeno fuzzy model development of batch biotechnological processes, *International Journal of Approximate Reasoning*, 26, 233, 2001.
36. Guillaume, S. and Charnomordic, B., Knowledge discovery for control purposes in food industry databases, *Fuzzy Sets and Systems*, 122, 487, 2001.
37. Rywotycki, R., Food frying process control system, *Journal of Food Engineering*, 59, 339, 2003.
38. Ioannou, I., Perrot, N., Curt, C., Mauris, G., and Trystram, G., Development of a control system using the fuzzy set theory applied to a browning process: a fuzzy symbolic approach for the measurement of product browning: development of a diagnostic model, part I, *Journal of Food Engineering*, 64, 497, 2004.
39. Ioannou, I., Perrot, N., Mauris, G., and Trystram, G., Development of a control system using the fuzzy set theory applied to a browning process: towards a control system of the browning process combining a diagnosis model and a decision model, part II, *Journal of Food Engineering*, 64, 507, 2004.
40. Lababidi, H. M. S. and Baker, C. G. J., Web-based expert system for food dryer selection, *Computers and Chemical Engineering*, 27, 997, 2003.
41. Baker, C. G. J. and Lababidi, H. M. S., Development of a fuzzy expert system for the selection of batch dryers for foodstuffs, *Drying Technology*, 18(1), 117, 2000.
42. Baker, C. G. J. and Lababidi, H. M. S., Developments in computer aided dryer selection, *Drying Technology*, 19(8), 1851, 2001.
43. Baker, C. G. J., Lababidi, H. M. S., and Masters, K., A fuzzy expert system for the selection of spray-drying equipment, *Drying Technology*, 22(1,2), 237, 2004.
44. Curt, C., Trystram, G., and Hossenlopp, J., Formalisation of at line human evaluations. Integration of human decision in the dry sausage ripening process, *Science des Aliments*, 21/6, 663, 2001.
45. Ali, Y. M. and Zhang, L., A methodology for fuzzy modeling of engineering systems, *Fuzzy Sets and Systems*, 118, 181, 2001.
46. Ross, T. J., *Fuzzy Logic with Engineering Applications*, 2nd ed., West Sussex, UK: Wiley, 2004.
47. Jang, J. S. R., Sun, C. T., and Mituzani, E., *Neuro-Fuzzy and Soft Computing: A Computational Approach to Learning and Machine Intelligence*, Upper Saddle Drive, NJ: Prentice-Hall, 1997.
48. Yager, R. and Filev, D., SLIDE: a simple adaptive defuzzification method, *IEEE Transactions on Fuzzy Systems*, 1, 69, 1993.
49. Berkan, R. C. and Trubatch, S. L., *Fuzzy Systems Design Principles, Building Fuzzy If–Then Rule Bases*, Piscataway, NJ: IEEE Press, 1997.
50. Hellendoom, H. and Thomas, C., Defuzzification in fuzzy controllers, *Intelligent and Fuzzy Systems*, 1, 109, 1993.
51. Linko, P., Uemura, K., and Eerikainen, T., Neural networks in fuzzy extrusion control, In *Food Engineering in a Computer Climate*, IChemE Symposium Series No. 126, EPCE Event No. 452. p. 401, IChemE, Rugby, UK, 1992.
52. Mathworks, *Fuzzy Logic Toolbox for Use with MATLAB, User's Guide, Version 2*, Natick, MA: The Mathworks, Inc., 2002.
53. Mathworks, *MATLAB: The Language of Technical Computing, Version 6*, Natick, MA: The Mathworks, Inc., 2002.
54. Bandemer, H. and Gottwald, S., *Fuzzy Sets, Fuzzy Logic, Fuzzy Methods with Applications*, West Sussex, UK: Wiley, 1995.
55. Hertz, J., Krogh, A., and Palmer, R. G., *Introduction to the Theory of Neural Computation, Lecture Notes Volume 1*, Redwood City, CA: Addison-Wesley, 1991.
56. Tambe, S. T., Kulkarni, B. D., and Deshpande, P. B., *Elements of Artificial Neural Networks, with Selected Applications in Chemical Engineering, and Chemical & Biological Sciences*, Louisville, KY: Simulation & Advanced Controls, 1996.
57. Schalkoff, R. J., *Artificial Neural Networks*, New York: McGraw-Hill, 1997.
58. Nguyen, H. T., Prasad, N. R., Walker, C. L., and Walker, E. A., *A First Course in Fuzzy and Neural Control*, Boca Raton, FL: Chapman & Hall/CRC, 2003.
59. Mitra, S. and Hayashi, Y., Neuro-fuzzy rule generation: survey in soft computing framework, *IEEE Transactions on Neural Networks*, 11, 748, 2000.

60. Duch, W., Adamczak, R., and Grabczewski, K., A new methodology of extraction, optimization and application of crisp and fuzzy logic rules, *IEEE Transactions on Neural Networks*, 12, 277, 2001.

61. Nauck, D. and Kruse, R., Neuro-fuzzy systems for function approximation, *Fuzzy Sets and Systems*, 101, 261, 1999.

62. Jang, J. S. R., ANFIS: adaptive-network-based fuzzy inference systems, *IEEE Transactions on Systems, Man, and Cybernetics*, 23, 665, 1993.

63. Forrest, S., Genetic algorithms: principles of natural selection applied to computation, *Science*, 261, 872, 1993.

64. Mitchell, M., *An Introduction to Genetic Algorithms*, Cambridge, MA: MIT Press, 1996.

65. Michalewicz, Z., *Genetic Algorithms + Data Structures = Evolution Programs*, 3rd ed., Heidelberg: Springer, 1996.

66. Banzhaf, W., Nordin, P., Keller, R. E., and Francone, F. D., *Genetic Programming: an Introduction: On the Automatic Evolution of Computer Programs and its Applications*, Morgan Kaufmann, Dpunkt: Springer, 1998.

67. Vose, M. D., *The Simple Genetic Algorithm*, Cambridge, MA: MIT Press, 1999.

PART III

Some Generic Modeling Techniques

CHAPTER 16

Monte Carlo Simulation

Kevin Cronin and James P. Gleeson

CONTENTS

16.1   INTRODUCTION

16.1.1   Probabilistic Modeling

Questions concerning model uncertainty arise in all scientific and engineering fields. The last few years have seen many new developments in the field of uncertainty analysis.[1] Decision makers are increasingly demanding some defensible representation of uncertainty surrounding the output of complicated mathematical models. Scientists, in turn, are putting more effort into the modeling and propagation of uncertainty.

In general, uncertainty analysis requires stochastic modeling environments. Historically, analysts have often developed models based on a single value or "point estimate" for each input variable, ignoring uncertainty. This, broadly speaking, is the deterministic approach to modeling or analysis. Often the point estimate is the mean value of the sample data. Uncertainty can cause significant discrepancies in predicted risk and can have a major influence on decisions based on the analysis. For example, using point estimates of the input variables, a point estimate of the output variable can be calculated. On the other hand, if the uncertainty in the input variables is incorporated into the analysis, a probability density function (PDF) (or its statistics) for the same output can be estimated. This is the probabilistic modeling approach, within which the Monte Carlo method is one tool.[2] The point estimate of the output is not necessarily the mean of its PDF and will generally not be, except for a linear input/output functional relationship. Clearly, a PDF is preferable to a point estimate, as it communicates more of the available information.

Modeling approaches can also be classified as to whether they are theoretical (analytical) or numerical (approximate). Because only a small fraction of problems lend themselves to exact or closed-form solutions, approximate methods play a central role both in research and engineering applications. The Monte Carlo method is one of the most universally used approximate numerical techniques, especially where stochastic effects are important, though it has also been used to examine deterministic problems.[3] The latter application includes the calculation of complex integrals, solution of equations, etc.

16.1.2   Uncertainty in Food and Bioprocess Engineering

Variability in food processing and bioprocessing arises from random fluctuations in environmental processing parameters such as temperatures, feed-rates, etc., and from dispersion of internal parameters such as heat and mass transfer coefficients, reaction constants, physical and thermal properties, etc.[4] The last of these is very significant because biomaterials, owing to their natural origin, inherently have a large dispersion in their physical and thermal properties. As a guide, the coefficient of variation in a broad spectrum of thermophysical properties of foods can be expected to lie within a band of 5–15%. Also, unlike variability in processing parameters (that can, to an extent, be ameliorated by good system design and control practice), this intrinsic dispersion is outside the control of the system designer. Therefore, an efficient and systematic method is required to evaluate uncertainties and to investigate the effects of the uncertainties in food process simulation.[5]

More particularly, the thermophysical properties of many biological products, as well as the external parameters such as initial and ambient temperature, surface heat transfer coefficient, etc., may change randomly both as a function of space and time coordinates.[6] For example, the chemical composition and physical structure of many agricultural materials are very heterogeneous and are a function of several factors such as harvest time and weather conditions, and the origin of the product. The thermophysical properties are affected by this heterogeneity and can vary inside the product as a function of the space coordinates. Other parameters, such as the temperature of the convective fluid flowing around the conducting solid, are intrinsically stochastic and may change in an unpredictable way during the heat-transfer process. For instance, the temperature inside a refrigerated room may fluctuate as a function of time as a consequence of unpredictable openings of refrigerator or oven doors, actions of the temperature control system, ambient conditions outside the refrigerated room, and changes of solar radiation flux. As a consequence, the temperature distribution inside the heated object is also random and can only be specified meaningfully by means of statistical characteristics such as its mean value, variance, and PDF. The Monte Carlo analysis technique is particularly suited to accommodate the intrinsically variable nature of the properties of foods and other bio-products and uncertainties in the manufacturing processes they undergo.[7]

16.1.3   Monte Carlo Simulation as a Probabilistic Modeling Tool

There is a wide range of methods available for analyzing uncertainties in model predictions due to variability in model input parameters. Probabilistic modeling approaches include transformation methods, perturbation methods (also known as *statistical differentials*), variance propagation analysis, sensitivity analysis, stochastic response surface modeling, Markov chains, Monte Carlo, etc. Chapter 10 includes a broader discussion of these techniques.

The term *Monte Carlo* is used to describe any approach to a problem where a probabilistic analogue to a given mathematical problem is setup and solved by stochastic sampling. This invariably involves the generation of many random numbers, so giving rise to the name. In direct Monte Carlo simulation, vectors of model inputs are obtained by sampling from known or assumed model input probability distributions. These distributions that characterize the variability in the input model variables, are typically based upon field measurements or other prior knowledge of the factor in question. The underlying deterministic model can vary in complexity from being a single algebraic equation to a series of algebraic or differential equations. Repeated model execution using these input vectors are then used to describe model outputs of interest in terms of a PDF. Each single model execution is known as a *run* or *trial*. The PDF of the output variable(s) can then be analyzed to determine statistics of interest including the mean, variance and higher moments of the results. With Monte Carlo simulation, stochastic components in the system being modeled can be incorporated, and the uncertainty associated with model predictions quantified.

The advantages of the Monte Carlo method are that (at least for direct Monte Carlo simulation) it is conceptually easy to understand and robust in operation. It is equally applicable as a stochastic solution technique to both phenomenological and empirical deterministic models of the underlying process. Although large numbers of iterations or simulations (10^3–10^6) must usually be carried out to generate statistically significant results, useful estimates of the uncertainties in model outputs can be obtained with only 50 or 100 model runs. The Monte Carlo method allows use of standard nonparametric statistical tests concerning confidence intervals. The single greatest disadvantage of the method is that it is computationally expensive.

Consider a deterministic model lying at the high end of the complexity range, particularly deriving from the finite element or finite difference technique. Typically, it will have a large numbers of input parameters, many equations and parameterizations, thousands of grid points and degrees of freedom and a high time resolution. Obviously, a large number of repeated iterations of the model is not a practicable proposition.[8] These comments concerning the drawbacks of the approach primarily apply to the so-called *direct Monte Carlo sampling method*, where straightforward random sampling from the entire distribution for each input variable is carried out. There are techniques available to reduce the number of model iterations that must be performed, which can help ameliorate this problem (see Section 16.3.3).

16.1.4   History of the Monte Carlo Method

The name and systematic development of the Monte Carlo method dates from about 1944.[9] There were, however, a number of isolated and undeveloped instances on earlier occasions. One of the first accepted occurrences of a Monte Carlo type problem is the Buffon's needle experiment that involves the repeated random throwing of a needle onto a board or a floor ruled with parallel straight lines where the value of π can be inferred from the number of intersections between needle and lines. A number of physicists, mathematicians and statisticians (including Lord Kelvin and W. S. Gosset) employed the method in a rudimentary form to tackle a variety of problems using the principles of numerical sampling. In most cases, this was to verify the more traditional analytical approach. In the beginning of the last century, the Monte Carlo method was used to examine the Boltzmann equation. In 1908, the statistician "Student" (the pseudonym of W. S. Gosset) used the Monte Carlo method to estimate the correlation coefficient in his t-distribution. Nonetheless, the real use of the Monte Carlo technique dates from work on the American atom bomb project of the Second World War as pioneered by von Neumann, Ulam, and Fermi. Von Neumann and Ulam coined the term "Monte Carlo" after the gambling casinos of the Mediterranean city. This particular work involved a direct simulation of the probabilistic issues associated with random neutron diffusion in fissile materials. In about 1948, Fermi, Metropolis, and Ulam obtained Monte Carlo estimates for the eigenvalues of the Schrödinger equation. Further important developments of the Monte Carlo method was given by Metropolis during the 1950s. Over the intervening years, the technique has grown in popularity due to the advent of widely available digital computers with ever-increasing computational power.

16.1.5   Chapter Outline

Because of the wide applicability of Monte Carlo simulations, the types of problems it can solve require some classification. In particular, the nature of the uncertainty or variability that is being investigated requires elucidation. The Monte Carlo method has long been used to simulate discrete random events; this is known as *event modeling*. Examples of such events are failure of equipment items or services, and this approach lies in the realm of operations research and system engineering. Event modeling finds applicability in the food industry as in many other industrial arenas.[10] However, this chapter will primarily focus on the simulation of continuous time processes where there is a continuous functional relationship (over time) between the input and output variables.

A straightforward statistical approach to the solution of random problems is the direct Monte Carlo method. In this method, a random sample of the stochastic input parameters is generated by the computer and the corresponding deterministic model is numerically solved. The solution is stored and the process is repeated a large number of times. In the end, the statistical characteristics of the output variables are estimated using classical inference techniques. The preponderance of this chapter will focus on direct simulation in Section 16.2. However, after outlining this method, some comment will be given on more involved procedures to handle complex variability where it arises. In particular, two major issues can complicate the analysis and require a more sophisticated approach:

- Nonindependent input random variables
- Random variables exhibiting noise

The principle of *variance reduction* is also quite complex and will be discussed with the above two issues in Section 16.3.

At the outset, it is necessary to appreciate the advantage of Monte Carlo simulation over simple limit analysis. The range of an output variable will be defined by its minimum and maximum values; these in turn, in most cases, can be found by inputting the extreme values (minimum or maximum, as appropriate) of the input variables into the deterministic model. However, this does not provide a realistic representation of the uncertainty in the output, as it is quite unlikely that all uncertain parameter values would take their most optimistic or most pessimistic values simultaneously.[11] Therefore, in addition to calculating the limits of the output variable, the Monte Carlo technique can calculate the likelihood of these limits occurring.

16.2   DIRECT MONTE CARLO SIMULATION

Generally, simulation based on Monte Carlo analysis is performed through the following six steps:

- The functional, deterministic model that relates the input and output variables is defined.
- Statistical analysis of the input variables to determine the PDF that will describe each variable.
- Generation of uniformly distributed random (or pseudo-random) numbers.
- Generation of random samples for each input variable.
- Repeated model executions until a meaningful level of statistical significance is achieved to assemble PDFs to describe the output variables.
- Analysis of the output variable to estimate important statistical quantities, such as the mean and variance, and to conduct sensitivity analysis.

Figure 16.1 displays in flowchart form the procedure that must be followed.

The intention here is not to review the theoretical basis of the Monte Carlo method; there are many excellent books available that satisfy that requirement.[12–14] Rather, this section outlines a practical "user's guide" as to how to carry out a basic direct Monte Carlo simulation in sequential steps. To aid understanding, a very simple deterministic model, representative of food processing, will be selected and solved by direct Monte Carlo simulation. Furthermore, the predictions of mean and variance in the output variable given by the Monte Carlo method will be compared to the theoretical solution for these quantities and to estimates provided by a quick, approximate method (Chapter of Bart Nicolai provides a fuller description of these latter methods). This will enable factors dealing with the accuracy and convergence of the Monte Carlo method to be highlighted. It should be noted that the theoretical and approximate solution are only possible when there is a known analytical function relating the output variables to the inputs.

The theory of functions of random variables[15] gives the expected value (i.e., the mean) of a function of a random variable, $y = f(x)$ from the equation:

$$E(y) = \mu_y = \int_{-\infty}^{\infty} f(x)g(x)\,dx,\qquad(16.1)$$

where $g(x)$ is the PDF of the input variable, x. The variance (square of the standard deviation) in y can then be calculated by

$$V(y) = \sigma_y^2 = E(y^2) - (E(y))^2\qquad(16.2)$$

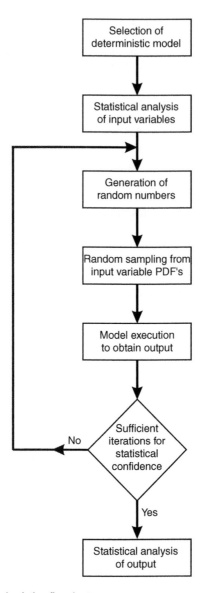

Figure 16.1 Direct Monte Carlo simulation flowchart.

Equation 16.1 and Equation 16.2 are appropriate for the case where there is a single input variable, x, and single output variable, y, though they can easily be generalized to the case of multiple input and output variables. Alternatively, the perturbation method can also be used to estimate the mean and variance in the output variable, given the mean and variance of the input variables.[16] Again, for a single input–output function, the mean and variance in the output variable can be approximated as:

$$\mu_y \approx f(\mu_x) \tag{16.3}$$

$$\sigma_y^2 \approx \left(\frac{\partial f}{\partial x}\right)^2_{x=\mu_x} \sigma_x^2 \tag{16.4}$$

where the derivative is evaluated at the mean value of the independent variables. For the special case where the function f is linear with respect to the input variables (and the input variables are independent), the preceding formulae hold exactly. Once the function f does not depart too severely from the condition of linearity, the input variables are independent and the magnitudes of the variances of the input variables remain low, then the statistical differential estimations for output mean and variance can provide a reasonably accurate prediction.

16.2.1 Selection of the Deterministic Model

The model may be a simple algebraic expression or a complicated set of equations and numerical procedures. It is only essential that the model is written as a mathematical function of some vector of arguments. Generally, a process model can be represented by a system of differential-algebraic equations as follows:

$$y = f(x_1, x_2, \ldots x_N, t) \tag{16.5}$$

where y is the output variable, x_i are the input variables, and t is time.

As previously mentioned, to illustrate the approach, a particularly simple model will be adopted; the first order decay of a variable with time

$$y = y_0 e^{-kt} \tag{16.6}$$

where y is the output variable of interest that changes with time t in the manner given by Equation 16.6. Note that this is the explicit solution of the first-order differential equation

$$\frac{dy}{dt} = -ky(t), \tag{16.7}$$

with initial condition $y(0) = y_0$. The constant k in this equation is known as the *rate constant* for the process. The inverse of k corresponds to the characteristic time of the system; the units of k will be the inverse of whatever unit the time is expressed in (whether it be seconds, minutes, hours, etc.). The input variables for this model will be y_0 and k. Because the process is stochastic, rather than interested in a single trajectory of the variable y against time, it is the transformation of the PDF that describes y with time that is of interest. This concept is illustrated in Figure 16.2.

16.2.2 Statistical Analysis of the Input Variables

To transform the deterministic model to a probabilistic tool, the input variables must be considered as random variables governed by probability distributions. There are a variety of probability distribution (or density) functions that can be used to represent the dispersion of the variables. The choice of the most appropriate probability distribution function for an uncertain variable depends on the characteristics of the variable and the amount of available information on the uncertainties of the variable. There are no restrictions on the shapes of the PDFs, although most studies make use of a few basic PDF shapes, such as normal, log-normal, uniform, or triangular. To select the most suitable probability distribution for the particular variable, statistical knowledge about its behavior is required.

16.2.2.1 Analysis of Experimental Data

Generally, information about the input variables is found from repeated experimental measurement. Alternatively, knowledge of the distribution of these variables is available from *a priori* considerations. From the experimentally available sample data of the input variables, summary

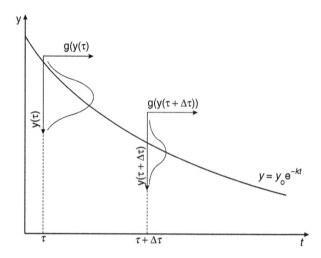

Figure 16.2 Stochastic process with evolving probability distribution.

statistics based on the statistical moments should be generated. The first three moments are usually sufficient. The first moment is the mean of the values. However, the mean of a probability distribution with very broad tails may result in a very poor estimate or a meaningless value. Other estimators such as median and mode can be used for such cases. Second moments, variance, and average deviation characterize width or variability around the mean value. The third moment, or skewness, characterizes the degree of asymmetry of a distribution around the mean. From the available data of the particular variable, the sample mean, x_m, standard deviation, s, and skewness, v, can be calculated as

$$x_m = \frac{\sum_{i=1}^{n} x_i}{n},$$

(16.8)

$$s = \sqrt{\frac{\sum_{i=1}^{n} (x_i - x_m)^2}{n-1}},$$

(16.9)

$$v = \frac{\sum_{i=1}^{n} (x_i - x_m)^3 / n - 1}{s^3}.$$

(16.10)

Note that n is the number of items (experimental measurements) in the sample.

In addition to calculating the sample statistics, a frequency histogram of the data from each variable should be constructed. This graphical technique is excellent in revealing the overall shape of the distribution, which can be matched to the PDF. The key issue here is the selection of the optimum number of intervals to break the data into (or equivalently the width of each interval). If too many intervals are selected (with too fine an interval width), the histogram will appear excessively fragmented, whereas if too few intervals are chosen, the shape of the histogram will approach that of an uneven rectangular block. Usually the best number of intervals is between 5 and 11.

Also, the input variables should be checked for independence. One informal, graphical procedure is to build scatter plots where the available values of one variable are plotted against

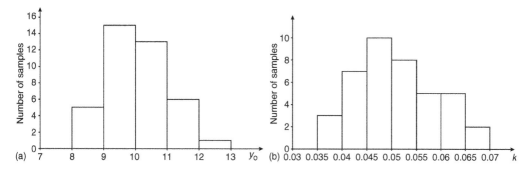

Figure 16.3 Frequency histograms of initial variables.

another to test for independence. Applying linear or non-linear regression, a correlation coefficient for each scatter plot can be obtained, and if the magnitude of this coefficient is close to zero, it is a strong indicator of independence between the variables. In many instances, theoretical considerations will show whether variables should or should not exhibit dependence.

To illustrate this step of the analysis, 40 individual measurements of the variables y_0 and k are assumed to have been taken. Furthermore, it is assumed that they are independent of each other. Frequency histograms showing the distribution in the initial value of y and in k are shown in Figure 16.3a and b, respectively. The distribution is slightly nonsymmetric about the mean and skewed to the right, as shown by the positive values for the skewness of the data. Sample statistics are summarized in Table 16.1.

From the sample statistics that quantify the input variables, the population parameters of these same variables can be estimated.[17] The experimental initial variable y_0 (sample size 40) has a mean value of 10 and a standard deviation of 0.99. From these sample statistics, $(x_m$ and $s)$, the population parameters μ and σ can thus be estimated as 10 and 1, respectively. The corresponding population parameters for the rate constant, k will be 0.05 and 0.008, respectively. Note the skewness of the underlying population of each variable is not required in this instance and is not estimated.

16.2.2.2 Selection of Probability Distribution Functions

In this subsection, the input random variables are assumed to be continuous, as opposed to discrete random variables; therefore, only continuous probability distributions are discussed.

Any probability distribution can be characterized by its parameters, which in turn can be estimated from the corresponding sample statistics. Generally, a probability distribution function has three parameters that can be geometrically interpreted as defining the location, scale, and shape of the distribution. A location parameter represents the position of the distribution on the x-axis by specifying an abscissa such as the minimum value or the average of the distribution. Changing the location parameter shifts the distribution left or right along the x-axis. The scale parameter

Table 16.1 Sample Statistics of Input Variables

Model Input Variables—Sample Statistics	y_0	k
Minimum	8.0	0.035
Maximum	12.3	0.066
Mean	10.0	0.05
Standard deviation	0.99	0.0079
Skewness	0.013	0.143

represents the width or dispersion of the distribution (such as the standard deviation in the normal distribution). Changing this parameter compresses or expands the distribution without changing its basic shape. The shape parameter represents the shape of distribution usually characterized by the skewness of the distribution. Note that not all distributions have a shape parameter and some have more than one (such as the beta distribution).

The first step in selecting a PDF to represent a given variable is to decide the general appropriateness of the candidate PDF based on the overall shape. In some cases, theoretical considerations can be used to select the correct PDF. Variables can be measured or shown to be capable of ranging from plus infinity to minus infinity, or alternatively to be capable of taking only positive values, or of being limited to within an upper or lower bound that are both positive. Analysis of the real problem can demonstrate that occurrences of the variable must be equiprobable within a certain range with no peak or alternatively, be unimodal, or indeed bimodal. Some variables must be, by definition, symmetric with respect to the modal value, whereas others from physical consideration must be right-skewed (maximum values of the distribution are much greater displaced from the mode than minimum values) or left-skewed (the reverse).

A brief overview of some of the potential PDFs that are available, their characteristics, and merits is given below. Table 16.2 lists the expression for the PDF for each distribution over a particular region of interest. In addition, each distribution is sketched in Figure 16.4 to provide an appreciation of its general shape. A more detailed treatment is available in Law and Kelton.[18]

- The uniform distribution is used when information about the dispersion in the variable of interest is poor and only the limiting values are known. The uniform distribution (also known as the *equiprobable distribution*) is defined by its minimum, x_{min}, and maximum, x_{max}, values. As it has no central tendency, the uncertainties result in broad distribution of the values of the output variables. Irrespective of the PDFs used to describe the input variables, the uniform distribution is always employed to initially generate the random numbers for the simulation that are subsequently used to generate the random input variables.
- The triangular distribution is used when the central and limiting values of a variable are known. It is defined by its minimum (x_{min}) mode (x_{mo}) and maximum (x_{max}) values. As with the uniform distribution, it is suitable when there is relatively little information about the actual distribution of the variable.

Table 16.2   Probability Density Functions of a Number of Common Probability Distributions

Probability Distribution	Probability Density Function
Uniform	$g(x) = \dfrac{1}{x_{max} - x_{min}} \qquad x_{min} \leq x \leq x_{max}$
Triangular	$g(x) = \dfrac{2(x - x_{min})}{(x_{max} - x_{min})(x_{mo} - x_{min})} \qquad x_{min} \leq x \leq x_{mo}$
Normal	$g(x) = \dfrac{1}{\sqrt{2\pi\sigma^2}} e^{-[(x-\mu)^2/2\sigma^2]} \qquad -\infty < x < \infty$
Log-normal	$g(x) = \dfrac{1}{x\sqrt{2\pi\sigma^2}} e^{-[(\ln x-\mu)^2/2\sigma^2]} \qquad 0 < x < \infty$
Exponential	$g(x) = \dfrac{1}{\beta} e^{-(x/\beta)} \qquad 0 \leq x < \infty$
Weibull	$g(x) = \dfrac{\alpha}{\beta} \left(\dfrac{x - \gamma}{\beta}\right)^{\alpha-1} e^{-((x-\gamma)/\beta)^\alpha} \qquad \gamma < x < \infty$

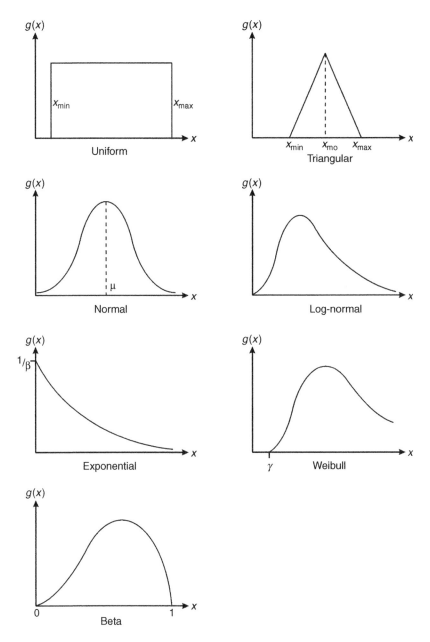

Figure 16.4 Shapes of some common probability density functions.

- The normal distribution (also known as the Gaussian distribution) is the most widely applicable distribution in statistics. Much mathematical analysis in the field of probability is based on (and restricted to) the normal distribution. The normal distribution is a symmetric distribution and is defined by its mean, μ and standard deviation, σ. Where experimental data, sampled from a variable, is unimodal and approximately symmetric (the "bell shaped" curve), the normal distribution can provide good results when representing the variable in question. The normal distribution has a range from plus infinity to minus infinity but can be applied to finite bounded data by the use of truncation.

- Mathematically, the log-normal distribution is a transformed variant of the normal distribution, though it is physically quite different. It fits the case where the data itself is not normally distributed but the (natural) log of the data is normally distributed. The log-normal distribution is a right-skewed distribution and is restricted to positive quantities with a range from zero to plus infinity. It is applicable where continuous random variables must take values greater than zero (if they occur), but have no practical upper limit. In particular, it is appropriate for environmental variables that are widely distributed and that tend to have a few large values. It is commonly employed to model particle size distributions, as they are transformed by processes such as prilling, grinding, and crystal growth.
- The exponential distribution is solely defined by its mean value, is right-skewed and has a range from zero to plus infinity. It is defined by a scale parameter, β, which must be positive. The parameter β is both the mean and standard deviation of the distribution. In the area of bioprocess engineering, it can be used to model the residence time of particles in a chamber or process.
- The Weibull distribution is defined by three parameters representing location (γ), scale (β), and shape (α). Similarly to the beta distribution, it is an artificial distribution, as opposed to one deriving from a particular probability model. It can take a variety of characteristic shapes depending upon the relative magnitudes of its parameters and hence is a multipurpose distribution that can be fitted to a large number of different types of random variables. It has a range from zero to plus infinity. The exponential distribution can be considered as a special case of the Weibull distribution.
- The beta distribution is commonly used to represent the uncertainty in the probability of occurrence of an event, because its range is limited between zero and one. It is defined in terms of the beta function and although an expression for its PDF is available, the equations are long and are not given in Table 16.2. The beta distribution is also very flexible in terms of the wide variety of shapes it can assume, including positively or negatively skewed, depending on the values of its parameters.

Many of the common PDFs that are available, such as the normal, log-normal, Weibull, etc., have ranges that extend to infinity, either on one side or both. Thus, if these are used to model variability, in some cases, physically unrealistic values of the variable can be generated by the probabilistic model. In such instances, truncated versions of the PDFs can be used where upper or lower (or both) cutoffs are applied. In the case of a normal distribution, these limits are generally selected to be plus or minus an integer number or half integer number of the standard deviation.

Once a PDF has been selected for a variable, its goodness-of-fit to the experimental data must be ascertained. The best-known test is the chi-square test, where in effect, the frequency distribution of the data from the experimental frequency histogram is compared with the expected theoretical distribution from the PDF. The greater the agreement (or the less the discrepancy) between the two, then the more likely it is that the chosen PDF is acceptable. Other tests include the more powerful Kolmogorov–Smirnov tests, which avoid some of the pitfalls of the chi-square method.[19] In some cases though, there will not be enough sample data to fit PDFs to the input variables using goodness-of-fit measures.

For the particular model used to illustrate this work, the distributions in both the initial value of the variable, y_0, and the rate constant k are finite bounded and slightly nonsymmetric about the mean (both skewed to the right). Both the Weibull and normal distributions are feasible to represent the data, and the above tests can be used to select the best fit. For simplicity, the normal distribution was chosen. It has theoretically an infinite range, which is obviously not physically possible for either variable, so outer cut off limits of ± 3 standard deviations are applied to the tails of the distribution. With these cutoff limits, the truncated normal distribution for y_0 has the PDF

$$g(y_0) = \frac{1.0027}{\sqrt{2\pi\sigma^2}} \exp - \left[\frac{(y_0 - \mu)^2}{2\sigma^2} \right]; \qquad \text{if } \mu - 3\sigma \leq y_0 \leq \mu + 3\sigma \tag{16.11}$$

$$g(y_0) = 0; \qquad\qquad\qquad\qquad\qquad \text{otherwise}$$

A similar expression can be produced for the rate constant.

16.2.3 Generation of Uniformly Distributed Random Numbers

The engine of the Monte Carlo method is some procedure to generate random numbers.[20] In practice, this is accomplished by using a computer to generate a sequence of pseudo-random numbers, based on some formula. The numbers generated must not have any obvious pattern. The numbers generated must all lie between zero and one (including these two numbers) and any number within this range must be capable of being generated. Thus, the random numbers correspond to the continuous uniform distribution on the interval [0,1]. They can then be employed to obtain random samples from any other PDF.

The search for methodologies to generate random numbers has exercised mathematicians over the last 60 years. The numerical procedures used to do so have become more complex to satisfy the stringent tests for randomness. The great majority of random-number generators in use today are linear congruential generators using the modulus function. Generally the modeler need not be concerned about the actual mechanics, and most simulation packages will return a random number using a standard function or key with a name such as RND, RAND, or RANDOM.

16.2.4 Sampling of Input Random Variables

After uniformly distributed random numbers have been generated, the next step is to generate random variates according to the selected input probability distribution function. Exactness, efficiency, and robustness are all issues that, in theory, should be considered when selecting the most appropriate algorithm. The main approaches are the inverse transform method, the composition technique, convolution and acceptance–rejection. All offer advantages and disadvantages and these are discussed comprehensively by Law and Kelton.[18] Again, the simulation software will generally perform this function.

As an example, if the variable in question is represented by the normal distribution, then the procedure is to initially select random variates from the standard normal distribution of mean zero and standard deviation one. These can then be converted to any particular normal distribution of mean μ and standard deviation σ. One traditional technique is the Box and Muller algorithm, which produces standard normal random variates in pairs.[21] Assuming u_1 and u_2 are two uniformly distributed random numbers, then z_1 and z_2 will be two standardized normal variates where

$$z_1 = [-2 \ln u_1]^{1/2} \cos(2\pi u_2), \tag{16.12}$$

$$z_2 = [-2 \ln u_1]^{1/2} \sin(2\pi u_2). \tag{16.13}$$

Two random variates from the particular normal distribution of interest (with parameters μ and σ) are then obtained by scaling as

$$x_1 = \mu + \sigma z_1, \tag{16.14}$$

$$x_2 = \mu + \sigma z_2. \tag{16.15}$$

For the particular illustrative model discussed here, cutoff limits of ± 3 standard deviations were applied to the tails of the experimental normal distributions; this Gaussian generator was modified to reject samples lying outside these limits, and they were not used in the computer simulation.

16.2.5   Generation of Output

The next step is the propagation of uncertainties through the model. This step comprises model calculation using the sampled values of the input random variables and storage of the calculation results. Each time the model is executed, values are sampled from the input distributions and each run or iteration of the model produces a unique final value for a given process time. Carrying out many runs means a large number of values for the output variable can be assembled. The statistics representing the output variable (mean, variance, etc.) will stochastically converge to the correct values with increasing number of iterations. The execution step can be terminated when the analyst is satisfied that the convergence has reached a satisfactory level.

Examining convergence in more detail, the main aim in Monte Carlo simulation is to obtain a small standard error in the final result. The standard error in the predicted mean value of the output variable is inversely proportional to the square root of the number of iterations, n (see Section 16.3.3.1). Thus, to reduce the standard error by a factor of, say 2, the number of iterations needs to be increased by 4. From heuristic considerations, the number of iterations necessary to ensure confidence in the results is assumed to occur when the output value stabilizes (within tolerance limits) and becomes independent of the number of trials. As the mean of the output is itself a stochastic variable, it will vary slightly even for a large number of trials. The t-statistic can be applied to estimate the necessary number of runs for a given level of accuracy. For relatively straightforward algebraic expressions, 10,000 or more simulations (i.e., functional evaluations) can be conducted, but for a high-level food process simulation model, this is generally not possible.

Alternatively, the results may be interpreted through statistical tolerance limits, which show that the output from direct Monte Carlo uncertainty analysis is valid even for a limited number of trials (for instance, 50 or 100). Tolerance limits differ from statistical confidence intervals in that the tolerance limits provide an interval within which at least a proportion, q, of the population lies, with probability, p or more, that the stated interval does indeed contain the proportion, q, of the population. Note that this procedure is valid only when the Monte Carlo simulations are based on direct random sampling and is not readily extended to the reduced-variance techniques discussed in Section 16.3.3.

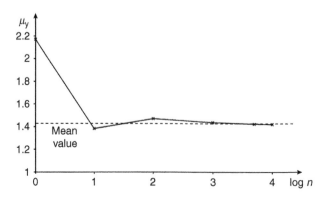

Figure 16.5   Stochastic convergence of output mean.

Table 16.3   Comparison of Model Output: Theoretical, Statistical Differentials, and Monte Carlo

Analysis Method	Output Variable Statistics	
	Mean (μ_y)	Standard Deviation (σ_y)
Theoretical	1.424	0.547
Statistical differentials	1.353	0.454
Monte Carlo simulation	1.421	0.493

16.2.6   Analysis of Output

For the particular model employed to illustrate this work, the output statistics of interest will be the mean and standard deviation in the variable y at a time of 40. The stochastic convergence of the mean towards its equilibrium value is displayed in Figure 16.5; the mean was evaluated after 1, 10, 100, 1,000, 5,000, and 10,000 iterations, respectively. Table 16.3 gives the predictions for the mean and standard deviation in the output variable y (at the time of 40). The mean and standard deviation (square root of the variance) as calculated by the theoretical approach (Equation 16.1 and Equation 16.2, respectively), statistical differentials (Equation 16.3 and Equation 16.4, respectively) and the Monte Carlo method (after 10,000 iterations) can be compared. A frequency histogram of the output variable y (at the time of 40) is illustrated in Figure 16.6.

As is evident, the Monte Carlo approach predicts a mean value very close to the theoretical value; the fractional error between the two is less than 0.25%. Increasing the number of iterations of the model that are conducted will in the long term reduce this error. The agreement between the Monte Carlo and theoretical approaches for the standard deviation is less satisfactory; here, the fractional error is just under 10%. This is not surprising because standard deviation (or variance) is always more difficult to estimate accurately than mean values. If a precise estimate of variance in the output variable is required, then generally a very large number of model iterations must be carried out.

The numerical output from any Monte Carlo simulation can be assembled in frequency histogram form. Further analysis can allow the distribution of the output variable to be described by various uncertainty displays such as cumulative distribution functions, PDFs, box plots, bar graphs, etc. Also, sensitivity analysis procedures can be used to analyse the output. It is usually the case that a model consists of a set of equations with m dependent or output variables and n independent variables plus input parameters. The sensitivity coefficient, S, can be defined as the ratio of the fractional change in

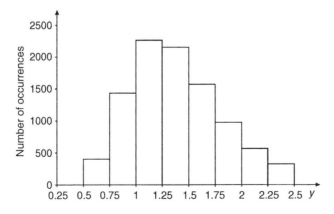

Figure 16.6   Frequency histogram of output mean.

an output variable to the corresponding fractional change in an input variable. As with any differential procedure, it is implied that the fractional changes are small (e.g., less than 10%).

16.3   ADVANCED MONTE CARLO ISSUES

16.3.1   Dependent Input Random Variables

For processes where the input random variables are not independent, then a simple or direct Monte Carlo simulation is invalid. The dependence between these variables must be taken into account using correlation methods. The issue of dependence between input parameters can be understood from thermal simulations. Product specific heat capacity and product density both contribute to the heating rate of an item, but there is a relationship (positively correlated) between the two quantities and from physical considerations, they are clearly not independent variables. Sometimes approximate analysis can simplify the problem. Assuming a strong positive correlation exists between the two variables, then one simple heuristic technique to capture their inter-dependence is to use the same random number when sampling from both.

16.3.1.1   Mathematical Basis

On a more mathematical basis, for processes where the input random variables are not independent, it is necessary to generate vectors of correlated random variables. Consider, for example, the generation of an N-dimensional vector $\mathbf{x} = (x_1, x_2, \ldots, x_N)^T$ from a normal distribution with given mean vector $\mathbf{x}_m$ and covariance matrix $\mathbf{C}$:

$$E((\mathbf{x} - \mathbf{x}_m)(\mathbf{x} - \mathbf{x}_m)^T) = \mathbf{C}. \tag{16.16}$$

Supposing that a vector $\mathbf{z}$ of uncorrelated, normal variables can be generated (e.g., by the Box–Muller algorithm, Section 16.2.4), it is possible to generate correlated random variables by taking linear combinations of the elements of $\mathbf{z}$:

$$\mathbf{x} = \mathbf{x}_m + \mathbf{L}\mathbf{z}, \tag{16.17}$$

where $\mathbf{L}$ is an $N \times N$ matrix. It is easy to show that if the elements of $\mathbf{z}$ have zero mean and unit variance, then the covariance matrix of $\mathbf{x}$ is

$$E((\mathbf{x} - \mathbf{x}_m)(\mathbf{x} - \mathbf{x}_m)^T) = \mathbf{L}\mathbf{L}^T. \tag{16.18}$$

Thus, the problem reduces to finding a matrix $\mathbf{L}$ that produces the desired covariance matrix, i.e., satisfying the equation

$$\mathbf{L}\mathbf{L}^T = \mathbf{C}. \tag{16.19}$$

This is a standard problem of matrix algebra, and may be solved by computing the Cholesky decomposition of $\mathbf{C}$ to yield the matrix $\mathbf{L}$. The vector of correlated normal variables is then given by $\mathbf{x} = \mathbf{x}_m + \mathbf{L}\mathbf{z}$. If the desired random variables have distributions other than the normal distribution, the generation of suitable correlated variables may not be so simple.

16.3.1.2   Example of Correlated Parameters

Suppose two correlated random variables are required from a joint normal probability distribution, for example y_0 and k for each realization (run) of the Monte Carlo simulations described in Section 16.2.1. If drying is the process of interest, then y_0 is the initial moisture content of a product and k is its drying rate. In some instances, the wetter an item is to begin with, the faster it will dry, thus giving rise to a dependence between the two variables. Let y_0 have mean μ_y, let k have mean μ_k, and define the variances and covariance (cross-correlation) as

$$E[(y_0 - \mu_y)^2] = \sigma_y^2, \tag{16.20}$$

$$E[(k - \mu_k)^2] = \sigma_k^2, \tag{16.21}$$

$$E[(y_0 - \mu_y)(k - \mu_k)] = \sigma_{yk}^2. \tag{16.22}$$

In the case where y_0 and k are independent, we have $\sigma_{yk}=0$, and the variables are easy to generate. However, when a correlation exists between the variables, the method described in Section 16.3.1.1 must be utilized. Here we describe the algorithm in a step-by-step fashion, in the special case of $N=2$ variables.

The Cholesy decomposition of the covariance matrix

$$\mathbf{C} = \begin{pmatrix} \sigma_y^2 & \sigma_{yk}^2 \\ \sigma_{yk}^2 & \sigma_k^2 \end{pmatrix} \tag{16.23}$$

yields the lower-triangular matrix

$$\mathbf{L} = \frac{1}{\sigma_y} \begin{pmatrix} \sigma_y^2 & 0 \\ \sigma_{yk}^2 & \sqrt{\sigma_y^2 \sigma_k^2 - \sigma_{yk}^4} \end{pmatrix}, \tag{16.24}$$

with $\mathbf{LL}^T = \mathbf{C}$. In each run, two independent normal variables, z_1 and z_2, of unit variance and zero mean are generated using, e.g., the Box–Muller algorithm. According to the algorithm presented in Section 16.3.1.1, the desired correlated variables y_0 and k are generated by setting

$$\begin{pmatrix} y_0 \\ k \end{pmatrix} = \begin{pmatrix} \mu_y \\ \mu_k \end{pmatrix} + \mathbf{L} \begin{pmatrix} z_1 \\ z_2 \end{pmatrix}. \tag{16.25}$$

Writing this vector equation in terms of components gives the recipe for generating correlated random variables using the independent variables z_1 and z_2:

$$y_0 = \mu_y + \sigma_y z_1 \tag{16.26}$$

$$k = \mu_k + \frac{\sigma_{yk}^2}{\sigma_y} z_1 + \frac{\sqrt{\sigma_y^2 \sigma_k^2 - \sigma_{yk}^4}}{\sigma_y} z_2. \tag{16.27}$$

The covariance, σ_{yk}^2 can be related to the correlation coefficient R_{yk} between the two variables y_0 and k using

$$\sigma_{yk}^2 = R_{yk} \sigma_y \sigma_k. \tag{16.28}$$

Hence, Equation 16.27 can also be written as

$$k = \mu_k + \sigma_k \left[R_{yk} z_1 + \sqrt{\left(1 - R_{yk}^2\right)} z_2 \right]. \tag{16.29}$$

Note that in the case of zero covariance $\left(\sigma_{yk}^2 = R_{yk} = 0\right)$, these equations reduce to the standard method for generating independent normal variables with given means and variances, compared to Equation 16.14 and Equation 16.15:

$$y_0 = \mu_y + \sigma_y z_1 \tag{16.30}$$

$$k = \mu_k + \sigma_k z_2. \tag{16.31}$$

16.3.2 Input Variables Exhibiting Noise

Another issue that must be considered is the nature of the randomness of the input variables and how they can dynamically vary in space and with time. In the direct Monte Carlo technique, input parameters are sampled or selected at the start of the run and held fixed for the duration of the particular iteration. Different values for these input variables will be sampled for the subsequent iteration, but remain constant within a given iteration. In many applications the input parameters may vary randomly with time, i.e., have a noise component.[22] In such circumstances, application of a simple Monte Carlo technique (taking the mean value of the signal over time) can produce model results that are very far removed from those observed in nature and also substantially different from analytical solutions. Although the mean of the Monte Carlo ensemble is close to the "correct" mean solution in the examples cited, the variance is too large. Even if the only goal of the modeling is to determine the mean result, then the large variance of the results means that many samples (model runs) are required for the sample mean to converge to the true mean, and if the aim is to investigate the unpredictability (or variability) of the model results, then the direct Monte Carlo method appears to be unreliable.

As an illustration of the above issue, consider the task of modeling the heating up behavior of a number of ostensibly identical items of discrete product in a sterilization chamber. Variability in a factor such as the surface heat-transfer coefficient can be incorporated in a number of ways. In the simplest implementation, the heat-transfer coefficient can be assumed to be randomly distributed between the different items, though constant (with respect to time and space) for each individual item. A more thorough treatment, though, would assume that in addition to varying from item to item, the heat-transfer coefficient will vary in a noisy fashion.

16.3.2.1 Mathematical Basis

The rigorous mathematical treatment of time-varying noise is based on the theory of stochastic differential equations. The fundamental concept is the Wiener process or "white noise" that describes very rapid unpredictable fluctuations in the input parameter. The mathematical idealization of white noise is rather unphysical, as it has infinite energy content due to the infinitely fast fluctuations. Moreover, the correct physical interpretation of the effects of white noise is not always clear, due to the so-called Ito-Stratonovich modeling dilemma.[23,24] To avoid these difficulties, various types of "colored" noise are commonly used in modeling applications. Coloured noise changes smoothly as a function of time, with a characteristic timescale of fluctuations known as the (de-)correlation time. Samples of colored noise may be generated by, for example, autoregressive random processes that consist of linear stochastic differential equations with white-noise forcing.[22]

In cases where the temporal variation of the parameters is important, a modified Monte Carlo method, in which the parameters are resampled from the underlying probability distribution at regular intervals through each model run (the *stochastic parameters* method), can produce more reasonable answers than the direct Monte Carlo method described above. However, the results from this method depend critically on the frequency with which the parameter resampling occurs. The variability in these parameters is continuous in nature, and parameter values separated by times shorter than the decorrelation timescale will tend to be similar. In modeling the effect of temporally varying parameters, it is important that the time scale of this variation is known, as well as the range of variability. It has been shown that for a parameter varying as an autoregressive process of order one, the resampling time of the stochastic parameter method should be approximately twice the decorrelation time to ensure that the Monte Carlo results are similar to those of the real system.[25]

The above approach to handle a factor varying in a random fashion with time can be extended to cases where a parameter exhibits random fluctuation in space as well as in time. In the example discussed before, it is possible that the surface heat-transfer coefficient of an object in the sterilization chamber could exhibit random spatial variation as well as the variation in time already discussed. Random fluctuations in space and time may be modeled as autoregressive random waves.[26] The output parameter is now the solution of a partial differential equation (or of a system of such equations), which typically requires solution by a finite element method. Standard spatial discretisation techniques such as the Galerkin method lead to systems of ordinary differential equations in time for the nodal values of the output variable(s). The parameters of these equations exhibit fluctuations in time, and so the methods discussed above may be applied to calculate the quantities of interest, e.g., the mean and variance of the temperature at each nodal point, by solving a matrix system.

16.3.2.2 Example of Parameters with Noise

As an example of a problem where the input parameters fluctuate randomly in time, consider the solution of the first order equation

$$\frac{dy}{dt} = -k(t)y(t), \tag{16.32}$$

with initial condition $y(0) = y_0$. Here, the rate of decay $k(t)$ is a random function of time; for clarity, y_0 is taken to be nonrandom. Assuming that $k(t)$ is a stationary random process, i.e., its statistical properties do not change over time, it can be characterized by, for example, its mean

$$E(k(t)) = \mu_k, \tag{16.33}$$

and by its correlation function:

$$E[(k(t) - \mu_k)(k(t') - \mu_k)] = R(t - t'). \tag{16.34}$$

If $k(t)$ is a *Gaussian* (normal) random process, then it is fully described by its mean and correlation function as defined above. The correlation function $R(\tau)$ measures how quickly the function decorrelates with time; generally, $R(\tau)$ decreases as the time difference τ increases, and when the value of $R(\tau)$ is close to zero, the values of $k(t)$ and $k(t + \tau)$ are almost uncorrelated (i.e., independent). The special case of white noise corresponds to infinitely fast changes in $k(t)$ (zero memory time), and so the correlation function decays instantly to zero:

$$R(\tau) = D'\delta(\tau)$$ (16.35)

where δ is the Dirac delta function. Colored noise contains smoother fluctuations than white noise, and so has a less singular correlation function; for instance, a first-order autoregressive (AR(1)) process (also known as an Ornstein–Uhlenbeck process) has a correlation function which decays exponentially over a timescale a^{-1}:

$$R(\tau) = De^{-a|\tau|}.$$ (16.36)

The solution of the differential Equation 16.32 is

$$y(t) = y_0 \exp\left[-\int_0^t k(s)ds\right],$$ (16.37)

which gives the unknown function $y(t)$ for each path or realization of the random process $k(t)$. The statistical properties of $y(t)$ may in general be calculated by Monte Carlo simulation but in this simple example it is possible to find the moments of $y(t)$ in closed form:

$$E(y^m(t)) = y_0^m \exp\left[-m\mu_k t + \frac{m^2}{2}\int_0^t\int_0^t R(s-s')ds'ds\right].$$ (16.38)

Thus, for example, the mean value of $y(t)$, when the decay rate $k(t)$ is a white noise process with the correlation function given in Equation 16.35, is

$$E(y(t)) = y_0 \exp\left[-\mu_k t + \frac{1}{2}D't\right].$$ (16.39)

Note that we have used the Stratonovich interpretation of white noise in this discussion; the Ito interpretation leads instead to the result

$$E(y(t))_{Ito} = y_0 \exp[-\mu_k t].$$ (16.40)

This ambiguity in the interpretation of white noise is a consequence of its unphysically rapid decorrelation—in essence, different regularizations of the delta function in Equation 16.35 can give different results. To avoid this problem, the case of colored noise with a decorrelation timescale a^{-1} may also be studied; using the correlation function of Equation 16.36 then yields for the moments of the process

$$E(y^m(t)) = y_0\exp[-m\mu_k t + m^2 Da^{-2}(at - 1 + e^{-at})].$$ (16.41)

Finally, a comparison is made to the direct Monte Carlo scenario where the decay rate k is a constant in each realization, but with the value of the constant chosen from a normal distribution. Because k does not vary in time in this case, the correlation function is a constant,

$$R(\tau) = D,$$ (16.42)

where D is simply the variance of the k values. In this case, Equation 16.38 gives the usual result for the moments

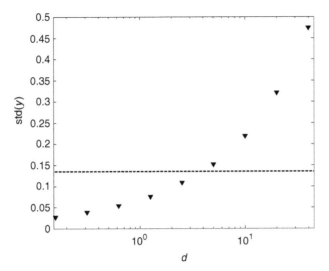

Figure 16.7 Standard deviation of the output variable as a function of the Monte Carlo resampling time.

$$E(y^m(t)) = y_0 \exp\left[-m\mu_k t + \frac{m^2}{2} Dt^2\right].$$ (16.43)

Note that the colored noise result (Equation 16.41) reduces to the constant-rate result in Equation 16.43 in the limit $a \to 0$; however, at large times $t \gg a^{-1}$, the behavior of Equation 16.41 resembles the random-walk growth of the white noise result in Equation 16.39. In the latter limit, the white noise intensity, D' is related to the colored noise variance by $D' = 2D/a$.

As an example, consider solving Equation 16.32 when $k(t)$ is a time-varying random process, e.g., an AR(1) process, and y_0 is deterministic. Taking values from Table 16.1 for the mean and standard deviation of k, and setting y_0 equal to 10, consider finding the standard deviation of y at time $t = 40$ by Monte Carlo simulation. The fluctuations in the value of k with time are described by a first-order autoregressive process, with correlation function given by Equation 16.36, and decorrelation timescale $a^{-1} = 2s^{-1}$ (for instance); note D is the variance of k. Because the value of k changes with time, the stochastic parameters method described in Section 16.3.2.1 to resample values of k at regular intervals during each run, must be used. The question of the best sampling frequency to use is answered in Figure 16.7, where the standard deviation of y (at time $t = 40$) found by resampling using various sampling intervals d is compared to the exact solution (dashed line) derived from Equation 16.41. As noted in Annan,[25] the exact solution for this AR(1) process is well approximated by choosing a sampling interval on the order of twice the decorrelation time, i.e., using a sampling interval d near 4.

16.3.3 Variance Reduction

Variance reduction techniques are available that can improve the efficiency of the Monte Carlo method by more than an order of magnitude. There are a number of such approaches with the importance sampling method being one of the more prominent.[27] The basic idea behind the importance sampling method is that certain values of the input random variables (or vectors) have more important impact on the quantities being estimated than others, and if these "important" values are sampled more frequently, i.e., sampled from a biased density function, the variance of the estimator can be reduced. The outputs from simulations are then weighted to correct the bias caused by sampling from the biased density function. The purpose of the importance sampling method is to obtain accurate estimates of output quantities with fewer samples than required in the direct Monte Carlo method. There are two major steps involved in the importance sampling method: the first is distortion of the original input process. Instead of taking samples from the original PDF, samples are taken from some other PDF, called importance density functions, such that some "important" regions of the sample space get more samples. The fundamental issue in implementing the importance sampling method is the choice of biased importance density functions. The second step is correction of the distortion by averaging the output from different samples (realizations) using weights that are related to the distortion, such that the mean of the quantity being estimated is preserved.

16.3.3.1 Mathematical Basis

To understand more about the mathematical basis of variance reduction strategies, it should be appreciated that techniques for the reduction of variance in Monte Carlo simulations are usually discussed in the context of multidimensional integration. Therefore, it is useful to first demonstrate the connection between simulation and integration problems. Consider Monte Carlo simulations with N random input parameters $x_1, x_2, \ldots x_N$, written for convenience as the N-dimensional column vector $\mathbf{x} = (x_1, x_2, \ldots, x_N)^T$. The goal of Monte Carlo is the calculation of the mean (average) value of some function $f(\mathbf{x})$ of the inputs, with the average being taken over the PDF of the random inputs, denoted $g(\mathbf{x})$. The function f is typically very complicated, requiring, for instance, the numerical solution of systems of differential equations, but for present purposes it may be accepted as given once the input vector $\mathbf{x}$ is chosen. The mean, or expectation value, of f may be written as the N-dimensional integral

$$E(f) = \int_\Omega f(\mathbf{x}) g(\mathbf{x})\, d\mathbf{x} \qquad (16.44)$$

with the domain of integration, Ω, depending on the distribution of the inputs; for example, if each input parameter is normally distributed, then Ω comprises all of the N-dimensional space. Note also that because (for instance) normally-distributed random variables $\mathbf{x}$ may be generated from vectors $\mathbf{u}$ of uniformly distributed variables (with finite domains) by the Box–Muller algorithm (for instance) discussed in Section 16.2.4, the integration domain may always be transformed to the N-dimensional unit (hyper-) cube $[0,1]^N$ by the inverse of the transformation $T:[0,1]^N \to \Omega$. Thus, the focus in the remainder of this section is on the N-dimensional integral

$$I = \int_{[0,1]^N} F(\mathbf{u})\, d\mathbf{u}, \qquad (16.45)$$

which gives the desired quantity $E(f)$ when F is related to f and g by the transformation T.

The integral I is over N dimensions, and is difficult to calculate accurately when N is large. As noted already, the direct Monte Carlo method uses n sample input vectors $\mathbf{u}_1, \mathbf{u}_2, \ldots, \mathbf{u}_n$ randomly chosen from uniform distributions to estimate the integral I as

$$F_m \pm n^{-1/2} s, \qquad (16.46)$$

where F_m and s are the sample mean and standard deviation, as in Equation 16.8 and Equation 16.9:

$$F_m \equiv \frac{1}{n} \sum_{i=1}^{n} F(\mathbf{u}_i) \quad s \equiv \sqrt{\frac{1}{n} \sum_{i=1}^{n} (F(\mathbf{u}_i) - F_m)^2}. \qquad (16.47)$$

The second term of Equation 16.46 is one standard deviation error estimate for the integral, not a rigorous bound. Also, as there is no guarantee that the error is normally distributed, the error term should be taken only as a rough indication of the possible error. Note that the error decreases as n increases at a rate proportional to $n^{-1/2}$. The goal of the various variance reduction methods is to choose the n sample points in such a way that the error term is reduced relative to that incurred by the direct Monte Carlo method in Equation 16.46.

16.3.3.2 Sampling Strategies

Returning to the importance sampling method, its application is motivated by the fact that choosing the sample vectors **u** from a uniform distribution may not be very efficient if the function $F(\mathbf{u})$ is sharply peaked in a small volume of the N-dimensional unit cube, but nearly zero everywhere else. In this case, many of the sample points will contribute very little information about the value of the integral. The importance sampling method is based upon the use of a biased density function (rather than a uniform density) for choosing the sample points, thus concentrating the sample points in the vicinity of the important (peak) values of the integrand. It can be shown that optimal reduction of variance occurs when the biased density function is chosen to be proportional to the magnitude of the integrand.[28] Of course, this requires a significant amount of prior information on the form of the integrand, and so in practice the importance sampling method is usually implemented in an adaptive fashion, with information from earlier samples used to estimate the optimal biased density function for later sample points.

The idea of *stratified sampling* is quite different from importance sampling. In the stratified sampling method, the integration domain is split into a number of nonoverlapping subdomains of equal volume, with the number of sample points in each subdomain chosen in a manner that leads to reduction of the overall variance. It can be shown that the optimal allocation of sample points is to have the number of sample points in each subdomain proportional to the standard deviation of the integrand over that subdomain. Like the importance sampling method, stratified sampling requires some prior knowledge about the behavior of the integrand. Note that stratified sampling concentrates sample points in regions of the domain where the variance of the integrand is largest, whereas importance sampling biases the sample points towards regions where the magnitude of the integrand is relatively large.

Other common strategies for reducing Monte Carlo variance attempt to ensure a uniform coverage of the integration domain—such methods are especially relevant when the number of sample points n is relatively small, or the dimension N of the integration domain is relatively large, so that the samples are sparsely distributed. The Latin hypercube sampling (LHS) method, for example, works by dividing each dimension into n segments, so that the whole domain is partitioned into n^N cells. The n cells containing the sample points are chosen as follows: initially one of the n^N cells is chosen at random to contain the first sample point. Next, all cells that share any of their parameters with the first cell (i.e., all cells in the same row, column, etc.) are eliminated. Another cell is then chosen at random from the remaining $(n-1)^N$ cells, its rows, columns, etc., are eliminated, and the process continues until only the final cell remains. The LHS leads to a sparse sample pattern that is the multidimensional analog of the two-dimensional Latin square (an $n \times n$ array of n symbols with exactly one symbol in each row and in each column). In effect, then, the LHS method forces the distribution of random samples to be more equally spread across the specified PDF, which is thought to be useful when large numbers of samples are not possible. The LHS method does not allow the application of standard nonparametric statistical tests, and it is often found to underestimate the total variance of the output parameters.

Quasi-random (or subrandom) sequences of numbers are used in quasi Monte Carlo methods. In such methods, the input parameters are not actually random variables, but are generated number-theoretically so that successive points fill gaps in the previously generated distribution. Thus quasi-random sequences fill the integration domain more uniformly than uncorrelated random points. Well-known examples of such sequences are associated with the names Halton and Sobol.[28]

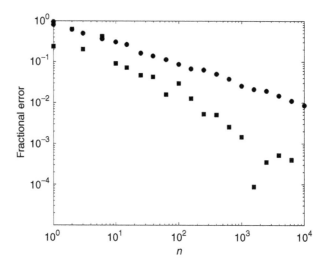

Figure 16.8 Fractional error in the calculation of the mean using direct Monte Carlo and the Halton sequence.

As a simple illustrative example of a variance reduction technique, consider the calculation of the mean value of y from Equation 16.6, with y_0 fixed (nonrandom), and with rate constant k chosen from a normal distribution with mean μ_k and variance D. The exact solution is given by Equation 16.43 with $m=1$, and may be calculated from the integral

$$E(y) = \frac{y_0 e^{-\mu_k t}}{\sqrt{2\pi D}} \int_{-\infty}^{\infty} \exp\left[-\sigma t - \frac{\sigma^2}{2D}\right] d\sigma. \qquad (16.48)$$

The integration domain may be transformed to $(0,1)$ by a change of variables, e.g., $\sigma = 1/\tan(\pi z)$. Figure 16.8 shows the fractional error of a Monte Carlo calculation of the mean of y, (circles) compared with a calculation using the quasi-random Halton sequence of base 2 (squares). Note the parameters used are $D=1$, $t=1$. The Monte Carlo error is averaged over 100 realizations, each using n random evaluation points; the fractional error decays as $n^{-1/2}$ in accordance with Equation 16.46. The Halton sequence is a deterministic sequence in which successive points "fill in" gaps left by previous points; it clearly converges more quickly to the exact value, at a rate on the order of n^{-1}.

16.4 APPLICATIONS IN FOOD AND BIOPROCESS ENGINEERING

16.4.1 Introduction

The realm of food and, more generally, bioprocessing, incorporates an enormous variety of operations, and the literature suggests that Monte Carlo analysis is applicable in a large number of instances. Some of the more prominent areas include the batch food sterilization process, broad food thermal processing, the packaging of foods and predictive microbiology.

Sterilization has proved to be one of the more popular processes to which the Monte Carlo method has been applied. The consequences of variability is critical here as the high levels of dispersion in the thermal diffusivity of food materials (with a coefficient of variation of up to 15%) causes a large variation in product temperature. As the thermal inactivation of microorganisms is highly dependent on the temperature, it is very possible to end up in a situation were some foods of the same batch are microbiologically safe, while others are not. Monte Carlo simulation has also been applied to the more general thermal processing of foods including the roasting, baking, frying and cooling operations. In a different context, it has also been employed to examine the food packaging question, specifically to determine the optimum strategy to form a pack of a given weight or weight range by assembling individual items of the product whose weight is randomly distributed. In the separate field of predictive microbiology, Monte Carlo modeling has also found wide application. Microbial contamination levels, concentration levels, risk of infection and transmission and relevant environmental factors such as temperature have been treated as random variables and used to solve a number of either mechanistic or statistical deterministic models to quantify the risk or danger of infection. Table 16.4 contains some of the more recent applications of the Monte Carlo method in food/bio processing. In Section 16.4.2 and Section 16.4.3, two detailed examples, taken from the literature of Monte Carlo simulation, will be examined.

16.4.2 Thermal Processing of Hazelnuts

The dry roasting of hazelnuts consists of loading the product into an oven and heating it for a certain time until the centre of the hazelnuts reach a target temperature.[36] Because of variability in the dimensional and thermal properties between individual hazelnuts, a distribution in center temperature is present at the end of the process. This dispersion in temperature will produce dispersion in the quality (color, texture) and safety (allergen destruction, aflatoxin content) of the hazelnuts. Therefore, if the process objective is that every hazelnut must achieve a minimum target temperature for the required safety, this implies that most of the hazelnuts that are smaller than the average size will reach temperatures greater than the target temperature with attendant consequences for quality. In the study, a deterministic model of heat transfer to the hazelnut was developed. The hazelnut was treated as a hollow sphere with a combined convective and radiative heat flux at its outer surface and heat diffusion through its body. The distribution in the physical and

Table 16.4 Applications of the Monte Carlo Method in Food/Bio Processing

Process	Deterministic Model	Random Variables	Output	Reference
Batch food sterilization	Heat transfer, sterilization kinetics	Initial temperature, heat-transfer coefficient, retort temperature	Product temperature and process lethality	29–33
Deep-fat frying	Percolation theory	Product structure	Degree of oil absorption	34, 35
Hazelnut roasting	Heat transfer	Product dimensions, thermal properties, oven temperature	Product temperature	36
Airborne contamination of food	Deposition	Settling velocity, bacteria count	Contamination level	37
Food preparation in a restaurant	Thermal inactivation	Infective dose	Risk of infection	38
Home cooking of hamburgers	Dose response	Presence and growth of organisms	Health risk of food consumption	39
Contamination of potable water	Destruction kinetics	Input contamination level, residence time, concentration	Safety of water	40
Shelf life of food	Growth and inactivation kinetics	Lag time	Microbial load	41
Food packaging	Food container/pack filling	Individual item weight	Optimum filling method	42–44

Table 16.5   Statistical Parameters of the Input Variables in Hazelnut Roasting

Variable	Units	Mean	Standard Deviation
Outside radius	mm	6.9	0.395
Inside radius	mm	3.69	0.57
Conductivity	W/m K	0.2	0.025
Specific heat	J/kg K	1994	188
Density	kg/m^3	875	17.3

thermal properties was experimentally measured, and these were treated as the input random variables. The objective of the work was to solve the deterministic model by the Monte Carlo approach and to predict the evolution of standard deviation in hazelnut center temperature versus time. These predictions were to be checked against experimental measurements of the variability in temperature.

The deterministic model to predict hazelnut center temperature (or more precisely, temperature on the inner surface) was a numerical, finite difference scheme with an explicit formulation. The largest time step that did not violate the stability criterion was selected and there were 100 nodes in the model. The input random variables to the model consisted of hazelnut outside radius, hazelnut hole internal radius, thermal conductivity, density, and specific heat. A further random variable was the location of the probe that was used to experimentally record the temperature. The random variables were assumed to be independent, normally distributed, and time-invariant. The only exception was conductivity and specific heat where a correlation existed. The mean and standard deviation in each of these variables was found by experimental sampling. Table 16.5 contains the magnitudes of the statistical parameters. Cutoff limits of ± 3 standard deviations were applied to the tails of the normal distribution to ensure the domain of the variables was finite bounded. Figure 16.9 displays the distribution, in frequency histogram form, of hazelnut outside radius and mass as found by experiment with a superimposed normal distribution curve. For each model run, each of the input variables was sampled and hazelnut temperature versus time was calculated. It was estimated that after 5,000 runs or iterations, mean final temperature would be calculated to within 0.3% of the actual value. From the 5,000 hazelnut temperature versus time profiles, the mean and standard deviation in temperature versus time were determined.

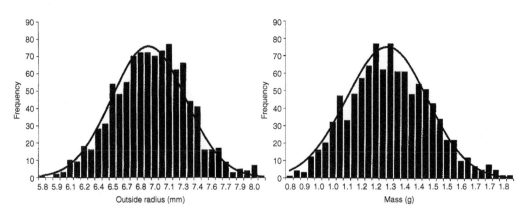

Figure 16.9   Distribution in hazelnut outside radius and mass.

16.4.3 Batch Thermal Processing of Packed Food in a Retort

In-pack thermal processing of foods in batch retort systems should deliver safe products throughout the retort.[32] Two main factors contribute to variability in product safety and quality during thermal processing: (1) nonuniformity in the process environment and, therefore, heat delivery to the product throughout the process chamber, and (2) variability in heat transfer within the product. The ideal thermal process should result in all the food items receiving the same thermal treatment as measured by the heating lethality (F_{0h}) value. The food product in the study consisted of a solid/liquid mixture within a pack. Because a theoretical analysis of heat transfer of such a system is excessively complex, an empirical deterministic model of heat transfer

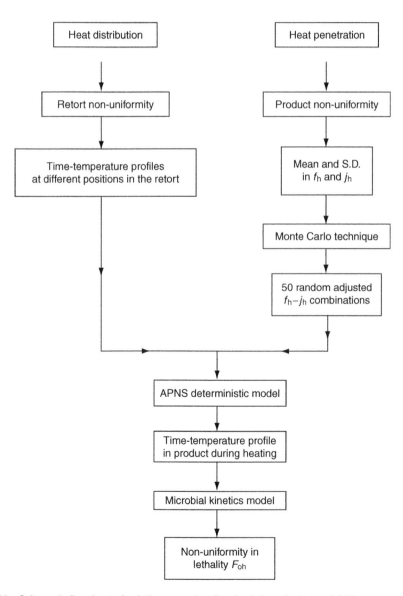

Figure 16.10 Schematic flowsheet of solution procedure for simulation of retort variability.

Table 16.6 Statistical Parameters of the Input and Output Variables in the Retorting of Green Beans

Parameter	Symbol	Units	Mean	Standard deviation
Heating-rate index	f_h	min	7.67	0.26
Heating lag factor	j_h	—	1.27	0.09
Thermal lethality	F_{0h}	min	8.35	1.48

was selected. The objective was to calculate the variability in the F_{0h} value that different containers of food would receive during treatment.

The empirical model of heat transfer required two parameters: a heating rate index, f_h, and a heating lag factor, j_h. From these two parameters, the time–temperature profile of the product during heating could be calculated based on the time–temperature profile in the retort. After the time–temperature profile of the food product was known, the heating lethality, F_{0h} was calculated using the standard D–z model of thermal inactivation kinetics (kinetic parameters for *Clostridium botulinum* were used). The input random variables to the deterministic model were the heating-rate index and heating lag factor; both were taken to be normally distributed. Variability in the processing conditions was limited to a spatial dispersion in the retort temperature and the surface heat-transfer coefficient to the product was assumed to be the same for all the packs. The mean and standard deviation in the two random variables was found from an experimental study. The solution procedure was that random numbers (normally distributed with mean zero and standard deviation of one) were generated and then used to sample values from the heating rate index and heating lag factor respectively. Lethality was then calculated with the deterministic model. Repeated simulations enabled frequency distributions of the lethality to be assembled. Figure 16.10 depicts a flowsheet of the solution procedure. In the first instance, both input variables were taken to be independent, and the sampling method described in Section 16.2.4 was employed. Subsequently, however, the dependence between them was taken into account to check the model's predictions, and the sampling method was as described in Section 16.3.1.2 of this chapter. It was shown that even where dependence does exist, ignoring it and treating the parameters as independent does not significantly affect the predictions of the approach. Typical coefficients of variation for the lethality values were found to range from 15 up to 63%, depending on the product being processed. The distribution in lethality when calculated at the same position in the retort was found to be normal, although the distribution in F_{0h} throughout the retort was non-normal. Table 16.6 contains the mean and standard deviation in the input variables (heating-rate index and heating lag factor) and the overall mean and standard deviation in the output variable (lethality) for the case where green beans were the product of interest in a water cascading retort.

ACKNOWLEDGMENTS

The second author acknowledges funding from Science Foundation Ireland under program number 02/IN.1/IM062.

GLOSSARY

Cumulative density function   The integral of the PDF and giving the probability that the random variable x, is less than a given value.

Deterministic model   A model or relationship that relates the input variables of the system under study to the output variables, in the absence of variability or uncertainty in the system.

Independent random variables   Two random variables, x and y, are independent where (knowledge of) the values of one variable, x does not change any of the probabilities associated with the values for y.

Monte Carlo modeling   Solution of a probabilistic model by random numerical sampling.

Noise   Where the random variable fluctuates (usually rapidly) with respect to time or a space variable.

Normal (Gaussian) distribution   The most important type of probability distribution has the characteristic shape of the cross section of a bell.

Probability density function   A form of the probability distribution that gives the probability of the variable, x having a value in the small interval, Δx.

Probability distribution   A function that associates each value of a random variable with the probability that it occurs.

Probabilistic (Stochastic) model   A model that relates input variables to output variables and incorporates uncertainty that is present in the system.

(Uniform) random number   The set of real numbers, lying between 0 and 1, that all have an equal probability of being selected for an experiment.

Random variable   For this work a physical variable (assumed continuous) whose magnitude is subject to random effects.

Standardized normal distribution   The form of the Normal distribution that describes a normally distributed variable having mean zero and a standard deviation of unity.

Statistically significant result   Where the Monte Carlo model has been run a sufficiently large number of times so the magnitude of the output statistics (mean, variance, etc.) is no longer influenced by chance variations.

Stochastic process   A process where the variables of interest change randomly with time and are described by a probabilistic model.

Uniform distribution   The probability distribution where each interval value of the random variable has an equal probability of occurring.

Variance reduction techniques   Any procedure to reduce the number of numerical samples that must be undertaken in order to calculate statistical parameters of the output variable by the Monte Carlo method.

NOMENCLATURE

a	Noise de-correlation frequency
D	Variance of colored noise
D'	Intensity of white noise
$f(x)$	Input/output functional relationship for single input variable x
$g(x)$	Probability density function of variable x
k	System rate constant (input variable)
n	Sample size
R	Correlation coefficient
$R(\tau)$	Noise correlation function
t	Time
x_i	Generic ith input variable
x_m	Mean value of sample
x_s	Standard deviation of sample
y	Generic output variable
y_0	Initial value of variable y (input variable)
z	Standardized normal variate

$\delta(\tau)$ Dirac delta function
μ Mean value of variable
σ_x Standard deviation of variable, x
σ_x^2 Variance of variable, x
σ_{xy}^2 Covariance of variables, x and y

REFERENCES

1. Torvi, H. and Hertzberg, T., Estimation of uncertainty in dynamic simulation results, *Computers and Chemical Engineering*, 21, 181–185, 1997.
2. Button, N. P. and Reilly, P. M., Uncertainty in incident rates for trucks carrying dangerous goods, *Accident Analysis and Prevention*, 32, 797–804, 2000.
3. Elishakoff, I., Essay on the role of the Monte Carlo method in stochastic mechanics, In *Monte Carlo Simulation*, Schueller, G. I. and Spanos, P. D., Eds., Lisse: A.A. Balkema Publishers, 2000.
4. Lee, K. L., Stochastic dynamic simulation of chemical processes with changing uncertainties, *Computers and Chemical Engineering*, 20, 557–562, 1996.
5. Van Impe, J. F., Bernaerts, K., Geeraerd, A. H., Poschet, F., and Versyck, K. J., Modeling and prediction in an uncertain environmemt, In *Food Process Modeling*, Tijskens, L. M. M., Hertog, M., and Nicolai, B. M., Eds., Cambridge, UK: Woodhead Publishing Limited, pp. 156–179, 2001.
6. Nicolai, B. M. and De Baerdemaeker, J., A variance propagation algorithm for the computation of heat conduction under stochastic conditions, *International Journal of Heat and Mass Transfer*, 42, 1513–1520, 1999.
7. Johns, W. R., Simulation of food processes with uncertain data, *Transactions IChemE, Part C*, 70, 59–68, 1992.
8. Hanna, S. R., Chang, J. C., and Fernau, J., Monte Carlo estimates of uncertainties in predictions by a photochemical grid model (uam-iv) due to uncertainties in input variables, *Atmospheric Environment*, 32(21), 3619–3628, 1998.
9. Hammersley, J. M. and Handscomb, D. C., *Monte Carlo Methods*, New York: Wiley, 1964.
10. Dubi, A., *Monte Carlo Applications in Systems Engineering*, New York: Wiley, 2000.
11. Cooke, R. M., Uncertainty modeling: examples and issues, *Safety Science*, 26, 49–60, 1997.
12. Rubinstein, R. Y., *Simulation and the Monte Carlo Method*, New York: Wiley, 1981.
13. Ross, S. M., *Simulation*, 2nd ed., San Diego, CA: Academic Press, 1997.
14. Ripley, B. D., *Stochastic Simulation*, New York: Wiley, 1987.
15. Cinlar, E., *Introduction to Stochastic Processes*, Engelwood Cliffs, NJ: Prentice-Hall, 1975.
16. Kempthorne, O. and Folks, L., *Probability, Statistics, and Data Analysis*, Ames, IA: Iowa State University Press, 1971.
17. Devore, J. and Peck, R., *Introductory Statistics*, 2nd ed., St. Paul, MN: West Publishing Company, 1994.
18. Law, A. M. and Kelton, W. D., *Simulation Modeling and Analysis*, 2nd ed., New York: McGraw-Hill, 1991.
19. Berthea, R. M. and Rhinehart, R. R., *Applied Engineering Statistics*, New York: Marcel Dekker, 1991.
20. Gentle, J. E., *Random Number Generation and Monte Carlo Methods*, New York: Springer, 1988.
21. Ross, S. M., *Introduction to Probability Models*, 6th ed., San Diego, CA: Academic Press, 1997.
22. Nicolai, B. M. and De Baerdemaeker, J., Simulation of heat transfer in foods with stochastic initial and boundary conditions, *Transactions IChemE, Part C*, 70, 78–82, 1992.
23. Van Kampen, N. G., *Stochastic Processes in Physics and Chemistry*, North-Holland: Amsterdam, 1992.
24. Gardiner, C. W., *Handbook of Stochastic Methods*, 2nd ed., Berlin: Springer, 1985.
25. Annan, J. D., Modeling under uncertainty: Monte Carlo methods for temporally varying parameters, *Ecological Modeling*, 136, 297–302, 2001.

26. Nicolai, B. M., Verboven, P., Scheerlinck, N., and De Baerdemaeker, J., Numerical analysis of the propagation of random parameter fluctuations in time and space during thermal food processes, *Journal of Food Engineering*, 38, 259–278, 1998.
27. Zhiming, L. and Dongxiao, Z., On importance sampling Monte Carlo approach to uncertainty analysis for flow and transport in porous media, *Advances in Water Resources*, 26, 1177–1188, 2003.
28. Press, W. H., Teukolsky, S. A., Vetterling, W. T., and Flannery, B. P., *Numerical Recipes in C++: The Art of Scientific Computing*, Cambridge: Cambridge University Press, 2004.
29. Lenz, M. and Lund, D. B., The Lethality-Fourier number method. Heating rate variations and lethality confidence intervals for forced-convection heated foods in containers, *Journal of Food Process Engineering*, 2, 227–271, 1979.
30. Lund, D., Statistical analysis of thermal process calculations, *Food Technology*, 3, 76–83, 1978.
31. Hayakawa, K., Wang, J., and De Massaguer, P., Simplified predictive equations for variability of thermal process lethality, *Journal of Food Process Engineering*, 19, 289–300, 1996.
32. Smout, C., Van Loey, A., and Hendrickx, M., Non-uniformity of lethality in retort processes based on heat distribution and heat penetration data, *Journal of Food Engineering*, 45, 103–110, 2000.
33. Varga, S., Oliveira, J., and Oliveira, F., Influence of the variability of processing factors on the F-value distribution in batch retorts, *Journal of Food Engineering*, 44, 155–161, 2000.
34. Moreira, R. and Barrufet, M., Spatial distribution of oil after deep fat frying of tortilla chips from a stochastic model, *Journal of Food Engineering*, 27, 279–290, 1996.
35. Rajkumar, V., Moreira, R., and Barrufet, M., Modeling the structural changes of tortilla chips during frying, *Journal of Food Engineering*, 60, 167–175, 2003.
36. Demir, A. D., Baucour, P., Cronin, K., and Abodayeh, K., Analysis of temperature variability during the thermal processing of hazelnuts, *Journal of Innovative Food Science & Emerging Technologies*, 4, 69–84, 2003.
37. Den Aantrekker, E., Beumer, R., Van Gerwen, S., Zwietering, M., Van Schothorsa, M., and Boom, M., Estimating the probability of recontamination via the air using Monte Carlo simulations, *International Journal of Food Microbiology*, 87, 1–15, 2003.
38. Bemrah, N., Bergis, H., Colmin, C., Beaufort, A., Millemann, Y., Dufour, B., Benet, J., Cerf, O., and Sanaa, M., Quantitative risk assessment of human Salmonellosis from the consumption of a turkey product in collective catering establishments, *International Journal of Food Microbiology*, 80, 17–30, 2003.
39. Cassin, M., Lammerding, A., Todd, E., Ross, W., and McColl, S., Quantitative risk assessment for Escherichia coli O157:H7 in ground beef hamburgers, *International Journal of Food Microbiology*, 41, 21–44, 1998.
40. Syposs, Z., Reichart, O., and Meszaros, L., Microbiological risk assessment in the beverage industry, *Food Control*, 16, 515–521, 2005.
41. Nicolai, B. M. and Van Impe, J. F., Predictive food microbiology: a probabilistic approach, *Mathematics and Computers in Simulation*, 42, 287–292, 1996.
42. Miller, W. M., Non-Parametric Estimation of automated weight-filling machinery for fresh citrus, *Journal of Food Engineering*, 5, 95–107, 1986.
43. Cronin, K., Fitzpatrick, J., and McCarthy, D., Packaging strategies to counteract weight variability in extruded food products, *Journal of Food Engineering*, 56, 353–360, 2003.
44. Cronin, K., A methodology for investigation into alternative packaging methods for column wrapped biscuits, *Journal of Food Engineering*, 39, 379–387, 1999.

CHAPTER 17

Dimensional Analysis

Law Chung Lim, Shyam S. Sablani, and Arun S. Mujumdar

CONTENTS

17.1 INTRODUCTION

Dimensional analysis is a mathematical tool that is used to reduce complex physical problems to the simplest forms before quantitative analysis and experimental investigation are carried out. The reduction number of variables uses the Buckingham theorem[1] as its central tool. In addition, it allows identification of dimensionless parameters even if the form of equation is still unknown.

Therefore, it reduces, by an order of magnitude, the number of experiments needed to generalize or empirically correlate a set of data.

Dimensional analysis, in its most primitive form, is used to check the validity of algebraic derivations or equations. Every term or quantity in a physically meaningful expression or equation has the same dimension, and they can be added and subtracted from one another. Both sides of any expression must have the same dimension.

In engineering, empirical results obtained from experiments are sometimes difficult to present in a readable form such as in the form of graphs. Here, dimensional analysis provides a way to choose the relevant data and to present them concisely. This, in turn, helps the development of theoretical modeling of the problem. Relationships between influencing factors can be determined, generalization of experimental data can be performed. It is useful for predicting performance of different systems.[2,3]

Its application in science and engineering is ubiquitous, including its use in food processing. It is an economical way to scale up processes as dimensional analysis reduces the degree of freedom of the physical problems to the minimum. Dimensional analysis is a useful tool in engineering experimentation and analysis, modeling, design, and scale up. This technique is extensively used in other fields, although the focus here is on engineering, scientific, and technological applications.

17.2　LIMITATIONS

One important limitation of dimensionless analysis is that it does not unravel the underlying physics or the nature of a physical phenomenon. Therefore, variables that affect or influence the phenomenon should be known prior to dimensional analysis. It should be noted that selection of variables is vital in ensuring a successful dimensional analysis. Therefore, it is important to know a priori the relevance of parameters to be included in such analysis. Sometimes, it may be necessary to carry out an iterative process by including or excluding certain parameters and correlating the resulting dimensionless groups using experimental data.

17.3　HOW TO OBTAIN DIMENSIONLESS NUMBERS

Dimensionless numbers can be derived by variables and parameters governing any process from a number of methods. Two most common methods in use are dimensionless analysis of the differential conservation equation and the Buckingham Π method. There are alternative avenues for the generation of dimensionless groups as well, but they are beyond the scopes of this concise chapter.

17.3.1　Buckingham's Π Theorem

Buckingham's first theorem states that the number of independent dimensionless numbers, m, that can be formed is equal to the total number of the physical quantities, n, minus the number of primary dimensions, r, that are used to express dimensional formulas of the physical quantities in question. In other words, every physical relationship between n physical quantities can be reduced to a relationship between m mutually independent dimensionless numbers.[1–3]

If a physical problem is expressed by n independent physical quantities, Q_i

$$f(Q_1, Q_2, Q_3, ..., Q_n) = 0,$$

then, according to the Buckingham's first theorem, this can also be expressed by $(n^N m$ independent dimensional numbers, π_i, as

$$f(\pi_1, \pi_2, \pi_3, \ldots, \pi_{n-m}) = 0.$$

Buckingham's second theorem states that each π group is a function of core variables (also known as *governing* or *repeating* variables) plus one of the remaining variables. Core variables are those variables that will most probably appear in all or most of the π groups. According to the second theorem, only m core variables are to be chosen. Core variables can be freely chosen by following the rules:

- A combination of the core variables must contain all basic dimensions.
- A combination of the core variables must not form a dimensionless group.
- The core variable does not necessarily appear in all π groups.
- The core variables that are chosen must be measurable in an experimental investigation that is of major interest to the designer.

It should be noted that if extra unimportant variables are introduced, then extra π groups will be formed. They will play small roles in influencing the physical behavior of the problem. However, if an important influential variable is missed, then a π group will be missing. Experimental analysis based on these results may miss significant behavioral changes. Therefore, it is very important to choose all influencing factors when a relevance list is made.

A relevance list consists of all influencing dimensional parameters. The parameters consist of a target quantity, geometric parameters, material parameters, and process-related parameters. In each case, only one target quantity is to be considered, and it is the only dependant parameter. The rest of the parameters in the list are independent of one another.[4,5]

Generally, the quantities and parameters are lished in a bracket { }. Semicolon separates the quantities and parameters into various categories which are mentioned below. Therefore, a reference list is normally written in the following form:

{target quantity; geometric parameters; material parameters; process-related parameters}

For an incompressible fluid of density ρ and viscosity μ which is flowing through a tube of length L at velocity v, the pressure drop over the tube length is ΔP. The relevance list is constructed based on the relevant physical quantities mentioned above.

In this example, the relevance list is

$$\{\Delta P; L; \rho, \mu; v, g\}.$$

Note that ΔP is the target quantity, L is the geometric parameter, P and μ are the material parameters, v and g are the process related parameters.

Fundamental dimensions that made up the quantities listed in the relevance list are length, time, and mass; therefore, r is 3, the number of physical quantities in question, n, is 6; consequently, the number of dimensionless groups, m, that can be formed is 3. Three variables are to be chosen as the core variables common to all three dimensionless groups. In this case, L, v, and ρ are selected. The three dimensionless groups (also known as π groups) are

$$\pi_1 = L^a v^b \rho^c g^1,$$
$$\pi_2 = L^d v^e \rho^f \Delta P^1,$$
$$\pi_3 = L^g v^h \rho^i \mu^1.$$

If expressed in terms of basic dimensions,

$$[\pi_1] = (L)^a (Lt^{-1})^b (ML^{-3})^c (Lt^{-2})^1,$$

$$[\pi_2] = (L)^d (Lt^{-1})^e (ML^{-3})^f (ML^{-1} t^{-2})^1,$$

$$[\pi_3] = (L)^g (Lt^{-1})^h (ML^{-3})^i (ML^{-1} t^{-1})^1.$$

To make the π groups dimensionless, the physical quantities must be raised to certain exponents, a, b, c, and so forth. To evaluate the exponents, let the dimension of the π groups be zero.

Π_1	$L^0 t^0 M^0 = L^{a+b-3c+1} t^{-b-2} M^c$
Π_2	$L^0 t^0 M^0 = L^{d+e-3f-1} t^{-e-2} M^{f+1}$
Π_3	$L^0 t^0 M^0 = L^{g+h-3i-1} t^{-h-1} M^{i+1}$

Next, these exponents are equated at both sides:

Π_1	For dimension L: $0 = a+b-3c+1$ For dimension t: $0 = -b-2$ For dimension M: $0 = c$ Hence, $c=0$; $b=-2$; $a=1$
Π_2	For dimension L: $0 = d+e-3f-1$ For dimension t: $0 = -e-2$ For dimension M: $0 = f+1$ Hence, $f=-1$; $e=-2$; $d=0$
Π_3	For dimension L: $0 = g+h-3i-1$ For dimension t: $0 = -h-1$ For dimension M: $0 = i+1$ Hence, $i=-1$; $h=-1$; $g=-1$

Substituting these exponents values into the π groups equations yield the following dimensionless groups:

$$\pi_1 = L^1 v^{-2} g^1 \text{ gives } \pi_1 = \frac{gL}{v^2} \text{ that is the reciprocal of the Froude number;}$$

$$\pi_2 = v^{-2} \rho^{-1} \Delta P^1 \text{ gives } \pi_2 = \frac{\Delta P}{\rho v^2} \text{ that is the Euler number;}$$

$$\pi_3 = L^{-1} v^{-1} \rho^{-1} \mu^1 \text{ gives } \pi_3 = \frac{\mu}{Lv\rho} \text{ that is the reciprocal of the Reynolds number.}$$

Another way to obtain dimensionless numbers in the Buckingham method is to use dimensional matrix to determine the dimensionless numbers. The columns of the dimensional matrix are assigned to the individual physical quantities and the rows are assigned to the exponent values.

The dimensional matrix is subdivided into a quadratic core matrix and a residual matrix.[5,6]

Dimension	L	v	ρ	g	ΔP	μ	
L	1	1	-3	1	-1	1	Quadratic core matrix (3 x 3)
							Residual matrix (3 x 3)
T	0	-1	0	-2	-2	-1	
M	0	0	1	0	1	1	
L+T	1	0	-3	-1	-3	-2	
-T	0	1	0	2	2	1	
M	0	0	1	0	1	1	
(L+T)+3M	1	0	0	-1	0	1	Identity matrix (3 x 3)
-T	0	1	0	2	2	1	
M	0	0	1	0	1	1	
	Core			Residual			

To obtain the dimensionless numbers, each quantity in the residual matrix forms the numerator of a fraction, whereas its denominator consists of the fillers from the core matrix (that has been transformed into identity matrix by mathematical manipulation) with exponents indicated in the residual matrix. The π groups are

$$\pi_1 = \frac{g}{L^{-1}v^2\rho^0} \text{ gives } \pi_1 = \frac{gL}{v^2} \text{ that is the reciprocal of the Froude number;}$$

$$\pi_2 = \frac{\Delta P}{L^0\rho^1v^2} \text{ gives } \pi_2 = \frac{\Delta P}{\rho v^2} \text{ that is the Euler number;}$$

$$\pi_3 = \frac{\mu}{L^1v^1\rho^1} \text{ gives } \pi_3 = \frac{\mu}{Lv\rho} \text{ that is the reciprocal of the Reynolds number.}$$

Both methods of the Buckingham theorem give a Froude number, Fr, an Euler number, Eu, and a Reynolds number, Re.

17.3.2 Dimensional Analysis of Governing Differential Equations

Differential equations are often derived from first principles to describe various transport phenomena. Essentially, they are equations of conservation of mass, momentum, species, and energy. Dimensional homogeneity requires that every term in a physically meaningful differential equation has the same units or dimensions. The ratio of one term in the equation to another term is necessarily dimensionless, and the ratio is known as dimensionless number (or group). Therefore, the interpretation of the dimensionless number is clear if one knows the physical meaning of each term in the equation.

For example, the x-component of the Navier–Stokes equation for laminar steady flow of an incompressible Newtonian fluid is

$$\vec{v}_x \frac{\partial \vec{v}_x}{\partial x} + \vec{v}_y \frac{\partial \vec{v}_y}{\partial y} + \vec{v}_z \frac{\partial \vec{v}_z}{\partial z} = g_x - \frac{\partial P}{\rho \partial x} + \frac{\mu}{\rho}\left(\frac{\partial^2 \vec{v}_x}{\partial x} + \frac{\partial^2 \vec{v}_y}{\partial y} + \frac{\partial^2 \vec{v}_z}{\partial z}\right) \quad (17.1)$$

Equation 17.1 can be expressed in terms of dimensional equality as Equation 17.2. Each term in Equation 17.2 has the dimension of $[L/t^2]$, viz. acceleration.

$$\left[\frac{\vec{v}^2}{L}\right] = [g] - \left[\frac{P}{\rho L}\right] + \left[\frac{\mu \vec{v}}{\rho L^2}\right] \tag{17.2}$$

The physical meaning of each term in Equation 17.2 is commonly given as

[inertia force] = [gravity force] − [pressure force] + [viscous force].

The equal sign here is not correct, however. This equation simply represents a balance between various forces acting on a fluid particle in the flow system.

There are six ratios that can be formed from the four terms shown in Equation 17.2, namely

$$\frac{[\vec{v}^2/L]}{[g]}, \frac{[\vec{v}^2/L]}{[P/\rho L]}, \frac{[\vec{v}^2/L]}{[\mu v/\rho L^2]}, \frac{[g]}{[P/\rho L]}, \frac{[g]}{[\mu \vec{v}/\rho L2]}, \text{ and, } \frac{[P/\rho L]}{[\mu \vec{v}/\rho L2]},$$

that gives the following dimensionless numbers:

$$\frac{\vec{v}^2}{gL}, \frac{\rho \vec{v}^2}{P}, \frac{L \vec{v} \rho}{\mu}, \frac{L \rho g}{P}, \frac{L^2 \rho g}{\mu \vec{v}}, \frac{PL}{\mu \vec{v}}.$$

Dimensionless numbers or their reciprocals derived this way have clear physical meaning. Some of the well-known dimensionless groups obtained from the Navier–Stokes equation are given in Table 17.1. Note that products of dimensionless groups as well as ratios of such groups raised to any integral power are also dimensionless. However, they have no physical meaning in general and are rarely used. It is important to note that the numerical values of some dimensionless numbers can lead to misleading interpretation of the physics involved, so care must be exercised in attributing significance to these numbers.

17.3.3 Dimensional Analysis on Transport Equations

The same method shown in the above example is performed on mass, heat, and momentum transport equations, respectively. Note that to obtain dimensionless heat- and mass-transfer coefficients, one needs to cast the convective boundary conditions in a special form so that the boundary

Table 17.1 Dimensionless Numbers Obtained from Dimensional Analysis of Navier–Stokes Equations

Dimensionless Number in Equation 17.4	Interpretation	Formula	Symbol	Dimensionless Group
$\vec{v}^2/gL$	Ratio of inertia force to gravity force	$\vec{v}^2/gL$	Fr	Froude number
$\rho \vec{v}^2/P$	Ratio of inertia force to pressure force	Reciprocal $= P/\rho \vec{v}^2$	Eu	Euler number
$L\vec{v}\rho/\mu$	Ratio of inertia force to viscous force	$L\vec{v}\rho/\mu$	Re	Reynolds number
$L\rho g/P$	Ratio of gravity force to pressure force	$P/L\rho g$	Eu·Fr	Product of Euler number and Froude number
$L^2\rho g/\mu \vec{v}$	Ratio of gravity force to viscous force	$L^2\rho g/\mu \vec{v}$	Re/Fr	Ratio of Reynolds number to Froude number
$PL/\mu \vec{v}^2$	Ratio of pressure force to viscous force	$PL/\mu \vec{v}$	Re·Eu	Product of Reynolds number and Euler number

condition has the same dimensions as those of the terms in the equation.

17.3.3.1 Mass-Transfer Equation

The mass transfer equation is given as

Rate of change of concentration

= Rate of change by convection + Rate of change by diffusion

+ Rate of change by homogenous chemical reaction

$$\frac{\partial c_A}{\partial t} = -\vec{v} \cdot \nabla c_A + D_{AB} \nabla^2 c_A + R_A. \tag{17.3}$$

Equation 17.3 can be expressed in terms of dimensional equality and it is given in Equation 17.4. Each term in Equation 17.4 has the dimension of $[N/L^3 t]$:

$$\left[\frac{c}{t}\right] = \left[\frac{\vec{v}c}{L}\right] + \left[\frac{D_{AB}c}{L^2}\right] + [R] \tag{17.4}$$

Likewise, there are six ratios that can be formed from the four terms in Equation 17.4.

$$\frac{L}{vt}, \frac{L^2}{D_{AB}t}, \frac{c}{Rt}, \frac{\vec{v}L}{D_{AB}}, \frac{\vec{v}c}{LR}, \frac{D_{AB}c}{L^2 R}$$

Table 17.2 shows the dimensionless numbers obtained from the mass transfer equation. The dimensionless numbers also relate to some well-known dimensionless groups used in engineering.

Every term in the mass-transfer equation can be divided with mass-transfer coefficient, $(k_L c/L)$ that has the dimension of $[N/L^3 t]$. The following four dimensionless numbers are obtained.

$$\frac{L}{k_L t}, \frac{\vec{v}}{k_L}, \frac{D_{AB}}{k_L L}, \frac{LR}{k_L c}$$

Table 17.3 shows the additional dimensionless numbers obtained from dimensional analysis of mass transfer equation by division of every term of the mass transfer (or species) equation with the mass transfer coefficient.

Table 17.2 Dimensionless Numbers Obtained from Dimensional Analysis of the Mass Transfer Equation

Dimensionless Number in Equation 17.4	Formula	Symbol	Dimensionless Group
$L/\vec{v}t$	Reciprocal $= vt/L$	Th	Thomson number
$L^2/D_{AB}t$	Reciprocal $= D_{AB}t/L^2$	Fo_m	Fourier number
c/Rt	Reciprocal $= Rt/c$	$Da_I \cdot Th$	Product of Damkohler number and Thomson number
$\vec{v}L/D_{AB}$	vL/D_{AB}	Pe_m	Peclet number
$\vec{v}c/LR$	Reciprocal $= LR/vc$	Da_I	Damkohler number
$D_{AB}c/L^2 R$	Reciprocal $= L^2 R/D_{AB}c$	Da_{II}	Damkohler number

Table 17.3   Additional Dimensionless Numbers from the Mass Transfer Equation and Convective Boundary Condition

Dimensionless Number in Equation 17.4	Formula	Symbol	Dimensionless Group
$L/k_L t$	Reciprocal $= k_K t/L$	$Fo \cdot Sh$	Product of Fourier number and Sherwood number
$\vec{v}/k_L$	Reciprocal $= k_L/\vec{v}$	Sh/Pe_m	Ratio of Sherwood number to Peclet number
$D_{AB}/k_L L$	Reciprocal $= k_L L/D_{AB}$	Sh	Sherwood number
$LR/k_L c$	$LR/K_L c$	$Pe_m/Sh \cdot Da_1$	—

17.3.3.2   Energy Equation

The energy transfer equation is given as

Rate of change of temperature

= Rate of change by convection + Rate of change by conduction

+ Rate of generation by viscous dissipation + Rate of generation by chemical reaction

+ Rate of generation by Joule heating

$$\frac{\partial T}{\partial t} = -\vec{v} \cdot \nabla T + \frac{k}{\rho C_p} \nabla^2 T + \frac{\mu \phi}{\rho C_p} + \frac{QR_A}{\rho C_p} + \frac{I^2}{\sigma_e \rho C_p}. \tag{17.5}$$

The last term in Equation 17.5 that is the rate of generation by Joule heating is omitted in the dimensional analysis performed below. Equation 17.6 in the dimensional equality of Equation 17.5. Each term in Equation 17.6 has the dimension of $[T/t]$.

$$\left[\frac{T}{t}\right] = \left[\frac{\vec{v}T}{L}\right] + \left[\frac{kT}{\rho C_p L^2}\right] + \left[\frac{\mu \vec{v}^2}{\rho C_p L^2 :}\right] + \left[\frac{QR}{\rho C_p}\right]. \tag{17.6}$$

Here the boundary condition involving heat transfer coefficient $(hT/\rho C_p L)$ is taken into account in the dimensional analysis. The division of terms in Equation 17.6 as well as the heat transfer coefficient yields the following dimensionless numbers:

$$\frac{L}{\vec{v}t}, \frac{\rho C_p L^2}{kt}, \frac{\rho C_p L^2 T}{\mu \vec{v}^2 t}, \frac{\rho C_p T}{QRt}, \frac{\rho C_p \vec{v}L}{k}, \frac{\rho C_p T}{\mu \vec{v}}, \frac{\rho C_p \vec{v}T}{LQR}, \frac{kT}{\mu \vec{v}^2}, \frac{kT}{L^2 QR}, \frac{\mu \vec{v}^2}{L^2 QR}, \frac{\rho C_p \vec{v}}{h}, \frac{\rho C_p \vec{v}}{hT}, \frac{k}{hL},$$
$$\frac{\mu \vec{v}^2}{LhT}, \frac{LQR}{hT}.$$

These dimensionless numbers include some well-known dimensionless groups that are frequently encountered in engineering; they are listed in Table 17.4.

Table 17.4 Dimensionless Numbers Obtained from Dimensional Analysis of the Differential Energy Conservation Equation

Dimensionless Number in Equation 17.4	Formula	Symbol	Dimensionless Group
$L/\bar{v}t$	Reciprocal = $\bar{v}t/L$	Th	Thomson number
$\rho C_p L^2/kt$	Reciprocal = $kt/\rho C_p L^2$	Fo	Fourier number
$\rho C_p L^2 T/\mu\bar{v}^2 t$	Reciprocal = $\mu\bar{v}^2 t/\rho C_p L^2 T$	Fo·Br	Product of Fourier number and Brinkman number
$\rho C_p T/QRt$	Reciprocal = $QRt/\rho C_p T$	Da$_{III}$·Th	Product of Damkohler number and Thomson number
$\rho C_p \bar{v}L/k$	$\rho C_p \bar{v}L/k$	Pe=Re·Pr	Product of Reynolds number and Prandtl number
$\rho C_p LT/\mu\bar{v}$	$\rho C_p LT/\mu\bar{v}$	Pe/Br	Ratio of Peclet number to Brinkman number
$\rho C_p \bar{v}T/LQR$	Reciprocal = $LQR/\rho C_p \bar{v}T$	Da$_{III}$	Damkohler number
$kT/\mu\bar{v}^2$	Reciprocal = $\mu\bar{v}^2/kT$	Br	Brinkman number
$kT/L^2 QR$	Reciprocal = $L^2 QR/kT$	Da$_{IV}$	Damkohler number
$\mu\bar{v}^2/L^2 QR$	$\mu\bar{v}^2/L^2 QR$	Br/Da$_{IV}$	Ratio of Brinkman number to Damkohler number
$\rho C_p L/ht$	Reciprocal = $ht/\rho C_p L$	Fo·Nu	Product of Fourier number and Nusselt number
$\rho C_p \bar{v}/h$	Reciprocal = $h/\rho C_p \bar{v}$	St=Nu/Pe	Stanton number
k/hL	Reciprocal = hL/k	Nu	Nusselt number
$\mu\bar{v}^2/LhT$	$\mu\bar{v}^2/LhT$	Br/Nu	Ratio of Brinkman number to Nusselt number
LQR/hT	LQR/hT	Da$_{III}$/St	Ratio of Damkohler number to Stanton number

17.3.3.3 Momentum-Transfer Equation

The momentum-transfer equation is given as

Rate of change of momentum

= Rate of change by convection + Rate of change by molecular viscous transfer

+ Rate of change due to pressure forces + Rate of change due to gravity forces

+ Rate of change due to magnetic forces

$$\frac{\partial \bar{v}}{\partial t} = -\bar{v}\cdot\nabla\bar{v} + \frac{\mu}{\rho}\nabla^2\bar{v} - \frac{\nabla P}{\rho} + g\left(\text{or} + g\frac{\Delta\rho}{\rho}\right) + \frac{\mu}{\rho}(I\times H_e). \qquad (17.7)$$

Note that the left side term expresses acceleration that is the rate of change of momentum per unit mass of the fluid.

Likewise, the last term in Equation 17.7 that is the rate of change because of magnetic forces is omitted in dimensional analysis for the momentum-transfer equation. Expressed Equation 17.7 in term of dimensional equality, it yields Equation 17.8. Each term in Equation 17.8 has the dimension of $[L/t^2]$.

$$\left[\frac{\bar{v}}{t}\right] = \left[\frac{\bar{v}^2}{L}\right] + \left[\frac{\mu\bar{v}}{\rho L^2}\right] + \left[\frac{P}{\rho L}\right] + [g]\left(\text{or}\left[g\frac{\Delta\rho}{\rho:}\right]\right). \qquad (17.8)$$

Here, the sample boundary conditions involving wall shear stress, $(\tau/\rho L)$, as well as surface tension, $(\sigma/\rho L^2)$, are taken into account in the dimensional analysis. The division of terms in

Table 17.5 Dimensionless Numbers Obtained from Dimensional Analysis of the Differential Momentum Conservation Equation

Dimensionless Number in Equation 17.4	Formula	Symbol	Dimensionless Group
$L/\bar{v}t$	Reciprocal $= \bar{v}t/L$	Th	Thomson number
$\rho L^2/\mu t$	$\rho L^2/\mu t$	Re/Th	Ratio of Reynolds number to Thomson number
$\rho\bar{v}L/Pt$	Reciprocal $= Pt/\rho\bar{v}L$	Eu·Th	Product of Euler number and Thomson number
$\bar{v}/gt$	$\bar{v}/gt$	Fr/Th	Ratio of Froude number to Thomson number
$\rho\bar{v}/gt\Delta\rho$	$\rho\bar{v}/gt\Delta\rho$	Re^2/Gr·Th	—
$\rho\bar{v}L/\mu$	$\rho\bar{v}L/\mu$	Re	Reynolds number
$\rho\bar{v}^2/P$	Reciprocal $= P/\rho\bar{v}^2$	Eu	Euler number
$\bar{v}^2/gL$	$\bar{v}^2/gL$	Fr	Froude number
$\rho\bar{v}^2/gL\Delta\rho$	$\rho\bar{v}^2/gL\Delta\rho$	Re/Gr	Ratio of Reynolds number square to Grash of number
$\mu\bar{v}/PL$	Reciprocal $= PL/\mu\bar{v}$	Re·Eu	Product of Reynolds number and Euler number
$\mu\bar{v}/\rho gL^2$	$\mu\bar{v}/\rho gL^2$	Fr/Re	Ratio of Froude number to Reynolds number
$\mu\bar{v}/gL^2\Delta\rho$	$\mu\bar{v}/gL^2\Delta\rho$	Re/Fr	Ratio of Reynolds number to Froude number
$P/\rho gL$	Reciprocal $= \rho gL/P$	Eu·Fr	Product of Euler number and Froude number
$P/gL\Delta\rho$	$P/gL\Delta\rho$	Eu·Re^2/GR	—
$\rho/\Delta\rho$	—	—	—
$\rho\bar{v}L/t\tau$	Reciprocal $= t\tau/\rho\bar{v}L$	Ne·Th	Product of Newton number and Thomson number
$\rho\bar{v}L^2/t\sigma$	$\rho\bar{v}L^2/t\sigma$	We/Th	Ratio of Weber number to Thomson number
$\rho\bar{v}^2/\tau$	Reciprocal $= \tau/\rho\bar{v}^2$	Ne	Newton number
$\rho\bar{v}^2L/\sigma$	$\rho\bar{v}^2L/\sigma$	We	Weber number
$\mu\bar{v}/L\tau$	Reciprocal $= L\tau/\mu\bar{v}$	Re·Ne	Product of Reynolds number and Newton number
$\mu\bar{v}/\sigma$	$\mu\bar{v}/\sigma$	Ca=We/Re	Capillary number
P/τ	P/τ	We/Ne	Ratio of Weber number to Newton number
PL/σ	PL/σ	Eu/We	Ratio of Euler number to Weber number
$\rho gL/\tau$	Reciprocal $= \tau/\rho gL$	Ne·Fr	Product of Newton number and Froude number
$\rho gL^2/\sigma$	$\rho gL^2/\sigma$	We/Fr	Ratio of Weber number to Froude number

Equation 17.8 as well as the two coefficients yield the following dimensionless numbers:

$$\frac{L}{\bar{v}t},\ \frac{\rho L^2}{\mu t},\ \frac{\rho\bar{v}L}{Pt},\ \frac{\bar{v}}{gt},\ \frac{\rho\bar{v}}{gt\Delta\rho},\ \frac{\rho\bar{v}L}{\mu},\ \frac{\rho\bar{v}^2}{P},\ \frac{\bar{v}^2}{gL},\ \frac{\rho\bar{v}^2}{gL\Delta\rho},\ \frac{\mu\bar{v}}{PL},\ \frac{\rho C_p\bar{v}}{h},\ \frac{\rho C_p\bar{v}}{hT},\ \frac{P}{\rho gL},$$

$$\frac{P}{gL\Delta\rho},\ \frac{\rho}{\Delta\rho},\ \frac{\rho\bar{v}L}{t\tau},\ \frac{\rho\bar{v}L^2}{t\sigma},\ \frac{\rho\bar{v}^2}{\tau},\ \frac{\rho v^2 L}{\sigma},\ \frac{\mu\bar{v}}{L\tau},\ \frac{\mu\bar{v}}{\sigma},\ \frac{P}{\tau},\ \frac{PL}{\sigma},\ \frac{\rho gL}{\tau},\ \frac{\rho gL^2}{\sigma},\ \frac{gL\Delta\rho}{\tau},$$

$$\frac{gL^2\Delta\rho}{\sigma}.$$

Table 17.5 lists the dimensionless numbers shown above, and these numbers can be related to some well-known dimensionless groups frequently used in engineering.

17.4 LIST OF DIMENSIONLESS NUMBERS

A list of dimensionless numbers that are frequently used in modeling or correlating mass, heat, and momentum transport processes as well as coupled processes is given in Table 17.6. Formulas and interpretation of the dimensionless numbers are given as well as the areas where these numbers are applied.

Table 17.6 Dimensionless Numbers Encountered in the Literature Dealing with Transport Processes

Group	Formula	Interpretation	Application
Archimedes number	$Ar = gL^3 \rho_f(\rho - \rho_f)/\mu^2$ g—gravitational acceleration (m/s^2) L—characteristic length (m) ρ—density of body (kg/m^3) ρ_f—density of fluid (kg/m^3) μ—viscosity (kg/s m)	Ratio of gravitational forces to viscous force, used to relate motion of fluids and particles due to density differences, as for fluidized beds	Momentum transfer (general); buoyancy, fluidization, and fluid motion due to density difference
Biot number	$Bi = (hL/k_s)$ h—overall heat transfer coefficient (W/m^2 · K) L—characteristic length (m) k_s—thermal conductivity of solid (W/m · K)	Ratio of the internal thermal resistance of a solid to the boundary layer (or surface film) thermal resistance Can also be regarded as ratio of conductive to convective heat resistance Biot number relates the heat transfer resistance inside and the surface of a solid $Bi > 1$ implies that the heat conduction inside the solid is slower than at its surface. Thus, temperature gradient inside the solid cannot be neglected	Unsteady state heat transfer
Mass transfer Biot number	$Bi_m = h_m L/D_{AB}$ h_m—overall mass transfer coefficient (m/s) L—characteristic length (m) D_{AB}—binary mass diffusion coefficient (m/s^2)	Ratio of mass transfer resistance in internal species to mass transfer resistance at boundary layer (interface) species Can also be regarded as ratio of diffusive to convective mass transfer resistance	Mass transfer between fluid and solid
Boltzmann number	$Bo = (\bar{v} c_p(sw))/\eta T$ $\bar{v}$— velocity (m/s) c_p—specific heat at constant pressure (J/kg · K) sw-specific weight (N/m^3) η—Stefan-Boltzmann constant T—absolute temperature (K)	Boltzmann number is a parameter of thermal radiation exchange that relates the enthalpy of gases and heat flow emitted at the surface	Simultaneous heat and momentum transfer
Bond number	$Bo = g(\rho_1 - \rho_v)L^2)/\sigma$ g—gravitational acceleration (m/s^2) ρ—density (kg/m^3) L—characteristic length (m) σ—surface tension, undisturbed surface tension (N/m)	Ratio of gravitational forces to surface tension forces	Momentum transfer (general); atomization, motion of bubbles and droplets

continued

Table 17.6 continued

Group	Formula	Interpretation	Application
Brinkman number	$Br = \mu\bar{v}^2/k\Delta T$ μ—viscosity (kg/s·m) $\bar{v}$—fluid velocity; local velocity (m/s) K—thermal conductivity (W/m·K) ΔT—temperature difference	Ratio of heat production by viscous dissipation to the heat transport by conduction	Heat transfer
Bulygin number	$Bu = (\lambda c_v P(T - T_0)/c)$ λ—latent heat of phase change (kJ/kg) c_v—specific vapor capacity (kJ/kg °C) P—pressure, local static pressure (Pa) T—temperature of medium/moist surface (wet-bulb temperature)/moving stream T_0—initial temperature/hot gas stream c—heat capacity of moist material (kJ/kg °C)	Ratio of heat of vaporization to sensible heat to bring liquid to boiling point Bulygin number represents high intensity heat and mass transfer during evaporation	Heat transfer during evaporation
Capillary number	$Ca = \mu\bar{v}/\sigma;$ $Ca = We/Re$ μ—viscosity (kg/s m) $\bar{v}$—fluid velocity; local velocity (m/s) σ—surface tension, undisturbed surface tension (N/m)	Ratio of viscous forces to surface tension forces Capillary number describes the flow of fluids through thin tubes (capillaries)	Momentum transfer (general); atomization and two-phase flow in beds of solids
Coefficient of friction	$C_f = \tau_s/(\rho\bar{v}^2/2)$ τ_s—shear stress (N/m²) ρ—density (kg/m³) $\bar{v}$—velocity (m/s)	Ratio of the force that maintains contact between an object and a surface and the frictional force that resists the motion of the object Can also be regarded as dimensionless surface shear stress	Momentum transfer
Dean number	$De = (d\bar{v}\rho/\mu)\sqrt{r/r_{eff}}$ d—diameter of pipe, particle, bubble, droplet, impeller, shaft, etc., (m) $\bar{v}$—velocity (m/s) ρ—density (kg/m³) μ—viscosity (kg/s m) r—radius of pipe, particle, bubble, droplet, impeller, shaft, etc., (m) r_{eff}—radius of curvature of bend	Ratio of centrifugal force to inertia force in fluid flow in a curve duct	Momentum transfer (general); flow in curved channels

Name	Formula/Symbols	Description	Application
Deborah number	$De = t_r/t_0$ t_r—relaxation time or reaction time (s) t_0—observation time (s) $De = (t_r D_{AB}/L^2)$ t_r—relaxation time or reaction time (s) D_{AB}—binary mass diffusion coefficient (m/s^2) L—characteristic length (m)	The first equation shows the ratio of relaxation time to observation time. The second equation shows the ratio of the relaxation time to the characteristics diffusion time. Deborah number indicates how "fluid" a material is; the smaller the Deborah number, the more fluid the material appears	Rheology
Eckert number	$Ec = \bar{v}^2/(c_p(T_s - T_\infty))$ $\bar{v}$—velocity (m/s) c_p—specific heat at constant pressure (J/kg·K) T—temperature (K)	Ratio of kinetic energy of the flow to the boundary layer enthalpy difference. Can also be regarded as ratio of temperature rise to temperature difference between wall and fluid at the boundary layer in adiabatic flow	Momentum and heat transfer (general); compressible flow
Euler number	$Eu = \Delta P/\rho\bar{v}^2$ ΔP—pressure drop due to friction (Pa) ρ—density (kg/m^3) $\bar{v}$—velocity (m/s)	Ratio of friction head to two velocity head. Euler number represents the relationship between the pressure drop due to friction and the inertia forces in a flow stream	Momentum transfer (general); fluid friction in conduits
Fedorov number	$Fe = d_e\{(4g\rho^2(\rho_p/\rho_g))/3\mu^2\}^{1/3}$ d_e—equivalent particle diameter (m) g—gravitational acceleration (m/s^2) ρ—density (kg/m^3) ρ_p—density of particle or droplet (kg/m^3) ρ_g—density of gas (kg/m^3) μ—viscosity (kg/s m)	Fedorov number represents the relationship between the flow of particles and gas in a fluidized bed	Fluidized beds
Fourier number	$Fo = \alpha t/L^2$; $\alpha = k/c_p\rho$ α—thermal diffusivity (m^2/s) t—time (s) L—characteristic length (m)	Ratio of the heat conduction rate to the rate of thermal energy storage in a solid. Can also be regarded as ratio of current time to time to reach steady state. Dimensionless time in temperature curves, used in explicit finite difference stability criterion. Fourier number characterizes the connection between the rate of change of temperature, physical properties, and the dimension of the product in the unsteady heat transfer, conduction for heat transfer	Heat transfer (general); unsteady state heat transfer
Mass transfer Fourier number	$Fo_m = D_{AB}t/L^2$ D_{AB}—binary mass diffusion coefficient (m/s^2) t—time (s)	Ratio of the species diffusion rate to the rate of species storage. Dimensionless time	Mass transfer (general); unsteady state mass transfer

continued

Table 17.6 continued

Group	Formula	Interpretation	Application
	L—characteristic length (m)	It characterizes the connection between the rate of change of temperature, physical properties, and the dimension of the product in the unsteady mass transfer, diffusion for mass transfer	
Friction factor	$f = \Delta P/((L/D)(\rho \bar{v}_f^2/2))$ ΔP—pressure drop (N/m^2) L—characteristic length (m) D—diameter (m) ρ—density (kg/m^3) $\bar{v}_f$—mass average fluid velocity (m/s)	Friction factor expresses the linear relationship between mean flow velocity and pressure gradient Can also be regarded as dimensionless pressure drop for internal flow	Internal flow (general); fluid friction in conduits
Froude number	$Fr = \bar{v}/gL$ $\bar{v}$—velocity (m/s) g—gravitational acceleration (m/s^2) L—characteristic length (m)	Ratio of inertia force and gravity force in homogenous fluid flow In fluid dynamics, the Froude number is the reciprocal of the square root of the Richardson number	Momentum transfer (general); open channel flow and wave and surface behavior
Galileo number	$Ga = L^3 g\rho^2/\mu^2$ L—characteristic length (m) g—gravitational acceleration (m/s^2) ρ—density (kg/m^3) μ—viscosity (kg/s·m)	Ratio of gravity force to viscous force Galileo number measures the force of molecular friction and the force of gravity in fluid flow, particularly for a viscous fluid	Momentum and heat transfer (general); viscous flow/circulation and thermal expansion calculations in particular
Gay Lussac number	$Ga = 1/\beta \Delta T$ β—coefficient of bulk expansion (K^{-1}) ΔT—liquid superheat temperature difference (K)		Thermal expansion processes
Graetz number	$Gz = \dot{m}c_p/kL$ $\dot{m}$—mass flow rate (kg/s) c_p—specific heat at constant pressure (kJ/kg·°C) k—thermal conductivity (W/m·K) L—characteristic length (m)	Ratio of thermal capacity fluid to convective heat transfer in forced convection of a fluid in streamline flow Equivalent to $\{(L/d)/(Re \cdot Pr)\}$ or $\{(L/d)/Pe\}$	Heat transfer (general); streamline flow, convection in laminar flow
Grashof number	$Gr = (g\beta(T_s - T_\infty)L^3)/V^2$; $v = \mu/\rho$ g—gravitational acceleration (m/s^2) β—volumetric thermal expansion coefficient (K^{-1}) T—temperature (K) L—characteristic length (m) v—kinetic viscosity (m^2/s)	Ratio of natural convection buoyancy force to viscous force acting on fluid in natural convection	Heat transfer (general); free convection

Colburn j factor (heat transfer)	$j_H = St \cdot Pr^{2/3}$ $j_H = (h/c_p\rho v)(c_p\mu/k)^{2/3}$ St—Stanton number Pr—Prandtl number	Dimensionless heat transfer coefficient	Heat transfer (general); free and forced convection
Colburn j factor (mass transfer)	$j_m = St_m Sc^{2/3}$ St_m—mass transfer Stanton number Sc—Schmidt number	Dimensionless mass transfer coefficient	Mass transfer (general)
Jakob number	$Ja = c_p(T_s - T_{sat})/h_{fg}$ c_p—specific heat at constant pressure (J/kg·K) T—temperature (K) λ—latent heat of vaporization (J/kg)	Ratio of sensible to latent heat energy absorbed during liquid-vapor phase change	Heat transfer (general); liquid-vapor phase change
Karman number	$Ka = (\rho d^3(-dP/dL))/\mu^2$ ρ—density (kg/m³) d—diameter of pipe, particle, bubble, droplet, impeller, shaft, etc., (m) P—pressure, local static pressure (Pa) L—characteristic length (m) μ—viscosity (kg/s·m)	Karmen number is a measure of stream turbulence in fluid flow	Momentum transfer (general); fluid friction in conduits
Kirpichev number (heat transfer)	$Ki_h = qL/K\Delta T$ q—heat flux (W/m²) L—characteristic length (m) k—thermal conductivity (W/m·K) ΔT—liquid superheat temperature difference (K)	Ratio of external heat transfer intensity to internal heat transfer intensity Kirpichev number relates the drying conditions for heat transfer	Heat transfer (general)
Kirpichev number (mass transfer)	$Ki_m = GL/D_{AB}\rho n$ G—mass low rate per unit area (kg/m²·s) L—characteristic length (m) D_{AB}—Binary mass diffusion coefficient (m²/s) ρ—density (kg/m³) n—specific mass constant (kg/kg)	Ratio of external mass transfer intensity to internal mass transfer intensity Kirpichev number relates the drying conditions for mass transfer	Mass transfer (general)
Knudsen number	$Kn = (L_{mfp}/L)$ L_{mfp}—length of mean free path (m) L—characteristic length (m)	Ratio of the molecular mean free path length to characteristics physical length Knudsen number represents the relationship of the mean free path of gaseous diffusion in drying beds in relation to the characteristic length dimension of the bed	Momentum and mass transfer (general); very low pressure gas flow
Kossovich number	$Ko = \lambda X/c_p\Delta T$ λ—latent heat of phase change (kJ/kg) X—moisture content (kg/kg)	Ratio of heat used for evaporation to heat used in raising temperature of body	Heat transfer (general); convective heat transfer during evaporation

continued

Table 17.6 continued

Group	Formula	Interpretation	Application
	c_p—specific heat (kJ/°C kg)		
	ΔT—liquid superheat temperature difference (K)		
Lebedov number	$Le = (\varepsilon b_t(T_{surr} - T_0))/c_v P_{\rho s}$	Ratio of the molar expansion flux to the molar vapor transfer flux for drying of porous materials	Drying of porous materials
	ε—voids		
	b_t—vapor expansion in capillaries (kg/m³K)		
	T_{surr}—temperature of surrounding medium (K)		
	T_0—initial temperature (K)		
	c_v—specific vapor capacity (kg/kg Pa)		
	P—pressure (Pa)		
	ρ_s—density of solids (kg/m³)		
Lewis number	$Le = \alpha/D_{AB}$; $\alpha = k/c_p\rho$; $Le = Sc/Pr$	Ratio of the thermal diffusivity and mass diffusivity of a material	Combined heat and mass transfer
	α—thermal diffusivity (m²/s)	Also known as the Lykov–Lewis number	
	D_{AB}—binary mass diffusion coefficient (m/s²)		
Miniovich number	$Mn = SR/\varepsilon$	Miniovich number relates the pore size and porosity of a product being dried	Drying
	S: particle area/particle volume (m⁻¹)		
	R: radius of pipe, pore, shaft, etc.; radius of curvature of bend (m)		
	ε: voidage or porosity [–]		
Newton number	$Ne = \tau/\rho\vec{v}^2$	Ratio of drag force to inertia force	Momentum transfer
	τ—torque or shear stress		
	ρ—density (kg/m³)		
	$\vec{v}$—fluid velocity (m/s)		
Nusselt number	$Ne = \tau/\rho\vec{v}^2$	Ratio of the total heat transfer to the conduction heat transfer in forced convection	Heat transfer (general); forced convection
	h—overall heat transfer coefficient (W/m²·K)		
	L—characteristic length (m)		
	k_f—thermal conductivity of fluid (W/m·K)	Dimensionless temperature gradient at the surface	
		It is used to calculate heat transfer coefficient, h	
Ostrogradsky number	$Os = q_v L^2/k\Delta T$	Ostrogradsky number relates the internal heating of a product the thermal properties of the medium (solid, liquid, gas)	Heat transfer
	q_v—strength of the internal heat source (W/m³)		
	L—characteristic length (m)		
	k—coefficient of thermal conductivity of medium (W/m K)		
	ΔT—temperature difference (K)		
Peclet number	$Pe = \vec{v}L/\alpha$; $\alpha = k/c_p\rho$; $Re \cdot Pr$	Ratio of bulk heat transfer to conductive heat transfer	Heat transfer (general); forced convection
	$\vec{v}$—fluid velocity (m/s)		

Name	Definition / symbols	Description	Application
Peclet number (mass transfer)	L—characteristic length (m); α—thermal diffusivity (m²/s); $Pe_m = vL/D_{AB}$; $Pe_m = Re\ Sc$; $\bar{v}$—fluid velocity (m/s); L—characteristic length (m); D_{AB}—binary mass diffusion coefficient (m²/s)	It is a dimensionless independent heat transfer parameter; Ratio of convective mass transfer to diffusive mass transfer; It is a dimensionless independent mass transfer parameter	Mass transfer
Prandtl number	$Pr = c_p\mu/k$; $Pr = (v/\alpha)$; c_p—specific heat at constant pressure (J/kg K); μ—viscosity (kg/s m); k—thermal conductivity (W/m·K); v—kinetic viscosity (m²/s)	Ratio of momentum diffusivity and thermal diffusivity; $Pr=1$ gives boundary layers of equal thickness; $Pr>1$ gives a thinner velocity boundary layer as momentum transfer is more rapid than heat transfer	Heat transfer (general); free and forced convection
Rayleigh number	$Ra = L^3\rho g\beta\Delta T/\mu\alpha$; $\alpha = k/c_p\rho$; L—characteristic length (m); ρ—density (kg/m³); g—gravitational acceleration (m/s²); β—volumetric thermal expansion coefficient (K^{-1}); T—temperature (K); μ—viscosity (kg/s·m); α—thermal diffusivity (m²/s)	Ratio of natural convective to diffusive heat/mass transport	Heat transfer (general); free convection
Reynolds number	$Re = \bar{v}L/v$; $Re = \rho\bar{v}L/\mu$; $\bar{v}$—velocity (m/s); L—characteristic length (m); V—kinetic viscosity (m²/s)	Ratio of inertia force and viscous force; This number provides a criterion for determining dynamic similarity	Momentum, heat, and mass transfer to account for dynamic similarity
Schmidt number (mass transfer Prandtl number)	$Sc = v/D_{AB}$; $v = (\mu/\rho)$; v—kinetic viscosity (m²/s); D_{AB}—binary mass diffusion coefficient (m²/s)	Ratio of the momentum and mass diffusivities	Mass transfer (general); diffusion in flowing systems
Sherwood number (mass transfer Nusselt number)	$Sh = h_m L/D_{AB}$; $Sh = j_m \cdot Re \cdot Sc^{1/3}$; h_m—overall mass transfer coefficient (m/s); L—characteristic length (m); D_{AB}—binary mass diffusion coefficient (m²/s)	Ratio of length scale to the diffusive boundary layer thickness; Sherwood number represents dimensionless concentration gradient at the surface; it is used to calculate mass transfer coefficient h_m	Mass transfer
Stanton number	$St = h/\bar{v}\rho c_p$; $St = Nu/Re\cdot Pr$; h—overall heat transfer coefficient (W/m² K); ρ—density (kg/m³); $\bar{v}$—velocity (m/s); c_p—specific heat at constant pressure (J/kg K)	Ratio of heat transfer to momentum transfer; Stanton number is a modified Nusselt number	Heat transfer (general); forced convection

continued

Table 17.6 continued

Group	Formula	Interpretation	Application
Mass transfer Stanton number	$St_m = h_m / \bar{v}$; $St_m = Sh/Re \cdot Sc$ h_m—overall mass transfer coefficient (m/s) $\bar{v}$—velocity (m/s)	Ratio of mass transfer to momentum transfer; it is a modified Sherwood number	Mass transfer (general); forced convection
Thomson number	$Th = \bar{v}t/L$ $\bar{v}$—velocity (m/s) L—characteristic length (m) t—time (s)	Ratio of convective transport to storage of quantity in question In mass transfer, it is the ratio of the rate of convective mass transfer to the rate of species storage In heat transfer, it is the ratio of the rate of bulk heat transfer (convective) to the rate of thermal energy storage In momentum transfer, it is the ratio of inertia force to species rate of change of velocity	Mass, heat and momentum transfer
Weber number	$We = \rho \bar{v}^2 L/\sigma$ ρ—density (kg/m³) $\bar{v}$—velocity (m/s) L—characteristic length (m) η—Stefan–Boltzmann constant	Ratio of inertia force to surface tension force	Momentum transfer (general), bubble/droplet formation, and breakage of liquid jets

This list is rather comprehensive and includes some groups found only in the Russian literature.
Source: From Hall, C. W., *Drying Technology*, 10(4), 1081–1095, 1992.

Table 17.7 Category of Dimensionless Numbers

Transport Processes	Dimensionless Numbers
Momentum transfer	Ar, Bo (Bond), Ca, C_f, De (Dean), Eu, f, Fr, Ga, Ka, Re, We
Mass transfer	Bi_m, Fo_m, j_m, Ki_m, Pe_m, Re, Sc, Sh, St_m
Heat transfer	Bi, Bu, Fo, Gz, Gr_L, J_H, Ja, Ki_h, Ko, Nu, Os, Pe_L, Pr, Ra, Re, St
Simultaneous heat and mass transfer	Bo (Boltzmann), Ec, Ga
Simultaneous mass and momentum transfer	Kn
Simultaneous heat and momentum transfer	Le, Lu
Specific area	
Fluidization	Ar, Fe
Drying	Kn, Le, Mn; Bi,[8,9] Di(Dincer)[8]
Atomization, bubbles, and droplets	Bo (Bond), Ca, We
Evaporation	Bu, Ja, Ko
Rheology	De (Deborah)
Compressible fluid	Ec
Thermal expansion	Ga
Fluid motion	Ar
Pneumatic transport (coarse grain materials)	Eu, f, Fr, Re[10]

The dimensionless numbers listed in Table 17.6 are categorized into three main transport processes, viz. heat, mass, and momentum transfers, and they are listed in Table 17.7. Dimensionless numbers that are frequently used in specific transport processes such as fluidization, drying, evaporation, and etc., are also listed in Table 17.7.

Table 17.8 lists some illustrative research findings on the application of dimensional analysis and the respective π space that consists of the dimensionless numbers of the process. Most textbooks of fluid mechanics, heat, or mass transfer provide detailed analysis of dimensional analysis and its application.

17.5 GENERALIZATION OF EXPERIMENTAL DATA TO OBTAIN EMPIRICAL CORRELATIONS

After the dimensionless numbers associated with a particular process in question are identified, experiments are designed and performed; the experiments are repeated for different values of the variables listed in the relevance list. Then the experimental result is analyzed and plotted on an $x-y$ plot. Analysis of the experimental results will show that certain variables have negligible influence and are irrelevant to the process in question. Such variables should then be removed form the relevance list and the dimensional analysis repeated. Including irrelevant parameters can distort the usefulness of dimensional analysis.

If experimental data is plotted on the graph lie and concentrated in a small band, then it indicates that an equation or an expression can be used to represent the relationship of the variables plotted on the graph. The equation obtained from the plot sometimes is valid only over the range of conditions experimentally tested. The function that formed the equation may consist of a dimensionless number or a function of several dimensionless numbers and parameters.

17.6 APPLICATIONS OF DIMENSIONAL ANALYSIS

Dimensional analysis has been applied to many areas, including areas related to food processing. In addition, many empirical equations and correlations that relate different dimensionless numbers in the respective π-space have been obtained. Here two examples are given to show in detail how the correlations and relationships between dimensionless groups are obtained from experimental findings.

Table 17.8 Some Research Findings on the Application of Dimensional Analysis in Areas Related to Food Processing

Operation	Relevance List {Target Quantity; Geometric Parameters; Material Properties; Process Parameters}	π-Space (References)
Mixing	{variation coefficient (mixing quality), χ; Drum diameter (D), drum length (L), mixing devise diameter (d) mean particle diameter (d_p), degree of fill of drum (ϕ); Effective axial dispersion coefficient (D_{eff}), particle density (ρ); Mixer rotational speed ($\bar{v}_\theta$), mixing time (θ), solid gravity ($g\rho$)}	Zlokarnik:[2] {χ, L/D, d/D, d_p/D, ϕ, $\theta\bar{v}_\theta$, Bo, Fr}
Drying	{Moisture ratio (MR); Film thickness (δ), film length (L); Gas density (ρ_g), solvent density (ρ_L), gas kinematic viscosity (v), gas heat capacity (C_{pg}), solvent heat capacity (C_{pl}), gas thermal diffusivity (α), solvent diffusivity (D_{AB}), solvent mass transfer coefficient (k_L), solvent vapor pressure (P_{vL}), Solvent heat of evaporation (ΔH); Gas throughput (F), gas pressure (P), gas temperature (T), drying time (t)}	Zlokarnik:[2] {MR; δ/L; ρ_L/ρ_g, C_{pL}/C_{pg}, P_{vL}/P, Sc, Sh, $k_L^2/\Delta H$, Pr; Fo, $\Delta H/C_{pL}T$, $P/\rho\Delta H$, Re}
Bubbling gas fluidized bed	{Target quantity; Bed diameter (D), Bed height (H) Fluid density (ρ_f), solid density (ρ_s), particle sphericity (ψ), fluid viscosity (μ), particle diameter (d) Superficial gas velocity (u_0), gravity acceleration (g)}	Zhang and Yang:[14] {gD/u_0^2, $\rho_s^2 g(\psi d)^4/\mu_f^2 D$} for $Re_p \leq 4$ {gD/u_0^2, $\rho_f D/\rho_s \psi_s d$} for $Re_p \geq 400$ {gD/u_0^2, $\rho_s^2 g(\psi d)^4/\mu_f^2 D$, $\rho_f D/\rho_s \psi_s d$} for $4 \leq Re_p \leq 400$
Circulating fluidized bed	{Target quantity; Column diameter (D); Particle diameter (d_p), particle density (ρ_p), gas density (ρ_g), gas viscosity (μ); External solids circulation flux (G_s), superficial gas velocity ($\bar{v}_g$), acceleration gravity (g)}	van der Meer, Thorpe, and Davidson:[15] d/D, ρ_p/ρ_g, $G_s/\rho_p u_0$, Re, Fr
Spouted bed	{Target quantity; Column diameter (D), bed height (H); Particle diameter (d_p), particle density (ρ_p), gas density (ρ_g), gas viscosity (μ), particle sphericity (ψ), bed voidage (ε), internal friction angle (ϕ); Superficial gas velocity (u_0), acceleration gravity (g)}	He, Lim, and Grace:[16] gd_p/u^2, $\rho_p d_p u_0/\mu$, ρ_g/ρ_p, H/d_p, D/d_p, ψ, ε, ϕ

17.6.1 Convective Heat-Transfer Coefficients in Cans

Thermal processing or canning is one of the most effective methods of food preservation and assurance of bacteriological safety. Since the early 1950s, agitation sterilization processing has been recognized as an effective method for achieving high quality foods. During agitation processing, the heat transfer to particulate liquids in cans is considerably more complex. Examples of particulate liquid cans are vegetable chunks in brine, fruit pieces in syrup or juices, meatballs in tomato sauce, etc. To establish a thermal processing schedule for such systems, experimental transient temperatures of liquid and particle center are needed. Theoretical models can also be used for the design, optimization, and validation of such systems. Overall heat transfer coefficient from heating medium to canned liquid (U) and liquid to particle heat transfer coefficient (h_{fp}) data can be used for the prediction of temperature profiles for liquid and solid particles besides relevant thermal and physical properties. The convective heat transfer coefficients associated with canned foods undergoing thermal processing are influenced by various operating conditions as well as liquid and particle properties.[11,12]

The convective heat-transfer coefficients (U and h_{fp}) are expressed in terms of the Nusselt number (Nu) that is a function of other dimensionless numbers, consisting of relevant properties of the liquid, particles, and system. Sablani et al.[12] developed dimensionless correlations for estimating convective heat transfer coefficients for cans with rotational processing. They used the data of experimental and mathematical study conducted by Sablani.[11] A summary of the range of operating and product parameters used in the determination of convective heat transfer coefficients is presented in Table 17.9. The experimental and mathematical procedure for estimation of the convective heat transfer coefficients in cans is described in Sablani et al.[12]

Experimental data obtained for U and h_{fp} were used to calculate the Nusselt number using the relationship $Nu = U$ (of h_{fp}) d_{cd}/k_l where d_{cd} and k_l are the characteristic dimension and thermal conductivity of the liquid, respectively. Other dimensionless numbers were calculated using the physical properties of liquid and particle (at average bulk temperature) and system (operating) parameters. The characteristic length $D_r + D_c$ (diameter of rotation + diameter of can) was used in the Nu based on overall heat-transfer coefficient, and equivalent particle diameter ($d_e = (6 \times$ volume of particle$/\pi)^{0.33}$) was used in the Nu based on liquid to particle heat transfer coefficient.[12]

Analysis of variance on the experimental data has shown that the convective heat transfer coefficients (U and h_{fp}) are influenced by rotational speed, liquid viscosity, particle size, shape, and concentration. Therefore, the Nusselt number (Nu) was modeled as a function of relevant dimensionless groups

Overall heat-transfer coefficient (U):

$$Nu = f\left(Re, Pr, Fr, Ar, \frac{\varepsilon}{100-\varepsilon}, \frac{D_e}{D_c}, \psi\right). \tag{17.9}$$

Liquid-to-particle heat-transfer coefficient (h_{fp}):

$$Nu = f\left(Re, Pr, \frac{k_p}{k_l}, \frac{\varepsilon}{100-\varepsilon}, \frac{D_e}{D_c}, \psi\right). \tag{17.10}$$

For the overall heat-transfer coefficient, a stepwise multiple regression analysis of experimental data on various factors, represented in dimensionless form, eliminated Froude number as nonsignificant ($P > 0.05$) in comparison with the other parameters. The following equation gave the best fit

Table 17.9 System and Product Parameters Used in the Determination of Convective Heat-Transfer Coefficients

System and Product Parameters	Symbol	Experimental Range
Heating medium temperature	T_R	110, 120, and 130°C
Diameter of rotation	D_r	0, 0.18, 0.38, and 0.54 m
Rotation speed	N	10, 15, and 20 rpm
Can liquids	Water and oil	
Particle type and shape	Nylon and sphere diameter (D)	0.01905, 0.02225, and 0.025 m
	Cube (L_{cu})	0.01905 m
	Cylinder ($L_{cyl} \times D_{cyl}$)	0.01905×0.01905 m
Particle concentration	ε	20, 30, and 40% (v/v)

$(R^2 = 0.99)$ for the experimental data for the overall heat transfer coefficient with multiple particles:

$$Nu = 0.71Re^{0.44}Pr^{0.36}\left(\frac{\varepsilon}{100-\varepsilon}\right)^{-0.37}\left(\frac{D_e}{D_c}\right)^{-0.11}\psi^{0.24}. \tag{17.11}$$

The above correlation (Equation 17.11) is valid for the Re in the range 1.7×10^4–5.4×10^5, Pr in the range of 2.6–90.7, ε in the range of 20–40% (v/v), the ratio of d_e/D_c in the range of 0.22–0.29, and ψ in the range of 0.806–1.

Regression analysis of the experimental data, obtained for liquid to particle heat transfer coefficient in the presence of multiple particles, gave the following correlation ($R^2 = 0.96$):

$$Nu = 0.167Re^{0.61}\left(\frac{k_p}{k_l}\right)^{1.98}\left(\frac{\varepsilon}{100-\varepsilon}\right)^{0067}\left(\frac{D_e}{D_c}\right)^{-0.70}\psi^{0.23} \tag{17.12}$$

The correlation (Equation 17.12) is valid for Re in the range of 28–1.55×10^3, k_p/k_l in the range from 0.56 to 2.24, d_e/D_c in the range from 0.22 to 0.29, ε in the range from 20 to 40% (v/v), and ψ in the range from 0.806 to 1.

17.6.2 Fastest Particle Flow in an Aseptic Processing System

In aseptic processing, the food is first heated in scraped surface heat exchangers (SSHE) and held for a pre-determined time in a hold tube, cooled quickly through a second set of SSHEs, filled, and aseptically sealed into sterile containers. Residence time distribution of particles and liquid to particle heat transfer coefficient are needed for process calculations. Residence time distribution of particles is critical because different particles take varying amounts of time to pass through the holding tube. The residence time of the fastest particle is required from the process safety point of view. Knowledge of the flow characteristics of viscous liquid and suspended food particles in SSHEs and holding tubes is essential to continuous aseptic processing of low acid liquid foods containing particulates.[13] Abdelrahim et al.[13] developed dimensionless correlations to describe the flow behavior of food particles (meat and carrot cubes) in the SSHE, holding tube, and the entire assembly of a pilot scale aseptic processing system.

Experimental study involved the determining of residence time of the fastest particle (meat or carrot) in holding tubes and the whole SSHE. The carrier fluid was starch solution of different concentration. The experimental conditions are listed in the Table 17.10. The details of the aseptic processing system and properties of carrier liquid and particles are described in Abdelrahim et al.[13]

Table 17.10   Different Experimental Conditions Used in the Particle Residence Time Study in the Aseptic Processing System

Parameters	Range
Carrier fluid	Thermo-flo starch (gelatinization temperature 140°C)
Concentration of starch	3, 4, 5, 6%
Density	1010, 1014, 1019, 1026 kg/m³
Flow rate	10, 15, 20, 25 kg/min
Particles	Carrots and meat
Particle size (carrot)	0.007, 0.016 m
Particle size (meat)	0.012, 0.019, 0.025 m
Particle concentration (carrot)	5%
Particle concentration (meat)	5%
Particle density (carrot)	1040 kg/m³
Particle density (meat)	1110 kg/m³

A particle moving in viscous liquid experiences a change of momentum equal to the sum of imposed forces: gravitational, buoyancy, drag, and fluid inertia.[13] The velocity of the particle can be described in the form of relative velocity (u_p/u_l), particle Froude number (Fr_p), or particle generalized Reynolds number (Re_p)

$$\frac{\vec{v}_p}{\vec{v}_l}, Fr_p, Re_p = f\left(Re_l, Fr_l, a, Ar_{l,p}, \frac{D_e}{D}\right), \tag{17.13}$$

where f represents a function of the various dimensionless numbers are $Re_p = v_p d_e \rho_l/\mu_{ap}$, $Re_l = v_l D \rho_l/\mu_{ap}$, $Fr_p = v_p^2/g d_e$, $Fr_l = v_l^2 g D$, $a = [\rho_p/\rho_l - 1]$, $Ar_p = a g d_e^3 \rho_l^2/\mu_{ap}^2$, $Ar_l = a g D^3 \rho_l^2/\mu_{ap}^2$.

In the above-defined dimensionless numbers, because of the non-Newtonian character of starch solutions (power-law liquids), the viscosity term is replaced by an apparent viscosity:

$$\mu_{ap} = \frac{2^{(n-3)} m \left[\frac{3n-1}{4n}\right]^n}{v_l^{(1-n)} D^{(n-1)}}. \tag{17.14}$$

The stepwise multiple regression of various dimensionless numbers (Equation 17.13) resulted in the following two equations that gave the best fit for the experimental data of particle Froude and Reynolds numbers in the SSHE (R^2 were 0.97 and 0.99, respectively):

$$Fr_p = 0.23 Fr_l^{0.60} \left(\frac{D_e}{D}\right)^{-0.48} Re_l^{0.50} Ar_l^{-0.24}. \tag{17.15}$$

$$Re_p = 2.69 Fr_l^{0.31} \left(\frac{D_e}{D}\right)^{-0.27} Re_l^{0.23} Ar_l^{-0.39}. \tag{17.16}$$

Both Froude and Reynolds numbers were influenced by the particle-to-tube diameter ratio, carrier fluid velocity, density, and viscosity incorporated in different dimensionless numbers. Similar types of correlations were developed for holding tube (i.e., without heating and cooling sections of the aseptic system):

$$Fr_p = 1.48 Fr_l^{0.70} \left(\frac{D_e}{D}\right)^{-0.65} Re_l^{0.15} Ar_l^{-0.08}, \tag{17.17}$$

$$Re_p = 6.65 Fr_l^{0.34} \left(\frac{D_e}{D}\right)^{-0.29} Re_l^{0.10} Ar_l^{-0.45}. \tag{17.18}$$

The particle Reynolds number showed a better fit ($R^2 = 0.99$) over the data compared with the Froude number correlation ($R^2 = 0.84$).

Here, $Re_p = v_p d_e \rho_l/\mu_{ap}$, $Re_l = v_l D \rho_l/\mu_{ap}$, $Fr_p = u_p^2/g d_e$, $Fr_l = v_l^2/g D$, $a = [\rho_p/\rho_l - 1]$, $Ar_p = a g d_e^3 \rho_l^2/\mu_{ap}^2$, $Ar_l = a g D^3 \rho_l^2/\rho_{ap}^2$.

Table 17.11 shows the correlations obtained from the generalization of experimental data of some processes related to food and bioprocessing.

17.7 SCALE-UP

Scaling up from laboratory scale to pilot plant or industrial scale is done by achieving similarity with the laboratory scale. There are three types of similarities: geometric similarity, where model and prototype have the same dimension scale ratio; kinematic similarity, where model and prototype have the same velocity scale ratio; dynamic similarity, where model and prototype have the

Table 17.11   Some Correlations Obtained from Dimensional Analysis

Processing Operation	Process Variables	Correlation and Reference
Canning (liquid only) end-over-end rotation	Rotation speed (N), diameter of can (D_c), density of liquid (ρ_l), height of can (H), viscosity of liquid (μ), volume of can (V_c), volume of can headspace (V_p), can angular velocity (ω), surface tension of liquid (σ), thermal conductivity of liquid (k_l)	Duquenoy:[17] $Nu = 17 \times 10^5$ $Re^{1.45} Pr^{1.19} We^{-0.551}$ $(D_c/2H)^{0.932} (V_p/V_c)^{0.628}$ $Nu = UD_c/2k_l$, $Re = 2\pi ND_c\rho_l L/(D_c+H)\mu$
Canning (liquid only) end-over-end rotation	Rotation speed (N), diameter of rotation (D_r), density of liquid (ρ_l), viscosity of liquid (μ), thermal conductivity of liquid (k_l)	Anantheswaran and Rao:[18] $Nu = 2.9 Re^{0.436} Pr^{0.287}$ $Nu = U(D_r+H)/k_l$, $Re = (D_r+H)^2 N\rho_l/\mu$
Canning (heat transfer to liquid in the presence of particles) axial rotation	Rotation speed (N), radius of reel (S), density of liquid (ρ_l), viscosity of liquid (μ), thermal conductivity of liquid (k_l)	Lenz and Lund:[19] $Nu = 115 + 15\ Re^{0.3} Pr^{0.08}$ (single particle in the can) $Nu = -33 + 53 Re^{0.28} Pr^{0.14} [d_s/S(1-\varepsilon)]^{0.46}$ $Nu = US/k_l$ $Re = S^2 N\rho_l/\mu$
Canning (heat transfer to liquid in the presence of particles) axial rotation	Rotation speed (N), diameter of can (D_c), density of liquid (ρ_l), height of can (H), viscosity of liquid (μ), can angular velocity (ω), particle concentration (ε), drag coefficient (C_D), thermal conductivity of liquid (k_l)	Deniston et al.:[20] $Nu = 1.87 \times 10^{-4} Re^{1.69} [((\rho_p - \rho_l)/C_D\rho_l)$ $((\omega^2 D_c + 2\ g)/\omega^2 D_c)(d_s/D_c)]^{0.530}$ $(\alpha_p/\omega D_c^2)^{0.126} [(1-\varepsilon)(H_{ce}/D_{ci})$ $(\omega D_c^2/\alpha_l)]^{-0.17}$ $Nu = UD_c/k_l$ $Re = \rho_l \omega D_c^2/2\mu$
Canning (heat transfer to particles) axial rotation	Rotation speed (N), density of liquid (ρ_l), viscosity of liquid (μ), thermal conductivity of liquid (k_l), equivalent diameter (d_e), particle sphericity (ψ)	Fernandez et al.:[21] $Nu = 2.7 \times 10^4 Re^{0.294} Pr^{0.33} \psi^{6.98}$ $Nu = h_{fp} d_e/k_l$ $Re = d_e 2_l^{Np}/\mu$
Aseptic processing (heat transfer to particles)	Relative velocity between liquid and particle (v_r), density of liquid (ρ_l), viscosity of liquid (μ, K, and n), thermal conductivity of liquid (k_l), particle diameter (d_p), specific heat of liquid (C_{pl}), volumetric thermal expansion coefficient of the liquid (β), average temperature of particle (T_{av})	Baptista et al.:[22] $Nu = Nu_s + 0.17 GRe^{0.71} GPr^{0.42} (d_p/d_t)^{0.28}$ $Nu_s = 2 + 0.025 Pr_s^{0.33} Gr^{0.5}$ $Nu = h_{fp} d_p/k_l$ $GRe = 8\rho_l v_r^{2-n} d_p^n/2^n K$ $((3n+1)/n)^n$ $GPr = C_K((3n+1)/n)^n 2^{n-3}/k_l$ $(V_r/d_p)^{1-n}$ $Gr = d_p^3 g\beta\rho_l^2(T_{av} - T_l)/2\mu_s$
Aseptic processing (heat transfer to particles): straight tube	Relative velocity between liquid and particle (v_r), density of liquid (ρ_l), viscosity of liquid (μ, K and n), thermal conductivity of liquid (k_l), particle diameter (d), specific heat of liquid (C_{pl}), volumetric thermal expansion coefficient of the liquid (β)	Ramaswamy and Zeareifard:[23] $Nu = 2 + 3.8 GRe^{0.479} GPr^{0.655}$ $(d/D)^{2.293} (V_p/V_s)^{0.514}$ $Nu = h_{fp}a/k_l$ $GRe = 8\rho_l v_r^{2-n} d_p^n/2^n K((3n+1)/n)^n$ $GPr = C_{pl} K((3n+1)/n)^n 2^{n-3}/k_l (V_r/d_p)^{1-n}$ $Gr = g\beta\rho_l^2 \Delta T d^3/\mu$
Aseptic processing (fastest particle velocity, v_p): Straight tube	Liquid velocity (v_l), density of liquid and particle (ρ_l and ρ_p), viscosity of liquid (μ, K, and n), particle and tube diameters (d_p, d_t), upward inclination of tube (I)	Baptista et al.:[24] Particle linear velocity (v_p): $V_p/V_l = 0.77 GRe^{0.053} Fr^{0.092} Ar^{0.011} \alpha^{-0.28}$ $(d_p/d_t)^{0.52}$ Particle angular velocity (ω): $\omega/v_l = 0.23 GRe^{0.33} Fr^{-0.25} Ar^{-0.083} \alpha^{0.50} (d_p/d_t)^{0.83}$ $GRe = 8\rho_l v_r^{2-n} d_p^n/2^n K((3n+1)/n)^n$ $Fr = v_l^2/gd_t$ $Ar = (\rho_l\rho_p - \rho_l) \sin(I)gd_p^3/\mu^2$
Mixing (propeller stirrer)	Stirrer speed ($\vec{v}_\theta$), mixing time (θ), height (H), diameter (D)	Zlokarnik:[25] $\vec{v}_\theta \theta \propto (H/D)^{0.85}$ $Re = 10^3$

continued

Table 17.11   continued

Processing Operation	Process Variables	Correlation and Reference
Heat transfer in bubbling beds	—	$\bar{v}_\theta \theta \propto (H/D)^{1.5}$ $Re = 10^4 - 10^5$ Molerus and Wirth:[26]
Liquid atomization	—	$Nu = 0.165(Ar/Pr)^{(1/3)}$ $10^5 \le Ar \le 10^8$ $Nu = 0.02469Ar^{0.4304}$ $Ar \ge 10^8$ Dahl and Muschelknautz:[27] $We = 4.5 \times 10^4 Oh^{1/6}$, here Oh is Ohnesorge number $(We^{1/2}/Re)$ Zlokarnik:[5] $We = 1.97 \times$ $10^4(\bar{v}\mu/\sigma)^{0.154}$

same force scale ratio. Sometimes, partial similarity is achieved when full similarity is not possible in scale-up analysis.[3] Similarity in scale up analysis can be achieved by ensuring that the correlations obtained from small scale equipment after dimensional analysis still hold for larger scale equipment. It is important to note that experiments must cover the full range of values of the relevant dimensionless groups encountered in the scaled-up version.

Scale up of many types of equipments, viz. spray dryer,[28] fluidized bed dryer,[29] spouted bed dryer,[30,31] rotary dryer,[32] pneumatic conveying dryer,[33] layer dryer,[34] mixer granulator,[35] and chemical reactor,[36] as well as industrial processes, viz. spray coating,[37] freeze drying,[38] fermentation,[39] agglomeration,[40] and fluidized beds[41] have been carried out and reported.

17.8   CONCLUDING REMARKS

Application of dimensional analysis is widespread. It is especially useful when a mathematical model is either not possible or not feasible. Certainly, food processing is one of the areas that can make use of dimensional analysis to perform experimental data analysis, model design, prototype testing, and equipment scaling up. In performing dimensional analysis, a relevance list consists of all influencing parameters in a physical problem is made, followed by the generation of π-space that consists of dimensionless numbers. Thereafter, experiments are to be carefully designed and performed. Generalization of experimental data gives relationships and correlations that relate the dimensionless numbers. Based on the relationships and correlations derived empirically, reliable scale up can be carried out.

NOTATION

Symbol	Quantity	Dimension	Unit
Basic physical quantities			
L	Length	$[L]$	m
t	Time	$[t]$	s
T	Temperature	$[T]$	K
M	Mass	$[M]$	kg
N	Quantity of matter	$[N]$	mol
Φ	Angle	$[\Phi]$	rad, deg

continued

continued

Symbol	Quantity	Dimension	Unit
Physical quantities			
A	Surface area	$[L^2]$	m^2
a	Density simplex	$[-]$	—
b_t	Vapor expansion in capillaries	$[ML^{-3}T^{-1}]$	$kg\,m^{-3}\,K^{-1}$
c_A	Molar concentration of component A	$[NL^{-3}]$	$mol\,m^{-3}$
c	Heat capacity per unit mass (incompressible substances)	$[L^2t^{-2}T^{-1}]$	$J\,kg^{-1}\,K^{-1}$
c_p	Heat capacity per unit mass at constant pressure	$[L^2t^{-2}T^{-1}]$	$J\,kg^{-1}\,K^{-1}$
c_v	Specific vapor capacity	$[Lt^2M^{-1}]$	$kg\,kg^{-1}\,Pa^{-1}$
D	Diameter	$[L]$	m
D_{AB}	Binary diffusion coefficient	$[L^2t^{-1}]$	$m^2\,s^{-1}$
D	Diameter	$[L]$	m
F	Volumetric flow rate	$[L^3t^{-1}]$	$m^3\,s^{-1}$
g	Gravity acceleration $=9.80665\ m\,s^{-2}$	$[Lt^{-2}]$	$m\,s^{-2}$
G	Mass flow rate per unit area	$[ML^{-2}t^{-1}]$	$Kg\,m^{-2}\,s^{-1}$
H	Height	$[L]$	m
H_e	Magnetic field strength	$[QL^{-2}t^{-1}]$	$C\,m^{-2}\,s^{-1}$
h	Heat-transfer coefficient	$[MT^{-1}t^{-3}]$	$W\,m^{-2}\,K^{-1}$
I	Electric charge flux	$[QL^{-2}t^{-1}]$	$C\,s^{-1}m^{-2}$
k	Thermal conductivity	$[MLT^{-1}t^{-3}]$	$W\,m^{-1}\,K^{-1}$
k_L	Mass-transfer coefficient	$[Lt^{-1}]$	$m\,s^{-1}$
L	Length	$[L]$	m
MR	Moisture ratio	—	—
$\dot{m}$	Mass flow rate	$[Mt^{-1}]$	$kg\,s^{-1}$
n	Specific mass constant	$[MM^{-1}]$	$kg\,kg^{-1}$
P	Pressure	$[ML^{-1}t^{-2}]$	Pa
P_v	Vapor pressure	$[ML^{-1}t^{-2}]$	Pa
Q	Heat released in chemical reaction per mole reacting	$[ML^2t^{-2}N^{-1}]$	$J\,mol^{-1}$
q	Heat flux	$[Mt^{-3}]$	$W\,m^{-2}$
R_A	Molar rate of generation of component A	$[Nt^{-1}L^{-3}]$	$mol\,s^{-1}\,m^{-3}$
R	Radius	$[L]$	m
r	Radius	$[L]$	m
S	Particle area/particle volume	$[L^{-1}]$	$m^2\,m^{-3}$
sw	Specific weight	$[Mt^{-2}L^{-2}]$	$N\,m^{-3}$
T	Temperature	$[T]$	°C, K
t	Time	$[t]$	s
U	Overall heat transfer coefficient	$[MT^{-1}t^{-3}]$	$W\,m^{-2}\,K^{-1}$
V	Volume	$[L^3]$	m^3
$\vec{v}$	Velocity	$[Lt^{-1}]$	$m\,s^{-1}$
$\vec{v}_\theta$	Rotational speed	$[t^{-1}]$	s^{-1}
ν	Kinetic viscosity	$[L^2t^{-1}]$	$m^2\,s^{-1}$
X	Moisture content	$[M\,M^{-1}]$	$kg\,kg^{-1}$
x, y, z	Cartesian coordinate	$[L]$	m
Physical Quantities (Greek Symbols)			
α	Thermal diffusivity	$[L^2t^{-1}]$	$m^2\,s^{-1}$
β	Volumetric thermal expansion coefficient	$[T^{-1}]$	K^{-1}
δ	Thickness	$[L]$	m
ε	Void fraction	—	—
η	Stefan–Boltzmann constant $=5.67\times10^{-8}\ Wm^{-2}K^{-4}$	$[Mt^{-3}T^{-4}]$	$W\,m^{-2}\,K^{-4}$
θ	Mixing time	$[t]$	s
λ	Latent heat of vaporization	$[L^2t^{-2}]$	$J\,kg^{-1}$

continued

continued

Symbol	Quantity	Dimension	Unit
μ	Viscosity	$[ML^{-1}t^{-1}]$	$N\,s\,m^{-2}$
ρ	Density	$[ML^{-3}]$	$kg\,m^{-3}$
σ	Surface tension	$[Mt^{-2}]$	$kg\,m^{-2}$
σ_e	Electrical conductivity	$[Q^2tL^{-3}M^{-1}]$	$C^2\,s\,m^{-3}\,kg^{-1}$
τ	Shear stress	$[ML^{-1}t^{-2}]$	$N\,m^{-2}$
ϕ	Viscous dissipation function	$[t^{-2}]$	s^{-2}
φ	Angle	$[\Phi]$	rad, deg (°)
χ	Mixing quality	—	—
ψ	Sphericity	—	—

Subscripts

A	Component A
c	Can
cd	Characteristic dimension
cy1	Cylinder
e	Equivalent
eff	Effective
f	Fluid
fg	Fluid-gas
fp	Fluid-particle
g	Gas
L	Solvent
l	Liquid
m	Mass transfer
0	Initial
p	Particle
r	Rotation
s	Solids
sat	Saturated
sph	Sphere
surr	Surrounding medium
V	Vapor
x	Direction in Cartesian coordinate
y	Direction in Cartesian coordinate
z	Direction in Cartesian coordinate
∞	Infinity

REFERENCES

1. Buckingham, E., On physically similar systems: Illustrations of the use of dimensional equation, *Physical Review*, 4, 345–376, 1921.
2. Barenblatt, G. I., *Scaling, Self-Similarity, and Intermediate Asymptotics*, Cambridge: Cambridge University Press, 1996.
3. Langhaar, H. L., *Dimensional Analysis and Theory of Models*, New York: Wiley, 1951.
4. Zlokarnik, M., Problems in the application of dimensional analysis and scale-up of mixing operations, *Chemical Engineering Science*, 53(17), 3023–3030, 1998.
5. Zlokarnik, M., *Scale-up in Chemical Engineering*, Weinheim: Wiley-VCH, 2002.
6. Kottowski, H. M. and Skouloudis, A. N., Potential and limits of dimensional analysis for the design of chemical facilities, *Journal of Loss Prevention in the Process Industries*, 4, 194–201, 1991.
7. Hall, C. W., Dimensionless numbers and groups for drying, *Drying Technology*, 10(4), 1081–1095, 1992.
8. Dincer, I. and Hussain, M. M., Development of a new bi–di correlation for solids drying, *International Journal of Heat and Mass Transfer*, 45, 3065–3069, 2002.
9. Rovedo, C. O., Suarez, C., and Viollaz, P., Analysis of moisture profiles, mass Biot number and driving forces during drying of potato slabs, *Journal of Food Engineering*, 36, 211–231, 1998.

10. Molerus, O. and Burschka, A., Pneumatic transport of coarse grained materials, *Chemical Engineering and Processing*, 34, 173–184, 1995.
11. Sablani, S. S., Heat transfer studies of canned liquid particle mixtures in cans subjected to end-over-end processing. PhD dissertation, McGill University, 1996.
12. Sablani, S. S., Ramaswamy, H. S., and Mujumdar, A. S., Dimensionless correlations for convective heat transfer to liquid and particles in cans subjected to end-over-end rotation, *Journal of Food Engineering*, 34, 453–472, 1997.
13. Abdelrahim, K. A., Ramaswamy, H. S., Grabowski, S., and Marcotte, M., Dimensionless correlations for the fastest particle flow in a pilot scale aseptic processing system, *Lebensmittel Wissenchaft und-Technologie*, 28, 43–49, 1995.
14. Zhang, M. C. and Yang, Y. K., On the scaling laws for bubbling gas fluidized bed dynamics, *Powder Technology*, 51, 159–165, 1987.
15. van der Meer, E. H., Thorpe, R. B., and Davidson, J. F., Dimensionless groups for practicable similarity of circulating fluidized beds, *Chemical Engineering Science*, 54, 5369–5376, 1999.
16. He, Y. L., Lim, C. J., and Grace, J. R., Scale up of spouted beds, *Chemical Engineering Science*, 52(2), 329–339, 1997.
17. Duquenoy, A., Heat transfer to canned liquids, Vol. 1, Eds., Linko, P., Malkki, Y., Olkku, J., and Larinkari, J., In *Food Processing Systems*, London: Applied Science, pp. 483–489, 1990.
18. Anantheswaran, R. C. and Rao, M. A., Heat transfer to model Newtonian liquid foods in cans during end over end rotation, *Journal of Food Engineering.*, 4, 1–19, 1985.
19. Lenz, M. K. and Lund, D. B., The lethality-Fourier number method. Heating rate variations and lethality confidence intervals for forced-convection heated foods in containers, *Journal of Food Process Engineering*, 2, 227–271, 1978.
20. Deniston, M. F., Hassan, B. H., and Merson, R. L., Heat transfer coefficients to liquids with food particles in axially rotating cans, *Journal of Food Science*, 52(962–966), 979, 1987.
21. Fernandez, C. L., Rao, M. A., Rajavasireddi, S. P., and Sastry, S. K., Particulate heat transfer to canned snap beans in Steritort, *Journal of Food Process Engineering*, 10, 183–198, 1988.
22. Baptista, P. N., Oliveira, F. A. R., Oliveira, J. C., and Sastry, S. K., Dimensionless analysis of fluid particle heat transfer coefficients, *Journal of Food Engineering*, 31, 199–218, 1997.
23. Ramaswamy, H. S. and Zareifard, M. R., Evaluation of factors influencing tube-flow fluid-to-particle heat transfer coefficient using a calorimetric technique, *Journal of Food Engineering*, 45, 127–138, 2000.
24. Baptista, P. N., Oliveira, F. A. R., Oliveira, J. C., and Sastry, S. K., Dimensionless analysis of the flow of spherical particles in two phase flow in straight tubes, *Journal of Food Engineering*, 31, 125–136, 1997.
25. Zlokarnik, M., Eignung von Rührern zum Homogenisieren von Flüssigkeitsgemischen (Homogenization of liquid mixtures), *Chemie Ingenieur Technik*, 39(9/10), 539–548, 1967.
26. Molerus, O. and Wirth, K. E., *Heat Transfer in Fluidized Beds*, London: Chapman & Hall, 1997.
27. Dahl, H. D. E. and Muschelknautz, E., Atomisation of liquids and suspensions with hollow cone nozzles, *Chemical Engineering Technology*, 15, 224–231, 1992.
28. Masters, K., Scale up of spray dryers, *Drying Technology*, 12(1–2), 235–257, 1994.
29. Bahu, R. E., Fluidized bed dryer scale up, *Drying Technology*, 12(1–2), 329–339, 1994.
30. Szentmarjay, T., Szalay, A., and Pallay, E., Scale up of the mechanically spouted bed dryer with inert particles, *Drying Technology*, 12(1–2), 341–350, 1994.
31. Passos, M. L., Mujumdar, A. S., and Massarani, G., Scale up of spouted bed dryers: criteria and applications, *Drying Technology*, 12(1–2), 351–391, 1994.
32. Papadakis, S. E., Langrish, T. A. G., Kemp, I. C., and Bahu, R. E., Scale up of cascading rotary dryers, *Drying Technology*, 12(1–2), 259–277, 1994.
33. Kemp, I., Scale up of pneumatic conveying dryers, *Drying Technology*, 12(1–2), 279–297, 1994.
34. Moyers, C. G., Scale up of layer dryers: A unified approach, *Drying Technology*, 12(1–2), 393–416, 1994.
35. Litster, J. D., Hapgood, K. P., Michaels, J. N., Sims, A., Roberts, M., and Kameneni, S. K., Scale-up of mixer granulators for effective liquid distribution, *Powder Technology*, 124, 272–280, 2002.
36. Donati, G. and Paludetto, R., Scale up of chemical reactors, *Catalysis Today*, 34, 483–533, 1997.

37. urtona, R. and Cheng, X. X., The scale-up of spray coating processes for granular solids and tablets, *Powder Technology*, 150, 78–85, 2005.
38. sinontides, S. C., Rajniak, P., Pham, D., Hunke, W. A., Placek, J., and Reynolds, S. D., Freeze drying—principles and practice for successful scale-up to manufacturing, *International Journal of Pharmaceutics*, 280, 1–16, 2004.
39. Thiry, M. and Cingolani, D., Optimizing scale-up fermentation processes, *Trends in Biotechnology*, 20(3), 103–105, 2002.
40. Mort, P. R., Scale-up of binder agglomeration processes, *Powder Technology*, 150, 86–103, 2005.
41. Matsen, J. M., Scale-up of fluidized bed processes: Principle and practice, *Powder Technology*, 88, 237–244, 1996.

CHAPTER **18**

Linear Programming

Eli Feinerman and Sam Saguy

CONTENTS

18.1   INTRODUCTION

Decision makers (such as consumers, producers, and policy makers) are often concerned with how to do things "best." Companies attempt to maximize profits or minimize costs subject to a variety of technological and source constraints, as well as legal regulations and demands. Consumers try to spend their free but limited income in a way that will maximize their utility from the consumption of the acquired goods and services. Policy makers or social planners try to allocate public funds and to design a set of regulations that will maximize the social welfare of the community. Analytical solutions for simple constrained optimization problems can sometimes be obtained via the classical methods of algebra and calculus. However, optimal solutions for more complex problems typically require the use of numerical algorithms.

Mathematical programming (MP) relates to the use of mathematical models for solving optimization problems. A typical MP model involves the selection of values for a limited number of decision variables (often called *activities*), focusing attention on a single objective function to be maximized (or minimized, depending on the context of the problem), subject to a finite set of constraints that limit the selection of the decision variables. Linear programming (LP) is the simplest and most widely used form of MP in which the objective and constraints are linear functions of the decision variables. In other words, LP is an optimization problem that involves maximizing or minimizing a linear objective function that includes several non-negative variables, the choice of which is subject to a set of linear constraints.

Compared to classical optimization methods, LP is a relatively new approach that grew out of troop supply problems arising during World War II. Mathematicians were looking for an approach that could make use of computers that were being developed at that time. The simplex algorithm[1] was developed by Dantzig in 1947. It is a simple recipe for solving LP problems of any size and is easily programmed on a computer. The extensive availability and widespread use of computers and their ever-growing computational power have turned LP into a broadly utilized tool that furnishes practical solutions for a spectrum of problems. Examples include: military analysis, production planning (especially for agricultural producers and oil refiners), transportation and shipping, new product development, efficient utilization of resources, and others. The food field, with its dynamic and spatial processes, is typically recognized as complex, highly interactive, nonlinear and spatially distributed. These properties make analysis, modeling, and even simulation a challenging task. However, in many cases, a linear framework is quite adequate, providing an accurate description, or a fairly close approximation, of the system at hand.[2] Several typical applications of LP in the food domain include processing, canning operations, livestock nutrition, design of a nutritionally adequate diet at the lowest cost, evaluating the economic value of the fortified product, formulation, etc. Therefore, LP furnishes a valuable and important tool for the food industry. Other problems defined as dynamic and nonlinear programming represent a different class of optimization and are not covered herein. A typical LP problem can have hundreds of variables and constraints. These large-scale problems can be solved in a practical amount of time due to recent advances in both computer power and algorithm efficiency. LP is commonly a relatively short-term tool, as the constraints normally change with market pricing, labor cost, varying requirements, etc. Therefore, LP needs frequent cost updates and repeatable verification of the optimal results.

For many years, the principal tool for the solution of LP models was the aforementioned simplex method.[1] Due to its broad applications, it will also be used as the main focus of this chapter. In the 1980s, another class of solution algorithm became competitive with the simplex method. This class of algorithms is called the *interior point method*[3] and some computer implementations of it are the most efficient means of solving very large LP models.

The rest of the chapter proceeds as follows: first, in Section 18.2, the main concepts and the general formulation of a LP problem are introduced by an illustration of its application via a simple hypothetical example. The example is also used to characterize and analyze the optimal solution. A graphical solution approach to LP problems and its application to a food-blending problem are presented in Section 18.3. Subsequently, the graphical approach is utilized to illustrate a few potential pathological situations of LP. Thereafter, in Section 18.4, a brief review of some of the vast amount of literature on LP applications in food processing and nutrition management is given. To further illustrate and highlight the utilization of computerized LP, two simple, albeit representative examples are presented that can also be used by the reader to verify the formulation and utilization of this straightforward yet powerful tool. Then, in Section 18.5, some of the currently available LP software is reviewed and a short introduction to the software associated with the development of new products is provided. The concluding remarks in Section 18.6 end the chapter.

18.2 LINEAR PROGRAMMING FORMULATION

18.2.1 Introduction

As indicated, the concept of MP relates to the use of mathematical models to solve optimization problems. A typical MP model involves the selection of values for a limited number of decision or control variables (often called *activities*). Attention will be focused on a single objective function to be maximized (or minimized, depending on the context of the problem), subject to a set of constraints that limit the selection of the decision variables. More specifically, the problem can be defined as choosing the level of n activities, denoted by $x_1, x_2,..., x_n$ that maximizes (or minimizes) an objective function, $F(x_1, x_2,..., x_n)$, subject to a set of m constraints. The activities can be summarized by a column vector:

$$\underline{X} = \begin{pmatrix} x_1 \\ x_2 \\ \vdots \\ x_n \end{pmatrix}.$$

The constraints are defined by a set of m functions of the decision variables, $g_1(\underline{X}) \le b_1$, $g_2(\underline{X}) \le b_2$, ..., $g_m(\underline{X}) \le b_m$ and by the requirement for non-negativity of the decision variables, (i.e., $\underline{X} \ge 0$). The coefficients $b_1, b_2,..., b_m$ are given and called *constraint constants* (also known as the *right-hand side* of the equation, or *RHS*). The LP problem is the most commonly applied form of MP, in which the objective function is linear and the constraints consist of linear equalities and/or inequalities.

18.2.2 General Formulation of a LP Problem

Linear programs have objective functions that are to be maximized or minimized, linear constraints that can be of three types (less than or equal to, equal to, and greater than or equal to), and non-negative decision variables (or activities). A constraint whose left-hand side (LHS) is less than or equal to (greater than or equal to) the constraint constant on the RHS is termed *maximum* (or *minimum*) *constraint*. A constraint whose LHS is equal to its RHS is called an *equality constraint*. A mathematical formulation of a standard form of a maximum LP problem with maximum constraints can be given as follows:

$$\underset{x}{\text{Max}} \{F = c_1 x_1 + c_2 x_2 + \cdots + c_n x_n\}$$

Subject to :

$$a_{11} x_1 + a_{12} x_2 + \cdots + a_{1n} x_n \leq b_1 \qquad (1)$$

$$a_{21} x_1 + a_{22} x_2 + \cdots + a_{2n} x_n \leq b_2 \qquad (2)$$

$$\vdots \qquad \qquad \vdots$$

$$a_{m1} x_1 + a_{m2} x_2 + \cdots + a_{mn} x_n \leq b_m \qquad (m)$$

$$x_j \geq 0, \quad j = 1,\ldots,n.$$

The coefficient c_j ($j=1,\ldots, n$) in the objective function represents the increase (if $c_j>0$) or decrease (if $c_j<0$) in the value of the objective function, F, per unit increase in x_j, b_i ($i=1,\ldots, m$) is the constant on the RHS of the ith constraints, and a_{ij} are the coefficients of the functional constraint equations, expressing the number of units from the constraint i consumed by one unit of activity j. Any non-negative vector X that satisfies all the constraints is called *a feasible solution* of the LP problem. A feasible solution that maximizes (or minimizes in the case of a minimization problem) the objective function is called an *optimal solution*.

18.2.3 Example 1: The Problem of a Hypothetical Juice Manufacturer: Underlying Assumptions of LP

Consider a manufacturer who owns a juice factory with a storage capacity of 400 m³ and produces apple, lemon, and cherry juice, stored in stainless steel containers, each of 1 m³. The production and marketing of one container of apple juice requires (per one month) 4 work-hours, $15 of capital investment, and 24 machine-hours. Similarly, one container of cherry juice requires 2.5 work-hours, $12 of capital, and 11 machine-hours; one container of lemon juice requires 3.5 work-hours, $12 of capital, and 20 machine-hours. Marketing obligations require that the number of containers of apple juice not exceed 40% of the number of containers of the other two juices. The net profits per container (denoted by $c_j, j=1,\ldots, 3$) of apple, cherry, and lemon juice are $354, $325, and $346, respectively. In addition to 400 m³ of storage capacity, the manufacturer has at its disposal, for the month under consideration, 2200 work-hours, $5000 available for capital investment, and 8500 machine-hours. Obviously, the manufacturer's objective is to maximize the total net profit from the plant. The first task, then, is to formulate the LP problem to establish the optimal feasible solution.

In this example, three ($n=3$) activities and four ($m=4$) constraints are identified. The activities are the three types of juices (apple, cherry, and lemon), measured in units of 1 m³ containers. Specifically, x_1, x_2, and x_3 are the numbers of containers of apple, cherry, and lemon juice, respectively. The constraints and the levels of the constraint constants are:

1. Storage capacity in cubic meters, b_1 ($=400$ m³)
2. Labor measured in work-hours, b_2 ($=2200$ work-hours)
3. Investment capital measured in dollars, b_3 ($=\$5000$)
4. Machine work measured in machine-hours, b_4 ($=8500$ machine-hours)

5. The juices' balance constraint measured in containers, b_5 ($=0$ containers, see explanation below)

The LP problem may now be formulated:

$$\text{Max}\{F = 354x_1 + 325x_2 + 346x_3\}$$
$$\underset{x}{\text{Subject to :}}$$

$1x_1 + 1x_2 + 1x_3 \quad\quad \leq 400$	(1)
$4x_1 + 2.5x_2 + 3.5x_3 \leq 2,200$	(2)
$15x_1 + 12x_2 + 12x_3 \leq 5,000$	(3)
$24x_1 + 11x_2 + 20x_3 \leq 8,500$	(4)
$1x_1 - 0.4x_2 - 0.4x_3 \leq 0$	(5)

$$x_j \geq 0, \quad j = 1,...,3.$$

The coefficients of the functional constraint equations are:

$$a_{11} = 1,...,a_{23} = 3.5,...,a_{32} = 12,...,a_{41} = 24,...,a_{53} = -0.4.$$

All the constraints are maximum constraints. The first four relate the total demand of the activities for the scarce resources (storage, labor, capital, and machine-hours) to the limited supply via the fundamental relation *demand* $\leq$ *supply*. The fifth (balance) constraint states that $x_1 \leq 0.4(x_2 + x_3)$. However, the RHS of a constraint should include only a constant and the LHS should include only the terms $a_{ij}x_j$. Thus, the term of the above inequality is arranged to obtain the fifth constraint.

Before proceeding, the major underlying assumptions of LP are briefly identified:

- *Boundedness*: The number of activities (n) and the number of constraints (m) is finite.
- *Fixedness*: At least one constraint has a nonzero RHS coefficient (b_i).
- *Certainty*: All the c_j, b_i, and a_{ij} coefficients in the model are constants that are assumed to be known with certainty.
- *Divisibility or continuity*: Every constrained resource and every activity (x_j) can be used in quantities that are fractional units.
- *Proportionality*: This property requires that the value of each term in the linear function be strictly proportional to the value of the activity in the term. This assumption asserts, for example, that if one unit of the apple juice activity requires 1 m³ of storage volume, 4 work-hours, $15 of capital, and 24 machine-hours, and its associated net profit is $354, then two units of apple juice activity will require 2 m³ of storage volume, 8 work-hours, $30 of capital, and 48 machine-hours, and its associated net profit will be $708.
- *Additivity*: This assumption asserts that if, in the current example, the production of two or more activities requires storage capacity, labor, capital, and machine work in amounts that may differ, the total demand for storage capacity, labor, capital and machine work is equal to the sum of the quantities demanded by all the activities.

The last two linear properties of proportionality and additivity preclude the use of a nonlinear objective function and nonlinear constraints. The additivity property prohibits cross-product terms (e.g., $10x_1x_2$), which might represent nonlinear interaction effects, for instance between apple juice and cherry in our example. The proportionality property, for example, requires that the total net profit associated with a specific activity always be directly proportional to the level of the activity; it follows that it is not possible to include a fixed start-up cost in the analysis.

18.2.4 Primary Analysis of the Optimal Solution, Concepts and Definitions

Algorithms for solving an LP problem are available via many computer programs, and will be discussed further on. At this stage, the optimal solution is presented and used to define and explain a few important concepts and results associated with this solution (optimal values are hereafter denoted by asterisks).

The activities' optimal levels in this example (i.e., the levels which maximize the objective function subject to the constraints) are: $x_1^* = 66.67$ containers of apple juice, $x_2^* = 0.0$ containers of lemon juice and $x_3^* = 333.33$ containers of cherry juice, implying that the optimal value of the objective function is: $F^* = 138,933.34 (=354x_1^* + 325x_2^* + 346x_3^*)$. It should be mentioned that the optimal activity levels are continuous. Therefore, in practice, the above results should be rounded off to the nearest integer value. Activities which are strictly positive in the optimal solution (x_1 and x_3 in the current example) are called *basic activities* (or *basic variables*). Activities that are equal to zero in the optimal solution (x_2 in the current example) are termed *nonbasic*.

18.2.5 Binding and Nonbinding Constraints, Shadow Prices

A specific constraint is *binding* (or *effective*) if, in the optimal solution, it is satisfied with the equality sign. A constraint is *nonbinding* (or *noneffective*) if it is satisfied with the inequality sign. In the current example, the first and third constraints are binding whereas the second, fourth, and fifth constraints are nonbinding:

$$1x_1^* + 1x_2^* + 1x_3^* \qquad = 400 \qquad\qquad (1)$$
$$4x_1^* + 2.5x_2^* + 3.5x_3^* = 1,433.33 < 2,200 \quad (2)$$
$$15x_1^* + 12x_2^* + 12x_3^* = 5000 \qquad\qquad (3)$$
$$24x_1^* + 11x_2^* + 20x_3^* = 8,266.66 < 8,500 \quad (4)$$
$$-1x_1^* + 0.4x_2^* + 0.4x_3^* = 66.67 > 0 \qquad\quad (5)$$

Generally speaking, the amounts of apple, lemon, and cherry juice that the manufacturer is able to produce are limited by the availability of the various resources and by the balance constraint, implying that (for given levels of the coefficients a_{ij} and c_j) the optimal levels of the activities are functions of the constraint constants summarized by the vector $\underline{b} = (b_1, b_2..., b_5)$: $x_1^* = x_1(\underline{b})$, $x_2^* = x_2(\underline{b})$, $x_3^* = x_3(\underline{b})$. The objective function, F, can also be expressed as a function of $\underline{b}$: $F^*(\underline{b}) = F(x_1(\underline{b}), x_2(\underline{b}), x_3(\underline{b}))$. The sensitivity of the optimal value of the objective function to variations in the constraint constants can be calculated via differentiation of F^* (with respect to b_i):

$$\lambda_i \equiv \frac{\partial F^*(\underline{b})}{\partial b_i}, \qquad i = 1, ..., 5.$$

The λ_i's are called *dual prices* or *shadow prices* of the constraints (in contrast to market prices that are visible to everybody). In addition to the optimal levels of activity, the shadow prices of the various constraints are obtained as by-products of the optimal solution.

It can be proven that the shadow price of a binding maximum constraint ($\leq$) is positive whereas the shadow price of a binding minimum constraint ($\geq$) is negative. The shadow price of a nonbinding constraint (whether it is a maximum or minimum constraint) is equal to zero. The shadow prices calculated in the current example are: $\lambda_1 = \$314.00$ per m^3, $\lambda_2 = \$0$ per work-hour, $\lambda_3 = \$3.67$ per dollar of capital investment, $\lambda_4 = \$0$ per machine-hour, and $\lambda_5 = \$0$ per container. The shadow price of a constraint can be interpreted as the marginal sacrifice that the planner must bear because of the presence of that constraint. In other words, if the ith constraint could be made less limiting by one unit, the optimal value of the objective function would increase by λ_i. Conversely, if the ith constraint becomes tighter by 1 unit, the optimal value of the objective function will decrease by λ_i. In the current example, the marginal contribution to the objective function of the last cubic meter of the limited storage capacity and the last dollar of the limited capital are given by $\lambda_1 = F^*(b_1 = 400/b_{-1}) - F^*(b_1 = 399/b_{-1}) = \$314/m^3$ and by $\lambda_3 = F^*(b_3 = 5000/b_{-3}) - F^*(b_3 = 4999/b_{-3}) = \$2.67/\text{capital}$, respectively. The term "$/b_{-i}$", $i = 1,3$ means that "all other constraint constants, except for the ith, are held fixed at their original level." On the other hand, constraints 2, 4, and 5 are not binding. For example, in the optimal solution, only 1433.33 work-hours out of the 2200 available to the manufacturer are used by the activities in the optimal solution. Namely, in the optimal solution there are 766.67 unused work-hours. Thus, if the available amount of work-hours is reduced by 1 unit, this will not affect the optimal value of the objective function:

$$\lambda_2 = F^*(b_2 = 2200/b_{-2}) - F^*(b_2 = 1999/b_{-2}) = \$0/\text{work day}.$$

18.2.6 Opportunity and Reduced Costs of the Activities

The opportunity cost of the jth activity, denoted by z_j, is the sacrifice of producing one additional unit of that activity resulting from the fact that alternative production opportunities (i.e., some other activities) must be forgone to satisfy the problem's constraint. In principal, the sacrifice can be either positive or negative. A positive sacrifice is to be avoided, whereas a negative sacrifice is welcome.

To illustrate the calculation of z_j, consider the apple juice enterprise. One unit of that activity requires $a_{11} = 1\,m^3$ of storage capacity, $a_{21} = 4$ work-hours, $a_{31} = \$15$ of capital, and $a_{41} = 24$ machine-hours. Note that $a_{51} = -1$ and that if the fifth constraint were binding, it would force the planner to increase the production of lemon juice and/or cherry juice to at least 0.4 containers above the optimal levels of these activities that would be obtained in the absence of the fifth constraint. This requirement, if binding, would reduce the value of the objective function and therefore $\lambda_5 \leq 0$. The opportunity cost of apple juice is equal to the sum of the input quantities required for one unit of the activity (apple juice) multiplied by the shadow prices of these inputs:

$$z_1 = \sum_{i=1}^{5} a_{i1}\lambda_i = 1(314) + 4(0) + 15\$(2.67) + 24(0) + -1(0) = \$354/m^3.$$

Similarly,

$$z_2 = \sum_{i=1}^{5} a_{i2}\lambda_i = \$346/\text{acre}; \quad \text{and} \quad z_3 = \sum_{i=1}^{5} a_{i3}\lambda_i = \$346/m^3.$$

The reduced costs of an activity, denoted by R_j, are defined by the difference $(z_j - c_j)$. In our example,

$$R_1 = 354 - 354 = 0, \qquad R_2 = 346 - 325 = 21, \quad \text{and} \quad R_3 = 346 - 346 = 0.$$

These results illustrate an additional important characteristic of the solution to the LP problem: the reduced cost of a basic activity $x_j > 0$ is equal to zero while the reduced cost of a nonbasic activity $x_j = 0$ is positive and represents a reduction in the optimal value of the objective function if the operation of one unit of a nonbasic activity were to be forced upon the planner. Lemon juice is a nonbasic activity in our example. If it were to be produced at a level of 1 unit (i.e., 1 container), the manufacturer would gain $c_2 = \$325$ on the one hand, and would lose $z_2 = \$346$ (forgone benefits), on the other. Because the loss exceeds the benefits, $x_2 = 0$ at the optimal solution. The lemon juice activity would be a candidate for becoming a basic activity if c_2 were to increase by at least $R_2 = 346 - 325 = \$21$ per container.

To further illustrate the relationships between the reduced costs of the activities and the shadow prices of the constraints, recall that in the optimal solution, $x_1^* = 66.67$, $x_2^* = 0$, and $x_3^* = 333.33$. By substituting these values into the constraints (as was illustrated above), only the first and third ones can be easily identified as binding, implying that $\lambda_1 > 0$, $\lambda_3 > 0$, $\lambda_2 = \lambda_4 = \lambda_5 = 0$. Two equations are required for calculating λ_1 and λ_3. Because x_1 and x_3 are basic activities, their associated reduced costs are equal to zero:

$$\left.\begin{array}{l} \underbrace{R_1 = 1(\lambda_1) + 15(\lambda_3)}_{z_1} \underbrace{-354}_{c_1} = 0 \\[2mm] \underbrace{R_3 = 1(\lambda_1) + 12(\lambda_3)}_{z_3} \underbrace{-346}_{c_3} = 0 \end{array}\right\} \Rightarrow \lambda_1 = 314, \ \lambda_3 = 2.67.$$

This section is concluded with an informal presentation of two important theorems of LP (the proofs are beyond the scope of this chapter; the reader is encouraged to consult available text-books[4-6]).

- If a LP problem has an optimal solution, $X^* = (x_1^*, x_2^*, \ldots, x_n^*)$, then the optimal value of the objective function, $F^* = \sum_{j=1}^{n} c_j x_j^* (=138,933.34$ in the example presented here), is equal to the sum of the constraint constants multiplied by their associated shadow prices: $W^* = \sum_{i=1}^{m} b_i \lambda_i (=400(314) + 2200(0) + 5000(2.67) + 8500(0) + 0(0) = F^*)$.

- The number of the basic (nonzero) activities in the optimal solution of a LP problem ($x_j^* > 0$) is no greater than the number of constraints (m) in that problem. This theorem is very important for model builders. If a specific LP model has $m = 50$ constraints and $n = 500$ activities, then 450 of those activities will be irrelevant to any given solution. Thus, this theorem may be viewed as a reminder of the importance of the row dimension in LP models.

18.3   GRAPHICAL SOLUTION APPROACH

18.3.1   Well-Behaved Problem

A simple way to solve an LP problem with only two activities is graphically. This is illustrated via a simple hypothetical example of maximizing a linear profit function, F, measured in dollars with $n = 2$ activities subject to $m = 3$ constraints:

$$\text{Max} \{F = 3x_1 + 2x_2\}$$
$$\underset{x}{\text{Subject to :}}$$

$10x_1 + 10x_2$	≤ 100	(1)
$100x_1 + 200x_2$	$\leq 1,600$	(2)
$10x_1 + 5x_2$	≤ 80	(3)

$$x_j \geq 0, \quad j = 1, 2.$$

Before graphing the above inequality constraints, consider a general formulation of $a_{i1}x_1 + a_{i2}x_2 \leq b_i$ and note that the equation $a_{i1}x_1 + a_{i2}x_2 = b_i$ represents all the combinations of x_1 and x_2 that satisfy the constraint with the above equality. To graph this equation on the (x_1, x_2) axes, it is convenient to rewrite it as:

$$x_2 = \underbrace{\frac{b_i}{a_{i2}}}_{\text{Intercept}} - \underbrace{\frac{a_{i1}}{a_{i2}}}_{\text{Slope}} x_1.$$

The linear equation is depicted in Figure 18.1.

The graph of the inequality constraint $a_{i1}x_1 + a_{i2}x_2 \leq b_i$, coupled with the non-negativity constraints $x_1 > 0$ and $x_2 > 0$, is represented by the shaded half-plane below the linear line in the first quadrant. The shaded area represents the *feasible region* of a LP problem with a single maximum constraint. This region contains all the combinations of x_1 and x_2 under which $x_1 > 0$, $x_2 > 0$ and $a_{i1}x_1 + a_{i2}x_2 \leq b_i$. If the constraint was a minimum constraint, i.e., $a_{i1}x_1 + a_{i2}x_2 \geq b_i$, then the feasible region would be the half-plane above the linear line in the first quadrant (depicted in Figure 18.2). The graph including all the constraints, (1), (2) and (3), of the above problem, is drawn in a similar way (Figure 18.3). The feasible region (the shaded polyhedron ABCDE in Figure 18.3), which contains all the pairs (x_1, x_2) that satisfy each of the three constraints (including the non-negativity constraints) simultaneously, is said to be the set of *feasible solutions* for the LP problem. The *optimal solution* of the LP problem (if it exists) is the feasible solution that maximizes the objective function $F = 3x_1 + 2x_2$.

To find the optimal solution, consider first an arbitrary level of profit, say F_0, and note that the equation $F_0 = 3x_1 + 2x_2$ or $x_2 = (F_0/2) - (3/2)x_1$ defines all the combinations of x_1 and x_2 that yield a profit of F_1 dollars, and is called an *iso-profit curve*.

The iso-profit curve for a profit of F_1 dollars is a straight line with an intercept of $F_1/2 (= F_1/c_2)$ and a slope of $3/2 (= c_1/c_2)$ (see Figure 18.4). Similarly, if one chooses another arbitrary level of profit, $F_2 > F_1$, another iso-profit curve, $x_2 = (F_2/2) - (3/2)x_1$, is obtained. The two iso-profit curves are parallel (i.e., they have the same slope, $3/2 (= c_1/c_2)$, but the second curve lies above the first one (i.e., the intercept $F_2/2$ is larger than the intercept $F_1/2$). In fact, there are an infinite number of iso-profit curves that are parallel to each other and higher profits occur as they move further from the origin (Figure 18.4).

All of the tools needed to determine the optimal solution are now in place. The map of the (parallel) iso-profit curves indicates what the total profits are, namely values of the objective functions, associated with various combinations of x_1 and x_2.

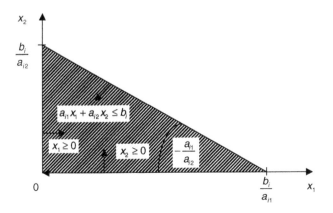

Figure 18.1 The feasible region with a single maximum constraint.

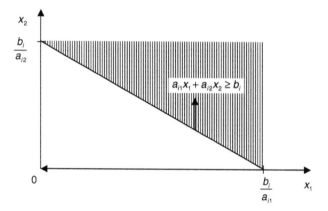

Figure 18.2 The feasible region with a single minimum constraint.

The feasible region defined by the constraints indicates which of the combinations is affordable. The planner's task is to put the two together and choose a combination of activities that will yield the highest affordable value of the objective function. Here, the feasible region is drawn together with a few iso-profit curves (Figure 18.5).

Of the three iso-profit curves, F_3 is preferred because it yields the highest profit; however, it is not feasible, nor is any other (x_1, x_2) combination that lies beyond the feasible region. A profit of F_1 dollars is not the highest feasible one. The planner's strategy is to keep moving to higher and higher iso-profit curves until the highest one that is still affordable is reached. The highest affordable profit, F_2, is obtained with the activity combination for which the iso-profit curve "touches" the corner point C of the feasible region. Note that C is the intersection of the boundary lines of constraints (1) and (3), namely, both constraints are satisfied with strict equality, implying:

$$\left. \begin{array}{ll} (1) & 10x_1 + 10x_2 = 100 \\ (3) & 10x_1 + 5x_2 = 80 \end{array} \right\} \Rightarrow x_1^* = 6, \ x_2^* = 4, \ F^* = 26.$$

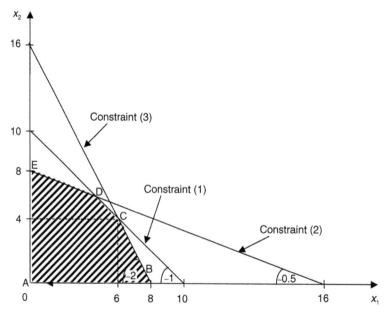

Figure 18.3 The feasible region with three maximum constraints.

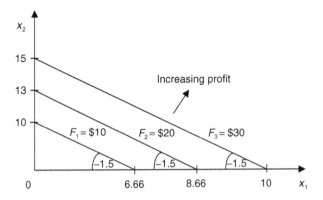

Figure 18.4 Iso-profit curves.

It is also worth noting that the second constraint is not binding at the optimal solution: $100x_1^* + 200x_2^* = 1400 < 1600$, implying that the shadow price of this constraint is zero, $\lambda_2 = 0$. To calculate the shadow prices of the two binding constraints, one should utilize the knowledge that the reduced costs associated with basic activities x_1 and x_2 are equal to zero:

$$\left. \begin{array}{l} R_1 = \underbrace{10(\lambda_1) + 10(\lambda_3)}_{z_1} - \underbrace{3}_{c_1} = 0 \\[2mm] R_2 = \underbrace{10(\lambda_1) + 5(\lambda_3)}_{z_2} - \underbrace{2}_{c_2} = 0 \end{array} \right\} \Rightarrow \lambda_1 = 0.10, \ \lambda_3 = 0.20$$

Also, note that

$$W^* = \sum_{i=1}^{3} b_i \lambda_i = \underbrace{100(0.1)}_{\lambda_1 b_1} + \underbrace{80(0.2)}_{\lambda_3 b_3} = 26 (= F^*).$$

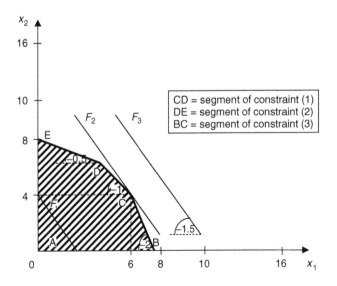

Figure 18.5 Optimal solution.

The choice of corner point C of the feasible region as the optimal point can be made by either using a ruler to move the iso-profit line away from the origin until the last point at which the ruler intersects the feasible region, or by comparing the absolute value of the slope of the iso-profit curve (c_1/c_2) to the absolute values of the slopes of the various segments of the feasible region (a_{i1}/a_{i2}). Specifically, in the current example, Figure 18.5 also indicates that:

If $[c_1/c_2 > 2] \rightarrow$ the optimal solution is obtained at the corner point B ($x_1^* = 8, x_2^* = 0$),

If $[2 \geq c_1/c_2 > 1] \rightarrow$ the optimal solution is obtained at vertex C ($x_1^* = 6, x_2^* = 4$),

If $[1 \geq c_1/c_2 > 0.5] \rightarrow$ the optimal solution is obtained at vertex D ($x_1^* = 4, x_2^* = 6$), and

If $[0.5 \geq c_1/c_2] \rightarrow$ the optimal solution is obtained at the corner point E ($x_1^* = 0, x_2^* = 8$).

These results can be generalized via the following theorem of LP (its proof is beyond the scope of the current chapter): if an optimal solution for a LP problem exists, then it must be obtained at a vertex or corner point of the feasible region. Furthermore, if the optimal solution exists at two adjacent vertices of the polygonal feasible region, then every point on the linear segment connecting the vertices is an optimum solution (if, for example, $c_1/c_2=1$, then all the points on segment CD of the feasible region yield the same maximum value of the objective function). In this case, the LP problem will have an infinite number of solutions. Because they all yield the same level of the objective function, we can choose the activity combination at one of the two vertices (either C or D in the example discussed here).

This important theorem indicates that the search for an optimal solution to a LP problem can be confined to the evaluation of the objective function at a finite set of vertex and corner points of the feasible region of the problem. Indeed, the simplex method, which is an efficient, successful and widely used algorithm to solve LP problems, searches for an optimal solution by proficiently iterating from one vertex or corner point to another, and the value of the objective function increases (in a maximum problem, or decreases in a minimum problem) with each iteration, until it converges to its optimal value. A presentation of the (relatively simple) simplex algorithm is beyond the scope of the current chapter and can be found elsewhere.[4,5] Next, an example of a minimum-optimization problem, also known as a *food-blending problem*, is discussed. This problem was chosen to highlight the principles; therefore, the complexity of a typical nutritional formulation was circumvented for the sake of simplicity.

18.3.2　Example 2: Food-Blending Problem

Consider a consumer who can purchase two food products to nourish his or her family. Being very familiar with recent nutritional requirements, he or she decides that at least 50, 150, and 100 units per week are required of nutritional values such as vitamin A, vitamin B1 and vitamin C, denoted A, B1, and C (no nutritional units are used for simplicity), respectively. The nutritional content and cost of 1 kg of each food product is depicted in Table 18.1.

The goal is to determine the least expensive product combination that will provide the nutritional requirements. To formulate the problem as an LP problem, let x_1 and x_2 be the number of

Table 18.1　Nutritional Composition of Food Products

	Nutritional Elements (units/kg)			
	A	B1	C	Cost ($/kg)
Food product 1	1.00	2.00	4.00	7.00
Food product 2	1.00	5.00	1.25	10.00

kilograms of products 1 and 2, respectively. The minimization problem is then given by:

$$\operatorname*{Min}_{x}\{C = 7x_1 + 10x_2\}$$

Subject to :

$1x_1 + 1x_2$	≥ 50	(1)
$2x_1 + 5x_2$	≥ 150	(2)
$4x_1 + 1.25x_2$	≥ 100	(3)

$$x_j \geq 0, j = 1,2.$$

In this case, the *lowest affordable iso-cost line* is sought. The optimal solution is presented in Figure 18.6, where the iso-cost curve, denoted IC_2, touches the shaded feasible region from below, at point C. Obviously, the (x_1, x_2) combinations along the iso-cost line IC_1 cost less than those along IC_2, but they are not feasible because they do not satisfy the minimum constraints.

Point C (Figure 18.6) is the intersection of the boundary lines of constraints (1) and (2); specifically, both constraints are satisfied with strict equality, implying:

$$\left.\begin{array}{ll} (1) & 1x_1 + 1x_2 = 50 \\ (2) & 2x_1 + 5x_2 = 150 \end{array}\right\} \Rightarrow x_1^* = 33.33 \text{ kg}, x_2^* = 16.67 \text{ kg}, C^* = 400 \text{ kg}.$$

The third constraint is not binding at the optimal solution: $4x_1^* + 1.25x_2^* = 154.17 > 100$, implying that its shadow price vanishes (i.e., $\lambda_3 = 0$). The shadow prices of the binding, minimum constraints (1) and (2) are negative, and represent the increase in value of the objective function (i.e., increase in costs, which is equivalent to *decrease* in free income to the consumer), associated with an increase of one unit in the minimum amount (the RHS constants of the constraints). To reduce possible confusion in calculating these shadow prices, this is illustrated in a manner consistent with that presented in the previous examples, where the objective function was maximized.

Towards this goal, it is worth noting that the objective function of the food management problem: $\operatorname*{Min}_X\{C = 7x_1 + 10x_2\}$ can be rewritten as: $\operatorname*{Max}_X\{-C = -7x_1 - 10x_2\}$. The two objective functions are completely equivalent and yield the same solution (the value of X that minimizes C is

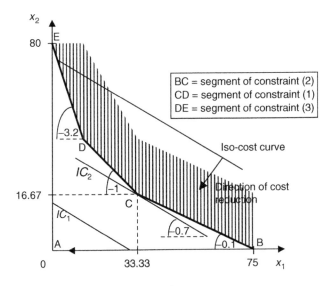

Figure 18.6 Optimal solution for the food-blending problem.

exactly identical to the value that maximizes $-C$). Recalling that in an LP maximization problem, the reduced costs associated with the nonzero basic activities are equal to zero and noting that $c_1 = -7$ and $c_2 = -10$:

$$\left.\begin{array}{l} \underbrace{R_1 = 1.(\lambda_1) + 2.(\lambda_2)}_{z_1} \underbrace{-(-7)}_{c_1} = 0 \\[2em] \underbrace{R_2 = 1.(\lambda_1) + 5.(\lambda_2)}_{z_2} \underbrace{-(-10)}_{c_2} = 0 \end{array}\right\} \Rightarrow \lambda_1 = -\$5/\text{Units of A}, \ \lambda_2 = -\$1/\text{Units of B}.$$

As was illustrated in the previous examples, here too:

$$W^* = \sum_{i=1}^{3} b_i \lambda_i = \underbrace{50.(-5)}_{\lambda_1 b_1} + \underbrace{150.(-1)}_{\lambda_2 b_2} = -400 \text{ dollars}(= -C^*).$$

18.3.3 Pathological Situations of LP

Selected limited typical geometric examples of pathological situations that occur in LP are outlined below.

18.3.3.1 Unbounded Solution

In this situation, there is no largest (or smallest) value of F, i.e., it is always possible to find values of (x_1, x_2) that will make F larger than any pre-assigned value. In other words, F can be made arbitrarily large. An example of a LP problem with an unbounded solution is:

$$\text{Max} \{F = 4x_1 + 3x_2\}$$
$$\underset{x}{}$$
$$\text{Subject to :}$$
$$-3x_1 + 2x_2 \leq 6 \qquad (1)$$
$$-1x_1 + 3x_2 \leq 18 \qquad (2)$$
$$x_1, x_2 \geq 0$$

The graphical presentation of this problem is depicted in Figure 18.7. These lines demonstrates that F can be made arbitrarily large.

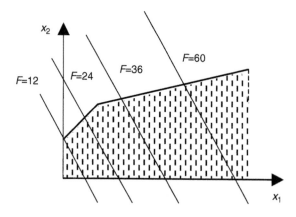

Figure 18.7 Unbounded solution.

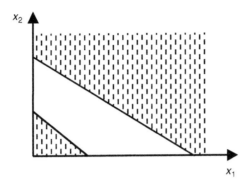

Figure 18.8 Infeasible solution.

18.3.3.2 Infeasible Solution

In this pathological situation, no feasible solution for the LP problem exists; specifically, a set of x_j's, $j = 1, \ldots, n$ that simultaneously satisfies all the constraints does not exist. This pathological situation commonly arises from an error in the formulation of the LP problem. An example of such a problem is:

$$\text{Max} \{F = 3x_1 + 5x_2\}$$
$$\underset{x}{\text{Subject to :}}$$

$$5x_1 + 5x_2 \leq 25 \qquad (1)$$
$$9x_1 + 13x_2 \geq 117 \qquad (2)$$
$$x_1, x_2 \geq 0$$

The graphical presentation of this problem is depicted in Figure 18.8, which clearly indicates that there is no feasible solution (i.e., there is no combination (x_1, x_2) that can be at the same time above the upper linear line and below the lower linear line).

18.4 TYPICAL APPLICATION

One of the many typical applications of LP is the formulation of foods delivering products that meet cost minimization while simultaneously meeting nutritional guidelines. Numerous other characteristic examples include: human diets,[7–11] economic value of fortified food supplements,[12] food security,[13] formulation of complementary infant and child foods,[14] food production and processing,[15–17] breadmaking,[18] accuracy and efficiency of estimating nutrient values in commercial food products,[19,20] general formulation,[21–23] and quality optimization.[24,25] These examples represent only the tip of the iceberg, highlighting both the vast applications and possibilities of the far-reaching capabilities that LP offers. In addition, LP deals with numerous other domains and topics. It has found practical application in almost all facets of business, from advertising to production planning, transportation, distribution, optimal control problems, integer programming, economic analysis, game theory, modern welfare economics, operations research, optimal allocation of resources, scheduling, shipping, telecommunication networks, oil refining, blending, and even stock and bond portfolio selection.

18.4.1 Nutritional Children's Formulation: Excel® Solver

Before listing some of the frequently utilized dedicated LP software, we would like to highlight a simple yet very powerful tool available on most PCs as a part of Excel (http://office.microsoft.com/en-us/FX010858001033.aspx). To illustrate its use, two typical examples have been chosen, both focusing on nutrition and food formulation. The overall objective here is to highlight the utilization of the computerized LP application by providing a simple yet representative example that can also be used by the reader to verify the formulation and application of this straightforward yet powerful tool.

Appendix A lists a complete description of how Excel Solver is used to derive the optimal solution for the composition of a nutritional children's product described previously by Briend et al.[11] This example highlights the principles behind formulating an LP problem, typical data required prior to attempting to utilize the Excel Solver, and data interpretation techniques. Due to its length and detailed nature, it is listed in the appendix.

18.4.2 Example 3: Reconstituted Juice-Blend Formulation

The objective in this example is to optimize (minimum cost) a formulation involving the blending of several juices while meeting marketing, taste, and nutrition constraints. The first step, as indicated for the previous example (Appendix A), is to list all the variables and to gather the database information on the ingredients. In this case, the data includes the solid concentration (^{0}Bx) of the concentrates and the juice, the minimum and maximum acidity of the concentrates, nutritional information (e.g., the amount of ascorbic acid), densities and prices ($/kg concentrate; this information should be frequently updated to accommodate fluctuations in the marketplace). Typical information is listed in Table 18.2. The first column lists the concentrations of the various juice concentrates; the second column lists the typical soluble solids of the reconstituted juice; the third and fourth columns provide the maximum and minimum acidity expected taking into account the natural variability of the various concentrates; the fifth column lists the ascorbic acid concentration; and the last two list the density and price per 1 kg of concentrate. Note that all values except those in the second column relate to the juice concentrate.

The LP goal is to determine the combination of juice concentrates that will cost the least, will meet all of the constraints, and will provide an ascorbic acid concentration that meets quality requirements. To formulate the problem as an LP problem, let $x_1, \ldots, x_{17}$ be kilograms of concentrated juice, flavor and water. The imposed constraints are as follows:

- A basis of reconstituted juice blend of 100 kg is chosen. This includes all the constituents (i.e., juice concentrates, flavors and water).
- Juice concentration after reconstitution of each fruit should be between 5 and 20%. The lower limit is imposed to ensure that the final blend includes a minimum concentration of each and every juice, thus complying with the product label that may specify it. The upper limit of 20% is imposed to ensure that no specific juice dominates the blend.
- Total concentration of the added flavors should be at least 0.35%, satisfying consumer preference. In addition, each flavor should be present at least 0.05%, ensuring an acceptable flavor combination.
- To provide the necessary sweetness, the final concentration of soluble solids in the reconstituted juice blend should be between 13.5 and 14.0 ^{0}Bx.
- Total acidity should be between 0.8 and 0.9%. This value is imposed to ensure an adequate acidity. In some cases, a constraint on the Brix-to-acidity ratio may be imposed.

Table 18.2 Typical Information on Juice Concentrate, Reconstituted Juice and Other Ingredients

Juice Concentrate	Conc. Soluble Solids (°Bx)	Juice Soluble Solids (°Bx)	TA[a] min (%)	TA max (%)	Ascorbic Acid (mg/100 g)	Density (g/mL)	Cost ($/kg Conc.)
Apple	70.0	10.50	1.60	2.03	3.42	1.350	1.127
Banana	23.0	23.00	0.60	0.70	95.74	1.096	0.862
Cherry	68.0	14.30	2.29	4.11	755.12	1.337	2.533
Concord	68.0	14.00	2.13	2.99	14.55	1.337	2.564
Cranberry	50.0	8.00	11.40	13.84	122.07	1.232	9.032
Lemon	38.0	6.00	32.00	33.00	157.29	1.232	1.684
Madera	68.0	14.00	1.54	1.96	221.70	1.337	1.239
Passion fruit	50.0	14.50	10.00	12.00	102.90	1.232	3.307
Pear	70.0	12.70	1.72	2.29	8.28	1.350	0.072
Pineapple	61.0	12.80	2.29	3.22	54.63	1.295	1.171
Peach	70.0	11.80	4.39	4.59	20.14	1.350	2.205
Pink grapefruit	68.0	14.00	2.65	3.16	136.00	1.337	1.596
Red raspberry	65.0	10.50	8.00	12.00	318.11	1.319	10.845
White grapefruit	68.0	14.00	0.68	1.19	214.38	1.337	1.151
Other ingredients							
Banana flavor							15.43
Cherry flavor							20.06
Raspberry flavor							17.64
Water						1.0	0.0

[a] TA, total acidity.

- The final ascorbic acid concentration should be at least 50 mg/100 g.
- The cost of the water used to reconstitute the juice concentrate is taken to be zero.

The minimization problem is formulated as:

$$Min_{\underline{x}}\{C = 1.127x_1 + 0.862x_2 + 2.564x_3 + \cdots + 17.637x_{17}\}$$

Subject to:

Juice content:	$0.70/0.105x_1 + 0.23/0.23x_2 + \cdots + 0.68/0.14x_{14}$	
	$+ x_{15} + x_{16} + x_{17} = 100$	(1)
Individual juice content (max):	$0.70/0.105x_1, 0.23/0.23x_2,..0.68/0.14x_{14} \leq 20$	(2)
Individual juice content (min):	$0.70/0.105x_1, 0.23/0.23x_2,..0.68/0.14x_{14} \geq 5$	(3)
Flavor content (max):	$x_{15} + x_{16} + x_{17} \leq 0.35$	(4)
Flavor content (min):	$x_{15}, x_{16}, x_{17} \geq 0.05$	(5)
Brix (min):	$0.70x_1 + 0.23x_2 + .. + 0.68x_{14} \geq 13.5$	(6)
Brix (max):	$0.70x_1 + 0.23x_2 + .. + 0.68x_{14} \leq 14.0$	(7)
Acidity (max):	$0.016x_1 + 0.006x_2 + .. + 0.0068x_{14} \leq 0.9$	(8)
Acidity (min):	$0.016x_1 + 0.006x_2 + .. + 0.0068x_{14} \geq 0.8$	(9)
Ascorbic acid (min):	$3.42x_1 + 95.74x_2 + .. + 214.38x_{14} \geq 50$	(10)

$x_j \geq 0, \quad j = 1,...,17$

The optimal solution found is $0.58/kg of juice blend and the other derived values are listed in Table 18.3.

The optimal solution gives the lowest cost of the juice-blend formulation at $0.58/kg. This solution meets all of the above constraints and simultaneously reduces the cost of the juice blend

Table 18.3 Optimal Solution for Least-Cost Juice Formulation

Juice Concentrate	Variable	Optimal Solution (kg Concentrate)	Constraint Status (for the Juice Blend)	Slack (for the Juice Blend)
Apple	x_1	0.75	Binding	0
Banana	x_2	8.56	Not binding	3.56
Cherry	x_3	2.73	Not binding	7.97
Concord	x_4	1.03	Binding	0
Cranberry	x_5	0.80	Binding	0
Lemon	x_6	0.79	Binding	0
Madera	x_7	1.67	Not binding	8.12
Passion fruit	x_8	1.45	Binding	0
Pear	x_9	0.90	Not binding	0
Pineapple	x_{10}	1.05	Not binding	0
Peach	x_{11}	0.84	Binding	0
Pink grapefruit	x_{12}	1.03	Binding	0
Red raspberry	x_{13}	0.81	Binding	0
White grapefruit	x_{14}	4.12	Binding	0
Banana flavor	x_{15}	0.25	Not binding	0.20
Cherry flavor	x_{16}	0.05	Binding	0
Raspberry flavor	x_{17}	0.05	Binding	0
Water		73.12	Not binding	
TA(%)		0.86	Not binding	0.06
Solid concentration (°Bx)		13.50	Binding	0
Ascorbic acid (mg/100 g)		50.00	Binding	0

to the minimum. Utilizing the density of the different concentrates as listed in Table 18.2 allows us to calculate the volume and cost of each aseptic carton (assuming that this is the juice blend's packaging), and the price per carton. This information is vital for guaranteeing profitable production.

Obviously, if the conditions in the marketplace were to change, the cost of each ingredient in Table 18.2 would need to be updated and the linear program rerun to determine the new formulation. Indeed, it is important to note that the prices of the concentrates do fluctuate quite often. Although it would produce the least-cost juice blend each and every time, frequent modification of the formulation by applying this technique may have an adverse effect on consumers, as changes could also result in different tastes and organoleptic characteristics. Therefore, every formulation change should be considered carefully, and sensory evaluations are highly recommended.

18.4.3 Example 4: Restaurant Management

"Greens" is a trendy, popular restaurant that uses only fresh vegetables for both its menu and its ingredients. All the vegetables (green beans, corn, tomatoes) are grown in the owner's 750 m^2 garden plot. One kilogram of beans requires 1 m^2 of land, 0.7 work-hours, and 1 m^3 of water per week. Similarly, 1 kg of corn requires 1 m^2 of land, 1 work-hour, and 0.5 m^3 of water, and 1 kg of tomatoes requires 2 m^2 of land, 1.1 work-hours, and 0.4 m^3 of water per week. The weekly growing cost per kg (including water and labor costs) of beans, corn, and tomatoes is $1.5, $1.1, and $2.0, respectively. The current staff can supply up to 600 work-hours per week and the weekly quota of irrigation water is 400 m^3.

These three vegetables are used to produce three dinner-menu items: sautéed vegetables (SV), vegetable soup (VS), and a vegetable pie (VP). The items SV, VS, and VP are listed on the menu at $7.75, $5.50, and $9.25 per large family-size serving, respectively. These relatively low prices are promotional. The preparation of SV, VS, and VP requires 0.1, 0.2, and 0.1 h of labor, and 1 h of the current team of workers costs the owner of the restaurant $2.5. Because the restaurant is so popular, all the menu items that are produced are consumed by its customers. One serving item of SV requires 0.5 kg of corn and 0.3 kg of tomatoes; one serving item of VS requires 0.6 kg of beans, 0.2 kg of corn, and 0.3 kg of tomatoes; and one serving item of VP requires 0.3 kg of beans, 0.3 kg of corn, and 0.4 kg of tomatoes. Based on past experience, the owner of the restaurant has asked his team to produce at least 50 serving items per week of VP and has requested that the total serving number of VS should be not less than half of the total serving numbers of the two other items.

The owner of the restaurant is seeking to develop a profit-maximizing production plan, which can be obviously done via LP. The formulation and solution of the problem are presented below. Start with a definition of the activities and constraints:

1. Activities (x_j's) and their associated income coefficients (c_j's):
   a. To grow 1 kg of beans, $c_1 = -\$1.5/\text{kg}$.
   b. To grow 1 kg of corn, $c_2 = -\$1.1/\text{kg}$.
   c. To grow 1 kg of tomatoes, $c_3 = -\$2.0/\text{kg}$.
   d. To produce one serving item of SV, $c_4 = \underbrace{7.75}_{\text{menue price}} \underbrace{-2.5 \text{x} 0.1}_{\text{labor costs}} = \$7.5/\text{item}$.
   e. To produce one serving item of VS, $c_5 = \underbrace{5.5}_{\text{menue price}} \underbrace{-2.5 \text{x} 0.2}_{\text{labor costs}} = \$5.0/\text{item}$.
   f. To produce one serving item of VP, $c_6 = \underbrace{9.25}_{\text{menue price}} \underbrace{-2.5 \text{x} 0.1}_{\text{labor costs}} = \$9.0/\text{item}$.

The level of x_1, x_2, and x_3, is measured in kilograms and the level of each of the last three x_4, x_5, and x_6, is measured in the relevant number of serving items.

The objective function of the LP problem is:

$$\text{Max } (\underbrace{F = -1.5x_1 - 1.1x_2 - 2x_3 + 7.5x_4 + 5x_5 + 9x_6}_{X}).$$

2. Constraints:
   a. Land constraint: $1x_1 + 1x_2 + 2x_3 \leq 750 \text{ m}^2$
   b. Labor constraint: $0.7x_1 + 1x_2 + 1.1x_3 + 0.1x_4 + 0.2x_5 + 0.1x_6 \leq 600 \text{ h}$
   c. Water constraint: $1x_1 + 0.5x_2 + 0.4x_3 \leq 400 \text{ m}^3$
   d. Beans: supply and demand:

$$1x_1 \underbrace{\geq 0x_4 + 0.6x_5 + 0.3x_6}_{\text{demand (kg)}} \Rightarrow -1x_1 + 0x_4 + 0.6x_5 + 0.3x_6 \leq 0$$
$$\underbrace{}_{\text{supply (kg)}}$$

   e. Corn: supply and demand:

$$1x_2 \underbrace{\geq 0.5x_4 + 0.2x_5 + 0.3x_6}_{\text{demand (kg)}} \Rightarrow -1x_2 + 0.5x_4 + 0.2x_5 + 0.3x_6 \leq 0,$$
$$\underbrace{}_{\text{supply (kg)}}$$

   f. Tomatoes: supply and demand:

$$1x_3 \underbrace{\geq 0.3x_4 + 0.3x_5 + 0.4x_6}_{\text{demand (kg)}} \Rightarrow -1x_3 + 0.3x_4 + 0.3x_5 + 0.4x_6 \leq 0,$$
$$\underbrace{}_{\text{supply (kg)}}$$

   g. Requested ratio between VS and the two other items:

$$x_5 \geq 0.5(x_4 + x_6) \Rightarrow 1x_5 - 0.5x_4 - 0.5x_6 \geq 0,$$

   h. Minimum SV: $1x_4 \geq 50$ serving items
   i. Non-negativity constraints: $x_j \geq 0, \quad j = 1,\dots,6$

The problem was solved via the Excel Solver and the results are summarized in Table 18.4 and Table 18.5.

The optimal value of the objective function, $F^* = \$3275/\text{week}$.

Note that $W^* = \sum_{i=1}^{8} \lambda_i b_i = 750(2.767) + 600(2.00) = \$3275/\text{week}(=F^*)$.

The optimal solution of the water quota is 52.23 m^3, constraints 3 and 8 are not utilized, and the optimal level of SV (175.16 serving items) exceeds the minimum level required (50) by 125.16 serving items. Thus, the shadow prices of these two constraints are equal to zero. An additional 1 m^2 of land will increase the restaurant's net profit (i.e., the value of the objective function) by $\lambda_1 = \$2.767$, and an additional work-hour will increase the net profits by $\lambda_2 = \$2.00$. Interpreting the meaning of the shadow prices of constraints 4, 5, and 6 is not trivial, and will be explained via constraint 4. Because the constraint is binding, it can be written as an equality: $-1x_1 + 0x_4 + 0.6x_5 + 0.3x_6 = b_4 = 0$. Increasing the level of b_4 by one unit means that total demand for beans, $0.6x_5 + 0.3x_6 = 176.75 \text{ kg}$, can be supplied by only $176.75 - 1 = 175.75 \text{ kg}$ of beans grown in the restaurant's garden plot. Thus, the shadow price $\lambda_4 = \$5.667/\text{kg}$ is the change in the value of

Table 18.4 Optimal Solution for the Restaurant Problem: Activities

	x_1 (kg)	x_2 (kg)	x_3 (kg)	x_4 (Serving Items)	x_5 (Serving Items)	x_6 (Serving Items)
Optimal value	176.75	187.90	192.68	175.16	191.08	207.01
Reduced costs	0	0	0	0	0	0

Table 18.5 Optimal Solution for the Restaurant Problem: Constraints

Constraint →	"1" (Land; m²)	"2" (Labor; Hours)	"3" (Water; m³)	"4" (Beans' Balance; m²)	"5" (Corn's Balance; m²)	"6" (Tomatoes' Balance; m²)	"7" (VS Relative to FV+VP)	"8" (Minimum FV; m²)
Constraint constant (bⱼ)	750	600	400	0	0	0	0	50
Surplus/slack	0	0	52.23	0	0	0	0	125.16
Shadow price (λⱼ)	2.767	2.000	0.000	5.667	5.867	9.733	-2.893	0.000

the objective function (i.e., the restaurant's net benefit) in response to an addition of 1 kg of beans to the total quantity produced in the garden plot. Similarly, the shadow prices λ_5 and λ_6 represent the contributions to net profits by an additional kilogram of corn and tomatoes, respectively. Constraint 7 is a minimum binding constraint and its negative shadow price (measured in $/kg) can be interpreted as the net profit that can be gained if the constraint is relaxed by one unit (i.e., if when $0.5(x_4+x_6)=191.08$ kg the restaurant will be "forced" to produce only 190.08 kg of VS, i.e., x_5 will be equal to 190.08 rather than 191.08).

This example is completed by assuming that the restaurant's owner can rent up to 50 m^2 of additional land from his neighbor at a weekly rate of $2 per 1 m^2. He can also hire up to 30 work-hours from a remote city at the relatively high cost (including transportation costs) of $5.5 per work-hour. Because the shadow prices of land and labor are positive, the owner should examine the profitability of the two alternatives by introducing them into the LP problem and solving it again. This can be accomplished by adding to the above problem (1) two additional activities, namely x_7, the number of m^2 rented from his neighbor and x_8, the number of work-hours hired from the remote city; and (2) two additional constraints, i.e., $1x_7 \leq b_9 = 50$ and $1x_8 \leq b_{10} = 30$. The income coefficients of (the new) activities 7 and 8 are $c_7 = -\$2/m^2$ and $c_8 = -\$3$/work hour, respectively. Recall that 1 work-hour of the current team of employees costs the restaurant owner $2.5. This cost was already taken into account in the calculation of the income coefficient c_1-c_7. To avoid counting it twice, $2.5 is deducted from the cost of hiring new work-hours to get $c_8 = -(5.5-2.5) = -\$3$. The extended LP problem can be formulated as:

$$\text{Max } \{F = -1.5x_1 - 1.1x_2 - 2x_3 + 7.5x_4 + 5x_5 + 9x_6 - 2x_7 - 3x_8\}$$
$$\underbrace{}_{X}$$

Subject to :

$1x_1 + 1x_2 + 2x_3 - 1x_7 \leq 750 \; m^2$ (1)
$0.7x_1 + 1x_2 + 1.1x_3 + 0.1x_4 + 0.2x_5 + 0.1x_6 - 1x_8 \leq 600 \; hours;$ (2)
$1x_1 + 0.5x_2 + 0.4x_3 \leq 400 \; m^3$ (3)
$-1x_1 + 0x_4 + 0.6x_5 + 0.3x_6 \leq 0$ (4)
$-1x_2 + 0.5x_4 + 0.2x_5 + 0.3x_6 \leq 0$ (5)
$-1x_3 + 0.3x_4 + 0.3x_5 + 0.4x \leq 0$ (6)
$1x_5 - 0.5x_4 - 0.5x_6 \geq 0$ (7)
$1x_4 \geq 50 \; \text{serving items}$ (8)
$1x_7 \leq 50$ (9)
$1x_8 \leq 30$ (10)
$x_j \geq 0, \quad j = 1, ..., 8.$

The results are summarized in Table 18.6 and Table 18.7.

It follows that hiring additional expensive work-hours is not profitable ($x_8 = 0$). On the other hand, the restaurant's owner can increase his total net profits by renting an additional ~ 18 m^2 of land ($x_7 = 18.14$). In this case, the value of the objective function will increase by $14/week ($= 3289 - 3275$).

Table 18.6 Optimal Solution for the Extended Restaurant Problem: Activities

	x_1 (kg)	x_2 (kg)	x_3 (kg)	x_4 (Items)	x_5 (Items)	x_6 (Items)	x_7 (m²)	x_8 (Work-Hours)
Optimal value	208.75	159.17	200.11	50.00	186.46	322.92	18.14	0.00
Reduced costs	0	0	0	0	0	0	0	0.092

The optimal value of the objective function $F^* = \$3289$/week.

Table 18.7 Optimal Solution for the Extended Restaurant Problem: Constraints

Constraint →	"1"	"2"	"3"	"4"	"5"	"6"	"7"	"8"	"9"	"10"
Constraint constant (b_i)	750	600	400	0	0	0	0	50	50	30
Surplus/slack	0	0	31.62	0	0	0	0	0	31.8	30.0
Shadow price (λ_i)	2.0	2.99	0	5.59	6.09	9.29	−2.96	−0.11	0	0

18.5 SOFTWARE

18.5.1 Computer Programs

There is an impressive amount of LP software currently available. A relatively recent OR/MS Today survey listed 44 different programs.[26] Because this topic is extremely active and companies are often merging, acquired, or cease to operate, the reader is strongly encouraged to consult the aforementioned survey that also includes a full description of the interfaces, platforms supported, maximum number of constraints, pricing information, algorithms, problem types, vendors, etc.

Another excellent resource is the Optimization Technology Center at Northwestern University and Argonne National Laboratory (http://www-fp.mcs.anl.gov/otc/Guide/SoftwareGuide/Categories/linearprog.html),that provides detailed information on available LP software and the unique features of each. The reader is encouraged to check their "Linear Programming Frequently Asked Questions" section, which furnishes explanations and makes specific recommendations (http://www-unix.mcs.anl.gov/otc/Guide/faq/linear-programming-faq.html#online_services). Typically utilized user-friendly software includes: LINDO (http://www.lindo.com) and Economics QM for Windows. QM for Windows is user-friendly Windows software available for quantitative methods, management science, and operations research. It features separate modules covering a spectrum of topics including integer programming, LP, mixed-integer programming, and others.

When teaching an introduction to optimization methods (including LP), GAMS (http://www.gams.com) or AMPL (http://www.ampl.com) are useful packages for the student. Moreover, free student versions (plus manuals) can be downloaded from their respective websites. In (and even outside of) operations research GAMS and AMPL are rapidly becoming the acceptable "standards" for the formulation of LP as well as NLP (nonlinear programming, which covers a very wide spectrum of cases) problems. In addition, there are several "web-based optimization solver" sites on the Internet where students and/or other practitioners can submit and solve LP and NLP problems. These usually accept AMPL and/or GAMS formats. The advantage of these sites is that students do not even need to install any software: they just need a PC with Internet access and a browser. These sites are also interesting because they include a library of examples that can also be useful for students.

To provide some idea of the relative performance of LP codes, a collection of benchmark results for optimization software was compiled (http://plato.la.asu.edu/bench.html). It includes tests of numerous simplex and interior-point codes for LP as well as branch-and-bound codes for linear integer programming; both commercial packages and downloadable, free implementations are included.

18.5.2 Product Development

New product development is a complex and time-consuming task that involves a plethora of activities and resources. Even after most decisions have been made as to the adequate concept, and with respect to consumer benefits, quality, sensory evaluation, etc., one of the more salient tasks when developing a new formulation is determining how to reduce costs to meet market pressure and economic constraints. Although there is a great variety of product-development software available,[27] this topic is not discussed here. However, to provide the reader with a short introduction to this important topic, several typical examples are listed. It is worth noting that the authors have no personal experience with these products and the information is based on available documentation found on the Internet and/or private communication with the developers.

18.5.2.1 *ProductVision*™

ProductVision eliminates the need for multiple systems (spreadsheets, databases, additional software packages) to maintain product recipe, nutritional, specification, and costing information. ProductVision gives the user a single source of information. With ProductVision, there is one database that holds all of the information concerning resources and formulas, and that same integrated system handles the automatic calculation of product properties, costs, nutrition "Facts" labels and ingredient statements (http://www.asdsoftware.com).

18.5.2.2 *TechWizard*™

TechWizard (The Owl Software, http://www.owlsoft.com) utilizes LP and is referred to as *goal-oriented formulation*.[28] One typical example of this software's application is least-cost formulation. The software unitizes current ingredient prices that are either entered into or retrieved from the ingredient database. The formula specifications are listed, including the goals for this formulation such as fat, total solids, sweetness, etc. LP and a combinatorial optimization algorithm are implemented to determine the best blend of ingredients for the various goals. The software is frequently utilized in food applications (L. G. Phillips, personal communication, 2004) such as formulating ice creams for mouthfeel and taste.[28]

18.5.2.3 *DevEX*®

In addition to many facets of product development, this software also includes formula-optimization tools (http://www.selerant.com).

18.6 CONCLUDING REMARKS

This chapter provides the rationale, fundamental aspects, and general principles underlying the utilization and application of LP. As indicated, LP offers a simple yet very effective algorithm for formulating and solving management problems. Although most food-processing applications related to engineering and kinetics are nonlinear in nature, LP provides a leading tool for formulation and cost reduction. LP furnishes a very economical and straightforward method that allows better utilization of resources, improving productivity and overall profitability. The formulation of an LP problem is not complicated and it enables a fast and efficient way of searching for an optimal solution. Overcoming the problem of nonlinear food systems is a topic that requires special attention and expertise. Nevertheless, LP is extremely versatile and should be considered due to its many potential benefits. The large variety of readily available software, as well as the possibility of using LP on Excel, which exists on most PCs, makes this tool paramount for product development, especially for improving and meeting nutritional requirements, as well as maximizing profitability.

ACKNOWLEDGMENTS

The authors would like to express their appreciation and gratitude to Ms. Connie Whippie for her assistance on the juice formulation example and Mr. Gilad Axelrad for his help with the formulation of some of the examples and with the drawing of the graphs.

18.A APPENDIX

18.A.1 Introduction

To use Excel to solve LP problems, the Solver add-in must be included. Typically, this feature is not installed by default when Excel is first set up on your hard disk. To add this facility to the Tools menu, you need to carry out the following steps (once only):

1. Select the menu option Tools|Add_Ins
2. From the dialogue box presented, check the box for Solver Add-In.
3. Click OK. You can now access the Solver option from the new Tools|Solver menu option.

To illustrate Excel Solver, a recent publication[11] was chosen both for its simplicity and because it can be implemented for numerous nutritional applications. Those who have used Excel previously can ignore this introductory example and delve directly into the juice formulation example.

18.A.2 Optimizing a Diet for Children

To utilize LP, the first stage is to create the necessary database outlining the various food compositions, costs, and other pertinent information in table form (Table 18.A.1): For each food item (variable), the cost (arbitrary units), energy, protein, calcium and iron are listed. This information is widely available through a variety of sources (e.g., USDA National Nutrient Database for Standard Reference, Release 16; http://www.nal.usda.gov/fnic/cgi-bin/nut_search.pl). Updated prices can be found in financial resources.

The constraints for the formulation of the children's diet should take into consideration the maximum allowed portion size per day. In our case, the maximum daily intake for lentil and liver is chosen to be 60 g. For maize flour and milk, a high arbitrary figure of 999 g is chosen, indicating that these are important and unrestricted components in the children's diet. In other words, while the constraints of lentils and liver may be binding, those of maize and fresh milk are not.

The goal is to determine the least expensive product combination that will fulfill the nutritional requirements. To formulate the problem as a LP one, let x_1, x_2, x_3, and x_4 be the grams of maize flour, fresh milk, cooked lentils and liver, respectively. The minimization problem is then

Table 18.A.1 Nutritional Composition and Arbitrary Prices

Ingredient	Maximum Allowed (g)	Price ($/kg)	Energy (kcal/100 g)	Protein (g/100 g)	Calcium (mg/100 g)	Iron (mg/100 g)
Maize Flour (x_1)	999	100	362	8.1	6	3.5
Fresh Milk (x_2)	999	150	66	3.2	115	0.1
Lentils (cooked) (x_3)	60	200	311	24.5	51	8.8
Liver (x_4)	60	300	157	18.1	14	8.5

Source: From Briend, A., Darmon, N., Ferguson, E., and Erhardt, J. G., *Journal of Pediatric Gastroenterology and Nutrition*, 36(1), 12–22, 2003.

given by:

$$Min_{\underline{x}}\{C = 100x_1 + 150x_2 + 200x_3 + 300x_4\}$$

Subject to :

Energy :	$362x_1 + 66x_2 + 311x_3 + 157x_4 = 746.0$	(1)
Protein :	$8.1x_1 + 3.2x_2 + 24.5x_3 + 18.1x_4 \geq 5.0$	(2)
Calcium :	$6x_1 + 115x_2 + 51x_3 + 14x_4 \geq 196.0$	(3)
Iron :	$3.5x_1 + 0.1x_2 + 8.8x_3 + 8.5x_4 \geq 11.8$	(4)
Weight :	$x_1, x_2 \leq 999; x_3, x_4 \leq 60$	(5)
	$x_j \geq 0, \quad j = 1,...,4.$	

This information is now transferred into Excel as shown below (Table 18.A.2):
The calculation table includes the basket ($x_1...x_4$; \$A\$12...\$A\$15) and an initial arbitrary value of 50 g for each variable (\$B\$12...\$B\$15), which apparently does not meet the imposed requirements (RHS; \$D\$21 to \$G\$21), i.e., at least 746 kcal, 5.0 g of protein, 196.0 mg of calcium and 11.8 mg of iron. The other values (\$C\$12...\$G\$15) are derived from the weight of the ingredient that needs to be optimized (\$B\$12...\$B\$15).
   Solving this problem requires the following steps:

1. Activate the solver.
2. Set the target cell (objective function, in this case minimum cost) to \$C\$18.
3. Set the cells that can be changed (variables x1... x4; \$B\$12...\$B\$15).
4. Set the constraints (Table 18.A.3):
   a. The weight of the ingredients to be less than or equal to the total weight allowed for the nutritional consideration (\$B\$12...\$B\$15 ≤ \$B\$4...\$B\$7)
   b. The amount of energy the formulation provides; this is a binding constraint (\$D\$18 = \$D\$21)

Table 18.A.2   Optimizing A Nutritional Children's Formulation Using Excel

Database

	Max (g)	Price ($)	Energy (kcal)	Protein (g)	Calcium (mg)	Iron (mg)
Maize flour (x1)	999	100	362	8.1	6	3.5
Fresh milk (x2)	999	150	66	3.2	115	0.1
Lentils (cooked) (x3)	60	200	311	24.5	51	8.8
Liver (x4)	60	300	157	18.1	14	8.5

Calculation

	Basket (g)	Price ($)	Energy (kcal)	Protein (g)	Calcium (mg)	Iron (mg)
Maize flour (x1)	50.0	5.0	181.0	4.1	3.0	1.8
Fresh milk (x2)	50.0	7.5	33.0	1.6	57.5	0.1
Lentils (cooked) (x3)	50.0	10.0	155.5	12.3	25.5	4.4
Liver (x4)	50.0	15.0	78.5	9.1	7.0	4.3
		Price ($)	Energy (kcal)	Protein (g)	Calcium (mg)	Iron (mg)
Total		37.5	448.0	27.0	93.0	10.5
Requirements (RHS)			746.0	5.0	196.0	11.8

Source: Briend, A., Darmon, N., Ferguson, E., Erhardt, J. G., *Journal of Pediatric Gastroenterology and Nutrition*, 36(1), 12–22, 2003.

c. The concentrations of protein, calcium and iron to be equal to or higher than those recommended (E18...G18 ≥ E21...G21)

5. The following Solver options are chosen: linear model, non-negative values of the variables, and automatic scaling.

Table 18.A.3 Solver Parameters

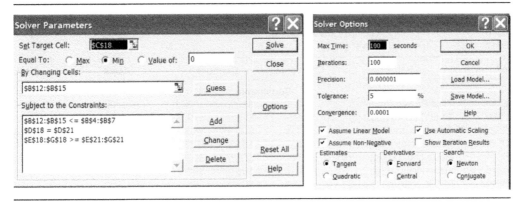

The tolerance option is only required for integer programs (IP): it allows the solver to use "near-integer" values, within the tolerance you specify, and this helps speed up the IP calculations. Checking the show iteration results box allows you to see each step of the calculation, but this may be prohibitive if the problem is complex. Automatic scaling is useful if there is a large difference in magnitude between the variables and the objective value. The bottom three options—estimates, derivatives, and search—affect the way the solver approaches finding a basic feasible solution, how the solver finds partial differentials of the objective and constraints, and how the solver decides which way to search for the next iteration. Essentially, the options affect how the solver uses memory and the number of calculations it makes. For most LP problems, they are best left as the default values. It is good practice to check the "Assume Linear Model" box, unless, of course, the model is not linear. This will ensure the correct result and quite importantly, provide the relevant sensitivity report.

The solution, listed in Table 18.A.4, shows a least-cost formulation of $51.9. It also shows that the binding-energy constraint has indeed been met. It demonstrates that both calcium and iron are also binding constraints, whereas protein is not, and the formulation provides almost six-fold more protein than required. An attempt to enforce a binding constraint on protein concentration will lead to a nonfeasible solution.

Table 18.A.4 Optimal Solution

Optimal Solution						
	Basket (g)	Price ($)	Energy (kcal)	Protein (g)	Calcium (mg)	Iron (mg)
Maize flour ($x1$)	118.6	11.9	429.4	9.6	7.1	4.2
Fresh milk ($x2$)	134.4	20.2	88.7	4.3	154.6	0.1
Lentils (cooked) ($x3$)	60.0	12.0	186.6	14.7	30.6	5.3
Liver ($x4$)	26.3	7.9	41.3	4.8	3.7	2.2
		Price ($)	Energy (kcal)	Protein (g)	Calcium (mg)	Iron (mg)
Total		51.9	746.0	33.4	196.0	11.8
Requirements (RHS)			746.0	5.0	196.0	11.8

Table 18.A.5 Excel Report

Cell	Name	Original Value	Final Value
	Target Cell (min)		
C20	Total Price ($)	—	51.9
	Adjustable Cells		
B14	Maize flour (x_1) (g)	50	118.6
B15	Fresh milk (x_2) (g)	50	134.4
B16	Lentils (cooked) (x_3) (g)	50	60.0
B17	Liver (x_4) (g)	50	26.3

Cell	Name	Cell Value	Formula	Status	Slack
		Constraints			
E20	Total Protein (g)	33.4	E20 > = E23	Not binding	28.4
F20	Total Calcium (mg)	196.0	F20 > = F23	Binding	0.0
G20	Total Iron (mg)	11.8	G20 > = G23	Binding	0.0
D20	Total Energy (kcal)	746.0	D20 = D23	Not binding	0.0
B14	Maize flour (x_1) (g)	118.6	B14 < = B5	Not Binding	880.4
B15	Fresh milk (x_2) (g)	134.4	B15 < = B6	Not binding	864.6
B16	Lentils (cooked) (x_3) (g)	60.0	B16 < = B7	Binding	0.0
B17	Liver (x_4) (g)	26.3	B17 < = B8	Not binding	33.7

Excel also provides a table that summarizes the solutions and highlights the binding constraints (Table 18.A.5). This report provides information on shadow values, reduced costs and the upper and lower limits for the decision variables and constraints.

GLOSSARY

Activities   The decision variables of an LP problem. One unit of a specific activity is fully characterized by its associated income coefficient and a series of coefficients expressing the number of units from each constraint required for the operation of (one unit of) the activity. Activities that are strictly positive in the optimal solution are called *basic activities* (or *basic variables*). Activities that are equal to zero in the optimal solution are termed *nonbasic*.

Algorithm   A complete procedure or method describing the steps to be taken for the solution of a problem.

Constraint   A linear relationship between the decision variables defining the structure of the LP problem. It can either be an equation or an inequality. Those that are satisfied with equality sign are called *binding* (or *effective*) constraints. Those that are satisfied with inequality sign are termed *nonbinding* (or *noneffective*) constraints.

Dual price (or shadow price)   A variable associated with a constraint in the LP problem. In a maximum LP problem the dual price of a binding maximum constraint ($\leq$) is *positive* while the dual price of a binding minimum constraint ($\geq$) is *negative*. The shadow price of a nonbinding constraint (whether it is a maximum or minimum constraint) is equal to zero.

Feasible solution   A solution for LP problem that satisfies all the linear constraints of the problem, including the non-negativity constraints of the activities.

Graphical solution approach   A simple way to solve a LP problem with only two activities.

Infeasible solution   A set of values for the variables of a given LP problem that violates one (or more) constraint(s).

Linear programming (LP)   An optimization problem that involves maximizing or minimizing of a *linear objective function* that includes several non-negative decision variables, the choice of which is subject to a set of *linear constraints*. LP is the simplest and most widely used form of MP. The LP problem is termed as a *maximizing, (minimizing), problem* if the objective function is to be maximized, (minimized). Several typical applications of LP in the food domain include processing, canning operations, livestock nutrition, design of a nutritionally adequate diet at the lowest cost, evaluating the economic value of the fortified product, formulation, etc.

LP software   A vast number of LP software is currently available. Typically utilized user-friendly software includes: LINDO and Economics QM for Windows. GAMS, AMPL and Excel are useful packages for students.

Mathematical programming (MP)   Relates to the use of mathematical models for solving optimization problems. A typical MP model involves the selection of values for a limited number of decision variables, focusing attention on a single objective function to be maximized (or minimized, depending on the context of the problem), subject to a finite set of constraints that limit the selection of the decision variables.

Opportunity cost   An economic criterion for the alternative costs associated with an activity in the LP problem. It is the sacrifice of producing one additional unit of the activity, resulting from the fact that alternative production opportunities (i.e., some other activities) must be forgone in order to satisfy the problem's constraints. In principal, the sacrifice can be either positive or negative.

Optimal solution   The feasible solution that optimizes (i.e., maximize or minimize) the value of the LP linear objective function.

Reduced cost   Defined as the difference between the opportunity costs and the income coefficient associated with the activity under consideration. The reduced cost of a basic activity is equal to zero while the reduced cost of a nonbasic activity is positive and, assuming a maximizing LP problem, represents a reduction in the optimal value of the objective function if the operation of one unit of a nonbasic activity is to be forced upon the optimal solution.

Simplex method   The principal and very successful algorithm for solving LP problems; invented primarily by G. B. Dantzig in 1947.

Unbounded solution   A feasible solution of an LP problem associated with an unbounded value of the objective function.

REFERENCES

1. Dantzig, G. B., *Linear Programming and Extensions*, Princeton, NJ: Princeton University Press, 1963.
2. Banga, J. R., Balsa-Canto, E., Moles, C. G., and Alonso, A. A., Improving food processing using modern optimization methods, *Trends in Food Science and Technology*, 14, 131–144, 2003.
3. Wright, S. J., *Primal-dual Interior-Point Methods*, Philadelphia, PA: Society for Industrial and Applied Mathematics, 1997.
4. Hadley, G., *Linear Programming*, Reading, MA: Addison-Wesley, 1962.
5. Thie, P. R., *An Introduction to Linear Programming and Game Theory*, 2nd ed., New York: Wiley, 1988.
6. Paris, Q., *An Economic Interpretation of Linear Programming*, Ames, IA: Iowa State University Press, 1991.
7. Owen-Smith, N., Evaluating optimal diet models for an African browsing ruminant, the kudu—How constraining are the assumed constraints?, *Evolution Ecology*, 7(5), 499–524, 1993.
8. Owen-Smith, N., Circularity in linear programming models of optimal diet, *Oecologia*, 108(2), 259–261, 1996.
9. Darmon, N., Ferguson, E. L., and Briend, A., A cost constraint alone has adverse effects on food selection and nutrient density: an analysis of human diets by linear programming, *Journal of Nutrition*, 132(12), 3764–3771, 2002.
10. Darmon, N., Ferguson, E. L., and Briend, A., Linear and nonlinear programming to optimize the nutrient density of a population's diet: an example based on diets of preschool children in rural Malawi, *American Journal of Clinical Nutrition*, 75, 245–253, 2002.

11. Briend, A., Darmon, N., Ferguson, E., and Erhardt, J. G., Linear programming: a mathematical tool for analyzing and optimizing children's diets during the complementary feeding period, *Journal of Pediatric Gastroenterology and Nutrition*, 36(1), 12–22, 2003.
12. Briend, A., Ferguson, E., and Darmon, N., Local food price analysis by linear programming: a new approach to assess the economic value of fortified food supplements, *Food and Nutrition Bulletin*, 22(2), 184–189, 2001.
13. Gladwin, C. H., Thomson, A. M., Peterson, J. S., and Anderson, A. S., Addressing food security in Africa via multiple livelihood strategies of women farmers, *Food Policy*, 26(2), 177–207, 2001.
14. Deliza, R., Sgarbieri, V. C., and Rosenthal, A., Formulation, nutritive-value and sensory evaluation of a new weaning food based on sweet corn (Nutrimaiz) dehydrated pulp, *Journal of Nutritional Science and Vitaminology*, 36(6), 587–597, 1990.
15. Chan, J. C. M., Hui, Y. Y., and Sculli, D., Computerization of production management information—a case-study in a food-processing factory, *Computers in Industry*, 18(1), 91–98, 1992.
16. Llewelyn, R. V. and Williams, J. R., Nonparametric analysis of technical, pure technical, and scale efficiencies for food crop production in East Java, Indonesia, *Agricultural Economics*, 15(2), 113–126, 1996.
17. Sharma, S. K., Mulvaney, S. J., and Rizvi, S. H., Product formulation and process optimization using linear programming, In *Food Process Engineering: Theory and Laboratory Experiments*, Wiley, New York, pp. 1–19, 2000.
18. Hayta, M. and Cakmakli, U., Optimization of wheat blending to produce breadmaking flour, *Journal of Food Process Engineering*, 24(3), 179–192, 2001.
19. Westrich, B. J., Buzzard, I. M., Gatewood, L. C., and McGovern, P. G., Accuracy and efficiency of estimating nutrient values in commercial food products using mathematical optimization, *Journal of Food Composition and Analysis*, 7(4), 223–239, 1994.
20. Westrich, B. J., Altmann, M. A., and Potthoff, S. J., Minnesota's nutrition coordinating center uses mathematical optimization to estimate food nutrient values, *Interfaces*, 28(5), 86–98, 1998.
21. Chen, J. S., Lee, C. M., and Crapo, C., Linear programming and response surface methodology to optimize surimi gel texture, *Journal of Food Science*, 58(3), 535–538, 1993.
22. Yoon, W. B., Park, J. W., and Kim, B. Y., Linear programming in blending various components of surimi seafood, *Journal of Food Science*, 62(3), 561–564, 1997.
23. Kupiec, B., Wardynska, K., and Kolczak, T., The optimisation of new meat product formulation, *Przemysl Spozywczy*, 53(2), 33–36, 1999.
24. Saguy, I., Constraints to quality optimization in aseptic processing, *Journal of Food Science*, 53(1), 306–307, 1988 see also p. 310
25. Buczkowski, J., Optimum utilization of raw material in meat factories, *Gospodarka Miesna*, 44(11), 20–23, 1992.
26. Fourer, R., Linear Programming survey, OR/MS Today 30(6), 2003, http://www.lionhrtpub.com/orms/surveys/LP/LP-survey.html (accessed on March 30,2005).
27. Giese, J., Product development software aids formulation, *Food Technology*, 57(10), 78–80, 2003.
28. Phillips, L. G. and Roland, A. M., Development of a software program for goal oriented formulation, Paper presented at IFT Annual Meeting, Chicago, IL, 1999.

Index

Printed and bound by CPI Group (UK) Ltd, Croydon, CR0 4YY

23/10/2024

01778251-0011